Human Molecular Genetics 2

Human Molecular Genetics 2

Second Edition

Tom Strachan

BSc PhD
Professor of Human Molecular Genetics
University of Newcastle
Newcastle-upon-Tyne, UK

and

Andrew P. Read

MA PhD FRCPath FMedSci
Professor of Human Genetics
University of Manchester
Manchester, UK

© BIOS Scientific Publishers Ltd, 1999

First published 1996
Second edition 1999

A CIP catalogue record for this book is available from the British Library.

ISBN 1 85996 202 5

BIOS Scientific Publishers Ltd
9 Newtec Place, Magdalen Road, Oxford OX4 1RE, UK
Tel. +44 (0) 1865 726286. Fax +44 (0) 1865 246823
World Wide Web home page: http://www.bios.co.uk/

Published in the United States of America, its dependent territories and Canada by John Wiley & Sons, Inc., by arrangement with BIOS Scientific Publishers Ltd.

Copies may be obtained from:

USA
John Wiley & Sons Inc.,
605 Third Avenue, New York,
NY 10158–0012, USA

Canada
John Wiley & Sons (Canada) Ltd,
22 Worcester Road, Rexdale,
Ontario M9W 1L1, Canada

Production Editor: Fran Kingston
Typeset by J&L Composition Ltd, Filey, North Yorkshire, UK
Printed by The Bath Press Ltd, Bath, UK

Contents

Chapter 8   Human gene expression   **169**

Chapter 9   Instability of the human genome: mutation and DNA repair   **209**

Chapter 10   Physical and transcript mapping   **241**

Chapter 19   Complex diseases: theory and results   445

Chapter 20   Studying human gene structure, expression and function using cultured cells and cell extracts   465

Chapter 21   Genetic manipulation of animals   491

Chapter 22   Gene therapy and other molecular genetic-based therapeutic approaches   515

Abbreviations

AAV	adeno-associated virus
ADA	adenosine deaminase
AFP	α-fetoprotein
AIDS	acquired immune deficiency syndrome
AMCA	aminomethylcoumarin acetic acid
APOE	apolipoprotein E
ARMS	amplification refractory mutation system
ARS	autonomously replicating sequence
ASO	allele-specific oligonucleotide
BAC	bacterial artificial chromosome
bp	base pair
BER	base excision repair
BMI	body mass index
BOR	branchio-oto-renal syndrome
CCD	charge coupled device
CCM	chemical cleavage of mismatches
cDNA	complementary DNA
CDR	complementarity determining region
CEN	centromere element
CEPH	Centre d'Études du Polymorphisme Humaine
CF	cystic fibrosis
CGH	comparative genome hybridization
CGRP	calcitonin gene-related peptide
CHLC	Cooperative Human Linkage Center
cM	centiMorgan
CMGT	chromosome-mediated gene transfer
CMPD	campomelic dysplasia
CMT	Charcot-Marie-Tooth disease
CNS	central nervous system
cR	centiRay
CRE	cAMP response element
CREB	CRE-binding protein
cRNA	complementary RNA
ddNTP	dideoxynucleoside triphosphate
DGGE	denaturing gradient gel electrophoresis
dHPLC	denaturing high performance liquid chromatography
DIRVISH	direct visual hybridization
DM	myotonic dystrophy
DMD	Duchenne muscular dystrophy
DNase	deoxyribonuclease
dNTP	deoxynucleoside triphosphate
DOP-PCR	degenerate oligonucleotide-primed PCR
DPE	downstream promoter element
DZ	dizygotic
EBV	Epstein–Barr virus
EMS	ethylmethylsulfonate
EMSA	electrophoretic mobility shift assay
ENU	ethylnitrosourea
ER	endoplasmic reticulum
ES cell	embryonic stem cell
EST	expressed sequence tag
EtBr	ethidium bromide
FAP	familial adenomatous polyposis
FH	familial hypercholesterolemia
FISH	fluorescence *in situ* hybridization
FITC	fluorescein isothiocyanate
FRAXA	fragile-X syndrome
FSHD	fascioscapulohumeral muscular dystrophy
GAT	gene augmentation therapy
GDB	Genome Data Base
GE	genome equivalents
GFP	green fluorescent protein
GM-CSF	granulocyte-macrophage colony stimulating factor
GPI	glycosylphosphatidyl inositol
GSS	Gerstmann–Sträussler–Scheinker syndrome
HD	Huntington disease
HDL	high density lipoprotein
HERV	human endogenous retrovirus
HIV	human immunodeficiency virus
HLA	human leukocyte antigen
HLH	helix–loop–helix
HMG	high mobility group
HNPCC	hereditary nonpolyposis colon cancer
HPRT	hypoxanthine phosphoribosyltransferase
HRR	haplotype relative risk
HSCR	Hirschprung disease
HSR	homogeneously staining region
HSV	herpes simplex virus

HTH	helix–turn–helix	Pu	purine
HUGO	The Human Genome Organization	QTL	quantitative trait locus
IBD	identical by descent	RACE	rapid amplification of cDNA ends
IBS	identical by state	RBS	ribosome-binding site
IDDM	insulin-dependent diabetes mellitus	RDA	representational difference analysis
IFN	interferon	rDNA	ribosomal DNA
Ig	immunoglobulin	RE	response element
IGF	insulin-like growth factor	RER	rough endoplasmic reticulum
IL	interleukin	REST	RE-1 silencing transcription factor
IRE	(i) iron-response element, or (ii) interspersed repeat element	RF	replicative form
		RFLP	restriction fragment length polymorphism
IRE-PCR	interspersed repeat element PCR	RNase	ribonuclease
IRP	island rescue polymerase chain reaction	RNP	ribonucleoprotein particles
kb	kilobase	rNTP	ribonucleoside triphosphate
LCR	locus control region	rRNA	ribosomal RNA
LDL	low density lipoprotein	RSP	restriction site polymorphism
LGS	Langer–Giedion syndrome	RT	reverse transcriptase
LHON	Leber's hereditary optic atrophy	RTLV	retrovirus-like element
LINE	long interspersed nuclear element	RT-PCR	reverse transcriptase PCR
LoH	loss of heterozygosity	SAGE	serial analysis of gene expression
LTR	long terminal repeat	SCA1	spinocerebellar ataxia type 1
mAb	monoclonal antibody	SINE	short interspersed nuclear element
MAGP	microfibril-associated glycoprotein	SKY	spectral karyotyping
Mb	megabase	snoRNA	small nucleolar RNA
MBP	myelin basic protein	SNP	single nucleotide polymorphism
MCS	multiple cloning site	snRNA	small nuclear RNA
MeCP	methylated CpG binding protein	snRNP	small nuclear ribonucleoprotein
MER	medium reiteration frequency	SRP	signal recognition particle
M-FISH	multiplex-FISH	SSCP	single-stranded conformation polymorphism
MIM	Mendelian Inheritance in Man		
MIN	minisatellite instability	STRP	short tandem repeat polymorphism
MODY	maturity-onset diabetes of the young	STS	sequence-tagged site
mRNA	messenger RNA	SV40	simian virus 40
mtDNA	mitochondrial DNA	SVAS	supravalvular aortic stenosis
MZ	monozygotic	TAF	TBP-associated factor
NER	nucleotide excision repair	TBP	TATA box-binding protein
NF	neurofibromatosis	TCR	T-cell receptor
NIDDM	noninsulin-dependent diabetes mellitus	TCS	Treacher Collins syndrome
NIH	National Institutes of Health (USA)	TDT	transmission disequilibrium test
NPL	nonparametric lod score	TF	transcription factor
NRSE	neural restrictive silencer element	THE	transposable human element
NRSF	neural restrictive silencer factor	TIL	tumor-infiltrating lymphocyte
OD	optical density	T_m	melting temperature
OLA	oligonucleotide ligation assay	TRITC	tetramethylrhodamine isothiocyanate
OMIM	on-line Mendelian Inheritance in Man	tRNA	transfer RNA
ORF	open reading frame	TS	tumor suppressor
PAC	P1–derived artificial chromosome	TTD	trichothiodystrophy
PAH	phenylalanine hydroxylase	UPD	uniparental disomy
PAR	pseudoautosomal region	UTR	untranslated region
PCR	polymerase chain reaction	UTS	untranslated sequence
PFD	polyostotic fibrous dysplasia	VNTR	variable number of tandem repeats
PFGE	pulsed-field gel electrophoresis	VPC	vector-producing cell
PIC	polymorphism information content	Xce	X-controlling element
PKD	polycystic kidney disease	X-gal	5-bromo, 4-chloro-3-indolyl β-D-galactopyranoside
PKU	phenylketonuria		
PNA	peptide nucleic acid	Xic	X-inactivation center
PTT	protein truncation test	YAC	yeast artificial chromosome

Preface to the first edition

The idea for this book grew from two earlier efforts, *The Human Genome* (TS; BIOS Scientific Publishers, 1992) and *Medical Genetics, an Illustrated Outline* (APR; Gower Medical Publishing, 1989). In these small books we tried to develop a treatment of human genetics based on understanding the structure and function of the normal human genome. Traditionally, textbooks of human genetics tended to start by considering meiosis and the way diseases segregate in pedigrees, whilst textbooks of molecular genetics rarely emphasized human topics. Until recently this was inevitable because so little was understood about the normal human genome. The Human Genome Project has changed all that, and the present book is an attempt to provide a comprehensive integrated study of human molecular genetics.

It would be hard to overstate the importance for biomedical science of the Human Genome Project. As the first 'big science' project in biology, it forms the focus for a mass collective effort by thousands of researchers world-wide to move our understanding of biology on to a new plane. Human molecular genetics not only forms the cutting edge of biomedical research, but at the same time it has immediate application to the diagnosis of disease and has great potential for treating disease. Thus it is of major interest to all students of biological science and medicine, and to a wide range of biomedical researchers.

Human molecular genetics is a large subject. We have tried to make it more digestible by organizing the text into clearly demarcated sections, using statement headings to define what the reader can find in each section, and identifying important new terms by bold typeface (most of which, apart from basic terms, can be found in the Glossary). The first section (Chapters 1–3) provides introductory-level material on DNA, chromosomes and pedigree patterns but the second section (Chapters 4–6), which describes general principles and applications of cloning and molecular hybridization, contains some advanced examples. The third section (Chapters 7–10) is the one which we believe truly distinguishes this book from the others, and provides a comprehensive guide to the structure, function, evolution and mutational instability of the human genome and human genes. It provides a solid base for relating the subsequent sections on mapping the human genome (Chapters 11–13), studying human genetic diseases (Chapters 14–18), and dissecting and manipulating genes (Chapters 19 and 20).

Currently, the pace of research in this field is extremely rapid, and new information and insights are pouring out of the laboratories at an almost unimaginable rate. At the time of writing, new partial human gene sequences (expressed sequence tags) are being added to the databases at a rate of 8000 per week. To cope with this flood, students need two tools: a good framework of principles and the ability to use modern informatics. One of our aims has been to encourage students to use the wonderful genetic resources available on the Internet (in particular, we have provided accession numbers for the OMIM online database for genetic diseases, rather than listing references).

Revolutionary times are nothing if not exciting. We have tried to convey the feel of fast-moving research, while providing a description in some depth of the techniques and data that are helping us to understand the evolution, nature and function of our genome. This book will have succeeded if readers finish it sharing our excitement and enthusiasm for the continuing voyage of discovery into our DNA. The journey is far from finished.

Tom Strachan and Andrew P. Read

Preface to the second edition

We were gratified by the favorable reception given to the first edition, and we thank the many people who wrote pointing out errors or making suggestions. Inevitably we could not incorporate every good suggestion into this new edition without an unacceptable increase in length. As we enter the new Millennium, the pace of change in human genetics continues to accelerate. As before, we expect readers to use the Internet as the main source of factual detail, and we have included a new preliminary section to reinforce this message. We have concentrated in our text on the principles rather than the details, but we hope that the comprehensive index (which now has a new Disease Index prefacing the standard index) will give readers quick access to the factual information that is used to illustrate the principles. Apart from general revision and updating, and some rearrangement of material, there are new chapters on gene expression and on analysis of complex diseases, reflecting the increasing importance of these topics in human molecular genetics. To assist teachers, a problem book linked to this text is in preparation, and all original illustrations are freely available in downloadable form from the BIOS website (http://www.bios.co.uk).

We thank Meryl Lusher for research assistance and help with proof reading; Jolene Blench, Rona Mayall, Leanne Morrison, and Margaret Weddle for secretarial assistance; Jenny Barrett, Nick Lemoine, Nalin Thakker, David Cooper and Andrew Wallace for commenting on drafts; many colleagues for providing material for illustrations; Applied Imaging and PE Biosystems for sponsorship of some color illustrations; Affymetrix, Inc. (Santa Clara, CA) for supplying several color images; and Fran Kingston, Lisa Mansell, Jonathan Ray and their team at BIOS Scientific Publishers for their hard work and understanding. Finally, we owe a great debt to our long-suffering wives and families and so thank you yet again, Meryl, Alex, James and Gilly for being so understanding.

Tom Strachan and Andrew P. Read

Before we start – genetic data and the Internet

This book is about the principles of human molecular genetics. Whether you are a student or a researcher, you will also need access to genetic data – to find where a gene maps, whether the gene has been cloned, what mutations have been found, what genes are contained in a particular chromosomal region, where and when a gene is expressed during embryonic development, etc. You need the most up-to-date information, and you usually need it quickly. The place to find all such genetic data is on the Internet.

All students of human molecular genetics should be using the Internet

The Internet is a world-wide system that links innumerable individual computers and local networks in universities, research laboratories and commercial and government organizations. It allows almost instant communication between computers anywhere in the world. Accessing the Internet requires appropriate hardware, software and permissions. The hardware needed is a network card inside your computer and, if you are using a telephone line, a modem. Software includes the basic programs to enable your computer to communicate across the network, and high-level tools such as a web browser program (Netscape Navigator™, Internet Explorer™, etc.). All these things are included as standard features in many personal computers.

Permission for network access is usually automatic within institutions; private users need to subscribe to an Internet Service Provider (advertised in computer magazines). Once everything is set up, surfing the net is extremely simple, but setting up requires computer expertise and local knowledge. Consult your institution's support service or local vendor. Many books are available to help the beginner use the Internet – choose the most recent available, because the system is evolving fast.

The World Wide Web is the primary way of using the Internet

The World Wide Web (WWW) is not a separate network, but a way of using the Internet through a web browser program. The clever design of these programs allows the user to move effortlessly from computer to computer, collecting and collating material of every sort, including text, data, images, video clips and sound. Individuals and organizations choose to make material located on their own computer accessible in this way, and Internet users access the material using their web browser. All Internet applications are increasingly being packaged as WWW applications (replacing older tools like Gopher), so that a web browser is now the standard tool for most Internet tasks. To the user, the web appears as a library and public notice board of unlimited size and with unlimited access – although there are also private areas, accessible (in theory) only to users who have been given a username and password.

An essential feature of the web, which exhilarates some people and worries others, is that access is completely open. There are no controls and no censorship, except within local institutions. Given the modest technology required, anybody can post anything on the web. Thus information on the web is very different from information in a peer-reviewed scientific journal, such as those we cite in chapter references. It may be good, poor, bad or even maliciously misleading. However, geneticists are exceedingly well served by many high quality databases, including those listed in *Box 1*. Gaining access to these should be a top priority for any serious student of human molecular genetics.

The key to the success of the Web is a text format known as hypertext. Hypertext documents look like ordinary text, and can be read, searched, edited and stored – for example almost all the illustrations in this book can be viewed and downloaded from the publishers' web site, http://www.bios.co.uk. But hypertext documents have one extra feature (*Figure 1*). Certain words or boxes in the

 **National Center for Biotechnology Information**

OMIM Home Page --
Online Mendelian Inheritance in Man

Welcome to OMIM(TM), Online Mendelian Inheritance in Man. This database is a catalog of human genes and genetic disorders authored and edited by Dr. Victor A. McKusick and his colleagues at Johns Hopkins and elsewhere, and developed for the World Wide Web by NCBI, the National Center for Biotechnology Information. The database contains textual information, pictures, and reference information. It also contains copious links to NCBI's Entrez database of MEDLINE articles and sequence information.

NEW The OMIM Morbid Map, a catalog of cytogenetic map locations organized by disease, is now available.

Browsing OMIM

- Search the OMIM Database
- Search the OMIM Gene Map
- Search the OMIM Morbid Map
- The OMIM numbering system
- How to create WWW links to OMIM
- View the OMIM Update Log
- OMIM Statistics
- Citing OMIM in the literature
- The OMIM Gene List

OMIM Allied Resources

- Entrez: the NCBI MEDLINE and GenBank retrieval system
- Human Gene Nomenclature Home Page
- The Davis Human/Mouse Homology Map
- Online Mendelian Inheritance in Animals (OMIA)
- The Alliance of Genetic Support Groups
- The Cardiff Human Gene Mutation Database (HGMD)
- The Jackson Laboratory: Courses, mouse resources, mouse databases
- RetNet: Genes causing Retinal Diseases
- MitoMap: the Emory University mitochondrial genome database
- HUM-MOLGEN: Courses, resources, databases, etc.
- Locus-specific mutation databases

OMIM is a trademark of the Johns Hopkins University.

For questions about the OMIM database in general,
Comments and questions? Send mail to the NCBI Help Desk

Credits : Brandon Brylawski

Figure 1: A typical hypertext document.

This is the home page of OMIM at the US National Center for Biotechnology Information (NCBI). Clicking on any of the blue items takes you to another web page. Some are within the NCBI computer system (e.g. 'Search the OMIM Database'); some are on computers elsewhere in the USA (e.g. 'MitoMap'); some take you to another continent (e.g. 'The Cardiff Human Gene Mutation Database'). The user does not need to know anything about the computer he or she is accessing through these links – clicking on the link opens up the page.

Box 1

Useful Internet starting points for human molecular genetics

Addresses often change as institutions upgrade their computers, individuals move etc. The addresses given here are ones that seem likely to remain stable for a few years, that have links to a good variety of useful sites and a likely commitment to keeping the links updated. Use them as starting points for building up your own list of bookmarked addresses.

The National Center for Biotechnology Information:
http://www.ncbi.nlm.nih.gov/
Has links to many genetic resources including GenBank, OMIM, Medline etc.

UK Human Genome Mapping Project Resource Centre:
http://www.hgmp.mrc.ac.uk/
Select 'Genome web sites' for many useful links.

US Human Genome Mapping Project:
http://www.ornl.gov/hgmis/
Another well-maintained site with lots of useful information and links.

Genome Database: http://www.gdb.org/
GDB was threatened with closure in 1998, but its future seems assured through a grant to the Bioinformatics Centre at the Hospital for Sick Children, Toronto.

Medline: http://www.ncbi.nlm.nih.gov/PubMed/

OMIM: USA: http://www.ncbi.nlm.nih.gov/Omim/
UK: http://www.hgmp.mrc.ac.uk/omim/

Search engine: http://www.yahoo.com
One of several indexing programs that you can use to find pages on a specified topic. See Lawrence and Giles (1999) for discussion of how well search engines perform.

document are highlighted in color. These are hypertext links. Clicking with a mouse on a link causes the browser program automatically to access another document. This might be on the same computer or on a computer thousands of miles away. This document in turn will have its own hyperlinks. Thus, users can follow a web of cross-references around the world, without having to know anything about the nature or location of the computers they are accessing.

Each web document has a unique identifier called a uniform resource locator (URL). URLs take the form http://filename (because the standard language used to communicate within the web is called the hypertext transmission protocol or http). For example the URL of the WWW server at the European Bioinformatics Institute is http://www.ebi.ac.uk/. Typing this into your web browser will call up the Home Page (the default document seen when connecting to a web server for the first time) seen in *Figure 2*. Each of the boxes is a hyperlink to another Web page. URL addresses can be complicated, but there is no need to memorize long lists of them. Your browser can store and recall ('bookmark') the URL of any

interesting page you find while wandering through the maze of hyperlinks.

Although in principle the web abolishes geography – you need never know whether the data you are using is held on a computer in Cambridge, UK, Cambridge, Massachussetts or Tokyo – in practice transmission bottlenecks over intercontinental lines can make some accesses infuriatingly slow. This problem can only get worse as the explosive growth of the Internet continues. It makes sense to access information locally wherever possible.

The World Wide Web gives access to most human genetic data

Primary genetic data is stored in sequence and mapping databases that are accessed through the Internet. Information published on paper is best located through Medline, either over the Internet or through local information services. For human genetics, a particularly useful database is OMIM, the database of human mendelian characters.

Comprehensive DNA and protein sequence databases cover all organisms

GenBank contains the totality of public DNA and protein sequence data. This vast and rapidly growing archive is maintained in identical copies (mirrors) at sites in the USA, UK and Japan. Divisions of GenBank include complete genome sequences of a growing list of organisms, sequences of individual genes, and of many anonymous clones. See Chapter 13 for more details. One use of the database is simply to retrieve the DNA or protein sequence of a known gene or gene product; but the main use is for researchers to check if a new sequence matches anything already in the database, and for analyzing sequences into families and evolutionary trees. This involves the researcher running programs on the central computer to search and analyze the database. Such accesses are normally free of charge to academic researchers, but require registration to obtain a user name and password.

OMIM is the standard database of human mendelian characters

OMIM, *Online Mendelian Inheritance in Man*, is a catalogue of the 6000 known human mendelian characters, created by Victor McKusick. It is updated continuously and is accessible at websites listed in *Box 1*. OMIM is the essential starting point for acquiring information on human mendelian characters, both pathological and

EMBL Outstation
European Bioinformatics Institute

The EMBL - European Bioinformatics Institute (EBI) is a centre for research and services in bioinformatics. The Institute manages databases of biological data including nucleic acid, protein sequences and macromolecular structures.

Information	The EBI main information pages about staff, site information, travel, jobs, conferences and seminars.
Databases	Access to SRS and the EMBL, TrEMBL, SWISS-PROT, RHdb, MSD databases, etc.
Tools	Fast sequence database searches and interactive analysis tools. e.g. Fasta3, WU-Blast2, ClustalW, etc.
Submissions	Information about how to submit data to databases maintained at the EBI, including EMBL using WEBIN.
Groups	Index of Groups and Team Leaders at the EBI and their projects, e.g. DALI, EDGP, CORBA, etc.
Publications	Lists of publications from the EBI.
FTP	The EBI databases and software public ftp archives.
Search EBI	The searchable collection of EBI's web pages; can be used to look up a person or a certain topic.

EMBL-European Bioinformatics Institute
Wellcome Trust Genome Campus, Hinxton,
Cambridge CB10 1SD, UK.

Figure 2: The home page of the World Wide Web server of the European Bioinformatics Institute.

This is a typical (good) home page, summarizing the content of the website and providing rapid links to every section, as well as to external sites.

non-pathological. Each character is given a 6–digit MIM number which is widely used to identify it in genetic literature – thus achondroplasia is MIM 100800 and Huntington disease is MIM 143100.

The first digit normally indicates the mode of inheritance:

1. autosomal dominant loci or phenotypes
2. autosomal recessive loci or phenotypes
3. X-linked loci or phenotypes
4. Y-linked loci or phenotypes
5. mitochondrial loci or phenotypes
6. autosomal loci or phenotypes added after 1994.

The database can be searched by OMIM number, name, author or any text item, and there are hyperlinks to mapping and sequence databases. OMIM entries usually include a discussion of the genetics of the condition and a list of leading references, but not much clinical description. The discussions accumulate over time and only some have been rewritten in the light of new knowledge, so that the earlier parts of an entry were often written without knowledge of the later parts. McKusick's catalogue is also available as a printed book updated every 2 years (McKusick, 1997), and as a CD-ROM.

We expect readers of this book to use OMIM or the printed equivalent, and in general we have not given other references for descriptions of diseases and the mutations causing them. When we list an OMIM number next to the name of a disease in the text, this means that the details referred to are available (or referenced) in OMIM. An added advantage, of course, is that OMIM will continue to be updated after this book is printed, and by using it you get the most up-to-date references.

Medline is the main way to locate published papers on a topic

Every paper published in thousands of biomedical journals is abstracted and indexed in the Medline database. This can be searched by author, title, keyword, etc., to produce a list of papers with a brief abstract of each. Many libraries subscribe to regular printed and CD-ROM issues of Medline, but it can also be accessed for free at various Internet sites (see *Box 1*), which also provide a useful range of hyperlinks. The PubMed version at NCBI is especially well designed.

Searching for a web page

One problem with listing addresses in a printed book is that they change quite often. *Box 1* lists one or two of the most useful current entry points into the Web; but if you find these addresses do not work, you may need to do a search to find a starting point. Public 'search engines' such as Yahoo (see *Box 1*) maintain searchable indexes of web pages. Your browser may have a built-in link to a search engine. Using this you can search for web documents on specific subjects. Once you have a starting point you can fan out through the hyperlinks and build up your own preferred set of bookmarked sites. On subsequent occasions you can access these sites from a list of bookmarks that your browser will maintain, without needing to know the address.

Other Internet applications allow private or semi-public exchange of data and ideas.

- The main form of private communication on the Internet is email. If you know somebody's email address you can send a message that can (in theory) be read only by that person. The email address directs the message to the recipient's local computer, where access to it requires a correct user name and password. Any computer file, including text, data, images, etc., can be sent by email. Email has many advantages over conventional mail, telephone or fax, but has the limitation that there are no public directories containing all email addresses.
- In moderated discussion groups, registered members receive and submit data, opinions, comments, etc., by email. A moderator controls what can be posted to the group.
- Bulletin boards are unmoderated discussion groups, open to anybody. They exist on every conceivable topic, including many relevant to this book. Many include a useful FAQ (frequently asked questions) section.

> Repeat message: all students of human molecular genetics should be using the Internet!

Reference

Lawrence S, Giles CL (1999) Accessibility of information in the web. *Nature*, **400**, 107–109.
McKusick VA (1997) *Mendelian Inheritance in Man*, 12th edn. Johns Hopkins University Press, Baltimore.

DNA structure and gene expression

1.1 Building blocks and chemical bonds in DNA, RNA and polypeptides

Molecular genetics is primarily concerned with the interrelationship between the information macromolecules DNA (deoxyribonucleic acid) and RNA (ribonucleic acid) and how these molecules are used to synthesize polypeptides, the basic component of all proteins. In some viruses RNA is the hereditary material, but in all cells genetic information is stored in DNA molecules. Selected regions of the cellular DNA molecules serve as templates for synthesizing RNA molecules. The great majority of the RNA molecules are in turn used to specify the synthesis of polypeptides, either directly or by assisting at different stages in gene expression. As the vast majority of gene expression is dedicated to polypeptide synthesis, proteins are the major functional endpoints of the DNA template and account for the majority of the dry weight of a cell. The term protein was derived from the Greek *proteios*, meaning 'of the first rank' and reflects the important roles of proteins in diverse cellular functions, as enzymes, receptors, storage proteins, transport proteins, transcription factors, signaling molecules, hormones, etc.

1.1.1 DNA, RNA and polypeptides are large polymers defined by a linear sequence of simple repeating units

In eukaryotes individual DNA molecules are found in the chromosomes of the nucleus and in mitochondria, and also in the chloroplasts of plant cells. They are large polymers, with a linear backbone of alternating sugar and phosphate residues. The sugar in DNA molecules is deoxyribose, a 5 carbon sugar, and successive sugar residues are linked by covalent phosphodiester bonds. Covalently attached to carbon atom number 1' (one *prime*) of each sugar residue is a nitrogenous base. Four types of base are found: adenine (A), cytosine (C), guanine

(G) and thymine (T) and they consist of heterocyclic rings of carbon and nitrogen atoms. They can be divided into two classes: purines (A and G) have two joined heterocyclic rings; pyrimidines (C and T) have a single such ring. A sugar with an attached base is called a nucleoside. A nucleoside with a phosphate group attached at carbon atom 5' or 3' constitutes a nucleotide, the basic repeat unit of a DNA strand (*Figures 1.1* and *1.2*). The composition of RNA molecules is similar to that of DNA molecules, but differs in that they contain ribose sugar residues in place of deoxyribose and uracil (U) instead of thymine (*Figures 1.1* and *1.2*).

Proteins are composed of one or more polypeptide molecules which may be modified by the addition of various carbohydrate side chains or other chemical groups. Like DNA and RNA, polypeptide molecules are polymers consisting of a linear sequence of repeating units, in this case **amino acids**. The latter consist of a positively charged amino group and a negatively charged carboxylic acid (carboxyl) group connected by a central carbon atom to which is attached an identifying side chain. The 20 different amino acids can be grouped into different classes depending on the nature of their side chains (*Figure 1.3*). Classification is based as follows:

- **Basic** amino acids carry a side chain with a net positive charge; an amino (NH_2) group or histidine ring in the side chain acquires a H^+ ion at physiological pH.
- **Acidic** amino acids carry a side chain with a net negative charge; a carboxyl group (COOH) in the side chain loses a H^+ ion at physiological pH to form COO^-.
- **Uncharged polar** amino acids are electrically neutral but carry side chains with polar chemical groups, which are distinguished by having fractional electronic charges [e.g. hydroxyl ($O^{\delta-}-H^{\delta+}$) groups, sulfhydryl ($S^{\delta-}-H^{\delta+}$) groups, etc.]. Like the basic and acidic amino acids, polar amino acids may react with other groups bearing electric charges.
- **Nonpolar neutral** amino acids are hydrophobic

(water-repelling). They often interact with one another and with other hydrophobic groups.

Polypeptides are formed by a condensation reaction between the amino group of one amino acid and the carboxyl group of the next to form a repeating backbone (–NH–CHR–CO–), where the R side chains differ from one amino acid to another (see *Figure 1.21*).

1.1.2 Covalent bonds confer stability on DNA, RNA and polypeptide molecules, but weaker noncovalent bonds are critically important in permitting intermolecular associations and in stabilizing structure

The stability of the nucleic acid and protein polymers is primarily dependent on the strong covalent bonds that connect the constituent atoms of their linear backbones. In addition to covalent bonds, a number of weak noncovalent bonds (see *Table 1.1*) are important in interactions between these molecules and between groups within a single nucleic acid or protein molecule. Typically, such noncovalent bonds are weaker than covalent bonds by a factor of more than 10. Unlike covalent bonds whose strength is determined only by the particular atoms involved, however, the strength of noncovalent bonds is also crucially dependent on their aqueous environment. The structure of water is particularly complex, with a rapidly changing network of noncovalent bonding occur-

ring between the individual H_2O molecules. The predominant force in this structure is the hydrogen bond, a weak electrostatic bond formed between a partially positive hydrogen atom and a partially negative atom which, in the case of water molecules, is an oxygen atom.

Charged molecules are highly soluble in water. Because of the phosphate charges present in their component nucleotides, both DNA and RNA are negatively charged (*polyanions*). Depending on their amino acid composition, proteins may carry a net positive charge (basic proteins) or a net negative charge (acidic proteins). The hydrogen bonding potential of water molecules means that molecules with polar groups (including DNA, RNA and proteins) can form multiple interactions with the water molecules, leading to their solubilization. Thus, even electrically neutral proteins are often readily soluble if they contain an appreciable number of charged or neutral polar amino acids. In contrast, membrane-bound proteins are often characterized by a high content of hydrophobic amino acids which are thermodynamically more stable in the hydrophobic environment of a lipid membrane.

Unlike covalent bonds which require considerable energy input to break them, noncovalent bonds are constantly being made and broken at physiological temperatures. As a result, they readily permit reversible and so *transient* molecular interactions, which are essential for biological function. In the case of nucleic acids and proteins, they play a whole variety of key roles, ensuring faithful replication of DNA, transcription of RNA, codon–anticodon recognition, etc. Although individually weak,

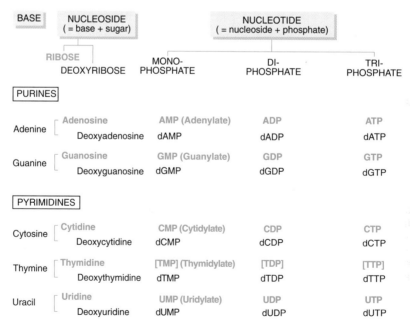

Figure 1.1: Common bases found in nucleic acids with corresponding nucleosides and nucleotides.

Ribonucleosides and ribonucleotides are shown in blue. Brackets denote nucleotides that are not normally found. Note that nucleotides with a single monophosphate are often described with names where the suffix -ine of the base is replaced by the suffix -ylate, as in the examples given for ribonucleotides.

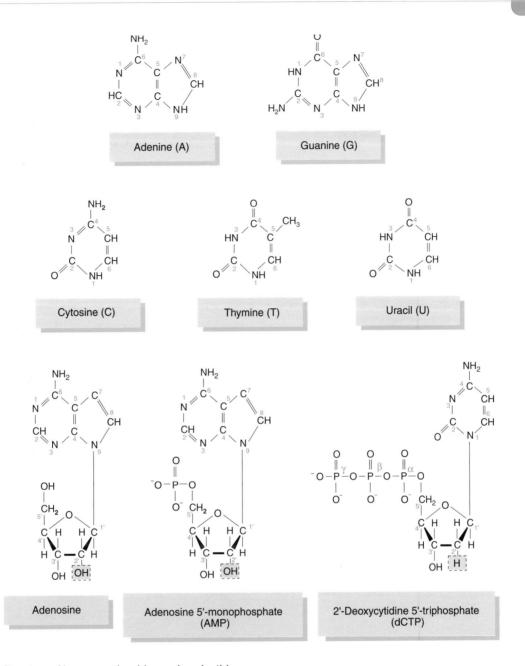

Figure 1.2: Structure of bases, nucleosides and nucleotides.

The bold lines at the bottom of the sugar rings are meant to indicate that the plane of the ring is set at an angle of 90° with respect to the plane of the corresponding base [i.e. if the plane of a base is represented as lying on the surface of the page, carbon atoms 2' and 3' of the sugar can be viewed as projecting upwards out of the page and the oxygen atom as projecting down below the surface of the page]. *Note* that numbering in deoxyribose and ribose sugars is confined to the five carbon atoms, numbered 1'–5', but the numbering of the bases includes both carbon and nitrogen atoms which occur within the heterocyclic rings. The highlighted hydroxyl and hydrogen atoms connected to carbon 2' indicate the essential difference between the ribose and deoxyribose sugar residues. Phosphate groups are denoted sequentially as α, β, γ, etc., according to proximity to the sugar ring.

the combined action of numerous noncovalent bonds can make large contributions to the stability of the structure (conformation) of these molecules and so can be crucially important for specifying the shape of a macro-molecule (see *Figure 1.7B* for the example of how intramolecular hydrogen bonding provides much of the shape of a transfer RNA molecule).

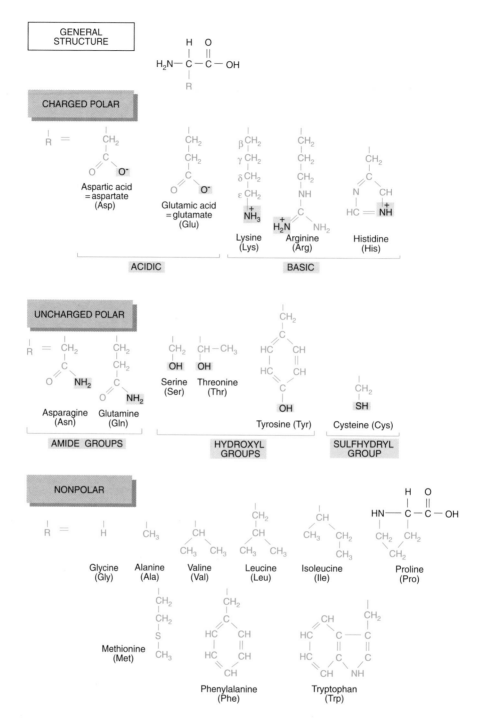

Figure 1.3: Structures of the amino acids.

Amino acids in a subclass (e.g. the acidic amino acids or the basic amino acids) are chemically very similar. Highlighted groups are polar chemical groups. The convention of numbering carbon atoms is to designate the central carbon atom as α and subsequent carbons of linear side chains as β, γ, δ, etc. (see figure for lysine side chain). Although, in general, polar amino acids are hydrophilic and nonpolar amino acids are hydrophobic, glycine (which has a very small side chain) and cysteine (whose sulfhydryl group is not as polar as a hydroxyl group) occupy intermediate positions on the hydrophilic–hydrophobic scale. *Note* that proline is unusual in that the side chain connects the nitrogen atom of the NH_2 group as well as the central carbon atom.

Table 1.1: Weak noncovalent bonding

Type of bond	Nature of bond
Hydrogen	Hydrogen bonds form when a *hydrogen atom* is sandwiched between two electron-attracting atoms, usually oxygen or nitrogen. See *Box 1.1* for examples of their importance in nucleic acid and protein structure and function.
Ionic	Ionic interactions occur between *charged groups*. They can be very strong in crystals but in an aqueous environment, the charged groups are shielded by H_2O molecules and other ions in solution and so are quite weak. Nevertheless, they can be very important in biological function, as in the case of enzyme–substrate recognition.
Van der Waals	*Any* two atoms which are very close to each other show a weak attractive bonding interaction due to their fluctuating electrical charges (van der Waals attraction) until they get extremely close, when they repel each other very strongly (van der Waals repulsion). Although individually very weak, van der Waals attractions can become important when there is a very good fit between the surfaces of two macromolecules.
Hydrophobic forces	Water is a polar molecule. When hydrophobic molecules or chemical groups are placed in an aqueous environ ment they are forced together to minimize their disruptive effects on the complex network of hydrogen bonding between water molecules. Hydrophobic groups which are forced together in this way are said to be held together by *hydrophobic bonds*, even though the basis of their attraction is due to a common repulsion by water molecules.

1.2　DNA structure and replication

1.2.1　*The structure of DNA is an antiparallel double helix*

The linear backbone of a DNA molecule and of an RNA molecule consists of alternating sugar residues and phosphate groups. In each case, the bond linking an individual sugar residue to the neighbouring sugar residues is a 3′, 5′-phosphodiester bond. This means that a phosphate group links carbon atom 3′ of a sugar to carbon atom 5′ of the neighbouring sugar (*Figure 1.4*).

Whereas the RNA molecules within a cell normally exist as single molecules, the structure of DNA is a double helix in which two DNA molecules (DNA strands) are held together by weak hydrogen bonds to form a DNA duplex. Hydrogen bonding occurs between laterally opposed bases, base pairs, of the two strands of the DNA duplex according to Watson–Crick rules: adenine (A) specifically binds to thymine (T) and cytosine (C) specifically binds to guanine (G) (*Figure 1.5*). As a result, the base composition of DNA from different cellular sources is not random: the amount of adenine equals that of thymine, and the amount of cytosine equals that of guanine. The base composition of DNA can therefore be

Box 1.1

Examples of the importance of hydrogen bonding in nucleic acids and proteins

Intermolecular hydrogen bonding in nucleic acids. This is important in permitting the formation of double-stranded nucleic acids as follows:

- *Double-stranded DNA.* The stability of the double helix is maintained by hydrogen bonding between A–T and C–G base pairs (Section 1.2.1 and *Figure 1.5*).
- *DNA–RNA duplexes.* These form naturally during RNA transcription and hydrogen bonding underpins the following types of base pairs: A–U; C–G and also A–T (bonding of A in the RNA strand with T in the DNA strand; Section 1.3.3 and *Figure 1.12*).
- *Double-stranded RNA.* The genomes of some viruses consist of RNA–RNA duplexes, but in addition transient RNA–RNA duplexes form in all cells during RNA processing and gene expression. RNA splicing requires recognition of exon–intron boundaries following hydrogen bonding between the unspliced RNA transcripts and various small nuclear RNA molecules

(Section 1.4.1). In addition, codon–anticodon recognition involves hydrogen bonding between two RNA molecules: mRNA and tRNA (Section 1.5.1 and *Figure 1.20*). Hydrogen bonding between RNA molecules involves not just A–U and C–G base-pairing but also G–U base pairs (*Table 1.5*).

Intramolecular hydrogen bonding in nucleic acids. This is important in providing secondary structure in both DNA and RNA molecules, as in the case of formation of hairpins in DNA and the complex arms of tRNA (Section 1.2.1 and *Figure 1.7*). In the latter case, note that base-pairing can include not just A–U and G–C but also G–U base pairs.

Intramolecular hydrogen bonding in proteins. Some fundamental units of protein secondary structure, such as the α-helix, the β-*pleated sheet* and the β-*turn* are largely defined by intrachain hydrogen bonding (Section 1.5.5 and *Figure 1.24*).

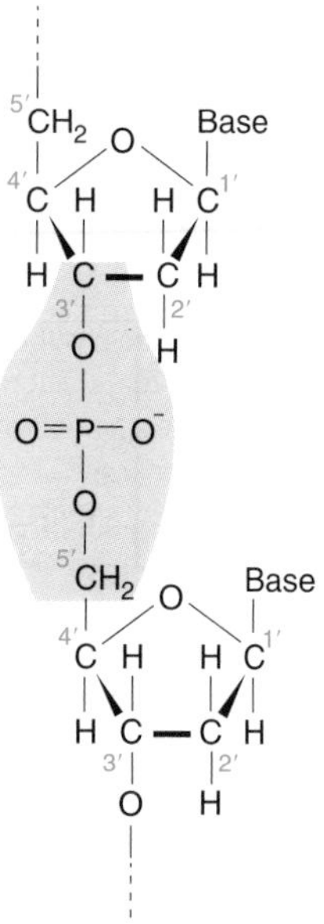

eukaryotic genome is of the B-DNA form in which each helical strand has a *pitch* (the distance occupied by a single turn of the helix) of 3.4 nm. As the phosphodiester bonds link carbon atoms number 3′ and number 5′ of successive sugar residues, one end of each DNA strand, the so-called 5′ end, will have a terminal sugar residue in which carbon atom number 5′ is not linked to a neighboring sugar residue (*Figure 1.6*). The other end is defined as the 3′ end because of a similar absence of phosphodiester bonding at carbon atom number 3′ of the terminal sugar residue. The two strands of a DNA duplex are said to be antiparallel because they always associate (**anneal**) in such a way that the 5′ → 3′ direction of one DNA strand is the opposite to that of its partner (*Figure 1.6*).

Genetic information is encoded by the linear sequence of bases in the DNA strands (the primary structure). Consequently, two DNA strands of a DNA duplex are said to have complementary sequences (or to exhibit base complementarity) and the sequence of bases of one DNA strand can readily be inferred if the DNA sequence of its complementary strand is already known. It is usual, therefore, to describe a DNA sequence by writing the sequence of bases of one strand only, and in the 5′ → 3′ direction. This is the direction of synthesis of new DNA molecules during DNA replication, and also of the nascent RNA strand produced during transcription (see below). However, when describing the sequence of a DNA region encompassing two neighboring bases (really a dinucleotide) on one DNA strand, it is usual to insert a 'p' to denote a connecting phosphodiester bond [e.g. CpG means that a cytidine is covalently linked to a neighboring guanosine *on the same DNA strand*, while a CG base pair means a cytosine on one DNA strand is hydrogen-bonded to a guanine *on the complementary strand* (*Figure 1.6*)].

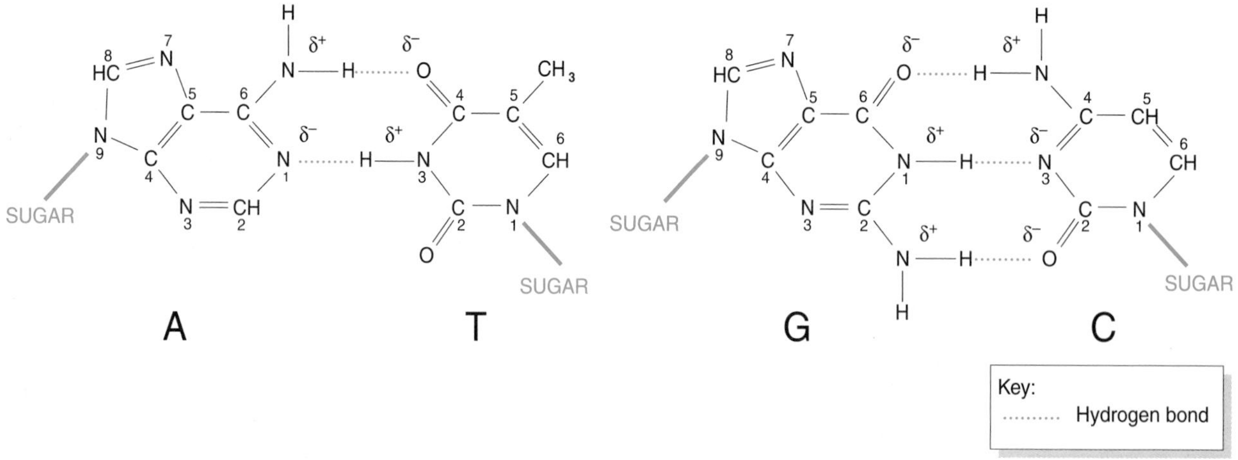

Figure 1.4: A 3′-5′ phosphodiester bond.

specified unambiguously by quoting its %GC (= %G + %C) composition. For example, if a source of DNA is quoted as being 42%GC, the base composition can be inferred to be: G, 21%; C, 21%; A, 29%; T, 29%.

DNA can adopt different types of helical structure. A-DNA and B-DNA are both right-handed helices (ones in which the helix spirals in a clockwise direction as it moves away from the observer). They have respectively 11 and 10 base pairs per turn. Z-DNA is a left-handed helix which has 12 base pairs per turn. Under physiological conditions, most of the DNA in a bacterial or

Figure 1.5: A–T base pairs have two connecting hydrogen bonds; G–C base pairs have three.

Fractional positive charges on hydrogen atoms and fractional negative charges on oxygen and nitrogen atoms are denoted by δ^+ and δ^-, respectively.

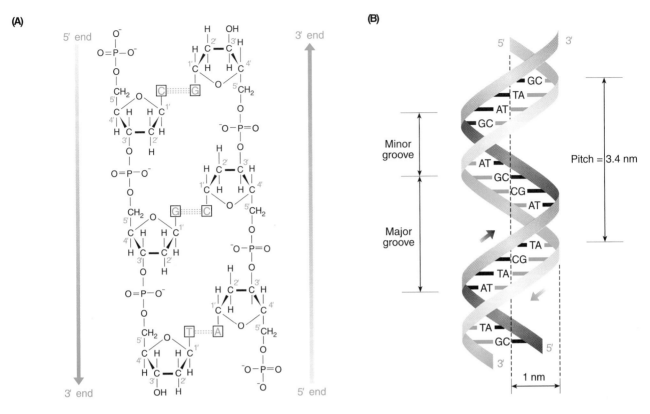

Figure 1.6: The structure of DNA is a double-stranded, antiparallel helix.

(**A**) Antiparallel nature of the two DNA strands. The two strands are antiparallel because they have opposite directions for linking of 3′ carbon atom to 5′ carbon atom. The structure shown is a double-stranded trinucleotide whose sequence can be represented as: 5′ pCpGpT–OH 3′ (DNA strand on left) / 5′ pApCpG–OH 3′ (DNA strand on right) (where p = phosphodiester bond and –OH = terminal OH group at 3′ end). This is normally abbreviated by deleting the 'p' *and* 'OH' symbols and giving the sequence on one strand only (e.g. the sequence could equally well be represented as 5′ CGT 3′ or 5′ ACG 3′). (**B**) The double helical structure of DNA. *Note* that the two strands are wound round each other to form a *plectonemic* coil. The *pitch* of each helix represents the distance occupied by a single turn and accommodates 10 nucleotides in B-DNA.

Intermolecular hydrogen bonding also permits RNA–DNA duplexes and double-stranded RNA formation which are important requirements for gene expression (see *Box 1.1*). In addition, hydrogen bonding can occur between bases within a single DNA or RNA molecule. Sequences having closely positioned complementary inverted repeats are prone to forming hairpin structures or loops which are stabilized by hydrogen bonding between bases at the neck of the loop (*Figure 1.7A*). Such structural constraints, which are additional to those imposed by the primary structure, contribute to the **secondary structure** of the molecule. Certain RNA molecules, such as transfer RNA (tRNA), show particularly high degrees of secondary structure (*Figure 1.7B*).

Note that in the case of hydrogen bonding in RNA–RNA duplexes and also in intramolecular hydrogen bonding within an RNA molecule, GU base pairs are occasionally found (*Figure 1.7B*). This form of base-pairing is not particularly stable, but does not significantly disrupt the RNA–RNA helix.

1.2.2 DNA replication is semiconservative and synthesis of DNA strands is semidiscontinuous

During the process of DNA synthesis (DNA replication), the two DNA strands of each chromosome are unwound by a helicase enzyme and each DNA strand directs the synthesis of a complementary DNA strand to generate two daughter DNA duplexes, each of which is identical to the parent molecule (*Figure 1.8*). As each daughter DNA duplex contains one strand from the parent molecule and one newly synthesized DNA strand, the replication process is described as **semiconservative**. The enzyme DNA polymerase catalyzes the synthesis of new DNA strands using the four deoxynucleoside triphosphates (dATP, dCTP, dGTP, dTTP) as nucleotide precursors.

DNA replication is initiated at specific points, which have been termed origins of replication. Starting from such an origin, the initiation of DNA replication results in a **replication fork**, where the parental DNA duplex bifurcates into two daughter DNA duplexes. As the two

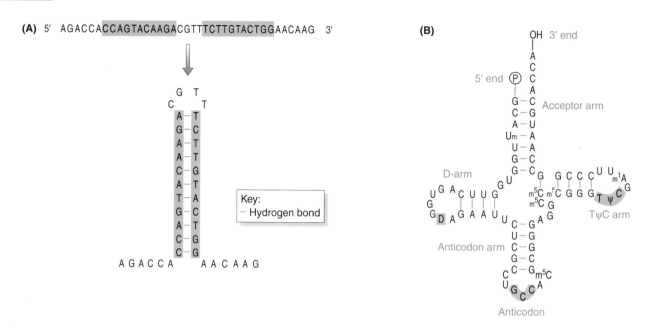

(A) 5′ AGACCACCAGTACAAGACGTTTCTTGTACTGGAACAAG 3′

(B)

Figure 1.7: Intramolecular hydrogen bonding in DNA and RNA.

(A) Formation of a double-stranded hairpin loop within a single DNA strand. The highlighted sequences in the DNA strand at the top represent inverted repeat sequences which can hydrogen-bond to form the hairpin structure below. **(B)** Transfer RNA (tRNA) has extensive secondary structure. The example shown is a human tRNA^{Glu} gene. *Note* the minor nucleosides: D,5,6-dihydrouridine; ψ, pseudouridine (5-ribosyl uracil); m^{5}C, 5-methylcytidine; m^{1}A, 1-methyladenosine. The cloverleaf structure is stabilized by extensive intramolecular hydrogen bonding, with conventional G–C and A–U base pairs, but also the occasional G–U base pair. Four arms are recognized: the acceptor arm is the one to which an amino acid can be attached (at the 3′ end); the TψC arm is defined by this trinucleotide; the D arm is named because it contains dihydrouridine residues; and the anticodon arm contains the anticodon trinucleotide in the center of the loop. The secondary structure of tRNAs is virtually invariant: there are always seven base pairs in the stem of the acceptor arm, five in the TψC arm, five in the anticodon arm, and three or four in the D arm.

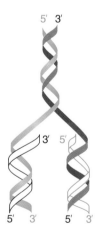

Figure 1.8: DNA replication is semiconservative.

The parental DNA duplex consists of two complementary, antiparallel DNA strands (solid colors), which unwind and then individually act as templates for the synthesis of new complementary, antiparallel DNA strands (outline colors). Each of the daughter DNA duplexes contains one parental DNA strand and one new DNA strand, forming a DNA duplex which is structurally identical to the parental DNA duplex. *Note* that this figure shows the result of DNA duplication but not the way the process works (for which, see *Figure 1.9*).

strands of the parental DNA duplex are antiparallel, but act individually as templates for the synthesis of a complementary antiparallel daughter strand, it follows that the two daughter strands must run in opposite directions [i.e. the direction of chain growth must be 5′ → 3′ for one daughter strand, the **leading strand**, but 3′ → 5′ for the other daughter strand, the **lagging strand** (*Figure 1.9*)].

The reactions catalyzed by DNA polymerases involve addition, to the free 3′ hydroxyl group of the growing DNA chain, of a dNMP moiety provided by a dNTP precursor [the two distal phosphate residues of the dNTP – that is, the β and γ residues (*Figure 1.2*) – are cleaved and the resulting pyrophosphate group (PP$_i$) discarded]. This requirement introduces an asymmetry into the DNA replication process: only the leading strand will have a free 3′ hydroxyl group at the point of bifurcation. This will permit sequential addition of nucleotides and continuous elongation in the same direction in which the replication fork moves. However, synthesis of the lagging strand has to be accomplished as a progressive series of small (typically 100–1000 nucleotides long) fragments, often referred to as Okazaki fragments. As only the leading strand is synthesized continuously, the synthesis of DNA strands is said to be semidiscontinuous. Each frag-

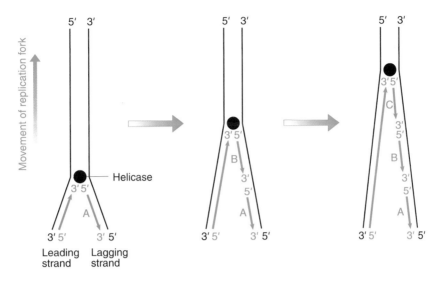

Figure 1.9: Asymmetry of strand synthesis during DNA replication.

Parental strands are shown in black, newly synthesized strands in blue. The helicase unwinds the DNA duplex to allow the individual strands to be replicated. As replication proceeds, the 5′ → 3′ direction of synthesis of the **leading strand** is the same as that in which the replication fork is moving, and so synthesis can be continuous. The 5′ → 3′ synthesis of the **lagging strand**, however, is in the direction opposite to that of movement of the replication fork. It needs to be synthesized in pieces (Okazaki fragments), first A, then B, then C which are subsequently sealed by the enzyme DNA ligase to form a continuous DNA strand. *Note* that synthesis of these fragments is initiated using an RNA primer. In eukaryotic cells, the leading and lagging strands are synthesized, respectively, by DNA polymerases δ and α (see *Table 1.2*).

ment of the lagging strand is synthesized in the 5′ → 3′ direction, which will be in the opposite direction to that in which the replication fork moves. Successively synthesized fragments are covalently joined at their ends using the enzyme DNA ligase so as to ensure chain growth in the direction of movement of the replication fork (*Figure 1.9*).

Five classes of mammalian DNA polymerases are known, including a polymerase that is dedicated to replication of the mitochondrial genome (*Table 1.2*). In individual mammalian chromosomes, DNA replication proceeds bidirectionally, to form replication bubbles from multiple initiation points. The distance between adjacent replication origins is about 50–300 kb, a distance which may be significant in chromosome structure (see Section 2.3.1). At different origins, DNA replication is initiated at different times in the S phase of the cell cycle (see Chapter 2) but eventually neighboring replication bubbles will

fuse (*Figure 1.10*). DNA replication is time-consuming: human cells in culture require about 8 hours to complete the process.

1.3 RNA transcription and gene expression

1.3.1 The flow of genetic information is almost exclusively one way: DNA → RNA → protein

The expression of genetic information in all cells is very largely a one-way system: DNA specifies the synthesis of RNA and RNA specifies the synthesis of polypeptides, which subsequently form proteins. Because of its universality, the DNA → RNA → polypeptide (protein) flow of genetic information has been described as the **central**

Table 1.2: The five classes of mammalian DNA polymerase

	Class				
	α	β	γ	δ	ε
Location	Nuclear	Nuclear	Mitochondrial	Nuclear	Nuclear
Function	Synthesis and priming of lagging strand	DNA repair	Replicates mitochondrial DNA	Synthesis of leading strand	DNA repair
3′ → 5′ exonuclease?	No	No	Yes	Yes	Yes

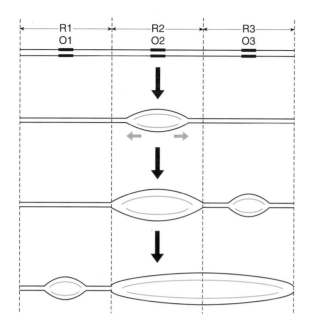

Figure 1.10: The chromosomes of complex organisms have multiple replication origins.

R1, R2 and R3 denote adjacent replication units (**replicons**) located on the same chromosome and have internally located origins of replication O1, O2 and O3, respectively. In this example, bidirectional replication is envisaged to proceed initially from O2, then O3 and finally O1. The bottom panel shows fusion of the replication bubbles initiated from O2 and O3, before the R1 replicon has completed replication.

dogma of molecular biology. The first step, the synthesis of RNA using a DNA-dependent RNA polymerase, is described as **transcription** and occurs in the nucleus of eukaryotic cells and, to a limited extent, in mitochondria and chloroplasts, the only other organelles which have a genetic capacity in addition to the nucleus (see *Figure 1.11*). The second step, polypeptide synthesis, is described as translation and occurs in ribosomes, large RNA–protein complexes which are found in the cytoplasm and also in mitochondria and chloroplasts. The RNA molecules which specify polypeptide are known as messenger RNA (mRNA). The expression of genetic information follows a colinearity principle: the linear sequence of nucleotides in DNA is decoded to give a linear sequence of nucleotides in RNA which can be decoded in turn in groups of three nucleotides (*codons*) to give a linear sequence of amino acids in the polypeptide product.

Although DNA is the hereditary material in all present-day cells, it is most likely that early in evolution RNA served that function. RNA molecules can, like DNA, undergo self-replication. However, the 2′ hydroxyl group on the ribose residues of RNA makes the sugar–phosphate bonds comparatively unstable chemically. In DNA the deoxyribose residues carry only hydrogen atoms at the 2′ position and so DNA is much more suited

than RNA to being a stable carrier of genetic information. Many different classes of present-day viruses nevertheless have a genome that consists of RNA, not DNA. Retroviruses such as HIV are a subclass of RNA viruses in which the RNA replicates via a DNA intermediate, using a **reverse transcriptase**, an RNA-dependent DNA polymerase. Recently it has become clear that eukaryotic cells, including mammalian cells, contain nonviral chromosomal DNA sequences which encode cellular reverse transcriptases (see Section 7.4.6 for examples in the human genome). Because some nonviral RNA sequences are known to act as templates for cellular DNA synthesis, the principle of unidirectional flow of genetic information is no longer strictly valid.

1.3.2 Only a small fraction of the DNA in complex organisms is expressed to give a protein or RNA product

Only a small proportion of all the DNA in cells is ever transcribed. According to their needs, different cells transcribe different segments of the DNA (**transcription units**) which are discrete units, spaced irregularly along the DNA sequence. However, the great majority of the cellular DNA is never transcribed in any cell.

Moreover, only a portion of the RNA made by transcription is translated into polypeptide. This is because:

■ *some transcription units are expressed to give an RNA molecule other than mRNA and so do not specify polypeptides directly.* Instead, the RNA products are mature molecules which can serve different functions, as in the case of ribosomal RNAs (rRNAs), tRNAs, and diverse small nuclear (sn) and cytoplasmic RNA molecules (see Section 7.2.2);

■ *the **primary transcript** (initial transcription product) of those transcription units which do encode polypeptides is subject to RNA processing events.* As a result, much of the initial RNA sequence is discarded to give a much smaller mRNA (Section 1.4.1);

■ *only a central part of the mature mRNA is translated*; sections of variable length at each end of the mRNA remain untranslated (Section 1.5.1).

In animal cells, DNA is found in both the nucleus and the mitochondria. The mitochondria have, however, only a small fraction of the total cellular DNA and a very limited number of genes (see Section 7.1.1); the vast majority of the DNA of a cell is located in the chromosomes of the nucleus.

The fraction of **coding DNA** in the genomes of complex eukaryotes is rather small. This is partly a result of the noncoding nature of much of the sequence within genes. Another reason is that a considerable fraction of the genome of complex eukaryotes contains repeated sequences which are nonfunctional or which are not transcribed into RNA. These include defective copies of functional genes (**pseudogenes** and gene fragments), and highly repetitive noncoding DNA.

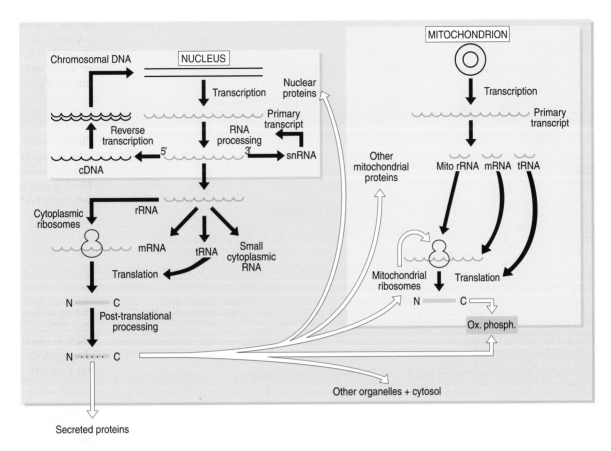

Figure 1.11: Gene expression in an animal cell.

Note that a small proportion of nuclear RNA molecules can be converted naturally to cDNA by virally encoded and cellular reverse transcriptases, and thereafter integrate into chromosomal DNA at diverse locations. *Note* also that the mitochondrion synthesizes its own rRNA and tRNA and a few proteins which are involved in the oxidative phosphorylation (Ox. phosph.) system. However, the proteins of mitochondrial ribosomes and the majority of the proteins in the mitochondrial oxidative phosphorylation system and other mitochondrial proteins are encoded by nuclear genes and translated on cytoplasmic ribosomes, before being imported into mitochondria.

1.3.3 Transcription is the process whereby genetic information in some DNA segments (genes) specifies the synthesis of RNA

RNA synthesis is accomplished using an RNA polymerase enzyme, with DNA as a template and ATP, CTP, GTP and UTP as RNA precursors. The RNA is synthesized as a single strand, with the direction of transcription being 5′ → 3′. Chain elongation occurs by adding the appropriate ribonucleoside monophosphate residue (AMP, CMP, GMP or UMP) to the free 3′ hydroxyl group at the 3′ end of the growing RNA chain. Such nucleotides are derived by splitting a pyrophosphate residue (PP$_i$) from the appropriate ribonucleoside triphosphate (rNTP) precursors. This means that the nucleotide at the extreme 5′ end *(the initiator nucleotide) will differ from all others in the chain by carrying a 5′ triphosphate group.*

Normally, only one of the two DNA strands acts as a template for RNA synthesis. During transcription, double-stranded DNA is unwound and the DNA strand which will act as a template for RNA synthesis forms a transient double-stranded RNA–DNA hybrid with the growing RNA chain. As the RNA transcript is complementary to this **template strand**, the transcript has the same 5′ → 3′ direction and base sequence (except that U replaces T) as the opposite, nontemplate strand of the double helix. For this reason the nontemplate strand is often called the **sense strand**, and the template strand is often called the **antisense strand** (*Figure 1.12*). In documenting gene sequences it is customary to show only the DNA sequence of the sense strand. Orientation of sequences relative to a gene sequence is commonly dictated by the sense strand and by the direction of transcription (e.g. the 5′ end of a gene refers to sequences at the 5′ end of the sense strand, and sequences **upstream** or **downstream** of a gene refer to sequences which flank the gene at the 5′ or 3′ ends, respectively, of the sense strand).

In eukaryotic cells, three different RNA polymerase molecules are required to synthesize the different classes of RNA (*Table 1.3*). The vast majority of cellular genes encode polypeptides and are transcribed by RNA

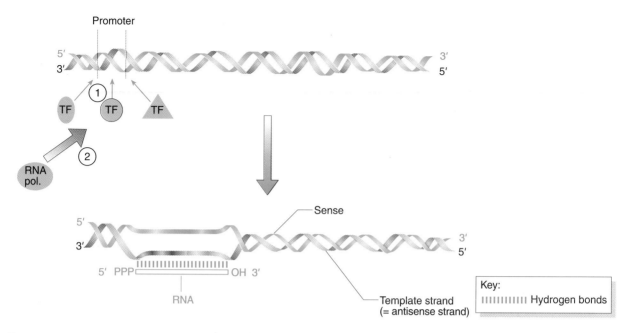

Figure 1.12: RNA is transcribed as a single strand which is complementary in base sequence to one strand (template strand) of a gene.

Various transcription factors (TF) are required to bind to a promoter sequence in the immediate vicinity of a gene ①, in order to subsequently position and guide the RNA polymerase that will transcribe the gene ②. Chain synthesis is initiated with a nucleoside triphosphate and chain elongation occurs by successive addition of nucleoside monophosphate residues provided by rNTPs to the 3′ OH. This means that the 5′ end will have a triphosphate group, which may subsequently undergo modification (e.g. by capping; see Section 1.4.2) and the 3′ end will have a free hydroxyl group. *Note* that the sequence of the RNA will normally be identical to the sense strand of the gene (except U replaces T) and complementary to the template strand.

polymerase II. Increasing importance, however, is being paid to genes which encode RNA as their mature product: functional RNA molecules are now known to play a wide variety of roles, and catalytic functions have been ascribed to some of them (see Section 7.2.2).

1.3.4 Transcription of eukaryotic genes requires interaction between cis-acting transcription elements and trans-acting transcription factors

Eukaryotic RNA polymerases cannot initiate transcription by themselves. Instead, combinations of short sequence elements in the immediate vicinity of a gene act as recognition signals for transcription factors to bind to the DNA in order to guide and activate the polymerase. A major group of such short sequence elements is often clustered upstream of the coding sequence of a gene, where they collectively constitute the **promoter**. After a number of general transcription factors bind to the promoter region, an RNA polymerase binds to the transcription factor complex and is activated to initiate the synthesis of RNA from a unique location. The transcription factors are said to be *trans-acting*, because they are synthesized by genes which are remotely located, and they migrate to their sites of

Table 1.3: The three classes of eukaryotic RNA polymerase

Class	Genes transcribed	Comments
I	28S rRNA; 18S rRNA; 5.8S rRNA	Localized in the nucleolus. A single primary transcript (45S rRNA) is cleaved to give the three rRNA classes listed
II	All genes that encode polypeptides; most snRNA genes	Polymerase II transcripts are unique in being subject to capping and polyadenylation
III	5S rRNA; tRNA genes; U6 snRNA; 7SL RNA; 7SK RNA; 7SM RNA	The promoter for some genes transcribed by RNA polymerase III (e.g. 5S rRNA, tRNA, 7SL RNA) is internal to the gene (see *Figure 1.13*) and for others (e.g. 7SK RNA) is located upstream

Table 1.4: Examples of *cis*-acting elements recognized by ubiquitous transcription factors

Cis element	DNA sequence is identical to, or a variant of	Associated *trans*-acting factors	Comments
GC box	GGGCGG	Spl	Spl factor is ubiquitous
TATA box	TATAAA	TFIID	TFIIA binds to the TFIID–TATA box complex to stabilize it
CAAT box	CCAAT	Many, e.g. C/EBP, CTF/NFI	Large family of *trans*-acting factors
CRE (cAMP response element)	GTGACGTA/CAA/G	CREB/ATF family, e.g. ATF-1	Genes activated in response to cAMP

action. In contrast, the promoter elements are ***cis-acting***; their function is limited to the DNA duplex on which they reside (*Table 1.4*).

In the case of genes which are actively transcribed by RNA polymerase II, either at a specific stage in the cell cycle (e.g. histones) or in specific cell types (e.g. β-globin), the promoter elements always include a TATA box, often TATAAA or a variant, at a position about 25 bp upstream (−25) from the transcriptional start site (*Figure 1.13*). Mutation at the TATA element does not prevent initiation of transcription, but does cause the startpoint of transcription to be displaced from the normal position.

The promoters of many other genes, including housekeeping genes, lack TATA boxes but often have a GC box, containing variants of the consensus sequence GGGCGG. Other common promoter elements include the CAAT box, often at about −80, which is usually the strongest determinant of promoter efficiency. Note, however, that the GC and CAAT boxes appear to be able to function in either orientation, although their sequences are asymmetrical (*Figure 1.13*). In addition to general upstream transcription elements which are recognized by ubiquitous transcription factors, more specific recognition elements are known which are recognized by tissue-restricted transcription factors (see *Table 8.2*).

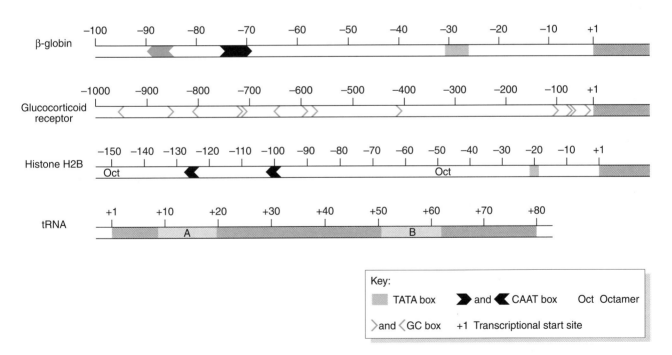

Figure 1.13: Eukaryotic promoters consist of a collection of conserved short sequence elements located at relatively constant distances from the transcription start site.

Alternative orientations for GC and CAAT box elements are indicated by chevron orientation: > = normal orientation; < = reverse orientation. The glucocorticoid receptor gene is unusual in possessing 13 upstream GC boxes (10 in the normal orientation; three in the reverse orientation). The tRNA genes are transcribed by RNA polymerase III and have an internal bipartite promoter comprising element A (usually within the nucleotides numbered +8 to +19 according to the standard tRNA nucleotide numbering system) and element B (usually between nucleotides +52 and +62). Specific transcription factors bind to these elements and then guide RNA polymerase III to start transcribing at +1.

Enhancers comprise groups of *cis*-acting short sequence elements, which can enhance the transcriptional activity of specific eukaryotic genes. However, unlike promoter elements whose positions relative to the transcriptional initiation site are relatively constant (*Figure 1.13*), enhancers are located a variable, and often considerable, distance from the transcriptional start site, and their function is independent of their orientation. They appear to bind gene regulatory proteins and, subsequently, the DNA between the promoter and enhancer loops out, allowing the proteins bound to the enhancer to interact with the transcription factors bound to the promoter, or with the RNA polymerase. **Silencers** are equivalent regulatory elements which can inhibit the transcriptional activity of specific genes. For further details, see *Box 8.2*.

1.3.5 *Tissue-specific gene expression involves selective activation of specific genes, and regions of transcriptionally active chromatin adopt an open conformation*

The DNA content of a specific type of eukaryotic cell, a myocyte for example, is virtually identical to that of a lymphocyte, hepatocyte or any other type of nucleated cell from the same organism. What makes the different cell types different is the pattern of genes which are expressed in the cell. Some cells, particularly brain cells, express a large number of different genes. In many other cell types, a large fraction of the genes is transcriptionally inactive. Clearly, the genes that are expressed are the ones which define the functions of the cell. Some of these functions are common ones which are essential for general cell functions and are specified by so-called **housekeeping genes**. The expression of other genes may be largely restricted to a specific cell type (tissue-specific gene expression). Note, however, that even in the case of genes which show considerable tissue specificity in expression, some gene transcripts occur at very low levels in all cell types (**illegitimate** or **ectopic transcription**).

The distinction between transcriptionally active and inactive regions of DNA in a cell is reflected in the structure of the associated chromatin. Transcriptionally inactive chromatin generally adopts a highly condensed conformation and is often associated with regions of the genome which undergo late replication during S phase of the cell cycle. It is associated with tight binding by the histone H1 molecule. By contrast, transcriptionally active DNA adopts a more open conformation and is often replicated early in S phase. It is marked by relatively weak binding by histone H1 molecules and extensive acetylation of the four types of nucleosomal histones, i.e. histones H2A, H2B, H3 and H4 (see Section 2.3.1). Additionally, in transcriptionally active chromatin the promoter regions of vertebrate genes are generally characterized by absence of methylated cytosines (see below). Transcription factors can displace nucleosomes, and so the open conformation of transcriptionally active **chro-**

matin can be distinguished experimentally because it also affords access to nucleases: at very low concentrations, the enzyme DNase I will digest long regions of nucleosome-free DNA. Although the regulatory regions may contain several sequence-specific binding proteins, the open chromatin structure is marked by the presence of **DNase I-hypersensitive sites** (see Section 8.5.2).

1.4   RNA processing

The RNA transcript of most eukaryotic genes undergoes a series of processing reactions. Often this involves removal of unwanted internal segments and rejoining of the remaining segments (RNA splicing). Additionally, in the case of RNA polymerase II transcripts, a specialized nucleotide linkage (7-methylguanosine triphosphate) is added to the 5′ end of the primary transcript (**capping**), and adenylate (AMP) residues are sequentially added to the 3′ end of mRNA to form a poly(A) tail (**polyadenylation**).

1.4.1 *RNA splicing ensures removal of intronic RNA sequences from the primary transcript and fusion of exonic RNA sequences*

The coding sequences of most vertebrate genes, both polypeptide-encoding genes and genes encoding RNA molecules other than mRNA, are split into segments (**exons**) which are separated by noncoding intervening sequences (**introns**). Transcription involves the production of RNA sequence complementary to the entire length of the gene, encompassing both exons and introns. Often, however, the RNA transcript undergoes RNA splicing, a series of processing reactions whereby the intronic RNA segments are snipped out and discarded and the exonic RNA segments are joined end-to-end (spliced) to give a shorter RNA product (*Figure 1.14*).

The mechanism of RNA splicing is dependent upon the identity of the nucleotide sequences at the exon/intron boundaries (splice junctions). In particular, it is critically dependent on what has been called the GT–AG rule: introns almost always start with GT (or really GU at the RNA level) and end with AG (see *Figure 1.15*).

Although the conserved GT (GU) and AG dinucleotides are crucially important for splicing, they are not by themselves sufficient to signal the presence of an intron. Comparisons of documented sequences have revealed that sequences adjacent to the GT and AG dinucleotides show a considerable degree of conservation (*Figure 1.15*). In addition, a third conserved intronic sequence that is known to be functionally important in splicing is the so-called **branch site** which is usually located very close to the end of the intron, at most 40 nucleotides before the terminal AG dinucleotide (*Figure 1.15*).

The splicing mechanism involves the following sequence:

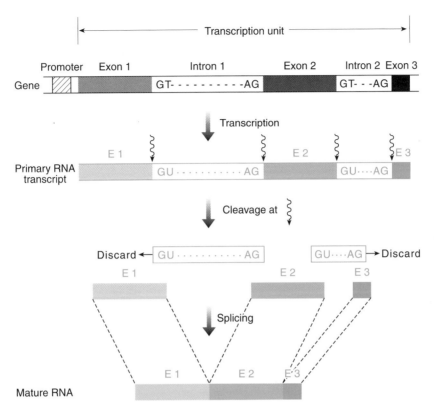

Figure 1.14: RNA splicing involves endonucleolytic cleavage and removal of intronic RNA segments and splicing of exonic RNA segments.

(i) cleavage at the 5′ splice junction;
(ii) nucleolytic attack by the terminal G nucleotide of the splice donor site at the invariant A of the branch site to form a lariat-shaped structure;
(iii) cleavage at the 3′ splice junction, leading to release of the intronic RNA as a lariat, and splicing of the exonic RNA segments (*Figure 1.16*).

The above reactions are mediated by a large RNA-protein complex, the **spliceosome**, which consists of five types of **snRNA** (small nuclear RNA) and more than 50 proteins (Stanley and Guthrie, 1998). Each of the snRNA molecules is attached to specific proteins to form snRNP particles and the specificity of the splicing reaction is established by RNA–RNA base-pairing between the RNA transcript and snRNA molecules. In the case of the vast majority of introns in genes which encode polypeptides, the five snRNA species are: U1, U2, U4, U5 and U6 snRNA. The 5′ terminus of the U1 snRNA has a sequence UACUUAC which base-pairs with the splice donor consensus (GUAAGUA). After the U1 snRNP has bound, U2 snRNA recognizes the branch site by a similar base-pairing reaction and, subsequently, interaction between U1 snRNP and U2 snRNP brings the two splice junctions close

Figure 1.15: Consensus sequences at the DNA level for the splice donor, splice acceptor and branch sites in introns of complex eukaryotes.

Highlighted nucleotides are almost invariant (but *note* that rare introns also exist where the conserved splice donor dinucleotide GT is replaced by AT and where the conserved splice acceptor dinucleotide AG is replaced by AC; see text). Other nucleotides represent the majority nucleotide found at this particular position. *Note* that in cases where pyrimidines (C/T or T/C) are preferred, no significance should be attached to which base comes first. For example, the consensus sequence of the branch site is written so as to highlight the similarity to the consensus branch site in yeast introns (TACTAAC), but the sequence given for the splice acceptor site does not signify a preference for C as opposed to T.

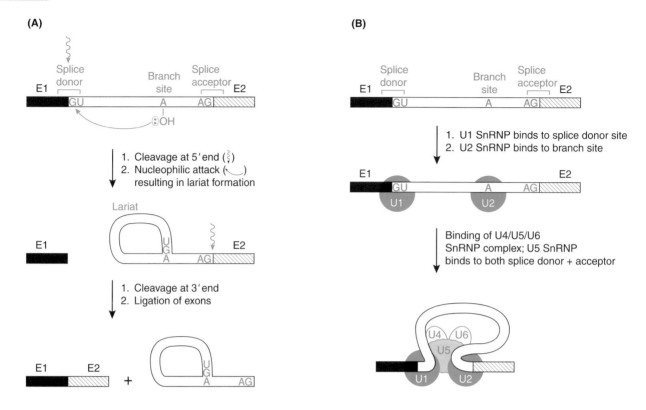

Figure 1.16: Mechanism of RNA splicing (GU–AG introns).

(**A**) Mechanism. See *Figure 1.15* for consensus sequences of splice donor, splice acceptor and branch site. The nucleophilic attack involves the 2′-hydroxyl group attached to the conserved A at the branch site and the G of the conserved GU at the start of the intron, and results in a new covalent bond linking these two nucleotides to give a branched structure (lariat). (**B**) Role of snRNPs. Small nuclear ribonucleoprotein particles (snRNPs) are part of the spliceosome. U1 snRNA has a terminal sequence complementary to the splice donor consensus and binds to it by *RNA–RNA base pairing*. After the U1 snRNP has bound, U2 snRNA recognizes the branch site by a similar base-pairing reaction. Interaction between the splice donor and splice acceptor junctions is stabilized by subsequent binding of a preformed multi-snRNP particle, containing U4, U5 and U6 snRNAs, with the U5 snRNP able to bind simultaneously to both splice donor and splice acceptor.

together. Thereafter a multi-snRNP particle, containing U4, U5 and U6 snRNAs, associates with the U1–U2 snRNP complex.

A second type of spliceosome is also known which processes a much rarer class of intron, the so-called AT–AC intron where the conserved GT and AG dinucleotides are replaced by AT and AC respectively. In this case U11 and U12 snRNA replace the functions of U1 and U2 (see Tarn and Steitz, 1997).

The spliceosome is envisaged to act in a processive manner: once a 5′ splice site is recognized, it scans the RNA sequence until it meets the next 3′ splice site (which would be signaled as a target by the branch site consensus sequence located just before it). However, the order in which intronic sequences are removed and their flanking exonic sequences spliced is not governed by their linear order in the RNA transcript; instead, the conformation of the RNA is thought to influence the accessibility of 5′ splice sites.

1.4.2 Specialized nucleotides are added to the 5′ and 3′ ends of most RNA polymerase II transcripts

In addition to RNA splicing, RNA polymerase II transcripts are subject to two additional RNA processing events.

Capping

This occurs shortly after transcription. In the case of primary transcripts which will be processed to give mRNA, a methylated nucleoside, 7-methylguanosine (m^7G) is linked to the first 5′ nucleotide of the RNA transcript by a special 5′–5′ phosphodiester bond. As this bond effectively bridges the 5′ carbon of the m^7G residue to the 5′ carbon of the first nucleotide, the 5′ end is said to be blocked or capped (*Figure 1.17*). Transcripts of the snRNA genes are also capped, but their caps may undergo additional modification. The cap has been envisaged to have several possible functions:

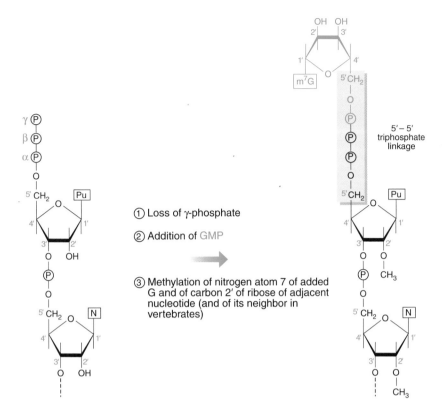

① Loss of γ-phosphate

② Addition of GMP

③ Methylation of nitrogen atom 7 of added G and of carbon 2' of ribose of adjacent nucleotide (and of its neighbor in vertebrates)

Figure 1.17: The 5' end of eukaryotic mRNA molecules is protected by a specialized nucleotide (capping).

After the original gamma phosphate of the terminal 5' nucleotide is removed, a new GMP residue is provided by a GTP precursor which forms a specialized 5'–5' triphosphate linkage with what was the terminal 5' nucleotide. Subsequent reactions lead to methylation of nitrogen atom 7 of the terminal G, and, in vertebrates, of the 2' carbon atom of the ribose of each of the two adjacent nucleotides. N, any nucleotide; Pu, purine.

- to protect the transcript from 5' → 3' exonuclease attack (decapped mRNA molecules are rapidly degraded);
- to facilitate transport from the nucleus to the cytoplasm;
- to facilitate RNA splicing;
- to play an important role in the attachment of the 40S subunit of the cytoplasmic ribosomes to the mRNA (see below).

Polyadenylation

Transcription by both RNA polymerase I and III is known to stop after the enzyme recognizes a specific transcription termination site. However, identifying possible termination sites for transcription by RNA polymerase II is difficult because the 3' ends of mRNA molecules are determined by a post-transcriptional cleavage reaction. The sequence AAUAAA is a major element that signals 3' cleavage for the vast majority of polymerase II transcripts (*transcripts from histone genes and snRNA genes are notable exceptions*). Cleavage occurs at a specific site located 15–30 nucleotides downstream of the AAUAAA element (*Figure 1.18*).

Following the cleavage point, transcription can continue for hundreds or thousands of nucleotides until termination occurs at one of several later sites. Once cleavage has occurred downstream of the AAUAAA element, about 200 adenylate (i.e. AMP) residues are sequentially added in mammalian cells by the enzyme poly(A) polymerase to form a poly(A) tail. The poly(A) tail has been envisaged to have several possible functions:

- to facilitate transport of the mRNA molecules to the cytoplasm;
- to stabilize at least some of the mRNA molecules in the cytoplasm [shortening of poly(A) tracts is associated with mRNA degradation, but some mRNA species (e.g. actin mRNA) remain stable with little or no poly(A)];
- it may facilitate translation by permitting enhanced recognition of the mRNA by the ribosomal machinery.

In the case of histone genes, which are unique in producing mRNA that does not become polyadenylated, termination of transcription also involves 3' cleavage of the primary transcript. This reaction is dependent on secondary structure in the RNA transcript, including a

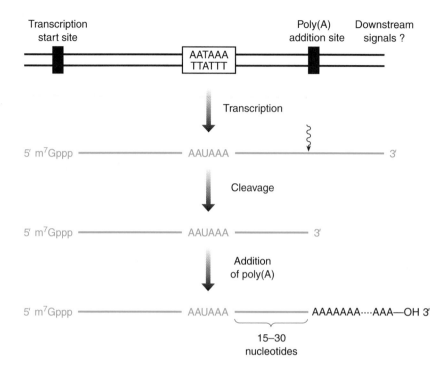

Figure 1.18: The 3′ end of most eukaryotic mRNA molecules is polyadenylated.

The end of transcription of RNA polymerase II transcripts is signaled by a 3′ cleavage in the transcribed RNA. In most mRNA species, this is achieved by an upstream AAUAAA signal in concert with, as yet, unidentified downstream signals. Cleavage occurs normally about 15–30 nucleotides downstream of the AAUAAA element and AMP residues are subsequently added by poly(A) polymerase to form a poly(A) tail. Histone mRNA undergoes a different 3′ cleavage reaction (see text).

conserved upstream hairpin sequence and a short down-stream sequence which base-pairs with a short sequence at the 5′ end of the U7 snRNA.

1.5 Translation, post-translational processing and protein structure

1.5.1 Translation is the process whereby mRNA is decoded on ribosomes to specify the synthesis of polypeptides

Following post-transcriptional processing, mRNA transcribed from genes in nuclear DNA migrates to the cytoplasm. Here it engages with ribosomes and other components to direct the synthesis of specific polypeptides. The mitochondria also have ribosomes and a limited capacity for protein synthesis (see Section 7.1.1).

Only the central segment of a typical eukaryotic mRNA molecule is translated to specify the synthesis of a polypeptide. The flanking sequences, **5′** and **3′ untranslated regions (5′ UTR; 3′ UTR)**, are originally copied from sequence derived from the 5′ and 3′ terminal exons and, like the 5′ cap and 3′ poly(A) tail, assist in binding and stabilizing the mRNA on the ribosomes where translation of the central segment occurs (*Figure 1.19*).

Ribosomes are large RNA–protein complexes composed of two subunits. In eukaryotes, cytoplasmic ribosomes have a large 60S subunit and a smaller 40S subunit (*S values* are a measure of how fast large molecular structures will sediment in the ultracentrifuge and are governed by both molecular mass and shape). The 60S subunit contains three types of rRNA molecule: 28S rRNA, 5.8S rRNA and 5S rRNA, and about 50 ribosomal proteins. The 40S subunit contains a single 18S rRNA and over 30 ribosomal proteins. Ribosomes provide a structural framework for polypeptide synthesis *in which the RNA components are predominantly responsible for the catalytic function of the ribosome*; the protein components are thought to enhance the function of the rRNA molecules, and a surprising number of them do not appear to be essential for ribosome function.

The assembly of a new polypeptide from its constituent amino acids is governed by a triplet genetic code. Successive groups of three nucleotides (**codons**) in the linear mRNA sequence are decoded sequentially in order to specify individual amino acids. The decoding process is mediated by a collection of tRNA molecules, to each of which a specific amino acid has been covalently bound (at the free 3′ hydroxyl group of the tRNA; see *Figure 1.20*) by a specific amino acyl tRNA synthetase. Different tRNA molecules bind different amino acids.

Each tRNA has a specific trinucleotide sequence, called the **anticodon**, at a crucially important site located in the center of one arm of the tRNA (*Figure 1.7B*). This site provides the necessary specificity to interpret the genetic

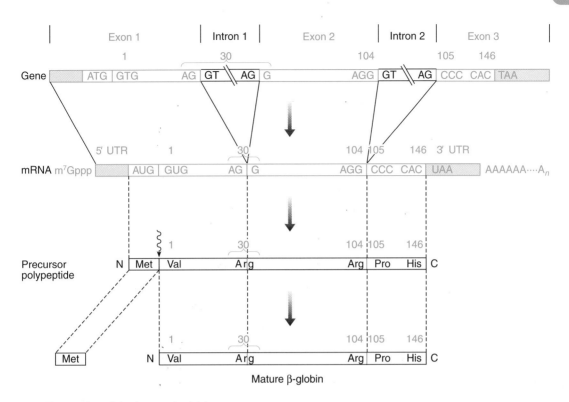

Figure 1.19: Expression of the human β-globin gene.

Exons 1 and 3 each contain noncoding sequences (shaded bars) at their extremities, which are transcribed and are present at the 5' and 3' ends of the β-globin mRNA, but are not translated to specify polypeptide synthesis. Such 5' and 3' untranslated regions (5' UTR and 3' UTR), however, are thought to be important in ensuring high efficiency of translation (see text). The stop codon UAA represents the first three nucleotides of the 3' untranslated region. *Note* that the initial translation product has 147 amino acids, but that the N-terminal methionine is removed by post-translational processing to generate the mature β-globin polypeptide. The first two bases of the codon specifying Arg30 are encoded by exon 1 and the third base is encoded by exon 2 (i.e. intron 1 separates the second and third bases of the codon, an example of a **phase 2 intron**; see *Box 14.3*). The second intron separates codons 104 and 105 and is an example of a **phase 0 intron**; see *Box 14.3*, and also *Figure 1.23* for an example of a **phase 1 intron**.

code: for an amino acid to be inserted in the growing polypeptide chain, the relevant codon of the mRNA molecule must be recognized via base-pairing with a suitably complementary anticodon of the appropriate tRNA molecule (*Figure 1.20*).

One model of translation envisages that the 40S ribosomal subunit initially recognizes the 5' cap via the participation of proteins that specifically bind to the cap. It then scans along the mRNA until it encounters the initiation codon, which is almost always AUG, specifying methionine (a few cases are known where ACG, CUG or GUG are used instead). Usually, though not always, the first AUG encountered will be the initiation codon. However, the AUG is recognized efficiently as an initiation codon only when it is embedded in a suitable sequence, the optimal being the sequence: GCC**Pu**CC*AUG***G**. The most important determinants in this sequence are the G following the AUG codon, and the purine, preferably A, preceding it by three nucleotides (Kozak, 1996). Subsequently, successive amino acids are incorporated into the growing polypeptide chain by a condensation reaction: the amino group of the incoming amino acid reacts with the carboxyl group

of the last amino acid to be incorporated, resulting in a peptide bond between successive residues (*Figure 1.21*). This is catalyzed by a peptidyl transferase activity which resides in the RNA component of the large ribosomal subunit.

1.5.2 The genetic code is degenerate

The genetic code is a three-letter code. There are four possible bases to choose from at each of the three base positions in a codon. There are, therefore, 64 ($=4^3$) possible codons, but only 20 different amino acids are specified. As a result, the genetic code is said to be *degenerate*: each amino acid is specified on average by about three different codons. Certain amino acids (such as methionine or tryptophan, in the case of the nuclear genetic code) are specified by only a single codon; others, including leucine and serine, are specified by six codons (*Figure 1.22*).

Although there are 64 codons, the corresponding number of tRNA molecules with different anticodons is less: just over 30 types of cytoplasmic tRNA and only 22 types of mitochondrial tRNA. The interpretation of all 64

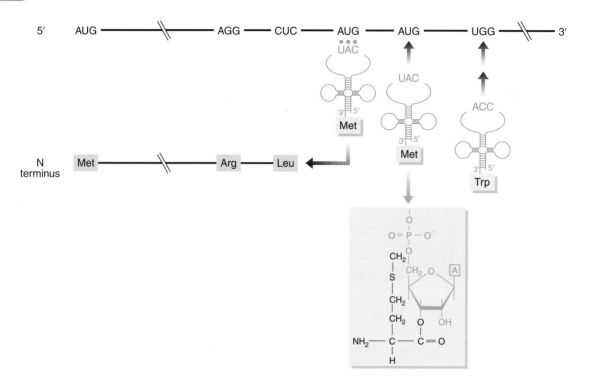

Figure 1.20: The genetic code is deciphered by codon–anticodon recognition.

The sequence of nucleotides in the mRNA sequence is interpreted from a translational start point (normally marked by the sequence AUG) and continues in the 5′ → 3′ direction until a stop codon is reached in that reading frame (see *Figure 1.19*). Each codon in the mRNA is recognized by the complementary anticodon sequence of a tRNA molecule to which a specific amino acid is covalently bonded to the adenosine at the 3′ end (see insert).

codons on both cytoplasmic and mitochondrial ribosomes is possible because the normal base-pairing rules are relaxed when it comes to codon–anticodon recognition. The wobble hypothesis states that pairing of codon and anticodon follows the normal A–U and G–C rules for the first two base positions in a codon, but that exceptional 'wobbles' occur at the third position and G–U base pairs are also admitted (*Table 1.5*).

Translation continues until a **termination codon** is

Peptide bond

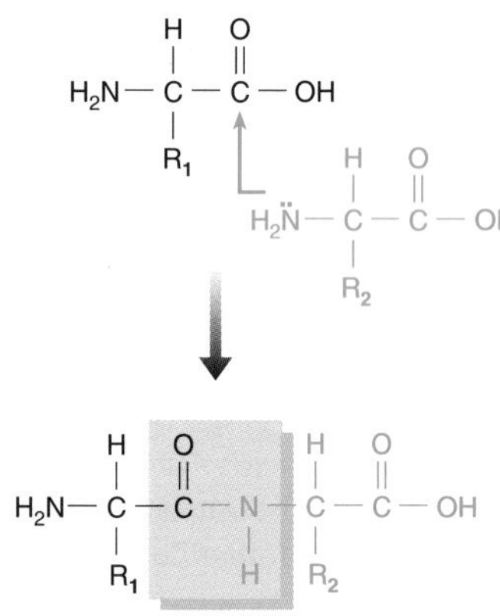

Figure 1.21: Polypeptides are synthesized by peptide bond formation between successive amino acids.

encountered (i.e. UAA, UAG or UGA in the case of nuclear-encoded mRNA; UAA, UAG, AGA or AGG in the case of mitochondrial-encoded mRNA; *Figure 1.22*). The backbone of the primary translation product will therefore have at one end a methionine with a free amino group (the N-terminal end) and at the other end an amino acid with a free carboxyl group (the C-terminal end). Note that although codons are translated in a specific translational reading frame, overlapping genes are occasionally found in eukaryotes, in which different translational reading frames are used (see *Figure 7.3* for an example).

The predominant step in the control of translation is ribosome binding. In addition to the 5′ cap, the 5′ UTR (often < 200 bp) and 3′ UTR (usually very much longer than the 5′ UTR) both play critical roles in mRNA recruitment for translation. Several *cis*-acting elements that are involved in this process have been characterized and, in addition, a few *trans*-acting factors which bind to these elements have been identified. It is possible that the 5′ and 3′ UTR sequences interact to enhance translation. The 3′ UTR has a key role in translational regulation and signals for controlling translation, mRNA stability and localization have all been found in this region (see Wickens *et al.*, 1997 and Section 8.2.5).

The mitochondrial protein synthesis machinery is similar to that found in cytoplasmic ribosomes. For example, mitochondrial ribosomes contain two subunits, a large

AAA ⎫
AAG ⎭ Lys   CAA ⎫ Gln   GAA ⎫ Glu   UAA ⎫ STOP
AAC ⎫   CAG ⎭   GAG ⎭   UAG ⎭
AAU ⎭ Asn   CAC ⎫ His   GAC ⎫ Asp   UAC ⎫ Tyr
            CAU ⎭   GAU ⎭   UAU ⎭

ACA ⎫
ACG ⎪ Thr   CCA ⎫   GCA ⎫   UCA ⎫
ACC ⎪   CCG ⎪ Pro   GCG ⎪ Ala   UCG ⎪ Ser
ACU ⎭   CCC ⎪   GCC ⎪   UCC ⎪
           CCU ⎭   GCU ⎭   UCU ⎭

|AGA| ⎰ Arg   CGA ⎫   GGA ⎫ |UGA| ⎰ STOP / *Trp*
|AGG| ⎱ STOP   CGG ⎪ Arg   GGG ⎪ Gly   UGG ⎱ Trp
AGC ⎫   CGC ⎪   GGC ⎪   UGC ⎫ Cys
AGU ⎭ Ser   CGU ⎭   GGU ⎭   UGU ⎭
     ⎰ Ile

|AUA| ⎰ *Met*   CUA ⎫   GUA ⎫   UUA ⎫ Leu
AUG ⎱ Met   CUG ⎪ Leu   GUG ⎪ Val   UUG ⎭
AUC ⎫ ⎰ Ile   CUC ⎪   GUC ⎪   UUC ⎫ Phe
AUU ⎭   CUU ⎭   GUU ⎭   UUU ⎭

Figure 1.22: The nuclear and mitochondrial genetic codes are similiar but not identical.

Blue boxes indicate the four codons which are interpreted differently in the nucleus and mitochondria of mammalian cells, with the mitochondrial interpretation given in blue. Thus, the mitochondrial code has 4 stop codons instead of three (UAA, UAG, AGA, AGG), two Trp codons instead of one (UGA, UGG), four Arg codons instead of six (CGA, CGC, CGG, CGU), two Met codons instead of one (AUA, AUG) and two Ile codons instead of three (AUC, AUU). *Note* that degeneracy of the genetic code most often involves the third base of the codon. Sometimes any base may be substituted (GGN = glycine, CCN = proline, etc., where N is any base). In other cases, any purine (Pu) or any pyrimidine (Py) will do (AAPu = lysine, AAPy = asparagine etc.).

subunit with a 23S rRNA and several proteins, and a small subunit with a 16S rRNA and several ribosomal proteins. However, the components are assembled from the products of both mitochondrial and nuclear genes: the RNA components are synthesized in the mitochondria (the two rRNA molecules and 22 types of tRNA and a few, usually about 10, types of mRNA; see Section 7.1.1 for a description of the human mitochondrial genome) whereas the proteins are encoded by nuclear genes which have been translated on cytoplasmic ribosomes and imported into the mitochondrion (RNA polymerase, 22 amino acyl tRNA synthetases and about 80 ribosomal proteins).

Table 1.5: Codon–anticodon pairing admits relaxed base-pairing (wobbles) at the third base position of codons

Base at 5′ end of tRNA anticodon	Base recognized at 3′ end of mRNA codon
A	U only
C	G only
G	C or U
U	A or G

1.5.3 Post-translational modifications of proteins are frequent and can involve addition of specific chemical groups to specific amino acids and cleavage of the primary translation product

Primary translation products often undergo a variety of modification reactions, involving the addition of chemical groups which are attached covalently to the polypeptide chain at the translational and post-translational levels. This can involve simple chemical modification (hydroxylation, phosphorylation, etc.) of the side chains of single amino acids or the addition of different types of carbohydrate or lipid groups (see *Table 1.6*).

Protein modification by addition of carbohydrate groups

Glycoproteins contain oligosaccharides which are covalently attached to the side chains of certain amino acids. Few proteins in the cytosol are *glycosylated*, that is have attached carbohydrate, and those that are carry a single sugar residue, *N*-acetylglucosamine, covalently linked to a serine or threonine residue. By contrast, those proteins which are secreted from cells or exported to lysosomes, the Golgi apparatus or the plasma membrane are glycosylated. Oligosaccharide components of glycoproteins are largely preformed and added *en bloc* to polypeptides. Two major types of glycosylation are recognized:

(i) **N-glycosylation** involves, in most cases, initial transfer of a common oligosaccharide sequence to the side chain NH_2 group of an Asn residue within the endoplasmic reticulum (ER; see *Table 1.6*). Subsequent trimming of residues and replacement with different monosaccharides occurs in the Golgi apparatus;

(ii) **O-glycosylation** – see *Table 1.6*.

Proteoglycans are proteins with attached *glycosaminoglycans* which usually contain disaccharide repeating units containing glucosamine or galactosamine. The most well characterized proteoglycans are components of the extracellular matrix.

Protein modification by addition of lipid groups

Some proteins, notably membrane proteins, are modified by the addition of fatty acyl or prenyl groups which typically serve as membrane anchors. Examples of fatty acyl groups include the myristoyl group which is a (C_{14}) lipid that is found attached to a glycine residue at the extreme N terminus, and enables the modified protein to interact with a membrane receptor or the lipid bilayer of the membrane. Another fatty acyl group which serves as a membrane anchor is the (C_{16}) palmitoyl group, which becomes attached to the S atom of cysteine residues.

Prenyl groups typically become attached to cysteine residues close to the C terminus, and include farnesyl (C_{15}) groups and geranylgeranyl (C_{20}) groups. Many proteins that participate in signal transduction and protein targeting contain either a farnesyl or a geranylgeranyl unit at their C terminus.

Table 1.6: Major types of modification of polypeptides

Type of modification (group added)	Target amino acids	Comments
Phosphorylation (PO_4^-)	Tyrosine, serine, threonine	Achieved by specific kinases. May be reversed by phosphatases
Methylation (CH_3)	Lysine	Achieved by methylases and undone by demethylases
Hydroxylation (OH)	Proline, lysine, aspartic acid	Hydroxyproline and hydroxylysine are particularly common in collagens
Acetylation (CH_3CO)	Lysine	Achieved by an acetylase and undone by deacetylase
Carboxylation (COOH)	Glutamate	Achieved by γ-carboxylase
N-glycosylation (complex carbohydrate)	Asparagine, usually in the sequence: **Asn**–X–Ser/Thr	Takes place initially in the endoplasmic reticulum; X is any amino acid other than proline
O-glycosylation (complex carbohydrate)	Serine, threonine, hydroxylysine	Takes place in the Golgi apparatus; less common than N-glycosylation
GPI (glycolipid)	Aspartate at C terminus	Serves to anchor protein to *outer* layer of plasma membrane
Myristoylation (C_{14} fatty acyl group)	Glycine at N terminus (see text)	Serves as membrane anchor
Palmitoylation (C_{16} fatty acyl group)	Cysteine to form S-palmitoyl link.	Serves as membrane anchor
Farnesylation (C_{15} prenyl group)	Cysteine at C terminus (see text)	Serves as membrane anchor
Geranylgeranylation (C_{20} prenyl group)	Cysteine at C terminus (see text)	Serves as membrane anchor

Anchoring of a protein to the outer layer of the plasma membrane uses a different mechanism: the attachment of a glycosylphosphatidyl inositol (GPI) group. This complex glycolipid group contains a fatty acyl group that serves as the membrane anchor which is linked successively to a glycerophosphate unit, an oligosaccharide unit and finally through a phosphoethanolamine unit to the C terminus of the protein. The entire protein except for the GPI anchor is located in the extracellular space.

Post-translational cleavage

The primary translation product may also undergo internal cleavage to generate a smaller mature product. Occasionally, the initiating methionine is cleaved from the primary translation product, as during the synthesis of β-globin (*Figure 1.19*). More substantial polypeptide cleavage is observed in the case of the maturation of many proteins, including plasma proteins, polypeptide hormones, neuropeptides, growth factors, etc. For example, all secreted polypeptides and also polypeptides which are transported across intracellular membranes (e.g. those synthesized in the cytoplasm and transported into the mitochondria) are synthesized initially as precursors, in which a **signal sequence** (sometimes called a **leader sequence**) at the N-terminal end acts as a recognition signal for transport across cellular membranes (see below). Thereafter the signal peptide is cleaved from the main polypeptide and degraded. Additionally, in some cases, a single mRNA molecule may specify more than one functional polypeptide chain as a result of proteolytic cleavage of a large precursor polypeptide (*Figure 1.23*).

1.5.4 Secretion of proteins and export to specific intracellular locations requires specific localization signals in the coding sequence or specific types of attached group

Proteins synthesized on mitochondrial ribosomes are required to function within the mitochondria. However, the numerous proteins that are synthesized on cytoplasmic ribosomes have diverse functions which may require them to be secreted from the cell where they were synthesized (as with hormones and other intercellular signaling molecules) or to be exported to specific intracellular locations, such as the nucleus (histones, DNA and RNA polymerases, transcription factors, RNA-processing proteins, etc.), the mitochondrion (mitochondrial ribosomal proteins, many respiratory chain components, etc.), peroxisomes, and so on. To do this, a specific localization signal needs to be embedded in the structure of the polypeptide so that it can be sent to the correct address. Usually, the localization signal takes the form of a short peptide sequence. Often, but not always, this constitutes a so-called **signal sequence** (or **leader sequence**) which is removed from the protein by a specialized signal peptidase once the sorting process has been achieved.

Signals for export to the endoplasmic reticulum and extracellular space

In the case of secreted proteins the signal peptide comprises the first 20 or so amino acids at the N-terminal end and always includes a substantial number of hydrophobic amino acids (see *Table 1.7*). The signal sequence is

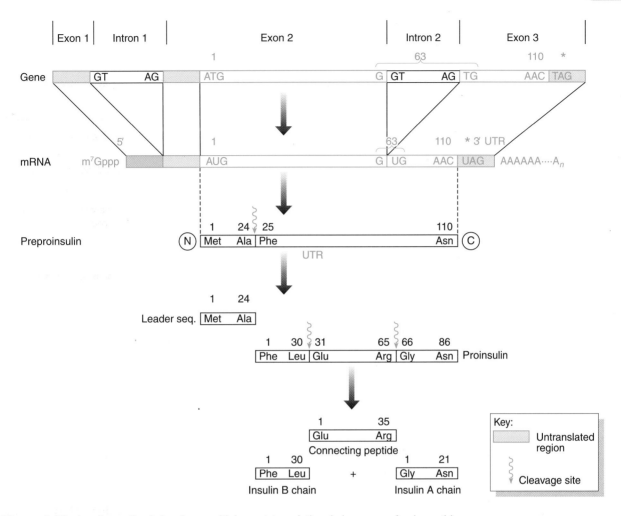

Figure 1.23: Insulin synthesis involves multiple post-translational cleavages of polypeptide precursors.

The first intron interrupts the 5′ untranslated region; the second intron interrupts base positions 1 and 2 of codon 63 and is classified as a phase I intron (see *Box 14.3*, and also *Figure 1.19* for other intron phases). The primary translation product, preproinsulin, has a leader sequence of 24 amino acids which is required for the protein to cross the cell membrane, and is thereafter discarded. The proinsulin precursor contains a central segment, the connecting peptide, which is thought to be important in maintaining the conformation of the A and B chain segments so that they can form disulfide bridges (see *Figure 1.25*).

guided to the ER by a signal recognition particle (SRP), an RNA–protein complex consisting of a small cytoplasmic RNA species, 7SL RNA, and six specific proteins. The SRP complex binds both the growing polypeptide chain and the ribosome and directs them to an SRP receptor protein on the cytosolic side surface of the rough endoplasmic reticulum (RER) membrane. Thereafter a polypeptide can pass into the lumen of the ER, destined for export from the cell, unless there are additional hydrophobic segments which stop the transfer process, as in the case of transmembrane proteins.

Other signals

Like the ER signal, an N-terminal signal sequence is required to traverse the mitochondrial membranes, and is subsequently cleaved. Typically, a mitochondrial signal peptide has, in addition to many hydrophobic amino acids, several positively charged amino acids, usually spaced at intervals of about four amino acids. This structure is thought to form an *amphipathic α-helix*, a helical structure with charged amino acids on one surface and hydrophobic amino acids on another (see below and *Figure 1.24*).

Nuclear localization signals can be located just about anywhere within the polypeptide sequence and typically consist of a stretch of four to eight positively charged amino acids, together with neighboring proline residues. Often, however, the signal is *bipartite*, and the positively charged amino acids are found in two blocks of two to four residues, separated by about 10 amino acids (*Table 1.7*). *Note* that some nuclear proteins lack any nuclear localization sequences

Table 1.7: Examples of protein localization sequences

Destination of protein	Location and form of signal	Examples
Endoplasmic reticulum and secretion from cell	N-terminal peptide of 20 or so amino acids; very hydrophobic	Human insulin – 24 amino acids, highly hydrophobic signal peptide: N–Met–Ala–Leu–Trp–Met–Arg–Leu–Leu–Pro–Leu–Leu–Ala–Leu–Leu–Ala–Leu–Trp–Gly–Pro–Asp–Pro–Ala–Ala–Ala
Mitochondria	N-terminal peptide; α-helix with positively charged residues on one face and hydrophobic ones on the other	Human mitochondrial aldehyde dehydrogenase – N-terminal 17 amino acids: N–Met–Leu–**Arg**–Ala–Ala–Ala–**Arg**–Phe–Gly–Pro–**Arg**–Leu–Gly–**Arg**–**Arg**–Leu–Leu
Nucleus	Internal sequence of amino acids; often a string of basic amino acids plus prolines; may be bipartite	SV40 T antigen – continuous: Pro–Pro–**Lys**–**Lys**–**Lys**–**Arg**–**Lys**–Val p53 – bipartite: **Lys**–**Arg**–Ala–Leu–Pro–Asn–Asn–Thr–Ser–Ser–Ser–Pro–Gln–Pro–**Lys**–**Lys**–**Lys**
Lysosome	Addition of mannose 6-phosphate residues	

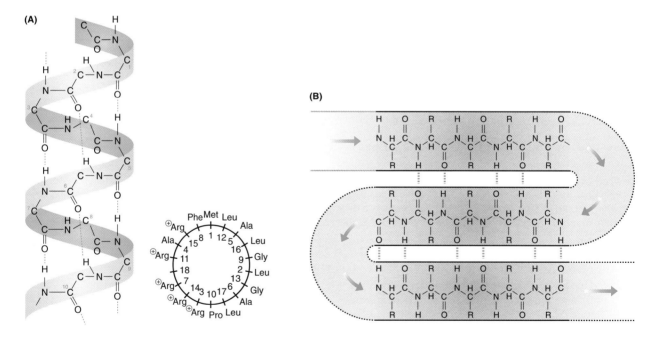

Figure 1.24: Regions of secondary structure in polypeptides are often dominated by intrachain hydrogen bonding.

(A) Structure of an α-helix. *Left:* only the backbone of the polypeptide is shown for clarity. The carbonyl (CO) oxygen of each peptide bond is hydrogen bonded to the hydrogen on the peptide bond amide group (NH) of the fourth amino acid away, so that the helix has 3.6 amino acids per turn. *Note* that for clarity some bonds have been omitted. The side chains of each amino acid are located on the outside of the helix and there is almost no free space within the helix. *Right:* charged amino acids and hydrophobic amino acids are located on different surfaces in an amphipathic α-helix. The sequence shown is that for the 17 amino acid long signal peptide sequence for the mitochondrial aldehyde dehydrogenase (see *Table 1.7*).
(B) Structure of a β-pleated sheet. *Note* that hydrogen bonding occurs between the CO oxygen and NH hydrogen atoms of peptide bonds on adjacent parallel segments of the polypeptide backbone. The example shows a case of bonding between antiparallel segments of the polypeptide backbone (antiparallel β-sheet) and the enforced abrupt change of direction between antiparallel segments is often accomplished using β-turns (see text). Arrows mark the direction from N terminus to C terminus. *Note* that parallel β-pleated sheets with the adjacent segments running in the same direction are also commonly found.

Table 1.8: Levels of protein structure

Level	Definition	Comment
Primary	The linear sequence of amino acids in a polypeptide	Can vary enormously in length from a small peptide to thousands of amino acids long
Secondary	The path that a polypeptide backbone follows in space	May vary locally, e.g. as α-helix or β-pleated sheet, etc.
Tertiary	The overall three-dimensional structure of a polypeptide	Can vary enormously, e.g. globular, rod-like, tube, coil, sheet, etc.
Quaternary	The overall structure of a multimeric protein, i.e. of a combination of protein subunits	Often stabilized by disulfide bridges and by binding to ligands, etc.

themselves but are transported into the nucleus with assistance from other nuclear proteins which have appropriate signals.

Lysosomal proteins are targeted to the lysosome by the addition of a mannose 6-phosphate residue which is added in the *cis*-compartment of the Golgi apparatus and is recognized by a receptor protein in the *trans*-compartment of the Golgi.

1.5.5 Protein structure is varied and complex and is not easily predicted from the amino acid sequence of polypeptides

Proteins are composed of one or more polypeptides, each of which can be subject to post-translational modification. They can interact with specific cofactors (for example, divalent cations such as Ca^{2+}, Fe^{2+}, Cu^{2+}, Zn^{2+} or small molecules which are required for functional enzyme activity, e.g. NAD^+) or ligands (any molecule which a protein specifically binds), each of which can be powerful influences on the conformation of the protein. At least

four different levels of structural organization have been distinguished for proteins (see *Table 1.8*).

Within a single polypeptide chain, there is ample scope for hydrogen bonding between different residues; irrespective of the side chains, the oxygen of a peptide bond's carbonyl (CO) group can hydrogen bond to the hydrogen of the NH group of another peptide bond. Fundamental structural units defined by hydrogen bonding between closely neighboring amino acid residues of a single polypeptide include:

■   *The α-helix*. This involves formation of a rigid cylinder. The structure is dominated by hydrogen bonding between the carbonyl oxygen of a peptide bond with the hydrogen atom of the amino nitrogen of a peptide bond located four amino acids away (see *Figure 1.24*). Note that the DNA-binding domains of transcription factors are usually α-helical (see Section 8.2.3). An amphipathic α-helix has charged residues on one surface and hydrophobic ones on another (*Figure 1.24*). Identical α-helices with a

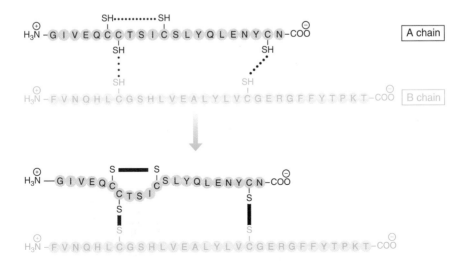

Figure 1.25: Intrachain and interchain disulfide bridges in human insulin.

The disulfide bridges (-S–S-) are formed by a condensation reaction between opposed sulfhydryl (-SH) groups of cysteine residues 6 and 11 of the A chain or between indicated residues of the different chains.

repeating arrangement of nonpolar side chains can coil round each other to form a particular stable structure called a coiled coil. Long rod-like coiled coils are found in many fibrous proteins, such as the α-keratin fibers of skin, hair and nails or fibrinogen of the blood clot.

■ *The β-pleated sheet*. This features hydrogen bond formation between opposed peptide bonds in parallel (often really antiparallel) segments of the same polypeptide chain (see *Figure 1.24*). β-pleated sheets form the core of most, but not all globular proteins.

■ *The β-turn*. Hydrogen bonding between the peptide bond CO group of amino acid residue *n* of a polypeptide with the peptide bond NH group of residue *n* + 3 results in a hairpin turn. By permitting the polypeptide to reverse direction abruptly, compact globular shapes can be achieved. β-turns are so named because they also often connect antiparallel strands in β-pleated sheets (see *Figure 1.24*).

More complex structural motifs consisting of combinations of the above structural modules constitute protein domains (compact regions of a protein formed by folding back of the primary structure so that elements of secondary structure can be stacked next to each other). Such domains often represent functional units involved in binding other molecules. In addition, covalent disulfide bridges are often formed between the sulfhydryl (−SH) groups of pairs of cysteine residues occurring within the same polypeptide chain or on different polypeptide chains (see *Figure 1.25*).

Clearly, the tertiary or quaternary structure of proteins is determined by the primary amino acid sequence. However although secondary structure motifs such as α-helices, β-pleated sheets and β-turns can be predicted by analyzing the primary sequence, the overall three-dimensional structure cannot, at present, be accurately predicted. In addition to the structural complexity of simple polypeptides, many proteins are organized as complex aggregates of multiple polypeptide subunits.

Further reading

Adams RPL, Knowler JT, Leader DP (1992) *Biochemistry of the Nucleic Acids*, 11th edn. Chapman & Hall, London.

Alberts B, Bray D, Lewis J, Raff M, Roberts K, Watson JD (1994) *Molecular Biology of the Cell*, 3rd edn. Garland Publishing, New York.

Carradine CR, Drew HR (1997) *Understanding DNA. The molecule and how it works*. Academic Press, London.

Lewin B (1997) *Genes VI*. Oxford University Press, Oxford.

Lodish H, Baltimore D, Berk A, Zipursky L, Matsudaira P, Darnell J (1995) *Molecular Cell Biology*, 3rd edn. Scientific American Books, New York.

References

Kozak M (1996) Interpreting cDNA sequences: some insights from studies on translation. *Mamm. Genome*, **7**, 563–574.

Stanley JP, Guthrie C (1998) Mechanical devices of the spliceosome: motors, clocks, springs and things. *Cell*, **92**, 315–326.

Tarn W-Y, Steitz J (1997) Pre-mRNA splicing: the discovery of a new spliceosome doubles the challenge. *Trends Biochem. Sci.*, **22**, 132–137.

Wickens M, Anderson P, Jackson RJ (1997) Life and death in the cytoplasm: messages from the 3′ end. *Curr. Biol.*, **7**, 220–232.

Chromosomes in cells

DNA functions in a context. Human DNA is structured into chromosomes that function within cells. Cells divide, differentiate and drive the development of the whole organism. Cell biology and development are huge subjects that we cannot possibly do justice to here, but we start with very brief overviews of these areas to help orient those students who come to human molecular genetics from a nonbiological direction. The main part of this chapter is an introduction to chromosomes – what they are, what they do, and how they can go wrong.

2.1 Organization and diversity of cells

All organisms more complex than viruses consist of cells, aqueous compartments bounded by membranes, which under restricted conditions are capable of existing independently. All cells are derived by cell division from other cells. Ultimately, there must be an unbroken chain of cells leading back to the first successful primordial cell that lived maybe 3.5 billion years ago. How that cell formed is an interesting question.

2.1.1 Prokaryotes and eukaryotes represent a fundamental division of living cells

All cellular organisms can be subdivided into two major classes, prokaryotes and eukaryotes, on the basis of the architecture of their cells (*Figure 2.1*).

Prokaryotes lack a defined nucleus and have a relatively simple internal organization. Under the electron microscope they appear relatively featureless. They comprise two kingdoms of life: eubacteria which include most of the bacteria; and the **archaea**, rather poorly understood organisms that superficially resemble bacteria and often grow in unusual environments, such as in acid hot springs, saturated brines, etc. The genome of a prokaryote typically consists of a single small circular chromosome in which the DNA is not packaged in any obviously organized way. Prokaryotes may be simple, but they are not primitive – they have been through far more generations of evolution than we have.

Eukaryotes have a much more complex intracellular organization with internal membranes, membrane-bound organelles including a nucleus, and a well-organized cytoskeleton. The general features are summarized briefly in *Box 2.1*. Eukaryotic cells have several linear chromosomes in their cell nuclei, in each of which a single very long DNA molecule is elaborately packaged by histone and other proteins. The number and DNA content of the chromosomes vary greatly between species (*Table 2.1*). In general the genome size tends to parallel the complexity of the organism, but there are many exceptions. Most mammals have much the same size genome. Humans do not have specially large genomes, while the cells of an onion and a lily contain respectively about five and 30 times as much DNA as a typical human cell. Eukaryotes are thought to have first apeared about 1.5 billion years ago.

2.1.2 Cell size and shape can vary enormously, but rates of diffusion fix some upper limits

Bacteria and some other simple organisms, such as the yeast *Saccharomyces cerevisiae*, consist of a single cell. Such cells are necessarily able to carry out all the functions that are required to sustain the organism. Multicellular organisms begin life as a single cell but then undergo repeated cell division, cell differentiation and cell turnover. They may end up containing huge numbers of cells. Cell differentiation ensures that individual multicellular organisms are composed of a variety of cell types that can vary greatly in size and shape.

Cells depend on diffusion to coordinate their metabolic activities. As they grow larger, the surface-to-volume ratio decreases. It is thought that the simple internal structure of prokaryotic cells limits their maximum size – typically bacterial cells are 1 μm in diameter. The complex internal membranes and compartmentalization of

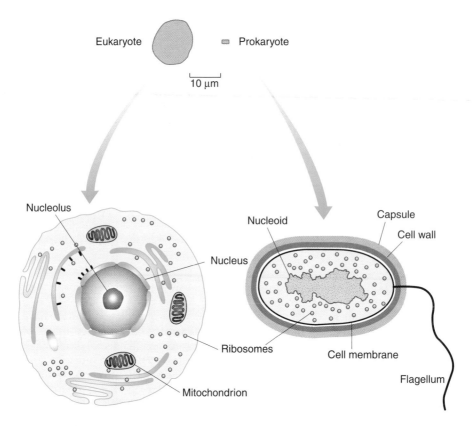

Figure 2.1: Prokaryotic and eukaryotic cell anatomy.

The top part of the figure shows a typical human cell and typical bacterium drawn to scale. The human cell is 10 µm in diameter and the bacterium is rod-shaped with dimensions 1 × 2 µm. Examples of some individual human cell types are described in *Box 2.3*. The lower drawings show the internal structures of typical animal and bacterial cells. Eukaryotic cells are characterized by their membrane-bound compartments, which are absent in prokaryotes. The bacterial DNA is contained in the structure called the nucleoid.

eukaryotic cells may be important in allowing them to grow larger. Nevertheless, metabolically active internal regions are seldom more than 15–25 µm from the cell surface, so that the limit of cell size is typically 30–50 µm. The average diameter of cells in a multicellular organism falls

Table 2.1: Variation in chromosome number and genome size

Species	Haploid chromosome number	Haploid genome size (Mb)
Saccharomyces cerevisiae (yeast)	16	14
Dictyostelium discoideum (slime mold)	7	70
Caenorhabditis elegans (nematode)	11/12	100
Drosophila melanogaster (fruit fly)	4	170
Gallus domesticus (chicken)	39	1200
Mus musculus (mouse)	20	3000
Xenopus laevis (toad)	18	3000
Homo sapiens (human)	23	3000
Zea mays (maize)	10	5000
Allium cepa (onion)	8	15000

within the range of 10–30 µm. Some individual human nerve cells can be as long as 1 metre, but the long projections are very thin. However, the ostrich egg warns us not to over-generalize.

2.1.3 Genetic content of cells: ploidy, the cell cycle and the life cycle

Most cells of humans are **diploid**. They contain two copies of the human genome. The DNA content and chromosome number of a genome are designated C and n respectively. For humans $C = 3.5 \times 10^{-12}$ g, approximately, and $n = 23$. The DNA content of diploid cells is $2C$ and they have $2n$ chromosomes. Almost all mammals are diploid, but among other organisms there are many examples of species that are normally **haploid** (n chromosomes, DNA content C), tetraploid ($4n$) or **polyploid**. **Triploidy** ($3n$) is less common because triploids have problems with meiosis (see below).

The diploid cells of our body are derived from the original diploid fertilized egg by repeated rounds of mitotic cell division. Each round can be summarized as one turn of the cell cycle (*Figure 2.2*). This comprises a short stage

Box 2.1

Anatomy of animal cells

Eukaryotic cells have complicated internal anatomies, with many internal membranes. Typically there is a nucleus and cytoplasm, the latter comprising various organelles, membranes and an aqueous compartment known as the cytosol. Different cells differ in their complement of organelles, but the following are typical of most animal cells. See the book by Alberts *et al.* (Further reading) for detail.

The nucleus: the repository of the genetic material. The nucleus contains almost all (typically 99.5%) the DNA of a cell, in the form of linear **chromosomes**. It is surrounded by a **nuclear envelope**, composed of two membranes separated by a narrow space and continuous with the endoplasmic reticulum (see below). Communication between the nucleus and the cytoplasm occurs via **nuclear pores**. Transport through the pores is selective and two-way. Within the nucleus, the chromosomes are arranged in a highly ordered way. The **nucleolus** is a region which contains numerous copies of genes encoding ribosomal RNA (that is, the short arms of chromosomes 13, 14, 15, 21 and 22 in human cells). These genes are continuously transcribed into rRNA and the resultant high RNA content gives the nucleolus a dense appearance in electron micrographs.

The mitochondria: the power-houses of aerobic eukaryotic cells. Mitochondria (and the chloroplasts of plants cells) are the only other eukaryotic organelles that contain DNA. The mitochondrial genome and genetic apparatus are described in Section 7.1. Most cells contain many copies of the mitochondrial DNA. Mutations can cause disease with an unusual pattern of inheritance (*Figure 3.4*). The role of mitochondria in the cell is to generate power by **oxidative phosphorylation**, a reaction where organic nutrients are oxidized by molecular oxygen and the chemical energy released is used to generate ATP. The mitochondrial **respiratory chain** is a series of five membrane-bound protein complexes, including various quinones, cytochromes and iron–sulfur proteins. Mitochondria have two membranes: a comparatively smooth outer membrane and a complex **inner mitochondrial membrane** which has a very large surface area because of numerous infoldings (*cristae*). The inner compartment, the **mitochondrial matrix**, is a very concentrated aqueous solution of many enzymes and chemical intermediates involved in energy metabolism.

Endoplasmic reticulum (ER): a major site of protein and lipid synthesis. The ER consists of flattened single membrane vesicles whose inner compartments, the **cisternae**, are interconnected to form channels throughout the cytoplasm. It is physically and functionally divided into two components. **Rough ER** is studded with ribosomes, whereas the **smooth ER** lacks any adhering ribosomes. Proteins synthesized on rough ER ribosomes cross the membrane of the ER and appear in the intracisternal space. From here, they are transported to the periphery of the cell where they will be incorporated into the plasma membrane, or secreted from the cell.

Golgi apparatus: machinery for secretion. The Golgi complex consists of flattened, single membrane vesicles which are often stacked. Its main function is in secretion of cell products, such as proteins, to the exterior, and in helping to form the plasma membrane and the membranes of lysosomes. Small vesicles containing secretory products (**secretory vacuoles**) are produced by pinching-off of the membranes.

Peroxisomes: organelles specializing in dangerous chemistry. Peroxisomes (microbodies) are small single-membrane vesicles containing a variety of enzymes that use molecular oxygen to oxidize their substrates and generate hydrogen peroxide. The hydrogen peroxide is used by a major peroxisomal enzyme, catalase, to oxidize a wide range of compounds. Reactive oxygen species like peroxide and the superoxide radical are highly toxic to cells and must be carefully contained.

Lysosomes: intracellular digestive organs. Lysosomes are small single-membraned vesicles, very diverse in form and size, that contain a cocktail of hydrolytic enzymes such as nucleases, proteases, lipases, glycosidases and phosphatases. Lysosomes digest materials brought into the cell by phagocytosis or pinocytosis, and degrade cell components during apoptosis.

The plasma membrane: guarded frontier of the cell. The plasma membrane is composed of a phospholipid bilayer. The hydrophobic lipid chains form the interior of the bilayer with hydrophilic phosphate groups on both sides in contact with the exterior of the cell and the cytoplasm. As well as providing a generally protective barrier, the plasma membrane has a variety of important roles:

- it is *selectively permeable*, regulating transport of a variety of ions and small molecules into and out of the cell. It contains active transport systems for ions such as Na^+, K^+ and Ca^{2+}, and for nutrients such as glucose and amino acids, as well as a number of enzymes;
- it contains a variety of integral membrane proteins or proteins anchored to one face of the membrane which play important roles in cell–cell signaling.
- the adhesive properties of the cell coat are specific and play an important general role in cell–cell recognition and tissue organization.

The cytosol: a highly concentrated aqueous solution. Cytosol makes up about half the volume of the cell and is the site of major metabolic activity including most protein synthesis and intermediary metabolism. As well as soluble components, the cytosol contains the cytoskeleton, a structure of microtubules, actin filaments and intermediate filaments that control the shape and movement of the cell. Microtubules act as tramways along which molecular motors (kinesin and dynein) move vesicles and organelles.

of cell division, the M phase (<u>mitosis</u>; *see Figure 2.10*) and a long intervening **interphase**. Interphase can be divided into S phase (DNA <u>synthesis</u>), G_1 phase (gap between M phase and S phase) and G_2 phase (gap between S phase and M phase). From anaphase of mitosis right through until DNA duplication in S phase, a chromosome of a diploid cell contains a single DNA double helix and the total DNA content is 2*C*. G_1 is the normal state of a cell,

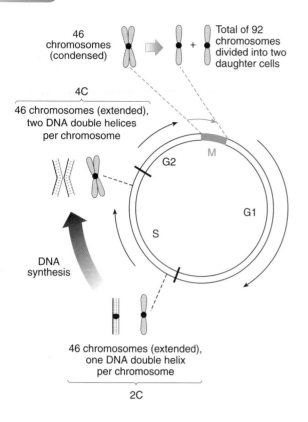

Figure 2.2: Human chromosomal DNA content during the cell cycle.

Interphase comprises G_1 + S + G_2. Chromosomes contain one DNA double helix from anaphase of mitosis right through until the DNA has duplicated in S phase. From this stage until the end of metaphase of mitosis, the chromosome consists of two chromatids each containing a DNA duplex, making two double helices per chromosome. The DNA content of a diploid cell before S phase is 2C (twice the DNA content of a haploid cell), while between S phase and mitosis it is 4C.

and the long-term end state of nondividing cells. Cells enter S phase only if they are committed to mitosis; nondividing cells remain in a modified G_1 stage, sometimes called G_0. The cell cycle diagram can give the impression that all the interesting action happens in S and M phases – but this is an illusion. A cell spends most of its life in G_0 or G_1 phase, and that is where the genome does most of its work.

A subset of the diploid body cells constitute the **germ line** (see *Figure 2.12*). These give rise to specialized diploid cells in the ovary and testis that can divide by meiosis to produce haploid gametes (sperm and egg). In humans ($n = 23$) each gamete contains 22 **autosomes** (nonsex chromosomes) plus one sex chromosome. In eggs the sex chromosome is always an X; in sperm it may be an X or a Y. After fertilization the zygote is diploid ($2n$) with the chromosome constitution 46,XX or 46,XY (*Figure 2.3*). The other cells of the body, apart from the germline, are known as **somatic cells**. Somatic cells have no input to succeeding generations – unless animal cloners intervene.

Although most somatic cells are diploid, there are exceptions. Some terminally differentiated cells, red blood cells for example, have no nuclei and others, such as keratinocytes, are devoid of organelles altogether. Such cells are nulliploid. Other cells are polyploid as a result of DNA replication without cell division (endomitosis). Regenerating cells of the liver and other tissues are naturally tetraploid because of endomitosis, while the giant megakaryocytes of the bone marrow usually contain 8C, 16C or 32C, and individually give rise to thousands of nulliploid platelet cells (*Figure 2.4A*). Other naturally occurring cells have multiple diploid nuclei as a result of cell fusion (*Figure 2.4B*).

2.2 Development

The development of any animal from a single fertilized egg cell is vastly complicated, but the early stages are common to all animals (*Box 2.2*), and at the molecular level development is controlled by a limited repertoire of developmental programs. All development depends on the basic processes of cell division, differentiation,

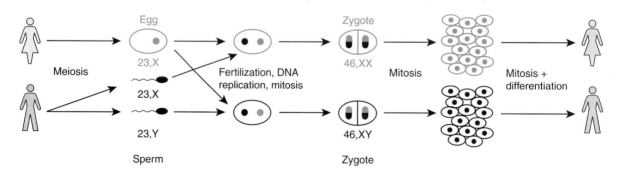

Figure 2.3: Human life, from a chromosomal viewpoint.

The haploid sperm and egg cells originate by meiosis from diploid precursors (see *Figure 2.12*). In the fertilized egg the sperm and egg chromosomes initially form separate male and female pronuclei. These combine during the first mitosis.

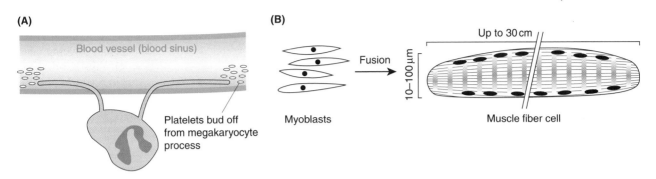

Figure 2.4: Some cells form by fragmentation or fusion of other cells.

(**A**) Platelets are formed by budding from a giant megakaryocyte. They have no nucleus. (**B**) Muscle cells are formed by fusion of large numbers of myoblast cells.

Box 2.2

A brief outline of animal development

This very brief summary concentrates on origins and cell lineages. We see that the earliest stages of human development are largely concerned with forming extraembryonic structures, and that the cells and tissues of a person are derived from only a very small part of the early zygote. See the book by Moore and Persaud (Further reading) for detail and illustrations.

Fertilization: formation of a unique individual
Sperm and egg cells are described in *Box 2.3*. After a sperm penetrates the zona pellucida its plasma membrane fuses with the membrane of the ovum and the head and tail, but not the body, of the sperm enter the egg. The egg completes meiosis II and expels the second polar body. Initially there are separate male and female pronuclei, each of which replicates its DNA once. The pronuclei then break down and the chromosomes are arranged for mitosis, which produces two diploid cells (the first cleavage division). By 3 days post-fertilization the **zygote** consists of a close-packed ball of 12–15 cells, the morula.

Blastocyst formation and implantation: the first appearance of recognizably different cell types
At about the 32 cell stage a fluid-filled cavity begins to form inside the morula. The zygote develops into a hollow ball (the blastocyst) with an outer layer of **trophectoderm** cells (which will give rise to extraembryonic membranes) and an off-center inner cell mass. The inner cell mass will contribute to both the extraembryonic membranes and the embryo itself. About 6 days after fertilization the blastocyst attaches to the uterine epithelium. Trophoblast cells proliferate rapidly and differentiate into an inner layer of cytotrophoblast and an outer syncytium (multinucleate cell layer) which starts to invade the connective tissue of the uterus. A cavity, the amniotic cavity, forms within the inner cell mass, enclosed by the amnion. The embryo, derived from part of the inner cell mass, now consists of a two-layered disc (hypoblast and epiblast) located between two fluid-filled cavities, the amniotic cavity and the yolk sac (derived from the blastocyst cavity). Hypoblast cells will form extraembryonic endoderm; epiblast cells will form all tissues of

the embryo in addition to the amnion and extraembryonic mesoderm.

Gastrulation: the origin of ectoderm, endoderm and mesoderm
Gastrulation takes place during the third week and is the beginning of morphogenesis. The orientation of the body is laid down and progenitor tissues for much of the body are formed. The embryo is converted into a structure of three germ layers, ectoderm, endoderm and mesoderm, all of which are derived from the epiblast. The **primitive streak** appears and then the **notochord**, defining the anteroposterior axis and head–tail orientation of the embryo. The notochord induces formation of overlying neuroectoderm that develops during the third and fourth weeks to form the neural tube (the progenitor of the brain, spinal cord and central nervous system), the neural crest (a highly versatile group of cells that form part of the peripheral nervous system, melanocytes, and some bone and muscle), the retina and other structures.

Organogenesis: the formation of organs, limbs, skeleton and other major structures of the body
During the fourth to eighth weeks of human development all the major organs and body systems are formed. In summary (but see Moore and Persaud, Further reading, for detail):

- **Ectoderm** gives rise to the central and peripheral nervous system; skin, hair and teeth; some bones of the head and face; some muscles; melanocytes; and some glands.
- **Mesoderm** gives rise to connective tissue; cartilage; most bone and muscle; blood vessels; and internal organs including the heart, kidneys, spleen, testes and ovaries.
- **Endoderm** gives rise to the epithelial lining of the gastrointestinal, respiratory and urinary tracts; and to most of the liver, pancreas, tonsils, thymus and thyroid and parathyroid glands. The primordial germ cells, progenitors of the sperm and eggs, are derived from endodermal cells of the yolk sac.

At the end of this period the embryo is described as a fetus and is for the first time recognizable as human.

morphogenesis and programed cell death (apoptosis). Differentiation is driven by gene switching: the difference between one cell type and another is primarily in the range of genes that are active in each cell. Morphogenesis, too, is ultimately driven by gene switching, as particular cells develop the capacity to respond to signals from neighboring cells by moving, dividing or dying. Apoptosis is an integral part of development: cells do not just happen to die, they have an inbuilt death program that is triggered in response to external or internal signals. All these developmental programs depend on cascades of signals and responses that have been remarkably highly conserved throughout the animal kingdom. Unraveling these programs is a major part of biological research. Probably the best introduction to how this is done, and why the results matter, is the book by Lawrence (Further reading).

2.2.1 Only a small percentage of the cells in the early embryo give rise to the mature organism

As shown in *Box 2.2*, the early stages of human development are largely concerned with establishing the future placenta and extraembryonic membranes, and implanting the conceptus in the uterine endometrium. About 2 weeks of development and many rounds of mitosis pass before the group of cells that will give rise to the embryo has any separate identity. Embryonic development proper begins with gastrulation in the third week after fertilization.

2.2.2 Cells differ in their potential to divide and differentiate

Multicellular animals begin life as a single cell following fertilization of an egg cell (*oocyte*) by a sperm cell. The fertilized egg proceeds to undergo a series of cell divisions. At the early stages of development, individual cells in the embryo are totipotent: each cell retains the capacity to differentiate into all the different types of cell in the body. As development proceeds cells become more *restricted* in their capacity to generate different types of descendant cells and are said to be pluripotent. Progenitor cells that can only develop into a single cell type are unipotent cells. The processes of cell differentiation lead to individual cells acquiring specialized forms and functions (*Box 2.3*).

Histology textbooks recognize about 210 different types of cell in the human body. Some, such as epithelial cells and fibroblasts, perform essentially the same role in a variety of organ systems. Others, such as hepatocytes,

---

Box 2.3

The diversity of human cells

Over 200 types of cells are described in histology textbooks. Here we illustrate the variety of size and form among common cell types.

Ovum – A large cell, 120 μm in diameter, surrounded by the zona pellucida. The first polar body, the product of meiosis I, lies under the zona. At ovulation, oocytes are in metaphase II. Meiosis II is not completed until after fertilization.

Sperm – 5 μm head containing highly compacted DNA; 5 μm cylindrical body with many mitochondria; 50 μm tail. At the front, the acrosome contains enzymes that help the sperm penetrate the zona pellucida. Semen typically contains 100 million sperm per ml.

Lymphocyte – small round cell, 6–8 μm diameter. Very little cytoplasm. Typically 5000 per μl of blood.

Erythrocyte – flat biconcave disk 7.2 μm diameter. Erythrocytes have no nucleus, mitochondria or ribosomes. Metabolism is entirely by glycolysis. Lifespan 120 days. Typically 5 million per μl of blood.

Megakaryocyte – large bone marrow cell 35–150 μm diameter. Irregular lobulated nucleus containing 8–32 genomes, formed by endomitosis. Megakaryocytes fragment to form thousands of platelets.

Platelet – 3–5 μm fragment of highly structured cytoplasm without nucleus. Lifespan 8 days. Typically 200 000 per μl of blood.

Epithelial cell – strongly adhesive cells that pack together to form epithelia with tight junctions (desmosomes) between cells. Some epithelial cells are specialized for ion transport, absorption or secretion.

Fibroblast – unspecialized cell of connective tissue capable of differentiating into cartilage, bone, fat and smooth muscle cells.

Hepatocyte – polyhedral cell 20–30 μm diameter, sometimes multinucleate. Rich in mitochondria and endoplasmic reticulum; contains lysosomes; may contain lipid droplets.

Macrophage – variable shape cell, wanders using pseudopodia. Specialized for engulfing particles by phagocytosis, contains many lysosomes to digest particles. Many macrophages fuse around large foreign bodies, forming giant tissue cells.

Muscle fiber cell – multinucleated cell made by fusion of myoblasts. 10–100 μm diameter, may be several cm long. Nuclei lie around periphery; most of interior is occupied by 1–2 μm myofibrils, with many mitochondria in between.

Neuron – very variable size and shape. Cell body 4–150 μm; usually many dendrites and one axon. One cell may have over 100 000 connections with other neurons. Axons of spinal cells innervating the feet are 1 m long.

Melanocyte – epithelial cell with long branched processes that lie between keratinocytes and pass packages of pigment (melanosomes) into them. Typically 1500 per mm^2 of skin (regardless of skin color).

Keratinocyte – mature keratinocytes are scale-like structures full of keratin and devoid of nucleus or any organelles.

may be restricted to an individual organ. Some mature *terminally differentiated* cells do not undergo cell division. Other cells (often distinguished by the suffix *-blast*, as in osteoblasts, chondroblasts, myoblasts, etc.) divide actively and act as precursors of terminally differentiated cells. In some cases, the precursor cells are also capable of undergoing self-renewal and are known as **stem cells**. Stem cells are an important target of gene therapy (Chapter 22). As an example, *Figure 2.5* illustrates the successive commitment of cells in the hemopoietic lineage.

2.2.3 X inactivation is a special feature of female development

Chromosomally-based sex-determination systems, like the human XX/XY system, create a problem for development. Having the wrong number of chromosomes almost always makes an organism develop abnormally, and yet the two sexes must develop normally with different chromosome constitutions. In humans and other mammals the solution is different for the X and the Y chromosomes. The Y chromosome contains very few genes, and these

are mostly genes governing male sexual function, so that females can get by perfectly well without a Y chromosome. The X chromosome, however, contains many genes that play vital roles in both sexes, and so some method of **dosage compensation** is required, to ensure that cells function normally with either one or two X chromosomes.

Mammals achieve X-chromosome dosage compensation by the mechanism of **X-inactivation**, often called **lyonization**, after Dr Mary Lyon who first suggested this mechanism (Lyon, 1999, Heard *et al., 1997; Figure 2.6*). Early in embryonic development (at late blastocyst stage in the mouse, and probably also in humans), cells somehow count their X chromosomes and then inactivate all but one of them. 46,XY cells (and also 45,X cells) leave their single X active; 46,XX cells (and also 47,XXY cells) inactivate one X, while a 47,XXX cell would inactivate two Xs. Thus, regardless of the number of X chromosomes in the karyotype, there is only one set of active X-linked genes in a cell. Males are constitutionally **hemizygous** for X chromosome genes (that is, they have only a single copy of each gene), while females become

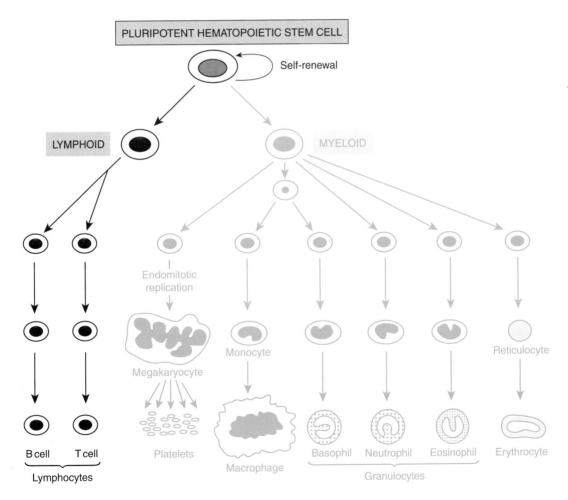

Figure 2.5: Commitment and differentiation in a cell lineage.

Blood cells are formed from pluripotent hematopoietic stem cells in the bone marrow.

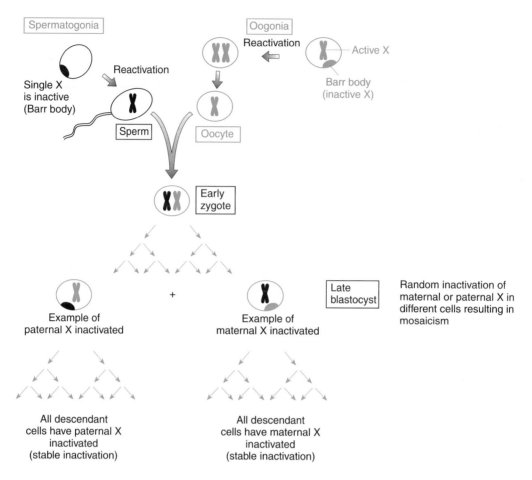

Figure 2.6: The process of X chromosome inactivation in mammals.

In the early XX female zygote, both X chromosomes are active but, around the time embryonic development begins, a choice is made randomly in each cell to inactivate either the paternal or the maternal X. The choice that a cell makes is preserved in all its descendants. An adult XX female has clonal populations of cells with the paternal or maternal X inactivated. The inactive X is reactivated in oocytes some time before meiosis. During spermatogenesis both the X and Y chromosomes are transiently inactivated. Adapted from Migeon (1994) *Trends Genet.*, **10**, pp. 230–235, with permission from Elsevier Trends Journals.

functionally hemizygous: they have only a single functional copy of each gene.

When chromosomes of a female cell are observed at metaphase of mitosis, the active and inactive Xs look the same – but this is because *all* chromosomes at metaphase of mitosis are condensed and largely inactive. After the end of cell division, the inactive X remains condensed while the other chromosomes decondense and resume transcriptional activity. In some cells the inactive X can be seen as a **Barr body** or **sex chromatin body** near the membrane of the interphase nucleus. This allows a simple but not very reliable method of sexing interphase cells.

Which X in a 46,XX cell is inactivated is random (with a few exceptions discussed later), so that in a female embryo, some cells will inactivate the paternal X and some the maternal X. Once the choice is made, it is remembered. When the cell divides, the daughter cells inactivate the same X as the mother cell. An adult female is a mosaic of clones derived from different embryonic cells. Within a clone, all the cells inactivate the same X,

but between clones the choice is random. If she happens to be a carrier of an X-linked recessive disease, this can have major implications (Section 3.1.2).

X inactivation is a fascinating and imperfectly understood phenomenon. One would like to know how the cell counts its X chromosomes, how the inactivation works, how it is perpetuated, and how it is reversed during gametogenesis (see *Figure 2.6*). We know that there is not blanket inactivation of all genes on the inactive X; certain regions (see Section 2.4.3), and certain genes in other regions, escape inactivation. The mechanisms are discussed in more detail in Section 8.5.2

2.3 Structure and function of chromosomes

Chromosomes as seen under the microscope and illustrated in textbooks are rather misleading. When we look at chromosomes in a dividing cell we see the genome of the cell largely switched off and packed up into neat bundles ready for cell division. The processes of cell division are fascinating in their own right, and errors in packaging or dividing up the genome have major medical consequences (Section 2.6). However, it is important to remember that the switched on, functioning interphase chromosome that controls cellular activities is a much more extended and diffuse structure than the metaphase chromosomes seen in *Figure 2.17*. Importantly, it comprises only a single chromatid and one DNA double helix, not the two-chromatid structure of mitotic chromosomes.

As functioning organelles, eukaryotic chromosomes seem to require only three classes of DNA sequence element: centromeres, telomeres and origins of replication. This simple requirement has been verified by the successful construction of artificial chromosomes in yeast: large foreign DNA fragments behave as autonomous chromosomes when ligated to short sequences that specify a functional centromere, two telomeres and a replication origin (*Figure 4.16*). Recently mammalian artificial chromosomes have been constructed on similar principles (Huxley, 1997; Schindelhauer, 1999).

2.3.1 Packaging of DNA into chromosomes requires multiple hierarchies of DNA folding

In the cell the structure of each chromosome is highly ordered (Manuelidis, 1990). Even in the interphase nucleus the 2 nm DNA double helix is subject to at least two levels of coiling (*Figure 2.7*).

■ The most fundamental unit of packaging is the **nucleosome**. This consists of a central core of eight **histone** proteins, small highly conserved basic proteins of 102–135 amino acids. Each core comprises two molecules each of histones H2A, H2B, H3 and H4, around which a stretch of 146 bp of double-stranded DNA is coiled in 1.75 turns. Adjacent nucleosomes are connected by a short length of spacer DNA. Electron micrographs of suitable preparations show a 'string of beads' appearance.

■ The string of beads, approximately 10 nm in diameter, is in turn coiled into a **chromatin fiber** of 30 nm diameter. The interphase chromosome seems to consist of these chromatin fibers, probably organized into long loops as described below.

During cell division, the chromosomes become ever more highly condensed. The DNA in a metaphase chromosome

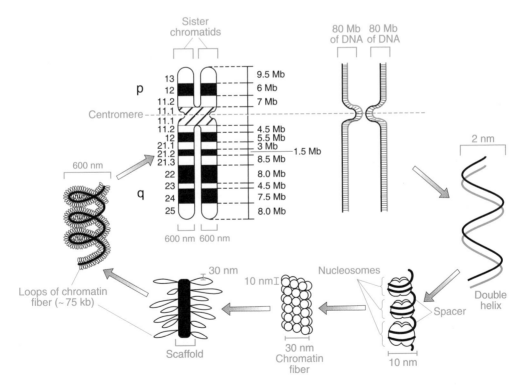

Figure 2.7: From DNA duplex to metaphase chromosome.

The figure shows human chromosome 17, as seen in a G-banded, 400 band preparation. The estimated packaging ratios (the degree of compaction of the linear DNA duplex) for human chromosomes are 1:6 for nucleosomes, 1:36 for the 30 nm fiber and >1:10 000 for the metaphase chromosome. Presently, it is uncertain whether the DNA at the centromere of the metaphase chromosome has been delayed in its replication unlike the rest of the chromatid, or whether full DNA replication has occurred in the S phase and the constriction at the centromere is due to some other cause.

is compacted to about 1/10 000 of its stretched-out length. Loops of the 30 nm chromatin fiber, containing 20–100 kb of DNA per loop, are attached to a central *scaffold*. This consists of nonhistone acidic proteins, notably topoisomerase II, an enzyme which has the interesting ability to pass one DNA double helix through another by cutting a gap and repairing it. Topoisomerase II and some other chromatin proteins are known to bind to AT-rich sequences, and the chromatin loops may be attached by stretches of several hundred base pairs of highly AT-rich (>65%) DNA (scaffold attachment regions). In the chromatids of a metaphase chromosome the loop–scaffold complex is compacted yet further by coiling (see *Figure 2.7*).

2.3.2 Chromosomes as functioning organelles: the centromere

Normal chromosomes have a single centromere that is seen under the microscope as the primary constriction, the region at which sister chromatids are joined. The centromere is essential for segregation during cell division. Chromosome fragments that lack a centromere (acentric fragments) do not become attached to the spindle, and so fail to be included in the nuclei of either of the daughter cells.

During late prophase of mitosis, a pair of kinetochores forms at each centromere, one attached to each sister chromatid. Multiple microtubules attach to each kinetochore, linking the centromere of a chromosome and the two spindle poles (see *Figure 2.11*). At anaphase, the kinetochore microtubules pull the two sister chromatids toward opposite poles of the spindle. Kinetochores play a central role in this process, by controlling assembly and disassembly of the attached microtubules and, through the presence of motor molecules, by ultimately driving chromosome movement.

Specific DNA sequences presumably specify the structure and function of centromeres. In simple eukaryotes, the sequences that specify centromere function are very short. For example, in the yeast *Saccharomyces cerevisiae* the centromere element (CEN) is about 110 bp long, comprising two highly conserved flanking elements of 9 bp and 11 bp and a central AT-rich segment of about 80–90 bp (*Figure 2.8*). The centromeres of such cells are interchangeable – a CEN fragment derived from one yeast chromosome can replace the centromere of another with no apparent consequence. In mammals, centromeres comprise hundreds of kilobases of repetitive DNA, some nonspecific and some chromosome-specific (Section 7.4.1).

2.3.3 Chromosomes as functioning organelles: origins of replication

The DNA in most diploid cells normally replicates only once per cell cycle. The initiation of replication is controlled by *cis*-acting sequences that lie close to the points at which DNA synthesis is initiated. Probably these are sites at which *trans*-acting proteins bind. Eukaryotic ori-

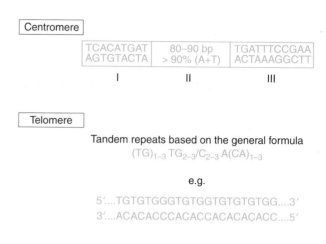

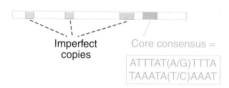

Figure 2.8: The functional elements of a yeast chromosome.

gins of replication have been most comprehensively studied in yeast, where the presence of a putative replication origin can be tested by a genetic assay. To test the ability of a random fragment of yeast DNA to promote autonomous replication, it is incorporated into a bacterial plasmid together with a yeast gene that is essential for growth of yeast cells. This construct is used to transform a mutant yeast that lacks the essential gene. The transformed cells can only form colonies if the plasmid can replicate in yeast cells. However, the bacterial replication origin in the plasmid does not function in yeast, therefore the few plasmids that transform at high efficiency must possess a sequence within the inserted yeast fragment that confers the ability to replicate extrachromosomally at high efficiency – that is an **autonomously replicating sequence (ARS)** element.

ARS elements are thought to derive from authentic origins of replication and, in some cases, this has been confirmed by mapping a specific ARS element to a specific chromosomal location and demonstrating that DNA replication is indeed initiated at this location. ARS elements extend for only about 50 bp and consist of an AT-rich region which contains a conserved core consensus and some imperfect copies of this sequence (*Figure 2.8*). In addition, the ARS elements contain a binding site for a transcription factor and a multiprotein complex is known to bind to the origin.

Mammalian replication origins have been much less well defined because of the absence of a genetic assay. Some initiation sites have been studied, but such studies have not been able to identify a unique origin of replication. This has led to speculation that replication can be initiated at multiple sites over regions tens of kilobases long. Mammalian artificial chromosomes seem to work without specific ARS sequences being provided. Computer analysis of regions encompassing several eukaryotic origins of replication, including some human and other mammalian examples, identified a consensus DNA sequence WAWTTDDWWWDHWGWHMAWTT where W = A or T; D = A or G or T; H = A or C or T; and M = A or C (Dobbs *et al.*, 1994).

2.3.4 Chromosomes as functioning organelles: the telomeres

Telomeres are specialized structures, comprising DNA and protein, which cap the ends of eukaryotic chromosomes. They have several likely functions:

- Maintaining the structural integrity of a chromosome. If a telomere is lost, the resulting chromosome end is unstable. It has a tendency either to fuse with the ends of other broken chromosomes, to be involved in recombination events or to be degraded. The loop structure of human telomeres (see below) means that natural chromosomes have no free DNA end.
- Ensuring complete replication of the extreme ends of chromosomes. During DNA replication, synthesis of the lagging strand is discontinuous and requires the presence of some DNA ahead of the sequence which is to be copied to serve as the template for an RNA primer (see *Figure 1.9*). However, at the extreme end of a linear molecule, there can never be such a template, and a different mechanism is required to solve the problem of replicating the ends of a linear DNA molecule (see below).
- Helping establish the three-dimensional architecture of the nucleus and/or chromosome pairing. Chromosome ends appear to be tethered to the nuclear membrane, suggesting that telomeres help position chromosomes.

Eukaryotic telomeres consist of a long array of tandem repeats. One DNA strand contains TG-rich sequences and terminates in the 3' end; the complementary strand is CA-rich. Unlike centromeres, the sequence of telomeres has been highly conserved in evolution – there is considerable similarity in the simple sequence repeat, for example TTGGGG (*Paramecium*), TAGGG (*Trypanosoma*) TTTAGGG (*Arabidopsis*) and TTAGGG (*Homo sapiens*) (see also *Figure 2.8*).

The problem of replicating the ends of a chromosome has been solved by extending the synthesis of the leading strand using a specialized enzyme, telomerase. This RNA–protein complex carries within its RNA component a short sequence which will act as a template to

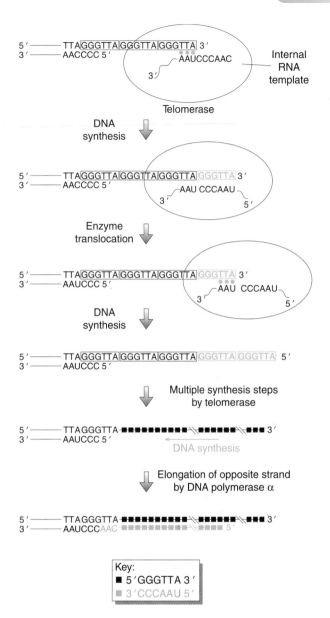

Figure 2.9: Telomerase extends the TG-rich strand of telomeres by DNA synthesis using an internal RNA template.

prime extended DNA synthesis of telomeric DNA sequences on the leading strand. Further extension of the leading strand provides the necessary template for DNA polymerase α to complete synthesis of the lagging strand (*Figure 2.9*). This mechanism leaves the telomere itself with a protruding 3' end. In mammalian chromosomes, the single-stranded end is believed to loop round and invade the double helix several kilobases proximally, producing a triple-stranded structure resembling the mitochondrial D-loop (*Figure 7.2*), which is stabilized by binding telomere-specific proteins (Greider, 1999). However, the actual nature of the telomere sequence may not be important. The telomere length is known to be highly variable and is subject to genetic control.

Just internal to the essential telomeric repeats,

eukaryotic chromosomes also have a more complex set of repeats called subtelomeric or telomere-associated repeats. Their sequences are not conserved in eukaryotes and their function is unknown.

2.3.5 Heterochromatin and euchromatin

In the interphase nucleus most of the chromatin (**euchromatin**) exists in an extended state, dispersed through the nucleus and staining diffusely. However, some chromatin remains highly condensed throughout the cell cycle and forms dark-staining regions (**heterochromatin**). Genes located in euchromatin may or may not be expressed, depending on the cell type and its metabolic requirements, but genes that are located within heterochromatin, either naturally or as the result of a chromosomal rearrangement, are very unlikely to be expressed. There are two classes of heterochromatin:

- **Constitutive heterochromatin** is always inactive and condensed. It consists largely of repetitive DNA and is found in and around the centromeres of chromosomes and in certain other regions (see *Figure 2.18*).
- **Facultative heterochromatin** can exist in either a genetically active (decondensed) or an inactive and condensed form, as in the case of mammalian X-chromosome inactivation (Section 2.2.3).

In euchromatin, the G bands (Section 2.5.2) partake of some of the properties of heterochromatin, but to a lesser degree. G band chromatin in metaphase chromosomes is more condensed than R band chromatin, and data on CpG island distribution (Section 7.1.2; *Figure 7.4*) show that G bands are relatively poor in genes. The subset of R bands that are revealed by T-banding have a particularly high density of genes. Section 1.3.5 discusses the different structures of chromatin in transcriptionally active and inactive chromosomal regions.

2.4 Mitosis and meiosis are the two types of cell division

2.4.1 Mitosis is the normal form of cell division

As a person develops from an embryo, through fetus and infant to an adult, cell divisions are needed to generate the large numbers of cells required. Additionally, many cells have a limited lifespan, so there is a continuous requirement to generate new cells in the adult. All these cell divisions occur by mitosis. Mitosis is the normal process of cell division, from cleavage of the zygote to death of the person. In the lifetime of a human there may be something like 10^{17} mitotic divisions (Section 9.2.1).

The M phase of the cell cycle (*Figure 2.2*) consists of the various stages of nuclear division (prophase, prometaphase, metaphase, anaphase and telophase of mitosis), and cell division (cytokinesis), which overlaps the final stages of mitosis (*Figure 2.10*). In preparation for cell division, the previously highly extended chromosomes contract and condense so that, by **metaphase** of mitosis, they are readily visible under the microscope. Even though the DNA was replicated some time previously, it is only at prometaphase that individual chromosomes can be seen to comprise two **sister chromatids**, attached at the centromere.

The mitotic spindle (*Figure 2.11*) is formed from tubulin-based microtubules and microtubule-associated proteins. **Polar fibers**, which extend from the two poles of the spindle towards the equator, develop at prophase while the nuclear membrane is still intact. Kinetochore fibers do not develop until prometaphase. These fibers attach to the kinetochore, a large multiprotein structure attached to the centromere of each chromatid, and extend in the direction of the spindle poles. The interaction between the different spindle fibers pulls the chromosomes towards the center, and by metaphase each chromosome is independently aligned on the equatorial plane (metaphase plate). Paternal and maternal homologs do not associate at all during mitosis. Following centromere division at anaphase, the spindle fibers pull the separated sister chromatids of each chromosome to opposite poles (*Figure 2.11*). The DNA of the two sister chromatids is identical, barring any errors in DNA replication. Thus the effect of mitosis is to generate daughter cells that contain precisely the same DNA sequences.

2.4.2 Meiosis is a specialized form of cell division giving rise to sperm and egg cells

Primordial germ cells migrate into the embryonic gonad and engage in repeated rounds of mitosis (many more in males than in females, which may be a significant factor in explaining sex differences in mutation rates – see *Figure 9.4*) to form oogonia in females and spermatogonia in males. Further growth and differentiation produces primary oocytes in the ovary and primary spermatocytes in the testis. These specialized diploid cells can undergo meiosis (*Figure 2.12*). Meiosis involves two successive cell divisions but only one round of DNA replication, so the products are haploid. In males, the product is four spermatozoa; in females, however, the cytoplasm divides unequally at each stage: the products of meiosis I (the first meiotic division) are a large secondary oocyte and a small cell (polar body). The secondary oocyte then gives rise to the large mature egg cell and a second polar body.

There are two crucial differences between mitosis and meiosis (*Table 2.2*).

- The products of mitosis are diploid; the products of meiosis are haploid.
- The products of mitosis are genetically identical; the products of meiosis are genetically different.

Mitosis involves a single turn of the cell cycle (*Figure 2.2*). The DNA is replicated in S phase and the two copies are divided exactly equally between the daughter cells in M

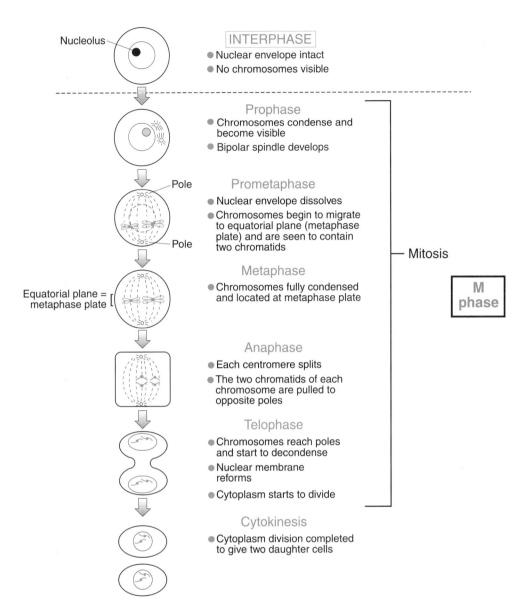

Figure 2.10: Cell division by mitosis.

phase. Meiosis is also preceded by one round of DNA synthesis, but then there are two cell divisions without intervening DNA synthesis, so that the products end up haploid. The second division of meiosis is identical to mitosis, but the first division has important differences whose purpose is to generate genetic diversity between the daughter cells. This is done by two mechanisms, independent assortment of paternal and maternal homologs (*Figure 2.13*), and recombination.

Independent assortment of paternal and maternal homologs

During meiosis I the maternal and paternal homologs of each chromosome pair form a **bivalent** by pairing together (synapsis) (*Figure 2.14*). Each chromosome consists of two sister chromatids following DNA replication,

so that the bivalent is a four-stranded structure at the metaphase plate. Spindle fibers then pull one complete chromosome (two chromatids) to either pole. However, for each of the 23 homologous pairs, the choice of which homolog enters which daughter cell is independent. This allows 2^{23} or about 8.4×10^6 possible combinations of parental chromosomes to be produced by one person.

Recombination

During prophase of meiosis I the synapsed homologs within each bivalent exchange segments in a random way. At the zygotene stage, each pair of homologs begins to form a synaptonemal complex consisting of the two chromosomes in close apposition, separated by a long linear protein core. Completion of this complex marks the start of the pachytene stage, which is when

(A)

(B)

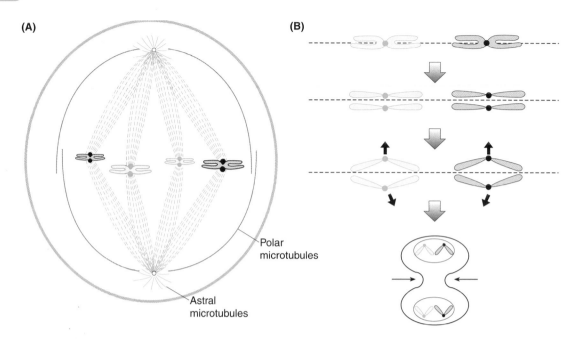

Polar
microtubules

Astral
microtubules

Figure 2.11: Mitosis: homologous chromosomes align independently on the metaphase plate and spindle fibers then pull the separated sister chromatids to opposite poles.

(**A**) At metaphase, paternal (black) and maternal (blue) homologs of each chromosome pair are independently aligned at the metaphase plate, and not associated with each other. Microtubules attached to the kinetochores link chromosomes to each of the poles. For clarity, only chromosomes 1 and 17 and a small fraction of the microtubules are shown. Other spindle microtubules include astral microtubules that radiate from each pole, and polar microtubules that form attachments linking the two poles. (**B**) At anaphase, the centromere of each of the 46 chromosomes duplicates and the two chromatids separate. Spindle fibers pull on the kinetochores of the centromeres, eventually pulling the two sister chromatids of each chromosome to opposite poles. At this stage (telophase) they become enclosed in a nuclear envelope. Subsequently the cell divides (cytokinesis).

recombination (or crossover) occurs. Crossing-over involves physical breakage of the double helix in one paternal and one maternal chromatid, and joining of maternal and paternal ends. Overall, the combination of recombination between homologs in prophase I plus independent assortment of homologs at anaphase I ensures that a single individual can produce an almost unlimited number of genetically different gametes.

The mechanism allowing alignment of the homologs is not understood. However, it is thought that such close apposition is required for recombination. Recombination nodules, very large multiprotein assemblies located at intervals on the synaptonemal complex, are thought to mediate the recombination events. The two homologs can be seen to be physically connected at specific points. Each such connection is described as a **chiasma** (plural **chiasmata**) and marks a crossover point. There are an average of 55 chiasmata in a male meiotic cell, and maybe 50% more in female meiosis. The genetic consequences of crossing over are considered in Chapter 11.

In addition to their role in recombination, chiasmata are thought to be essential for correct chromosome segregation at meiosis I. By holding the maternal and paternal homologs of each chromosome pair together on the spindle until anaphase I (*Figures 2.14* and *2.15*), they have a role that is analogous to that of the centromeres in

mitosis and meiosis II. There is genetic evidence that children with wrong numbers of chromosomes are often the product of gametes where a bivalent lacked crossovers.

Meiosis II appears identical to mitosis, except that there are only 23 chromosomes instead of 46. Each chromosome consists of two chromatids, and these are separated in anaphase II. However, there is one difference. The sister chromatids of a mitotic chromosome are identical, being copies of each other. The two chromatids of a chromosome in meiosis II may be genetically different as a result of crossovers in meiosis I (*Figure 2.15*).

2.4.3 X–Y pairing and the pseudoautosomal regions

In female meiosis, each chromosome has a fully homologous partner, and the two Xs synapse and cross over just like any other pair of homologs. In male meiosis there is a problem. The human X and Y sex chromosomes are very different from one another. Nevertheless, they do pair in prophase I in males, thus ensuring that at anaphase I each daughter cell receives one sex chromosome, either the X or the Y. X–Y pairing is end-to-end rather than along the whole length, and it is made possible by a 2.6 Mb region of homology between the X and Y chromosomes at the tips of their short arms. Pairing is sustained by an obliga-

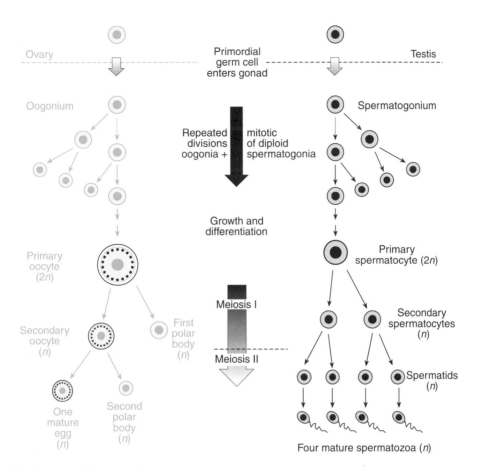

Figure 2.12: Development of the germ line.

The germ line develops by repeated mitotic division of diploid cells, culminating in production of primary oocytes and primary spermatocytes. These diploid cells can undergo meiosis. Meiosis involves two cell divisions but only one round of DNA replication, so the products are haploid. In humans, primary oocytes enter meiosis I during fetal life but then arrest at the prophase stage right through to puberty or later. During this time, the primary oocytes complete their growth phase, acquiring an outer jelly coat, cortical granules, ribosomes, mRNA, yolk, etc. After puberty, one oocyte a month completes meiosis. Sperm are produced continuously from puberty onwards.

Table 2.2: Mitosis and meiosis compared

	Mitosis	Meiosis
Location	All tissues	Only in testis and ovary
Products	Diploid somatic cells	Haploid sperm and egg cells
DNA replication and cell division	Normally one round of replication per cell division	Only one round of replication but two cell divisions
Extent of prophase	Short (~30 min in human cells)	Meiosis I is long and complex; can take years to complete
Pairing of homologs	None	Yes (in meiosis I)
Recombination	Rare and abnormal	Normally at least once in each chromosome arm
Relationship between daughter cells	Genetically identical	Different (recombination and independent assortment of homologs)

tory crossover in this region. Genes in the pairing segment have some interesting properties:

- they are present as homologous copies on the X and Y chromosomes;
- they are not subject to X-inactivation (as expected since each sex has two copies);

- because of the crossing over, alleles at these loci do not show the normal X-linked or Y-linked patterns of inheritance, but segregate like autosomal alleles.

Because of this behavior, this region is known as the major pseudoautosomal region. A second smaller pseudoautosomal region of 320 kb is located at the tips of

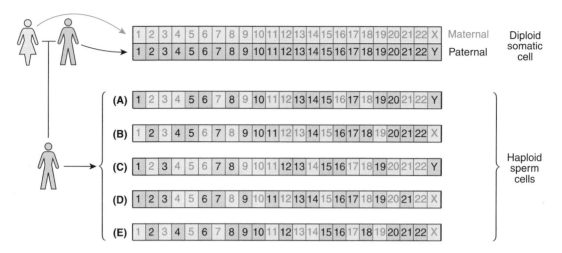

Figure 2.13: Meiosis: independent assortment of maternal and paternal homologs at meiosis I produces the first level of genetic diversity.

There are 2^{23} or 8.4 million different ways of picking one chromosome from each of the 23 pairs in a diploid cell. Gametes A–E show just five of the possible combinations of maternal and paternal chromosomes. This diagram ignores recombination, which introduces a second level of genetic diversity by ensuring that each individual chromosome passed on is a mixture of maternal and paternal sequences.

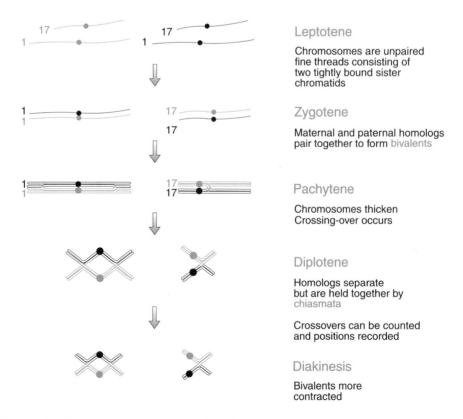

Figure 2.14: Meiosis: the five stages of prophase in meiosis I.

Two representative pairs of homologs are shown. There are two crossovers in the chromosome 1 bivalent and one in the chromosome 17 bivalent. For the sake of clarity, the two crossovers on chromosome 1 involve the same two chromatids. In reality the number of crossovers is likely to be higher, and multiple crossovers may involve three or even all four chromatids in a bivalent, as shown in *Figure 11.2*.

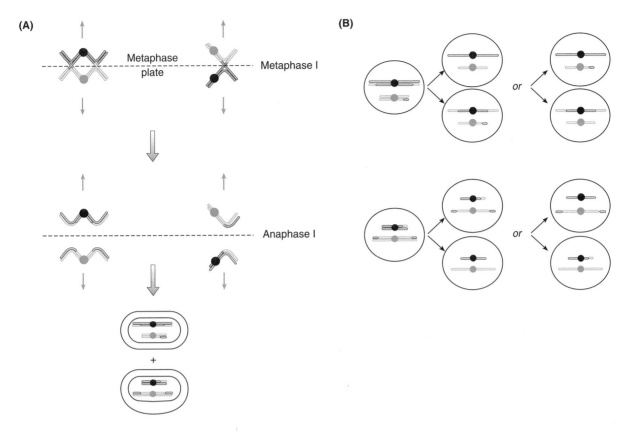

Figure 2.15: Meiosis: from metaphase I to the gametes.

The figure follows on from *Figure 2.14*. (**A**) From metaphase to cell division in meiosis I. The figure shows one possible segregation pattern of the two bivalents. (**B**) Meiosis II. Although the two sister chromatids of each chromosome will segregate to different daughter cells, the different chromosomes behave independently and so many combinations are possible, as shown.

the long arms of both chromosomes, but pairing and crossing-over in this minor pseudoautosomal region is not an obligatory feature of male meiosis.

2.5 Studying human chromosomes

2.5.1 Mitotic chromosomes can be seen in any dividing cell, but most conveniently in lymphocytes; meiotic chromosomes are hard to study in humans

Chromosomes can only be seen in dividing cells, and obtaining dividing cells directly from the human body is difficult. Bone marrow is a possible source, but it is much easier all round to take an accessible source of nondividing cells and culture them in the laboratory. Blood is the material of choice – most people don't mind giving a few millilitres, and the T lymphocytes in blood can be easily induced to divide by treatment with lectins such as phytohemagglutinin. Other common sources include fibroblasts grown from skin biopsies, and (for prenatal

diagnosis) chorionic villi or fetal cells shed into the amniotic fluid.

Although chromosomes were described accurately in some organisms as early as the 1880s, for many decades all attempts to prepare spreads of human chromosomes produced a tangle that defied analysis. The key to getting analyzable spreads was a new technique, growing cells in liquid suspension and treating them with hypotonic saline to make them swell. This allowed the first good quality preparations to be made in 1956. White cells from blood are put into a rich culture medium laced with phytohemagglutinin and allowed to grow for 48–72 hours, by which time they should be dividing freely. Nevertheless, because M phase occupies only a small part of the cell cycle, few cells will be actually dividing at any one time. The mitotic index (proportion of cells in mitosis) is increased by treating the culture with a spindle disrupting agent such as colcemid. Cells reach M phase of the cycle, but are unable to leave it, and so cells accumulate in metaphase of mitosis. Often it is preferable to study prometaphase chromosomes, which are less contracted and so show more detail. Cell cultures can be prevented from cycling by thymidine starvation; when the block is released the cells progress through the cycle synchronously. By trial and error, the time after release can be determined when a good proportion of cells are in the desired prometaphase stage.

Meiosis can only be studied in testicular or ovarian samples. Female meiosis is especially difficult, as it is active only in fetal ovaries, whereas male meiosis can be studied in a testicular biopsy from any post-pubertal male who is willing to give one. The results of meiosis can be studied by analyzing chromosomes from sperm, although the methodology for this is cumbersome. Meiotic analysis is used for some investigations of male infertility.

2.5.2 Chromosomes are identified by their size, centromere position and banding pattern

Until the 1970s chromosomes were identified on the basis of their size and the position of the centromeres. This allowed chromosomes to be classified into groups (*Table 2.3*) but not unambiguously identified. The introduction of banding techniques (*Box 2.4*) finally allowed each chromosome to be identified, as well as permitting more accurate definition of translocation breakpoints, subchromosomal deletions, etc. Banding resolution can be increased by using more elongated chromosomes, for example chromosomes from prometaphase or earlier, rather than metaphase. Typical high-resolution banding procedures for human chromosomes can resolve a total of 400, 550 or 850 bands (*Figures 2.16, 2.18*).

The chromosome constitution is described by a **karyotype** that states the total number of chromosomes and the sex chromosome constitution. Human females and males are 46,XX and 46,XY respectively. When there is a chromosomal abnormality the karyotype also describes the type of abnormality and the chromosome bands or subbands affected. See *Box 2.5* for details of chromosome nomenclature. Chromosomes are displayed as a **karyo-**

gram (often loosely described as a karyotype). Karyograms such as *Figure 2.17* are prepared by cutting up a photograph of the spread, matching up homologous chromosomes and sticking them back down on a card – or

---

Chromosome banding

G-banding – the chromosomes are subjected to controlled digestion with trypsin before staining with Giemsa, a DNA-binding chemical dye. Dark bands are known as G bands. Pale bands are G negative (*Figures 2.17, 2.18*).

Q-banding – the chromosomes are stained with a fluorescent dye which binds preferentially to AT-rich DNA, such as Quinacrine, DAPI (4′,6–diamidino-2–phenylindole) or Hoechst 33258, and viewed by UV fluorescence. Fluorescing bands are called Q bands and mark the same chromosomal segments as G bands.

R-banding – is essentially the reverse of the G-banding pattern. The chromosomes are heat-denatured in saline before being stained with Giemsa. The heat treatment denatures AT-rich DNA, and R bands are Q negative. The same pattern can be produced by binding GC-specific dyes such as chromomycin A_3, olivomycin or mithramycin.

T-banding – identifies a subset of the R bands which are especially concentrated at the telomeres. The T bands are the most intensely staining of the R bands and are visualized by employing either a particularly severe heat treatment of the chromosomes prior to staining with Giemsa, or a combination of dyes and fluorochromes.

C-banding – is thought to demonstrate constitutive heterochromatin, mainly at the centromeres. The chromosomes are typically exposed to denaturation with a saturated solution of barium hydroxide, prior to Giemsa staining.

Table 2.3: Human chromosome groups

Group	Chromosomes	Description
A	1–3	Largest; 1 and 3 are metacentric but 2 is submetacentric
B	4,5	Large; submetacentric with two arms very different in size
C	6–12,X	Medium size; submetacentric
D	13–15	Medium size; acrocentric with satellites
E	16–18	Small; 16 is metacentric but 17 and 18 are submetacentric
F	19,20	Small; metacentric
G	21,22,Y	Small; acrocentric, with satellites on 21 and 22 but not on the Y

Autosomes are numbered from largest to smallest, except that chromosome 21 is smaller than chromosome 22.

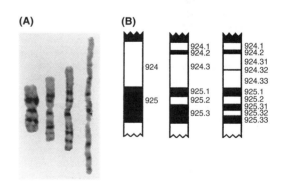

Figure 2.16: Different chromosome banding resolutions can resolve bands, sub-bands and sub-sub-bands.

(**A**) G-banded chromosome 1 at different banding resolutions. (**B**) Numbering of bands, sub-bands, and sub-sub-bands. Reproduced from Wolstenholme (1992) in *Human Cytogenetics: a Practical Approach*, vol. 1, 2nd edn, IRL Press. By permission of Oxford University Press.

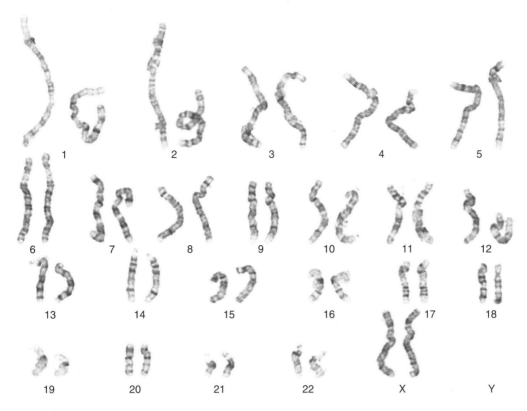

Figure 2.17: G-banded prometaphase karyogram of mitotic chromosomes from lymphocytes of a normal female.

Compare with the idealized ideograms in *Figure 2.18*. Overall lengths of metaphase chromosomes range between 2 and 10 μm; the DNA of the cell, if stretched out, would be about 2 m long. Courtesy of Dr Sue Hamilton, St. Mary's Hospital, Manchester.

Chromosome nomenclature

The International System for Human Cytogenetic Nomenclature (ISCN) is fixed by the Standing Committee on Human Cytogenetic Nomenclature (see Mitelman, 1995). The basic terminology for banded chromosomes was decided at a meeting in Paris in 1971, and is often referred to as the Paris nomenclature.

Short arm locations are labeled p (*petit*) and long arms q (*queue*). Each chromosome arm is divided into regions labeled p1, p2, p3 etc., and q1, q2, q3, etc., counting outwards from the centromere. Regions are delimited by specific landmarks, which are consistent and distinct morphological features, such as the ends of the chromosome arms, the centromere and certain bands. Regions are divided into bands labeled p11 (one–one, not eleven!), p12, p13, etc., sub-bands labeled p11.1, p11.2, etc., and sub-sub-bands e.g. p11.21, p11.22, etc., in each case counting outwards from the centromere (*Figures 2.16* and *2.18*).

The centromere is designated 'cen' and the telomere 'ter'. *Proximal* Xq means the segment of the long arm of the X that is closest to the centromere, while *distal* 2p means the portion of the short arm of chromosome 2 that is most distant from the centromere, and therefore closest to the telomere.

nowadays more often by getting an image analysis computer to do the job.

Chromosome banding picks out structural organization on a 1–10 Mb scale

Various treatments involving denaturation and/or enzymatic digestion, followed by incorporation of a DNA-specific dye, can cause human and other mitotic chromosomes to stain as a series of light and dark bands (*Box 2.4, Figure 2.18*; see Craig and Bickmore, 1993). Banding patterns are interesting (as well as being useful to cytogeneticists) because they provide evidence of some sort of structure over 1–10 Mb regions. The banding patterns correlate with other properties. Regions that stain as dark G bands replicate late in S phase of the cell cycle and contain more condensed chromatin, while R bands (light G bands) generally replicate early in S phase, and have less condensed chromatin. Genes are mostly concentrated in the R bands, while the later replicating, more condensed G-band DNA is less active transcriptionally. There are also differences in the types of dispersed repeat elements found in G and R bands (Sections 7.4.5 and 7.4.6).

Bands similar to G bands can be produced by staining with quinacrine, which preferentially binds to AT-rich DNA, while the R-banding pattern can be elicited using chromomycin, which preferentially binds GC-rich DNA.

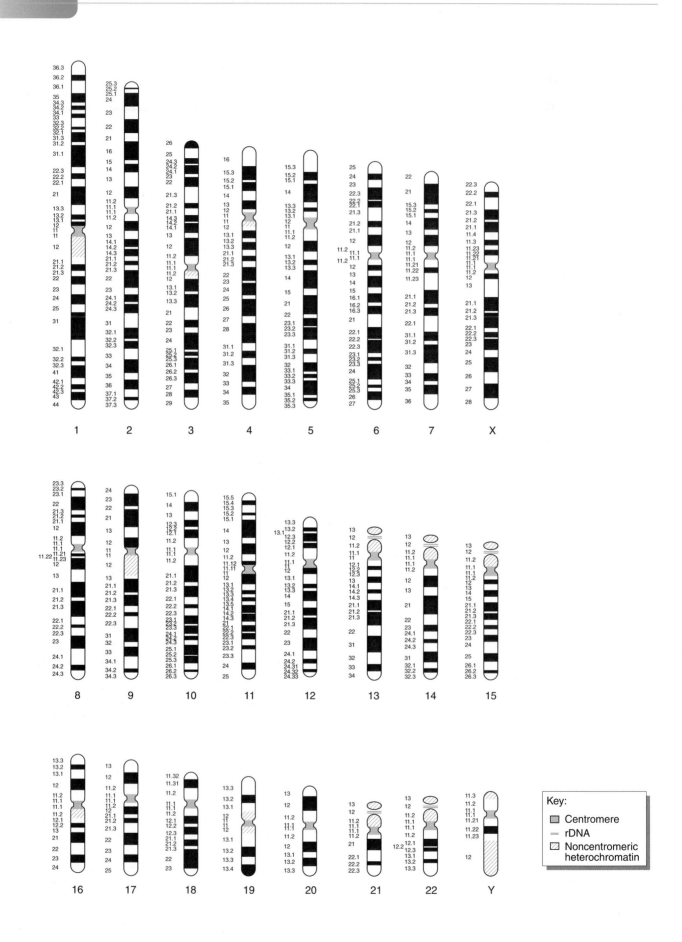

Key:

■ Centromere

— rDNA

▨ Noncentromeric heterochromatin

Figure 2.18: Banding pattern of human chromosomes.

This is a compilation of the best banding patterns that might be seen on each chromosome, and not a picture of how chromosomes appear in any one cell under the microscope. Chromosomes are numbered in order of size, except that 21 is actually smaller than 22. Arrays of repeated ribosomal DNA genes on the short arms of the acrocentric chromosomes 13, 14, 15, 21 and 22 often appear as thin stalks carrying knobs of chromatin (satellites). Heterochromatin occurs at centromeres, on much of the Y chromosome long arm, at secondary constrictions on 1q, 9q and 16q, and on the short arms of the acrocentric chromosomes.

However, the AT content of human G band DNA is only a few per cent higher than R band DNA. Saitoh and Laemmli (1994) suggested the difference depends on differences in the scaffold–loop structure (see *Figure 2.7*). Chromatin loops are thought to attach to the chromosome scaffold at special scaffold attachment regions (SARs). According to the Saitoh and Laemmli model, there are more SARs per unit length of DNA in G bands than in R bands. G bands have smaller loops and a tighter 'queue' of SARs along the scaffold, so that there are more SARs per unit length of chromosome, leading to stronger staining with AT-selective stains like Giemsa.

2.6 Chromosome abnormalities

Chromosome abnormalities might be defined as changes resulting in a visible alteration of the chromosomes. How much can be seen depends on the technique used. The smallest loss or gain of material visible by traditional methods on standard cytogenetic preparations is about 4 megabases of DNA. However, fluorescence *in situ* hybridization (FISH, Section 10.1.4) allows much smaller changes to be seen; the development of molecular cytogenetics has removed any clear dividing line between changes described as chromosomal abnormalities and changes thought of as molecular or DNA defects. An alternative definition of a chromosomal abnormality is an abnormality produced by specifically chromosomal mechanisms. Most chromosomal aberrations are produced by misrepair of broken chromosomes, by improper recombination or by malsegregation of chromosomes during mitosis or meiosis.

2.6.1 Types of chromosomal abnormality

A chromosomal abnormality may be present in all cells of the body (constitutional abnormality), or may be present in only certain cells or tissues (somatic or acquired abnormality). Constitutional abnormalities must have been present very early in development, most likely the result of an abnormal sperm or egg, or maybe abnormal fertilization or an abnormal event in the early embryo. By contrast, an individual with a somatic abnormality is a **mosaic** (see *Figure 3.9*), containing cells with two different chromosome constitutions, with both cell types deriving from the same zygote. Chromosomal abnormalities, whether constitutional or somatic, mostly fall into two categories: numerical and structural abnormalities (*Box 2.6*). Occasionally, abnormalities have been identified in which chromosomes have the correct number and structure, but represent unequal contributions from the two parents (Section 2.6.4). Perhaps unexpectedly, correct parental origin matters.

Box 2.6

Nomenclature of chromosome abnormalities

Numerical abnormalities
Triploidy	69,XXX, 69,XXY, 69,XYY
Trisomy	e.g. 47,XX, +21
Monosomy	e.g. 45,X
Mosaicism	e.g. 47,XXX / 46,XX

Structural abnormalities:
Deletion	e.g. 46,XY, del(4) (p16.3)[a] 46,XX, del(5) (q13q33)[a]
Inversion	e.g. 46,XY, inv(11) (p11p15)
Duplication	e.g. 46,XX, dup(1) (q22q25)
Insertion	e.g. 46,XX, ins(2) (p13q21q31)[b]
Ring	e.g. 46,XY, r(7) (p22q36)
Marker	e.g. 47,XX, +mar[c]
Translocation, reciprocal	e.g. 46,XX, t(2;6) (q35;p21.3)[d]
Translocation, Robertsonian	e.g. 45,XY, der(14;21) (q10;q10)[e] 46,XX, der(14;21) (q10;q10), +21[f]

Notes:
[a] Terminal deletion (breakpoint at 4p16.3) and interstitial deletion (5q13-q33).
[b] A rearrangement of one copy of chromosome 2 by insertion of segment 2q21-q31 into a breakpoint at 2p13.
[c] Karyotype of a cell that contains an extra unidentified chromosome (a *marker*).
[d] A balanced reciprocal translocation with breakpoints in 2q35 and 6p21.3.
[e] A balanced carrier of a 14;21 Robertsonian translocation. q10 is not really a chromosome band, but indicates the centromere. der means derivative.
[f] Translocation Down syndrome; a patient with one normal chromosome 14, a Robertsonian translocation 14;21 chromosome and two normal copies of chromosome 21.

This is a short nomenclature; a more complicated nomenclature is defined by the ISCN that allows complete description of any chromosome abnormality – see Further reading.

2.6.2   Numerical chromosomal abnormalities involve gain or loss of complete chromosomes

Three classes of numerical chromosomal abnormalities can be distinguished: polyploidy, aneuploidy and mixoploidy.

Polyploidy

Between 1 and 3% of recognized human pregnancies are **triploid**. The most usual cause is two sperm fertilizing a single egg (dispermy); sometimes the cause is a diploid gamete (*Figure 2.19*). Triploids very seldom survive to term, and the condition is not compatible with life. Tetraploidy is much rarer and always lethal. It is usually due to failure to complete the first zygotic division: the DNA has replicated to give a content of 4C, but cell division has not then taken place as normal. Although constitutional polyploidy is rare and lethal, all normal people have some polyploid cells (Section 2.1.3).

Aneuploidy

Euploidy means having complete chromosome sets (*n*, 2*n*, 3*n*, etc.). **Aneuploidy** is the opposite, that is, one or more individual chromosomes extra or missing from a euploid set. **Trisomy** means having three copies of a particular chromosome in an otherwise diploid cell, for example trisomy 21 (47,XX or XY, +21) in Down syndrome. Monosomy is the corresponding lack of a chromosome, for example monosomy X (45,X) in Turner syndrome. Cancer cells often show extreme aneuploidy, with multiple chromosomal abnormalities (*Figure 18.6*). Aneuploid cells arise through two main mechanisms:

- **Nondisjunction**: failure of paired chromosomes to separate (disjoin) in anaphase of meiosis I, or failure

of sister chromatids to disjoin at either meiosis II or at mitosis. Nondisjunction in meiosis produces gametes with 22 or 24 chromosomes, which after fertilization by a normal gamete make a trisomic or monosomic zygote. Nondisjunction in mitosis produces a mosaic.

- **Anaphase lag**: failure of a chromosome or chromatid to be incorporated into one of the daughter nuclei following cell division, as a result of delayed movement (lagging) during anaphase. Chromosomes that do not enter a daughter cell nucleus are lost.

Mixoploidy

Mixoploidy includes **mosaicism** (an individual possesses two or more genetically different cell lines all derived from a single zygote) and **chimerism** (an individual has two or more genetically different cell lines originating from different zygotes – see *Figure 3.9*). Abnormalities that would be lethal in constitutional form may be compatible with life in mosaics.

Aneuploidy mosaics are common. For example, mosaicism resulting in a proportion of normal cells and a proportion of aneuploid (e.g. trisomic) cells can be ascribed to nondisjunction or chromosome lag occurring in one of the mitotic divisions of the early embryo (any monosomic cells that are formed usually die out). Polyploidy mosaics (e.g. human diploid/triploid mosaics) are occasionally found. As gain or loss of a haploid set of chromosomes by mitotic nondisjunction is most unlikely, human diploid/triploid mosaics most probably arise by fusion of the second polar body with one of the cleavage nuclei of a normal diploid zygote.

Clinical consequences of numerical abnormalities

Having the wrong number of chromosomes has serious, usually lethal, consequences (*Table 2.4*). Even though the extra chromosome 21 in a man with Down syndrome is a perfectly normal chromosome, inherited from a normal parent, its presence causes multiple congenital abnormalities. Autosomal monosomies have even more catastrophic consequences than trisomies. These abnormalities must be the consequence of an imbalance in the levels of gene products encoded on different chromosomes. Normal development and function depend on innumerable interactions between gene products, including many that are encoded on different chromosomes. Altering the relative numbers of chromosomes will affect these interactions.

Having the wrong number of sex chromosomes has far fewer ill effects than having the wrong number of any autosome. 47,XXX and 47,XYY people often function within the normal range; 47,XXY men have relatively minor problems compared to people with any autosomal trisomy, and even monosomy, in 45,X women, has remarkably few major consequences. In fact, since normal people can have either one or two X chromosomes, and either no or one Y, there must be special mechanisms that allow normal function with variable numbers of sex chro-

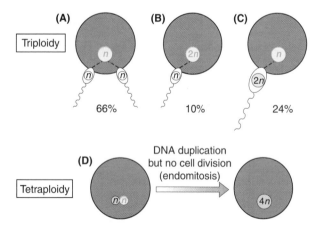

Figure 2.19: Origins of triploidy and tetraploidy.

About two-thirds of human triploids arise by fertilization of a single egg by two sperm (**A**). Other causes are a diploid egg (**B**) or sperm (**C**). Most human triploids abort spontaneously; very rarely they survive to term, but not beyond. Tetraploidy (**D**) results from failure of the first mitotic division after fertilization, and is incompatible with development.

Table 2.4: Consequences of numerical chromosomal abnormalities

Polyploidy		
Triploidy	(69,XXX, XXY or XYY)	1–3% of all conceptions; almost never liveborn; do not survive
Aneuploidy		
Autosomes	**nullisomy** (missing a pair of homologs)	Preimplantation lethal
	monosomy (one chromosome missing)	Embryonic lethal
	trisomy (one extra chromosome)	Usually embryonic or fetal lethal
		Trisomy 13 (Patau syndrome) and trisomy 18 (Edwards syndrome) may survive to term
		Trisomy 21 (Down syndrome) may survive to age 40 or longer
Sex chromosomes		
	XXX, XXY, XYY	Relatively minor problems, normal lifespan
	45,X	Turner syndrome – 99% abort spontaneously; survivors are of normal intelligence but infertile and show minor physical signs

mosomes. In the case of the Y chromosome, this is because it carries very few genes, whose only important function is to determine male sex. For the X chromosome, the special mechanism of lyonization (Section 2.2.3) controls the level of X-encoded gene products independently of the number of X chromosomes present in the cell.

Autosomal monosomy is invariably lethal at the earliest stage of embryonic life. On every chromosome there are probably a few genes where a halving of the level of the gene product is incompatible with development. Also, while such a halving is not obviously pathogenic for most genes (Section 16.4.3), it may have minor effects, and the combination of hundreds or thousands of these minor effects could be enough to disrupt normal development of the embryo. Trisomies make a smaller change than monosomies in relative levels of gene products, and their effects are somewhat less. Trisomic embryos survive longer than monosomic ones, and trisomies 13, 18 and 21 are compatible with survival until birth. Interestingly, these three chromosomes seem to be relatively poor in genes (Section 7.1.2). It is not so obvious why triploidy is lethal in humans and other animals. With three copies of every autosome, the dosage of autosomal genes is balanced and should not cause problems. Triploids are always sterile because triplets of chromosomes cannot pair and segregate correctly in meiosis, but many triploid plants are in all other respects healthy and vigorous. The lethality in animals is probably explained by imbalance between products encoded on the X chromosome and autosomes, which lyonization is unable to compensate.

2.6.3 Structural chromosomal abnormalities result from misrepair of chromosome breaks or from malfunction of the recombination system

Chromosome breaks occur either as a result of damage to DNA (by radiation or chemicals, for example) or as part of the mechanism of recombination. In G_2 phase of the cell cycle (*Figure 2.2*) chromosomes consist of two chromatids. Breaks occurring at this stage are manifest as chromatid breaks, affecting only one of the two sister chromatids. Breaks occurring in G_1 phase, if not repaired before S phase, appear later as chromosome breaks, affecting both chromatids. Cells have enzyme systems that recognize and if possible repair broken chromosome ends. Repair can be either by joining two broken ends together, or by capping a broken end with a telomere. Cell cycle checkpoint mechanisms (Section 18.7.3) normally prevent cells with unrepaired chromosome breaks from entering mitosis; if the damage cannot be repaired, the cell commits suicide (apoptosis).

Structural abnormalities arise when breaks are repaired incorrectly. Provided there are no free broken ends, the cell cycle checkpoints can be negotiated satisfactorily, but in fact the wrong broken ends may have been joined together. Any resulting chromosome that has no centromere (acentric) or two centromeres (dicentric) will not segregate stably in mitosis, and will eventually be lost. Chromosomes with a single centromere can be stably propagated through successive rounds of mitosis, even if they are structurally abnormal. Meiotic recombination between mispaired chromosomes is a common cause of translocations, especially in spermatogenesis. *Table 2.5* summarizes the main stable structural abnormalities and *Figures 2.20* and *2.21* illustrate how they arise.

An additional rare class of structural abnormality not shown in *Table 2.5* are **isochromosomes**. These are symmetrical chromosomes consisting of either two long arms or two short arms of a particular chromosome. They are believed to arise from an abnormal U-type exchange between sister chromatids just next to the centromere of a chromosome. Isochromosomes are rare except for i(Xq); i(21q) are an occasional cause of Down syndrome.

Structural chromosomal abnormalities are *balanced* if there is no net gain or loss of chromosomal material, and *unbalanced* if there is net gain or loss. In general, balanced abnormalities (inversions, balanced translocations) have

Table 2.5: Structural abnormalities resulting from misrepair of chromosome breaks or recombination between nonhomologous chromosomes

	One chromosome involved	Two chromosomes involved
One break	Terminal deletion (healed by adding telomere)	—
Two breaks	Interstitial deletion; Inversion; Ring chromosome (*Figure 2.20*) Duplication or deletion by unequal sister-chromatid exchange (*Figure 9.7*)	Reciprocal translocation (*Figure 2.21*) Robertsonian translocation (*Figure 2.21*) Duplication or deletion by unequal recombination (*Figure 9.7*)
Three breaks	Various rearrangements, e.g. inversion with deletion, intrachromosomal insertion	Interchromosomal insertion (direct or inverted)

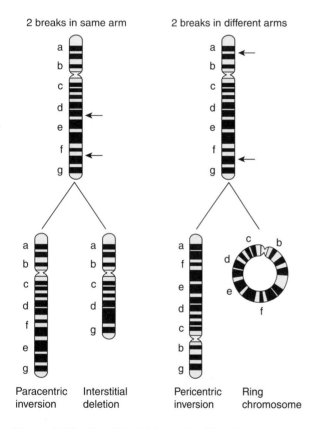

Paracentric inversion   Interstitial deletion   Pericentric inversion   Ring chromosome

Figure 2.20: Possible stable results of two breaks on a single chromosome.

no effect on the phenotype, although there are important exceptions to this:

- a chromosome break may disrupt an important gene;
- the break may affect expression of a gene even though it does not disrupt the coding sequence. It may separate a gene from a control element, or it may put the gene in an inappropriate chromatin environment, for example translocating a normally active gene into heterochromatin;

- balanced X-autosome translocations cause problems with X-inactivation (see *Figure 15.9*).

Robertsonian translocations are sometimes called **centric fusions**, but this is misleading because in fact the breaks are in the proximal short arms. The translocation chromosome is really dicentric, but because the two centromeres are very close together they function as one, and the chromosome segregates regularly. The distal parts of the two short arms are lost as an acentric fragment. Short arms of acrocentric chromosomes contain only arrays of repeated ribosomal RNA genes, and the loss of two short arms has no phenotypic effect. Because there is no phenotypic effect, Robertsonian translocations are regarded as balanced, even though in fact some material has been lost.

Unbalanced abnormalities can arise directly, through deletion or, rarely, duplication, or indirectly by malsegregation of chromosomes during meiosis in a carrier of a balanced abnormality. Carriers of balanced structural abnormalities can run into trouble during meiosis, if the structures of homologous pairs of chromosomes do not correspond:

- A carrier of a balanced reciprocal translocation can produce gametes that after fertilization give rise to an entirely normal child, a phenotypically normal balanced carrier, or various unbalanced karyotypes that always combine monosomy for part of one of the chromosomes with trisomy for part of the other (*Figure 2.22*). It is not possible to make general statements about the relative frequencies of these outcomes. The size of any unbalanced segments depends on the position of the breakpoints. If the unbalanced segments are large, the fetus will probably abort spontaneously; unbalance for smaller segments may result in liveborn abnormal babies.
- A carrier of a balanced Robertsonian translocation can produce gametes that after fertilization give rise to an entirely normal child, a phenotypically normal balanced carrier, or a conceptus with full trisomy or full monosomy for one of the chromosomes involved (*Figure 2.23*).
- A carrier of a pericentric inversion may produce

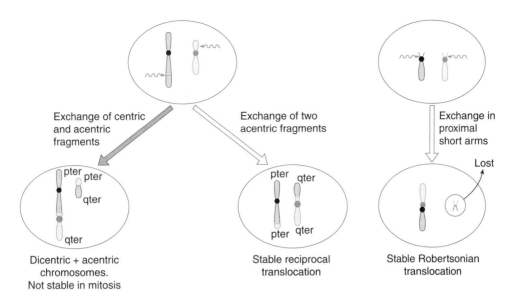

Figure 2.21: Origins of translocations.

Dicentric and acentric chromosomes are not stable through mitosis. Robertsonian translocations are produced by exchanges between the proximal short arms of the acrocentric chromosomes 13, 14, 15, 21 and 22. Both centromeres are present, but they function as one and the chromosome is stable. The small acentric fragment is lost, but this has no pathological consequences because it contains only repeated rDNA sequences, which are also present on the other acrocentric chromosomes.

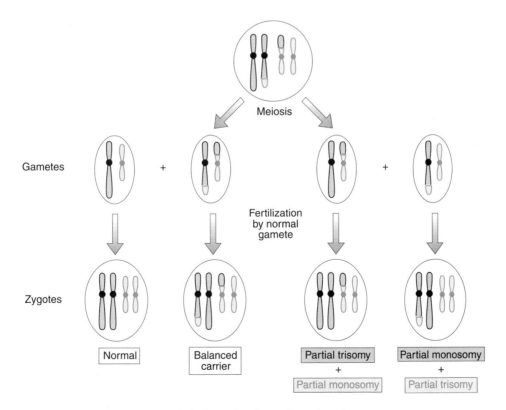

Figure 2.22: Results of meiosis in a carrier of a balanced reciprocal translocation.

Other modes of segregation are also possible, for example 3:1 segregation. The relative frequency of each possible gamete is not readily predicted. The risk of a carrier having a child with each of the possible outcomes depends on its frequency in the gametes and also on the likelihood of a conceptus with that abnormality developing to term. See the book by Gardner and Sutherland (Further reading) for discussion.

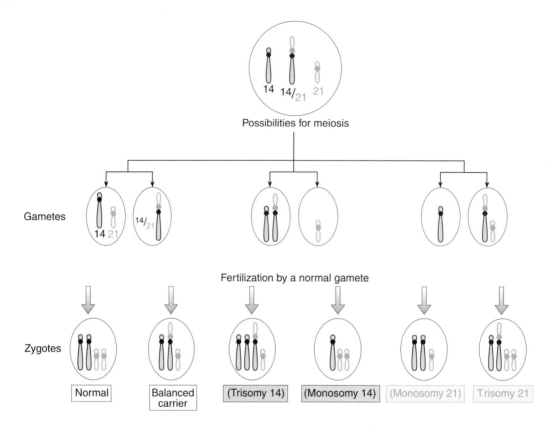

Figure 2.23: Results of meiosis in a carrier of a Robertsonian translocation.

Carriers are asymptomatic but often produce unbalanced gametes which can result in a monosomic or trisomic zygote. The bracketed monosomic and trisomic zygotes in this example would not develop to term.

unbalanced offspring because when the inverted and noninverted homologs pair they form a loop so that matching segments pair along the whole length of the chromosomes. If a crossover occurs within the loop, the result is a chromosome carying an unbalanced deletion and duplication. Paracentric chromosome inversions form similar loops, but any crossover within the loop generates an acentric or dicentric chromosome, which is unlikely to survive. For details of meiosis in carriers of inversions, see the book by Gardner and Sutherland (Further reading) or any other cytogenetics text.

2.6.4 Apparently normal chromosomal complements may be pathogenic if they have the wrong parental origin

The rare abnormalities described below demonstrate that it is not enough to have the correct number and structure of chromosomes; they must also have the correct parental origin. 46,XX conceptuses in which both genomes originate from the same parent (uniparental diploidy) never develop correctly. For some individual chromosomes, having both homologs derived from the same parent (**uniparental disomy**) also causes abnormality. A small

number of genes are imprinted with their parental origin (Section 8.5) and are expressed differently according to the origin. It is assumed that the abnormalities of uniparental disomy and uniparental diploidy are caused by abnormal expression of such imprinted genes.

Uniparental diploidy is seen in hydatidiform moles, abnormal conceptuses with a 46,XX karyotype of exclusively paternal origin. Molar pregnancies show widespread hyperplasia of the trophoblast but no fetal parts, and have a significant risk of transformation into choriocarcinoma. Genetic marker studies show that most moles are homozygous at all loci, indicating that they arose by chromosome doubling from a single sperm. Ovarian teratomas are the result of maternal uniparental diploidy. These rare benign tumors of the ovary consist of disorganized embryonic tissues, without extraembryonic membranes. They arise by activation of an unovulated oocyte.

Uniparental disomy (UPD), affecting a single pair of homologs, goes undiagnosed if the result is not abnormal, but is detected for chromosomes for which it produces characteristic syndromes (see *Box 16.6*). UPD can be isodisomy, where both homologs are identical, or heterodisomy, where they are derived from both homologs in one parent. The usual cause is thought to be trisomy rescue: a conceptus that is trisomic and would otherwise die, occasionally loses one chromosome by mitotic nondisjunction or anaphase lag from a totipotent cell. The euploid progeny of this cell form the embryo, while all the aneuploid cells die. If each of the three copies has an equal chance of being lost, there will be a two in three chance of a single chromosome loss leading to the normal chromosome constitution and a one in three chance of uniparental disomy (either paternal or maternal). Uniparental isodisomy may possibly arise by selection pressure on a monosomic embryo to achieve euploidy by selective duplication of the monosomic chromosome.

Further reading

Alberts B, Bray D, Lewis J, Raff M, Roberts K, Watson JD (1994) *Molecular Biology of the Cell*, 3rd edn. Garland Publishing, New York. *A comprehensive text on cell biology.*

Gardner RJM, Sutherland GR (1996) *Chromosome Abnormalities and Genetic Counseling*, 2nd edn. OUP, Oxford. *A thorough introduction to the nature, origin and consequences of human chromosomal abnormalities.*

Lawrence PA (1992) *The Making of a Fly.* Blackwell, Oxford. *Some of the detail is out of date, but the overall picture is superbly presented.*

Moore KL, Persaud TVN (1998) *The Developing Human: Clinically Oriented Embryology*, 6th edn. WB Saunders, Philadelphia. *Excellently illustrated description of development, week by week.*

Rooney DE, Czepulkowski BH (1992) (eds) *Human Cytogenetics: a Practical Approach*, Vols 1 and 2. IRL Press, Oxford. *Detailed laboratory protocols.*

Therman E, Susman M (1992) *Human Chromosomes: Structure, Behavior and Effects*, 3rd edn. Springer, New York. *An excellent compact introduction; emphasis on scientific bases rather than clinical implications.*

Tyler-Smith C, Willard HF (1993) Mammalian chromosome structure. *Curr. Opin. Genet. Dev.*, **3**, 390–397.

References

Craig JM, Bickmore WA. (1993) Chromosome bands – flavours to savour. *Bioessays*, **15**, 349–354.

Dobbs DL, Shaiu W-L, Benbow RM (1994) Modular sequence elements associated with origin regions in eukaryotic chromosomal DNA. *Nucleic Acids Res.*, **22**, 2479–2489.

Greider CW (1999) Telomeres do D-loop–T-loop. *Cell*, **97**, 419–422.

Heard E, Clerc P, Avner P (1997) X-chromosome inactivation in mammals. *Annu. Rev. Genet.*, **31**, 571–610.

Huxley C (1997) Mammalian artificial chromosomes and chromosome transgenics. *Trends Genet.*, **13**, 345–347.

Lyon MF (1999) X-chromosome inactivation. *Curr. Biol.*, **9**, R235–R237.

Manuelidis L (1990) A view of interphase chromosomes. *Science*, **250**, 1533–1540.

Migeon BR (1994) X-chromosome inactivation: molecular mechanisms and genetic consequences. *Trends Genet.*, **10**, 230–235.

Mitelman F (ed.) (1995) *An International System for Human Cytogenetic Nomenclature* (ISCN). Karger, Basel.

Saitoh Y, Laemmli UK (1994) Metaphase chromosome structure: bands arise from a differential folding path of the highly AT-rich scaffold. *Cell*, **76**, 609–622.

Schindelhauer D (1999) Construction of mammalian artificial chromosomes: prospects for defining an optimal centromere. *BioEssays*, **21**, 76–83.

Wolstenholme J (1992) In: *Human Cytogenetics: a Practical Approach* (DE Rooney, BH Czepulkowski eds), Vol. 1, 2nd edn, pp. 1–30. IRL Press, Oxford.

Genes in pedigrees

3

3.1 Mendelian pedigree patterns

The simplest genetic characters are those whose presence or absence depends on the **genotype** at a single **locus**. That is not to say that the character itself is programmed by only one pair of genes – expression of any human character is likely to require a large number of genes and environmental factors. However, sometimes a particular genotype at one locus is both necessary and sufficient for the character to be expressed, given the normal genetic and environmental background of the organism. Such characters are called **mendelian**. In humans over 10 000 mendelian characters are known. As described in the *Introduction*, the essential starting point for acquiring information on any human mendelian character, whether pathological or non-pathological, is the OMIM Internet database.

3.1.1 Dominance and recessiveness are properties of characters, not genes

A character is **dominant** if it is manifest in the **heterozygote** and **recessive** if not. Note that dominance and recessiveness are properties of characters, not genes. Thus sickle cell anemia is recessive because only HbS **homozygotes** manifest it. Heterozygotes for the same gene show sickling trait, which is therefore a dominant character. Most human dominant syndromes are known only in heterozygotes. Sometimes homozygotes have been described, born from matings of two heterozygous affected people, and often the homozygotes are much more severely affected. Examples are achondroplasia (short-limbed dwarfism) and Type 1 Waardenburg syndrome (deafness with pigmentary abnormalities). Nevertheless we describe achondroplasia and Waardenburg syndrome as dominant because these terms describe **phenotypes** seen in heterozygotes. In experimental organisms, where this uncertainty does not exist, geneticists tend to use the term **semidominant** when the heterozygote has an intermediate phenotype, reserving 'dominant' for conditions where the homozygote is indistinguishable from the heterozygote – Huntington disease (adult-onset progressive neurological deterioration) for example. The question of dominance has been well reviewed by Wilkie (1994). Males are **hemizygous** for loci on the X and Y chromosomes, where they have only a single copy of each gene, so the question of dominance or recessiveness does not arise in males for X- or Y-linked characters.

3.1.2 There are five basic mendelian pedigree patterns

Figure 3.1 shows the symbols used for drawing pedigrees. Mendelian characters may be determined by loci on an autosome or on the X or Y sex chromosomes. Autosomal characters in both sexes and X-linked characters in females can be dominant or recessive. Nobody has two genetically different Y chromosomes (in the rare XYY

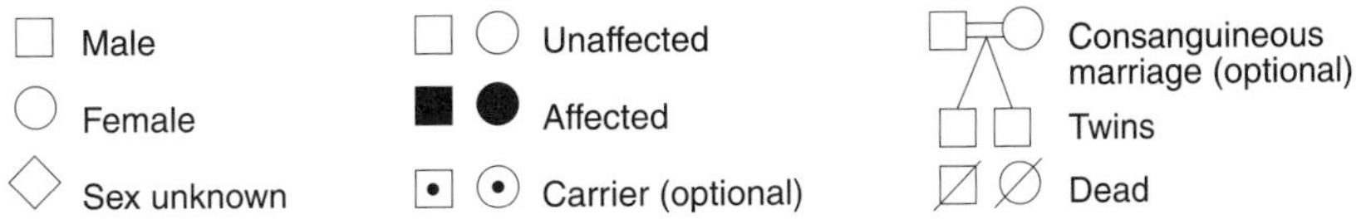

Figure 3.1: Main symbols used in pedigrees.

Generations are usually labeled in Roman numerals, and individuals within each generation in Arabic numerals; III-7 or III$_7$ is the seventh person from the left (unless explicitly numbered otherwise) in generation III. An arrow → can be used to indicate the proband or propositus (female: proposita) through whom the family was ascertained.

males, the two Y chromosomes are duplicates). Thus there are five archetypal mendelian pedigree patterns (*Figure 3.2*; *Box 3.1*). Special considerations apply to X- and Y-linked conditions as described below, so that in practice the important mendelian pedigree patterns are autosomal dominant, autosomal recessive and X-linked (dominant or recessive). These basic patterns are subject to various complications discussed in Section 3.2, and illustrated in *Figure 3.5*.

X-inactivation (lyonization) blurs the distinction between dominant and recessive X-linked conditions

Carriers of 'recessive' X-linked conditions often manifest some signs, while compared to affected males, heterozygotes for 'dominant' conditions are usually more mildly and variably affected. This is a consequence of X-inactivation. As described in Chapter 2 (Section 2.2.3) mammals compensate for the unequal numbers of X

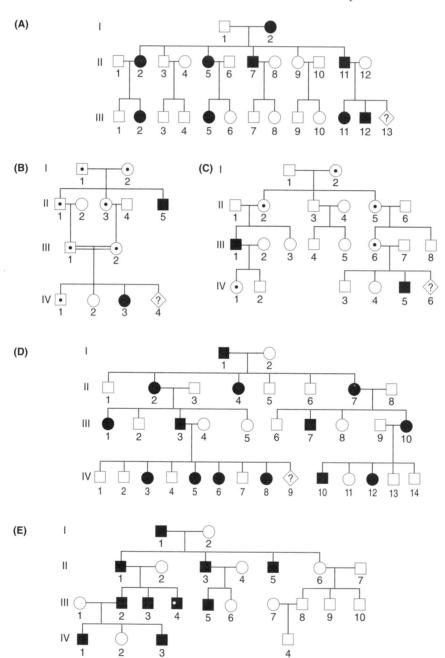

Figure 3.2: Basic mendelian pedigree patterns.

(**A**) Autosomal dominant; (**B**) autosomal recessive; (**C**) X-linked recessive; (**D**) X-linked dominant; (**E**) Y-linked. The risk for the individuals marked with a query are (A) 1 in 2, (B) 1 in 4, (C) 1 in 2 males or 1 in 4 of all offspring, (D) negligibly low for males, 100% for females. See Section 3.2 and *Figure 3.5* for complications to these basic patterns.

Box 3.1

Mendelian pedigree patterns

Autosomal dominant inheritance
(Figure 3.2A)

- An affected person usually has at least one affected parent (for exceptions see *Figure 3.5*).
- Affects either sex.
- Transmitted by either sex.
- A child of an affected × unaffected mating has a 50% chance of being affected (this assumes the affected parent is heterozygous, which is usually true for rare conditions).

Autosomal recessive inheritance
(Figure 3.2B)

- Affected people are usually born to unaffected parents.
- Parents of affected people are usually asymptomatic carriers.
- There is an increased incidence of parental consanguinity.
- Affects either sex.
- After the birth of an affected child, each subsequent child has a 25% chance of being affected.

X-linked recessive inheritance
(Figure 3.2C)

- Affects mainly males.
- Affected males are usually born to unaffected parents; the mother is normally an asymptomatic carrier and may have affected male relatives.
- Females may be affected if the father is affected and the mother is a carrier, or occasionally as a result of non-random X-inactivation (Section 3.1.2).
- There is no male-to-male transmission in the pedigree (but matings of an affected male and carrier female can give the appearance of male to male transmission, see *Figure 3.5G*).

X-linked dominant inheritance
(Figure 3.2D)

- Affects either sex, but more females than males.
- Females are often more mildly and more variably affected than males.
- The child of an affected female, regardless of its sex, has a 50% chance of being affected.
- For an affected male, all his daughters but none of his sons are affected.

Y-linked inheritance *(Figure 3.2E)*

- Affects only males.
- Affected males always have an affected father (unless there is a new mutation).
- All sons of an affected man are affected.

tion takes place early in embryonic life, and once a cell has chosen which X to inactivate, that choice is transmitted clonally to all its daughter cells.

A female heterozygous for an X-linked condition (dominant or recessive), is a mosaic (see Section 3.2.6). Each cell expresses either the normal or the abnormal allele, but not both. Where the phenotype depends on a circulating product, as in hemophilia (failure of blood to clot), there is an averaging effect between the normal and abnormal cells. Female carriers have an intermediate phenotype, and are usually clinically unaffected but biochemically abnormal. Where the phenotype is a localized property of individual cells, as in hypohidrotic ectodermal dysplasia (MIM 305100: missing sweat glands, abnormal teeth and hair) female carriers show patches of normal and abnormal tissue. Occasional **manifesting heterozygotes** are seen for X-linked recessive conditions: these women may be quite severely affected because by bad luck most cells in some critical tissue have inactivated the normal X.

There are probably no Y-linked diseases

Although a few Y-linked characters have been described, no Y-linked diseases are known, apart from disorders of male sexual function. Conceivably such a disease may exist undiscovered, but this is unlikely for two reasons. First, the pedigree pattern would be strikingly noticeable, especially in societies that trace family through the male line, yet they have not been noted (claims for 'porcupine men' are dubious, see MIM 146600). Second, the Y-chromosome cannot carry any genes whose function is important for health, because females are perfectly normal without any Y-linked genes. Thus any Y-linked genes must code either for non-essential characters or for male-specific functions, and defects are unlikely to cause diseases apart from defects of male sexual function. Genes present as functional copies on both the Y and the X might prove an exception to this argument.

3.1.3 The mode of inheritance can rarely be defined unambiguously in a single pedigree

Given the limited size of human families, it is rarely possible to be completely certain of the mode of inheritance of a character simply by inspecting a single pedigree. In experimental animals one would set up a test cross and check for a 1 in 2 or 1 in 4 ratio. In human pedigrees the proportion of affected children is not a very reliable indicator. Mostly this is because the numbers are too small, but in addition, the way in which the family was ascertained can bias the ratio of affected to unaffected children observed. For recessive conditions, the proportion of affected children often seems to be greater than 1 in 4. This is because families are normally ascertained when they have an affected child; families where both parents are carriers but, by good fortune, nobody is affected are systematically missed. These biases of ascertainment, and the ways of correcting them, are discussed in Section 19.4.

For many of the rarer conditions, the stated mode of

chromosomes in male and female cells by permanently inactivating all but one X chromosome in each cell. XY males keep their single X active, whilst XX females inactivate one X (chosen at random) in each cell. Inactiva-

inheritance is no more than an informed guess. Assigning modes of inheritance is important, because that is the basis of the risk estimates used in genetic counseling. However, it is important to recognize that the modes of inheritance are often working hypotheses rather than established fact. OMIM uses an asterisk to denote entries with relatively well-established modes of inheritance. Only when a cloned copy of the gene is available can the inheritance be determined with certainty.

3.1.4 One gene – one enzyme does not imply one gene – one syndrome

Pedigree patterns provide the essential entry point into human genetics, but they are only a starting point for defining genes. It would be a serious error to imagine that the 10 000 or so known mendelian characters define 10 000 DNA coding sequences. This would be an unjustified extension of the one gene – one enzyme hypothesis of Beadle and Tatum. Back in the 1940s this hypothesis allowed a major leap forward in understanding how genes determine phenotypes. Since then it has been extended – some genes encode nontranslated RNAs, some proteins are not enzymes, and many proteins contain several separately encoded polypeptide chains. But even with these extensions, Beadle and Tatum's hypothesis cannot be used to imply a one-to-one correspondence between entries in the OMIM catalogue and entries in Genbank, the DNA sequence database.

The genes of classical genetics are abstract entities. Any character that is determined at a single chromosomal location will segregate in a mendelian pattern – but the determinant may not be a gene in the molecular geneticist's sense of the word. Fascio-scapulo-humeral muscular dystrophy (MIM 158900: severe but nonlethal weakness of certain muscle groups) is associated with small deletions of sequences at 4q35, but nobody (at the time of writing) has managed to find a protein-coding sequence at that location, despite intensive searching and sequencing. The 'gene' for Charcot-Marie-Tooth disease type 1A (MIM 118220: motor and sensory neuropathy) turned out to be a 1.5 Mb tandem duplication on chromosome

17p11.2 (Section 16.6.2). These examples are unusual; most OMIM entries probably do describe the consequences of mutations affecting a single transcription unit, but because of locus and allelic heterogeneity, there is still no one-to-one correspondence with Genbank entries.

Locus heterogeneity is common in syndromes that result from failure of a complex pathway

Profound congenital hearing loss is often genetic, and when genetic it is usually autosomal recessive. However, when two people with autosomal recessive profound hearing loss marry, as they often do, the children usually have normal hearing (*Figure 3.3*). This is an example of **complementation** (*Box 3.2*). The children will have normal hearing whenever the parents carry mutations in different genes. Diseases and developmental defects represent the failure of a pathway. It is easy to see that many different genes would be needed to construct so exquisite a machine as the cochlear hair cell, and a defect in any of

---

Box 3.2

The complementation test to discover whether two recessive characters are determined by allelic genes

Parental cross	$a_1a_1 \times a_2a_2$	$aaBB \times AAbb$
	|	|
	a_1a_2	AaBb
Offspring	mutant	wild type
Phenotype	**one locus**	**two loci**

Animals homozygous for the two characters are crossed and the phenotype of the offspring observed. If both animals carry mutations at the same locus the progeny will not have a wild-type allele, and so will be phenotypically abnormal. If there are two different loci the progeny are heterozygous for each of the two recessive characters, and therefore phenotypically normal. Very occasionally alleles at the same locus can complement each other (interallelic complementation).

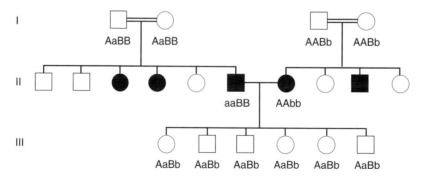

Figure 3.3: Complementation: parents with autosomal recessive profound hearing loss often have children with normal hearing.

II$_6$ and II$_7$ are offspring of unaffected but consanguineous parents, and each has affected sibs, making it likely that each has autosomal recessive hearing loss. All their children are unaffected, showing that II$_6$ and II$_7$ have nonallelic mutations.

those genes could lead to deafness. Such **locus hetero-geneity** is only to be expected in conditions like deafness, blindness or mental retardation, where a rather general pathway has failed; but even with more specific patholo-gies, multiple loci are very frequent. A striking example is Usher syndrome, an autosomal recessive combination of hearing loss and retinitis pigmentosa, which can be caused by mutations at eight or more unlinked loci (Smith *et al.*, 1994). OMIM has separate entries for known examples of locus heterogeneity (defined by linkage or mutation analysis), but there must be many undetected examples still contained within single entries.

Allelic series are a cause of clinical heterogeneity

Sometimes several apparently distinct human pheno-types turn out to be all caused by different allelic muta-tions at the same locus. The difference may be one of degree – mutations that partially inactivate the dys-trophin gene produce Becker muscular dystrophy, while mutations that completely inactivate the same gene pro-duce the similar but more severe Duchenne muscular dystrophy (lethal muscle wasting). Sometimes the differ-ence is qualitative – inactivation of the androgen receptor gene causes androgen insensitivity (MIM 313700; 46,XY embryos develop as females), but expansion of a run of glutamine codons within the same gene causes a very different disease, spinobulbar muscular atrophy or Kennedy disease (MIM 313200). These and other genotype-phenotype correlations are discussed in more depth in Chapter 16.

3.1.5 Mitochondrial inheritance gives a recognizable matrilinear pedigree pattern

In addition to the mutations in genes carried on the nuclear chromosomes, mitochondrial mutations are a significant cause of human genetic disease. The mitochon-drial genome (see *Figure 7.2*) is small but highly mutable compared to nuclear DNA, probably because mito-chondrial DNA replication is more error-prone and the number of replications is much higher. Mitochondrially-encoded diseases have two unusual features, **matrilineal inheritance** and frequent heteroplasmy (Wallace, 1994).

Inheritance is matrilineal, because sperm do not con-tribute mitochondria to the zygote (this assertion rests on limited evidence; however, paternally-derived mitochon-drial variants are not detected in children). Thus a mito-chondrially inherited condition can affect both sexes, but is passed on only by affected mothers (*Figure 3.4A*), giv-ing a recognizable pedigree pattern.

Cells contain many mitochondrial genomes. In some patients with a mitochondrial disease, every mitochon-drial genome carries the causative mutation (**homo-plasmy**), but in other cases a mixed population of normal and mutant genomes is seen within each cell (**hetero-plasmy**). Unlike nuclear genetic mosaicism, which must arise post-zygotically (Section 3.2.6), mitochondrial het-eroplasmy can be transmitted from heteroplasmic mother to heteroplasmic child. In such cases the proportion of abnormal mitochondrial genomes can vary remark-ably between the mother and child, suggesting that a

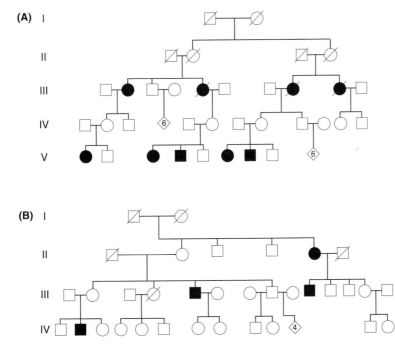

Figure 3.4: Pedigrees of mitochondrial diseases.

(**A**) A typical pedigree pattern, showing mitochondrially-determined hearing loss (family reported by Prezant *et al.*, 1993). (**B**) The atypical pattern of Leber's hereditary optic atrophy, which affects mainly males (family reported by Sweeney *et al.*, 1992).

surprisingly small number of maternal mitochondrial DNA molecules give rise to all the mitochondrial DNA of the child (see *Box 9.3*).

One of the best known mitochondrial diseases shows an unusual mode of inheritance: Leber's hereditary optic atrophy (LHON, MIM 535000: sudden and irreversible loss of sight) is associated with various mitochondrial mutations and is inherited matrilinearly, but unexpectedly, almost all affected patients are male (*Figure 3.4B*). Possibly LHON requires both a mitochondrial and an X-linked mutation, but attempts to demonstrate an X-linked susceptibility have not been successful, so the reason for the male excess remains unknown (Riordan-Eva and Harding, 1995). The complicated molecular pathology of mitochondrial diseases is discussed in Chapter 16.

3.2 Complications to the basic pedigree patterns

In real life various complications often disguise a basic mendelian pattern. *Figure 3.5* shows a number of common complications.

3.2.1 Common recessive conditions can give a pseudo-dominant pedigree pattern

If a character is common in the population, there is a high chance that it may be brought into the pedigree independently by two or more people. A common recessive character like blood group O may be seen in successive generations because of repeated marriages of group O people with heterozygotes. This produces a pattern resembling dominant inheritance (*Figure 3.5A*). Thus the classic pedigree patterns are best seen with rare characters.

3.2.2 Failure of a dominant condition to manifest is called nonpenetrance

With dominant conditions, **nonpenetrance** is a frequent complication. The **penetrance** of a character, for a given genotype, is defined as the probability that a person who has the genotype will manifest the character. By definition, a dominant character is manifest in a heterozygous person, and so should show 100% penetrance. Nevertheless, many human characters, while generally showing dominant inheritance, occasionally skip a generation. In *Figure 3.5B*, II_2 has an affected parent and an affected child, and almost certainly carries the mutant gene, but is phenotypically normal. This would be described as a case of non-penetrance.

There is no mystery about nonpenetrance; indeed, 100% penetrance is the more surprising phenomenon. Very often the presence or absence of a character depends, in the main and in normal circumstances, on the genotype at one locus, but an unusual genetic background, a particular lifestyle or maybe just chance means

that the occasional person may fail to manifest the character. Nonpenetrance is a major pitfall in genetic counseling. It would be an unwise counselor who, knowing the condition in *Figure 3.5B* was dominant and seeing III_7 was free of signs, told her that she had no risk of having affected children. One of the jobs of genetic counselors is to know the usual degree of penetrance of each dominant syndrome.

Frequently, of course, a character depends on many factors and does not show a mendelian pedigree pattern even if entirely genetic. There is a continuum of characters from fully penetrant mendelian to multifactorial (Section 3.4.2; *Figure 3.10*), with increasing influence of other genetic loci and/or the environment. No logical break separates imperfectly penetrant mendelian from multifactorial characters; it is a question of which is the most useful description to apply.

Late-onset diseases show age-related penetrance

A particularly important case of reduced penetrance is seen with late-onset diseases. Genetic conditions are, of course, not necessarily congenital (present at birth). The genotype is fixed at conception, but the phenotype may not manifest until adult life. In such cases the penetrance is age-related. Huntington disease is a well-known example (*Figure 3.6*). Delayed onset might be caused by slow accumulation of a noxious substance, by slow tissue death or by inability to repair some form of environmental damage. Hereditary cancers are caused by a second mutation affecting a cell of a person who already carries one mutation in a tumor suppressor gene (Chapter 18). Depending on the disease, the penetrance may become 100% if the person lives long enough, or there may be people who carry the gene but who will never develop symptoms no matter how long they live. Age-of-onset curves such as *Figure 3.6* are important tools in genetic counseling, because they enable the geneticist to estimate the chance that an at-risk but asymptomatic person will subsequently develop the disease.

3.2.3 Many conditions show variable expression

Related to nonpenetrance is the **variable expression** frequently seen in dominant conditions. *Figure 3.5C* shows an example from a family with Waardenburg syndrome. Different family members show different features of the syndrome. The cause is the same as with nonpenetrance: other genes, environmental factors or pure chance have some influence on development of the symptoms. Nonpenetrance and variable expression are typically problems with dominant, rather than recessive, characters. Partly this reflects the difficulty of spotting nonpenetrant cases in a typical recessive pedigree. However, as a general rule, recessive conditions are less variable than dominant ones, probably because the phenotype of a heterozygote involves a balance between the effects of the two alleles, so that the outcome is likely to be more

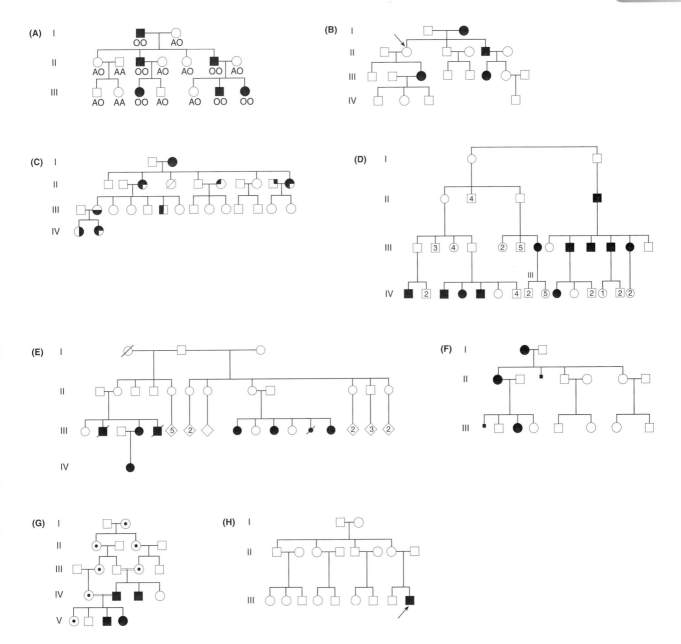

Figure 3.5: Complications to the basic mendelian patterns.

(**A**) A common recessive, such as blood group O, can give the appearance of a dominant pattern. (**B**) Autosomal dominant inheritance with nonpenetrance in II$_2$. (**C**) Autosomal dominant inheritance with variable expression: in this family with Waardenburg syndrome, shading of 1st quadrant = hearing loss; 2nd quadrant = different colored eyes; 3rd quadrant = white forelock; 4th quadrant = premature graying of hair. (**D**) Genetic imprinting: in this family autosomal dominant glomus tumors manifest only when the gene is inherited from the father (family reported by Heutink *et al.*, 1992). (**E**) Genetic imprinting: in this family autosomal dominant Beckwith–Wiedemann syndrome manifests only when the gene is inherited from the mother (family reported by Viljoen and Ramesar, 1992). (**F**) X-linked dominant incontinentia pigmenti. Affected males abort spontaneously (small squares). (**G**) An X-linked recessive pedigree where inbreeding gives an affected female and apparent male-to-male transmission. (**H**) A new autosomal dominant mutation, mimicking an autosomal or X-linked recessive pattern.

sensitive to outside influence than the phenotype of a homozygote. However, both nonpenetrance and variable expression are occasionally seen in recessive conditions.

These complications are much more conspicuous in humans than in plants or other animals, because laboratory animals and crop plants are far more genetically uni-

form than humans. What we see in human genetics is typical of a wild population. Nevertheless, mouse geneticists are familiar with the way expression of a mutant can change when it is bred onto a different genetic background, and understand its importance when studying mouse models of human diseases.

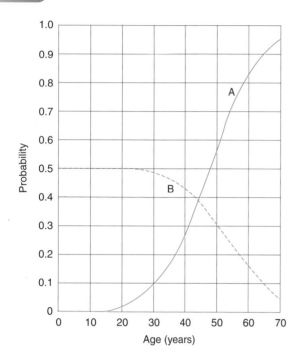

Figure 3.6: Age of onset curve for Huntington disease.

Curve A: probability that an individual carrying the disease gene will have developed symptoms by a given age. Curve B: risk that a healthy child of an affected parent carries the disease gene at a given age. Reproduced from Harper (1998) *Genetic Counselling*, 5th edn, with permission from Butterworth-Heinemann Ltd.

Anticipation is a special type of variable expression

Anticipation describes the tendency of some variable dominant conditions to become more severe in successive generations. Until recently, most geneticists were skeptical that this ever really happened. The problem is that true anticipation is very easily mimicked by random variations in severity. A family comes to clinical attention when a severely affected child is born. Investigating the history, the geneticist notes that one of the parents is affected, but only mildly. This looks like anticipation, but may actually be just a bias of ascertainment. Had the parent been severely affected, he or she would most likely never have become a parent, and had the child been mildly affected, the family would not have come to notice. Given the lack of any plausible mechanism for anticipation, and the statistical problems of demonstrating it in the face of these biases, most geneticists were unwilling to consider anticipation seriously until molecular developments obliged them to do so.

Anticipation suddenly became respectable, even fashionable, with the discovery of unstable expanding trinucleotide repeats in Fragile-X syndrome (MIM 309550: mental retardation with various physical signs), and later in myotonic dystrophy (MIM 160900: a very variable multisystem disease with characteristic muscular dysfunction) and Huntington disease (see *Box 16.7*). Severity or age of onset of these diseases correlates with the repeat length, and the repeat length tends to grow as the gene is transmitted down the generations. Thus these conditions show true anticipation. Now once again we see claims for anticipation being made for many diseases, and it is important to bear in mind that the old objection about bias of ascertainment remains valid. To be credible, a claim of anticipation requires careful statistical backing, and not just anecdotal evidence.

3.2.4 For imprinted genes, expression depends on parental origin

Certain human characters are autosomal dominant and transmitted by parents of either sex, but they manifest only when inherited from a parent of one particular sex. For example there are families with autosomal dominant glomus tumors that are expressed only in people who inherit the gene from their father (*Figure 3.5D*), while Beckwith–Wiedemann syndrome (MIM 130650: exomphalos, macroglossia, overgrowth) is sometimes dominant but expressed only by people who inherit it from their mother (*Figure 3.5E*). These parental sex effects are evidence of **imprinting**, a poorly understood phenomenon whereby certain genes are somehow marked (imprinted) with their parental origin. The many questions that surround the mechanism and evolutionary purpose of imprinting are discussed in Chapter 7 and a particularly striking clinical example is described in *Box 16.6*.

3.2.5 Male lethality may complicate X-linked pedigrees

For some X-linked dominant conditions, absence of the normal allele is lethal before birth. Thus affected males are not born, and we see a condition that affects only females, who pass it on to half their daughters but none of their sons (*Figure 3.5F*). There may be a history of miscarriages, but families are rarely big enough to prove that the number of sons is only half the number of daughters. An example is incontinentia pigmenti (MIM 308310: linear skin defects following defined patterns known as Blaschko's lines, often accompanied by neurological or skeletal problems).

3.2.6 New mutations often complicate pedigree interpretation, and can lead to mosaicism

Many cases of severe genetic disease are the result of fresh mutations, striking without warning in a family with no previous history of the disease. People with severe genetic diseases seldom reproduce, so they do not pass on their mutant genes. On the assumption that, averaged over time, new mutations exactly replace the disease genes lost through natural selection, there is a simple relationship (described in Section 3.3) between the rate at which natural selection is removing disadvantageous genes, the rate at which new mutation is creating them, and their frequency in the population. The general mechanisms that affect the population frequency of alleles are discussed in Section 9.2.3.

This mutation-selection dynamic has different effects on pedigrees, depending on the mode of inheritance. Autosomal recessive pedigrees are not significantly affected – any new mutations probably happened many generations ago, and we can safely assume that the parents of an affected child are both carriers. For dominant conditions however the turnover of disease genes is much faster, because they are constantly exposed to selection. A fully penetrant lethal dominant would necessarily always occur by fresh mutation, and the parents would never be affected (an example is thanatophoric dysplasia, MIM 187600: severe shortening of long bones and abnormal fusion of cranial sutures). People with nonlethal but severe dominant conditions often have unaffected parents and no previous family history of the condition. Serious X-linked recessives also show a significant proportion of fresh mutations, because the gene is exposed to natural selection whenever it is in a male.

When a normal couple with no relevant family history have a child with severe abnormalities (*Figure 3.5H*), deciding the mode of inheritance and recurrence risk can be very difficult: the problem might be autosomal recessive, autosomal dominant with a new mutation, X-linked recessive (if the child is male) or nongenetic. A further complication is introduced by germinal mosaicism (see below).

Mosaics have two (or more) genetically different cell lines

We have seen that in serious autosomal dominant and X-linked diseases, where affected people have few or no children, the disease genes are maintained in the population by recurrent mutation. A common assumption is that an entirely normal person produces a single mutant gamete. However, this is not necessarily what happens. Unless there is something special about the mutational process, such that it can happen only during gametogenesis, mutations may arise at any time during post-zygotic life. Post-zygotic mutations produce **mosaics** with two (or more) genetically distinct cell lines. The older literature on human mosaicism refers only to chromosomal mosaicism, because that was the only type of mosaicism that could be detected before DNA analysis was developed, but mosaicism for single gene mutations is at least as frequent and important.

Mosaicism can affect somatic and/or germ line tissues. Post-zygotic mutations are not merely frequent, they are inevitable. Human mutation rates are typically 10^{-7} per gene per cell generation, and our bodies contain perhaps 10^{13} cells. It follows that every one of us must be a mosaic for innumerable genetic diseases. Indeed, as Professor John Edwards memorably remarked, a normal man may well produce the whole of the OMIM catalogue in every ejaculate. This should cause no anxiety. If a cell in your finger mutates to the Huntington disease genotype, or a cell in your ear picks up a cystic fibrosis mutation, there

are absolutely no consequences for you or your family. Only if a somatic mutation results in the emergence of a substantial clone of mutant cells is there a risk to the whole organism. This can happen in two ways:

(i) The mutation causes abnormal proliferation of a cell that would normally replicate little or not at all, thus generating a clone of mutant cells. This, of course, is how cancer happens, and this whole topic is discussed in detail in Chapter 18.

(ii) the mutation occurs in an early embryo, affecting a cell which is the progenitor of a significant fraction of the whole organism. In that case the mosaic individual may show clinical signs of disease.

Mutations occurring in a parent's germ line can cause *de novo* inherited disease in a child. When an early germ-line mutation has produced a person who harbors a large clone of mutant germ-line cells (**germinal**, or **gonadal**, **mosaicism**), a normal couple with no previous family history may produce more than one child with the same serious dominant disease. The pedigree mimicks recessive inheritance. Even if the correct mode of inheritance is realized, it is very difficult to calculate a recurrence risk to use in counseling the parents. Usually an empiric risk (Section 3.4.4) is quoted. *Figure 3.7* shows an example of

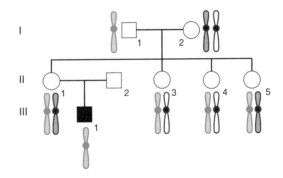

Figure 3.7: A new mutation in X-linked recessive Duchenne muscular dystrophy.

The three grandparental X chromosomes were distinguished using genetic markers, and are shown in blue, gray and white (ignoring recombination). III_1 has the grandpaternal X, which has acquired a mutation at some point in the pedigree. There are four possible points at which this could have happened:

- If III_1 carries a new mutation, the recurrence risk for all family members is very low.
- If II_1 is a germinal mosaic, there is a significant risk (but hard to quantify) for her future children, but not for her sisters.
- If II_1 was the result of a single mutant sperm, she has the standard recurrence risk for X-linked recessives, but her sisters are free of risk.
- If I_1 was a germinal mosaic all the sisters have a significant risk, which is hard to quantify.

the uncertainty that mosaicism introduces into counseling, in this case in an X-linked disease.

Molecular studies can be a great help in these cases. Sometimes it is possible to demonstrate directly that a normal father is producing a proportion of mutant sperm (*Figure 3.8*). Direct testing of the germ line is not possible in women, but other accessible tissues such as fibroblasts or hair roots can be examined for evidence of mosaicism. A negative result on somatic tissues does not rule out germ line mosaicism, but a positive result, in conjunction with an affected child, proves it.

Chimeras contain cells from two separate zygotes in a single organism

Mosaics are presumed (though rarely proved) to derive from a single fertilized egg. **Chimeras** on the other hand are the result of fusion of two zygotes into a single embryo (the reverse of twinning), or alternatively of limited colonization of one twin by cells from a nonidentical co-twin (*Figure 3.9*). Chimerism is proved by the presence in pooled tissue samples of too many parental alleles at several loci (if just one locus were involved, one would suspect mosaicism for a single mutation). Blood-grouping centers occasionally discover chimeras among normal donors, and some intersex patients turn out to be XX/XY chimeras. Strain *et al.* (1998) describe a remarkable case of a 46,XY/46,XX boy whose 46,XX cell line is parthenogenetic, derived by diploidization of a haploid maternal cell with no paternal contribution.

3.3 Factors affecting gene frequencies

3.3.1 There can be a simple relation between gene frequencies and genotype frequencies

A thought experiment: picking genes from the gene pool

Over a whole population there may be many different alleles at a particular locus, although each individual person has just two alleles, which may be identical or different. We can imagine a **gene pool**, consisting of all alleles at the *A* locus in the population. The **gene frequency** of allele A_1 is the proportion of all *A* alleles in the gene pool which are A_1. Consider two alleles, A_1 and A_2 at the *A* locus. Let their gene frequencies be p and q respectively (p and q are each between 0 and 1) Let us perform a thought experiment:

- Pick an allele at random from the gene pool. There is a chance p that it is A_1 and a chance q that it is A_2.
- Pick a second allele at random. Again the chance of picking A_1 is p and the chance of picking A_2 is q (we assume the gene pool is sufficiently large that removing the first allele has not significantly changed the gene frequencies of the remaining alleles).
- The chance that both alleles were A_1 is p^2.

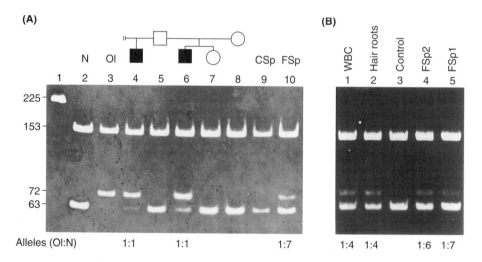

Figure 3.8: Germinal mosaicism in autosomal dominant osteogenesis imperfecta.

The father, though phenotypically normal, carries a mutation in the *COL1A1* gene, demonstrable by PCR amplification of sperm. The normal allele gives the 63 bp band and the mutant allele the 72 bp band. Both affected sons are heterozygous with a 1 : 1 ratio of bands (the intensities in the picture are not equal because of unequal amplification). The blood sample from the father gives only the normal band (panel A lane 5) but a sperm sample (panel A lane 10) contains both alleles with a 1 : 7 ratio of mutant to normal. A sperm sample from a normal control (panel A lane 9) gives only the normal band, as expected. Panel B shows the ratio of mutant to normal alleles observed in various samples from the father. FSp, sperm of father; CSp, control sperm; WBC, white blood cells. Reproduced from Cohn *et al.* (1990) *Am. J. Hum. Genet.* **46**, with permission from The University of Chicago Press.

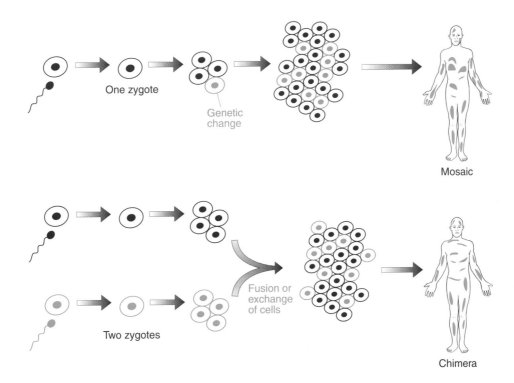

Figure 3.9: Mosaics and chimeras.

Mosaics have two or more genetically different cell lines derived from a single zygote. The genetic change indicated may be a gene mutation, a numerical or structural chromosomal change, or in the special case of lyonization, X-inactivation. A chimera is derived from two zygotes, which are usually both normal but genetically distinct.

■ The chance that both alleles were A_2 is q^2.

■ The chance that the first allele was A_1 and the second A_2 is pq. The chance that the first was A_2 and the second A_1 is qp. Overall, the chance of picking one A_1 and one A_2 allele is 2pq.

The Hardy–Weinberg distribution

If we pick a person at random from the population, this is equivalent to picking two genes at random from the gene pool. The chance the person is A_1A_1 is p^2, the chance they are A_1A_2 is 2pq, and the chance they are A_2A_2 is q^2. This simple relationship between gene frequencies and genotype frequencies (the **Hardy–Weinberg distribution**, see *Box 3.3*) holds whenever a person's two genes are drawn independently and at random from the gene pool. A_1 and A_2 may be the only alleles at the locus (in which case p + q = 1) or there may be other alleles and other genotypes (p + q < 1). For X-linked loci males, being hemizygous (only one allele) are A_1 or A_2 with frequencies p and q respectively, while females can be A_1A_1, A_1A_2 or A_2A_2 (see *Box 3.3*).

Limitations of the Hardy–Weinberg distribution

These simple calculations break down if the underlying assumption, that a person's two genes are picked independently from the gene pool, is violated. In particular, there is a problem if there has not been random mating. **Assortative mating** can take several forms, but the most

generally important is **inbreeding**. If you marry a relative you are marrying somebody whose genes resemble your own. This increases the likelihood of your children being homozygous and decreases the likelihood that they will be heterozygous. Rare recessive conditions are strongly associated with parental consanguinity, and Hardy–Weinberg calculations that ignore this will overestimate the carrier frequency in the population at large.

Use of the Hardy–Weinberg distribution in genetic counseling

Gene frequencies or genotype frequencies are essential inputs into many forms of genetic analysis, such as linkage analysis (Section 11.3) and segregation analysis

---

Box 3.3

Hardy–Weinberg equilibrium genotype frequencies for allele frequencies p (A_1) and q (A_2)

	Autosomal locus			X-linked locus			
				Males		Females	
Genotype	A_1A_1	A_1A_2	A_2A_2	A_1 A_2		A_1A_1 A_1A_2 A_2A_2	
Frequency	p^2	2pq	q^2	p q		p^2 2pq q^2	

Note that these genotype frequencies will be seen whether or not A_1 and A_2 are the only alleles at the locus.

The Hardy–Weinberg distribution can be used (with caution) to calculate carrier frequencies and simple risks for counseling

An autosomal recessive condition affects 1 newborn in 10 000. What is the expected frequency of carriers?

Phenotypes	Unaffected		Affected
Genotypes	AA	Aa	aa
Frequencies	p^2	$2pq$	$q^2 = 1/10\,000$

q^2 is 10^{-4}, and therefore $q = 10^{-2}$ or 1/100.
1 in 100 genes at the A locus are a, 99/100 are A.
The carrier frequency, $2pq$, is $2 \times 99/100 \times 1/100$, very nearly 1 in 50.

If a parent of a child affected by the above condition remarries, what is the risk of producing an affected child in the new marriage?

To produce an affected child, both parents must be carriers, and the risk is then 1 in 4. Thus the overall risk is:

(the parent's carrier risk)	×	(the new spouse's carrier risk)	× 1/4
= 1	×	1/50	× 1/4
= 1/200			

This assumes there is no family history of the same disease in the new spouse's family.

X-linked red–green color blindness affects 1 in 12 British males; what proportion of females will be carriers and what proportion will be affected?

	Males		Females		
Genotypes	A_1	A_2	$A_1 A_1$	$A_1 A_2$	$A_2 A_2$
Frequencies:	p	q = 1/12	p^2	$2pq$	q^2

$q = 1/12$, therefore $p = 11/12$
$2pq = 2 \times 1/12 \times 11/12 = 22/144$
$q^2 = 1$ in 144. Thus this single-locus model predicts that 15% of females will be carriers and 0.7% will be affected.

(Section 19.4), and they have a particular importance in calculating genetic risks. *Box 3.4* gives examples.

3.3.2 Genotype frequencies can be used (with caution) to calculate mutation rates

Mutant genes are being created by fresh mutation and being removed by natural selection (Section 9.2.3). For a given level of selection we can calculate the mutation rate that would be required to replace the genes lost by selection. If we assume that there is an equilibrium in the population between the rates of loss and of replacement, the calculation tells us the present mutation rate. We can define the **coefficient of selection (s)** as the relative chance of reproductive failure of a genotype due to selec-

tion (the fittest type in the population has s = 0, a genetic lethal has s = 1).

■ For an autosomal recessive condition, a proportion q^2 of the population are affected. The loss of disease genes each generation is sq^2. This is balanced by mutation at the rate of $\mu(1 - q^2)$ where μ is the mutation rate per gene per generation. At equilibrium $sq^2 = \mu(1 - q^2)$, or approximately (if q is small) $\mathbf{\mu = sq^2}$.

■ For a rare autosomal dominant condition homozygotes are excessively rare. Heterozygotes occur with frequency $2pq$ (frequency of disease gene = p). Only half the genes lost through their reproductive failure are the disease allele, so the rate of gene loss is very nearly sp. Again this is balanced by a rate of new mutation of μq^2, which is approximately μ if q is almost 1. Thus $\mathbf{\mu = sp}$.

■ For an X-linked recessive disease the rate of gene loss through affected males is sq. This is balanced by a mutation rate 3μ, since all X chromosomes in the population are available for mutation, but only the one third of X chromosomes which are in males are exposed to selection. Thus $\mathbf{\mu = sq/3}$.

These results are summarized in *Box 3.5*. Estimates derived using them can be compared with the general expectation, from studies in many organisms, that mutation rates are typically 10^{-5}–10^{-7} per gene per generation.

Heterozygote advantage can be much more important than recurrent mutation for determining the frequency of a recessive disease

The formula $\mu = sq^2$ gives an unexpectedly high mutation rate for some autosomal recessive conditions. Consider cystic fibrosis (CF), for example. Until very recently, virtually nobody with CF lived long enough to reproduce, therefore s = 1. CF affects about one birth in 2000 in the UK. Thus $q^2 = 1/2000$, and the formula gives $\mu = 5 \times 10^{-4}$. This would be a strikingly high mutation rate for any gene, but there is evidence that new CF mutations are in fact very rare. This follows from the uneven ethnic distribution of CF and the existence of strong linkage disequilibrium (Section 12.4.1).

Mutation-selection equilibrium

Autosomal recessive condition*	$\mu = sq^2$	or $\mu = F(1 - f)$
Autosomal dominant condition	$\mu = sp$	or $\mu = \frac{1}{2}F(1 - f)$
X-linked recessive condition	$\mu = sq/3$	or $\mu = \frac{1}{3}F(1 - f)$

μ	=	mutation rate per gene per generation
p, q	=	gene frequencies
s	=	coefficient of selection
f	=	biological fitness = 1 − s
F	=	frequency of condition in the population

* This formula gives a seriously wrong estimate of the mutation rate if there is heterozygote advantage, see *Box 3.6*.

Selection in favor of heterozygotes for cystic fibrosis

For CF, the disease frequency in the UK is about one in 2000 births.

Phenotypes	Unaffected		Affected
Genotypes	AA	Aa	aa
Frequencies	p^2	$2pq$	$q^2 = 1/2000$

q^2 is 5×10^{-4}, therefore $q = 0.022$ and $p = 1 - q = 0.978$

$p/q = 0.978 / 0.022 = 43.72 = s_2/s_1$

If $s_2 = 1$ (affected homozygotes never reproduce), $s_1 = 0.023$

The present CF gene frequency will be maintained, even without fresh mutations, if Aa heterozygotes have on average 2.3% more surviving children than AA homozygotes.

The missing factor is **heterozygote advantage**. CF carriers have, or had in the past, some reproductive advantage over normal homozygotes. There has been much debate over what this advantage might be. The CF gene encodes a membrane chloride channel, which is required by *Salmonella typhi* for it to enter epithelial cells, so maybe heterozygotes are relatively resistant to typhoid fever (Pier *et al.*, 1998). Whatever the cause of the heterozygote advantage, if s_1 and s_2 are the coefficients of selection against the AA and aa genotypes respectively, then an equilibrium is established (without recurrent mutation) when the ratio of the gene frequencies of A and a, p/q, is s_2/s_1. *Box 3.6* illustrates the calculation for cystic fibrosis, and shows that a heterozygote advantage too small to observe in population surveys can have a major effect on gene frequencies.

It is worth remembering that the medically important mendelian diseases are those that are both common and serious. They must all have some or other special trick to remain common in the face of selection. The trick may be an exceptionally high mutation rate (Duchenne muscular dystrophy), or propagation of non-pathological premutations (Fragile X), or onset of symptoms after reproductive age (Huntington disease) – but for common serious recessive conditions it is most often heterozygote advantage.

3.4 Nonmendelian characters

3.4.1 Research into simple and complex traits has long defined two separate traditions within human genetics

By the time Mendel's work was rediscovered in 1900, a rival school of genetics was well established in the UK and elsewhere. Francis Galton, the remarkable and eccentric cousin of Charles Darwin, devoted much of his vast talent to systematizing the study of human variation. Starting with an article on *Hereditary Talent and Character*

published the same year, 1865, as Mendel's paper (and expanded in 1869 to a book, *Hereditary Genius*), he spent many years investigating family resemblances.

Galton was devoted to quantifying observations and applying statistical analysis. His Anthropometric Laboratory, established in 1884, recorded from his subjects (who paid him threepence for the privilege) their weight, sitting and standing height, arm span, breathing capacity, strength of pull and of squeeze, force of blow, reaction time, keenness of sight and hearing, color discrimination and judgements of length. Except for color blindness, these are quantitative, continuously variable characters. In one of the first applications of statistics he compared physical attributes of parents and children, and established the degree of correlation between relatives. By 1900 he had established a large body of knowledge about the inheritance of such attributes, and a tradition (**biometrics**) of their investigation.

A historical controversy

When Mendel's work was rediscovered, a controversy arose. The claims of the mendelians, championed by Bateson, were resisted by biometricians. Biometricians allowed that mendelian genes might explain a few rare abnormalities or curious quirks, but pointed out that most of the characters likely to be important in evolution (body size, build, strength, skill in catching prey or finding food) were **continuous** or **quantitative characters** and not amenable to mendelian analysis. You cannot define their inheritance by drawing pedigrees and marking in the affected people, because we all have these characters, only to different degrees. Mendelian analysis requires **dichotomous characters** (characters like extra fingers, that you either have or don't have). The controversy ran on, heatedly at times, until 1918. That year saw a seminal paper by RA Fisher demonstrating that continuous characters governed by a large number of independent mendelian factors (**polygenic characters**) would display precisely the quantitative variation and family correlations described by the biometricians. Later Falconer extended this model to cover dichotomous characters (see Section 19.3).

Two traditions in human genetics

In principle Fisher's description of polygenic inheritance unified genetics. This was indeed generally true for the genetics of experimental organisms or farm animals. In human genetics, however, studies of mendelian and quantitative characters tended to continue as separate traditions, and until very recently few investigators felt at home in both worlds. The spectacular advances of 1970–1990 were entirely in mendelian genetics, whilst investigation of nonmendelian characters remained largely limited to statistical studies of family resemblances. Geneticists from the mendelian tradition were often reluctant to get involved in these studies, partly because of the complex statistical methodology and no doubt also because of a feeling that they were a poor investment of research effort compared to mapping and cloning genes for mendelian characters. Also many studies concerned sensitive areas of behavioral genetics such as the heritability of IQ, where violent controversies and a distastefully confrontational style of argument often reigned.

3.4.2 Multifactorial nonmendelian characters can be oligogenic or polygenic

The further away a character is from the primary gene action, the less likely is it to show a simple mendelian pedigree pattern. DNA sequence variants are virtually always cleanly mendelian – which is their major attraction as genetic markers. Protein variants (electrophoretic mobility or enzyme activity) are usually mendelian but can depend on more than one locus because of post-translational modification (Section 1.5.3). The failure or malfunction of a developmental pathway that results in a birth defect is likely to involve a complex balance of factors. Thus the common birth defects (cleft palate, congenital dislocation of the hip, congenital heart disease, etc.) are rarely mendelian. Behavioral traits like IQ test performance or schizophrenia are still less likely to be mendelian. This does not however mean that they may not be genetic, either partly or entirely.

Nonmendelian characters may depend on two, three or many genetic loci, with greater or smaller contributions from environmental factors (*Figure 3.10*). We use multi-

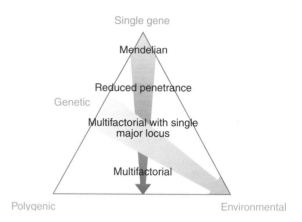

Figure 3.10: The spectrum of human characters.

Few characters are purely mendelian, purely polygenic or purely environmental. Most depend on some mix of major and minor genetic determinants, together with environmental influences. The mix of factors determining any given character could be represented by a point located somewhere within the triangle.

factorial here as a catch-all term covering all these possibilities. More specifically, the genetic determination may involve a small number of loci (**oligogenic**) or many loci each of individually small effect (**polygenic**); or there may be a single major locus with a multifactorial back-

Figure 3.11: Multifactorial determination of a disease or malformation.

The angels and devils can represent any combination of genetic and environmental factors. Adding an extra devil or removing an angel can tip the balance, without that particular factor being *the* cause of the disease. Courtesy of Professor RW Smithells.

ground. For dichotomous characters the underlying loci are envisaged as susceptibility genes, while for quantitative characters they are seen as **quantitative trait loci** (QTLs). *Figure 3.11* presents quite a useful way of thinking about the role of individual factors in multifactorial malformations or diseases.

3.4.3 The new synthesis uses mendelian markers to analyze nonmendelian phenotypes

Recent developments have finally brought together the study of mendelian and complex human phenotypes. Automation is allowing genetic analysis and sequencing on a scale scarcely imagined ten years ago. This has had two consequences. Most human genes are now identified, at least as expressed sequence tags (ESTs), so that molecular geneticists are looking for fresh fields to conquer. At the same time, marker studies can now be done on a scale that is probably large enough to deliver the statistical power needed to detect individual quantitative trait loci and susceptibility loci. Given the overwhelming preponderance of nonmendelian conditions in human disease, molecular dissection of complex phenotypes is widely seen as the next frontier in medical genetics.

There are not two sorts of genes, mendelian genes and polygenes. It is possible however that there are two sorts of mutations. The sequence variants that confer susceptibility to polygenic disease may not, for the most part, be the gross mutations seen in mendelian conditions, which typically totally inactivate a gene (Chapter 16). They are likely to include more subtle variants that slightly modify the expression of a gene. These will exist as common nonpathogenic variants in the healthy population, that cause problems only when combined with a whole series of similar variants at other loci. Genetics of nonmendelian characters is discussed in detail in Chapter 19.

3.4.4 Counseling in nonmendelian conditions uses empiric risks

In genetic counseling for nonmendelian conditions, risks are not derived from polygenic theory; they are **empiric risks** obtained through population surveys (see, for example, *Table 19.4*). This is fundamentally different from mendelian conditions, where the risks come from theory. The effect of family history is also quite different. If a couple are both carriers of cystic fibrosis, the risk of their next child being affected is 1 in 4. This remains true regardless how many affected or normal children they have already produced. If they have had a baby with neural tube defect, the recurrence risk is about 1 in 25 in the UK – but if they have already had two affected babies, the recurrence risk is about 1 in 12. It is not that having a second affected baby has caused their recurrence risk to increase, but it has enabled us to recognize them as a couple who always had been at particularly high risk. A cynic would say it involves the counselor being wise after the event, but the practice accords with our understanding based on threshold theory (Section 19.3), as well as with epidemiological data, and it is the best we can offer in an imperfect state of knowledge.

Further Reading

Cavalli-Sforza L, Bodmer WF (1971) *Genetics of Human Populations*. Freeman, San Francisco. *Although some parts are very out of date, it remains an excellent general textbook of human population genetics.*

Forrest DW (1974) *Francis Galton: the life and work of a Victorian genius*. Elek, London.

McKusick VA. (1997) *Mendelian Inheritance in Man*, 12th edn. Johns Hopkins University Press, Baltimore. *The print version of the OMIM database.*

References

Cohn DH, Starman BJ, Blumberg B, Byers PH (1990) Recurrence of lethal osteogenesis imperfecta due to parental mosaicism for a dominant mutation in a human type I collagen gene (*COL1A1*). *Am. J. Hum. Genet.*, **46**, 599.

Fisher RA (1918) The correlation between relatives under the supposition of mendelian inheritance. *Trans. Roy. Soc.*, **52**, 399–433.

Harper PS (1998) *Genetic Counselling*, 5th edn. Butterworth-Heinemann, Oxford.

Heutink P, van der Mey AG, Sandkujl LA *et al.* (1992) A gene subject to genomic imprinting and responsible for hereditary paragangliomas maps to chromosome 11q23 qter. *Hum. Mol. Genet.*, **1**, 7–10.

Pier GB, Grout M, Zaidi T *et al.* (1998) *Salmonella typhi* uses CFTR to enter intestinal epithelial cells. *Nature*, **393**, 79–82.

Prezant TR, Agapian JV, Bowman MC *et al.* (1993) Mitochondrial ribosomal RNA mutation associated with both antibiotic induced and non syndromic deafness. *Nature Genet.*, **4**, 289–294.

Riordan-Eva P, Harding AE (1995) Leber's hereditary optic atrophy: the clinical relevance of different mitochondrial DNA mutations. *J. Med. Genet,.* **32**, 81–87.

Smith RJH, Berlin CI, Hejtmancik JF *et al.* (1994) Clinical diagnosis of the Usher syndromes. *Am. J. Med. Genet.,* **50**, 32–38.

Strain L, Dean JC, Hamilton MP, Bonthron DT (1998) A true hermaphrodite chimera resulting from embryo amalgamation after in vitro fertilization. *N. Engl. J. Med.,* **338**, 166–169.

Sweeney MG, Davis MB, Lashwood A, Brockington M, Toscano A, Harding AE (1992) Evidence against an X-linked locus close to DXS7 determining visual loss susceptibility in British and Italian families with Leber hereditary optic neuropathy. *Am. J. Hum. Genet.,* **51**, 741–748.

Van der Meulen MA, van der Meulen MJP, te Meerman GJ (1995) Recurrence risk for germinal mosaics revisited. *J. Med. Genet.,* **32**, 102–104.

Viljoen D, Ramesar R (1992) Evidence for paternal imprinting in familial Beckwith–Wiedemann syndrome. *J. Med. Genet.,* **29**, 221–225.

Wallace DC (1994) Mitochondrial DNA sequence variation in human evolution and disease. *Proc. Natl Acad. Sci. USA,* **91**, 8739–8746.

Wilkie AOM (1994) The molecular basis of dominance. *J. Med. Genet.,* **31**, 89–98.

Cell-based DNA cloning

4

4.1 Fundamentals of DNA technology and the importance of DNA cloning

The fundamentals of current DNA technology are very largely based on two quite different approaches to studying specific DNA sequences within a complex DNA population (see *Figure 4.1*):

■ **DNA cloning.** The desired fragment must be *selectively amplified* so that it is purified essentially to homogeneity. Thereafter, its structure and function can be comprehensively studied, for example by DNA sequencing, *in vitro* expression studies, etc.,

and various manipulations can be achieved to change its structure by *in vitro* mutagenesis.

■ **Molecular hybridization.** The fragment of interest is not amplified, but instead is *specifically detected* within a complex mixture of many different sequences. Its chromosomal location can be determined in this way and some information can be gained regarding its structure. If expressed, the sequence of interest can be detected within a complex RNA or cDNA population from specific cells, enabling comprehensive analysis of its expression patterns.

Before DNA cloning, our knowledge of DNA was extremely limited. DNA cloning technology changed all

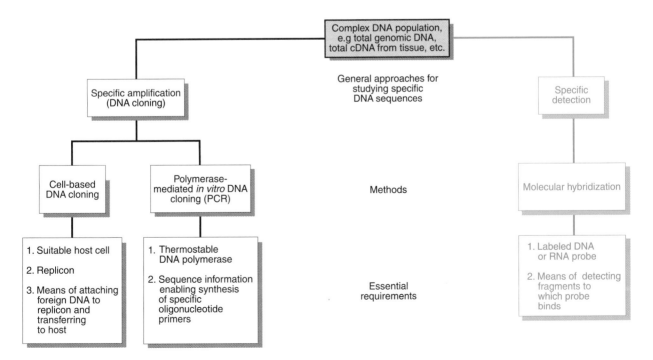

Figure 4.1: General approaches for studying specific DNA sequences in complex DNA populations.

that and revolutionized the study of genetics. Why was DNA cloning such an important technological advance? One important consideration is the tremendous size and complexity of DNA sequences (compared to, say, protein sequences). Individual nuclear DNA molecules contain hundreds of millions of nucleotides. When DNA is isolated from cells using standard methods, these huge molecules are fragmented by shear forces, generating complex mixtures of still very large DNA fragments (typically 50–100 kb long).

Given the above, how can relatively homogeneous DNA populations be prepared from such a complex starting mixture. In the case of DNA from human cells and a wide variety of complex eukaryotic cells, one early approach had been to separate different classes of DNA fragments according to their base composition. The DNA preparation was submitted to ultracentrifugation in equilibrium density gradients (e.g. in CsCl density gradients). When this was achieved, the DNA was fractionated into a major band (the bulk DNA) and several minor **satellite DNA** bands. The satellite DNA species have different buoyant densities to the bulk DNA because they have unusual sequences and their base composition is different to the majority of the DNA in cells (these properties in turn reflect the involvement of satellite DNA in specific aspects of chromosome structure and function (see Section 2.3.2). Although valuable and interesting, the purified satellite DNAs were, however, a minor component of the genome and did not contain genes.

What DNA cloning offered was a general method for studying *any* DNA sequence.

4.1.1 DNA cloning is a general method of selectively amplifying DNA sequences to generate homogenous DNA populations

In order to study genes, methods had to be developed to purify them. Because mammalian genomes are complex, any specific gene or DNA fragment of interest normally represents only a tiny fraction of the total DNA in a cell. For example, the β-globin gene comprises only 0.00005% of the 3300 megabases (Mb) of human genomic DNA, and even the massive 2.5 Mb dystrophin gene, the largest gene that has been identified, accounts for only about 0.08% of human genomic DNA.

One way of enriching for gene sequences is to isolate total RNA, or poly(A)$^+$ messenger RNA (mRNA) from suitable cells and convert this to **complementary DNA (cDNA)** using the enzyme **reverse transcriptase**. In some cases, this can result in a profound enrichment for specific exonic DNA sequences when the relevant genes are known to be expressed at very high levels in a specific cell type. In most cases, however, the desired gene sequences still represent only a tiny proportion of the total cDNA population.

In order to have a general method of studying a specific DNA sequence within a complex DNA population, the

technology of DNA cloning was developed. The essential characteristic of DNA cloning is that the desired DNA fragments must be *selectively amplified* in some way, resulting in a programmed large increase in copy number of selected DNA sequences. In practice, this involves multiple rounds of DNA replication catalyzed by a DNA polymerase acting on one or more types of template DNA molecule. Essentially two different DNA cloning approaches are used (see *Figure 4.1*):

- **Cell-based DNA cloning.** This was the first form of DNA cloning to be developed, and is an *in vivo* cloning method. The first step in this approach involves attaching foreign DNA fragments *in vitro* to DNA sequences which are capable of independent replication. The recombinant DNA fragments are then transferred into suitable host cells where they can be propagated selectively. In the past, the term DNA cloning has been used exclusively to signify this particular approach.

- **Cell-free DNA cloning.** The polymerase chain reaction (PCR) is a newer form of DNA cloning which is enzyme mediated and is conducted entirely *in vitro*. This approach is described in detail in Chapter 6.

4.2 Principles of cell-based DNA cloning

4.2.1 Cell-based DNA cloning requires attachment in vitro of DNA fragments to purified replicons and propagation in suitable host cells

The essence of cell-based DNA cloning involves four steps (*Figure 4.2*):

(i) *Construction of recombinant DNA molecules* by *in vitro* covalent attachment (**ligation**) of the desired DNA fragments (**target DNA**) to a replicon (any sequence capable of independent DNA replication). This step is facilitated by cutting the target DNA and replicon molecules with specific restriction endonucleases before joining the different DNA fragments using the enzyme DNA ligase.

(ii) *Transformation.* The recombinant DNA molecules are transferred into host cells (often bacterial or yeast cells) in which the chosen replicon can undergo DNA replication independently of the host cell chromosome(s).

(iii) *Selective propagation of cell clones* involves two stages. Initially the transformed cells are plated out by spreading on an agar surface in order to encourage the growth of *well-separated* cell colonies. These are cell clones (populations of identical cells all descended from a single cell). Subsequently, *individual* colonies can be picked from a plate and the cells can be further expanded in liquid culture.

(iv) *Isolation of recombinant DNA clones* by harvesting

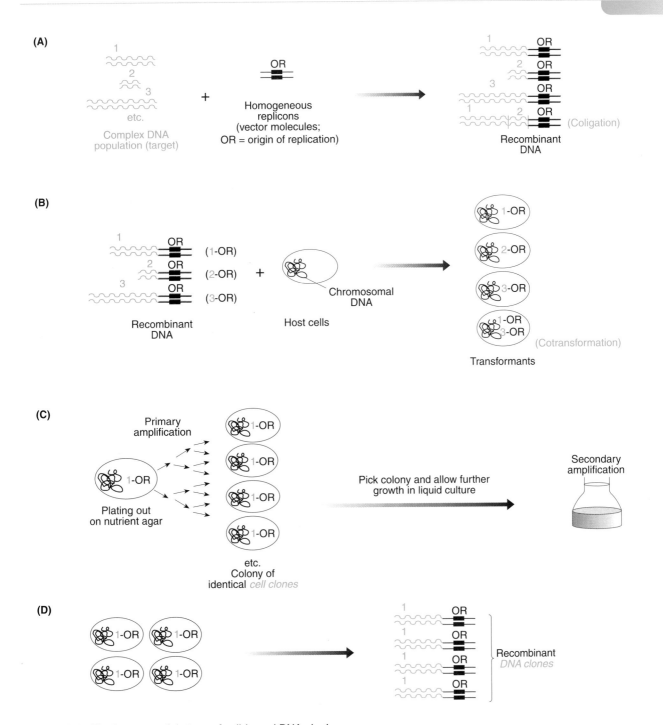

Figure 4.2: The four essential steps of cell-based DNA cloning.

(A) Formation of recombinant DNA. *Note* that, in addition to simple vector–target ligation products, coligation events may occur whereby two unrelated target DNA sequences may be ligated in a single product (e.g. sequences 1 plus 2 in the bottom example). **(B) Transformation.** This is a key step in DNA cloning because cells normally take up only one foreign DNA molecule. *Note* that occasionally, however, cotransformation events are observed, such as the illustrated cell at the bottom which has been transformed by two different DNA molecules (a recombinant molecule containing sequence 1 and a recombinant molecule containing sequence 3). **(C) Amplification** to produce numerous cell clones. After plating out the transformed cells, *individual* clone colonies can be separated on a dish and then *individually picked* into a secondary amplification step to ensure clone homogeneity. **(D) Isolation of recombinant DNA clones.**

expanded cell cultures and selectively isolating the recombinant DNA.

Replicons and host cells

Cell cloning requires that the foreign DNA fragments which are introduced into a host cell must be able to replicate. If not, the foreign DNA would soon be diluted out as the host cell undergoes many rounds of cell division. However, foreign DNA fragments will generally lack an origin of replication that will function in the host cell. They require, therefore, to be attached to an independent replicon so that their replication is controlled by the replicon's origin of replication.

In principle, the necessary replicon could be provided in two ways. One possibility is to introduce the fragments into the host cells in such a way that in each cell one, or a very few, fragments integrate into a host cell chromosome and are propagated under the control of a host cell replicon. This is the way that retroviruses, for example, integrate their DNA into host cell chromosomes. However, as a general method for cloning, this approach suffers from numerous disadvantages, including the difficulty in retrieving the inserted DNA. Instead, cell-based cloning very largely relies on a different approach: the foreign DNA fragments are attached *in vitro* to a purified replicon and the resulting hybrid molecules are transferred into host cells, where they replicate *independently of the host cell chromosomes*. Because the foreign DNA fragments (target DNA) can be viewed as passengers of the replicon, replicons used for cloning are described as **vector** molecules. Notice that, although the resulting replication of the introduced DNA fragments is independent of the host chromosomes, the vector may have an origin of replication that originates from either a natural extrachromosomal replicon or, in some cases, a chromosomal replicon (as in the case of yeast artificial chromosomes, see Section 4.3.4).

Although some DNA cloning systems involve human and other mammalian cells as hosts, the great bulk of cell-based DNA cloning has used modified bacterial or fungal host cells. The former cells are widely used because of their capacity for rapid cell division. Bacterial cell hosts have a single circular double-stranded chromosome with a single origin of replication. Replication of the host chromosome subsequently triggers cell division so that each of the two resulting daughter cells contains a single chromosome like their parent cell (i.e. the copy number is maintained at one copy per cell). However, the replication of extrachromosomal replicons is not constrained in this way: many such replicons go through several cycles of replication during the cell cycle and can reach high copy numbers.

Extrachromosomal replicons as vector molecules

Two basic types of extrachromosomal replicons are found in bacterial cells:

- **Plasmids** are small circular double-stranded DNA molecules which individually contain very few genes. Their existence is intracellular, being vertically distributed to daughter cells following host cell division, but they can be transferred horizontally to neighboring cells during bacterial conjugation. Natural examples include plasmids which carry the sex factor (F) and those which carry drug-resistance genes.
- **Bacteriophages** are viruses which infect bacterial cells. DNA-containing bacteriophages often have genomes containing double-stranded DNA which may be circular or linear. Unlike plasmids, they can exist extracellularly. The mature virus particle (**virion**) has its genome encased in a protein coat so as to facilitate adsorption and entry into a new host cell.

In order for naturally occurring replicons to be used as vector molecules for cell-based DNA cloning, various modifications need to be made. Similarly, the host cells that are used for cloning are specialized cells whose genotype has been selected to optimize their use in DNA cloning. Typically, cloning systems are designed to ensure that joining of the foreign DNA fragment occurs at a unique location in the vector molecule. Additionally, they have in-built selection systems so that cells which contain the relevant vector molecule can be specifically selected. In many cases, there are additional screening systems to ensure detection and propagation of cells containing recombinant DNA (see Section 4.2.5).

4.2.2 Restriction endonucleases enable the target DNA to be cut into pieces of manageable size and facilitate ligation to similarly cut vector molecules

A major boost to the development of cell-based DNA cloning was the discovery and exploitation of type II restriction endonucleases, enzymes which normally cleave DNA whenever a small, specific recognition sequence, usually 4–8 base pairs (bp) long occurs (see *Box 4.1* and *Table 4.1*).

The recognition sequences for the vast majority of type II restriction endonucleases are normally **palindromes**, that is the sequence of bases is the same on both strands when read in the $5' \rightarrow 3'$ direction, as a result of a twofold axis of symmetry. In some cases, the cleavage points occur exactly on the axis of symmetry, giving products (restriction fragments) which are **blunt-ended** (*Figure 4.3*). In most cases, however, the cleavage points do not fall on the symmetry axis, so that the resulting restriction fragments possess so-called 5' overhangs or 3' overhangs.

Overhanging ends generated by cleavage with a restriction nuclease are often described as **sticky ends** or **cohesive termini** because the two overhanging ends of each fragment are complementary in base sequence, and will have a tendency to associate with each other, or with any other similarly complementary overhang, by forming base pairs. Different fragments with the same sequences in their overhanging ends can be generated by: (i) cutting with the same restriction nuclease; (ii) cutting with different restriction endonucleases that happen to recognize

Box 4.1

Restriction endonucleases and modification–restriction systems

Bacteriophages that are liberated from a bacterium of a particular strain can infect other bacteria of the same strain but not those of a different strain. This is because the phage DNA has the same modification pattern as the DNA of bacterial strains it can infect; the phage is 'restricted' to that strain of bacteria. The restriction is not an absolute one: some phages can escape restriction and can acquire the modification pattern of the new host.

The basis of modification–restriction systems is now known to involve two types of enzyme activity:

(i) a sequence-specific DNA methylase activity provides the basis of the modification pattern;

(ii) a sequence-specific restriction endonuclease activity underpins the restriction phenomenon by cleaving phage DNA whose modification (methylation) pattern is different from that of the host cell DNA.

The bacterial strain possesses a *DNA methylase activity with the same sequence specificity as the corresponding restriction endonuclease activity*. As a result, cellular restriction endonucleases will not cleave the appropriately methylated host cell DNA but may cleave incoming phage DNA, if not methylated appropriately. *Note*, however, that some plasmids and bacteriophages possess genes for modification and restriction systems so that their presence in a bacterium determines its specificity.

the same target sequence (**isoschizomers**); or (iii) by cutting with enzymes which have different recognition sequences but happen to produce compatible sticky ends, for example *Bam*HI and *Mbo*I (*Figure 4.3*).

The termini of restriction fragments which have the same type of overhanging ends can associate in a variety of different ways, either intramolecularly (cyclization), or between molecules to form linear **concatemers** or circular compound molecules. Intermolecular reactions occur most readily at high DNA concentrations. At very low DNA concentrations, however, individual termini on different molecules have less opportunity of making contact with each other, and intramolecular cyclization is favored.

Because the overhanging ends generated by restriction endonucleases are very short (typically four nucleotides or less), hydrogen bonding between complementary overhanging ends provides a rather weak contact between two molecules, and can only be maintained at low temperatures. However, it does facilitate subsequent covalent bonding between the two associated molecules (DNA ligation). This is performed using the enzyme DNA ligase. Ligation of blunt-ended fragments is also possible, although less efficient than sticky end ligation.

Generally, ligation reactions are designed to promote the formation of recombinant DNA (by ligating target DNA to vector DNA), although vector cyclization, vector–vector concatemers and target DNA–target DNA ligation are also possible (see *Figure 4.4*). To achieve this, the vector molecules are often treated so as to prevent or minimize their ability to undergo cyclization. There are two common ways of achieving this.

Table 4.1: Restriction endonucleases

Enzyme	Source	Sequence cut	Average expected fragment size (kb) in human DNA[a]
*Alu*I	*Arthrobacter luteus*	AGCT	0.3
*Hae*III	*Hemophilus aegyptus*	GGCC	0.6
*Taq*I	*Thermus aquaticus*	TCGA	1.4
*Mnl*I	*Moraxella nonliquefaciens*	CCTC/GAGG	0.4
*Hind*III	*Hemophilus influenzae* Rd	AAGCTT	3.1
*Eco*RI	*Escherichia coli* R factor	GAATTC	3.1
*Bam*HI	*Bacillus amyloliquefaciens* H	GGATCC	7.0
*Pst*I	*Providencia stuartii*	CTGCAG	7.0
*Mst*I	*Microcoleus* species	CCTNAGG[c]	7.0
*Sma*I	*Serratia marcescens*	CCCGGG	78
*Bss*HII	*Bacillus stearothermophilus*	GCGCGC	390[b]
*Not*I	*Norcadia otitidis-caviarum*	GCGGCCGC	9766[b]

[a] Assuming 40% G + C, and a CpG frequency 20% of that expected.
[b] Observed average sizes are often lower than these estimates.
[c] N = A, C, G or T.

Note: Names are normally derived from the first letter of the genus and the first two letters of the species name, e.g. *Pst*I is the first restriction nuclease to have been isolated from *Providencia stuartii*. *Mnl*I is an example of an enzyme whose recognition sequence is not palindromic. So-called rare-cutters often have recognition sequences containing one or more CpG dinucleotides and cut vertebrate DNA comparatively infrequently.

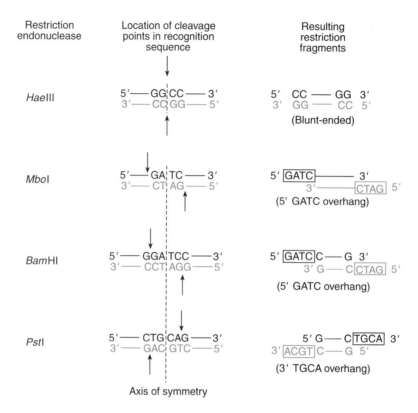

Figure 4.3: Restriction endonucleases can generate blunt-ended fragments or fragments with 5′ or 3′ overhanging ends.

■ *Cutting of vectors with two different restriction endo-nucleases.* Often, vector molecules have multiple unique restriction sites, in which foreign DNA can be cloned, occurring in a short segment of the molecule. It is often convenient, therefore, to cut the vector with two restriction endonucleases which do not produce complementary overhanging ends (e.g. *Eco*RI and *Bam*HI), and remove the small vector fragment between the sites, resulting in a vector molecule whose two ends cannot religate. However, if target DNA is cut with the same enzyme combination, recombinant DNA can easily be formed.

■ *Vector dephosphorylation.* During DNA ligation *in vitro*, the enzyme DNA ligase will catalyze the formation of a phosphodiester bond only if one nucleotide contains a 5′ phosphate group and the other contains a 3′ hydroxyl group. If the 5′ phosphate groups at both ends of the vector DNA are removed by treatment with alkaline phosphatase, the tendency for the vector DNA to recircularize will therefore be minimized. A foreign DNA insert can, however, provide 5′-terminal phosphates which can then be joined to the 3′ hydroxyl groups provided by the vector. This method, therefore, increases the frequency of cells containing recombinant DNA.

4.2.3 Introducing recombinant DNA into recipient cells provides a method for fractionating a complex starting DNA population

The plasma membrane of cells is selectively permeable and does not normally admit large molecules such as long DNA fragments. However, cells can be treated in certain ways (e.g. by exposure to certain high ionic strength salts, short electric shocks, etc.) so that the permeability properties of the plasma membranes are altered. As a result, a fraction of the cells become **competent**, that is capable of taking up foreign DNA from the extracellular environment. Only a small percentage of the cells will take up the foreign DNA (DNA transformation). However, those that do will often take up only a single molecule (which can, however, subsequently replicate many times within a cell). *This is the basis of the critical fractionation step in cell-based DNA cloning*; the population of transformed cells can be thought of as a sorting office in which the complex mixture of DNA fragments is sorted by depositing individual DNA molecules into individual recipient cells (*Figure 4.5*).

Because circular DNA (even nicked circular DNA) transforms much more efficiently than linear DNA, most of the cell transformants will contain cyclized products rather than linear recombinant DNA concatemers and, if

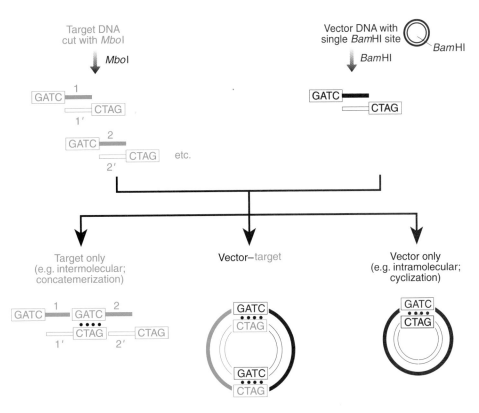

Figure 4.4: Cohesive termini can associate intramolecularly and intermolecularly.

Note that only some of the possible outcomes are shown. For example, vector molecules may also form intermolecular concatemers, multimers can undergo cyclization and co-ligation events can involve two different target sequences being included with a vector molecule in the same recombinant DNA molecule (see *Figure 4.2A*). The tendency towards cyclization of individual molecules is more pronounced when the DNA is at low concentration and the chance of collision between different molecules with complementary sticky ends is reduced.

an effort has been made to suppress vector cyclization (e.g. by dephosphorylation), most of the transformants will contain recombinant DNA. Note, however, that cotransformation events (the occurrence of more than one type of introduced DNA molecule within a cell clone, see *Figure 4.2B*) may be comparatively common in some cloning systems.

The transformed cells are allowed to multiply. In the case of cloning using plasmid vectors and a bacterial cell host, a solution containing the transformed cells is simply spread over the surface of nutrient agar in a petri dish (plating out). This usually results in the formation of bacterial colonies which consist of cell clones (identical progeny of a single ancestral cell). Picking an individual colony into a tube for subsequent growth in liquid culture permits a secondary expansion in the number of cells which can be scaled up to provide very large yields of cell clones, all identical to an ancestral single cell (*Figure 4.5*). If the original cell contained a single type of foreign DNA fragment attached to a replicon, then so will the descendants, resulting in a huge amplification in the amount of the specific foreign fragment.

Expanded cultures representing cell clones derived

from a single cell can then be processed to recover the recombinant DNA. To do this, the cells are lysed, and the DNA is extracted and purified using procedures that result in recovery of the recombinant DNA. Such procedures take advantage of differences between the recombinant DNA and the host chromosomal DNA. In the case of bacterial cells, the bacterial chromosome is circular double-stranded DNA, like any plasmids containing introduced foreign DNA. However, the chromosomal DNA is relatively very large (~4.6 Mb) compared with most recombinant DNA molecules (often only a few kilobases long). During cell lysis and subsequent extraction procedures, the very large bacterial chromosomal DNA, but not the small recombinant plasmid DNA, will undergo nicking and shearing, generating linear DNA fragments with free ends. This difference can be exploited by subjecting the isolated DNA to a denaturation step, often as a result of exposure to alkaline pH in the 12.0–12.5 range. Following this treatment, the linearized host cell DNA readily denatures, but the strands of covalently closed circular (CCC) plasmid DNA are unable to separate. After normal conditions have been restored, the two strands of the CCC DNA rapidly re-align in

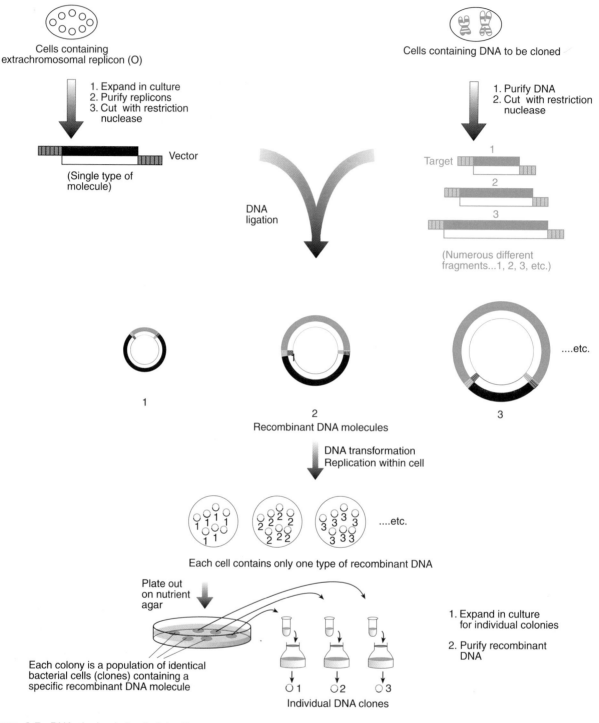

Figure 4.5: DNA cloning in bacterial cells.

The example illustrates cloning of genomic DNA but could equally be applied to cloning cDNA (see *Figure 4.8*).

perfect register to form native superhelical molecules or so-called supercoiled DNA (this higher order form of twisting occurs in any CCC DNA because the tension introduced by twisting of the double helix cannot be relaxed, unlike in DNA with one or two free ends where relaxation is possible by rotation of a free end; *Figure 4.6*).

The denatured host cell DNA precipitates out of solution, leaving behind the CCC plasmid DNA.

If required, further purification is possible by column chromatography. Alternatively, equilibrium density gradient centrifugation (isopycnic centrifugation) is used: the partially purified DNA is centrifuged to equilibrium

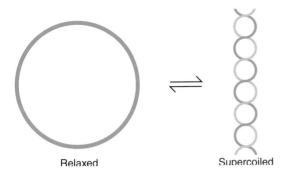

Relaxed   Supercoiled

Figure 4.6: Covalently closed circular (CCC) DNA and DNA supercoiling.

The double-stranded CCC DNA is shown on left in schematic form (no attempt is made to show the double helical structure). Unless one of the two DNA strands becomes nicked, the twisting of the double helix cannot be relaxed and the tension induced causes spontaneous formation of a supercoiled structure shown on the right. Nicking of either DNA strand, however, relieves the tension by permitting rotation at the free end.

in a solution of cesium chloride containing ethidium bromide (EtBr). EtBr binds by *intercalating* between the base pairs, thereby causing the DNA helix to unwind. Unlike chromosomal DNA, a CCC plasmid DNA has no free ends and can only unwind to a limited extent, which limits the amount of EtBr it can bind. EtBr–DNA complexes are denser when they contain less EtBr, so CCC plasmid DNA will band at a lower position in the cesium chloride gradient than chromosomal DNA and plasmid circles that are open, enabling separation of the recombinant DNA from host cell DNA. The resulting recombinant DNA molecules will normally be identical to each other (representing a single target DNA fragment) and are referred to as DNA clones.

4.2.4 DNA libraries are a comprehensive set of DNA clones representing a complex starting DNA population

The first attempts at cloning human DNA fragments in bacterial cells concentrated on target sequences which were highly abundant in a particular starting DNA population. For example, most human cells contain much the same complex collection of DNA sequences in the nucleus, but the mRNA populations of different cell types can be quite different. Although the mRNA population in each cell is complex, some cells are particularly devoted to synthesizing a specific type of protein and so their mRNA populations have a few predominant mRNA species (e.g. much of the mRNA made in erythrocytes is α and β globin mRNA). The enzyme **reverse transcriptase** (RNA-dependent DNA polymerase) can be used to make a DNA copy that is complementary in base sequence to the mRNA, so-called cDNA. Hence, cDNA from erythrocytes is greatly enriched in globin cDNA, facilitating its isolation.

Modern DNA cloning approaches offer the possibility of making comprehensive collections of DNA clones (**DNA libraries**) from extremely complex starting DNA populations (such as total human genomic DNA). This approach enables DNA sequences which are very rare in the starting population to be represented in a library of DNA clones, whence they can be isolated individually by selecting a suitable host cell colony and amplifying it. Two basic varieties of this method have popularly been undertaken: the construction of genomic DNA libraries and cDNA libraries.

Newly constructed libraries are said to be *unamplified*, although this is a misleading term because the initially transformed cells have been amplified to form separated cell colonies. Often, cell colony formation is allowed to proceed on top of membranes that are overlaid on to the surface of nutrient agar in sterile culture dishes. Copies of the library can then be made by replica plating on to a similar sized membrane prior to overlaying on to a nutrient agar surface and colony growth. More recently, individually picked cell colonies have been spotted in gridded arrays on to suitable membranes or into the wells of microtiter dishes where they can be stored for long periods at −70°C in the presence of a cell-stabilizing medium such as glycerol. For multiple distribution, amplified libraries are required. The cells from representative primary filters are washed off into cell culture medium, diluted and stabilized by the presence of glycerol, or some alternative stabilizing agent. Individual aliquots can then be plated out at a later stage to regenerate the library. This additional amplification step, however, may result in distortion of the original representation of cell clones because during the amplification stage there may be differential rates of growth of different colonies.

Genomic DNA libraries

In the case of complex eukaryotes, such as mammals, all nucleated cells have essentially the same DNA content, and it is often convenient to prepare a genomic library from easily accessible cells, such as blood cells. The starting material is genomic DNA which has been fragmented in some way, usually by digesting with a restriction endonuclease. Typically, the genomic DNA is digested with a 4 bp cutter such as *Mbo*I which recognizes the sequence GATC. This sequence will occur about every 275 bp on average in human genomic DNA, so that there will be few DNA sequences which lack a recognition site for this enzyme. Clearly, complete digestion of the starting DNA with this enzyme would produce very small fragments. Instead, the extent of enzymic cleavage is deliberately minimized: under conditions of partial restriction digestion (low enzyme concentration, short time of incubation, etc.), cleavage will occur at only a small number of the potential restriction sites. This clearly has the benefit of being able to produce large fragments for cloning. Importantly, it also allows random fragmentation of the DNA. Thus, for a specific sequence location, the pattern of cutting will be different on different copies of the same starting DNA sequence (*Figure 4.7*). Such random fragmentation ensures that the library will contain as much representation as possible of the starting DNA. Additionally, it has the advantage that it results in clones with overlapping inserts. As a result, after characterization of the insert of one clone, attempts can be made to access clones from the same general region by identifying those with inserts showing some similarities to that of the original clone (see Section 10.3.2).

The complexity (number of independent DNA clones) of a genomic DNA library can be defined in terms of genome equivalents (GE). A genome equivalent of 1, a so-called one-fold library, is obtained when the number of independent clones = genome size/average insert size. For example, for a human genomic DNA library that has an average insert size of 40 kb, 1 GE = 3000 Mb/40 kb = 75 000 independent clones. A library such as this which has 300 000 clones is sometimes termed a 4-fold library because it has 4 GE. Because of sampling variation, however, the number of GEs must be considerably greater than 1 to have a high chance of including any particular sequence within that library. Consequently, attempts are normally made to prepare libraries with several GEs.

Subgenomic DNA libraries have also been a valuable resource: the ability to fractionate individual human chromosomes or specific chromosome bands has permitted construction of chromosome-specific libraries (see Section 10.1.5) and chromosome band-specific libraries, obtained following chromosome dissection (see Section 15.3.5).

cDNA libraries

As gene expression can vary in different cells and at different stages of development, the starting material for making cDNA libraries is usually total RNA from a specific tissue or specific developmental stage of embryogenesis. From this, poly(A)$^+$ mRNA can be selected by specific binding to a complementary single-stranded oligonucleotide or polynucleotide which is bound to a solid matrix. For example, chromatography columns containing poly(U) bound to sepharose or oligo(dT) bound to cellulose have been widely used: the poly(A)$^+$ mRNA selectively binds to the poly(U) or oligo(dT) components and subsequently can be eluted using buffers of high ionic strength to disrupt the hydrogen bonding. The isolated poly(A)$^+$ mRNA can then be converted, using reverse transcriptase, to a double-stranded cDNA copy. To assist cloning, oligonucleotide linkers which contain suitable restriction sites are ligated to each end of the cDNA (see *Figure 4.8*; see also *Box 4.2* for a general description of oligonucleotide linkers).

4.2.5 Recombinant screening is often achieved by insertional inactivation of a marker gene

An essential requirement for cell-based DNA cloning systems is a method of detecting cells containing the appropriate vector molecule, and the subset which contains recombinant DNA.

Screening for cells transformed by vector molecules

Identification of cells containing the vector molecule requires engineering or selection of the vector molecule to contain a suitable marker gene whose expression pro-

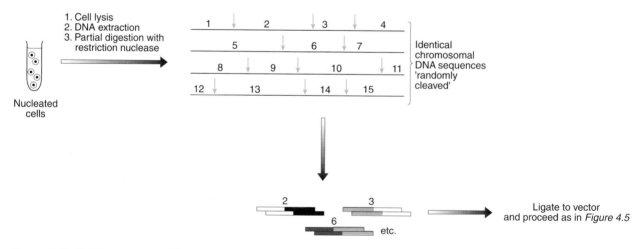

Figure 4.7: Making a genomic DNA library.

All nucleated cells of an individual will have the same genomic DNA content so that any easily accessible cells (e.g. blood cells) can be used as source material. Because DNA is extracted from numerous cells with identical DNA molecules, the isolated DNA will contain large numbers of identical DNA sequences. However, partial digestion with a restriction endonuclease will cleave the DNA at only a small subset of the available restriction sites, and the pattern will differ between individual molecules, resulting in almost random cleavage. This will generate a series of restriction fragments which, if they derive from the same locus, may share some common DNA sequence (e.g. fragment 6 partially overlaps fragments 2 and 3, as shown, and also fragments 9 and 10, and 13 and 14).

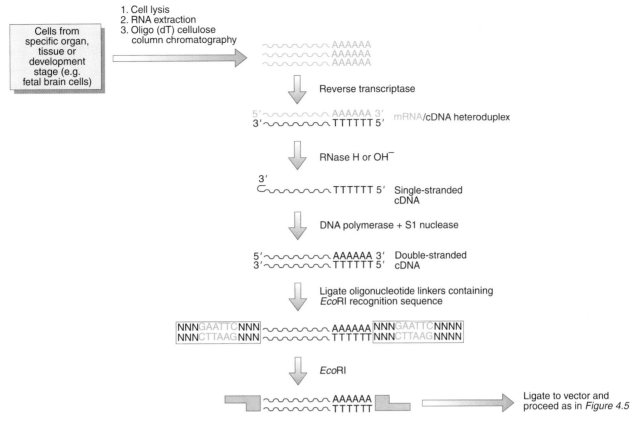

Figure 4.8: Making a cDNA library.

The reverse transcriptase step often uses an oligo(dT) primer to prime synthesis of the cDNA strand. More recently, mixtures of random oligonucleotide primers have been used instead to provide a more normal representation of sequences. RNase H will specifically digest RNA that is bound to DNA in an RNA–DNA hybrid. The 3′ end of the resulting single-stranded cDNA has a tendency to loop back to form a short hairpin. This can be used to prime second strand synthesis by DNA polymerase and the resulting short loop connecting the two strands can then be cleaved by the S1 nuclease which specifically cleaves regions of DNA that are single stranded.

vides a means of identifying cells containing it. Two popularly used marker gene systems are based on:

■ **antibiotic resistance genes.** A host cell strain is chosen that is sensitive to a particular antibiotic, often ampicillin, tetracycline or chloramphenicol. The corresponding vector has been engineered to contain a gene which confers resistance to the antibiotic. After transformation, cells are plated on agar containing the antibiotic to rescue cells transformed by the vector.

■ **β-galactosidase gene complementation.** The host cell is a mutant which contains a fragment of the β-galactosidase gene but cannot make any functional β-galactosidase. The vector is engineered to contain a different fragment of the β-galactosidase gene. After transformation by the vector, functional complementation occurs resulting in active β-galactosidase which can be assayed by acting on a colorless substance, Xgal (5-bromo, 4-chloro, 3-indolyl β-D-galactopyranoside), to make a blue product.

Recombinant screening

Identification of cells containing recombinant DNA (vector molecules with inserts) is often accomplished by **insertional inactivation** of the marker gene. The vector molecule is designed to have a *multiple cloning site* (*polylinker*) located within the marker gene. The number of nucleotides in the polylinker sequence containing the multiple cloning sites is a multiple of three, so that its insertion in the marker gene does not result in a shift in the translation reading frame. As a result of maintaining the reading frame, and the very small size of the polylinker, the activity of the marker gene is maintained despite the insertion of the polylinker. However, if the vector then contains recombinant DNA inserted in the middle of the polylinker, the resulting large insertion causes loss of expression of the marker gene. In the case of a marker β-galactosidase gene, insertional inactivation means that cells containing recombinant DNA are colorless in the presence of Xgal, while cells containing nonrecombinant vector are blue.

Another common screening system involves the use of **suppressor tRNA** genes, mutant tRNA genes which can

Box 4.2

Oligonucleotide linkers

For many applications in molecular genetics it is necessary to covalently attach (by DNA ligation) a synthetic double-stranded oligonucleotide (an oligonucleotide linker) to a target DNA molecule in order to endow the target DNA molecule with some desired property. Often, for example, oligonucleotide linkers are attached in order to provide a unique restriction site that will, following digestion with the appropriate restriction nuclease, allow subsequent ligation of the target DNA to a vector DNA molecule as in *Figure 4.8*. In other cases, it is desirable to attach the linkers to DNA fragments of unknown sequence in order to permit amplification of the target DNA molecules using PCR with linker-specific primers (see *Figure 6.12*). In order to prepare the double-stranded linkers, two complementary single-stranded oligonucleotides with the desired sequences are individually synthesized by standard methods, then allowed to anneal to form a double-stranded linker before being ligated to the target DNA molecule. Note that essentially the same result can also be achieved by PCR using primers whose sequences are specific for the target DNA at the 3′ ends but which carry additional sequence at the 5′ end that endows the target sequence with the desired property (see Section 6.4.2 and *Figure 6.20A* for an example).

Box 4.3

Nonsense suppressor mutations

	Glu		**Amber**
Codon	5′ GAG 3′		5′ UAG 3′
	• • •		• • •
Anticodon of:	3′ CUC 5′	→	3′ AUC 5′
	tRNA^Glu	→	tRNA^Glu*
			(amber suppressor)

	Glu		**Ochre**
Codon	5′ GAA 3′		5′ UAA 3′
	• • •		• • •
Anticodon of:	3′ CUU 5′	→	3′ AUU 5′
	tRNA^Glu	→	tRNA^Glu*
			(ochre suppressor)

Base changes in the anticodon of a tRNA may enable it to insert an amino acid in response to a stop codon. The tRNA^Glu at top carries a CUC anticodon which recognizes the codon GAG. Mutation of the tRNA gene can produce a mutant tRNA^Glu* which has a C → A change at the 3′ base in the anticodon. This mutant tRNA can now recognize the amber stop codon UAG and by inserting a glutamate it *suppresses* the amber stop signal. The bottom example illustrates a similar method of generating another mutant tRNA which can suppress the effect of an ochre stop codon.

recognize a normal termination codon signal and, in response, introduce an amino acid in the polypeptide chain. This is possible because the mutation in such tRNA genes (often mutated tRNA^Glu or tRNA^Tyr genes) results in an altered anticodon sequence that is complementary to one of the normal termination codons: UAA (ochre), UAG (amber) or UGA (opal) (see *Box 4.3*). In such systems, the host cell often carries a defective marker gene that is designed to have a premature stop codon (i.e. an amber, ochre or opal mutation) resulting in a phenotype that can be easily scored. If the vector carries a suppressor tRNA gene that suppresses the effect of this termination codon, the marker gene activity will be recovered and the wild-type phenotype will be restored. However, if the suppressor tRNA gene is inactivated by insertional inactivation, a mutant phenotype will again be produced (see *Figure 4.16* for an example of one such system).

If there is already available a DNA probe that is closely related to the desired recombinant DNA, colonies containing the latter can be detected directly by colony hybridization (see *Figure 5.18*).

4.2.6 Initial characterization of recombinant DNA clones involves restriction mapping

Restriction mapping of DNA clones involves cutting the DNA with one or more of a series of different restriction nucleases and separating the resulting fragments according to size by agarose gel electrophoresis. Because of the conspicuous deficiency in the CpG dinucleotide in verte-

brate genomes (see *Box 8.5*), recognition sites that are GC-rich will occur comparatively less frequently than expected in vertebrate DNA, but will not be so rare in bacterial cell DNA (see *Table 4.1*).

Double digests (cleavage by two different enzymes) and **partial digests** (reduced digestion so that not every cleavage site is actually cut) help in relating the different restriction fragments to each other. The resulting information can be used to construct a **restriction map**, a linear map of the relative positions of recognition sites for a variety of restriction endonucleases (*Figure 4.9*). If the restriction maps of two independently isolated DNA fragments show extensive sharing of restriction sites, it is highly likely that the two fragments contain overlapping DNA sequences or are closely related members of a repeated DNA sequence family. Restriction mapping of a recombinant DNA clone immediately provides details of the length of the insert DNA and some information on the location of unique restriction sites which may be useful for subcloning purposes (where fragments of a recombinant DNA are cloned into a different vector molecule, often so as to study some aspect of its structure or function).

4.3 Vector systems for cloning different sizes of DNA fragments

Cell-based DNA cloning has been used widely as a tool for producing quantities of pure DNA for physical characterization and functional studies of individual genes, gene clusters or other DNA sequences of interest. How-

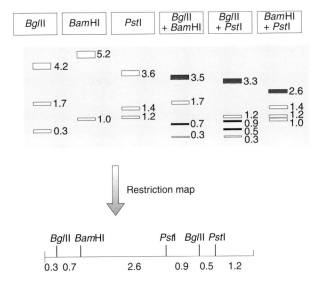

Figure 4.9: Generating a restriction map.

The size patterns from double digests provide information on the relative locations of restriction sites. The example shows size fractionation by agarose gel electrophoresis of restriction fragments following incubation of a 6.2 kb DNA fragment with the indicated enzymes. New bands in the double digests (i.e. not found in the original single digests) are indicated by black boxes. In the *Bgl*II + *Bam*HI double digest, the original 1.7 kb and 0.3 kb bands from the *Bgl*II digest alone are maintained, suggesting that these fragments do not have a *Bam*HI site, while the 4.2 kb *Bgl*II fragment is replaced by 3.5 kb and 0.7 kb fragments, suggesting that there is a *Bam*HI site within 0.7 kb from one end of the 4.2 kb *Bgl*II fragment. Similarly, in the *Bam*HI + *Pst* I double digest, the 1.4 kb and 1.2 kb fragments seen in the *Pst*I digest alone are maintained, suggesting that they lack a *Bam*HI site, while the 3.6 kb *Pst*I fragment is replaced by a 2.6 kb + 1.0 kb fragment, as a result of possession of an internal *Bam*HI site located 1.0 kb from one end. By comparing all three patterns of double digestion, the restriction map at the bottom can be deduced. *Note* that restriction mapping is often helped by the use of partial digests and also by end-labeling (Section 5.1.1).

ever, the size of different DNA sequences of interest can vary enormously (e.g. human gene sizes are known to vary between 0.1 kb and 2.5 Mb). The first cell-based cloning systems to be developed could clone only rather small DNA fragments. Recently, however, there have been rapid developments in cloning systems that permit cloning of very large DNA fragments.

4.3.1 Plasmid vectors provide a simple way of cloning small DNA fragments in bacterial (and simple eukaryotic) cells

In order to adapt natural plasmid molecules as cloning vectors, several modifications are normally made:

■ Insertion of a multiple cloning site polylinker. This is a short (~30 bp) synthetic sequence which contains

unique restriction sites for a variety of common restriction nucleases (pre-existing restriction sites for these enzymes will be deleted from the plasmid if necessary to ensure the presence of unique cloning sites).

■ Insertion of an antibiotic resistance gene. The host cells that are used must naturally be sensitive to the antibiotic in question so that any vector molecule which transforms a host cell can confer antibiotic resistance. By plating transformed cells on a medium containing the antibiotic, only those cells that have been transformed by vector molecules survive.

■ Insertion of a selection system for screening for recombinants. Typically this involves arranging for the multiple cloning site polylinker to be inserted into an expressible gene or gene fragment within the plasmid (see Section 4.2.5).

The plasmid vector pUC19 contains a polylinker with unique cloning sites for multiple restriction nucleases and an ampicillin resistance gene to permit identification of transformed cells (*Figure 4.10*). In addition, selection for recombinants is achieved by insertional inactivation of a component of the β-galactosidase gene, a complementary portion of this gene being provided by using a specially modified *E. coli* host cell.

4.3.2 Lambda and cosmid vectors provide an efficient means of cloning moderately large DNA fragments in bacterial cells

The major disadvantage of plasmid vectors is that their capacity for accepting large DNA fragments is severely limited: most inserts are a few kilobases in length and inserts larger than 5–10 kb are very rare. Additionally, standard methods of transformation of bacterial cells with plasmid vectors are relatively inefficient. To address these difficulties, attention was focused at an early stage on the possibility of using bacteriophage lambda as a cloning vector. The wild-type λ virus particle (virion) contains a genome of close to 50 kb of linear double-stranded DNA packaged within a protein coat and has evolved a highly efficient mechanism of infecting *E. coli* cells. After the λ virion attaches to the bacterial cell, the coat protein is discarded and the λ DNA is injected into the cell. At the extreme termini of the λ DNA are overhanging 5′ ends which are 12 nucleotides long and complementary in base sequence. Because these large 5′ overhangs can base-pair, they are effectively sticky ends, similar to, but more cohesive than, the small sticky ends generated by some restriction nucleases (see Section 4.2.2). Such cohesive properties are recognized in the name given to this sequence – the **cos** sequence. Once inside the bacterial cell, the cos sequences base-pair, and sealing of nicks by cellular ligases results in the formation of a double-stranded circular DNA. Thereafter the λ DNA can enter one of two alternative pathways (*Figure 4.11*):

■ **The lytic cycle.** The λ DNA replicates, initially bi-directionally, and subsequently by a rolling circle

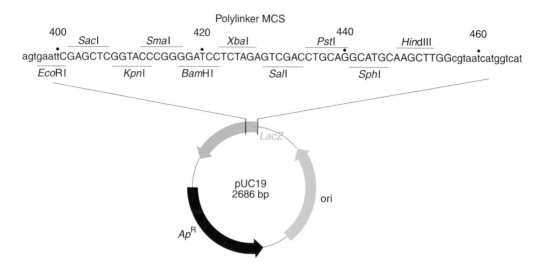

Figure 4.10: Map of plasmid vector pUC19.

The origin of replication (ori) was derived originally from a ColE1-like plasmid, pMB1. The ampicillin resistance gene (Ap^R) was derived originally from the plasmid RSF 2124 and permits selection for cells containing the vector molecule. A portion of the *lacZ* gene is included and is expressed to give an amino-terminal fragment of β-galactosidase. This is complemented by a mutant *lacZ* gene in the host cell: the products of the vector and host cell *lacZ* sequences, although individually inactive, can associate to form a functional product. The 54 bp polylinker multiple cloning site (capital letters) is inserted into the vector *lacZ* (lower case letters) component in such a way as to preserve the reading frame and functional expression. However, cloning of an insert into the multiple cloning site (MCS) will cause insertional inactivation, and absence of β-galactosidase activity.

model which generates linear multimers of the unit length. Coat proteins are synthesized and the λ multimers are snipped at the cos sites to generate unit lengths of λ genome which are packaged within the protein coats. Some of the λ gene products lyse the host cell, allowing the virions to escape and infect new cells.

■ **The lysogenic state**. The λ genome possesses a gene *att* which has a homolog in the *E. coli* chromosome. Apposition of the two *att* genes can result in recombination between the λ and *E. coli* genomes and subsequent integration of the λ DNA within the *E. coli* chromosome. In this state, the λ DNA is described as a **provirus** and the host cell as a lysogen because, although the λ DNA can remain stably integrated for long periods, it has the capacity for excision from the host chromosome and entry into the lytic cycle (*Figure 4.11*). Genes required for lysogenic function are located in a central segment of the λ genome (*Figure 4.12*).

The decision to enter the lytic cycle or the lysogenic state is controlled by two regulatory genes, *cI* and *cro*. These two genes are mutually antagonistic: in the lytic state the *cro* protein dominates, leading to repression of *cI*, whereas in the lysogenic state the *cI* repressor dominates and suppresses transcription of other λ genes including *cro*. In normally growing host cells, the lysogenic state is favored and the λ genome replicates along with the host chromosomal DNA. Damage to host cells favors a transition to the lytic cycle, enabling the virus to escape the damaged cell and infect new cells.

In order to design suitable cloning vectors based on λ, it was necessary to design a system whereby foreign DNA could be attached to the λ replicon *in vitro* and for the resultant recombinant DNA to be able to transform *E. coli* cells at high efficiency. The latter requirement was achieved by developing an *in vitro* packaging system which mimicked the way in which wild-type λ DNA is packaged in a protein coat, resulting in high infection efficiency (*Figure 4.13*).

Several major types of cloning vector that have been developed by modifying phage λ, or utilizing the size selection imposed by cos sequences, are described in the following sections.

Replacement λ vectors

Only DNA molecules from 37 to 52 kb in length can be stably packaged into the λ particle. The central segment of the λ genome contains genes that are required for the lysogenic cycle but are not essential for lytic function. As a result, it can be removed and replaced by a foreign DNA fragment. Using this strategy, it is possible to clone foreign DNA up to 23 kb in length, and such vectors are normally used for making genomic DNA libraries.

Insertion λ vectors

Lambda vectors used for making cDNA libraries do not require a large insert capacity (most cDNAs are <5 kb long). Design of insertion vectors often involves modification of the λ genome to permit insertional cloning into the *cI* gene.

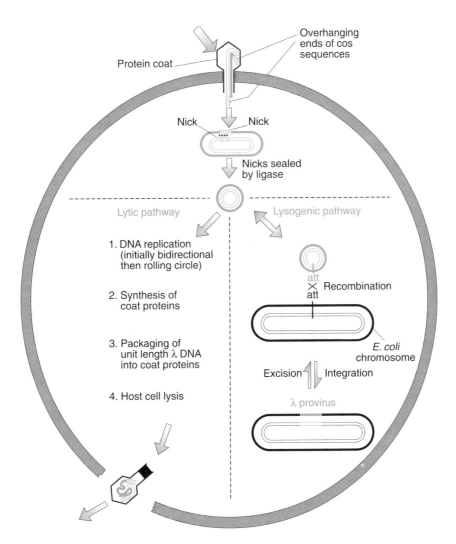

Figure 4.11: Phage λ can enter both lytic and lysogenic pathways.

Cosmid vectors

Cosmid vectors contain <u>cos</u> sequences inserted into a small plasmid vector. Large (~30–44 kb) foreign DNA fragments can be cloned using such vectors in an *in vitro* packaging reaction because the total size of the cosmid vector is usually only about 8 kb (see *Figure 4.14*).

4.3.3 Large DNA fragments can be cloned in bacterial cells using vectors based on bacteriophage P1 and F factor plasmids

Because the human and other mammalian genomes are so large, and because many individual genes can be very large (see *Figure 7.7*), there was a need for the development of new cloning vectors that could accept large DNA inserts. A number of such vectors have been developed recently (see *Table 4.2*) and have found immediate uses in general physical mapping of genomes and in permitting the characterization and expression of large genes or gene complexes. Examples of such vectors are discussed below.

Bacterial artificial chromosome (BAC) vectors

Many vectors which are popularly used for DNA cloning in bacterial cells contain high to medium copy number replicons. The advantage of vectors which contain such replicons is the high yield of DNA they afford: each cell in which a vector molecule is propagated will have several to multiple copies of the vector molecule, depending on the replication efficiency of its replicon. However, an important disadvantage is that such vectors often show structural instability of inserts, resulting in deletion or rearrangement of portions of the cloned DNA. Such instability is particularly common in the case of DNA inserts of eukaryotic origin where repetitive sequences occur frequently and, as a result, it is difficult to clone and maintain intact large DNA in bacterial cells.

In order to overcome this limitation, attention has recently been focused on vectors based on low copy number replicons, such as the *E. coli* fertility plasmid, the F-factor. This plasmid contains two genes, *parA* and *parB*, which maintain the copy number of the F factor at 1–2 per *E. coli* cell. Vectors based on the F factor system are able to

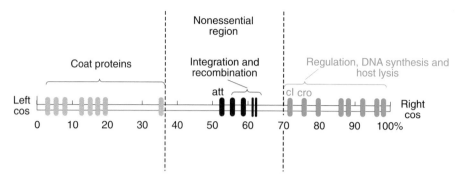

Figure 4.12: Map of the λ genome, showing positions of genes (vertical bars).

In λ replacement vectors, the nonessential region is removed by restriction endonuclease digestion, leaving a left λ arm and a right λ arm. A foreign DNA fragment can be ligated to the two arms in place of the original 'stuffer' fragment, providing maximal insert sizes of over 20 kb.

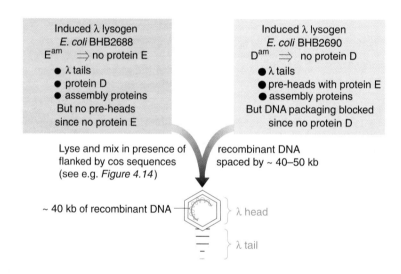

Figure 4.13: *In vitro* DNA packaging in a phage λ protein coat can be performed using a mixed lysate of two mutated λ lysogens.

Normal *in vivo* packaging of λ DNA involves first making pre-heads, structures composed of the major capsid protein encoded by gene E. A unit length of λ DNA is inserted in the pre-head, the unit length being prepared by cleavage at neighboring cos sites. A minor capsid protein D is then inserted in the pre-heads to complete head maturation, and the products of other genes serve as assembly proteins, ensuring joining of the completed tails to the completed heads. A defect in producing protein E, resulting from an amber mutation introduced into gene E (E^{am}), prevents pre-heads being formed by BHB2688. An amber mutation in gene D (D^{am}) prevents maturation of the pre-heads, with enclosed DNA, into complete heads. The components of the BHB2688/BHB2690 mixed lysate, however, complement each other's deficiency and provide all the products for correct packaging.

accept large foreign DNA fragments (>300 kb). The resulting recombinants can be transferred with considerable efficiency into bacterial cells using electroporation (a method of exposing cells to high voltages in order to relax the selective permeability of their plasma membranes). However, because the resulting **bacterial artificial chromosomes (BACs)** contain a low copy number replicon, only very low yields of recombinant DNA can be recovered from the host cells (Shizuya *et al.*, 1992).

Bacteriophage P1 vectors and PACs
Certain bacteriophages have relatively large genomes, thereby affording the potential for developing vectors that can accommodate large foreign DNA fragments. One such is bacteriophage P1 which, like phage λ, packages its genome in a protein coat, and 110–115 kb of linear DNA is packaged in the P1 protein coat. P1 cloning vectors have therefore been designed in which components of P1 are included in a circular plasmid. The P1 plasmid vector

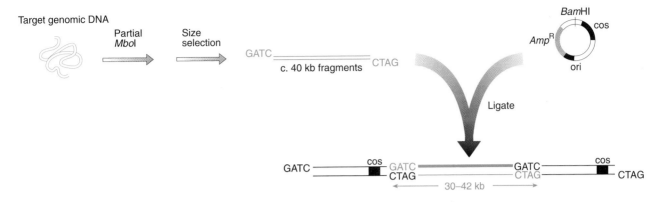

Figure 4.14: Ligation to cleaved cosmid vector molecules can produce vector–target concatemers, resulting in a large exogenous DNA fragment flanked by cos sequences.

can be cleaved to generate two vector arms to which up to 100 kb of foreign DNA can be ligated and packaged into a P1 protein coat *in vitro*. The recombinant P1 phage can be allowed to adsorb to a suitable host, following which the recombinant P1 DNA is injected into the cell, circularizes and can be amplified (Sternberg, 1992) (see *Figure 4.15*).

An improvement on the size range of inserts accepted by the basic P1 cloning system has been the use of bacteriophage T4 *in vitro* packaging systems with P1 vectors which enables the recovery of inserts up to 122 kb in size. More recently, features of the P1 and F-factor systems have been combined to produce P1-derived artificial chromosome (PAC) cloning systems (Iouannou *et al.*, 1994).

4.3.4 Yeast artificial chromosomes (YACs) enable cloning of megabase fragments

The most popularly used system for cloning very large DNA fragments involves the construction of **yeast artificial chromosomes** (**YACs**; see Burke et al., 1987; Schlessinger, 1990). Cloning in yeast cells offers, in principle, some advantages over cloning in bacterial cells. Certain eukaryotic sequences, notably those with repeated sequence organizations, are difficult, or impos-

sible, to propagate in bacterial cells which do not have such types of DNA organization, but would be anticipated to be tolerated in yeast cells which are eukaryotic cells. However, the main advantage offered by YACs has been the ability to clone very large DNA fragments. Such a development proceeded from the realization that the great bulk of the DNA in a chromosome is not required for normal chromosome function. As detailed in Sections 2.3.2–2.3.4, the essential functional components of chromosomes are:

(i) **Centromeres**, required for disjunction of sister chromatids in mitosis and of homologous chromosomes at the first meiotic division.

(ii) **Telomeres**, required for complete replication of linear molecules and for protection of the ends of the chromosome from nuclease attack.

(iii) **Autonomous replicating sequence** (**ARS**) elements, required for autonomous replication of the chromosomal DNA. They are thought to act as specific replication origins.

In each case, the DNA segment necessary for functional activity *in vivo* in yeast is limited to at most a few hundred base pairs of DNA (*Figure 2.8*). As a result, it became possible to envisage a novel cloning system based on the use of chromosomal replicons (ARS elements) as an alternative to the ubiquitous use of cloning vectors based on extrachromosomal replicons (those found in plasmids and bacteriophages).

To make a YAC, it is simply necessary to combine four short sequences that can function in yeast cells: two telomeres, one centromere and one ARS element, together with a suitably sized foreign DNA fragment to give a linear DNA molecule in which the telomere sequences are correctly positioned at the termini (*Figure 4.16*). The resulting construct cannot be transfected directly into yeast cells. Instead, yeast cells have to be treated in such a way as to remove the external cell walls. The resulting yeast spheroplasts can accept exogenous fragments but are osmotically unstable and need to be embedded in agar. The overall transformation efficiency is very low

Table 4.2: Sizes of inserted DNA commonly obtained with different cloning vectors

Cloning vector	Size of insert
Standard high copy number plasmid vectors	0–10 kb
Bacteriophage λ insertion vectors	0–10 kb
Bacteriophage λ replacement vectors	9–23 kb
Cosmid vectors	30–44 kb
Bacteriophage P1	70–100 kb
PAC (P1 artificial chromosome) vectors	130–150 kb
BAC (bacterial artificial chromosome) vectors	up to 300 kb
YAC (yeast artificial chromosome) vectors	0.2–2.0 Mb

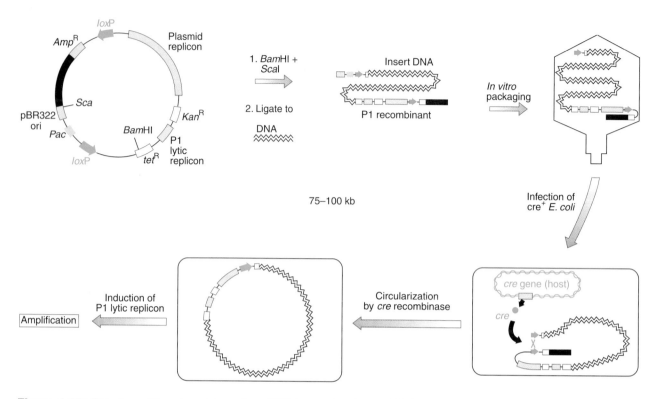

Figure 4.15: The phage P1 vector system allows DNA fragments of up to 100 kb to be cloned.

The P1 plasmid vector Ad10 incorporates various elements of the P1 genome, notably *pac*, the P1 packaging site, and two *lox*P sites, which are the sites naturally recognized by the phage recombinase, the product of the *cre* gene. The vector is digested so as to generate two arms, a short arm and a long arm to which 85–100 kb size-selected foreign DNA fragments are ligated. Packaging of the recombinant DNA occurs *in vitro* using P1 packaging extracts: pacase cleaves the recombinant DNA at the *pac* site and then works with other components to insert the DNA (maximum of 115 kb) into phage heads. Tail proteins are attached and the recombinant phage is allowed to adsorb to a *cre*⁺ strain of *E. coli*. The host cell *cre* product acts on the *lox*P sites so as to produce a circular plasmid which is maintained at low copy number (by the plasmid replicon) but can be amplified by inducing the P1 lytic replicon.

and the yield of cloned DNA is low (about one copy per cell). Nevertheless, the capacity to clone large exogenous DNA fragments (up to 2 Mb) has made YACs a vital tool in physical mapping (see Section 10.3).

4.4 Cloning systems for preparing single-stranded DNA and for studying gene expression

4.4.1 Single-stranded DNA clones for use in DNA sequencing are generally obtained using M13 or phagemid vectors

DNA templates provided in the form of isolated single-stranded DNA are preferred for conventional DNA sequencing because the sequences obtained are clearer and easier to read (Section 6.3). Although single-stranded DNA templates can be prepared by PCR-based methods (see Section 6.3.3), single-stranded recombinant DNA clones are usually used as templates. They are

obtained using vectors based on certain bacteriophages which naturally assume a single-stranded DNA form at some stage in their life cycle. Because the vector sequence is already known, it is often convenient to use a single vector-specific sequencing primer which is complementary to a sequence in the vector adjacent to the cloning site.

M13 vectors

M13 is one of a group of filamentous bacteriophages (including the fd and f1 phages) which can infect certain strains of *E. coli*. The genomes of these phages consist of single-stranded circular DNA molecules about 6.4 kb long, which are very highly related to each other in DNA sequence and enclosed in a protein coat, forming a long filamentous structure. After adsorption to the bacterium, the phage genome enters the bacterial cell where it is converted from the single-stranded form to a double-stranded form, the replicative form (RF). The latter serves as a template for making numerous copies of the genome and, after a certain time, a phage-encoded product switches DNA synthesis towards production of single

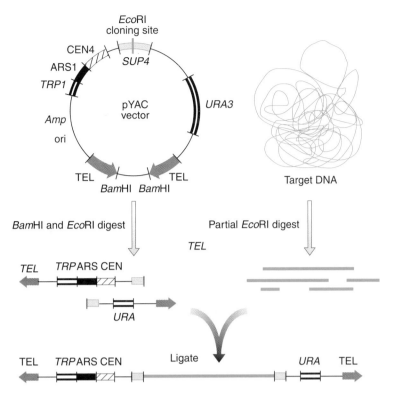

Figure 4.16: Making YACs.

Vector DNA sequences include: CEN1, centromere sequence; TEL, telomere sequences; ARS1, autonomous replicating sequence; *Amp*, gene conferring ampicillin-resistance; ori, origin of replication for propagation in an *E. coli* host. The vector is used with a specialized yeast host cell, AB1380, which is red colored because it carries an ochre mutation in a gene, *ade-2*, involved in adenine metabolism, resulting in accumulation of a red pigment. However, the vector carries the *SUP4* gene, a suppressor tRNA gene (Section 4.2.5) which overcomes the effect of the *ade-2* ochre mutation and restores wild-type activity, resulting in colorless colonies. The host cells are also designed to have recessive *trp1* and *ura3* alleles which can be complemented by the corresponding *TRP1* and *URA3* alleles in the vector, providing a selection system for identifying cells containing the YAC vector. Cloning of a foreign DNA fragment into the *SUP4* gene causes insertional inactivation of the suppressor gene function, restoring the mutant (blue color) phenotype.

strands. The latter migrate to the cell membrane where they are enclosed in a protein coat, and hundreds of mature phage particles are extruded from the infected cell without cell lysis. M13 vectors are based upon the double-stranded RF form and have a multiple cloning site for generating double-stranded recombinant DNA circles. The latter can be transfected into suitable strains of *E. coli*. After a certain period, phage particles are harvested and stripped of their protein coats to release single-stranded recombinant DNA for direct use as templates in DNA sequencing reactions (*Figure 4.17A*).

Phagemid vectors
Phagemid vectors are plasmids which have been artificially manipulated so as to contain a small segment of the genome of a filamentous phage, such as M13, fd or f1. The selected phage sequences contain all the *cis*-acting elements required for DNA replication and assembly into phage particles. They permit successful cloning of inserts several kilobases long (unlike M13 vectors in which such inserts tend to be unstable). Following transformation of

a suitable *E. coli* strain with a recombinant phagemid, the bacterial cells are *superinfected* with a filamentous **helper phage**, such as f1, which is required to provide the coat protein. Phage particles secreted from the superinfected cells will be a mixture of helper phage and recombinant phagemids (*Figure 4.17B*). The mixed single-stranded DNA population can be used directly for DNA sequencing because the primer for initiating DNA strand synthesis is designed to bind specifically to a sequence of the phagemid vector adjacent to the cloning site. Commonly used phagemid vectors include the pEMBL series of plasmids and the pBluescript family (see *Figure 4.17B*).

4.4.2 Expression cloning involves cloning gene sequences using a cloning system which is designed to permit expression of the gene

For many purposes, DNA cloning is geared simply towards amplifying the introduced DNA to obtain sufficient quantities for a variety of subsequent structural and

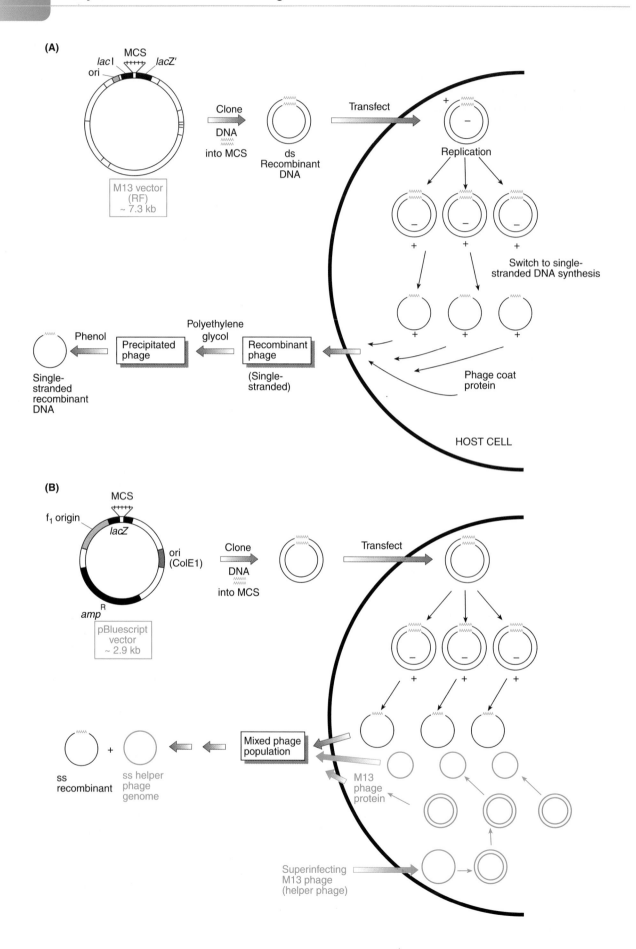

functional studies. However, in the case of gene sequences, there are many circumstances where instead of just amplifying and propagating the cloned DNA, it is desirable to be able to express the gene in some way (**expression cloning**). In each case, appropriate expression signals need to be provided by the cloning system. Expression cloning can be conducted using PCR-based systems, but is often conducted using cell-based cloning systems. A wide variety of cloning systems can be used, depending on:

- **Type of expression product.** For some purposes, it may be sufficient to be able to obtain an RNA product. Examples include generation of antisense RNA probes (**riboprobes** – see *Figure 5.4*) for use in tissue *in situ* hybridization studies, or generation of antisense RNA for inhibiting or destroying expression of specific genes, either in functional studies or for therapeutic purposes (Section 22.3.2). In many cases, however, a protein product is desired.

- **Type of environment.** Sometimes, it may be sufficient to express the product *in vitro*. Often, however, it may be desirable to be able to express the product in a particular cellular system which may be a highly defined prokaryotic or eukaryotic cell line.

- **Purpose of expression system.** The expression system may be designed simply for investigating expression. In some cases, however, the purpose may be to retrieve large quantities of an expression product, as in the need to generate large quantities of a specific protein to assist subsequent crystallography studies or attempts to raise specific antibodies against the protein.

A large variety of expression cloning vectors have been designed to be used in different host cell systems ranging from bacterial cells to mammalian cells, with specific vectors being engineered to be useful in specific host cell types.

cDNA expression libraries for immunological screening

Expression of eukaryotic proteins in bacterial cells can also be used as a target system for screening with antibodies to identify previously uncharacterized genes. In such cases, even although there may be no information concerning the coding sequence of the gene in question, partial purification of a protein product may have allowed a specific antibody to be raised. Using vectors such as λgt11, it is possible to prepare a cDNA library from a desired eukaryotic tissue in which the bacterial clones can express proteins from their eukaryotic inserts. Filters containing individual bacterial colonies (colony filters, see *Figure 5.18*) can be screened by exposure to the antibody (see *Figure 4.18*). Positively reacting bacteria can then be propagated to isolate the cDNA clone, and the isolated cDNA clone can, in turn, be used to screen a genomic library to identify the cognate gene.

Large scale expression of eukaryotic polypeptides in bacterial cells

In this case, the inserted eukaryotic cDNA is simply a provider of genetic information required to specify a polypeptide; the control of expression is provided externally. Expression of important eukaryotic genes in bacterial cells can provide an endless bulk supply of important, medically relevant compounds or of proteins for basic follow-up research studies. As the production of very large amounts of a protein (which are often foreign to the host cell) can be detrimental to host cell growth, expression systems are often designed with inducible promoters. In such cases, cells with the required recombinant DNA can be selected and grown up in large quantities without expression of the foreign gene. Expression of the inserted gene can be switched on by exposure to an inducing agent and cells can be harvested shortly thereafter. One major problem in some cases, however, concerns the importance of post-translational processing in the production of many proteins. Because of differences between such systems in prokaryotes and eukaryotes, eukaryotic proteins produced by expression of cDNA clones in bacterial cells may be unstable, or show limited or no biological activity.

An important application of this approach is designed to permit the construction of an antibody specific for a eukaryotic polypeptide where none have previously been available. In these cases, the object is to prepare **fusion proteins** consisting of an amino terminal peptide encoded by a portion of a bacterial gene, such as the *E. coli lacZ* gene, attached to the desired eukaryotic polypeptide (see *Box 20.2*).

Figure 4.17 (Opposite): Producing single-stranded recombinant DNA using M13 and phagemid vectors.

(A) M13 vectors. M13 vectors are replicative (RF) forms of M13 derivatives containing a nonfunctional component of the *lacZ* β-galactosidase system which can be complemented in function by the presence of a complementary *lacZ* component in the *E. coli* JM series. The double-stranded M13 recombinant DNA enters the normal cycle of DNA replication to generate numerous copies of the genome, prior to a switch to production of single-stranded DNA (+ strand only). Mature recombinant phage exit from the cell without lysis. **(B)** Phagemid vectors. The pBluescript series of plasmid vectors contain two origins of replication: a normal one from *Col*E1 and a second from phage f1 which, in the presence of a filamentous phage genome, will specify production of single-stranded DNA. Superinfection of transformed cells with M13 phage results in two types of phage-like particles released from the cells: the original superinfecting phage and the plasmid recombinants within a phage protein coat. Sequencing primers specific for the phagemid vector are used to obtain unambiguous sequences. Abbreviation: MCS, multiple cloning site.

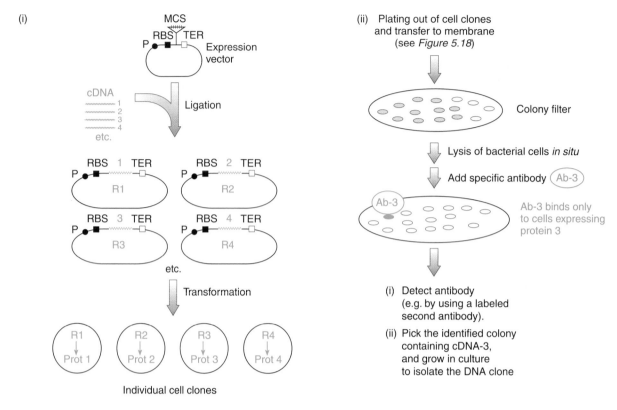

Figure 4.18: Antibody screening of a cDNA expression library.

To permit expression in a bacterial host, the vector requires suitable expression sequences flanking the cDNA insert, including a strong, usually inducible, bacterial or phage promoter (P) plus a ribosome-binding site (RBS) upstream of the multiple cloning site (MCS). In addition, downstream transcriptional terminator sites (TER) are required to prevent readthrough from disrupting plasmid replication. cDNA cloning using such vectors results in a library of bacterial clones containing recombinant molecules (R1, R2, R3, R4 in figure) able to express the insert to give a protein (prot1, prot2, etc.). Individual bacterial colonies can be separated by plating out and the colonies transferred to colony filters (see *Figure 5.18*). The bacterial cells are lysed *in situ* by treatment with lysozyme and washed to remove cell debris. Exposure to a suitable antibody can result in specific antibody binding to a colony expressing the protein of interest, and detection of bound antibody. The identified colony containing the cDNA of interest is grown in culture to enable isolation of the desired cDNA clone.

Expressing eukaryotic genes in eukaryotic cells

The biological properties of many eukaryotic proteins synthesized in bacteria may not be very representative of the native molecules because of incorrect or inefficient protein folding and post-translational processing. As a result, considerable effort has been developed to express mammalian proteins in eukaryotic and preferably mammalian cells. Major expression systems include:

■ those that are designed to produce transient or stable

expression of the transfected DNA. The most widely used of the transient expression systems is the monkey *COS* cell system which was derived by transformation of simian kidney CV-1 cells with a replication origin-defective SV40 genome (Gluzman, 1981);

■ those that involve the use of viral expression vectors, including vectors derived from SV40, adenovirus, vaccinia virus, retroviruses and the insect baculovirus. The latter expression system is popularly used for bulk preparation of a specific eukaryotic protein.

Further reading

Berger SL, Kimmel AR (1987) *Guide to Molecular Cloning Techniques. Methods in Enzymology*, Vol. 152. Academic Press, San Diego.

Glover D, Hames BD (eds) (1995) *DNA Cloning: a Practical Approach. Core Techniques*, Vol. 1. IRL Press, Oxford.

Glover D, Hames BD (eds) (1995) *DNA Cloning: a Practical Approach. Expression Systems*, Vol. 2. IRL Press, Oxford.

Old RW, Primrose SB (1994) *Principles of Gene Manipulation. An Introduction to Genetic Engineering*, 5th edn. Blackwell Scientific Publications, Oxford.

Sambrook J, Fritsch EF, Maniatis T (1989) *Molecular Cloning: a Laboratory Manual*, 2nd edn. Cold Spring Harbor Laboratory Press, Cold Spring Harbor, NY.

References

Gluzman Y (1981) SV40-transformed simian cells support the replication of early SV40 mutants. *Cell*, **23**, 175–182.

Iouannou PA, Amemiya CT, Garnes J, Kroisel PM, Shizuya H, Chen C, Batzer MA, de Jong P (1994) A new bacteriophage P1-derived vector for the propagation of large human DNA fragments. *Nature Genet.*, **6**, 84–89.

Schlessinger D (1990) Yeast artificial chromosomes: tools for mapping and analysis of complex genomes. *Trends Genet.*, **6**, 248–258.

Shizuya H, Birren B, Kim U-J, Mancino V, Slepak T, Tachiiri Y, Simon M (1992) Cloning and stable maintenance of 300 kilobase pair fragments of human DNA in *Escherichia coli* using an F-factor based vector. *Proc. Natl Acad. Sci. USA*, **89**, 8794–8797.

Sternberg N (1992) Cloning high molecular weight DNA fragments by the bacteriophage P1 system. *Trends Genet.*, **8**, 11–16.

Nucleic acid hybridization assays

5

Nucleic acid hybridization is a fundamental tool in molecular genetics which takes advantage of the ability of individual single-stranded nucleic acid molecules to form double-stranded molecules (that is, to hybridize to each other). For this to happen, the interacting single-stranded molecules must have a sufficiently high degree of base complementarity (see Section 1.2.1). Standard nucleic acid hybridization assays involve using a labeled nucleic acid **probe** to identify related DNA or RNA molecules (that is, ones with a significantly high degree of sequence similarity) within a complex mixture of unlabeled nucleic acid molecules, the target nucleic acid.

5.1 Preparation of nucleic acid probes

In standard nucleic acid hybridization assays the probe is labeled in some way. Nucleic acid probes may be made as single-stranded or double-stranded molecules (see *Figure 5.1*), but the working probe must be in the form of single strands.

Conventional DNA probes are isolated by cell-based DNA cloning or by PCR. In the former case, the starting DNA may range in size from 0.1 kb to hundreds of

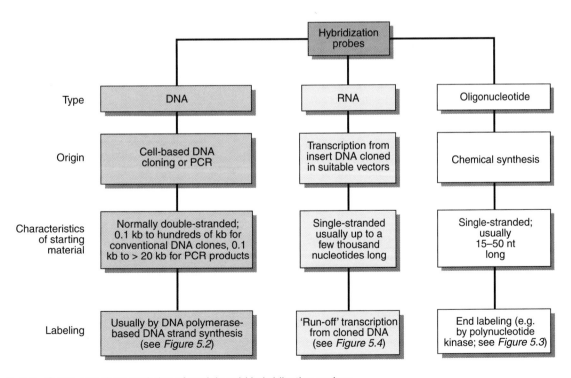

Figure 5.1: Origin and characteristics of nucleic acid hybridization probes.

kilobases in length and is usually (but not always) originally double-stranded. PCR-derived DNA probes have often been less than 10 kb long and are usually, but not always, originally double-stranded. Conventional DNA probes are usually labeled by incorporating labeled dNTPs during an *in vitro* DNA synthesis reaction (see Section 5.1.1).

RNA probes can conveniently be generated from DNA which has been cloned in a specialized plasmid vector (Melton *et al.*, 1984). Such vectors normally contain a phage promoter sequence immediately adjacent to the multiple cloning site. An RNA synthesis reaction is employed using the relevant phage RNA polymerase and the four rNTPs, at least one of which is labeled. Specific labeled RNA transcripts can then be generated from the cloned insert (see Section 5.1.1).

Oligonucleotide probes are short (typically 15–50 nucleotides) single-stranded pieces of DNA made by chemical synthesis: mononucleotides are added, one at a time, to a starting mononucleotide, conventionally the 3' end nucleotide, which is bound to a solid support. Generally, oligonucleotide probes are designed with a specific sequence chosen in response to prior information about the target DNA. Sometimes, however, oligonucleotide probes are used which are **degenerate** in sequence. Typically this involves parallel syntheses of a set of oligonucleotides which are identical at certain nucleotide positions but different at others. Oligonucleotide probes are often labeled by incorporating a ^{32}P atom or other labeled group at the 5' end (see next section).

5.1.1 DNA and RNA can conveniently be labeled in vitro by incorporation of nucleotides (or nucleotide components) containing a labeled atom or chemical group

Although, in principle, DNA and RNA can be labeled *in vivo*, by supplying labeled deoxynucleotides to tissue culture cells, this procedure is of limited general use; it has been restricted largely to preparing labeled viral DNA from virus-infected cells, and studying RNA processing events. A much more versatile method involves *in vitro* labeling: the purified DNA, RNA or oligonucleotide is labeled *in vitro* by using a suitable enzyme to incorporate labeled nucleotides. Two major types of procedure have been widely used:

- **Labeling of new strands during *in vitro* DNA or RNA synthesis.** In this type of procedure, DNA or RNA polymerase is used to make labeled DNA or RNA copies of a starting DNA. The *in vitro* DNA or RNA synthesis reaction requires that at least one of the four nucleotide precursors carries a labeled group. Labeling of DNA by *in vitro* DNA synthesis is normally accomplished using one of three methods: nick-translation (this section); random primed labeling (this section); or PCR-mediated labeling (see Section 6.1.1). Labeling of RNA generally is carried

out using an *in vitro* transcription system (see further on in this section).

- **End-labeling.** This type of procedure involves addition of a labeled group to one or a few terminal nucleotides. It is less widely used, but is useful for a number of procedures, including labeling of single-stranded oligonucleotides (see below) and restriction mapping. Inevitably, because only one or a very few labeled groups are incorporated, the specific activity (the amount of radioactivity incorporated divided by the total mass) of the labeled DNA is much less than that for probes in which there has been incorporation of several labeled nucleotides along the length of the DNA.

Labeling DNA by nick translation

The nick-translation procedure involves introducing single-strand breaks (*nicks*) in the DNA, leaving exposed 3' hydroxyl termini and 5' phosphate termini. The nicking can be achieved by adding a suitable endonuclease such as pancreatic deoxyribonuclease I (DNase I). The exposed nick can then serve as a start point for introducing new nucleotides at the 3' hydroxyl side of the nick using the DNA polymerase activity of *E. coli* DNA polymerase I at the same time as existing nucleotides are removed from the other side of the nick by the 5' → 3' exonuclease activity of the same enzyme. As a result, the nick will be moved progressively along the DNA ('translated') in the 5' → 3' direction (see *Figure 5.2A*). If the reaction is carried out at a relatively low temperature (about 15° C), the reaction proceeds no further than one complete renewal of the existing nucleotide sequence. Although there is no net DNA synthesis at these temperatures, the synthesis reaction allows the incorporation of labeled nucleotides in place of the previously existing unlabeled ones.

Random primed DNA labeling

The random primed DNA labeling method (sometimes known as oligolabeling) (Feinberg and Vogelstein, 1983) is based on hybridization of a mixture of all possible hexanucleotides: the starting DNA is denatured and then cooled slowly so that the individual hexanucleotides can bind to suitably complementary sequences within the DNA strands. Synthesis of new complementary DNA strands is primed by the bound hexanucleotides and is catalyzed by the Klenow subunit of DNA polymerase I (which contains the polymerase activity in the absence of associated exonuclease activities). DNA synthesis occurs in the presence of the four dNTPs, at least one of which has a labeled group (see *Figure 5.2B*). This method produces labeled DNAs of high specific activity. Because all sequence combinations are represented in the hexanucleotide mixture, binding of primer to template DNA occurs in a random manner, and labeling is uniform across the length of the DNA.

End-labeling of DNA

Single-stranded oligonucleotides are usually end-labeled

(A)

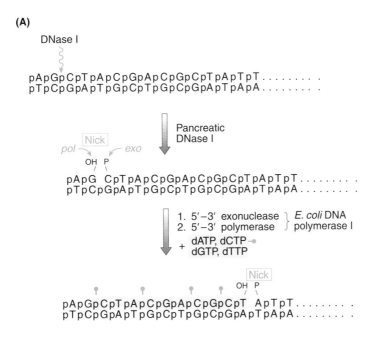

(B)

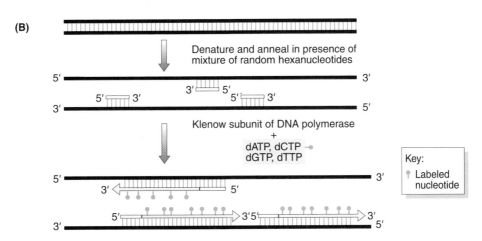

Figure 5.2: DNA labeling by *in vitro* DNA strand synthesis.

(A) Nick translation. Pancreatic DNase I introduces single-stranded nicks by cleaving internal phosphodiester bonds (p), generating a 5′ phosphate group and a 3′ hydroxyl terminus. Addition of the multisubunit enzyme *E. coli* DNA polymerase I contributes two enzyme activities: (i) a 5′ → 3′ exonuclease attacks the exposed 5′ termini of a nick and sequentially removes nucleotides in the 5′ → 3′ direction; (ii) a DNA polymerase adds new nucleotides to the exposed 3′ hydroxyl group, continuing in the 5′ → 3′ direction, thereby replacing nucleotides removed by the exonuclease and causing lateral displacement (translation) of the nick. (B) Random primed labeling. The Klenow subunit of *E. coli* DNA polymerase I can synthesize new radiolabeled DNA strands using as a template separated strands of DNA, and random hexanucleotide primers.

using polynucleotide kinase (kinase end-labeling). Typically, the label is provided in the form of a ^{32}P at the γ-phosphate position of ATP and the polynucleotide kinase catalyses an exchange reaction with the 5′-terminal phosphates (see *Figure 5.3A*). The same procedure can also be used for labeling double-stranded DNA. In this case, fragments carrying label at one end only can then be gen-

erated by cleavage at an internal restriction site, generating two differently sized fragments which can be separated by gel electrophoresis and purified.

Larger DNA fragments can be end-labeled by various alternative methods. **Fill-in end-labeling** (*Figure 5.3B*) is one popular approach, and uses the *Klenow subunit* of *E. coli* DNA polymerase. Again, fragments carrying label

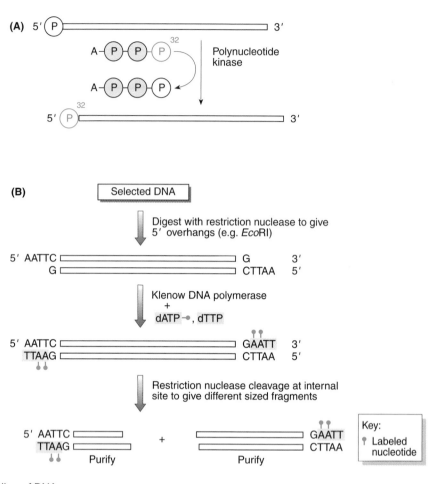

Figure 5.3: End-labeling of DNA.

(A) Kinase end-labeling of oligonucleotides. The 5'-terminal phosphate of the oligonucleotide is replaced in an exchange reaction by the ^{32}P-labeled γ-phosphate of [γ-^{32}P]ATP. The same procedure can be used to label the two 5' termini of double-stranded DNA. **(B)** Fill-in end-labeling by Klenow. The DNA of interest is cleaved with a suitable restriction nuclease to generate 5' overhangs. The overhangs act as a primer for Klenow DNA polymerase to incorporate labeled nucleotides complementary to the overhang. Fragments labeled at one end only can be generated by internal cleavage with a suitable restriction site to generate two differently sized fragments which can easily be size-fractionated.

at one end only can be generated by restriction cleavage and size fractionation. An alternative PCR-based method is primer-mediated 5' end-labeling (see Section 6.1.1).

Labeling of RNA

The preparation of labeled RNA probes (**riboprobes**) is most easily achieved by *in vitro* transcription of insert DNA cloned in a suitable plasmid vector. The vector is designed so that adjacent to the multiple cloning site is a phage promoter sequence, which can be recognized by the corresponding phage RNA polymerase. For example, the plasmid vector pSP64 contains the bacteriophage SP6 promoter sequence immediately adjacent to a multiple cloning site (see *Figure 5.4*). The SP6 RNA polymerase can then be used to initiate transcription from a specific start point in the SP6 promoter sequence, transcribing through any DNA sequence that has been inserted into the multiple cloning site. By using a mix of NTPs, at least one of

which is labeled, high specific activity radiolabeled transcripts can be generated (*Figure 5.4*). Bacteriophage T3 and T7 promoter/RNA polymerase systems are also used commonly for generating riboprobes. Labeled sense and antisense riboprobes can be generated from any gene cloned in such vectors (the gene can be cloned in either of the two orientations) and are widely used in tissue *in situ* hybridization (Section 5.3.4).

5.1.2 Nucleic acids can be labeled by isotopic and nonisotopic methods

Isotopic labeling and detection

Traditionally, labeling of nucleic acids has been conducted by incorporating nucleotides containing radioisotopes. Such radiolabeled probes contain nucleotides with a radioisotope (often ^{32}P, ^{33}P, ^{35}S or ^{3}H), which can be detected specifically in solution or, much more com-

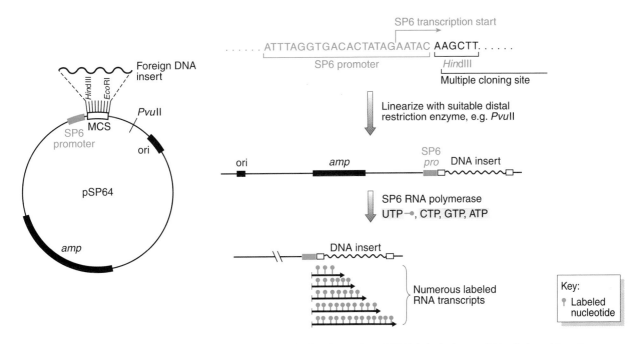

Figure 5.4: Riboprobes are generated by run-off transcription from cloned DNA inserts in specialized plasmid vectors.

The plasmid vector pSP64 contains a promoter sequence for phage SP6 RNA polymerase linked to the multiple cloning site (MCS) in addition to an origin of replication (ori) and ampicillin resistance gene (*amp*). After cloning a suitable DNA fragment in one of the 11 unique restriction sites of the MCS, the purified recombinant DNA is linearized by cutting with a restriction enzyme at a unique restriction site just distal to the insert DNA (*Pvu* II in this example). Thereafter labeled insert-specific RNA transcripts can be generated using SP6 RNA polymerase and a cocktail of NTPs, at least one of which is labeled (UTP in this case).

monly, within a solid specimen (autoradiography – see *Box 5.1*).

The intensity of an autoradiographic signal is dependent on the intensity of the radiation emitted by the radioisotope, and the time of exposure, which may often be long (one or more days, or even weeks in some applications). ^{32}P has been used widely in Southern blot hybridization, dot-blot hybridization, colony and plaque hybridization (see below) because it emits high energy β-particles which afford a high degree of sensitivity of detection. It has the disadvantage, however, that it is relatively unstable (see *Table 5.1*). Additionally, its high energy β-particle emission can be a disadvantage under circumstances when fine physical resolution is required to interpret the resulting image unambiguously. For this reason, radionuclides which provide less energetic

β-particle radiation have been preferred in certain procedures, for example ^{35}S-labeled and ^{33}P-labeled nucleotides for DNA sequencing and tissue *in situ* hybridization, and ^{3}H-labeled nucleotides for chromosome *in situ* hybridization. ^{35}S and ^{33}P have moderate half-lives while ^{3}H has a very long half-life. However, the latter isotope is disadvantaged by its comparatively low energy β-particle emission which necessitates very long exposure times.

^{32}P-labeled and ^{33}P-labeled nucleotides used in DNA strand synthesis labeling reactions have the radioisotope at the α-phosphate position, because the β- and γ-phosphates from dNTP precursors are not incorporated into the growing DNA chain. Kinase-mediated end-labeling, however, uses [γ-^{32}P]ATP (see *Figure 5.3A*). In the case of ^{35}S-labeled nucleotides which are incorporated during the synthesis of DNA or RNA strands, the NTP or dNTP carries a ^{35}S isotope in place of the O^{-} of the α-phosphate group. ^{3}H-labeled nucleotides carry the radioisotope at several positions. Specific detection of molecules carrying a radioisotope is most often performed by autoradiography (see *Box 5.1*).

Nonisotopic labeling and detection
Nonisotopic labeling systems involve the use of nonradioactive probes. Although developed only comparatively recently, they are becoming increasingly popular and are finding increasing applications in a variety of

Table 5.1: Characteristics of radioisotopes commonly used for labeling DNA and RNA probes

Radioisotope	Half-life	Decay type	Energy of emission
^{3}H	12.4 years	β^{-}	0.019 MeV
^{32}P	14.3 days	β^{-}	1.710 MeV
^{33}P	25.5 days	β^{-}	0.248 MeV
^{35}S	87.4 days	β^{-}	0.167 MeV

Principles of autoradiography

Autoradiography is a procedure for localizing and recording a radiolabeled compound within a solid sample, which involves the production of an image in a photographic emulsion. In molecular genetic applications, the solid sample often consists of size-fractionated DNA or protein samples that are embedded within a dried gel, fixed to the surface of a dried nylon membrane or nitrocellulose filter, or located within fixed chromatin or tissue samples mounted on a glass slide. The photographic emulsions consist of silver halide crystals in suspension in a clear gelatinous phase. Following passage through the emulsion of a β-particle or a γ-ray emitted by a radionuclide, the Ag^+ ions are converted to Ag atoms. The resulting latent image can then be converted to a visible image once the image is developed, an amplification process in which entire silver halide crystals are reduced to give metallic silver. The fixing process results in removal of any unexposed silver halide crystals, giving an autoradiographic image which provides a two-dimensional representation of the distribution of the radiolabel in the original sample.

Direct autoradiography involves placing the sample in intimate contact with an X-ray film, a plastic sheet with a coating of photographic emulsion; the radioactive emissions from the sample produce dark areas on the developed film. This method is best suited to detection of weak to medium strength β-emitting radionuclides (e.g. ^3H, ^{35}S, etc.). However, it is not suited to high energy β-particles (e.g. from ^{32}P): such emissions pass through the film, resulting in the wasting of the majority of the energy. Indirect autoradiography is a modification in which the emitted energy is converted to light by a suitable chemical (scintillator or fluor). One popular approach uses intensifying screens, sheets of a solid inorganic scintillator which are placed behind the film in the case of samples emitting high energy radiation, such as ^{32}P. Those emissions which pass through the photographic emulsion are absorbed by the screen and converted to light. By effectively superimposing a photographic emission upon the direct autoradiographic emission, the image is intensified.

(*Figure 5.6*). The reporter molecules on modified nucleotides need to protrude sufficiently far from the nucleic acid backbone to facilitate their detection by the affinity molecule and so a long carbon atom *spacer* is required to separate the nucleotide from the reporter group.

Two indirect nonisotopic labeling systems are widely used:

- The **biotin–streptavidin** system utilizes the extremely high affinity of two ligands: biotin (a naturally occurring vitamin) which acts as the reporter, and the bacterial protein streptavidin, which is the affinity molecule. Biotin and streptavidin bind together extemely tightly with an affinity constant of 10^{-14}, one of the strongest known in biology. Biotinylated probes can be made easily by including a suitable biotinylated nucleotide in the labeling reaction (see *Figure 5.7*).

- **Digoxigenin** is a plant steroid (obtained from *Digitalis* plants) to which a specific antibody has been raised. The digoxigenin-specific antibody permits detection of nucleic acid molecules which have incorporated nucleotides containing the digoxigenin reporter group (see *Figure 5.7*).

A variety of different marker groups or molecules can be conjugated to affinity molecules such as streptavidin or the digoxigenin-specific antibody. They include various fluorophores (see *Box 5.2*), or enzymes such as alkaline phosphatase and peroxidase which can permit detection via colorimetric assays or chemical luminescence assays, etc.

5.2 Principles of nucleic acid hybridization

5.2.1 Nucleic acid hybridization is a method for identifying closely related nucleic acid molecules within two populations, a complex target population and a comparatively homogeneous probe population

Definition of nucleic acid hybridization

Nucleic acid hybridization involves mixing single strands of two sources of nucleic acids, a **probe** which typically consists of a homogeneous population of *identified molecules* (e.g. cloned DNA or chemically synthesized oligonucleotides) and a **target** which typically consists of a *complex, heterogeneous population* of nucleic acid molecules. If either the probe or the target is initially double-stranded, the individual strands must be separated, generally by heating or by alkaline treatment. After mixing single strands of probe with single strands of target, strands with complementary base sequences can be allowed to reassociate. Complementary probe strands can reanneal to form *homoduplexes*, as can complementary tar-

different areas (Kricka, 1992). Two types of non-radioactive labeling are conducted:

- **Direct nonisotopic labeling**, where a nucleotide which contains the label that will be detected is incorporated. Often such systems involve incorporation of modified nucleotides containing a **fluorophore** (*Figure 5.5A*), a chemical group which can fluoresce when exposed to light of a certain wavelength (fluorescence labeling – see *Box 5.2*).

- **Indirect nonisotopic labeling**, usually featuring the chemical coupling of a modified **reporter molecule** to a nucleotide precursor. After incorporation into DNA, the reporter groups can be specifically bound by an affinity molecule, a protein or other ligand which has a very high affinity for the reporter group. Conjugated to the latter is a **marker** molecule or group which can be detected in a suitable assay

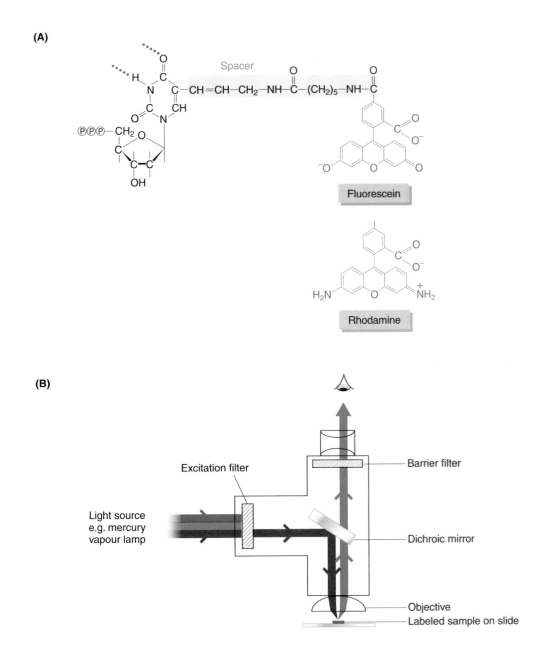

Figure 5.5: Fluorescence microscopy and structure of common fluorophores.

(**A**) Structure of fluorophores. The example on top shows fluorescein-dUTP. The fluorescein group is linked to the 5′ carbon atom of the uridine by a spacer group so that when the modified nucleotide is incorporated into DNA, the fluorescein group is readily accessible. Below is the structure of rhodamine from which a variety of fluorophores have been derived.
(**B**) Fluorescence microscopy. The excitation filter is a color barrier filter which in this example is selected to let through only blue light. The transmitted blue light is of an appropriate wavelength to be reflected by the dichroic (beam-splitting) mirror onto the labeled sample which then fluoresces and emits light of a longer wavelength, green light in this case. The longer wavelength of the emitted green light means that it passes straight through the dichroic mirror. The light subsequently passes through a second color barrier filter which blocks unwanted fluorescent signals, leaving the desired green fluorescence emission to pass through to the eyepiece of the microscope. A second beam-splitting device can also permit the light to be recorded in a CCD camera.

get DNA strands. However, it is the annealing of a probe DNA strand and a complementary target DNA strand to form a labeled probe–target heteroduplex that defines the usefulness of a nucleic acid hybridization assay. The

rationale of the hybridization assay is to use the identified probe to query the target DNA by identifying fragments in the complex target which may be related in sequence to the probe (*Figure 5.8*).

Box 5.2

Fluorescence labeling and detection systems

Fluorescence labeling of nucleic acids was developed in the 1980s and has proved to be extremely valuable in many different applications including chromosome *in situ* hybridization, tissue *in situ* hybridization and automated DNA sequencing. The fluorescent labels can be used in direct labeling of nucleic acids by incorporating a modified nucleotide (often 2′ deoxyuridine 5′ triphosphate) containing an appropriate **fluorophore**, a chemical group which fluoresces when exposed to a specific wavelength of light. Popular fluorophores used in direct labeling include fluorescein, a pale green fluorescent dye, rhodamine, a red fluorescent dye and amino methyl coumarin, a blue fluorescent dye (see *Figure 5.5A*). Alternatively, indirect labeling systems can be used whereby modified nucleotides containing a reporter group (such as biotin or digoxigenin) are incorporated into the nucleic acid (see *Figure 5.7*) and then the reporter group is specifically bound by an affinity molecule (such as streptavidin or a digoxigenin-specific antibody) to which is attached a fluorophore e.g. amino methylcoumarin acetic acid (AMCA), fluorescein isothiocyanate (FITC) or other fluorescein derivatives, and tetramethylrhodamine isothiocyanate (TRITC) or other rhodamine derivatives.

Detection of the fluorophore in direct or indirect labeling systems is accomplished by passing a beam of light from a suitable light source (e.g. a mercury vapor lamp in fluorescence microscopy, an argon laser in automated DNA sequencing) through an appropriate color filter (excitation filter) designed to transmit light at the desired excitation wavelength. In fluorescence microscopy (*Figure 5.5B*) this light is reflected onto the fluorescently labeled sample on a microscope slide using a dichroic mirror, one which reflects light of certain wavelengths while allowing light of other wavelengths to pass straight through. The light then excites the fluorophore to fluoresce and as it does so, it emits light at a slightly longer wavelength, the emission wavelength. The light emitted by the fluorophore passes back up and straight through the dichroic mirror, through an appropriate barrier filter and is then transmitted into the eye piece of the microscope, and can also be captured using a suitable charged coupled device (CCD) camera. Maximum emission and excitation wavelengths for common fluorophores are as indicated below.

Fluorophore	Color	Excitation max. (nm)	Emission max. (nm)
AMCA	Blue	399	446
Fluorescein	Green	494	523
CY3	Red	552	565
Rhodamine	Red	555	580
Texas Red	Red	590	615

Melting temperature and hybridization stringency

Denaturation of double-stranded probe DNA is generally achieved by heating a solution of the labeled DNA to a temperature which disrupts the hydrogen bonds that hold the two complementary DNA strands together. The energy required to separate two perfectly complementary DNA strands is dependent on a number of factors, notably:

■ **strand length** – long homoduplexes contain a large number of hydrogen bonds and require more energy to separate them; because the labeling procedure typically results in short DNA probes, this effect is negligible above an original length (i.e. prior to labeling) of 500 bp;

■ **base composition** – because GC base pairs have one more hydrogen bond than AT base pairs (see *Figure 1.2*), strands with a high % GC composition are more difficult to separate than those with a low % GC composition;

■ **chemical environment** – the presence of monovalent cations (e.g. Na^+ ions) stabilizes the duplex, whereas chemical denaturants such as formamide and urea destabilize the duplex by chemically disrupting the hydrogen bonds.

A useful measure of the stability of a nucleic acid duplex is the melting temperature (T_m). This is the temperature corresponding to the midpoint in the observed transition from double-stranded to single-stranded form. Conveniently, this transition can be followed by measuring the optical density of the DNA. The bases of the nucleic acids absorb 260 nm ultraviolet (UV) light strongly. However, the adsorption by double-stranded DNA is considerably less than that of the free nucleotides. This difference, the so-called hypochromic effect, is due to interactions between the electron systems of adjacent bases, arising from the way in which adjacent bases are stacked in parallel in a double helix. If duplex DNA is gradually heated, therefore, there will be an increase in the light absorbed at 260 nm (the optical density$_{260}$ or OD_{260}) towards the value characteristic of the free bases. The temperature at which there is a midpoint in the optical density shift is then taken as the T_m (see *Figure 5.9*).

For mammalian genomes, with a base composition of about 40% GC, the DNA denatures with a T_m of about 87°C under approximately physiological conditions. The T_m of perfect hybrids formed by DNA, RNA or oligonucleotide probes can be determined according to the formulae in *Table 5.2*. Often, hybridization conditions are chosen so as to promote heteroduplex formation and the hybridization temperature is often as much as 25°C below the T_m. However, after the hybridization and removal of excess probe, hybridization washes may be conducted under more stringent conditions so as to disrupt all duplexes other than those between very closely related sequences. Probe–target heteroduplexes are most stable thermodynamically when the region of duplex formation contains perfect base matching. Mismatches between the two strands of a heteroduplex reduce the T_m: for normal DNA probes, each 1% of mismatching reduces the T_m by approximately 1°C. Although probe–target heteroduplexes are usually not as stable as reannealed probe

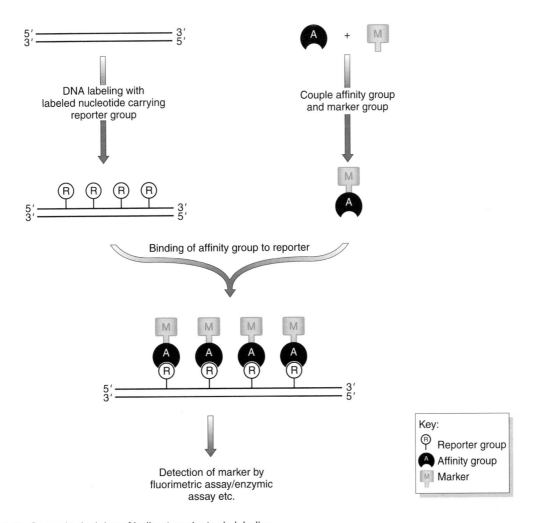

Figure 5.6: General principles of indirect nonisotopic labeling.

The protein recognizing the reporter group is often a specific antibody, as in the digoxigenin system, or any other ligand that has a very high affinity for a specific group, such as streptavidin in the case of using biotin as the reporter (see *Figure 5.7*). The marker can be detected in various ways. If it carries a specific fluorescent dye, it can be detected in a fluorimetric assay. Alternatively, it can be an enzyme such as alkaline phosphatase which can be coupled to an enzyme assay, yielding a product that can be measured colorimetrically.

homoduplexes, a considerable degree of mismatching can be tolerated if the overall region of base complementarity is long (>100 bp; see *Figure 5.10*).

Increasing the concentration of NaCl and reducing the temperature reduces the **hybridization stringency**, and enhances the stability of mismatched heteroduplexes. This means that comparatively diverged members of a multigene family or other repetitive DNA family can be identified by hybridization using a specific family member as a probe. Additionally, a gene sequence from one species can be used as a probe to identify homologs in other comparatively diverged species, provided the sequence is reasonably conserved during evolution (see *Figure 10.21* and *Box 20.1*).

Conditions can also be chosen to maximize hybridization stringency (e.g. lowering the concentration of NaCl

and increasing the temperature), so as to encourage dissociation (denaturation) of mismatched heteroduplexes. If the region of base complementarity is small, as with oligonucleotide probes (typically 15–20 nucleotides), hybridization conditions can be chosen such that a single mismatch renders a heteroduplex unstable (see Section 5.3.1).

5.2.2 The kinetics of DNA reassociation are defined by the product of DNA concentration and time (C_ot)

When double-stranded DNA is denatured, for example by heat, and the complementary single strands are then allowed to reassociate to form double-stranded DNA, the speed at which the complementary strands reassociate will depend on the starting concentration of the DNA. If

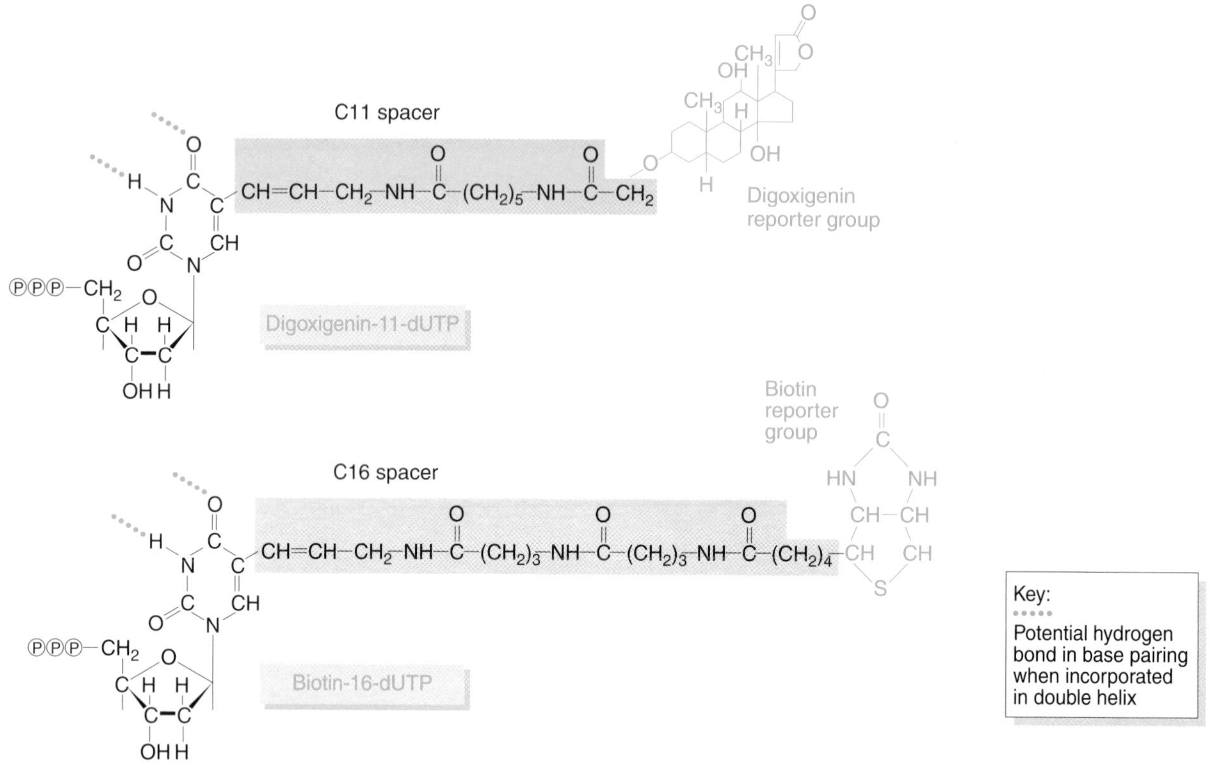

Figure 5.7: Structure of digoxigenin- and biotin-modified nucleotides.

Note that the digoxigenin and biotin groups in these examples are linked to the 5′ carbon atom of the uridine of dUTP by spacer groups consisting respectively of a total of 11 carbon atoms (digoxigenin-11-UTP) or 16 carbon atoms (biotin-16-dUTP). The digoxigenin and biotin groups are **reporter groups**: after incorporation into a nucleic acid they are bound by specific ligands containing an attached marker such as a fluorophore.

there is a high concentration of the complementary DNA sequences, the time taken for any one single-stranded DNA molecule to find a complementary strand and form a duplex will be reduced. **Reassociation kinetics** is the term used to measure the speed at which complementary single-stranded molecules are able to find each other and form duplexes. It is determined by two major parameters: the starting concentration (C_o) of the specific DNA sequence in moles of nucleotides per liter and the reaction time (t) in seconds. Since the rate of reassociation is proportional to C_o and to t, the $C_o t$ value (often loosely referred to as the cot value) is a useful measure. The $C_o t$ value will also vary depending on the temperature of reassociation and the concentration of monovalent cations. As a result, it is usual to use fixed reference values: a reassociation temperature of 65°C and a Na$^+$ concentration of 0.3 M NaCl.

Most hybridization assays use an excess of target nucleic acid over probe in order to encourage probe–target formation. This is so because the probe is usually *homogenous*, often consisting of a single type of cloned DNA molecule or RNA molecule, but the target nucleic acid is typically *heterogeneous*, comprising for example genomic DNA or total cellular RNA. In the latter case the

concentration of any one sequence may be very low, thereby causing the rate of reassociation to be slow. For example, if a Southern blot uses a cloned β-globin gene as a probe to identify complementary sequences in human genomic DNA, the latter will be present in very low concentration (the β-globin gene is an example of a single copy sequence and in this case represents only 0.00005% of human genomic DNA). It is therefore necessary to use several micrograms of target DNA to drive the reaction. By contrast, certain other sequences are highly repeated in genomic DNA (see Section 7.3), and this greatly elevated DNA concentration results in a comparatively rapid reassociation time.

Because the amount of target nucleic acid bound by a probe depends on the copy number of the recognized sequence, hybridization signal intensity is proportional to the copy number of the recognized sequence. Single copy genes give weak hybridization signals, highly repetitive DNA sequences give very strong signals. If a particular probe is heterogeneous and contains a low copy sequence of interest, such as a specific gene, mixed with a highly abundant DNA repeat, the weak hybridization signal obtained with the former will be completely masked by the strong repetitive DNA hybridization signal. This

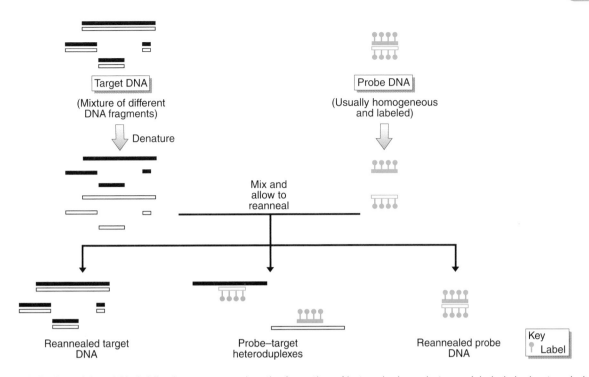

Figure 5.8: A nucleic acid hybridization assay requires the formation of heteroduplexes between labeled single-stranded nucleic acid probes and complementary sequences within a target nucleic acid.

The probe is envisaged to be strongly related in sequence to a central segment of one of the many types of nucleic acid molecule in the target. Mixing of denatured probe and denatured target will result in reannealed probe–probe homoduplexes (bottom right) and target–target homoduplexes (bottom left) but also in **heteroduplexes** formed between probe DNA and any target DNA molecules that are significantly related in sequence (bottom centre). If a method is available for removing the probe DNA that is not bound to target DNA, the heteroduplexes can easily be identified by methods that can detect the label.

effect can, however, be overcome by competition hybridization (see *Box 5.3*).

5.2.3 A wide variety of nucleic acid hybridization assays can be used

Early experiments in nucleic acid hybridization utilized solution hybridization, involving mixing of aqueous

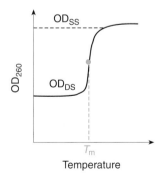

Figure 5.9: Denaturation of DNA results in an increase in optical density.

OD_{SS} and OD_{DS} indicate the optical density of single-stranded and double-stranded DNA respectively. The difference between them represents the hypochromic effect (see text).

solutions of probe and target nucleic acids. However, the very low concentration of single copy sequences in complex genomes meant that reassociation times were inevitably slow. One widely used way of increasing the reassociation speed is to artificially increase the overall DNA concentration in aqueous solution by abstracting water molecules (e.g. by adding high concentrations of polyethylene glycol).

An alternative to solution hybridization which facilitated detection of reassociated molecules involved immobilizing the target DNA on a solid support, such as a membrane made of nitrocellulose or nylon, to both of which single-stranded DNA binds readily. Attachment of labeled probe to the immobilized target DNA can then be followed by removing the solution containing unbound probe DNA, extensive washing and drying in preparation for detection.

This is the basis of the standard nucleic acid hybridization assays currently in use. More recently, however, reverse hybridization assays have become more popular. In these cases, the probe population is unlabeled and fixed to the solid support, while the target nucleic acid is labeled and present in aqueous solution. Note, therefore, that probe and target are not primarily distinguished by which is the labeled and which is the unlabeled population. Instead, the important consideration is that the target DNA should be the complex imperfectly understood

Table 5.2: Equations for calculating T_m

Hybrids	T_m (°C)
DNA–DNA	$81.5 + 16.6 \; (\log_{10}[Na^+]^a) + 0.41 \; (\%GC^b) - 500/L^c$
DNA–RNA or RNA–RNA	$79.8 + 18.5 \; (\log_{10}[Na^+]^a) + 0.58 \; (\%GC^b) + 11.8 \; (\%GC^b)^2 - 820/L^c$
oligo-DNA or oligo-RNA[d]	For <20 nucleotides: $\quad 2 \; (l_n)$ For 20–35 nucleotides: $\quad 22 + 1.46 \; (l_n)$

[a]Or for other monovalent cation, but only accurate in the 0.01–0.4 M range.
[b]Only accurate for %GC in the 30% to 75% range.
[c]L = length of duplex in base pairs.
[d]Oligo, oligonucleotide; l_n, effective length of primer = $2 \times$ (no. of G + C) + (no. of A + T).

Note that for each 1% formamide, the T_m is reduced by about 0.6°C, while the presence of 6 M urea reduces the T_m by about 30°C

population which the probe (whose molecular identity is known) attempts to query. Depending on the nature and form of the probe and target, a very wide variety of nucleic acid hybridization assays can be devised (*Box 5.4*).

Box 5.3

Competition hybridization and Cot-1 DNA

Competition (or **suppression**) **hybridization** involves blocking a potentially strong repetitive DNA signal which can be obtained when using a complex DNA probe. The labeled probe DNA is denatured and allowed to reassociate in the presence of unlabeled total genomic DNA in solution, or preferably a fraction that is enriched for highly repetitive DNA sequences. In either case, the highly repetitive DNA within the unlabeled DNA is present in large excess over the repetitive elements in the labeled probe. As a result, such sequences will readily associate with complementary strands of the repetitive sequences within the labeled probe, thereby effectively blocking their hybridization to target sequences.

Instead of using total genomic DNA as a blocking agent in hybridization, it is more effective to use a fraction of total genomic DNA that is enriched for highly repetitive DNA sequences, such as the *Alu*, LINE-1 and *THE* repeats of human DNA. For human DNA, and other mammalian DNA where the genome size is much the same as that of the human genome, the latter usually involves preparing a fraction of DNA known as **Cot-1 DNA** (i.e. DNA with a cot value of 1.0). Total purified human genomic DNA is sonicated to an average length of about 400 bp, denatured by heating, then allowed to renature in 0.3 M NaCl at 65°C at a starting concentration of x moles of nucleotides per liter for a time of t seconds, where $xt = 1.0$. For example, since 1 mole of nucleotide = 330 g on average, then a starting DNA concentration of 1 mg ml^{-1} (3 mmol l^{-1}) and a renaturation time of about 5.5 min (330 s) will produce a cot value of about 1.0 in mole nucleotides s^{-1} l^{-1}.

5.3 Nucleic acid hybridization assays using cloned DNA probes to screen uncloned nucleic acid populations

Numerous applications in molecular genetics involve taking an individual DNA clone and using it as a hybridization probe to screen for the presence of related sequences within a complex target of uncloned DNA or RNA. Sometimes the assay is restricted to simply checking for presence or absence of sequences related to the probe. In other

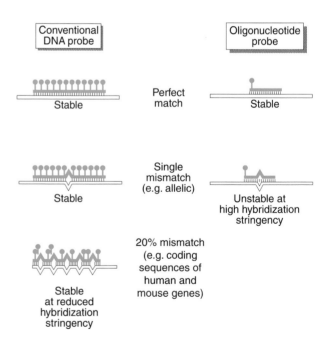

Figure 5.10: Nucleic acid hybridization can identify target sequences that are considerably diverged from a conventional DNA (or RNA) probe, or that are identical to an oligonucleotide probe.

Box 5.4

Standard and reverse nucleic acid hybridization assays

Standard assays	Labeled probe in solution	Unlabeled target bound to solid support
Dot-blot (*Figure 5.11*)	Any labeled DNA or RNA but often an oligonucleotide	Complex DNA or RNA population; not size-fractionated, but spotted directly onto membrane
Southern blot (*Figures 5.12, 5.16*)	Any	Often complex genomic DNA (but may be individual DNA clones); digested with restriction nuclease and size fractionated, then transferred to membrane
Northern blot (*Figure 5.13*)	Any	Complex RNA population (e.g. total cellular RNA or poly A$^+$ RNA) which has been size-fractionated, then transferred to membrane
Chromosome *in situ* hybridization (Section 5.3.4; *Figure 10.5*)	Usually a labeled genomic clone	DNA within chromosomes (often metaphase) of lysed cells on a microscope slide
Tissue *in situ* hybridization (*Figure 5.17*)	Usually a labeled antisense riboprobe or oligonucleotide	RNA within cells of fixed tissue sections on a microscope slide
Colony blot (*Figure 5.18*)	Any	Cell colonies separated after plating out on agar then transferred to membrane
Plaque lift (Section 5.4.1)	Any	Phage-infected bacterial colonies separated after plating out on agar and transferred to membrane
Gridded clone hybridization assay (*Figure 5.19*)	Any	Clones robotically spotted onto membrane in geometric arrays

Reverse assays	Labeled target in solution	Unlabeled probes bound to solid support
Reverse dot-blot	Complex DNA	Oligonucleotides spotted onto membrane
DNA microarray (*Figures 5.20A and 20.6A*)	Complex DNA	DNA clones robotically spotted onto microscope slide
Oligonucleotide microarray (*Figures 5.20B and 20.6B*)	Complex DNA	Oligonucleotides synthesized on glass

cases, useful information can be obtained regarding the size of the complementary sequences, their subchromosomal location or their locations within specific tissues or groups of cells.

5.3.1 Dot-blot hybridization is a rapid screening method which often employs allele-specific oligonucleotide (ASO) probes to discriminate between alleles differing at a single nucleotide position

The general procedure of dot-blotting involves taking an aqueous solution of target DNA, for example total human genomic DNA, and simply spotting it on to a nitrocellulose or nylon membrane then allowing it to dry. The variant technique of slot-blotting involves pipetting the DNA through an individual slot in a suitable template. In both methods the target DNA sequences are denatured, either by previously exposing to heat, or by exposure of

the filter containing them to alkali. The denatured target DNA sequences now immobilized on the membrane are exposed to a solution containing single-stranded labeled probe sequences. After allowing sufficient time for probe–target heteroduplex formation, the probe solution is decanted, and the membrane is washed to remove excess probe that may have become nonspecifically bound to the filter. It is then dried and exposed to an autoradiographic film.

A useful application of dot-blotting involves distinguishing between alleles that differ by even a single nucleotide substitution. To do this **allele-specific oligonucleotide (ASO)** probes are constructed from sequences spanning the variant nucleotide site. ASO probes are typically 15–20 nucleotides long and are normally employed under hybridization conditions at which the DNA duplex between probe and target is stable *only if there is perfect base complementarity between them*: a single mismatch between probe and target sequence is sufficient to render the short heteroduplex unstable (*Figure 5.10*). Typically, this involves designing the oligonucleotides so that the single nucleotide difference between alleles occurs in a central segment of the oligonucleotide sequence, thereby maximizing the thermodynamic instability of a mismatched duplex. Such discrimination can be employed for a variety of research and diagnostic purposes. Although ASOs can be used in conventional Southern blot hybridization (see below), it is more convenient to use them in dot-blot assays (see *Figure 5.11*).

Another method of ASO dot blotting uses a reverse dot-blotting approach. This means that the oligonucleotide probes are not labeled and are fixed on a filter or membrane whereas the target DNA is labeled and provided in solution. Positive binding of labeled target DNA to a specific oligonucleotide on the membrane is taken to mean that the target has that specific sequence. This approach, and related DNA microarray

methods, have many diagnostic applications (see Section 17.1.4).

5.3.2 Southern and Northern blot hybridizations detect target DNA and RNA fragments that have been size-fractionated by gel electrophoresis

Southern blot hybridization

In this procedure, the target DNA is digested with one or more restriction endonucleases, size-fractionated by agarose gel electrophoresis, denatured and transferred to a nitrocellulose or nylon membrane for hybridization (*Figure 5.12*). During the electrophoresis, DNA fragments, which are negatively charged because of the phosphate groups, are repelled from the negative electrode towards the positive electrode, and sieved through the porous gel. Smaller DNA fragments move faster. For fragments

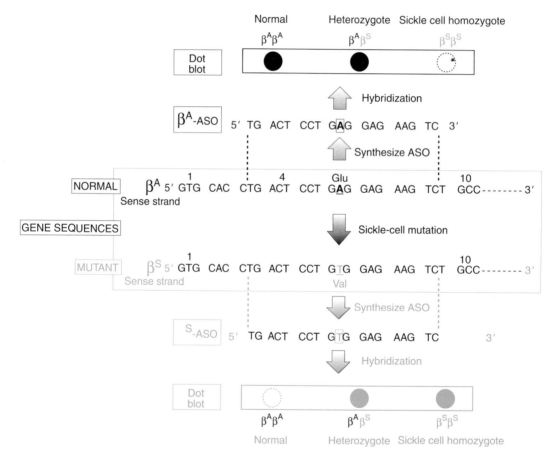

Figure 5.11: Allele-specific oligonucleotide (ASO) dot-blot hybridization can identify individuals with the sickle cell mutation.

The schematic dot-blot at top shows the result of probing with an ASO specific for the normal β-globin allele (β^A-ASO; shown immediately below). The results are positive (filled circle) for normal individuals and for heterozygotes but negative for sickle cell homozygotes (dashed, unfilled circle). The dot-blot at the bottom shows the result of probing with an ASO specific for the sickle cell β-globin allele (β^S-ASO; shown immediately above), and in this case the results are positive for the sickle cell homozygotes and heterozygotes but negative for normal individuals. The β^A-ASO and β^S-ASO were designed to be 19 nucleotides long in this case chosen from codons 3 to 9 of respectively the sense β^A and β^S globin gene sequences surrounding the sickle cell mutation site. The latter is a single nucleotide substitution (A → T) at codon 6 in the β-globin gene, resulting in a GAG (Glu) → GTG (Val) substitution (see middle sequences).

between 0.1 and 20 kb long, the migration speed depends on fragment length, but scarcely at all on the base composition. Thus, fragments in this size range are fractionated by size in a conventional agarose gel electrophoresis system. To achieve efficient size-fractionation of large fragments (40 kb to several megabases), a more specialized system is required, such as a pulsed-field gel electrophoresis apparatus (see Section 10.2.2).

Following electrophoresis, the test DNA fragments are denatured in strong alkali. As agarose gels are fragile, and the DNA in them can diffuse within the gel, it is usual to transfer the denatured DNA fragments by blotting on to a durable nitrocellulose or nylon membrane, to which single-stranded DNA binds readily. The individual DNA fragments become immobilized on the membrane at positions which are a faithful record of the size separation achieved by agarose gel electrophoresis. Subsequently, the immobilized single-stranded target DNA sequences are allowed to associate with labeled single-stranded probe DNA. The probe will bind only to related DNA sequences in the target DNA, and their position on the membrane can be related back to the original gel in order to estimate their size.

An important application of Southern blot hybridization in mammalian genetics is the ability to identify a given DNA probe as a member of a repetitive DNA family. Many important mammalian genes belong to multigene families, and many other DNA sequences show varying degrees of repetition. Once a newly isolated probe is demonstrated to be related to other uncharacterized sequences, attempts can then be made to isolate the other members of the family by screening genomic DNA libraries. Additionally, screening can also be conducted on genomic DNA samples from different species to identify interspecific related sequences. An important route to identifying coding DNA involves identifying sequences that are highly conserved in evolution (Section 10.4.1).

Northern blot hybridization

Northern blot hybridization is a variant of Southern blotting in which the target nucleic acid is RNA instead of DNA. A principal use of this method is to obtain information on the expression patterns of specific genes. Once a gene has been cloned, it can be used as a probe and hybridized against a Northern blot containing, in

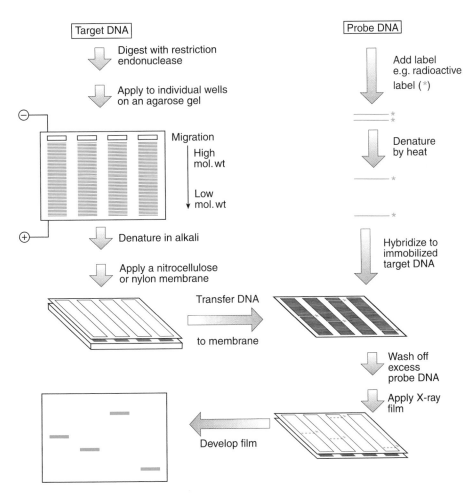

Figure 5.12: Southern blot hybridization detects target DNA fragments that have been size-fractionated by gel electrophoresis. See *Figure 5.16* for a specific application.

different lanes, samples of RNA isolated from a variety of different tissues (see *Figure 5.13*). The data obtained can provide information on the range of cell types in which the gene is expressed, and the relative abundance of transcripts. Additionally, by revealing transcripts of different sizes, it may provide evidence for the use of alternative promoters, splice sites or polyadenylation sites.

5.3.3 Southern blot hybridization permits restriction mapping and assay of RFLPs and moderately small scale mutations

Southern blot hybridization has been used extensively in molecular genetic studies as a means of genomic restriction mapping: a labeled DNA probe from one genome can be used to infer the structure of related sequences in the same or different genomes. Because the genomic DNA samples are fractionated by separation of restriction fragments according to size, mutations which alter a restriction site, and significantly large insertions or deletions occurring between neighboring restriction sites, can be typed. Such mutations will result in altered restriction fragment lengths, that is **restriction fragment length polymorphisms (RFLPs)**.

Direct detection of pathogenic point mutations by restriction mapping

Very occasionally, a pathogenic mutation directly abolishes or creates a restriction site, enabling direct screening for the pathogenic mutation. For example, the sickle cell mutation is a single nucleotide substitution (A → T) at codon 6 in the β-globin gene, which causes a missense mutation (Glu → Val), and at the same time abolishes an *Mst*II restriction site which spans codons 5 to 7. The nearest flanking restriction sites for *Mst*II, located 1.2 kb upstream in the 5′-flanking region and 0.2 kb downstream at the 3′ end of the first intron, are well conserved. Consequently, a β-globin DNA probe can differentiate the normal β^A-globin and the mutant β^S-globin alleles in *Mst*II-digested human DNA: the former exhibits 1.2 kb and 0.2 kb *Mst*II fragments, whereas the sickle cell allele exhibits a 1.4 kb *Mst*II fragment (*Figure 5.14*).

Detection of conventional RFLPs

The great majority of mutations are not associated with disease; instead, they often occur within noncoding DNA sequences. As a large number of recognition sequences are known for type II restriction endonucleases, many point mutation polymorphisms will be characterized by alleles which possess or lack a recognition site for a specific restriction endonuclease and therefore display restriction site polymorphism (RSP). Accordingly, individual RSPs normally have two detectable alleles (one lacking and one possessing the specific restriction site). RSPs can be assayed by digesting genomic DNA samples with the relevant restriction endonuclease and identifying specific restriction fragments whose lengths are characteristic of the two alleles, so-called RFLPs (*Figure 5.15*).

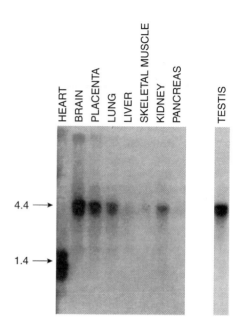

Figure 5.13: Northern blot hybridization is used to evaluate the gross expression patterns of a gene.

Northern blotting involves size-fractionation of samples of total RNA [or purified poly(A)^+ mRNA], transfer to a membrane and hybridization with a suitable labeled nucleic acid probe. The example shows the use of a labeled cDNA probe from the *FMR1* (Fragile-X mental retardation syndrome) gene. Highest levels are detected in the brain and testis (4.4 kb), with decreasing expression in the placenta, lung and kidney respectively. Multiple smaller transcripts are present in the heart. Reproduced from Hinds *et al.* (1993) *Nature Genetics*, **3**, pp. 36–43, with permission from Nature America Inc.

VNTR-based RFLPs and DNA fingerprinting

DNA probes can also be used to monitor **VNTR polymorphisms** where alleles differ by a variable number of tandem repeats. To do this, genomic DNA samples are digested with a restriction endonuclease which recognizes well-conserved restriction sites flanking a specific VNTR locus. The resulting restriction fragments are separated according to size on agarose gels, transferred to a suitable membrane and hybridized with a probe representing a unique sequence from the corresponding locus. The resulting pattern of locus-specific RFLPs does not reflect RSP: instead, the differences in sizes of the restriction fragments represent integral numbers of the tandemly repeated unit.

Although the term VNTR could, in theory, encompass a wide range of repeat lengths, in practice the term is usually reserved for moderately large arrays of a repeat unit which is typically in the 5–64 bp region (so-called hypervariable minisatellite DNA, distinguishing it from simple tandem repeat polymorphism (where the repeat unit length is from 1 to 4 bp; i.e. microsatellite DNA) and

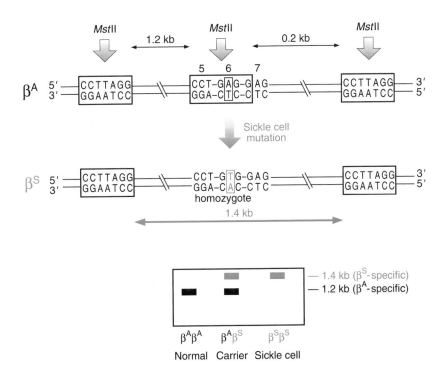

Figure 5.14: The sickle cell mutation destroys an *Mst*II site and generates a disease-specific RFLP.

The *Mst*II restriction nuclease recognizes the sequence CCTNAGG where N = A, C, G or T. A restriction site for *Mst*II is found in the normal β^A-globin allele but is destroyed by the sickle cell mutation. The nearest flanking *Mst*II sites are located, respectively, 1.2 kb upstream in the 5′-flanking region of the β-globin gene and 0.2 kb downstream at the 3′ end of the first intron. Conservation of these flanking sites results in the β^A-associated (1.2 kb + 0.2 kb) *Mst*II RFLP and the sickle cell-associated 1.4 kb *Mst*II RFLP.

tandem repeat polymorphism associated with very large arrays of satellite DNA.

If the VNTR locus is a member of a repeated DNA family, the use of a VNTR repeat probe, rather than a unique flanking probe, will produce a complex polymorphic pattern. For example, hypervariable minisatellite DNA clones have been used as probes against Southern blots of appropriately digested genomic DNA. Cross-hybridization of such probes with the other members of this highly repeated DNA family results in a pattern of hybridization bands representing the summed contributions of two alleles at each of many hypervariable loci scattered throughout the genome. Consequently, the overall polymorphism of the multilocus hybridization

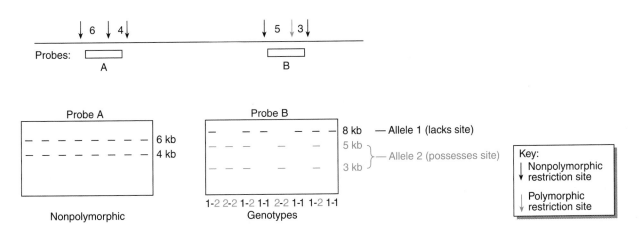

Figure 5.15: Assay of conventional (RSP-based) RFLPs.

patterns is uniquely high. Because it permits distinction between any two individuals who are not identical twins, probing with hypervariable minisatellites has been termed **DNA fingerprinting** (see Sections 17.4.2–17.4.4).

Detection of gene deletions by restriction mapping

Certain diseases are associated with a high frequency of deletion of all or part of a gene. If a partial restriction map has been established for the gene under investigation, deletions can be screened by Southern blot hybridization using an appropriate intragenic DNA probe. If the deletion is a small one, for example a few hundred base pairs, it is often apparent as a consistent reduction in size of normal restriction fragments in the gene. An individual who is homozygous for this mutation, or is a heterozygote with one normal allele and another with a small deletion, can easily be identified by detecting the aberrant size restriction fragments.

Large deletions will lead to absence of specific restriction fragments. Homozygous deletion of large DNA segments can easily be detected as complete absence of appropriate restriction fragments associated with the gene. If, however, an individual is heterozygous for a relatively large gene deletion, the deletion may still be detected by demonstrating comparatively reduced intensity of specific gene fragments. For example, patients with 21-hydroxylase deficiency often have deletions of about 30 kb of the 21-hydroxylase/*C4* gene cluster. Such pathological deletions eliminate the functional 21–hydroxylase gene, *CYP21*, and an adjacent *C4B* gene, leaving the related *CYP21P* pseudogene and *C4A* genes. Patients with homozygous deletions will show absence of diagnostic restriction fragments associated with *CYP21* and *C4B*, while carriers of the deletion will show a 2:1 ratio of *CYP21P:CYP21* and of *C4A:C4B* (Collier *et al.*, 1989; *Figure 5.16*).

5.3.4 In situ *hybridization usually involves hybridizing a nucleic acid probe to the denatured DNA of a chromosome preparation or the RNA of a tissue section fixed on a glass slide*

Chromosome in situ *hybridization*

A simple procedure for mapping genes and other DNA sequences is to hybridize a suitable labeled DNA probe against chromosomal DNA that has been denatured *in*

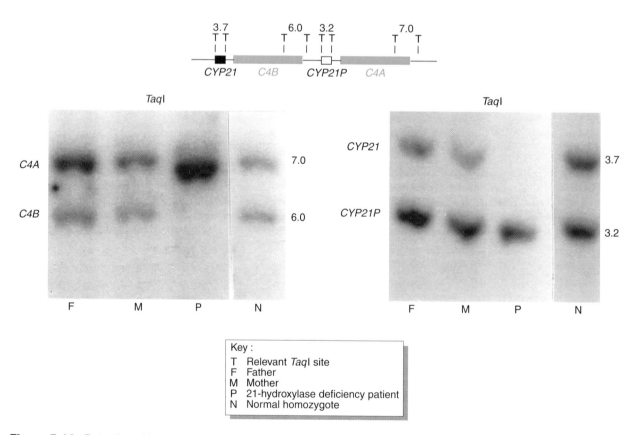

Figure 5.16: Detection of heterozygous and homozygous gene deletions associated with 21-hydroxylase deficiency.

Numbers indicate the size in kilobases of indicated *Taq*I restriction fragments. Southern blots of *Taq*I-digested genomic DNA from family members were hybridized with a complement *C4* gene probe (recognizing the duplicated *C4A* and *C4B* genes equally) (left panel) or with a 21-hydroxylase (*CYP21*) gene probe (right panel).The father and mother are heterozygotes for a deletion spanning *CYP21* and *C4B*, and the patient lacks *CYP21* and *C4B* sequences.

situ. To do this, an air-dried microscope slide preparation of metaphase or prometaphase chromosomes is made, usually from peripheral blood lymphocytes or lymphoblastoid cell lines. Treatment with RNase and proteinase K results in partially purified chromosomal DNA, which is denatured by exposure to formamide. The denatured DNA is then available for *in situ* hybridization with an added solution containing a labeled nucleic acid probe, overlaid with a coverslip. Depending on the particular technique that is used, chromosome banding of the chromosomes can be arranged either before or after the hybridization step. As a result, the signal obtained after removal of excess probe can be correlated with the chromosome band pattern in order to identify a map location for the DNA sequences recognized by the probe. Chromosome *in situ* hybridization has been revolutionized by the use of **fluorescence *in situ* hybridization (FISH)** techniques (see Section 10.1.4).

Tissue *in situ hybridization*

In this procedure, a labeled probe is hybridized against RNA in tissue sections (Wilkinson, 1998). Tissue sections are made from either paraffin-embedded or frozen tissue using a cryostat, and then mounted on to glass slides. A hybridization mix including the probe is applied to the section on the slide and covered with a glass coverslip. Typically, the hybridization mix has formamide at a concentration of 50% in order to reduce the hybridization temperature and minimize evaporation problems.

Although double-stranded cDNAs have been used as probes, single-stranded complementary RNA probes (**riboprobes***) are preferred: the sensitivity of initially single-stranded probes is generally higher than that of double-stranded probes, presumably because a proportion of the denatured double-stranded probe renatures to form probe homoduplexes. cRNA riboprobes that are complementary to the mRNA of a gene are known as antisense riboprobes and can be obtained by cloning a gene in the reverse orientation in a suitable vector such as pSP64 (see *Figure 5.4*). In such cases, the phage polymerase will synthesize labeled transcripts from the opposite DNA strand to that which is normally transcribed *in vivo*. Useful controls for such reactions include sense riboprobes which should not hybridize to mRNA except in rare occurrences where both DNA strands of a gene are transcribed.

Labeling of probes is performed using either selected radioisotopes, notably ^{35}S, or by nonisotopic labeling. In the former case, the hybridized probe is visualized using autoradiographic procedures. The localization of the silver grains is often visualized using only dark-field microscopy (direct light is not allowed to reach the objective; instead, the illuminating rays of light are directed from the side so that only scattered light enters the microscopic lenses and the signal appears as an illuminated object against a black background). However, bright-field microscopy (where the image is obtained by direct transmission of light through the sample) provides better signal detection (see *Figure 5.17*). Fluorescence labeling is a

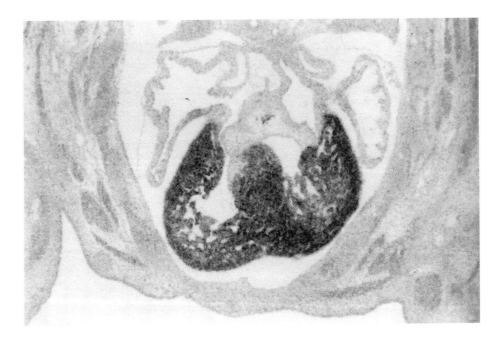

Figure 5.17: Tissue *in situ* hybridization.

The example shows the pattern of hybridization produced using a ^{35}S-labeled β-myosin heavy chain antisense riboprobe against a transverse section of tissue from a 13 day embryonic mouse. The dark areas represent strong labeling, notably in the ventricles of the heart. Kindly supplied by Dr David Wilson, University of Newcastle upon Tyne.

popular nonisotopic labeling approach and detection is accomplished by fluorescence microscopy (see *Box 5.2, Figure 5.5A*).

5.4 Nucleic acid hybridization assays using cloned target DNA, and microarray hybridization technology

Some of the technologies described in the preceding section (e.g. Southern blot hybridization and dot-blot hybridization) are also used to study cloned DNA as well as uncloned DNA. The techniques described in the next two sections, however, are dedicated to analysing cloned DNA. In addition, the very recently developed and very powerful microarray technologies are described in a third section.

5.4.1 Colony blot and plaque lift hybridization are methods for screening separated bacterial colonies or plaques following bacteriophage infection of bacteria

As described in Section 4.2.5 colonies of bacteria or other suitable host cells which contain recombinant DNA can generally be selected or identified by the ability of the insert to inactivate a marker vector gene (e.g. β-galactosidase, or an antibiotic-resistance gene). A specific approach can be used when the desired recombinant DNA contains a DNA sequence that is closely related to an available nucleic acid probe. In the case of bacterial cells used to propagate plasmid recombinants, the cell colonies are allowed to grow on an agar surface and then transferred by surface contact to a nitrocellulose or nylon membrane, a process known as colony blotting (see *Figure 5.18*). Alternatively the cell mixture is spread out on a

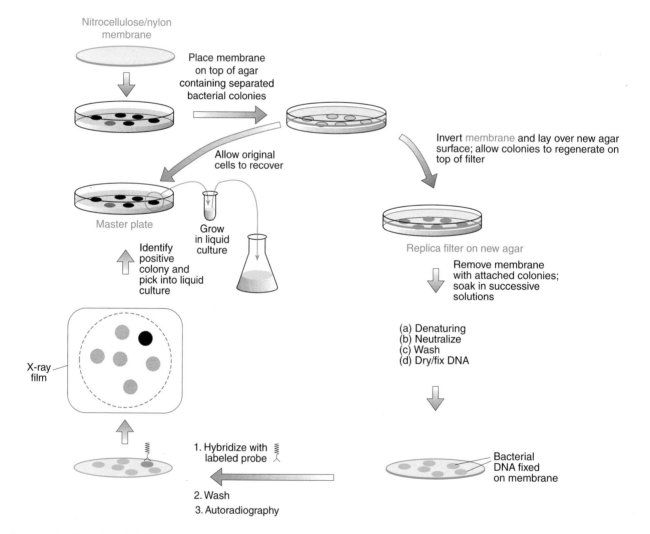

Figure 5.18: Colony hybridization involves replicating colonies on to a durable membrane prior to hybridization with a labeled nucleic acid probe.

This method is popularly used to identify colonies containing recombinant DNA, should a suitable labeled probe be available.

nitrocellulose or nylon membrane placed on top of a nutrient agar surface, and colonies are allowed to form directly on top of the membrane. In either approach, the membrane is then exposed to alkali to denature the DNA prior to hybridizing with a labeled nucleic acid probe.

After hybridization, the probe solution is removed, and the filter is washed extensively, dried and submitted to autoradiography using X-ray film. The position of strong radioactive signals is related back to a master plate containing the original pattern of colonies, in order to identify colonies containing DNA related to the probe. These can then be individually picked and amplified in culture prior to DNA extraction and purification of the recombinant DNA.

A similar process is possible when using phage vectors. The plaques which are formed following lysis of bacterial cells by phage will contain residual phage particles. A nitrocellulose or nylon membrane is placed on top of the agar plate in the same way as above, and when removed from the plate will constitute a faithful copy of the phage material in the plaques, a so called plaque-lift. Subsequent processing of the filter is identical to the scheme in *Figure 5.18*.

5.4.2 Gridded high density arrays of transformed cell clones or DNA clones have greatly increased the efficiency of DNA library screening

Once it became possible to create complex DNA libraries, more efficient methods of clone screening were required. Rather than simply plate out cell colonies on a cell culture dish and transfer them to a membrane in standard colony blotting, it was preferable to pick individual colonies and transfer them onto large membranes in the format of a high density gridded array. The process was enormously simplified by the application of robotic gridding devices which could perform the necessary spotting automatically by pipetting from clones arranged in microtiter dishes into pre-determined linear coordinates on a membrane. The resulting high density clone filters permitted rapid and efficient identification library screening (see *Figure 5.19*) and could be copied and distributed to numerous laboratories throughout the world.

5.4.3 DNA microarray technology has enormously extended the power of nucleic acid hybridization

Recently developed **DNA microarrays ('DNA chips')** have provided a scale-up in hybridization assay technology because of their huge capacity for miniaturization and automation (Schena *et al.*, 1998). Although reminiscent of the filter-based arrays, microarray construction involves quite different procedures. The surfaces involved are glass rather than porous membranes and the microarrays can be divided into two main classes according to their method of construction:

- Microarrays of DNA clones delivered by microspotting. Here the DNA clones have been prepared in

advance and then printed onto the surface of a microscope slide (*Figure 5.20A*). See *Figure 20.6A* for an application.

- Microarrays of oligonucleotides synthesized *in situ*. This approach has been pioneered by the company Affymetrix, Inc., in Santa Clara, CA, and typically involves a combination of photolithography technology from the semiconductor industry with the chemistry of oligonucleotide synthesis (*Figure 5.20B*). See *Figure 20.6B* for an application.

As in the case of reverse dot-blotting (Section 5.3.1), the DNA microarray technologies employ a reverse nucleic acid hybridization approach: the probes consist of unlabeled DNA fixed to a solid support (the arrays of DNA or oligonucleotides) and the target is labeled and in solution. Although the technology for establishing DNA microarrays was only developed in the last few years, already there have been numerous important applications and their impact on future biomedical research and diagnostic approaches is expected to be profound (see *Box 5.5*).

Figure 5.19: Gridded clone hybridization filters have facilitated physical mapping of the human genome.

The figure illustrates an autoradiograph of a membrane containing human YAC clones (i.e. total DNA from individual yeast clones containing human YACs). The membrane contains a total of 17 664 clones which had been gridded in arrays of a unit grid of 6 × 6 clones. The hybridization signals include weak signals from all clones by using a ^{35}S-labeled probe of total yeast DNA plus strongly hybridizing signals obtained with a ^{32}P-labeled unique sex chromosome probe (*DXYS646*). Original photo from Dr Mark Ross, Sanger Centre, Cambridge. Reprinted, with permission, from Ross and Stanton, *Current Protocols in Human Genetics*, Vol. 1. Copyright © 1995 John Wiley & Sons, Inc.

(A)

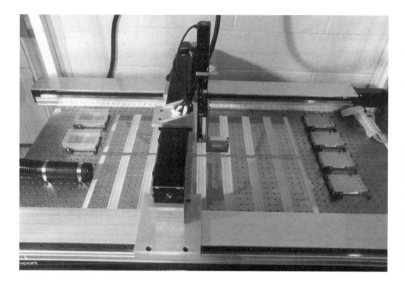

(B)

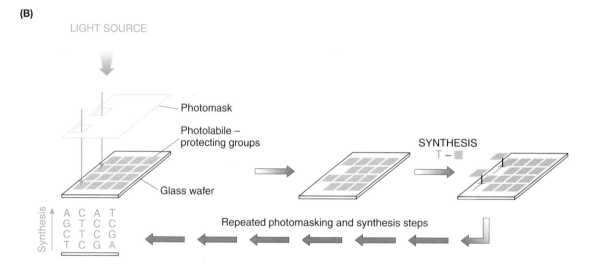

Figure 5.20: Construction of DNA and oligonucleotide microarrays.

(**A**) Robotic spotting for construction of DNA microarrays. Left, a microarray robot, with a table configuration which contains 160 slides with four microtiter plates, two wash stations and the dryer. Right, a laser scanner showing the optical table, power supplies for the lasers and photomultiplier tube cooling, the Ludi stage and lenses (see Cheung *et al.*, 1999 for more details). The microspotting of samples by robots can be performed by physical contact between spotting pins and the solid surface (of a microscope slide) or by an ink-jetting approach as is used in standard printing (the sample is loaded into a miniature nozzle equipped with a piezoelectric fitting and an electric current is used to expel a precise amount of liquid from the jet onto the substrate). Images kindly supplied by Aldo Massimi, Raju Kucherlapati and Geoffrey Childs at the Albert Einstein College of Medicine, New York. Reprinted from Cheung *et al.* (1999) *Nature Genet.*, **21** (suppl.), pp. 15–19, with permission from Nature America, Inc. (**B**) Construction of an oligonucleotide microarray by combining photolithography and *in situ* synthesis of oligonucleotides. Oligonucleotides are synthesized *in situ* in sequential steps starting from a 3′ mononucleotide which is anchored to the surface of a glass wafer. The photolithography entails modifying the glass wafer with photolabile protecting groups which can be eliminated when exposed to light and the use of carefully constructed photomasks which allow light to pass through onto carefully selected spatial coordinates. For those areas of the wafer which receive light passing through the photomask, the removal of the photolabile protective groups permits a new synthesis step. In this example thymidine is shown being coupled together with a protective photolabile group.

Box 5.5

Evolution and applications of DNA microarrays ('DNA chips')

Traditionally, molecular genetic analyses have focused on intensive studies of one or a few genes at a time. As the genome projects began to fulfil the promise of identifying huge numbers of genes, serious thought was given to the need to develop technologies that could permit large-scale gene screening and gene analyses, and the possibility of single pass whole genome analyses. Hybridization assays using high-density gridded arrays of clones on nylon membranes provided a technological step in the right direction, but this technology is limited in scope by the nature of the matrix supporting the clones. In order to permit whole genome analyses involving simultaneous screening of tens or hundreds of thousands of DNA clones or oligonucleotides, new technologies were required.

Two comparatively recent innovations made possible the new microarray technologies. One was the use of nonporous solid supports such as glass, which is much more amenable to miniaturization and fluorescence-based detection. Protocols developed by Professor Pat Brown and colleagues at Stanford permit robotic spotting of about 10 000 cDNA clones onto a microscope slide and hybridization with a double-labeled probe (Schena *et al.*,1995, 1998). The second innovation was high-density spatial synthesis of oligonucleotides on glass wafers using the photolithographic masking techniques which are used in the semiconductor industry (Fodor *et al.*, 1991). By 1998, Affymetrix, the leading manufacturer of such arrays, were able to produce chips containing more than 300 000 oligonucleotides within a 1.28×1.28 cm array with experimental versions exceeding one million such probes per array (Lipschutz *et al.*, 1999). The latter approach is more versatile in its applications (see below) and has the key advantage that the oligonucleotides can be synthesized at will, allowing chips to be manufactured directly from sequence databases.

Already, there have been numerous applications of the microarray technologies (Schena *et al.*, 1998; see also the Chipping Forecast (various authors), 1999, in Further reading) but the two major subsets of applications are large-scale screening of gene expression (at the RNA level) and of DNA variation.

- **Expression screening.** The focus of most current microarray-based studies is the monitoring of RNA expression levels which can be done by using either cDNA clone microarrays or gene-specific oligonucleotide microarrays (see *Figure 20.6* and Section 20.2.2). In the case of some organisms whose sequences have been completely established, such as that of the yeast *Saccharomyces cerevisiae*, whole genome expression screening has been possible and the technology is being developed to permit whole human genome expression screening in the near future (see Section 13.4.1).

- **Screening of DNA variation.** Oligonucleotide microarrays are required for this general purpose and several applications have been devised. The resequencing of the human mitochondrial genome by using DNA microarray was a successful test of the power of this technology for assessing large-scale sequence variation in individuals (see Section 6.3.4). There is also huge potential for assaying for mutations in known disease genes, as recently exemplified in the case of the breast cancer susceptibility gene, *BRCA1* (Section 17.1.4). In addition, there have been vigorous efforts to identify and catalog human single nucleotide polymorphism (SNP) markers (Section 13.2.5).

This technology is a very dynamic one and is currently spawning a variety of derivative technologies including the development of protein and antibody microarrays and cell microarrays. The interested reader is advised to consult recent reviews in the primary genetics and biotechnology journals.

Further reading

Berger SL, Kimmel AR (1987) *Guide to Molecular Cloning Techniques, Methods in Enzymology*, Vol. 152. Academic Press, San Diego, CA.

Hames BD, Higgins SJ (1985) *Genes Probes: A Practical Approach*. IRL Press, Oxford.

Old RW, Primrose SB (1994) *Principles of Gene Manipulation: an Introduction to Genetic Engineering*, 5th edn. Blackwell Scientific Publications, Oxford.

Sambrook J, Fritsch EF, Maniatis T (1989) *Molecular Cloning: a Laboratory Manual*, 2nd edn. Cold Spring Harbor Laboratory Press, Cold Spring Harbor, NY.

Wilkinson, D (1998) In Situ *Hybridization: a Practical Approach*. 2nd edn. IRL Press, Oxford.

Various authors (1999) The Chipping Forecast. *Nature Genet.,* **21** (suppl.), 1–60.

References

Cheung VG, Morley M, Aguilar F, Massimi A, Kucherlapati R, Childs G (1999) Making and reading microarrays. *Nature Genet.,* **21** (suppl.), 15–19.

Collier S, Sinnott PJ, Dyer PA, Price DA, Harris R, Strachan T (1989) Pulsed field gel electrophoresis identifies a high degree of variability in the number of tandem 21-hydroxylase and complement C4 gene repeats in 21-hydroxylase deficiency. *EMBO J.,* **8**, 1393–1402.

Feinberg AP, Vogelstein B (1983) A technique for radiolabelling DNA restriction endonuclease fragments to high specific activity. *Anal. Biochem.,* **132**, 6–13.

Fodor SP, Read JL, Pirrung MC, Stryer L, Lu AT, Solas D (1991) Light-directed, spatially addressable parallel chemical synthesis. *Science,* **251**, 767–773.

Hinds HL, Ashley CT, Sutcliffe JS *et al.* (1993) Tissue specific expression of FMR-1 provides evidence for a functional role in fragile X syndrome. *Nature Genet.,* **3**, 36–43.

Kricka LJ (1992) *Nonisotopic DNA Probing Techniques*. Academic Press, San Diego, CA.

Lipschutz RJ, Fodor SPA, Gingeras TR, Lockhart DJ (1999) High density synthetic oligonucleotide arrays. *Nature Genet.,* **21** (suppl.), 20–24.

Melton DA, Krieg PA, Rebagliati MR, Maniatis T, Zinn K, Green MR (1984) Efficient *in vitro* synthesis of biologically active RNA and RNA hybridization probes from plasmids containing a bacteriophage SP6 promoter. *Nucleic Acids Res.,* **12**, 7035-7056.

Ross MT, Stanton VPJ (1995) Screening large-insert libraries by hybridization. In: *Current Protocols in Human Genetics* (NJ Dracopoli *et al.*, eds), vol. 1, pp. 5.6.1–5.6.30. John Wiley & Sons, Inc., New York.

Schena M, Shalon D, Davis RW, Brown PO (1995) Quantitative monitoring of gene expression patterns with a complementary DNA microarray. *Science,* **270**, 467–470.

Schena M, Heller RA, Theriault TP, Konrad K, Lachenmeier E, Davis RW (1998) Microarrays: biotechnology's discovery platform for functional genomics. *Trends Biotechnol.,* **16**, 301–306.

Wilkinson, D (1998) In Situ *Hybridization: a Practical Approach*, 2nd edn. IRL Press, Oxford.

PCR, DNA sequencing and *in vitro* mutagenesis

6

6.1 Basic features of PCR

The polymerase chain reaction (PCR) has revolutionized molecular genetics by permitting rapid cloning and analysis of DNA. Since the first reports describing this new technology in the mid 1980s, there have been numerous applications in both basic and clinical research. Two other fundamental technologies are DNA sequencing and *in vitro* mutagenesis, both of which can be accomplished using PCR-based and non PCR-based methods.

6.1.1 PCR is a cell-free method of DNA cloning

The standard PCR reaction: selective DNA amplification

PCR is a rapid and versatile *in vitro* method for amplifying defined target DNA sequences present within a source of DNA. Usually, the method is designed to permit *selective amplification* of a specific target DNA sequence(s) within a heterogeneous collection of DNA sequences (e.g. total genomic DNA or a complex cDNA population). To permit such selective amplification, some prior DNA sequence information from the target sequences is required. This information is used to design two oligonucleotide primers (**amplimers**) which are specific for the target sequence and which are often about 15–25 nucleotides long. After the primers are added to denatured template DNA, they bind specifically to complementary DNA sequences at the target site. In the presence of a suitably heat-stable DNA polymerase and DNA precursors (the four deoxynucleoside triphosphates, dATP, dCTP, dGTP and dTTP), they initiate the synthesis of new DNA strands which are complementary to the individual DNA strands of the target DNA segment, and which will overlap each other (*Figure 6.1*).

The PCR is a chain reaction because newly synthesized DNA strands will act as templates for further DNA synthesis in subsequent cycles. After about 25 cycles of DNA synthesis, the products of the PCR will include, in addition to the starting DNA, about 10^5 copies of the specific target sequence, an amount which is easily visualized as a discrete band of a specific size when submitted to agarose gel electrophoresis. A heat-stable DNA polymerase is used because the reaction involves sequential cycles composed of three steps:

(i)   Denaturation, typically at about 93–95°C for human genomic DNA.
(ii) Reannealing at temperatures usually from about 50°C to 70°C depending on the T_m (see Section 5.2.1) of the expected duplex (the annealing temperature is typically about 5°C below the calculated T_m).
(iii) DNA synthesis, typically at about 70–75°C.

Suitably heat-stable DNA polymerases have been obtained from microorganisms whose natural habitat is hot springs. For example, the widely used *Taq* DNA polymerase is obtained from *Thermus aquaticus* and is thermostable up to 94°C, with an optimum working temperature of 80°C.

Specificity of amplification and primer design

The specificity of amplification depends on the extent to which the primers can recognize and bind to sequences other than the intended target DNA sequences. For complex DNA sources, such as total genomic DNA from a mammalian cell, it is often sufficient to design two primers about 20 nucleotides long. This is because the chance of an accidental perfect match elsewhere in the genome for either one of the primers is extremely low, and for both sequences to occur by chance in close proximity in the specified direction is normally exceedingly low. Although conditions are usually chosen to ensure that only strongly matched primer–target duplexes are stable, spurious amplification products can nevertheless be observed. This can happen if one or both chosen primer sequences contain part of a repetitive DNA sequence, and primers are usually designed to avoid matching to known repetitive DNA sequences, including large runs of a single nucleotide (*Figure 6.2*).

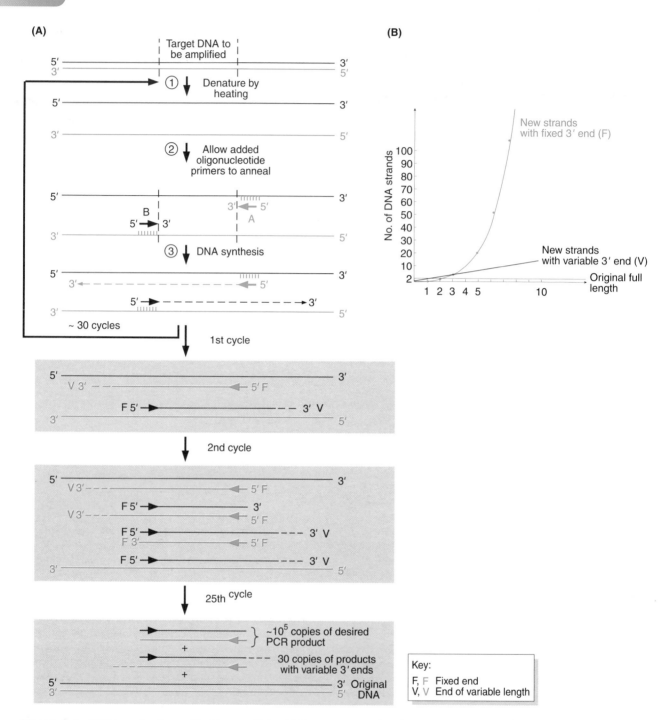

Figure 6.1: PCR is an *in vitro* method for amplifying DNA sequences using defined oligonucleotide primers.

Oligonucleotide primers A and B are complementary to DNA sequences located on opposite DNA strands and flanking the region to be amplified. Annealed primers are incorporated into the newly synthesized DNA strands. The first cycle will result in two new DNA strands whose 5′ end is fixed by the position of the oligonucleotide primer but whose 3′ end is variable ('ragged' 3′ ends). The two new strands can serve in turn as templates for synthesis of complementary strands of the desired length (the 5′ ends are defined by the primer and the 3′ ends are fixed because synthesis cannot proceed past the terminus of the opposing primer). After a few cycles, the desired fixed length product begins to predominate.

Length	Usually about 20 nt for target sequences in complex genomic DNA; can be much less if target DNA is less complex
Base composition	Substantial tandem repeats of one or more nucleotides to be avoided. Overall %GC plus length to be chosen so that the T_m of each oligonucleotide (*Table 5.2*) should be equal or nearly identical
Secondary structure	Avoid sequences prone to secondary structure which could form hairpins etc. (see *Figure 1.7A*)
3' end	Base complementarity of the two bases at the extreme 3' end of the two primers to be avoided. Otherwise primer dimers can result, reducing amplification efficiency

Figure 6.2: PCR primer design.

Accidental matching at the 3' end of the primer is critically important: spurious products may derive from substantially mismatched primer–target duplexes unless the 3' end of the primer shows perfect matching. Several strategies can be adopted to optimize reaction specificity:

- **Nested primers**. The products of an initial amplification reaction are diluted and used as the target DNA source for a second reaction in which a different set of primers is used, corresponding to sequences located close, but internal, to those used in the first reaction.
- **Hot-start PCR**. Mixing of all PCR reagents prior to an initial heat denaturation step allows more opportunity for nonspecific binding of primer sequences. To reduce this possibility, one or more components of the PCR are physically separated until the first denaturation step. A popular approach is to use a specially formulated wax bead designed to fit snugly within a PCR reaction tube. The reaction components minus the enzyme and reaction buffer are added to the tube followed by the molten wax bead which floats on top and then solidifies on cooling. The thermostable polymerase is then added with buffer. At the initial denaturation step the wax melts again and rises to the surface causing all the reaction components to come into contact with each other.
- **Touch-down PCR**. Most thermal cyclers can be programed to perform runs in which the annealing temperature is lowered incrementally during the PCR cycling from an initial value above the expected

T_m to a value below the T_m. By keeping the stringency of hybridization initially very high, the formation of spurious products is discouraged, allowing the expected sequence to predominate.

DNA labeling by PCR

The standard PCR reaction can be modified to permit incorporation of labeled nucleotides. Two methods are commonly used:

- **Standard PCR-based DNA labeling**. The PCR reaction is modified to include one or more labeled nucleotide precursors which become incorporated into the PCR product throughout its length.
- **Primer-mediated 5' end labeling**. PCR is conducted using a primer in which a labeled group is attached to the 5' end. As PCR proceeds the primer with its 5' end-label is incorporated into the PCR product. This method is often used with fluorophore labels during DNA sequencing (see legend to *Figure 6.18*) and is an example of a general PCR mutagenesis method known as **5' add-on mutagenesis** which has many applications (see Section 6.4.2 and *Figure 6.20A*).

6.1.2 The major advantages of PCR as a cloning method are its rapidity, sensitivity and robustness

Because of its simplicity, PCR is a popular technique with a wide range of applications which depend on essentially three major advantages of the method.

Speed and ease of use

DNA cloning by PCR can be performed in a few hours, using relatively unsophisticated equipment. Typically, a PCR reaction consists of 30 cycles containing a denaturation, synthesis and reannealing step, with an individual cycle typically taking 3–5 min in an automated thermal cycler. This compares favorably with the time required for cell-based DNA cloning, which may take weeks. Clearly, some time is also required for designing and synthesizing oligonucleotide primers, but this has been simplified by the availability of computer software for primer design and rapid commercial synthesis of custom oligonucleotides. Once the conditions for a reaction have been tested, the reaction can then be repeated simply.

Sensitivity

PCR is capable of amplifying sequences from minute amounts of target DNA, even the DNA from a single cell (Li *et al.*, 1988). Such exquisite sensitivity has afforded new methods of studying molecular pathogenesis and has found numerous applications in forensic science, in diagnosis, in genetic linkage analysis using single-sperm typing and in molecular paleontology studies, where samples may contain minute numbers of cells. However, the extreme sensitivity of the method means that great care has to be taken to avoid contamination of the sample under investigation by external DNA, such as from minute amounts of cells from the operator.

Robustness

PCR can permit amplification of specific sequences from material in which the DNA is badly degraded or embedded in a medium from which conventional DNA isolation is problematic. As a result, it is again very suitable for molecular anthropology and paleontology studies, for example the analysis of DNA recovered from archaeological remains. It has also been used successfully to amplify DNA from formalin-fixed tissue samples, which has important applications in molecular pathology and, in some cases, genetic linkage studies.

6.1.3 The major disadvantages of PCR are the general requirement for prior target sequence information, short size and limiting amounts of product, and infidelity of DNA replication

Despite its huge popularity, PCR has certain limitations as a method for selectively cloning specific DNA sequences.

Need for target DNA sequence information

In order to construct specific oligonucleotide primers that permit selective amplification of a particular DNA sequence, some prior sequence information is necessary. This normally means that the DNA region of interest has been partly characterized previously, often following cell-based DNA cloning. However, a variety of techniques have been developed that reduce or even exclude the need for prior DNA sequence information concerning the target DNA, when certain aims are to be met. For example, previously uncharacterized DNA sequences can sometimes be cloned using PCR with degenerate oligonucleotides if they are members of a gene or repetitive DNA family at least one of whose members has previously been characterized. In some cases, PCR can be used effectively without any prior sequence information concerning the target DNA to permit *indiscriminate amplification* of DNA sequences from a source of DNA that is present in extemely limited quantities (Section 6.2.4). Therefore, although PCR can be applied to ensure whole genome amplification, it does not have the advantage of cell-based DNA cloning in offering a way of separating the individual DNA clones comprising a genomic DNA library.

Short size and limiting amounts of PCR product

A clear disadvantage of PCR as a DNA cloning method has been the size range of the DNA sequences that can be cloned. Unlike cell-based DNA cloning where the size of cloned DNA sequences can approach 2 Mb (Section 4.3.4), reported DNA sequences cloned by PCR have typically been in the 0.1–5 kb size range, often at the lower end of this scale. Although small segments of DNA can usually be amplified easily by PCR, it becomes increasingly more difficult to obtain efficient amplification as the desired

product length increases. Recently, however, conditions have been identified for effective amplification of longer targets, including a 42-kb product from the bacteriophage λ genome. Often, the conditions for long range PCR involve a combination of modifications to standard conditions with a two-polymerase system. This provides optimal levels of DNA polymerase and $3' \rightarrow 5'$ exonuclease activity which serves as a proofreading mechanism (see *Box 6.1*).

The amount of PCR product obtained in a single reaction is also much more limited than the amount that can be obtained using cell-based cloning where scale-up of the volumes of cell cultures is possible. The efficiency of a PCR reaction will vary from template to template and according to various factors that are required to optimize the reaction but typically only comparatively small amounts of product are achieved.

Infidelity of DNA replication

Cell-based DNA cloning involves DNA replication *in vivo*, which is associated with a very high fidelity of copying because of proofreading mechanisms (see *Box 6.1*). However, when DNA is replicated *in vitro* the copying error rate is considerably greater. Of the heat-stable DNA polymerases required for PCR, the most widely used is *Taq* DNA polymerase derived from *T. aquaticus*. This DNA polymerase, however, has no associated $3' \rightarrow 5'$

Box 6.1

Proofreading by DNA polymerase-associated $3' \rightarrow 5'$ exonuclease activity

The fidelity of DNA replication *in vivo* is extremely high: during replication of mammalian genomes, for example, only one base in about 3×10^9 is copied incorrectly. Misincorporation occurs at a low frequency, dependent on the relative free energies of correctly and incorrectly paired bases. Very minor changes in helix geometry can stabilize G–T base pairs (with two hydrogen bonds; note the frequent occurrence of G–U base-pairing in RNA; Section 1.2.1). *In vivo* copying normally shows an error rate much lower than these thermodynamic limitations would imply. This is achieved by proofreading mechanisms, one of which is a common property of DNA polymerases. Unlike RNA polymerases, DNA polymerases *absolutely* require the 3'-hydroxyl end of a base-paired primer strand as a substrate for chain extension. Additionally, DNA molecules with a mismatched nucleotide at the 3' end of the primer strand are not effective templates for DNA synthesis. Many DNA polymerases, including that of *E. coli*, contain an integral $3' \rightarrow 5'$ exonuclease activity. When an incorrect base is inserted during DNA synthesis, DNA synthesis does not proceed. Instead, the $3' \rightarrow 5'$ exonuclease activity removes one nucleotide at a time from the 3' hydroxyl terminus until a correctly base-paired terminus is obtained, enabling DNA synthesis to proceed again. As a result, DNA polymerases are usually self-correcting, removing errors made by the DNA polymerase activity during DNA synthesis.

exonuclease to confer a proofreading function, and the error rate due to base misincorporation during DNA replication is rather high: for a 1 kb sequence that has undergone 20 effective cycles of duplication, approximately 40% of the new DNA strands synthesized by PCR using this enzyme will contain an incorrect nucleotide resulting from a copying error. This means that, even if the PCR reaction involves amplification of a single DNA sequence, the final product will be a mixture of extremely similar, but not identical DNA sequences.

Despite the errors due to replication *in vitro*, DNA sequencing of the total PCR product may give the correct sequence. This is because, although individual DNA strands in the PCR product often contain incorrect bases, the incorporation of incorrect bases is essentially random. As a result, *for each base position*, the contribution of one incorrect base on one or more strands is overwhelmed by the contributions from the huge majority of strands which will have the correct sequence. What it does mean, however, is that further analysis of the product may be difficult. If the PCR product is to be cloned in cells (e.g. to facilitate DNA sequencing or to permit functional studies in a cell-based expression system), transformation selects for a single molecule, and the cell clones chosen to be amplified will contain identical molecules, each the same as a single starting molecule which may well have the incorrect DNA sequence because of a copying error during PCR amplification. As a result, several individual clones may need to be sequenced in order to determine the correct (consensus) sequence, before selecting one with the authentic sequence for subsequent experiments.

More recently, the problem of infidelity of DNA replication during the PCR reaction has been considerably reduced by using alternative heat-stable DNA polymerases which have associated 3′–5′ exonuclease activity. For example, the *Pyrococcus furiosus* DNA polymerase is becoming more widely used because of the proofreading conferred by its associated 3′–5′ exonuclease activity (Cline *et al.*, 1996). The resulting PCR product has a much lower level of mutations introduced by copying errors: for a 1 kb segment of DNA that has undergone 20 effective cycles of duplication, about 3.5% of the DNA strands in the product carry an altered base.

6.1.4 Cell-based cloning of PCR amplification products is often required to permit subsequent structural and functional studies

The amount of material that can be cloned in a single PCR reaction is limited, and it is time-consuming and expensive to repeat the same PCR reaction many times to achieve large quantities of the desired DNA. In addition, the PCR product may not be in a suitable form that will permit some subsequent studies. As a result, it is often convenient to clone the PCR product in a cell-based cloning system in order to obtain large quantities of the desired DNA and to permit a variety of analyses. As

described in the previous section, it is important to verify that the sequence of the cloned product is representative of the original PCR product.

Various plasmid cloning systems are used to propagate PCR-cloned DNA in bacterial cells. Once cloned, the insert can be cut out using suitable restriction nucleases and transferred into other plasmids which may have specialized usages in permitting expression to give an RNA product, or to provide large quantities of a protein, etc. Several thermostable polymerases including *Taq* DNA polymerase have a terminal deoxynucleotidyl transferase activity which selectively modifies PCR-generated fragments by adding a single nucleotide, generally adenine, to the 3′ ends of amplified DNA fragments. The resulting overhangs can make it difficult to clone PCR products and a variety of approaches are commonly used to facilitate cloning, including the use of vectors with overhanging T residues in their cloning site polylinker and the use of 'polishing' enzymes such as T4 polymerase or *Pfu* polymerase which can remove the overhanging single nucleotides (*Figure 6.3*).

6.2 Applications of PCR

Although PCR was first developed only a decade and a half ago, the simplicity and the versatility of the technique have ensured that it is among the most ubiquitous of molecular genetic methodologies, with a wide range of general applications (*Figure 6.4*).

6.2.1 PCR enables rapid amplification of template DNA for screening of uncharacterized mutations

Because of its rapidity and simplicity, PCR is ideally suited to providing numerous DNA templates for mutation screening. Partial DNA sequences, at the genomic or the cDNA level, from a gene associated with disease, or some other interesting phenotype, immediately enable gene-specific PCR reactions to be designed. Amplification of the appropriate gene segment then enables rapid testing for the presence of associated mutations in large numbers of individuals. By contrast, cell-based DNA cloning of the gene from numerous different individuals is far too slow and labor-intensive to be considered as a serious alternative.

Typically, the identification of exon–intron boundaries and sequencing of the ends of introns of a gene of interest offers the possibility of genomic mutation screening. Individual exon-specific amplification reactions are developed by designing primers which recognize intronic sequences located close to the exon–intron boundary (*Figure 6.5A*). The resulting PCR products are then analyzed by rapid mutation-screening methods, in which the optimal size for mutation screening is usually about 200 bp (see Section 15.5.1). Conveniently, the average size of a human exon is about 180 bp but, in the case of very large exons, it is usual to design a series of primers to

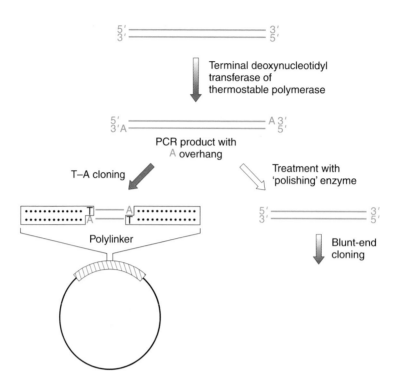

Figure 6.3: Cloning of PCR products in bacterial cells.

PCR products frequently have an overhanging adenosine at their 3′ ends (see text). The T–A cloning system has a polylinker system with complementary thymine overhangs to facilitate cloning. An alternative is to trim back the adenine overhangs using a suitable 'polishing' enzyme, which leaves the fragment blunt-ended.

generate overlapping exonic products. PCR can also quickly provide amplified cDNA sequences for mutation screening. Such cDNA mutation screening may be the only way in which mutations can be screened if the exon–intron organization of a gene has not been established. To do this, mRNA is isolated from a convenient source of tissue, such as blood cells, converted into cDNA using reverse transcriptase and the cDNA is used as a template for a PCR reaction. This version of the standard genomic PCR reaction is consequently often referred to as **RT-PCR** (**reverse transcriptase-PCR**; *Figure 6.5B*). Clearly, the method is ideally suited to genes expressed at high levels in easily accessible cells, such as blood cells. However, as a result of low level **ectopic transcription** of genes in all tissues, it has also been applied to transcript analysis of genes which are not significantly expressed in blood cells, such as the dystrophin (*DMD*) gene (Chelly *et al.*, 1989).

6.2.2 PCR permits rapid genotyping for polymorphic markers

Restriction site polymorphisms (RSPs) result in alleles possessing or lacking a specific restriction site. Such polymorphisms can be typed using Southern blot hybridization. A DNA probe representing the locus is hybridized against genomic DNA samples that have been digested with the appropriate restriction enzyme and size-fractionated by agarose gel electrophoresis. The resulting RFLPs have two alleles corresponding to the presence or absence of the restriction site (Section 5.3.3). As a convenient alternative to RFLPs, PCR can type RSPs by simply designing primers using sequences which flank the polymorphic restriction site, amplifying from genomic DNA, then cutting the PCR product with the appropriate restriction enzyme and separating the fragments by agarose gel electrophoresis (*Figure 6.6*).

Short tandem repeat polymorphisms (STRPs), also called microsatellite markers, consist of a short sequence, typically from one to four nucleotides long, that is tandemly repeated several times, and often characterized by many alleles. For example, $(CA)_n/(TG)_n$ repeats are often polymorphic when n exceeds 12, and have been widely used as polymorphic markers in the human genome (see below). Increasingly, however, trinucleotide and tetranucleotide marker polymorphisms are being typed. In each case the STRPs can be typed conveniently by PCR. Primers are designed from sequences known to flank a specific STRP locus, permitting PCR amplification of alleles whose sizes differ by integral repeat units (*Figure 6.7*). The PCR products can then be size-fractionated by polyacrylamide gel electrophoresis. The PCR normally includes a radioactive or fluorescent nucleotide precursor which becomes incorporated into

Typing genetic markers	RFLPs (see *Figure 6.6*); STRPs (see *Figures 6.7* and *6.8*)
DNA templates for mutation screening	Genomic mutation screening and RT-PCR (see *Figure 6.5*)
Detecting point mutations	Mutations changing restriction site – same principle as *Figure 6.6*. Other mutations by allele-specific amplification (ARMS; *Figure 6.9*)
cDNA cloning	From amino acid sequence by DOP-PCR (*Figure 6.11*); cloning of the ends of cDNA by RACE (*Figure 20.1*); making cDNA libraries from limiting amounts of material
Genomic DNA cloning	Cloning of new members of a DNA family by DOP-PCR (Section 6.2.4). Single cell PCR and whole genome amplification or subgenomic amplification (e.g. microdissected chromosome bands) by DOP-PCR or linker-primed PCR (*Figure 6.12*)
Genome walking†	Inverse PCR (see *Figure 10.15*); bubble linker (vectorette) PCR (see *Figure 10.16*), *Alu*-PCR (see *Figure 10.18*)
DNA sequencing	Double-stranded DNA for direct sequencing or for conventional cloning then sequencing, cycle sequencing (*Figure 6.18*)
In vitro mutagenesis	Using 5′ add-on mutagenesis to create a recombinant PCR product (see *Figure 6.20A*). Using mispaired primers to change a single predetermined nucleotide (see *Figure 6.20B*)
Gene expression studies	RT-PCR (Section 20.2.4); differential display (*Figure 20.7*)

Figure 6.4: PCR has numerous general applications.

The figure illustrates general applications. Specific applications are described in separate chapters. †Genome walking means accessing uncharacterized DNA starting from a neighboring characterized sequence.

the small PCR products and facilitates their detection. To ensure adequate size fractionation of alleles, the PCR products are denatured prior to electrophoresis. An example of the use of a CA repeat marker is shown in *Figure 6.8*.

6.2.3 A wide variety of PCR-based methods can be used to assay for known mutations

PCR is a very rapid and valuable tool for detecting pathogenic mutations and other mutations of interest. The examples below illustrate some popular methods.

Allelic discrimination by size or susceptibility to restriction enzyme

Small insertions or deletions (such as the three nucleo-

tide deletion in the common cystic fibrosis (CFTR) allele, F508del) can be simply detected by designing primers from regions closely flanking the mutation site and distinguishing the normal and mutant alleles by size on polyacrylamide or agarose gels. If the mutation changes a restriction site, mutant and normal alleles can be distinguished by amplifying across the mutant site and digesting the PCR product with relevant restriction endonuclease, exactly as in *Figure 6.6*.

Allelic discrimination by susceptibility to an artificially introduced restriction site

Even if the mutation does not result in a restriction site difference, it may be possible to exploit the difference between normal and mutant alleles by amplification-created restriction site PCR. This is a form of mismatched primer mutagenesis (see Section 6.4.2) in

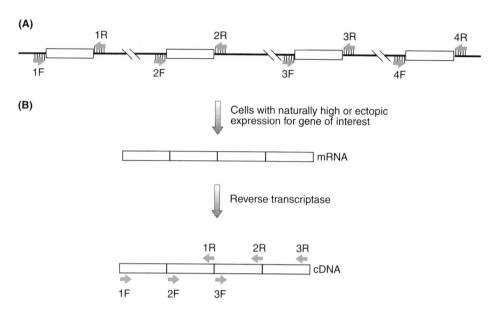

Figure 6.5: PCR products for gene mutation screening are obtained from genomic DNA using intron-specific primers flanking exons or by RT-PCR.

(**A**) Genomic DNA. Exons 1-4 can be amplified separately from genomic DNA using pairs of intron-specific primers 1F + 1R, 2F + 2R, etc. (**B**) RT-PCR. This relies on at least some mRNA being present in easily accessible cells such as blood cells, permitting conversion to cDNA. The cDNA can then be used as a template for pairs of exon-specific primers (1F+1R, 2F+2R, etc.) to generate overlapping DNA fragments.

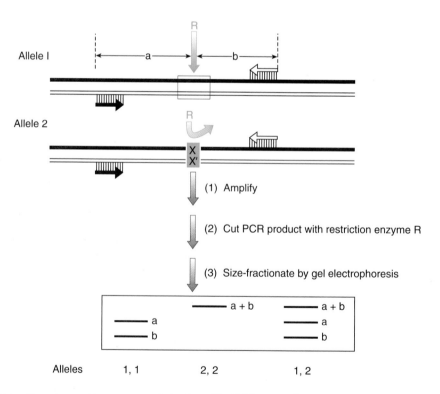

Figure 6.6: Restriction site polymorphisms can easily be typed by PCR as an alternative to laborious RFLP assays.

Alleles 1 and 2 are distinguished by a polymorphism which alters the nucleotide sequence of a specific restriction site for restriction nuclease R: allele 1 possesses the site, but allele 2 has an altered nucleotide(s) X, X' and so lacks it. PCR primers can be designed simply from sequences flanking the restriction site to produce a short product. Digestion of the PCR product with enzyme R and size-fractionation can result in simple typing for the two alleles.

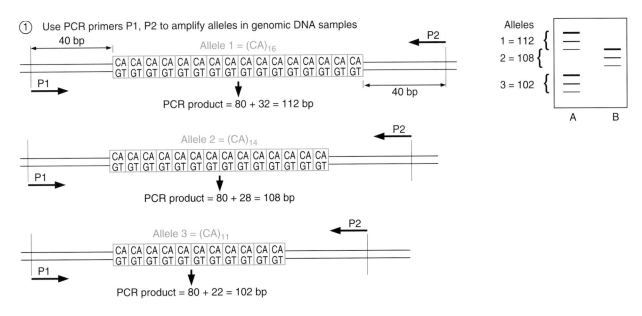

① Use PCR primers P1, P2 to amplify alleles in genomic DNA samples

② Denature PCR products and size-fractionate by polyacrylamide gel electrophoresis

③ Autoradiography

Figure 6.7: PCR can be used to type short tandem repeat polymorphisms (STRPs).

The example illustrates typing of a (CA)/(TG) dinucleotide repeat polymorphism which has three alleles as a result of variation in the number of the (CA) repeats. On the autoradiograph each allele is represented by a major upper band and two minor 'shadow bands' (see *Figure 6.8*). Individuals A and B have genotypes (in brackets) as follows: A (1,3); B (2,2).

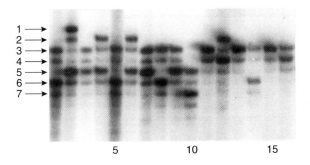

Figure 6.8: Example of typing for a CA repeat.

The example illustrated shows typing of members of a large family with the (CA)/(TG) marker *D17S800*. Arrows to the left mark the top (main) band seen in different alleles 1-7. *Note* that individual alleles show a strong upper band followed by two lower 'shadow bands', one of intermediate intensity immediately underneath the strong upper band, and one that is very faint and is located immediately below the first shadow band. For the indicated individuals, the genotypes (in brackets) are as follows: 1 (3,6); 2 (1,5); 3 (3,5); 4 (2,5); 5 (3,6); 6 (2,5); 7 (3,5); 8 (3,6); 9 (3,5); 10 (5,7); 11 (3,3); 12 (2,4); 13 (3,3); 14 (3,6); 15 (3,3); 16 (3,4). *Note* that in the latter case, the middle band is particularly intense because it contains both the main band for allele 4 plus the major shadow band for allele 3. Slipped strand mispairing (see Section 9.3.1) is thought to be the major mechanism responsible for producing shadow bands at tandem dinucleotide repeats (Hauge and Litt, 1993).

which a primer is deliberately designed from sequence immediately adjacent to, but not encompassing, the restriction site. The primer is deliberately designed to have a mismatched nucleotide which together with the sequence of the mutant site creates a restriction site not present in normal alleles (see *Figure 17.2* for a specific example).

Allele-specific PCR (ARMS test)

Oligonucleotide primers can be designed so as to discriminate between target DNA sequences that differ by a single nucleotide in the region of interest. This is a form of allele-specific PCR, the PCR equivalent of the allele-specific hybridization which is possible with ASO probes (Section 5.3.1). In the case of allele-specific hybridization, alternative ASO probes are designed to have differences in a central segment of the sequence (to maximize thermodynamic instability of mismatched duplexes). However, in the case of allele-specific PCR, ASO primers are designed to differ at the nucleotide that occurs *at the extreme 3' terminus*. This is so because the DNA synthesis step in a PCR reaction is crucially dependent on correct base-pairing at the 3' end (*Figure 6.9*). This method can be used to type specific alleles at a polymorphic locus, but has found particular use as a method for detecting a specific pathogenic mutation, the so-called amplification refractory mutation system (ARMS; Newton *et al.*, 1989).

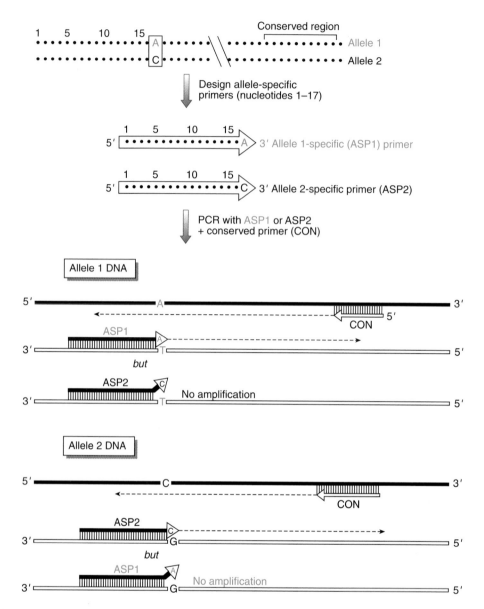

Figure 6.9: Correct base-pairing at the 3′ end of PCR primers is the basis of allele-specific PCR.

The allele-specific oligonucleotide primers ASP1 and ASP2 are designed to be identical to the sequence of the two alleles over a region preceding the position of the variant nucleotide, *up to and terminating in the variant nucleotide itself*. ASP1 will bind perfectly to the complementary strand of the allele 1 sequence, permitting amplification with the conserved primer. However, the 3′-terminal C of the ASP2 primer mismatches with the T of the allele 1 sequence, making amplification impossible. Similarly ASP2 can bind perfectly to allele 2 and initiate amplification, unlike ASP1.

Mutation detection using the 5′ → 3′ exonuclease activity of *Taq* DNA polymerase (TaqMan™ assay)

Taq polymerase does not possess a proofreading 3′ → 5′ exonuclease activity but does possess a 5′ → 3′ exonuclease activity. This property can be exploited to facilitate detection of specific alleles (Holland *et al.*, 1991; Lee *et al.*, 1993). Such an assay involves hybridization of three primers, the third primer being intended to bind just downstream of one of the conventional primers which should be allele-specific. The additional primer carries a

blocking group at the 3′ terminal nucleotide so that it cannot prime new DNA synthesis and at its 5′ end carries a labeled group. In modern versions of the assay, the label is a fluorogenic group and the third primer also carries a quencher group (see *Figure 6.10*). If the upstream primer which is bound to the same strand is able to prime successfully, *Taq* DNA polymerase will extend a new DNA strand until it encounters the third primer in which case its 5′ → 3′ exonuclease will degrade the primer causing release of separate nucleotides containing the dye and the quencher, and an observable increase in fluorescence.

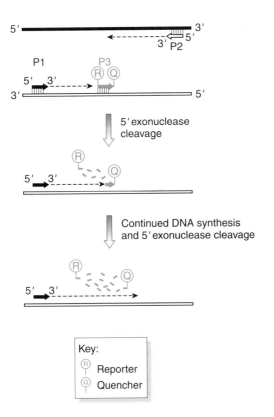

Key:
Ⓡ Reporter
Ⓠ Quencher

Figure 6.10: The TaqMan™ 5′ exonuclease assay.

In addition to two conventional PCR primers, P1 and P2, which are specific for the target sequence, a third primer, P3, is designed to bind specifically to a site on the target sequence downstream of the P1 binding site. P3 is labeled with two fluorophores, a reporter dye (R) is attached at the 5′ end, and a quencher dye (D), which has a different emission wavelength to the reporter dye, is attached at its 3′ end. Because its 3′ end is blocked, primer P3 cannot by itself prime any new DNA synthesis. During the PCR reaction, *Taq* DNA polymerase synthesizes a new DNA strand primed by P1 and as the enzyme approaches P3, its 5′ → 3′ exonuclease activity processively degrades the P3 primer from its 5′ end. The end result is that the nascent DNA strand extends beyond the P3 binding site and the reporter and quencher dyes are no longer bound to the same molecule. As the reporter dye is no longer in close proximity to the quencher, the resulting increase in reporter emission intensity is easily detected.

6.2.4 Degenerate oligonucleotide primers and primers specific for ligated linker sequences permit co-amplification of sequence families, or even indiscriminate amplification

DOP-PCR (degenerate oligonucleotide-primed PCR) is a form of PCR which is deliberately designed to permit possible amplification of several products. The two primers may be *partially degenerate* oligonucleotides, composed of panels of oligonucleotide sequences that have the same base at certain nucleotide positions, but are dif-

ferent at others. As a result, there may be comparatively many primer binding sites in the source DNA. This provides a means of searching for a new or uncharacterized DNA sequence that belongs to a family of related sequences either within or between species. *Note:* the use of such primers also provides a way of cloning a gene when only a limited portion of amino acid sequence is known for the product (*Figure 6.11*).

DOP-PCR can also be used to permit comparatively indiscriminate amplification of target DNA. Primer sequences with random sequences can bind to numerous locations in the template DNA and permit a form of **whole-genome amplification** (Zhang *et al.*, 1992; Cheung and Nelson, 1996). This can be advantageous where the amount of starting DNA may be limiting (as in the case of extracts from ancient DNA samples, microdissected chromosome bands, single cell typing, etc.), and PCR amplification of essentially all sequences increases the amount of DNA for study.

Linker-primed PCR (ligation adaptor PCR)

Another way of enabling amplification of essentially all DNA sequences in a complex DNA mixture involves first ligating a known sequence to all fragments. To do this, the target DNA population is digested with a suitable restriction endonuclease, and double-stranded oligonucleotide **linkers** (also called **adaptors**) with a suitable overhanging end are ligated to the ends of target DNA fragments. Amplification is then performed using oligonucleotide primers which are specific for the linker sequences. In this way, all fragments of the DNA source which are flanked by linker oligonucleotides can be amplified (*Figure 6.12*).

6.2.5 Anchored PCR uses a target-specific primer and a universal primer for amplifying sequences adjacent to a known sequence

It is often desirable to be able to amplify previously uncharacterized DNA sequences that neighbor a known DNA sequence, either at the genomic or cDNA level. To do this a form of **anchored PCR** is used (see *Figure 6.13*). One of the primers is specific for the target sequence and the second primer is specific for a common sequence that can be introduced in different ways, such as by using a linker–primer method as described in the previous section, or by using primers that are modified at the 5′ end so as to introduce a novel sequence.

6.3 DNA sequencing

6.3.1 DNA sequencing usually involves enzymatic DNA synthesis in the presence of base-specific dideoxynucleotide chain terminators

Formerly, chemical DNA sequencing methods were often employed, using base-specific chemical modification and

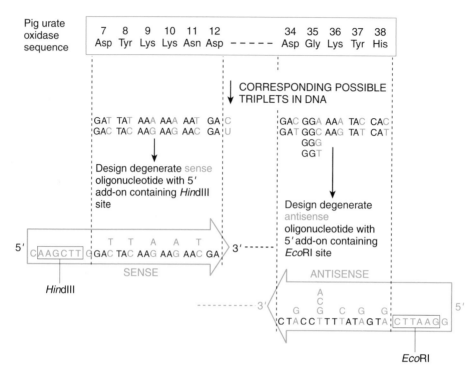

Figure 6.11: DOP-PCR can permit cDNA cloning using degenerate oligonucleotides.

The figure illustrates cloning of a cDNA for porcine urate oxidase using degenerate oligonucleotides corresponding to a known amino acid sequence. The sense primer was constructed to correspond to the codons 7–11 plus the first two bases of codon 12, and the antisense primer corresponded to codons 34–38 (Lee *et al.*, 1988). The amino acid sequences chosen for constructing primers were selected on the basis of their high content of amino acids which were specified by only two codons (Asp, Tyr, Lys, Asn, His, see *Figure 1.22*). The primers have 5′ extensions containing recognition sequences for restriction nucleases, in order to facilitate subsequent cell-based cloning.

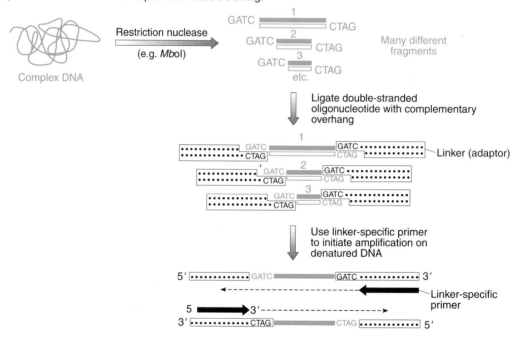

Figure 6.12: Linker-primed PCR permits indiscriminate amplification of DNA sequences in a complex target DNA.

The linker (adaptor) molecule is a double-stranded oligonucleotide formed by ligating two single-stranded oligonucleotides which are complementary in sequence except that one possesses a 5′ overhang compatible with a restriction nuclease overhang (in this case, the 5′ GATC overhang produced by *Mbo*I). After ligation of the linker to the target restriction fragments, a linker-specific primer can result in amplification of all fragments by binding to two flanking linker molecules.

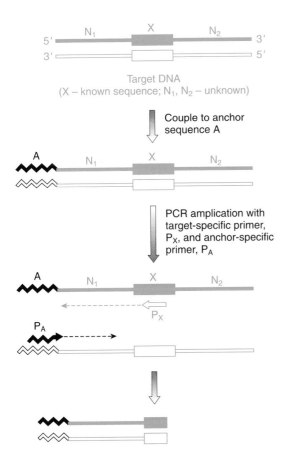

Figure 6.13: 'Genome walking' by anchored PCR.

The target may be a complex source of DNA comprised of many fragments to which an anchor sequence is attached, for example a double-stranded oligonucleotide linker. The idea is to use a primer specific for the anchor sequence and one specific for a known sequence X to be able to rescue fragments containing sequence X and so gain access to previously unidentified sequences adjacent to X. In this example the anchored sequence is shown only on the left hand side for clarity and permits amplification of the previously characterized N1 sequence adjacent to known sequence X. A variety of derivative methods have been devised, such as bubble-linker PCR (*Figure 10.16*).

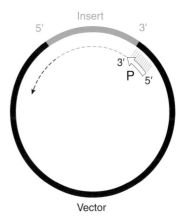

Figure 6.14: A universal sequencing primer can be used to sequence many different template DNAs.

DNA templates for DNA sequencing are often single-stranded recombinant DNA molecules. Different clones will often contain different inserts within the same vector molecule. As a result, a universal sequencing primer (P) can be designed to be complementary to a short vector sequence located next to the cloning site(s), allowing sequencing of different insert DNAs.

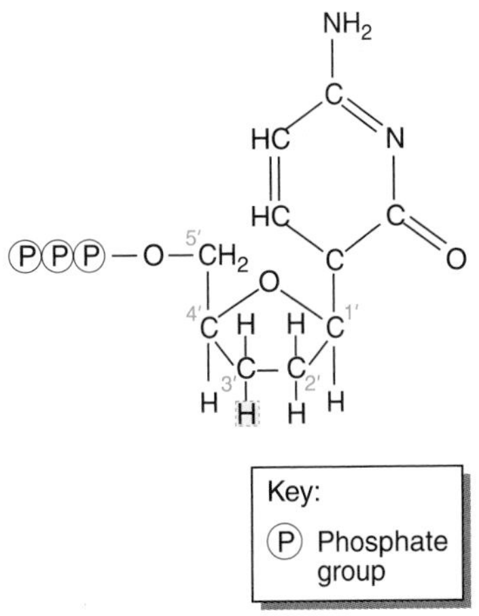

Figure 6.15: Structure of a dideoxynucleotide, 2', 3' dideoxy CTP.

Note that the hydroxyl group which is attached to carbon 3' in normal nucleotides (see *Figure 1.2*) is replaced by a hydrogen atom.

subsequent cleavage of the DNA. Currently, however, the vast majority of DNA sequencing is carried out using an enzymatic method: the DNA to be sequenced is provided in a single-stranded form from which DNA polymerase synthesizes new complementary DNA strands. Usually, the single-stranded DNA template is obtained using a cloning system which permits recovery of single-stranded recombinant DNA, as with M13 or phagemid cloning systems (Section 4.4.1 and *Figure 4.17*). The subsequent DNA sequencing reactions involve DNA synthesis using one or more labeled nucleotides and a universal sequencing primer that is complementary to the vector sequence flanking the cloning site (*Figure 6.14*).

In addition to the normal nucleotide precursors, DNA synthesis is carried out in the presence of base-specific **dideoxynucleotides (ddNTPs)**. The latter are analogs of the normal dNTPs but differ in that they lack a hydroxyl group at the 3' carbon position as well as the 2' carbon (*Figure 6.15*). A dideoxynucleotide can be incorporated into the growing DNA chain by forming a phosphodiester bond between its 5' carbon atom and the 3' carbon of the previously incorporated nucleotide. However, since ddNTPs lack a 3' hydroxyl group, any ddNTP that is incorporated into a growing DNA chain cannot

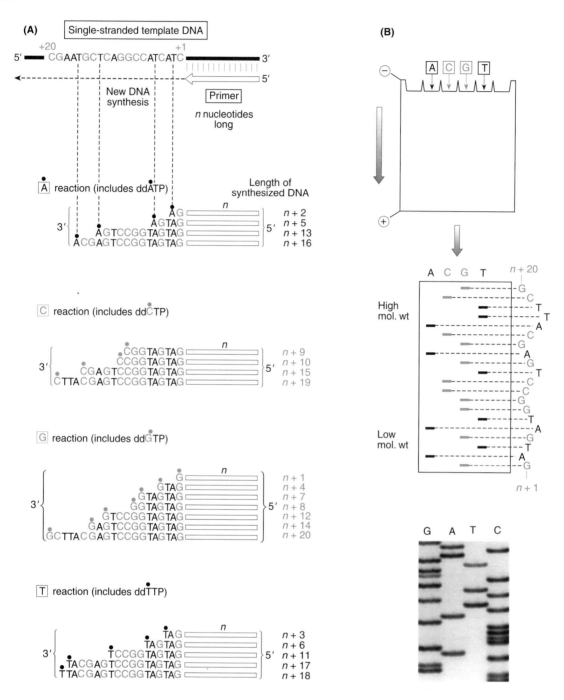

Figure 6.16: Dideoxy DNA sequencing relies on synthesizing new DNA strands from a single-stranded DNA template and random incorporation of a base-specific dideoxynucleotide to terminate chain synthesis.

(**A**) Principle of dideoxy sequencing. The sequencing primer binds specifically to a region 3′ of the desired DNA sequence and primes synthesis of a complementary DNA strand in the indicated direction. Four parallel base-specific reactions are carried out, each with all four dNTPs and with one ddNTP. Competition for incorporation into the growing DNA chain between a ddNTP and its normal dNTP analog results in a population of fragments of different lengths. The fragments will have a common 5′ end (defined by the sequencing primer) but variable 3′ ends, depending on where a dideoxynucleotide (shown with a filled circle above) has been inserted. For example, in the A-specific reaction chain, extension occurs until a ddA nucleotide (shown as A with a filled black circle above) is incorporated. This will lead to a population of DNA fragments of lengths $n + 2$, $n + 5$, $n + 13$, $n + 16$ nucleotides, etc. (**B**) Conventional DNA sequencing. This generally involves using a radioactively labeled nucleotide and size-fractionation of the products of the four reactions in separate wells of a polyacrylamide gel. The dried gel is submitted to autoradiography, allowing the sequence of the complementary strand to be read (from bottom to top). The bottom panel illustrates a practical example, in this case a sequence within the gene for type II neurofibromatosis.

participate in phosphodiester bonding at its 3′ carbon atom, thereby causing abrupt termination of chain synthesis.

Four parallel base-specific reactions are conducted using a mix of all four dNTPs and also a small proportion of one of the four ddNTPs. By setting the concentration of the ddNTP to be very much lower than that of its normal dNTP analog, chain termination will occur randomly at one of the many positions containing the base in question. Each reaction is therefore a *partial reaction*: chain termination occurs randomly at one of the possible bases *in any one DNA strand*. However, the DNA to be sequenced in a DNA sequencing reaction is a *population* of (usually) identical molecules. As a result, each one of the four base-specific reactions will generate *a collection of labeled DNA fragments of different sizes*, with *a common 5′ end but variable 3′ ends* (the common 5′ end is defined by the sequencing primer and the 3′ ends which terminate with the chosen ddNTP are variable because the insertion of the dideoxynucleotide occurs randomly at one of the many different positions that will accept that specific base – *Figure 6.16A*).

Fragments that differ in size by even a single nucleotide can be separated on a denaturing polyacrylamide gel. The differently sized fragments can be detected by incorporating labeled groups into the reaction products, either by incorporating labeled nucleotides or by using a primer with a labeled group. The sequence can then be read off by reading from the bottom of the gel to the top, a direction that gives the 5′ → 3′ sequence of the complementary strand of the provided DNA template (see *Figure 6.16B*).

6.3.2 DNA sequencing is increasingly being conducted using fluorescent labeling systems and automated detection systems

Traditional dideoxy sequencing methods have employed radioisotope labeling: the dNTP mix contains a proportion of radiolabeled nucleotides which are incorporated within the growing DNA chains. Following electrophoresis, the gel is dried and an autoradiographic film is placed in contact with the dried gel. After a suitable exposure time, the film is developed, giving a characteristic pattern of dark bands (*Figure 6.16B*). ^{32}P-labeled nucleotides are not very suitable for this purpose: the high energy β-radiation causes considerable scattering of the signal, leading to diffuse bands. Instead, ^{35}S- or ^{33}P-labeled nucleotides have been used.

Large-scale DNA sequencing efforts are dependent on improving efficiency by partial automation of the technologies involved. One major improvement in recent years has been the development of automated procedures for fluorescent DNA sequencing (Wilson *et al.*, 1990). These procedures generally use primers or dideoxynucleotides to which are attached fluorophores (chemical groups capable of fluorescing – see Section 5.1.2). During electrophoresis, a monitor detects and records the fluorescence signal as the DNA passes through a fixed point in the gel (*Figure 6.17A*). The use of different fluorophores in the four base-specific reactions means that, unlike conventional DNA sequencing, all four reactions can be loaded into a single lane. The output is in the form of intensity profiles for each of the differently colored fluorophores (*Figure 6.17B*), but the information is simultaneously stored electronically. This precludes transcription errors when an interpreted sequence is typed by hand into a computer file. Recent advances in technology mean that the accuracy of DNA sequencing using automated methods is acceptably high.

6.3.3 PCR-amplified products are often used for DNA sequencing

Cycle sequencing

Double-stranded DNA templates can be used in standard dideoxy sequencing by denaturing the DNA prior to binding the oligonucleotide primer. However, the quality of sequences from initially double-stranded DNA templates is often poor. Cycle sequencing, also called linear amplification sequencing, is a kind of PCR sequencing approach which overcomes this problem. Like the standard PCR reaction, it uses a thermostable DNA polymerase and a temperature cycling format of denaturation, annealing and DNA synthesis. The difference is that cycle sequencing employs only one primer and includes a ddNTP chain terminator in the reaction. The use of only a single primer means that unlike the exponential increase in product during standard PCR reactions, the product accumulates *linearly* (see *Figure 6.18*). Because the product accumulates during the reaction, and because of the high temperature at which the sequencing reactions are carried out, and the multiple heat denaturation steps, small amounts of double-stranded plasmids, cosmids, λDNA and PCR products may be sequenced reliably without a separate heat denaturation step.

6.3.4 DNA microarray technology permits an alternative approach to DNA sequencing

DNA sequencing can be accomplished by hybridization of the target DNA to a series of oligonucleotides of known sequence, usually about 7–8 nucleotides long. If the hybridization conditions are specific, it is possible to check which oligonucleotides are positive by hybridization, feed the results into a computer and use a program to look for sequence overlaps in order to establish the required DNA sequence. DNA microarrays have permitted sequencing by hybridization to oligonucleotides on a large scale (Southern, 1996) and in a test system, the sequence of human mtDNA previously first determined in 1981 was recently re-sequenced by DNA microarray hybridization. This type of technology is increasing in importance for assessing sequence variation over at least modest lengths of DNA and diagnostic applications in mutation analysis are proliferating (Hacia, 1999; Section 17.1.4).

(A)

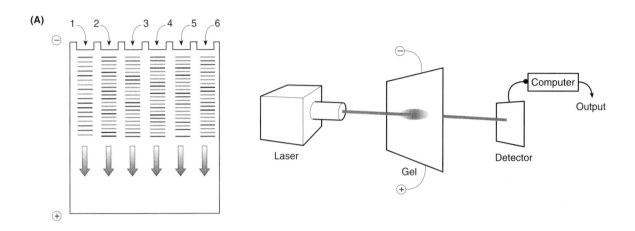

(B)

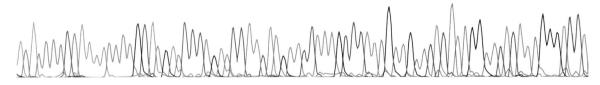

Figure 6.17: Automated DNA sequencing using fluorescent primers.

(**A**) Principles of automated DNA sequencing. Automated DNA sequencing involves loading all four reaction products into single lanes of the electrophoresis gel and capture of sequence data during the electrophoresis run. Four separate fluorescent dyes are used as labels for the base-specific reactions (the label can be incorporated by being attached to a base-specific ddNTP, or by being attached to the primer and having four sets of primers corresponding to the four reactions). During the electrophoresis run, a laser beam is focused at a specific constant position on the gel. As the individual DNA fragments migrate past this position, the laser causes the dyes to fluoresce. Maximum fluorescence occurs at different wavelengths for the four dyes, and the information is recorded electronically and the interpreted sequence is stored in a computer database. (**B**) Example of DNA sequence output. This shows a typical output of sequence data from an AB1377 automated DNA sequencer as a succession of dye-specific (and therefore base-specific) intensity profiles. The example illustrated represents sequencing of the end of a BAC clone from chromosome 3q26.3. Data provided by Dr Emma Tonkin, University of Newcastle upon Tyne. Figure kindly sponsored by PE Biosystems, a PE Corporation Business.

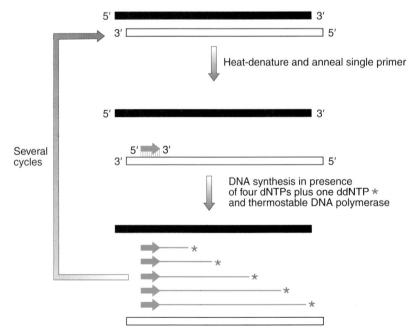

Figure 6.18: Cycle sequencing involves linear amplification using a single primer to initiate DNA synthesis.

Cycle sequencing using the dideoxynucleotide method involves setting up four parallel DNA sequencing reactions in which DNA synthesis occurs, using a mix of all four dNTPs plus one of the four ddNTPs. The reactions resemble PCR reactions because they involve the same thermocycling format as PCR. Since only a single primer is used, the product accumulates in a linear fashion, rather than exponentially as in PCR. In this example, label is introduced at the 3′ terminal end of a DNA strand when a labeled ddNTP is introduced. However, an alternative is to use a primer carrying a labeled group at its 5′ end (primer-mediated 5′ end-labeling).

6.4 *In vitro* site-specific mutagenesis

Mutagenesis is a fundamentally important DNA technology which seeks to change the base sequence of DNA and test its effect on gene or DNA function. The mutagenesis can be conducted *in vivo* (in studies of model organisms, or cultured cells) or *in vitro* and the mutagenesis can be directed to a specific site in a pre-determined way (**site-directed mutagenesis**), or can be random. In the case of *in vivo* mutagenesis, for example, gene targeting offers exquisite site-directed mutagenesis within living cells (Section 21.3.1) while exposure of male mice to high levels of a powerful mutagen such as ethylnitrosurea (ENU) and subsequent mating of the mice offers a form of random mutagenesis which can be important in generating new mutants (Section 21.4.1).

In vitro **mutagenesis** can involve essentially random approaches to mutagenesis, which may be valuable in producing libraries of new mutants. In addition, if a gene has been cloned and a functional assay of the product is available, it is also very useful to be able to employ a form of *in vitro* mutagenesis which results in alteration of a specific amino acid or small component of the gene product in a predetermined way.

6.4.1 *Oligonucleotide mismatch mutagenesis is a popular method of introducing a predetermined single nucleotide change into a cloned gene*

Many *in vitro* assays of gene function wish to gain information on the importance of individual amino acids in the encoded polypeptide. This may be relevant when attempting to assess whether a particular missense mutation found in a known disease gene is pathogenic, or just generally in trying to evaluate the contribution of a specific amino acid to the biological function of a protein. A popular general approach involves cloning the gene or cDNA into an M13 or phagemid vector which permits recovery of single-stranded recombinant DNA (Section 4.4.1). A mutagenic oligonucleotide primer is then designed whose sequence is perfectly complementary to the gene sequence in the region to be mutated, but with a single difference: at the intended mutation site it bears a base that is complementary to the desired mutant nucleotide rather than the original. The mutagenic oligonucleotide is then allowed to prime new DNA synthesis to create a complementary full-length sequence containing the desired mutation. The newly formed heteroduplex is used to transform cells, and the desired mutant genes can be identified by screening for the mutation (see *Figure 6.19*).

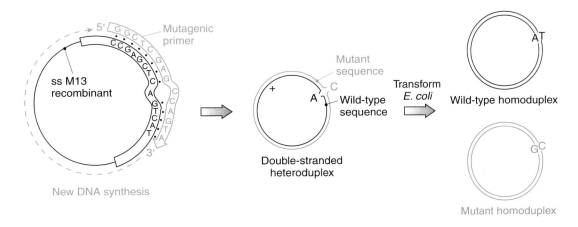

Figure 6.19: Oligonucleotide mismatch mutagenesis can create a desired point mutation at a unique predetermined site within a cloned DNA molecule.

The figure illustrates only one of many different methods of cell-based oligonucleotide mismatch mutagenesis (for alternative PCR-based site-directed mutagenesis, see Section 6.4.2). The example illustrates the use of a mutagenic oligonucleotide to direct a single nucleotide substitution in a gene. The gene is cloned into M13 in order to generate a single-stranded recombinant DNA (Section 4.4.1). An oligonucleotide primer is designed to be complementary in sequence to a portion of the gene sequence encompassing the nucleotide to be mutated (A) and containing the desired noncomplementary base at that position (C, not T). Despite the internal mismatch, annealing of the mutagenic primer is possible, and second strand synthesis can be extended by DNA polymerase and the gap sealed by DNA ligase. The resulting heteroduplex can be transformed into *E. coli*, whereupon two populations of recombinants can be recovered: wild-type and mutant homoduplexes. The latter can be identified by molecular hybridization (by using the mutagenic primer as an allele-specific oligonucleotide probe; see *Figure 5.11*) or by PCR-based allele-specific amplification methods (see *Figure 6.9*).

Other small-scale mutations can also be introduced in addition to single nucleotide substitutions. For example, it is possible to introduce a three-nucleotide deletion that will result in removal of a single amino acid from the encoded polypeptide, or an insertion that adds a new amino acid. Provided the mutagenic oligonucleotide is long enough, it will be able to bind specifically to the gene template even if there is a considerable central mismatch. Still larger mutations can be introduced by using cassette mutagenesis in which case a specific region of the original sequence of the original gene is deleted and replaced by oligonucleotide cassettes (Bedwell *et al.*, 1989).

6.4.2 PCR can be used to couple desired sequences or chemical groups to a target sequence and to produce specific pre-determined mutations in DNA sequences

In addition to long-established nonPCR based methods, site-directed mutagenesis by PCR has become increas-

ingly popular and various strategies have been devised to enable base substitutions, deletions and insertions (see below and Newton and Graham, 1997). In addition to producing specific predetermined mutations in a target DNA, a form of mutagenesis known as 5' add-on mutagenesis permits addition of a desired sequence or chemical group in much the same way as can be achieved using ligation of oligonucleotide linkers (see *Box 4.2*).

5' Add-on mutagenesis

This is a commonly used practice in which a new sequence or chemical group is added to the 5' end of a PCR product by designing primers which have the desired specific sequence for the 3' part of the primer while the 5' part of the primer contains the novel sequence or a sequence with an attached chemical group. The extra 5' sequence does not participate in the first annealing step of the PCR reaction (only the 3' part of the primer is specific for the target sequence), but it subsequently becomes incorporated into the amplified prod-

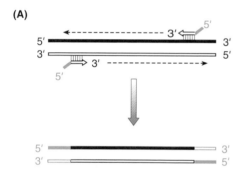

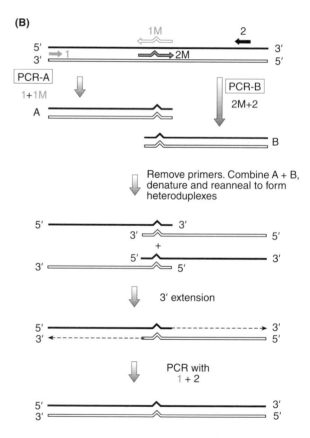

Figure 6.20: PCR mutagenesis.

(**A**) 5' add-on mutagenesis. Primers can be modified at the 5' end to introduce, for example, a labeled group (*Figure 10.24*), a sequence containing a suitable restriction site (*Figure 20.12*) or a phage promoter to drive gene expression. (**B**) Site-specific mutagenesis. The mutagenesis shown can result in an amplified product with a specific pre-determined mutation located in a central segment. PCR reactions A and B are envisaged as amplifying overlapping segments of DNA containing an introduced mutation (by deliberate base mismatching using a mutant primer – 1M or 2M). After the two products are combined, denatured and allowed to reanneal, the DNA polymerase can extend the 3' end of heteroduplexes with recessed 3' ends. Thereafter, a full length product with the introduced mutation in a central segment can be amplified by using the outer primers 1 and 2 only.

uct, thereby generating a recombinant product (*Figure 6.20A*). Various popular alternatives for the extra 5′ sequence include: (i) a suitable restriction site which may facilitate subsequent cell-based DNA cloning; (ii) a functional component, e.g. a promoter sequence for driving expression (see *Figure 17.9* for an example); a modified nucleotide containing a reporter group or labeled group, such as a biotinylated nucleotide (see *Figure 10.24* for an example) or fluorophore.

Mismatched primer mutagenesis

The primer is designed to be only partially complementary to the target site but in such a way that it will still bind specifically to the target. Inevitably this means that the mutation is introduced close to the extreme end of the PCR product. As described in Section 6.2.3 this approach may be exploited to introduce an artificial diagnostic restriction site that permits screening for a known mutation. Mutations can also be introduced at any point within a chosen sequence using mismatched primers. Two mutagenic reactions are designed in which the two separate PCR products have partially overlapping sequences containing the mutation. The denatured products are combined to generate a larger product with the mutation in a more central location (Higuchi, 1990; *Figure 6.20B*).

Further reading

Ehrlich HA (1989) *PCR Technology. Principles and Applications for DNA Amplification*. Stockton Press, New York.

Ehrlich HA, Gelfand D, Sninsky JJ (1991) Recent advances in the polymerase chain reaction. *Science*, **252**, 1643–1651.

Innis MA, Gelfand DH, Sninsky JJ, White TJ (1990) *PCR Protocols. A Guide to Methods and Applications*. Academic Press, San Diego, CA.

Ling MM, Robinson BH (1997) Approaches to DNA mutagenesis: an overview. *Analyt. Biochem.*, **254**, 157–178.

McPherson MJ, Taylor GR, Quirke P (1991) *PCR: a Practical Approach*. IRL Press, Oxford.

Newton CR, Graham A (1997) *PCR*, 2nd edn. BIOS Scientific Publishers, Oxford.

References

Bedwell DM, Strobel SA, Yun K, Jongeward GD, Emr SD (1989) Sequence and structural requirements of a mitochondrial protein import signal defined by saturation cassette mutagenesis. *Mol. Cell Biol.*, **9**, 1014–1025.

Chelly J, Concordet JP, Kaplan JC, Kahn A (1989) Illegitimate transcription: transcription of any gene in any cell type. *Proc. Natl Acad. Sci. USA*, **86**, 2617–2621.

Cheng S, Fockler C, Barnes WM, Higuchi R (1994) Effective amplification of long targets from cloned inserts and human genomic DNA. *Proc. Natl Acad. Sci. USA*, **91**, 5695–5699.

Cheung VG, Nelson SF (1996) Whole genome amplification using a degenerate oligonucleotide primer allows hundreds of genotypes to be performed on less than one nanogram of genomic DNA. *Proc. Natl Acad. Sci. USA*, **93**, 14676–14679.

Cline J, Braman JC, Hogrefe HH (1996) PCR fidelity of *Pfu* DNA polymerase and other thermostable DNA polymerases. *Nucleic Acids Res.*, **24**, 3546–3551.

Hacia JG (1999) Resequencing and mutational analysis using oligonucleotide microarrays. *Nature Genet.*, **21** (suppl.), 42–47.

Hauge Y, Litt M (1993) A study of the origin of 'shadow bands' seen when typing dinucleotide repeat polymorphisms by the PCR. *Hum. Mol. Genet.*, **2**, 411–415.

Higuchi R (1990) Recombinant PCR. In: *PCR Protocols. A Guide to Methods and Applications* (MA Innis, DH Gelfand, JJ Sninsky, TJ White, eds), pp. 177–183. Academic Press, San Diego, CA.

Holland PM, Abramson RD, Watson R, Gelfand DH (1991) Detection of specific polymerase chain reaction product by utilizing the 5′ → 3′ exonuclease activity of *Thermus aquaticus*. *Proc. Natl Acad. Sci. USA*, **88**, 7276–7280.

Lee CC, Wu X, Gibbs RA, Cook RG, Muzny DM, Caskey CT (1988) Generation of cDNA probes directed by amino acid sequence: cloning of urate oxidase. *Science*, **239**, 1288–1291.

Lee LG, Connell CR, Bloch W (1993) Allelic discrimination by nick-translation PCR with fluorogenic probes. *Nucleic Acids Res.*, **21**, 3761–3766.

Li H, Gyllenstein UB, Cui X, Saiki RK, Ehrlich H, Arnheim N (1988) Amplification and analysis of DNA sequences in single human sperm and diploid cells. *Nature*, **335**, 414–417.

Mead DA, Pey NK, Herrnstadt C, Marcil RA, Smith LM (1991) A universal method for the direct cloning of PCR amplified nucleic acid. *Biotechnology*, **9**, 657–663.

Newton CR, Graham A (1997) *PCR*, 2nd edn, pp. 75–84. BIOS Scientific Publishers, Oxford.

Newton CR, Graham A, Heptinstall LE, Powell SJ, Summers C, Kalsheker N, Smith JC, Markham AF (1989) Analysis of any point mutation in DNA. The amplification refractory mutation system (ARMS). *Nucleic Acids Res.*, **17**, 2503–2516.

Southern EM (1996) DNA chips: analysing sequence by hybridization to oligonucleotides on a large scale. *Trends Genet.,* **12**, 110–115.

Wilson RK, Chen C, Avdalovic N, Burns J, Hood L (1990) Development of an automated procedure for fluorescent DNA sequencing. *Genomics*, **6**, 626–634.

Zhang L, Cui X, Schmitt K, Hubert R, Navidi W, Arnheim N (1992) Whole genome amplification from a single cell: implications for genetic analysis. *Proc. Natl Acad. Sci. USA,* **89**, 5847–5851.

Organization of the human genome

7

7.1 General organization of the human genome

The human genome is the term used to describe the total genetic information (DNA content) in human cells. It really comprises two genomes: a complex nuclear genome which accounts for 99.9995% of the total genetic information, and a simple mitochondrial genome which accounts for the remaining 0.0005% (*Figure 7.1*). The nuclear genome provides the great bulk of essential genetic information, most of which specifies polypeptide synthesis on cytoplasmic ribosomes. Mitochondria possess their own ribosomes and the few polypeptide-encoding genes in the mitochondrial genome produce mRNAs which are translated on the mitochondrial ribosomes. However, the mitochondrial genome specifies only a very small portion of the specific mitochondrial functions; the bulk of the mitochondrial polypeptides are encoded by nuclear genes and are synthesized on cytoplasmic ribosomes, before being imported into the mitochondria (see *Figure 1.11*).

Like other complex genomes, a sizeable component of the human genome is made up of noncoding DNA. In addition, the human genome is representative of mammalian genomes and other complex genomes in having a considerable amount of repetitive DNA, including both noncoding repetitive DNA and multiple copy genes and gene fragments.

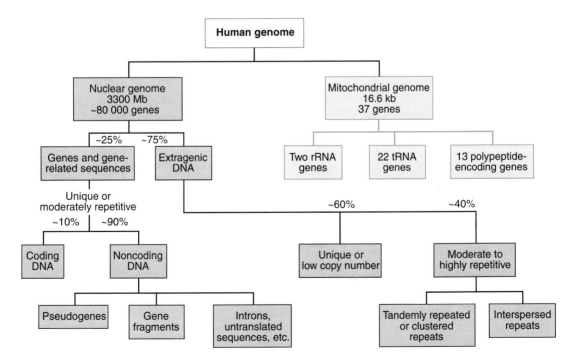

Figure 7.1: Organization of the human genome.

7.1.1 The mitochondrial genome consists of a small circular DNA duplex which is densely packed with genetic information

General structure and inheritance of the mitochondrial genome

The human mitochondrial genome is defined by a single type of circular double-stranded DNA whose complete nucleotide sequence has been established (Anderson *et al.*, 1981). It is 16 569 bp in length and is 44% (G + C). The two DNA strands have significantly different base compositions: the heavy (H) strand is rich in guanines, the light (L) strand is rich in cytosines. Although the mitochondrial DNA is principally double-stranded, a small section is defined by a triple DNA strand structure. This is because a short segment of the heavy strand is replicated for a second time, giving a structure known as 7S DNA (see *Figure 7.2* and Clayton, 1992 for a general

review of transcription and replication of animal mitochondrial DNAs). Human cells usually contain thousands of copies of the double-stranded mitochondrial DNA molecule. Accordingly, although a single mitochondrial DNA duplex has only about 1/8000 as much DNA as an average sized chromosome, the total mitochondrial DNA complement can account for up to about 0.5% of the DNA in a nucleated somatic cell.

During zygote formation, a sperm cell contributes its nuclear genome, but not its mitochondrial genome, to the egg cell. Consequently, the mitochondrial genome of the zygote is determined exclusively by that originally found in the unfertilized egg. The mitochondrial genome is therefore maternally inherited: males and females both inherit their mitochondria from their mother but males cannot transmit their mitochondria to subsequent generations. Thus mitochondrially encoded genes or DNA variants give the pedigree pattern shown in *Figure 3.4*. During mitotic cell division, the mitochondrial DNA

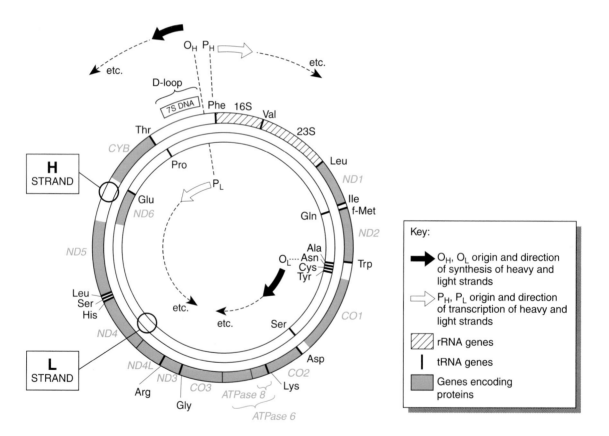

Figure 7.2: The human mitochondrial genome.

The **D loop** is marked by a triple-stranded structure and encompasses a duplicated short region of the heavy strand, *7S DNA*. Transcription of the heavy (H) strand actually originates from two closely spaced promoters located in the D loop region, which for the sake of clarity are grouped as P$_H$. Transcription from these promoters runs clockwise round the circle; transcription from the light strand promoter P$_L$ runs anticlockwise. In both cases, the resulting large transcripts are then cleaved to generate RNAs for individual genes. The high coding sequence density results from absence of introns in all genes and close apposition of genes, including one case of overlapping genes: the *ATPase 8* gene partly overlaps the *ATPase 6* gene (see *Figure 7.3*). Other polypeptide-encoding genes specify seven NADH dehydrogenase subunits (*ND4L* and *ND1–ND6*); three cytochrome *c* oxidase subunits (*CO1–CO3*) and cytochrome *b* (*CYB*).

molecules of the dividing cell segregate in a purely random way to the two daughter cells.

Mitochondrial genes

The human mitochondrial genome contains 37 genes: 28 are encoded by the heavy strand, and nine by the light strand (*Figure 7.2*). Of the 37 genes, a total of 24 specify a mature RNA product: 22 mitochondrial tRNA molecules and two mitochondrial rRNA molecules, a 23S rRNA (a component of the large subunit of mitochondrial ribosomes) and a 16S rRNA (a component of the small subunit of the mitochondrial ribosomes). The remaining 13 genes encode polypeptides which are synthesized on mitochondrial ribosomes.

Each of the 13 polypeptides encoded by the mitochondrial genome is a subunit of one of the mitochondrial respiratory complexes, the multichain enzymes of oxidative phosphorylation which are engaged in the production of ATP. Note, however, that there are a total of about 100 different polypeptide subunits in the mitochondrial oxidative phosphorylation system, and so the vast majority are encoded by nuclear genes (see *Box 7.1*). All other mitochondrial proteins, including numerous enzymes, transport proteins, structural proteins etc., are encoded by the nuclear genome and are translated on cytoplasmic ribosomes before being imported into the mitochondria (see *Figure 1.11*).

The mitochondrial genetic code

The mitochondrial genetic code (which is used to decipher only 13 different mitochondrial mRNAs on mitochondrial ribosomes) differs slightly from the nuclear genetic code (which specifies perhaps about 70 000–80 000 different mRNAs on cytoplasmic ribosomes). The mitochondrial genome encodes all the ribosomal RNA and tRNA molecules it needs for synthesizing proteins but relies on nuclear-encoded genes to provide all other components (such as the protein components of mitochondrial ribosomes, amino acyl tRNA synthetases, etc.).

As there are only 22 different types of human mitochondrial tRNA, individual tRNA molecules need to be able to interpret several different codons. Eight of the 22 tRNA molecules have anticodons which are each able to recognize families of four codons differing only at the third base, and 14 recognize pairs of codons which are identical at the first two base positions and share either a purine or a pyrimidine at the third base. Between them, therefore, the 22 mitochondrial tRNA molecules can recognize a total of 60 codons [(8 × 4) + (14 × 2)]. The remaining four codons, UAG, UAA, AGA and AGG cannot be recognized by mitochondrial tRNA and act as stop codons (see *Figure 1.22*).

In addition to their differences in genetic capacity and different genetic codes, the mitochondrial and nuclear genomes differ in many other aspects of their organization and expression (*Table 7.1*).

Coding and noncoding DNA

Unlike its nuclear counterpart, the human mitochondrial genome is extremely compact: approximately 93% of the DNA sequence represents coding sequence. All 37 mitochondrial genes lack introns and they are tightly packed (on average one per 0.45 kb). The coding sequences of some genes (notably those encoding the sixth and eighth subunits of the mitochondrial ATPase) show some overlap (*Figures 7.2* and *7.3*) and, in most other cases, the

---

Box 7.1

The limited autonomy of the mitochondrial genome

	Encoded by mitochondrial genome	Encoded by nuclear genome
Components of oxidative phosphorylation system	13 subunits	>80 subunits
I NADH dehydrogenase	7 subunits	>41 subunits
II Succinate CoQ reductase	0 subunits	4 subunits
III Cytochrome b–c1 complex	1 subunit	10 subunits
IV Cytochrome c oxidase complex	3 subunits	10 subunits
V ATP synthase complex	2 subunits	14 subunits
Components of protein synthesis apparatus	24	~80
tRNA components	22 tRNAs	None
rRNA components	2 rRNAs	None
Ribosomal proteins	None	~80
Other mitochondrial proteins	None	All, e.g. mitochondrial DNA polymerase, RNA polymerase plus numerous other enzymes, structural and transport proteins, etc.

Table 7.1: The human nuclear and mitochondrial genomes

	Nuclear genome	Mitochondrial genome
Size	3300 Mb	16.6 kb
No. of different DNA molecules	23 (in XX) or 24 (in XY) cells, all linear	One circular DNA molecule
Total no. of DNA molecules per cell	23 in haploid cells; 46 in diploid cells	Several thousand
Associated protein	Several classes of histone and nonhistone protein	Largely free of protein
Number of genes	~65 000–80 000	37
Gene density	~1/40 kb	1/0.45 kb
Repetitive DNA	Large fraction, see *Figure 7.1*.	Very little
Transcription	The great bulk of genes are transcribed individually	Continuous transcription of multiple genes
Introns	Found in most genes	Absent
% of coding DNA	~3%	~93%
Codon usage	See *Figure 1.22*	See *Figure 1.22*
Recombination	At least once for each pair of homologs at meiosis	Not evident
Inheritance	Mendelian for sequences on X and autosomes; paternal for sequences on Y	Exclusively maternal

coding sequences of neighboring genes are contiguous or separated by one or two noncoding bases. Some genes even lack termination codons; to overcome this deficiency, UAA codons have to be introduced at the post-transcriptional level (Anderson *et al.*, 1981; see legend to *Figure 7.3*).

The only significant region lacking any known coding DNA is the **displacement (D) loop region**. This is the region in which a triple-stranded DNA structure is generated by synthesizing an additional short piece of the H-strand DNA, known as 7S DNA (see *Figure 7.2*). The replication of both the H and L strands is unidirectional and starts at specific origins. In the former case, the origin is in the D loop and only after about two-thirds of the daughter H strand has been synthesized (by using the L

strand as a template and displacing the old H strand) does the origin for L strand replication become exposed. Thereafter, replication of the L strand proceeds in the opposite direction, using the H strand as a template (*Figure 7.2*). The D loop also contains the predominant promoter for transcription of both the H and L strands. Unlike transcription of nuclear genes, in which individual genes are almost always transcribed separately using individual promoters, transcription of the mitochondrial DNA starts from the promoters in the D loop region and continues, in opposing directions for the two different strands, round the circle to generate large multigenic transcripts (see *Figure 7.2*). The mature RNAs are subsequently generated by cleavage of the multigenic transcripts.

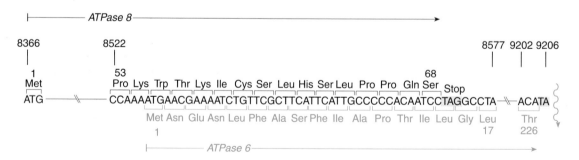

Figure 7.3: The genes for mitochondrial ATPase subunits 6 and 8 are partially overlapping and translated in different reading frames.

Note that the overlapping genes share a common sense strand, the H strand. Coding sequence coordinates are as follows: *ATPase* subunit 8, 8366–8569; *ATPase* subunit 6, 8527–9204. The C terminus of the *ATPase 6* subunit gene is defined by the post-transcriptional introduction of a UAA codon: following transcription the RNA is cleaved after position 9206 and polyadenylated, resulting in a UAA codon where the first two nucleotides are derived ultimately from the TA at positions 9205–9206 and the third nucleotide is the first A of the poly(A) tail. Other human genes are known to be overlapping but are often transcribed from opposite strands.

7.1.2 The nuclear genome is distributed between 24 different types of DNA duplex which show considerable regional variation in base composition and gene density

Size and banding patterns of human chromosomes

The nucleus of a human cell contains more than 99% of the cellular DNA. The nuclear genome is distributed between 24 different types of linear double-stranded DNA molecule, each of which has histones and other nonhistone proteins bound to it, constituting a chromosome. The 24 different chromosomes (22 types of autosome and two sex chromosomes, X and Y) can easily be differentiated by chromosome banding techniques (see *Figure 2.18*), and have been classified into groups largely according to size and, to some extent, centromere position (see *Table 2.3*). In addition to the primary constriction (centromere) present on each chromosome, the long arms of chromosomes 1, 9 and 16 possess so-called secondary constrictions (light staining, apparently uncoiled chromosomal regions) which, like the centromeres, are composed of **constitutive heterochromatin** (see Section 2.3.5). By comparison with the size of a mitochondrial DNA molecule, an average size human chromosome has an enormous amount of DNA, approximately 130 Mb on average, but varying between approximately 50 and 260 Mb (*Table 7.2*). In a 550 band metaphase chromosome preparation (see *Figure 2.18*), an average band corresponds to about 6 Mb of DNA.

Base composition in the human nuclear genome

Since the entire nucleotide sequence of the human mitochondrial genome is known, its precise base composition is known. The sequence of the human nuclear genome is still being established (and is not expected to be finished before 2003), but current estimates suggest a figure of about 42% GC. However, the proportion of specific combinations of nucleotides can vary considerably. Like other vertebrate nuclear genomes, for example, the human nuclear genome has a conspicuous shortage of the dinucleotide CpG (that is, neighboring cytosine and guanine residues on the same DNA strand in the 5' → 3' direction). Taking the average figure of 42% GC, the individual base frequencies are : C = G = 0.21, and so the expected frequency for the dinucleotide CpG is $(0.21)^2 = 0.0441$. However, the observed frequency of the CpG dinucleotide is approximately one-fifth of this (see Bird, 1986).

In vertebrate DNA, cytosine residues occurring in CpG dinucleotides are targets for methylation at carbon atom 5. Only about 3% of the cytosines in human DNA are methylated, but most that are methylated are found in the CpG dinucleotide, producing 5-methylcytosine. Over evolutionarily long periods of time, 5-methylcytosine spontaneously deaminates to give thymine and so CpG is continuously being depleted and replaced by TpG (or CpA on the complementary strand). Despite the overall background, certain small regions of DNA noted for their transcriptional activity are characterized by the expected CpG density (**CpG islands**; see *Box 8.5*).

Base composition can also show regional subchromosomal variation. For example, human telomeres are defined by numerous repeats of a 50% GC sequence, TTAGGG. Large tracts of condensed heterochromatin found at centromeric regions and in defined noncentromeric regions of many chromosomes (see *Figure 2.18*) are composed of specialized repetitive DNA whose sequence can vary considerably in base composition from the bulk DNA. Such variation is the basis for the ability of equilibrium density ultracentrifugation to fractionate the total DNA into subclasses (**satellite DNA** – see Section 7.4.1). The alternating pale and light staining chromosome bands also differ in a number of features, with dark G bands being characterized by a lower GC content than the pale staining bands (*Table 7.3*).

Gene density in the human nuclear genome

The total number of genes in the human genome has been estimated to be about 70 000–80 000 (see Section 7.2.1). As all but 37 of these genes are located in the nuclear genome, this gives a rough estimate of about 3000 genes per chromosome. However, gene density can vary substantially between chromosomal regions and also between whole chromosomes. For example, heterochromatic regions are known to be very largely composed of repetitive noncoding DNA, and the centromeres and large regions of the Y chromosome, in particular, are notably devoid of genes.

Recently, insight into gene distribution along the lengths of the different chromosomes has been obtained by hybridizing purified CpG island fractions of the genome (which are associated with perhaps about 56% of human genes; Antequara and Bird, 1993) to

Table 7.2: DNA content of human chromosomes[a]

Chromosome	Amount of DNA (Mb)	Chromosome	Amount of DNA (Mb)
1	263	13	114
2	255	14	109
3	214	15	106
4	203	16	98
5	194	17	92
6	183	18	85
7	171	19	67
8	155	20	72
9	145	21	50
10	144	22	56
11	144	X	164
12	143	Y	59

[a] The DNA content is given for chromosomes prior to entering the S (DNA replication) phase of cell division (see *Figure 2.2*). Data abstracted from electronic reference 1.

Table 7.3: Properties of chromosome bands seen with standard Giemsa staining

Dark bands (G bands)	Pale bands (correspond to R bands – see *Box 2.4*)
Stain strongly with dyes that bind preferentially to AT-rich regions, such as Giemsa and Quinacrine	Stain weakly with Giemsa and Quinacrine
May be comparatively AT-rich	May be comparatively GC-rich
DNase insensitive	DNase sensitive
Condense early during the cell cycle but replicate late	Condense late during cell cycle but replicate early
Gene poor. Genes may be large because exons are often separated by very large introns	Gene rich. Genes are comparatively small because of close clustering of exons
LINE rich, but may be poor in *Alu* repeats	LINE poor, but may be enriched in *Alu* repeats

metaphase chromosomes (Craig and Bickmore, 1994). On this basis, it is clear that gene density is high in sub-telomeric regions and that some chromosomes (e.g. 19 and 22) are gene rich while others (e.g. 4 and 18) are gene poor (*Figure 7.4*).

Differences between pale and dark G bands, which are, respectively, gene rich and gene poor, are illustrated by the contrast between the human leukocyte antigen (HLA) complex and the dystrophin gene (*DMD*) regions. The former is located in the pale G band, 6p21.3: at the time of writing, the most intensively investigated region of the HLA complex, the class III region, had 70 genes in 0.9 Mb of DNA, giving a density of one per 13 kb. By contrast, a full 2.4 Mb of DNA in the dark G band region, Xp21, appears to be devoted exclusively to the dystrophin gene (*Figure 7.5*).

7.2 Organization and distribution of human genes

7.2.1 The nuclear genome contains about 65 000-80 000 genes but only about 3% of the genome represents coding sequences

The number of genes in the human genome has been the subject of much speculation; while the small mitochondrial genome is known to have precisely 37 genes, the number in the nuclear genome remains unknown. Theoretical calculations based on the *mutational load* that a genome can tolerate and observed average mutation rates of human genes (~10^{-5} per gene per generation) suggest an upper limit of about 100 000. A variety of different approaches have been used to obtain more precise estimates of the total gene number. Three approaches have suggested a best estimate of about 65 000-80 000 genes:

■ **Genomic sequencing.** Extrapolation from sequencing of large chromosomal regions may suggest that there are about 70 000 genes (Fields *et al.*, 1994). This is based on the observation that gene-rich regions have an average gene density of close to one per

20 kb, but gene-poor regions have a much lower density, say one-tenth of this density, and that the genome is split 50:50 into gene-rich and gene-poor regions.

■ **CpG island number.** Restriction enzyme analysis using the methylation-sensitive enzyme *Hpa*II suggests that the total number of CpG islands (see *Box 8.5*) in the human genome is 45 000 (Antequara and Bird, 1993). Using an estimate that approximately 56% of genes are associated with CpG islands, these authors have suggested a total of about 80 000 human genes.

■ **EST analysis.** Large-scale random sequencing of cDNA clones provides so-called **expressed sequence tags** (ESTs, see Section 13.2.3). Comparison of known human EST sequences with a large set of different human genomic coding DNA sequences listed in sequence databases has suggested a figure of about 65 000 human genes (Fields *et al.*, 1994).

The above values suggest that the genes in the nuclear genome represent about 99.95% of the total number of cellular genes. If the average size of a human nuclear gene, including introns, is taken to be about 10–15 kb, this would mean that if the genes did not show overlaps, the total nuclear DNA occupied by genes would be about 70 000 × (10–15) kb or about 700–1050 Mb which corresponds roughly to about 25–35% of the genome. As the vast majority of nuclear genes encode polypeptides and the coding sequence required for an average size human polypeptide is taken to be about 500–600 codons, that is 1.5–1.8 kb, only about 3% of the nuclear genome (80–100 Mb of the 3300 Mb) would be expected to have a coding function.

7.2.2 RNA-encoding gene families often have numerous family members

While the great majority of human genes are expected to encode polypeptides, a significant minority encode mature RNA molecules of diverse function. The mitochondrial genome is exceptional in that 65% (24/37) of the genes encode RNA but even in the case of the nuclear genome about 5% of the genes, perhaps 3000–4000 genes

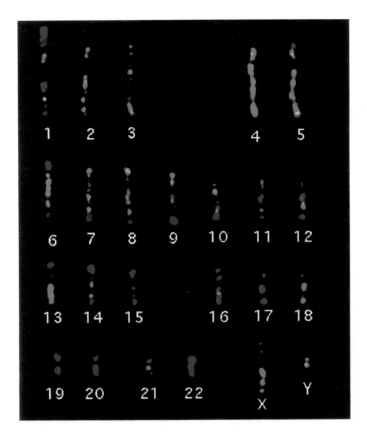

Figure 7.4: Clustering of CpG islands in the human genome.

The diagram represents FISH of a CpG island fraction from human DNA to human metaphase chromosomes (Craig and Bickmore, 1994). The Texas Red signal is derived from the CpG island probe, while the fluorescein isothiocyanate (FITC) green signal represents late replicating regions (which are mostly transcriptionally inactive), recognized by incorporation of bromodeoxyuridine (BrdU). Black regions represent overlap of the signals, indicating hybridization of the CpG island fraction to late replicating DNA. There is no counterstain, so that early replicating regions of the genome which do not have high densities of CpG islands are invisible, as are centromeres (where the anti-BrdU cannot get access). In addition to the rDNA clusters on the short arms of chromosomes 13–15, 21 and 22, high CpG island density is found on chromosomes 1, 9, 15–17, 19, 20 and 22. Adapted from Craig and Bickmore (1994) *Nature Genetics*, **7**, pp. 376–381, with permission from Nature America Inc.

in all, are expected to encode RNA molecules. In common with other cellular genomes a considerable variety of genes in the human genome are devoted to making mature RNA molecules which assist in the general process of gene expression. Some, notably rRNA and tRNA are involved in translation of mRNA. In addition, many other RNA families are involved in reactions leading to maturation not only of mRNA but also of rRNA, tRNA and other RNA species, involving both cleavage reactions and base-specific modification reactions. In addition, several other RNAs have diverse functions (*Table 7.4*).

Ribosomal RNA (rRNA) genes
There are multiple rRNA genes. In addition to the two mitochondrial rRNA molecules, the 28S, 18S and 5.8S cytoplasmic rRNAs are encoded by a single **transcription unit** (see *Figure 8.1*) which is tandemly repeated about 250 times, comprising five clusters of about 50 tandem repeats located on the short arms of human chromosomes 13, 14, 15, 21 and 22. In addition, the 5S cytoplasmic

rRNA is encoded by several hundred gene copies in at least three clusters on the long arm of chromosome 1. The major rationale for the repetition of cytoplasmic rRNA genes is likely to be based on gene dosage: by having a comparatively large number of these genes, the cell can satisfy the huge demand for cytoplasmic ribosomes needed for protein synthesis.

Transfer RNA (tRNA) genes
These belong to a very large dispersed gene family, comprising more than 40 different subfamilies each with several members which encode the different species of cytoplasmic tRNA. In addition to multiple copies of genes specifying the individual cytoplasmic tRNA molecules, there are several defective gene copies (*pseudogenes*).

Small nuclear RNA (snRNA) genes
A heterogeneous collection of several hundred **small nuclear RNA** species are encoded by a large dispersed

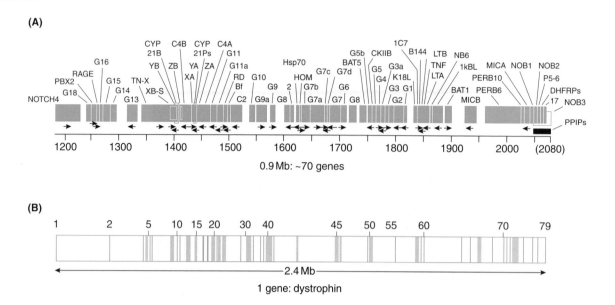

Figure 7.5: Contrasting gene densities in the HLA region and the dystrophin gene (DMD) region.

For the sake of clarity, only a 900 kb segment from the class III region of the 4 Mb HLA cluster is shown. *Note* that the great bulk of the genes in the HLA region have multiple exons (not shown) and this region is characterized by a very high density of exons in marked contrast to the dystrophin gene region, where there is a single gene with ~80 exons. The very high gene density in the HLA region is partly due to the presence of several overlapping genes (as indicated by internal/external boxes, e.g. at the 1400 kb position).

family of genes. Many of the snRNA species are uridine-rich and are named accordingly, e.g. U3 snRNA means the third uridine-rich small nuclear RNA to be classified. Individual species of RNA are associated with specific proteins to form ribonucleoprotein particles (RNPs). Some are known to be important in RNA splicing. A large subfamily of perhaps about 200 genes are present in the nucleolus, and have been termed small nucleolar RNA (snoRNA). They have important roles in specific cleavage reactions and base-specific modifications during maturation of ribosomal RNA (see Smith and Steitz, 1997).

Other RNA genes

Additional RNA genes encode functionally diverse products, including the 7SL RNA component of the signal recognition particle which is required for protein export and the RNA component of telomerase, the enzyme required to synthesize DNA at the telomeres (Section 2.3.4). More recently, evidence has been obtained suggesting that certain RNA genes encode products that are important in gene regulation. An important example is the *XIST* gene. This gene is thought to be the major gene involved in initiating the process of X chromosome inactivation, being expressed exclusively from inactivated X chromosomes. No long open reading frames can be identified, and gene function is thought to be carried out through an RNA product by a mechanism that remains obscure (see Section 8.5.6).

In addition, several RNA genes have been found at a variety of chromosomal regions that are known to be **imprinted** (imprinted genes are normally expressed from a maternally inherited copy or a paternally inherited copy, but not both; see Section 8.5.4). For example, the *H19* gene contains five exons and is expressed to give a polyadenylated cytoplasmic RNA which does not however associate with ribosomes. It shows a restricted pattern of expression during early development (fetal and neonatal liver, visceral endoderm and fetal gut) and is imprinted, since only the maternally inherited allele is expressed. Its functional significance is, however, unclear.

7.2.3 Functionally similar genes are occasionally clustered in the human genome, but are more often dispersed over different chromosomes

As seen in the previous section, some families of RNA genes are clustered. In the case of polypeptide-encoding gene families, some genes encoding identical or functionally related products are clustered, but often they are dispersed on several chromosomes.

Functionally identical genes

A very few human polypeptides are known to be encoded by two or more identical gene copies. Often, these are encoded by recently duplicated genes in a gene cluster, as in the case of the duplicated α-globin genes (see Section 7.3.4). In addition, some genes on different

Table 7.4: Functional diversity of RNA

Class of RNA	Examples	Function
A. RNA classes involved in assisting general gene expression		
Ribosomal RNA (rRNA)	28S rRNA	Component of large cytoplasmic ribosomal subunit
	5.8S rRNA	Component of large cytoplasmic ribosomal subunit
	5S rRNA	Component of large cytoplasmic ribosomal subunit
	18S rRNA	Component of small cytoplasmic ribosomal subunit
	23S rRNA	Component of large mitochondrial ribosomal subunit
	16S rRNA	Component of small mitochondrial ribosomal subunit
Transfer RNA (tRNA)	>40 different cytoplasmic tRNA; 22 types of mitochondrial tRNA	Binding to codons in mitochondrial or nuclear-encoded mRNA
Small nuclear RNA (snRNA)	Many, including	
	U1 snRNA	Component of major spliceosome
	U2 snRNA	Component of major spliceosome
	U4 snRNA	Component of major spliceosome
	U5 snRNA	Component of major and minor spliceosome
	U6 snRNA	Component of major spliceosome
	U4acat snRNA	Component of minor spliceosome
	U6acat snRNA	Component of minor spliceosome
	U11snRNA	Component of minor spliceosome
	U12 snRNA	Component of minor spliceosome
	U7 snRNA	Histone mRNA transcriptional termination
Small nucleolar RNA (snoRNA)	About 200 types, including	
	U3 snoRNA	rRNA processing
	U8 snoRNA	rRNA processing
	various box C/D snoRNAs	Site-specific methylation of the 2′ OH group of rRNA
	various box H/ACA snoRNAs	Site-specific rRNA modification by formation of pseudouridine.
B. Other RNA classes		
	7SL RNA	Component of signal recognition particle for transporting proteins (see Section 1.5.4)
	7SK RNA	Function uncertain
	Telomerase RNA	Component of telomerase (Section 2.3.4)
	XIST RNA	Regulatory gene imposing X-chromosome inactivation (Section 7.2.2)
	H19 RNA	Imprinted gene, function unclear (Section 7.2.2)
	SRA RNA	Encodes a steroid receptor coactivator

chromosomes encode identical polypeptides. Examples include members of histone gene subfamilies. As mentioned in Section 2.3.1, histones can be classified into five groups in terms of structure: H1 (the *linker histone*) and the four *core histones*, H2A, H2B, H3 and H4. In addition histone genes can be classified into three groups according to expression: (i) replication-dependent (restricted to the S phase of the cell cycle); (ii) replication-independent (expressed at a low level throughout the cell cycle to give so-called replacement histones); (iii) tissue-specific, e.g. the H1t and H3t genes are expressed exlusively in the testis. There appears to be a total of 61 human histone genes which comprise several subfamilies (Albig and Doenecke, 1997; electronic reference 2). Most of the histone genes are found in two multifamily clusters on the short arm of chromosome 6, but genes on several

other human chromosomes can specify identical copies of a particular histone subtype (*Figure 7.6*).

Functionally similar genes

A large fraction of human genes are members of gene families where individual genes are closely related but not identical in sequence. In many such cases the genes are clustered and have arisen by tandem gene duplication, as in the case of the different members of each of the α-globin and β-globin gene clusters (see Section 7.3.4). Genes which encode clearly related products but which are located on different chromosomes are generally less related, as in the case of the α-globin and β-globin genes. However, in the case of the *HOX* homeobox gene family which consists of clusters of approximately 10 genes on each of four chromosomes, individual genes on different chromosomes may be more related to each other than they are to members of the same gene cluster (Section 14.2.2 and *Figure 14.5*).

In addition to the above, genes encoding closely related tissue-specific isoforms, or subcellular compartment-specific isozymes are often nonsyntenic (i.e. located on different chromosomes; see *Table 7.5*).

Functionally related genes

Some genes encode products which may not be so closely related in structure, but are clearly functionally related. The products may be subunits of the same protein or macromolecular structure, components of the same metabolic or developmental pathway, or may be required to specifically bind to each other as in the case of ligands and their relevant receptors. In almost all such cases, the genes are not clustered and are usually found on different chromosomes (see *Table 7.5* for some examples).

7.2.4 Human genes show enormous variation in size and internal organization

Size diversity

Genes in simple organisms such as bacteria are comparatively similar in size, and usually very short. By contrast, complex organisms such as mammals show wide variation in gene size, a feature found especially in human genes which can vary in length from hundreds of nucleotides to several megabases (*Figure 7.7*). The enormous size of some human genes means that transcription can be time-consuming. For example, the human dystrophin gene requires about 16 hours to be transcribed, and transcripts undergo splicing before transcription is completed (Tennyson *et al.*, 1995).

As one would expect, there is a direct correlation between the size of a gene and the size of its product, but there are some striking anomalies. For example,

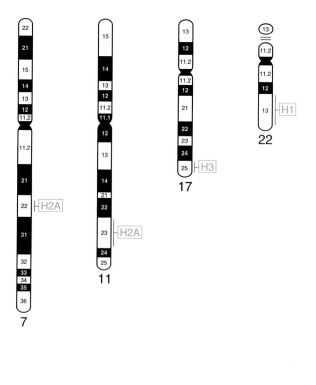

Figure 7.6: Chromosomal distribution of the human histone gene family.

Eleven clusters comprising a total of about 60 histone genes are distributed over seven human chromosomes. The two clusters on 6p contain the great majority of histone genes. Other clusters contain only one or two of the histone gene subtypes. *Note* that identical histones can be specified by genes on different chromosomes.

Table 7.5: Distribution of genes encoding functionally related products

Genes which encode	Organization	Examples
The same product	Often clustered but may also be on different chromosomes	The two α-globin genes on 11p; genes encoding rDNA (*Figure 8.1*) and histones (*Figure 7.6*).
Tissue-specific protein isoforms or isozymes	Sometimes clustered; sometimes nonsyntenic	Clustering of pancreatic and salivary amylase genes (1p21); nonsynteny of α-actin genes expressed in skeletal (1p) and cardiac (15q) muscle
Isozymes specific for different subcellular compartments	Usually nonsyntenic	Cytoplasmic (c) and mitochondrial (m) isozymes for various enzymes e.g. aldehyde dehydrogenase: *ALDH1*, *ALDH2* on 9q and 12q, respectively; aconitase: *ACO1*, *ACO2* on 9p and 22q respectively; thymidine kinase: *TK1*, *TK2* on 17q and 16 respectively.
Enzymes in the same metabolic pathway	Usually nonsyntenic	Genes encoding enzymes in steroidogenesis steroid 11-hydroxylase 8q, steroid 17-hydroxylase 10 steroid 21-hydroxylase 6p
Subunits of the same protein or enzyme	Usually nonsyntenic	Hemoglobin: α-chain – 16p; β chain – 11p; collagens: α(1)I chain – 7q; α(2)I chain – 17q; ferritin: heavy chain – 11q; light chain – 22q; class I HLA: heavy chain – 6p; light chain – 15q; immunoglobulins: heavy chain – 14q; light chain – 2p or 22q
Ligand plus associated receptor	Usually nonsyntenic	Genes encoding insulin, interferons and their receptors insulin *INS*-11p, but insulin receptor *INSR*-19p interferon α *IFNA* – 9p, and receptor, *IFNAR* – 21q interferon β *IFNB* – 9p, and receptor, *IFNBR* – 21q interferon γ *IFNG* – 12q, and receptor, *IFNGR* – 18

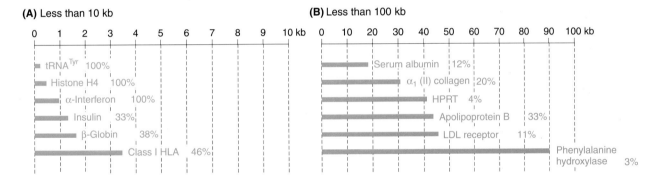

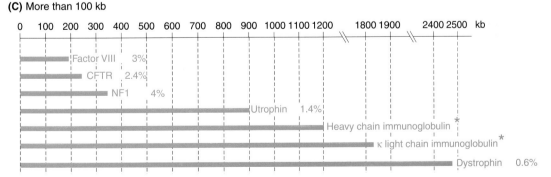

Figure 7.7: Human genes vary enormously in size and exon content.

Exon content is shown as a percentage of the lengths of indicated genes. *Note* the generally inverse relationship between gene length and percentage of exon content. Asterisks emphasize that the lengths given for the indicated Ig heavy chain and light chain loci correspond to the germline organizations. Immunoglobulin and T-cell receptor genes have unique organizations, requiring cell-specific somatic rearrangements in order to be expressed in B or T lymphocytes respectively (see Section 8.6). Abbreviations: CFTR, cystic fibrosis transmembrane regulator; HPRT, hypoxanthine phosphoribosyl transferase; NF1, neurofibromatosis type 1.

apolipoprotein B has 4563 amino acids and is encoded by a 45 kb gene while the dystrophin gene is 2.4 Mb in length and encodes a product in muscle cells of 3685 amino acids.

Diversity in internal organization

There is an inverse correlation between gene size and the proportion of the gene length which is expressed at the RNA level (*Figure 7.7*). A very small minority of human genes lack introns and are generally very small genes (see *Table 7.6* for examples). For those that do possess introns, the exon content as a percentage of gene length tends to be very small in large genes. This does not arise because exons in large genes are smaller than those in small genes: the average exon size in human genes is about 200 bp and, although very large exons are known (see *Box 7.2*), exon size is comparatively independent of gene length (*Table 7.7*). Instead, the explanation is due to the huge variation in intron lengths: large genes tend to have very large introns (*Table 7.7*). The relationship between gene and intron length is not, however, without anomalies: the human

Table 7.6: Examples of human genes with uninterrupted coding sequences

All 37 mitochondrial genes

Histone genes

Many genes encoding small RNA, e.g. most tRNA genes

Various neurotransmitter and hormone receptor genes, e.g. dopamine D1 and D5 receptors, 5-HT$_{1B}$ serotonin receptor, angiotensin II type 1 receptor, formyl peptide receptor, bradykinin B2 receptor, α2 adrenergic receptor

Autosomal processed copies of intron-containing X-linked genes
Typically have testis-specific expression patterns, e.g. *PGK2* (phosphoglycerate kinase), *GK* (glycerol kinase), *MYCL2* (myc family member), *PDHA2* (pyruvate dehydrogenase E1a), *GLUD2* (glutamate dehydrogenase)

Others, e.g. IFN-α, thrombomodulin, *SRY* and many *SOX* (*SRY* HMG box-related) genes, *XIST*, neurogenin genes

---

Box 7.2

Human gene organization

Gene number	
Total genome	Mitochondrial genome: 37 (Section 7.1.1). Nuclear genome: 65 000-80 000.
Chromosome	Average of about 3000; but dependent on chromosome length and also on chromosome type (see *Figure 7.4*).
Chromosome band	Average of about 130 per band in a 550-band chromosome preparation.
Gene density	Averages of about one per 0.45 kb in the mitochondrial genome one per 40–45 kb in the nuclear genome.
Gene size	Average 10–15 kb, but enormous variation (see *Figure 7.7*).
Intergenic distance	On average, 25–30 kb in nuclear genome.
Exon number	Generally correlated with gene length, and shows wide variation from small genes with a single exon (see *Table 7.6*) to large genes with numerous exons, e.g. the dystrophin gene (*DMD*) has 79 exons.
Exon size	Average size about 200 bp with comparatively little length variation, although coding sequence exons are a bit shorter on average and exons containing 3′ UTR sequences are considerably longer (Zhang, 1998). Some exceptionally long exons have been reported e.g. exon 26 of the apoB gene (*APOB*), 7.6 kb exon 15 of the adenomatous polyposis coli gene (*APC*), 6.5 kb exon 11 of the *BRCA1* breast cancer gene, 3.4 kb.
Intron size	Enormous variation. Strong direct correlation with gene size: small genes tend to have small introns and large genes tend to have large introns. Examples of typical intron sizes are as follows: β-globin gene (*HBB*; 1.6 kb) 0.5 kb myoglobin gene (*MB*; 10.4 kb) 4.7 kb dystrophin gene (*DMD*; 2.5 Mb) 30.0 kb
mRNA size	Average of about 2.5 kb, but considerable variation
5′ UTR	Average of about 0.1 kb (see Zhang, 1998)
coding DNA	Average of about 1.5–1.8 kb (500-600 codons)
3′ UTR	Average of about 0.6 kb (see Zhang, 1998) but this is likely to be an underestimate because of underreporting of genes with long 3′ UTRs.

Table 7.7: Average sizes of exons and introns in human genes

Gene product	Size of gene (kb)	Number of exons	Average size of exon (bp)	Average size of intron (bp)
tRNA^{tyr}	0.1	2	50	20
Insulin	1.4	3	155	480
β-Globin	1.6	3	150	490
Class I HLA	3.5	8	187	260
Serum albumin	18	14	137	1100
Type VII collagen	31	118	77	190
Complement C3	41	29	122	900
Phenylalanine hydroxylase	90	26	96	3500
Factor VIII	186	26	375	7100
CFTR (cystic fibrosis)	250	27	227	9100
Dystrophin	2400	79	180	30 000

type 7 collagen gene (*COL7A1*) is an intermediate size gene (31 kb) but has a total of 118 exons and an average intron size of only 188 bp (Christiano *et al.*, 1994). The extraordinary number of exons in *COL7A1* is not even matched by the number of exons in the giant 2.4 Mb dystrophin (*DMD*) gene (*Table 7.7*).

7.2.5 Rare examples of overlapping genes and genes within genes are known in the human genome

Partially overlapping genes

The genes of simple organisms are generally more clustered than those in complex organisms. The average gene density in the human genome is about one per 40–45 kb of DNA. Assuming a mean size of, say, 10–15 kb, human genes should be separated by about 30 kb of nongenic DNA on average. By contrast, average gene densities in simple organisms are very much higher: roughly one per 1, 2 and 5 kb, respectively, for *E. coli*, *Saccharomyces cerevisiae* and *Caenorhabditis elegans*. Simple genomes such as those of certain phages and bacteria often show examples of partially overlapping genes which use different reading frames, sometimes from a common sense strand. The human mitochondrial genome is another example of a simple genome packed with genetic information and it too has an example of such overlapping genes (see *Figure 7.3*).

Reported occurrences of overlapping genes in the complex nuclear genomes of mammals are rare and, where they do occur, the overlapping genes are often transcribed from the two different DNA strands. As noted in Section 7.1.2, the degree of gene clustering in the nuclear genome is largely dependent on the chromosomal region, and in regions of high density occasional examples of overlapping genes have been noted. For example, the class III region of the HLA complex at 6p21.3 has an average gene density of about one gene per 13 kb, and is known to contain several examples of overlapping genes (*Figure 7.5*).

Genes within genes

The small nucleolar RNA (snoRNA) genes are unusual in that the majority of them are located within other genes, often ones which encode a ribosome-associated protein or a nucleolar protein. Possibly this arrangement has been maintained to permit coordinate production of protein and RNA components of the ribosome (Tycowski *et al.*, 1993).

In addition to the snoRNA genes there are a few examples of other genes being located within the introns of larger genes, and in some cases the internal genes as well as the host genes are known to encode polypeptides. Three illustrative examples are:

■ *The neurofibromatosis type I (NF1) gene.* Intron 27 of the *NF1* gene spans about 40 kb and contains three small genes, each with two exons which are transcribed from the opposite strand to that used for the *NF1* gene (Viskochil *et al.*, 1991; see *Figure 7.8*).

■ *The factor VIII gene.* Intron 22 of the blood clotting factor VIII gene (*F8C*) contains a CpG island from which two internal genes, *F8A* and *F8B* are transcribed in opposite directions (Levinson *et al.*, 1992). *F8A* is transcribed from the opposite strand to that used by the factor VIII gene. *F8B* is transcribed in the same direction as the factor VIII gene to give a short mRNA containing a new exon spliced on to exons 23–26 of the factor VIII gene (see *Figure 9.20*).

■ *The retinoblastoma susceptibility gene RB1.* Intron 17 of this gene is 72 kb long and contains a G protein-coupled receptor gene, U16, which is actively transcribed from the opposite strand (see *Figure 7.19*).

7.3 Human multigene families and repetitive coding DNA

DNA sequences in the nuclear diploid genome usually exist as two allelic copies (on paternal and maternal homologous chromosomes). In addition to this degree of

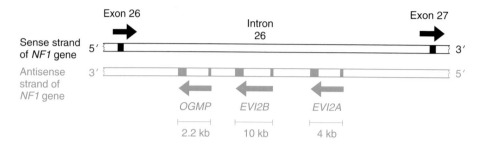

Figure 7.8: Genes within genes: intron 26 of the gene for neurofibromatosis type I (NF1) contains three internal genes each with two exons.

Note that the three internal genes are transcribed from the opposing strand to that used for transcription of the *NF1* gene. Genes are: *OGMP*, oligodendrocyte myelin glycoprotein; *EVI2A* and *EVI2B*, human homologs of murine genes thought to be involved in leukemogenesis, and located at ecotropic viral integration sites.

repetition, about 40% of the human nuclear genome in both haploid and diploid cells is composed of sets of closely related nonallelic DNA sequences (DNA sequence families or repetitive DNA). Within the considerable variety of different repetitive DNA sequences are DNA sequence families whose individual members include functional genes (**multigene families**), and also many examples of nongenic repetitive DNA sequence families.

7.3.1 The reassociation kinetics of human DNA suggest three broad classes of DNA sequence

Reassociation kinetics (Section 5.2.2) first suggested that complex genomes, such as the human genome, comprise different sequence classes on the basis of the **copy number**. Typically, human DNA is randomly sheared (e.g. by sonication) to give fragments whose average size is about 500 bp and the sheared DNA is denatured by heating to separate the complementary strands of each fragment. Thereafter the DNA is cooled to a temperature of about 20–30°C below the melting temperature, T_m (which marks the mid-point of the transition between the double-stranded and single-stranded states of DNA heated in solution). The cooled DNA renatures but the rate of reassociation depends not only on time (t) but also on the initial concentration (C_o) of that sequence (i.e. the $C_o t$ value).

The above type of analysis has suggested that the human genome consists of three broad sequence components:

- *Single copy*, or at least very low copy number, DNA (60%) – reassociates very slowly. A single strand from a single copy sequence will require some considerable time to find a complementary partner strand, given that the vast majority of DNA fragments are unrelated to it.
- *Moderately repetitive* (30%) – intermediate speed of reassociation.
- *Highly repetitive* (10%) – reassociates very rapidly. There are numerous copies of the same sequence and

the chances of quickly finding complementary partners within the mass of different fragments are high.

7.3.2 Members of DNA sequence families can be identified by a variety of different approaches.

The operational definition of a DNA sequence family is the comparatively high level of DNA sequence similarity (**sequence homology**) between whole family members, or components of the family members. Members of a DNA sequence family can be identified and actively sought by a variety of methods:

- *DNA sequencing* – Allows direct calculations on the degree of sequence relatedness of family members.
- *DNA hybridization and cloning.* A probe from a gene family member typically gives a complex band pattern when hybridized against a Southern blot of genomic DNAs. Individual family members can then be cloned by screening genomic DNA libraries.
- *PCR cloning* – Permits identification of novel family members by designing **degenerate primers** corresponding to highly conserved nucleotide or amino acid sequences.

When two members of a repetitive DNA sequence family exhibit a high degree of sequence homology, a recent common evolutionary origin is indicated. As detailed in the following sections, DNA sequence families show considerable variation in the number of different repeat unit members in the family, the size of the repeating unit, chromosomal location, mode of repetition and capacity for expression.

7.3.3 Human gene families vary in the overall sequence relatedness of different family members and the extent to which particularly conserved subgenic sequences define the family

A large percentage of actively expressed human genes are members of families of DNA sequences which show a high degree of sequence similarity. However, the extent of sequence sharing and the organization of family members can vary widely. Some family members may be nonfunctional (**pseudogenes** and gene fragments – see below) and rapidly accumulate sequence differences, leading to marked sequence divergence.

Classical gene families

Classical gene families are distinguished by members which exhibit a high degree of sequence homology over most of the gene length or, at least, the coding DNA component, a feature which automatically identifies such sequences as being closely related evolutionarily as well as functionally. In some cases, such as the individual rRNA gene families and individual histone gene families, there is an extremely high degree of sequence similarity between family members. Many other large gene families show a high degree of sequence similarity between family members.

Gene families encoding products with large, highly conserved domains

In some gene families there is particularly pronounced homology within specific strongly conserved regions of the genes; the corresponding sequence similarity between the remaining portion of the coding sequence in the different genes may be quite low. Often such families encode transcription factors that play important roles in early development, and the conserved sequence encodes a protein domain which is required to bind specifically to the DNA of selected target genes (see *Table 7.8*).

Gene families encoding products with very short conserved amino acid motifs

The members of some gene families may not be very obviously related at the DNA sequence level, but nevertheless encode gene products that are characterized by a common general function and the presence of very short conserved sequence motifs. Examples, some of which are illustrated in *Figure 7.9*, include:

■ the DEAD box gene family – contains several different genes whose products appear to function as RNA

helicases and are characterized by the presence of eight short amino acid sequence motifs, including the DEAD box (the sequence Asp–Glu–Ala–Asp as represented by the one-letter amino acid code);

■ the WD repeat gene family – gene products have a regulatory function, but there is considerable diversity in function. Typically characterized by tandem repeats with a central core of fixed length and containing small conserved amino acid motifs, including the WD (tryptophan–aspartate) sequence);

■ the ankyrin repeat gene family – wide functional diversity but often involved in protein–protein interactions. Characterised by tandem repeats of a 33 amino acid sequence characterized by the presence of select amino acids at only a few positions;

■ the LIM domain gene family – encode a characteristic cysteine-rich 56 amino acid domain most likely involved in protein–protein interactions.

Gene superfamilies

In some types of gene family, the genes encode products that are known to be functionally related in a general sense, and show only very weak sequence homology over a large segment, without very significant conserved amino acid motifs. Instead, there may be some evidence for general common structural features. Such genes, which appear to be evolutionarily related but more distantly than those in a classical or conserved domain/motif gene family, have been considered to be members of a gene superfamily. For example, in addition to the immunoglobulin gene family, other related genes such as the *HLA* genes, TCR genes, *T4* and *T8* genes are known to encode products with an immune system function and a domain structure that resembles that of immunoglobulins. Although, therefore, the level of sequence homology between such genes may be very low, the similarities in function and general domain structure have suggested the existence of a so-called Ig superfamily, in which there appears to be a distant common evolutionary relationship (*Figure 7.10*; see also *Figure 14.16* for the globin superfamily).

Table 7.8: Examples of human genes with sequence motifs which encode highly conserved domains

Gene family	Number of genes	Sequence motif/domain
Homeobox genes	30 *HOX* genes (see *Figure 14.5*) plus ~60 orphan homeobox genes	*Homeobox* specifies a *homeodomain* of ~60 amino acids. A wide variety of different subclasses have been defined
PAX genes	9	*Paired box* encodes a *paired domain* of ~130 amino acids; PAX genes often have in addition a type of homeodomain known as a *paired-*type homeodomain
SOX genes	~15	*SRY-like HMG box* which encodes a domain of ~70 amino acids
TBX genes	~15	*T-Box* which encodes a domain of ~170 amino acids
Forkhead domain genes	~15	The *forkhead domain* is about 110 amino acids long
POU domain genes	~15	The *POU domain* is ~150 amino acids long

(A) DEAD box

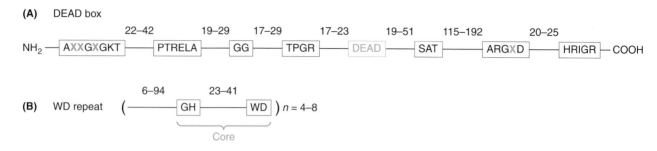

(B) WD repeat

Figure 7.9: Some gene families are defined by functionally related gene products bearing very short conserved amino acid motifs: consensus motifs for the DEAD box and WD repeat families.

(A) Motifs in the DEAD box family. This gene family encodes products implicated in cellular processes involving alteration of RNA secondary structure, such as translation initiation and splicing. Eight very highly conserved amino acid motifs are evident, including the DEAD box (Asp–Glu–Ala–Asp). Numbers refer to frequently found size ranges for intervening amino acid sequences (see Schmid and Linder, 1992). X = any amino acid. See inside front cover for the one-letter amino acid code. **(B)** WD repeat family. This gene family encodes products that are involved in a variety of regulatory functions, such as regulation of cell division, transcription, transmembrane signaling, mRNA modification, etc. The gene products are characterized by between four and eight tandem repeats containing a core sequence of fixed length (from 27–45 amino acids, terminating in the dipeptide WD, i.e. Trp–Asp) preceded by a unit whose length can vary between repeats (see Neer *et al.*, 1994).

7.3.4 Human gene families can occur as closely clustered genes at specific subchromosomal locations, or as widely dispersed genes

A wide variety of human gene families have been identified and show considerable variation in both the organization of the genes and the extent to which individual genes within the gene family are related in terms of sequence and function. In terms of gene organization, two basic arrangements can be discerned: families where there is evidence of close gene clustering and families that are dispersed over several different chromosomal locations. This classification is rather arbitrary, however, since several gene families consist of multiple gene clusters at different chromosomal locations (*Table 7.9*).

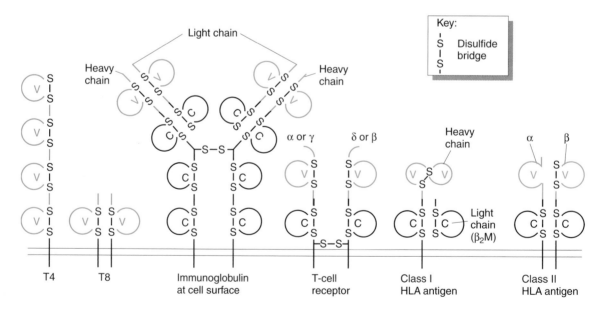

Figure 7.10: Members of the Ig superfamily are surface proteins with similar types of domain structure.

Most members of the Ig superfamily are dimers consisting of extracellular variable domains (V) located at the N-terminal ends and constant (C) domains, located at the C-terminal (membrane-proximal) ends. The light chain of class I HLA antigens has a single constant domain and does not span the membrane. It associates with the transmembrane heavy chain which has two variable and one constant domain, giving an overall structure similar to that of the class II HLA antigens.

Table 7.9: Examples of clustered and interspersed multigene families

Family	Copy no.	Organization location	Chromosome
A. Clustered gene families			
Single cluster gene families			
Growth hormone gene cluster	5	Clustered within 67 kb; one conventional pseudogene	17q22–24
α-Globin gene cluster	7	Clustered over ~50 kb (see *Figure 8.23*)	
Class I HLA heavy chain genes	~20	Clustered over 2 Mb (see *Figure 8.4*)	
Multiple cluster gene families			
HOX genes	38	Organized in four clusters (*Figure 9.5*)	2p, 7, 12,17
Histone gene family	61	Modest-sized clusters at a few locations; two large clusters on chromosome 6	(see *Figure 7.6*)
Olfactory receptor gene family	~1000	About 25 large clusters scattered throughout the genome	Many
B. Interspersed gene families			
Aldolase	5	Three functional genes and two pseudogenes on five different chromosomes	Many
PAX	9	All nine genes are expressed	Many
NF1 (Neurofibromatosis type I)	>12	One functional gene at 17q11.2; others are defective non-processed DNA copies	Mostly pericentromeric
Ferritin heavy chain	>15	One functional gene known on chromosome 11; most are processed pseudogenes	Many
Glyceraldehyde 3-phosphate dehydrogenase	>18	One functional gene on 12p; many processed pseudogenes	Many
Actin	>20	Four functional genes and many processed pseudogenes	Many

Gene families organized in a single cluster

Genes in an individual gene cluster are thought to arise by **tandem gene duplication** events. Different organizations are evident:

- **Tandem gene organization.** This arrangement is exemplified by the organisation of genes in individual rRNA gene clusters (see Section 7.2.2). The genes are highly related to each other in terms of both sequence and function, although certain family members may be nonfunctional (**pseudogenes**, see Section 7.3.5 below).
- **Close clustering.** The individual genes may be more physically separate than in a tandemly repeated cluster but nevertheless may be closely clustered, even to the extent of being subject to a common regulatory mechanism (**locus control region** – see the example of the α- and β-globin gene clusters in *Figures 7.11* and *8.23*. Again the individual genes usually show a high degree of sequence and functional identity to each other, but many family members may be pseudogenes.
- **Compound clusters.** In other clustered gene families, however, the physical relationship between genes in a cluster may be less close and a cluster of related genes may also contain within it genes that are unrelated in sequence and function, a compound gene cluster. For example, the HLA complex on 6p21.3 is dominated by families of genes which encode class I and II HLA antigens and various serum complement factors, but individual family members may be separated by functionally unrelated genes such as members of the steroid 21-hydroxylase gene family, etc. The gene organization in the latter case shows evidence of an ancestral duplication event that involved a segment containing different genes that are unrelated in function and sequence (see *Figure 7.5*).

Gene families organized in multiple gene clusters

Some gene families are organized in clusters distributed over two or more chromosomal locations. Again different organizations are evident, with some families showing very high similarity between genes on different clusters, while others are marked by comparatively low sequence homology between genes on different clusters.

- **High cluster similarity.** Here sequence homology between gene family members on different chromosomes may be very high as in the case of the different rRNA gene clusters on the short arms of chromosomes 13, 14, 15, 21 and 22. Another useful example is the olfactory receptor gene family which encodes a diverse repertoire of receptors which allow us to discriminate thousands of different odors. The family consists of perhaps 1000 genes, making it the largest in the human genome, and is organized in large clusters at more than 25 different chromosomal locations (Rouquier *et al.*, 1998).
- **Low cluster similarity.** Often sequence homology is greater within a cluster than between clusters. For example, the globin gene family includes genes at

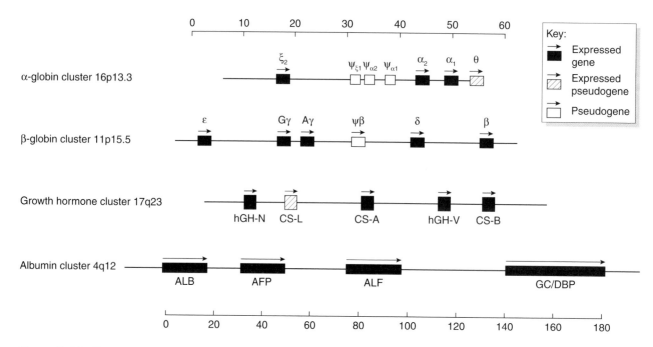

Figure 7.11: Examples of human clustered gene families.

Genes in a cluster are closely related in sequence and are often transcribed from the same DNA strand.

three locations: the α-globin cluster on 16p, the β-globin cluster on 11p and the myoglobin gene at 22q and although all globin genes are clearly related, those within a cluster are more related to each other than they are to genes in one of the other clusters (see *Figure 14.16).*

Interspersed gene families

Some gene families show no obvious physical relationship between family members, which are dispersed as solitary genes at two or more different chromosomal locations. The family members may show considerable sequence divergence unless their dispersion has been a relatively recent event, or there has been considerable selection pressure to maintain sequence conservation. The following examples illustrate some of the many different types of organization.

■   *Families expected to originate from two genomes.* Current ideas on mitochondrial origins have envisaged that the original genome was derived from a prokaryote but that many of the genes were subsequently transferred to the nuclear genome. Some families of nuclear genes encode cytoplasm-specific and mitochondrial-specific isoforms for certain enzymes and other key metabolic products (see *Table 7.5* for some examples).

■   *Families originating from ancient genome duplication or gene duplication events.* This class of interspersed gene families typically contains only a few members, as in the case of the *PAX* gene family, and appear to have evolved by a combination of gene duplication and/or genome duplication events over a long

period of evolutionary time. Typically, all or most of the genes are functional, and may individually encode highly related products.

■   *Families originating largely by retrotransposition events.* Some gene families have expanded comparatively recently in evolutionary terms by a process whereby RNA transcribed from one or a small number of functional genes is converted by cellular reverse transcriptase into natural cDNA which then becomes integrated elsewhere in the chromosomes. Most such copies are nonfunctional (see next section).

7.3.5 Pseudogenes, truncated gene copies and gene fragments are commonly found in multigene families

Families of RNA genes or polypeptide-encoding genes are frequently characterized by defective copies of essentially all of the gene or its coding sequence (**pseudogenes**), or a portion of it, in some cases a single exon (gene fragments). A large variety of different classes are found (see *Box 7.3*). The following examples are meant to be illustrative of the types of defective gene copies found in different types of gene family.

Nonprocessed pseudogenes in a gene cluster

Individual gene clusters are typically characterized by the presence of defective gene copies which have been copied *at the level of genomic DNA.* This means that the defective gene copies can contain copies of the exons, introns and promoter regions of the functional genes (nonprocessed pseudogenes). Classical examples are the α-globin and β-globin clusters (see *Figure 7.11*). Only one member of the α-globin gene family is a nonprocessed pseudogene, but three of the seven gene copies in the β-globin gene cluster are. Another β-globin gene family member, *HBQ1*, is likely to be an **expressed pseudogene**: it encodes a type of globin polypeptide, θ-globin, but there is no evidence that the latter is ever incorporated into a haemoglobin molecule and so θ-globin is likely to lack any function (Clegg, 1987)

Truncated genes and gene fragments in a gene cluster

The class I HLA gene family at 6p21.3 is a classical example of a gene cluster which is characterized by nonprocessed pseudogenes, truncated gene copies and gene fragments. Although the number of class I *HLA* genes can vary on different chromosome 6s, comprehensive analysis of one of these identified 17 family members clustered over about 2 Mb (Geraghty *et al.*, 1992; see

Figure 7.12). Six of the genes are known to be expressed, although the precise functions of some of these are still not clearly understood. The remaining members are clearly defective. Four are conventional full-length pseudogenes, but another five represent truncated gene copies (lacking the 5′ end in four cases and the 3′ end in the other case) and two are fragments which contain a small component of the gene sequence, even a single exon (see *Figure 7.12*).

Nonprocessed pseudogenes in an interspersed gene family

For several polypeptide-encoding genes, defective copies containing intronic sequence have been found elsewhere in the genome. Two illustrative examples are:

- *The NF1 (neurofibromatosis type 1) gene.* This gene is located close to the chromosome 17 centromere at 17q11.2 and has at least 11 nonprocessed pseudogene or gene fragment copies, nine of which are located at pericentromeric regions on seven different chromo-

---

Box 7.3

Pseudogenes and gene fragments

The processes that give rise to gene families often result in the formation of nonfunctional copies of a gene or a fragment of a gene, either a **pseudogene** (a nonfunctional copy of most or all of a gene, or at least its coding DNA), or truncated genes and gene fragments (nonfunctional copies of a segment of the gene). Various different classes of such sequences exist.

- **Nonprocessed (conventional) pseudogenes**. These are nonfunctional copies of the *genomic* DNA sequence of a gene, and so contain sequences corresponding to exons and introns, and often to flanking sequences. Nonprocessed pseudogenes can often be recognized by the presence of inappropriate termination codons in sequences corresponding to exons of functional gene homologs. Such pseudogenes are common in clustered gene families and are thought to arise by tandem gene duplication events (see *Figure 14.2*).

- **Expressed nonprocessed pseudogene**. Nonprocessed pseudogenes originate from gene duplication events. Immediately after a gene duplication event, both gene copies will be functional, but thereafter one copy may pick up deleterious mutations and eventually lose its original function and even the capacity to be expressed (see *Figure 14.2*). Early in this process, therefore, there may be a transition stage where the gene is no longer functional but continues to be expressed at the RNA level, and possibly in some cases even at the polypeptide level. For example, the θ-globin gene (*HBQ1*, see *Figure 7.11*) is known to be expressed, but there is no evidence that a θ-globin polypeptide becomes incorporated in a functional hemoglobin (Clegg, 1987). The human chorionic somatomammotropin-like gene *CS-L* may be another example.

- **Processed pseudogenes**. Processed pseudogenes are nonfunctional copies of the *exonic sequences* of an

active gene and are often found in interspersed gene families. They usually contain at one end an oligo(dA)/(dT) sequence and are thought to arise by integration into chromosomes of a natural complementary DNA sequence generated by a reverse transcriptase from an RNA transcript (see *Figure 7.13*). Processed pseudogenes derived from genes transcribed by RNA polymerase II (including all genes encoding polypeptides are normally incapable of expression because they lack promoter sequences. Processed pseudogenes derived from genes transcribed by RNA polymerase III, such as the *Alu* repeat, can reach very high copy numbers (see *Table 7.12*).

- **Expressed processed pseudogenes**. A processed pseudogene may originate by integrating into a chromosomal DNA site which just happens, by chance, to be adjacent to a promoter which can drive expression of the processed gene copy. In some cases, selection pressure may insure that the expressed processed gene copy continues to be expressed. A variety of such processed genes are known to have testis-specific expression patterns (see the example of the pyruvate dehydrogenase gene, *PDHA2*, Section 14.3.3).

  Another class of expressed processed pseudogene results from processed copies of genes transcribed by RNA polymerase III, which are unusual because the promoter is internal to the gene (see *Figure 8.3*). As the processed copy in this case contains promoter elements, it may be expressed (see the example of expressed *Alu* sequences, Section 7.4.5).

- **Truncated genes and gene fragments**. Sometimes genomic sequences closely resemble a small component of a functional gene [e.g. a 5′ fragment or a 3′ fragment (truncated genes), or a very small segment of the gene, even a single exon from a multiexon gene (gene fragment)]. These sequences are often found in clustered gene families (see *Figure 7.12*) and are likely to have originated by unequal crossover or unequal sister chromatid exchanges.

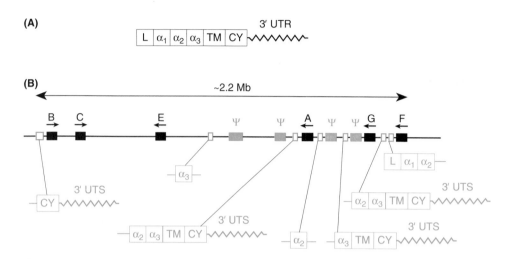

Figure 7.12: Clustered gene families often contain nonprocessed pseudogenes and truncated genes or gene fragments: example of the class I HLA gene family.

(**A**) Structure of a class I HLA heavy chain mRNA. The full-length mRNA contains a polypeptide-encoding sequence. Blocks represent different domains as follows: L, leader sequence; α_1, α_2, α_3 extracellular domains; TM, transmembrane sequence; CY, cytoplasmic tail and a 3'-untranslated sequence (3'-UTR). The three extracellular domains α_1–α_3 are each encoded essentially by a single exon. The very small 5'-UTR is not shown. (**B**) The class I HLA heavy chain gene cluster. The cluster is located at 6p21.3 and comprises about 20 genes. They include six expressed genes (black), four full-length nonprocessed pseudogenes (filled blue blocks) and a variety of nonfunctional truncated genes or gene fragments (open blue blocks). Some of the latter are truncated at the 5' end (e.g. the one next to *HLA-B*), some are truncated at the 3' end (e.g. the one next to *HLA-F*) and some contain single exons (e.g. the one next to *HLA-E*).

somes (Regnier *et al.*, 1997). Characterization of the chromosome 15-specific *NF1* gene copies revealed that they contain copies of a segment of the *NF1* gene spanning exons 8 and 27 and including the intron sequences. This type of gene duplication may be a result of what has been called pericentromeric plasticity.

■ *The* PKD1 *(adult polycystic kidney disease) gene.* This gene consists of 46 exons spanning 50 kb at 16p13.3. A truncated 5' gene copy comprising approximately 70% of the gene and encompassing exons 1 to 34 and all the intervening introns has been faithfully replicated at least three times and inserted into a more proximal location at 16p13.1 (the European Polycystic Kidney Disease Consortium, 1994).

An apparently related case concerns the *CFTR* (cystic fibrosis) gene at 7q31.2. which has 24 exons spanning 250 kb. Copies of a ~30 kb sequence encompassing exon 9 and flanking intron sequences are distributed at a large number of chromosomal locations. However, in this case a likely explanation proposed by Rozmahel *et al.* (1997) is a retrotransposition model. Transcription from the sense strand of the *CFTR* gene was envisaged to generate an antisense transcript encompassing the exon 9 sequence and flanking intron regions. The transcript could then have been converted to cDNA by cellular reverse transcriptase prior to integration at several sites in the genome.

Processed pseudogenes in an interspersed polypeptide-encoding gene family

Interspersed gene families frequently have several defective gene copies which have been copied *at the cDNA level* (processed pseudogenes; see *Table 7.9B* for examples*). This form of retrotransposition is carried out by cellular reverse transcriptases which transcribe mRNA into natural cDNA which can then integrate into chromosomal DNA at sites of temporary breakage (see *Figure 7.13* for one possible mechanism).

Although processed pseudogenes are typically not expressed, several examples are known where natural cDNA copies of a gene appear to be expressed, often in a testis-specific fashion (see Section 14.3.3). Often the functional intron-containing gene locus in such families is located on the X chromosome (see Section 14.3.3).

Processed pseudogenes in an RNA-encoding gene family

Although the size of some interspersed polypeptide-encoding gene families testifies to the success of retrotransposition as a mechanism for generating processed gene copies, the really successful (in terms of high copy number) retrotranspositions have been performed from RNA polymerase III transcripts. For example, the Alu repeat family (see Section 7.4.5) is considered to have arisen as processed pseudogenes copied from the 7SL RNA gene. Genes such as this which are transcribed by RNA polymerase III often contain an internal promoter

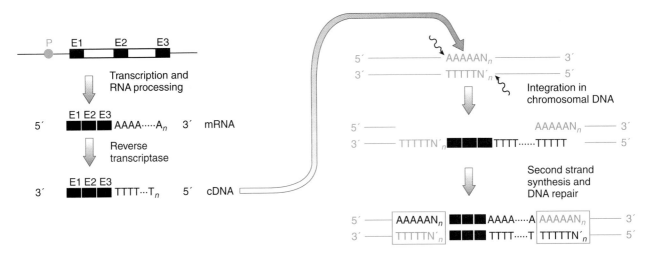

Figure 7.13: Processed pseudogenes originate by reverse transcription from RNA transcripts.

The reverse transcriptase function could be provided by LINE-1 (*Kpn*) repeats (see *Figure 7.18*). The model for integration shown in the figure is only one of several possibilities (see Vanin, 1985). This model envisages integration at staggered breaks (indicated by curly arrows) in A-rich sequences. If the A-rich sequence is included in a 5′ overhang, it could form a hybrid with the distal end of the poly(T) of the cDNA, facilitating second strand synthesis. Because of the staggered breaks during integration, the inserted sequence will be flanked by short direct repeats (boxed sequences). E1–E3 represent exons. P = promoter (shown here as for genes transcribed by RNA polymerase II, but note that genes transcribed by RNA polymerase III often contain internal promoters, see *Figure 8.3*). Some processed gene copies may possibly be functional, such as in the case of some autosomal processed copies of X-linked genes which conserve an important biological function during spermatogenesis (see Section 14.3.3).

which facilitates the expression of newly transposed copies (see Section 8.2.1 and *Figure 8.3*).

7.3.6 The coding sequences of many human genes contain repeated sequence motifs

In addition to the types of gene sequence duplication discussed in the previous sections, many human genes, like other eukaryotic genes, contain intragenic repeated sequences. Repeated sequences in coding DNA may involve different forms of repeated structure, including comparatively large sequences encoding protein domains or small sequence motifs. Different modes of repetition can be seen but tandem repetition is rather common.

■ Tandem repetition of **microsatellite sequences** (short sequence motifs – see Section 7.4.3) is common and may simply reflect statistical expected frequencies for certain base compositions. For example, certain genes which have a very high % GC often have long runs of single cytosines or guanines in the coding sequence. Such sequences are comparatively unstable and are prone to single nucleotide deletion or insertion events which are often pathogenic. A variety of genes are known to have long tracts of certain trinucleotides in their coding sequences and runs of the CAG trinucleotides in the coding sequences of several genes have been found to be

capable of expansion causing pathogenesis (see *Box 16.7*).

■ Tandem repetition of sequences encoding known or assumed protein domains is quite common, and may be functionally advantageous in some cases by providing a more available biological target. In some cases, the sequence homology between the repeats can be very high; in other cases it may be rather low (see *Table 7.10*).

7.4 Extragenic repeated DNA sequences and transposable elements

The human nuclear genome, like that of other complex eukaryotes, contains a large amount of highly repeated DNA sequence families which are largely transcriptionally inactive. A wide variety of different repeats are known and dedicated repeat sequence databases have been established (see electronic reference 3). Like multigene families, noncoding repetitive DNA shows two major types of organization: tandemly repeated and interspersed.

Tandemly repeated noncoding DNA

Such families are defined by **blocks** (or arrays) of tandemly repeated DNA sequences. Individual arrays can occur at a few or many different chromosomal locations. Depending on the average size of the arrays of

Table 7.10: Examples of intragenic repetitive coding DNA (see also Box 16.7)

Gene product	Size of encoded repeat in amino acids	No. of copies	Nucleotide sequence homology between copies
Ubiquitin (*UbB* and *UbC* genes)	76	3 (UbB) 9 (UbC)	High homology
Involucrin	10	59	High homology for central 39 repeats
Apolipoprotein (a)	114	37	High homology; 24 of the repeats are identical in sequence
Plasminogen	~75–80	5	Low homology but conserved protein domains
Collagen	18	57	Low homology but conserved amino acid motifs based on $(Gly–X–Y)_6$
Serum albumin	195	3	Low homology
Proline-rich protein genes	16–21	5	Low homology
Tropomyosin α-chain	42	7	Low homology
Immunoglobulin ε-chain, C region	108	4	Low homology
Dystrophin	109	24	Low homology

repeat units, highly repetitive noncoding DNA belonging to this class can be grouped into three subclasses: **satellite**, **minisatellite** and **microsatellite DNA**.

In addition to these three major subclasses (which are detailed in the following three sections), a fourth class has recently been recognized and described as megasatellite or macrosatellite DNA. Despite the name, this type of DNA is characterized by array lengths which can be comparatively modest compared to some satellite DNA arrays. Instead, the prefix mega- has been used to emphasize the large size of the repeating unit which can be several kilobases long. This class is exemplified by the RS447 megasatellite which consists of about 60 tandem copies of a novel 4.7 kb repeat on 4p15, plus another array of several copies on distal 8p (Gondo *et al.*, 1998). Array lengths can be highly polymorphic and the RS447 repeat is reported to contain a putative open reading frame of 1590 bp (Gondo *et al.*, 1998). See *Table 7.11* for some other examples.

Interspersed repetitive noncoding DNA
The individual repeat units are not clustered, but are dispersed at numerous locations in the genome, and together account for perhaps one third of the DNA in the human genome (Smit, 1996). Most of the DNA families belonging to this class contain some members that are capable of undergoing retrotransposition (i.e. transposition through an RNA intermediate).

The chromosomal locations of different types of tandemly repeated DNA can show a very restricted or highly dispersed pattern, whereas different classes of interspersed repeat DNA can show preferential location within different types of chromosome bands (see below and *Figure 7.14*).

7.4.1 Satellite DNA is composed of very long arrays of tandem repeats which can be separated from bulk DNA by buoyant density gradient centrifugation

Human satellite DNA is comprised of very large arrays of tandemly repeated DNA with the repeat unit being a simple or moderately complex sequence (*Table 7.11*; see Singer, 1982a). Repeated DNA of this type is not transcribed and accounts for the bulk of the heterochromatic regions of the genome, being notably found in the vicinity of the centromeres (pericentromeric heterochromatin). The base composition, and therefore density, of such DNA regions is dictated by the base composition of their constituent short repeat units and may diverge substantially from the overall base composition of bulk cellular DNA.

Isolation by buoyant density gradient centrifugation
When DNA is isolated from human cells by conventional methods, it is subject to mechanical shearing. Fragments are generated from the bulk DNA (with a base composition of ~42% GC) and fragments from the satellite DNA regions which may have a similar or different base composition. If the base composition is significantly different, satellite DNA sequences can be separated from the bulk DNA by buoyant density gradient centrifugation. Following centrifugation, they appear as minor (or **satellite**) bands of different buoyant density from a major band which represents bulk DNA. Typically, human DNA is complexed with Ag^+ ions and then fractionated in buoyant density gradients containing cesium sulfate, whereupon three satellite bands are identified at

Table 7.11: Major classes of tandemly repeated human DNA

Class	Size of repeat	Major chromosomal location(s)
'Megasatellite' DNA (blocks of hundreds of kb in some cases)	several kb	Various locations on selected chromosomes
RS447	4.7 kb	~50–70 copies on 4p15 plus several copies on distal 8p
untitled	2.5 kb	~400 copies on 4q31 and 19q13
untitled	3.0 kb	~50 copies on the X chromosome
Satellite DNA (blocks often from 100 kb to several Mb in length)	5–171 bp	Especially at centromeres
α (alphoid DNA)	171 bp	Centromeric heterochromatin of all chromosomes
β (*Sau3* A family)	68 bp	Centromeric heterochromatin of 1, 9, 13, 14, 15, 21, 22 and Y
Satellite 1 (AT-rich)	25–48 bp	Centromeric heterochromatin of most chromosomes and other heterochromatic regions
Satellites 2 and 3	5 bp	Most, possibly all, chromosomes
Minisatellite DNA (blocks often within the 0.1–20 kb range)	6–64 bp	At or close to telomeres of all chromosomes
telomeric family	6 bp	All telomeres
hypervariable family	9–64 bp	All chromosomes, often near telomeres
Microsatellite DNA (blocks often less than 150 bp)	1–4 bp	Dispersed throughout all chromosomes

different densities: satellite $1 - 1.687$ gcm^{-3}; satellite $2 - 1.693$ gcm^{-3}; satellite $3 - 1.697$ gcm^{-3}.

Each of these satellite classes includes a number of different tandemly repeated DNA sequence families (satellite subfamilies), some of which are shared between different classes. DNA sequence analysis has revealed that some of the repetitive DNA families in the satellites are based on very simple repeat units. For example, both satellite 2 and

satellite 3 contain sequence arrays which are based on tandem repetition of the sequence ATTCC. Additionally, restriction mapping has revealed satellite subfamilies which show additional higher order repeat units superimposed on the small basic repeat units. Such subfamilies are thought to arise as a result of subsequent amplification of a unit which is larger than the initial basic repeat unit and contains some diverged units (*Figure 7.15*).

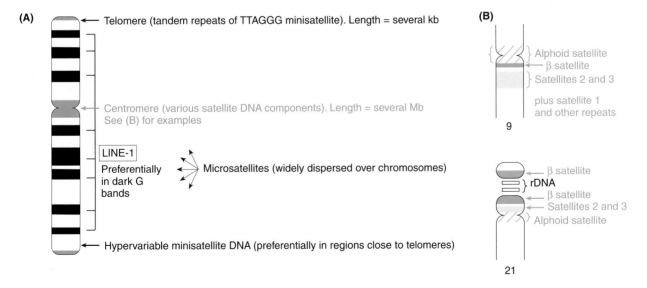

Figure 7.14: Chromosomal location of major repetitive DNA classes.

(**A**) General overview. *Note* the restricted locations of certain types of tandemly repeated DNA, such as satellite DNAs which are found in heterochromatin (notably at the centromeres) and minisatellite DNAs which are often found at telomeres or close to them. (**B**) Satellite DNA organization at centromeres. The locations of different classes of satellite DNA are shown for chromosome 9 and for chromosome 21 (one of the five examples of an autosomal acrocentric chromosome). The illustration in this case is redrawn from Tyler-Smith and Willard (1993) *Curr. Opin. Genet. Dev.*, **3**, 390–397, with permission from Current Biology Ltd.

Alphoid DNA and centromeric heterochromatin

Other types of satellite DNA sequence cannot easily be resolved by density gradient centrifugation. They were first identified by digestion of genomic DNA with a restriction endonuclease which typically has a single recognition site in the basic repeat unit. In addition to the basic repeat unit size (monomer), such enzymes will produce a characteristic pattern of multimers of the unit length because of occasional random loss of the restriction site in some of the repeats (Singer, 1982a). Alpha satellite (or alphoid DNA) constitutes the bulk of the centromeric heterochromatin and accounts for about 3–5% of the DNA of each chromosome. It is characterized by tandem repeats of a basic mean length of 171 bp, although higher order units are also seen. The sequence divergence between individual members of the alphoid DNA family can be so high that it is possible to isolate chromosome-specific subfamilies for each of the human chromosomes (Choo *et al.*, 1991).

The function of satellite DNA remains unclear (see Csink and Henikoff, 1998). The centromeric DNA of human chromosomes largely consists of various families of satellite DNA (see *Figure 7.14B*). Of these, only the α-satellite is known to be present on all chromosomes, and its repeat units often contain a binding site for a specific centromere protein, CENP-B. Recently cloned α-satellite arrays have been shown to seed *de novo* centromeres in human cells, indicating that α-satellite plays an important role in centromere function (Grimes and Cook, 1998).

7.4.2 Minisatellite DNA is composed of moderately sized arrays of tandem repeats and is often located at or close to telomeres

Minisatellite DNA comprises a collection of moderately sized arrays of tandemly repeated DNA sequences which are dispersed over considerable portions of the nuclear genome (*Table 7.11*). Like satellite DNA sequences, they are not normally transcribed (but see below).

Hypervariable minisatellite DNA

Hypervariable minisatellite DNA sequences are highly polymorphic and are organized in over 1000 arrays (0.1–20 kb long) of short tandem repeats (Jeffreys, 1987). The repeat units in different hypervariable arrays vary considerably in size, but share a common core sequence, GGGCAGGAXG (where X = any nucleotide), which is similar in size and in G content to the *chi* sequence, a signal for generalized recombination in *E. coli*. While many of the arrays are found near the telomeres, several hypervariable minisatellite DNA sequences occur at other chromosomal locations. The great majority of hypervariable minisatellite DNA sequences are not transcribed, except for elements occuring within noncoding intragenic sequences. Some, however, are expressed. For example, the *MUC1* locus on 1q is an expressed hypervariable minisatellite locus. It encodes a glycoprotein found in several epithelial tissues and body fluids which is highly polymorphic as a result of extensive variation in the number of minisatellite-encoded repeats (Swallow *et al.*, 1987).

The significance of hypervariable minisatellite DNA is not clear, although it has been reported to be a 'hotspot' for homologous recombination in human cells (Wahls *et al.*, 1990). Nevertheless it has found many applications. Various individual loci have been characterized and used as genetic markers, although the preferential localization in subtelomeric regions has limited their use for genome-wide linkage studies. A major application has been in **DNA fingerprinting**, in which a single DNA probe which contains the common core sequence can

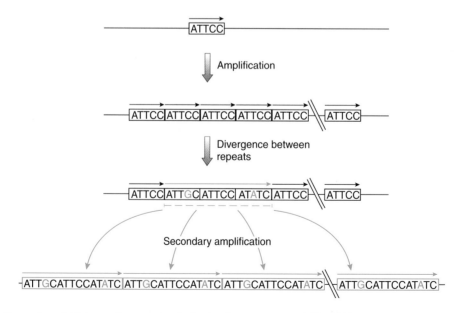

Figure 7.15: Formation of higher order repeat units in simple sequence satellite DNA.

Secondary amplification is envisaged in this case to involve a 15 bp repeat unit comprising three diverged 5 bp repeats.

hybridize simultaneously to multiple minisatellite DNA loci on all chromosomes, resulting in a complex individual-specific hybridization pattern (see Section 17.4.1).

Telomeric DNA

Another major family of minisatellite DNA sequences is found at the termini of chromosomes, the telomeres. The principal constituent of telomeric DNA is 10–15 kb of tandem hexanucleotide repeat units, especially TTAGGG, which are added by a specialized enzyme, telomerase (see *Figure 2.9*). By acting as buffers to protect the ends of the chromosomes from degradation and loss and by providing a mechanism for replicating the ends of the linear DNA of chromosomes, these simple repeats are directly responsible for telomere function (see *Figure 2.9* and Section 2.3.4).

7.4.3 Microsatellite DNA is defined by the presence of short arrays of tandem simple repeat units and is dispersed throughout the human genome

Microsatellite DNA families include small arrays of tandem repeats which are simple in sequence (often 1–4 bp) and are interspersed throughout the genome. Of the mononucleotide repeats, runs of A and of T are very common (see *Figure 7.19*) and together account for about 10 Mb, or 0.3% of the nuclear genome. By contrast, runs of G and of C are very much rarer. In the case of dinucleotide repeats, arrays of CA repeats (TG repeats on the complementary strand) are very common, accounting for 0.5% of the genome, and are often highly polymorphic (see Section 11.2.3). CT/AG repeats are also common, occurring on average once every 50 kb and accounting for 0.2% of the genome, but CG/GC repeats are very rare. This is so because C residues which are flanked at their 3′ end by a G residue (i.e. CpG) are prone to methylation and subsequent deamination, resulting in TpG (or CpA on the opposite strand, see *Box 8.5*). Trinucleotide and tetranucleotide tandem repeats are comparatively rare, but are often highly polymorphic and increasingly have been investigated to develop highly polymorphic markers.

The significance of microsatellite DNA is not known. Alternating purine–pyrimidine repeats, such as tandem repeats of the dinucleotide pair CA/TG, are capable of adopting an altered DNA conformation, Z-DNA, *in vitro*, but there is little evidence that they do so in the cell. Although microsatellite DNA has generally been identified in intergenic DNA or within the introns of genes, a few examples have been recorded within the coding sequences of genes. Tandem repeats of three nucleotides in coding DNA may be sites that are prone to pathogenic expansions (see Section 9.5.2 and *Box 16.7*).

7.4.4 Highly repeated interspersed DNA families contain a small percentage of actively transposing DNA elements

Two major classes of mammalian interspersed repetitive DNA families have been discerned on the basis of repeat unit length (Singer, 1982b): SINEs and LINEs.

SINEs (short interspersed nuclear elements)

The most conspicuous human SINE is the *Alu* repeat family (so called because of early attempts at characterizing the sequence using the restriction nuclease *Alu*I). The *Alu* repeat contains an internal RNA polymerase III promoter sequence. It has attained a very high copy number in the human genome (*Table 7.12*) and appears to have originated by retrotransposition from the 7SL RNA gene (see Section 7.4.5). The *Alu* repeat is primate-specific but other mammals have similar types of sequence derived from the 7SL RNA gene such as the B1 family in mouse.

Unlike the *Alu* repeat, another major human SINE family is not restricted to primates, with copies being found in marsupials and monotremes. In accordance with its distribution this family has been termed the MIR (mammalian-wide interspersed repeat) family (see *Table 7.12*).

LINEs (long interspersed nuclear elements)

Human LINEs are exemplified by the *LINE-1* or *L1* element (also called the *Kpn* repeat because of early attempts at characterizing this family using the restriction nuclease *Kpn*I; see Section 7.4.6). The LINE-1 element is also found in other mammals such as the mouse.

In addition to the human *Alu* and LINE-1 repeat families, there are many smaller families, including the THE-1 (transposable human element family), many MER (medium reiteration frequency) families and families of human endogenous retroviruses (HERV) or retrovirus-like elements (RTLV) – see Lower *et al.*, 1996 and *Table 7.12*.

Some members of the interspersed repeat families have been considered as transposable elements, unstable DNA elements which can migrate to different regions of the genome (*Figure 7.16*). Rare examples are known of human DNA sequences which appear to have been copied by a DNA-mediated transposition event (**transposons**) and a variety of sequences have been found in the human genome which resemble known DNA transposons such as the *mariner* transposon and others (Smit and Riggs, 1996). However, the latter are not thought to be actively transposing. Instead, the great majority of human transposable elements undergo retrotransposition, that is their RNA transcripts can be converted within the cell to a complementary DNA form which can reinsert back into chromosomal DNA at a variety of different locations (see *Figure 7.13*). Three classes of mammalian sequence are known to be able to transpose through an RNA intermediate (see *Box 7.4*). In each case, the transposition event involves duplication of a very short sequence at the target site, causing the transposed sequence to be flanked by short repeats (see *Figures 7.13* and *7.17*).

7.4.5 The Alu repeat occurs about once every 3 kb in the human genome and includes examples that are transcribed

The *Alu* repeat is the most abundant sequence in the human genome, with a copy number of about 1 000 000 (see Deininger, 1989; Smit, 1996). *Alu* repeats have a

Table 7.12: Major classes and families of interspersed human repetitive DNA (adapted from Smit, 1996)

Class[a]	Family[a]	Size of repeat unit	No. of copies	Percentage of genome
SINE	*Alu* family	Full length ~0.3 kb	~1 000 000	~ 7.0%
	MIR families	Average size ~0.13 kb	~400 000	~1.7%
LINE	LINE-1 (Kpn) family	Full length is 6.1 kb, but average size ~0.8 kb	~200 000–500 000	~5–12%
	LINE-2 family	Average size ~0.25 kb	~270 000	~2.1%
LTR	HERV	Average size ~1.3 kb	~50 000	~1.3%
	Others	Average size ~0.5 kb	~200 000	~3.3%
DNA transposon	Mariner & other families	Varies; perhaps average size = 0.25 kb	~200 000	~1.6%
Others	Various	Perhaps average size of about 0.4 kb	~60 000	~0.8%

[a]See text.

relatively high GC content and, although dispersed mainly throughout the euchromatic regions of the genome, have been reported to be preferentially located in R chromosome bands (Korenberg and Rykowski, 1988). The latter correspond to the pale bands seen when using standard Giemsa staining (see *Box 2.4*) and represent the most transcriptionally active regions of the genome.

The full-length *Alu* repeat is about 280 bp long and is usually flanked by short (often 6–18 bp) *direct repeats* (i.e. the repeats are in the same orientation). The typical *Alu* sequence is a tandemly repeated dimer, with the repeats sharing an approximately 120 bp sequence followed by a short sequence which is rich in A residues on one strand and T residues on the complementary strand.

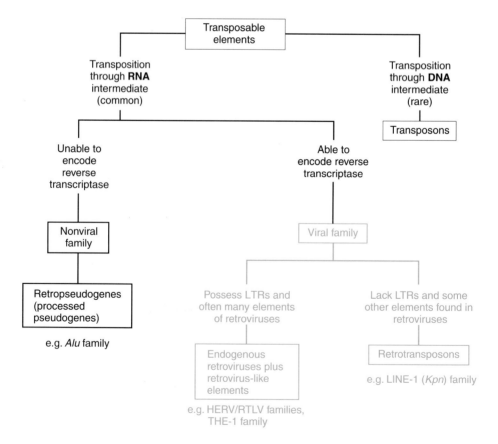

Figure 7.16: Human transposable elements.

Only a small proportion of members of any of the above families may be capable of transposing; many have lost such capacity by acquiring inactivating mutations and many are short truncated copies. See *Figures 7.17* and *7.18* for the typical structures of some human transposable elements. *Note* that retropseudogenes can transpose by using cellular reverse transcriptase provided by other sequences, e.g. LINE-1 elements.

Classes of mammalian sequence which undergo transposition through an RNA intermediate

Endogenous retroviruses are sequences which resemble retroviruses but which cannot infect new cells and are therefore restricted to one genome. They include sequences which have the flanking *long terminal repeats (LTRs)* of retroviruses and other elements found in fully functional retroviruses including sequences which can encode a functional reverse transcriptase. Truncated and degraded retrovirus-like elements (RTLVs) are comparatively common, and can include solitary LTRs.

Retrotransposons (sometimes abbreviated as **retroposons**) lack LTRs and often other elements of retroviruses (e.g. the *env* gene of retroviruses which encodes coat proteins). They encode a reverse transcriptase and so are capable of independent transposition. They contain an $(A)_n/(T)_n$ sequence at one end. Examples include the human and mouse LINE-1 elements.

Processed pseudogenes (retropseudogenes) lack a reverse transcriptase and so are incapable of independent transposition, relying instead on the reverse transcriptase activity of other elements. They contain an $(A)_n/(T)_n$ sequence at one end. This class includes two groups:

- low copy number processed pseudogenes derived from genes transcribed by RNA polymerase II (i.e. polypeptide-encoding genes). *Note* that some rare examples of processed copies of genes may possibly be functional (see Section 14.3.3);
- high copy number mammalian SINEs, such as the human *Alu* and the mouse B1 repeat families, are derived from genes transcribed by RNA polymerase III using an internal promoter (see Section 8.2.1 and *Figure 8.3*).

However, there is asymmetry between the tandem repeats: one repeat unit contains an internal 32-bp sequence lacking in the other (*Figure 7.18*). Monomers, containing only one of the two tandem repeats, and various truncated versions of dimers and monomers are also common.

The two repeated units of the *Alu* sequence show a striking resemblance to the sequence for 7SL RNA, a component of the signal recognition particle, which facilitates transport of proteins across the membrane of the endoplasmic reticulum. Because of this and the observation of the A-rich regions and the flanking direct repeats, it has been widely assumed that the *Alu* sequence has been propagated by retrotransposition from 7SL RNA, and therefore represents a processed 7SL RNA pseudogene (*Figure 14.25*). Certainly, transposition by *Alu* sequences is known to occur [presumably as a result of *trans*-acting cellular reverse transcriptases such as those encoded by LINE-1 (*Kpn*) elements (see below), and may occasionally cause clinical problems (Section 9.5.6)]. Possibly, the very high copy number achieved by this processed pseudogene is related to the presence of a promoter sequence in the 7SL RNA sequence (the 7SL RNA gene, like tRNA genes is transcribed by RNA polymerase III from an internal promoter, see *Section 8.2.1*). By contrast, processed pseudogenes from RNA polymerase II transcripts lack promoter sequences and their only chance of expression is if the integration event places them next to a functional promoter sequence.

Currently, the function of the *Alu* sequence, if any, is unknown, although several roles have been considered (see Schmid, 1998). Although the average expected frequency is one copy per 3 kb, high density clustering of *Alu* repeats is known to occur in certain regions. Because of their ubiquity, *Alu* sequences have been considered to promote unequal recombination, a mechanism which while occasionally causing disease, may be evolutionarily advantageous in promoting gene duplication (Section 9.3.2). Although conspicuously absent from coding

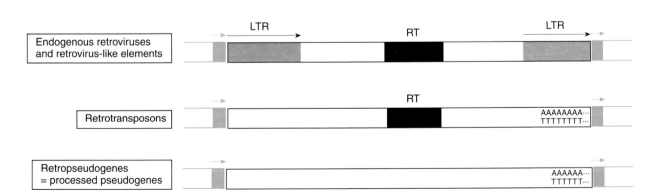

Figure 7.17: Classes of mammalian transposable elements which undergo transposition through an RNA intermediate.

Blue blocks flanking the sequences represent short repeats of a sequence originally present at the target site that was duplicated during the integration process (see *Figure 7.13*). Abbreviations: LTR, long terminal repeat; RT, reverse transcriptase.

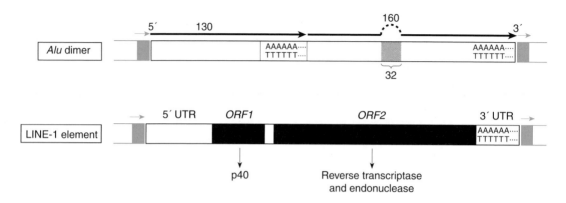

Figure 7.18: Structures of full-length Alu and LINE-1 repeats.

The consensus standard *Alu* dimer is shown with two similar repeats terminating in an $(A)_n$ /$(T)_n$ like sequence. They have different sizes because of the insertion of a 32 bp element within the larger repeat. *Alu* monomers also exist in the human genome, as do various truncated copies of both monomers and dimers. The consensus full-length LINE-1 element is 6.1 kb long but most LINE-1 elements are truncated and the average size is very much smaller. ORF1, ORF2, open reading frames 1 and 2.

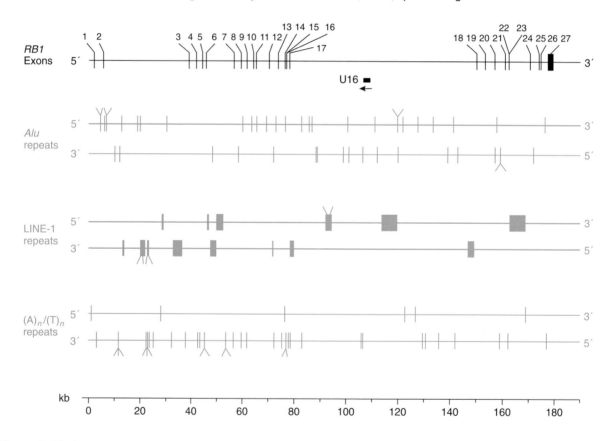

Figure 7.19: Location of Alu, LINE-1 and $(A)_n$ /$(T)_n$ repeats within the human retinoblastoma susceptibility gene.

The entire sequence of the 180 kb human retinoblastoma susceptibility gene, *RB1*, has been determined, enabling identification within the gene of many examples of abundant DNA repeats. Note that the 72 kb intron 17 contains a G protein-coupled receptor gene, *U16*, which is actively transcribed in the opposite direction to the *RB1* gene. The top line ($5' \rightarrow 3'$) of each pair shows the repeat elements orientated in the sense direction; the bottom line ($3' \rightarrow 5'$) shows them in the antisense orientation. There are 46 *Alu* repeats corresponding to the expected frequency of one per 4 kb. There is a particularly high frequency of LINE-1 elements in this gene (one per 11 kb), but only two of the elements approach the full 6.1 kb length. The $(A)_n$ /$(T)_n$ sequences ($n = 12$ or greater) which are indicated are only those not found within the sequences of the interspersed repeats. No examples were found for $(C)_n$ /$(G)_n$ for $n = 12$ or greater. The frequencies of dinucleotide repeats with $n = 6$ or greater were as follows: $(CG)_n$/$(GC)_n$ 0; $(AT)_n$ 1; $(CT)_n$ /$(AG)_n$ 2; $(CA)_n$/$(TG)_n$ 4 (three of which are polymorphic). Adapted from Toguchida *et al.* (1993) *Genomics*, **17**, 535–543, with permission from Academic Press Inc.

sequences, *Alu* sequences are often found in noncoding intragenic locations, notably in introns and occasionally in untranslated sequences (see *Figure 7.19*). Consequently, they are often represented in the primary transcript RNA from genes encoding polypeptides, and occasionally in mRNA. The *Alu* sequence can also be transcribed from its internal promoter by RNA polymerase III *in vitro*, and *in vivo* transcription of some *Alu* sequences can result in the accumulation of a small cytoplasmic RNA which can be specifically bound by two cytoplasmic signal recognition particle proteins, SRP9 and SRP14 (Chang *et al.*, 1994).

7.4.6 The LINE-1 (Kpn repeat) includes examples that appear to encode a reverse transcriptase and are actively transposing

The human LINE-1 (L1, *Kpn* repeat) family is expected to consist of over 100 000 and perhaps as many as 400 000–500 000 interspersed repeats (Smit, 1996). Of these, multiple members are known to be actively transposing (Sassaman *et al.*, 1997). The full-length consensus element is 6.1 kb long and has two **open reading frames (ORFs)**. ORF1 is located close to one end (conventionally known as the 5′ end) of the consensus element and encodes a protein of unknown function, p40 (so called because its molecular weight is ~40 kDa). ORF2 encodes a protein with an endonuclease domain and also a reverse transcriptase domain (see Sassaman *et al.*, 1997; Kazazian and Moran, 1998). The full-length consensus element contains an internal promoter within a region of untranslated DNA preceding ORF1 (conventionally called the 5′-UTR) while at the other end there is an $(A)_n/(T)_n$ sequence, often described as the 3′ poly(A) tail. As in the case of other elements that transpose, the LINE-1 elements are flanked by short duplicated repeats (*Figure 7.18*).

The full-length LINE-1 element is comparatively rare and most repeats are truncated at the 5′ end, resulting in a population that is heterogeneous in length but sharing a common 3′ end with the poly(A) tail. LINE-1 elements are primarily located in euchromatic regions but show an inverse relationship with *Alu* repeats by appearing to be located preferentially in the dark G bands (Giemsa positive) of metaphase chromosomes. Like the *Alu* repeats, they are conspicuously absent from coding sequences but may be found in intragenic noncoding sequences (*Figure 7.19*). As a result, they may be represented in the primary RNA transcript of large genes, but they are conspicuously absent from coding sequences.

Further reading

Gardiner K (1995) Human genome organization. *Curr. Opin. Genet. Dev.*, **5**, 315–322.
Schmid C (1996) Alu: structure, origin, evolution, significance and function of one-tenth of human DNA. *Prog. Nucl. Acid Res. Mol. Biol.*, **53**, 283–319.

Electronic References (e-Refs)

1. The Genome Monitoring Table at http://www.ebi.ac.uk/~sterk/genome-MOT
2. The Histone Sequence Database at http://genome.nhgri.nih.gov/histones/
3. The Repeat Sequence Database at http://www.girinst.org/

References Q15

Albig W, Doenecke D (1997) The human histone gene cluster at the D6S105 locus. *Hum. Genet.*, **101**, 284–294.
Anderson S *et al.* (1981) Sequence and organization of the human mitochondrial genome. *Nature*, **290**, 457–465.
Antequara F, Bird A (1993) Number of CpG islands and genes in human and mouse. *Proc. Natl Acad. Sci. USA*, **90**, 11995–11999.
Bird A (1986) CpG islands and the function of DNA methylation. *Nature*, **321**, 209–213.
Chang D-Y, Nelson B, Bilyeu T, Hsu K, Darlington GJ, Maraia RJ (1994) A human Alu-RNA binding protein whose expression is associated with accumulation of small cytoplasmic Alu RNA. *Mol. Cell. Biol.*, **14**, 3949–3959.
Choo KH, Vissel B, Nagy A, Earle E, Kalitsis P (1991) A survey of the genomic distribution of alpha satellite DNA on all the human chromosomes, and derivation of a new consensus sequence. *Nucleic Acids Res.*, **19**, 1179–1182.
Christiano A, Hoffman GG, Chung-Honet LC, Lee S, Cheng W, Uitto J, Greenspan DS (1994) Structural organization of the human type VII collagen gene (*COL7A1*), composed of more exons than any previously characterized gene. *Genomics*, **21**, 169–179.

Clayton DA (1992) Transcription and replication of animal mitochondrial DNAs. *Int. Rev. Cytol.*, **141**, 217–232.

Clegg JB (1987) Can the product of the theta gene be a real globin? *Nature*, **329**, 465–467.

Craig JM, Bickmore WA (1994) The distribution of CpG islands in mammalian chromosomes. *Nature Genet.*, **7**, 376–381.

Csink AK, Henikoff S (1998) Something from nothing: the evolution and utility of satellite repeats. *Trends Genet.*, **14**, 200–204.

Deininger P (1989) In: *Mobile DNA* (DE Berg, MM Howe, eds), pp. 619–636. American Society for Microbiology, Washington, DC.

European Polycystic Kidney Disease Consortium (1994) The polycystic kidney disease 1 gene encodes a 14 kb transcript and lies within a duplicated region on chromosome 16. *Cell*, **77**, 881–894.

Fields C, Adams MD, White O, Venter JC (1994) How many genes in the human genome? *Nature Genet.*, **7**, 345–346.

Geraghty DE, Koller BH, Hansen JA, Orr HT (1992) Examination of four HLA class I pseudogenes. Common events in the evolution of HLA genes and pseudogenes. *J. Immunol.*, **149**, 1934–1936.

Gondo Y, Okada T, Matsuyama N, Saitoh Y, Yanagisawa Y, Ikeda J-E (1998) Human megasatellite DNA RS447: copy-number polymorphisms and interspecies conservation. *Genomics*, **54**, 39–49.

Grimes B, Cooke H (1998) Engineering mammalian chromosomes. *Hum. Mol. Genet.*, **7**, 1635–1640.

Hewitt JE, Lyle R, Clark LN, Valleley EM, Wright TJ, Wijmenga C, van Deutckom JCT, Francis F, Sharpe PT, Hofker M, Frants RR, Williamson R (1994) Analysis of the tandem repeat locus D4Z4 associated with fascioscapulohumeral muscular dystrophy. *Hum Molec Genet*, **3**, 1287–1295.

Jeffreys AJ (1987) Highly variable minisatellites and DNA fingerprints. *Biochem. Soc. Trans.*, **15**, 309–317.

Kazazian HH, Moran JV (1998) The impact of L1 retrotransposons on the human genome. *Nature Genet.*, **19**, 19–24.

Korenberg JR, Rykaowski MC (1988) Human genome organization: Alu, lines, and the molecular structure of metaphase chromosome bands. *Cell*, **53**, 391–400.

Levinson B, Kenwrick S, Gamel P, Fisher K, Gitschier J (1992) Evidence for a third transcript from the human factor VIII gene. *Genomics*, **14**, 585–589.

Lower R, Lower J, Kurth R (1996) The viruses in all of us: characteristics and biological significance of human endogenous retrovirus sequences. *Proc. Natl Acad. Sci. USA*, **93**, 5177–5184.

Neer EJ, Schmidt CJ, Nambudripad R, Smith TF (1994) The ancient regulatory-protein family of WD-repeat proteins. *Nature*, **371**, 297–300.

Rouquier et al. (1998) Distribution of olfactory receptor genes in the human genome. *Nature Genet*, **18**, 243–250.

Rozmahel R, Heng HH, Duncan AM, Shi XM, Rommens JM, Tsui LC (1997) Amplification of CFTR exon 9 sequences to multiple locations in the human genome. *Genomics*, **45**, 554–561.

Sassaman DM et al. (1997) Many human L1 elements are capable of retrotransposition. *Nature Genet.*, **16**, 37–43.

Schmid CW (1998) Does SINE evolution preclude Alu function? *Nucleic Acids Res.*, **26**, 4541–4550.

Schmid SR, Linder P (1992) D-E-A-D protein family of putative RNA helicases. *Mol. Microbiol.*, **6**, 283–292.

Smit AF (1996) The origin of interspersed repeats in the human genome. *Curr. Opin. Genet. Dev.*, **6**, 743–748.

Smit AF, Riggs AD (1996) Tiggers and DNA transposon fossils in the human genome. *Proc. Natl Acad. Sci. USA*, **93**, 1443–1448.

Smith CM, Steitz JA (1997) Sno storm in the nucleolus: new roles for myriad small RNPs. *Cell*, **89**, 669–672.

Swallow DM, Gendler S, Griffiths B, Corney G, Taylor-Papadimitrou J, Bramwell ME (1987) The human tumour-associated epithelial mucins are coded by an expressed hypervariable gene locus PUM. *Nature*, **328**, 82–84.

Tennyson CN, Klamut HJ, Worton RG (1995) The human dystrophin gene requires 16 hours to be transcribed and is co-transcriptionally spliced. *Nature Genet.*, **9**, 184–190.

Toguchida J, McGee TL, Paterson JC, Eagle JR, Tucker S, Yandell DW, Dryda TP (1993) Complete genomic sequence of the human retinoblastoma susceptibility gene. *Genomics*, **17**, 535–543.

Tycowski KT, Shu M-D, Steitz JA (1993) A small nucleolar RNA is processed from an intron of the human gene encoding ribosomal protein S3. *Genes Dev.*, **7**, 1176–1190.

Tyler-Smith C, Willard HF (1993) Mammalian chromosome structure. *Curr. Opin. Genet. Dev.*, **3**, 390–397.

Viskochil D et al. (1991) The gene encoding the oligodendrocyte-myelin glycoprotein is embedded within the neurofibromatosis type 1 gene. *Molec. Cell Biol.*, **11**, 906–912.

Wahls WP, Wallace LJ, Moore PJ (1990) Hypervariable minisatellite DNA is a hotspot for homologous recombination in human cells. *Cell*, **60**, 95–103.

Zhang MQ (1998) Statistical features of human exons and their flanking regions. *Hum. Mol. Genet.*, **7**, 919–932.

Human gene expression

8.1 An overview of gene expression in human cells

The control mechanisms used to regulate human gene expression are fundamentally similar to those found in other mammals, and generally resemble those in eukaryotes in general. Although much more complex than equivalent mechanisms in organisms with small genomes, many of the same basic principles apply and as in other eukaryotes, a major level at which gene expression is controlled is the initiation of transcription. Mammals are particularly complex multicellular organisms and so it is perhaps unsurprising that there are some gene control mechanisms which are not used in bacteria or in some other eukaryotes. Various regulation mechanisms are required to maintain many different facets of mammalian gene expression, both at the spatial and temporal levels (*Box 8.1*). Although simplistic, it is convenient to consider three broad levels at which gene regulation can operate.

Transcriptional regulation of gene expression
We have long been accustomed to the idea that a primary control of gene regulation in eukaryotes occurs at the level of initiation of transcription. Regulation of expression can occur through the core promoter of a gene, at the level of recruitment and processivity of the relevant RNA polymerase. Expression of genes is initiated by the binding of transcription factors to the promoter. Basal levels of transcription can be modulated by binding of protein factors to other regulatory regions occurring in the sequences flanking the gene or sometimes within introns of the gene.

Post-transcriptional regulation of gene expression
This category overlaps with the previous section since it includes mechanisms operating at the level of RNA processing, such as RNA splicing which may more accurately be considered as co-transcriptional rather than post-transcriptional (Steinmetz, 1997). In addition to RNA processing, other levels at which control of gene expression can be exerted include: mRNA transport, translation, mRNA stability, protein processing, protein targeting, protein stability, etc.

A surprising variety of mechanisms are employed at the level of RNA processing with single alleles in an individual often able to generate a variety of different gene products (**isoforms**). The occurrence of these and other mechanisms has required a more flexible definition of the term *gene* than has formerly been used (see below). Several mechanisms are involved in regulating gene expression at the level of translation and an increasing number of regulatory sequences have been identified in the 5' and especially 3' untranslated regions of mRNA. Control of gene expression at the level of protein processing, targeting and stability has been shown in certain systems. For example, activation of some peptide hormones such as insulin requires post-translational cleavage from precursor forms (see *Figure 1.23*).

Epigenetic mechanisms and long range control of gene expression
In addition to genetic factors, additional factors which can be transmitted to progeny cells following cell division *but which are not directly attributable to the DNA sequence* are described as **epigenetic**. DNA methylation is an epigenetic mechanism which plays an important part in mammalian gene control, acting as a general method of maintaining repression of transcription. In addition, a variety of other mechanisms which affect the chromatin environment of a gene and hence its capacity for gene expression are known to operate in mammalian cells. In some cases, the mechanisms ensure that within a cell only one of the two parentally inherited alleles is normally expressed, even although the nucleotide sequence of the allele which is not expressed may be identical to the one which is expressed.

Table 8.1 provides an overview of the different types of mechanism known to be involved in regulating expression of human genes.

Spatial and temporal restriction of gene expression in mammalian cells

The regulation of gene expression in human and mammalian cells is exerted at different levels in order to achieve restricted expression at a variety of spatial and temporal levels.

Spatial restriction of gene expression

Some genes require to be expressed in essentially all types of nucleated cells because they encode a key product that is required to fulfil a general function in all cells, e.g. protein synthesis and energy production. Such genes are often termed **housekeeping genes**. Many human and mammalian genes, however, show much more restricted tissue-specific gene expression patterns. Spatial restriction of gene expression can occur at different levels:

- *Multiple organ/tissue pattern.* Different types of multisystem expression are found. In some cases, a gene may be performing a similar type of role in different organ systems. A variety of genes which play a crucial role in early development may be involved in regulating target genes in several different organ systems, e.g. the sonic hedgehog gene which is expressed in various parts of the developing nervous system, in developing limbs, and elsewhere. In other cases a gene can encode different variants (**isoforms**) in different tissues by using tissue-specific promoters or tissue-specific alternative splicing. In some cases these may have different functions (Section 8.3.2).
- *Specific tissue, cell lineage or cell type.* Some genes have a function which is appropriate for a particular cell type or cell lineage, as in the case of the β-globin gene which is expressed in erythroid cells.
- *Individual cells.* Some genes produce different products in *individual* cells that are of the same cell type. For example, individual B lymphocytes express different antibody molecules, the T-cell receptors are different in individual T lymphocytes, and individual olfactory neurons produce different olfactory receptors. Another example concerns cells where one of the two alleles of a gene is normally repressed. It may consistently be the paternal allele or the maternal allele (genomic imprinting; see Section 8.5.4), or one allele may be repressed independent of the parent of origin (Section 8.5.3).
- *Intracellular distribution.* The proteins of different genes are transported to different intracellular (or extracellular) locations. In some cases, different isoforms of the same gene may be sent to different intracellular locations (Section 8.3.2). In addition, gene control mechanisms are required to send mRNA for some genes to different intracellular localizations (Section 8.2.5).

Temporal restriction of gene expression

- *Developmental stage.* At the very earliest stages of development transcription does not occur; instead cells rely on previously synthesized RNA. Later in development some genes may be expressed transiently at specific stages. Some gene families contain members that are expressed at different developmental stages as in the case of globin genes (see *Figure 8.23*).
- *Differentiation stage.* As cells undergo development and differentiate, their genomes are modified resulting in altered gene expression patterns. In some terminally differentiated cells, transcription does not occur. The genome modifications that result in the progression to a nucleated adult somatic cell used to be thought to be irreversible until the cloning of Dolly the sheep (Section 21.5.2).
- *Cell cycle stage.* Some genes are only expressed at specific times in the cell cycle. For example, many histone genes are expressed only at the S (DNA synthesis) phase.
- *Inducible expression.* Some genes are activated in response to environmental cues or extracellular signaling from other cells (Section 8.2.4). Such gene expression is easily reversed if the inducing factor is removed.

8.2 Control of gene expression by binding of *trans*-acting protein factors to *cis*-acting regulatory sequences in DNA and RNA

A common molecular basis for much of the control of gene expression (whether it occurs at the level of initiation of transcription, RNA processing, translation or RNA transport) is the binding of protein factors to regulatory nucleic acid sequences. The latter can be DNA sequences found in the vicinity of the gene or even within it, or RNA transcript sequences at the level of precursor RNA or mRNA. As the protein factors engaged in regulating gene expression are themselves encoded by distantly located genes, they are required to migrate to their site of action, and so are called *trans*-acting factors. In contrast, the regulatory sequences to which they bind to are on the same DNA or RNA molecule as the gene or RNA transcript that is being regulated. Such sequences are said to be *cis*-acting.

Control by DNA–protein binding

A major control of gene expression in eukaryotic cells is exerted at the level of *initiation* of transcription where three different types of RNA polymerase are known to transcribe different classes of genes (see *Table 1.3*). All three types of RNA polymerase are large enzymes, consisting of 8–14 subunits, and in each case the polymerase is recruited to transcribe a gene following binding of proteins (transcription factors) to specific regulatory DNA sequences within the gene or in its vicinity. Chromatin is a highly organized and densely packed structure which

Table 8.1: Overview of the regulation of gene expression in human cells

Selective expression mechanism	Examples
Transcriptional	
Binding of tissue-specific transcription factors to cis-acting elements of a single gene	See *Table 8.2*
Direct binding of hormones, growth factors or intermediates to response elements in inducible transcription elements	cAMP response elements, steroid hormone response elements etc. (see *Table 8.4* and *Figure 8.9*)
Use of alternative promoters in a single gene	See *Figure 8.13* for dystrophin gene and *Figure 8.20* for *Dnmt1* gene. Also applies to many other genes
Post-transcriptional	
Alternative splicing	Section 8.3.2 and *Figure 8.14*
Alternative polyadenylation	Section 8.3.2 and *Figure 8.15*
Tissue-specific RNA editing	Section 8.3.3 and *Figure 8.16*
Translational control mechanisms	Section 8.2.5 and *Figure 8.12*
Epigenetic mechanisms and long-range control of gene expression by chromatin structure	
Allelic exclusion	DNA rearrangements in B and T lymphocytes which produce cell-specific immunoglobulins and T-cell receptors (Section 8.6.2) Imprinting of certain genes (Section 8.5.4)
X chromosome inactivation	Inactivation by the *XIST* gene product of many genes on the one X chromosome on which it is expressed in female cells (Section 8.5.6) Random allelic exclusion by unknown mechanisms e.g. *IL-2, IL-4, PAX5*, etc (Section 8.5.3 and *Box 8.6*)
Long-range control by chromatin structure	Classic position effects (e.g. possibly in the case of fascioscapulohumeral dystrophy (Section 8.5.1) Suppression of gene expression by chromatin domains (e.g. of the *PAX6* gene in aniridia and the *SOX9* gene in campomelic dysplasia (Section 8.5.1) Competition for enhancers or silencers (e.g. in globin expression; see Section 8.5.2) and the imprinting of *H19* and *IGF2* genes (Section 8.5.4)
Cell position-dependent, short-range signaling	Section 8.4.1

does not easily afford access to RNA polymerases and so transcription factors are required to help activate it to give a more open structure that will enable transcription to take place.

Control by RNA–protein binding

In addition to transcription factors, RNA-binding proteins are used to regulate gene expression. The best-studied examples involve binding to regulatory sequences in the untranslated sequences of mRNA, permitting translational control of gene expression. In addition, specific RNA–protein binding interactions are expected to be involved in the control of gene expression at the level of differential RNA processing too, as in the case of binding of SR and HnRNP proteins to pre-mRNA in order to modulate the choice of exons in splicing. The latter mechanisms are considered separately in Section 8.3 to illustrate the tremendous complexity of expression mechanisms that can be used to decode single genes, and

the significance of the large numbers of isoforms that can be produced as a result.

8.2.1 Ubiquitous transcription factors are required for transcription of RNA polymerases I and III

RNA polymerases I and III in eukaryotic cells are dedicated to transcribing genes to give RNA molecules (rRNA, tRNA, etc.) which assist in expression of the polypeptide-encoding genes. The transcribed genes are housekeeping genes since rRNA and tRNA are required in essentially all cells to assist in protein synthesis. As a result, ubiquitous transcription factors are required to assist RNA polymerases I and III.

Transcription by RNA polymerase I

RNA polymerase I is confined to the nucleolus and is devoted to transcription of the 18S, 5.8S and 28S rRNA genes. The latter are consecutively organized on a common 13 kb transcription unit (*Figure 8.1*). A compound unit of the 13-kb transcription unit and an adjacent 27 kb nontranscribed spacer is tandemly repeated about 50–60 times on the short arms of each of the five human acrocentric chromosomes, at the **nucleolar organizer regions** (see *Figure 2.18*). The resulting five clusters of rRNA genes, each about 2 Mb long, are referred to as ribosomal DNA or rDNA.

Initiation of transcription of the 28S, 5.8S and 18S rRNA genes is initiated following binding of two transcription factors to a core promoter element at the transcription initiation site and an upstream control element located over 100 nucleotides upstream. One of the transcription factors, UBF (upstream binding factor), is a homodimer and its identical subunits may bind first to the core promoter element and upstream control element, bringing them together so that they can be bound by the second factor, SL1 (selectivity factor 1; known in mouse as TFI-1B; *Figure 8.2*). The bound transcription factors subsequently recruit RNA polymerase I to form an initiation complex.

The primary transcript expressed from the single 13 kb transcription unit is a 45S precursor rRNA which undergoes a variety of cleavage reactions and base-specific modifications (carried out by a large number of different types of **small nucleolar RNA (snoRNA)**) to generate the mature 28S, 5.8S and 18S rRNA species (see *Figure 8.1*). Thus, these genes differ from the vast majority of nuclear genes, which are individually transcribed. Instead, rDNA transcription resembles mtDNA transcription (see Section 7.1.1 and *Figure 7.2*): both result in multigenic transcripts which yield functionally related products. This unusual use of polygenic primary transcripts is no different in principle, however, from the way in which a single primary translation product is occasionally cleaved to generate two or more functionally related polypeptides (see the example of human insulin in *Figure 1.23*).

Transcription by RNA polymerase III

RNA polymerase III is also involved in transcription of a variety of housekeeping genes, encoding various small stable RNA molecules such as 5S rRNA, tRNA molecules, 7SL RNA and some of the snRNA molecules needed for RNA splicing. These genes are characterized by promoters that lie *within* the coding sequence of the gene, rather than upstream of it. In tRNA genes, the promoter is bipartite, consisting of two well conserved sequences, the A box and the B box, while in the 5S rRNA gene, a single promoter element is present, the C box. In each case, transcription by RNA polymerase III is thought to proceed by binding of ubiquitous transcription factors to the promoter elements, followed by subsequent binding of other factors and finally recruitment of the polymerase (*Figure 8.3*).

8.2.2 Transcription of polypeptide-encoding genes often requires complex sets of cis-acting transcriptional control sequences and tissue-specific transcription factors

RNA polymerase II is responsible for transcribing all genes which encode polypeptides and also certain species

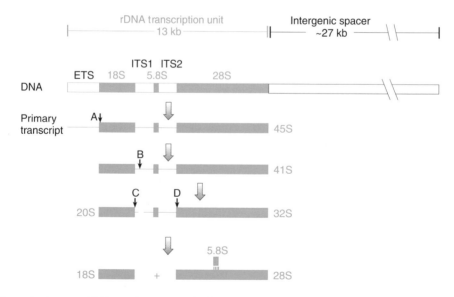

Figure 8.1: The major human rRNA species are synthesized by cleavage from a common 13 kb transcription unit which is part of a 40 kb tandemly repeated unit.

Small arrows indicated by letters A–D signify positions of endonuclease cleavage of RNA precursors. Cleavage of the 41S precursor at B generates two products: 20S + 32S. Following cleavage of the 32S precursor at D, and excision of the small 5.8S rRNA, hydrogen bonding takes place between the 5.8S rRNA and a complementary central segment of the 28S rRNA. The approximately 6 kb of RNA sequence originating from the external and internal transcribed spacer units (ETS, ITS1 and ITS2) are degraded in the nucleus. S is the sedimentation coefficient, a measure of size.

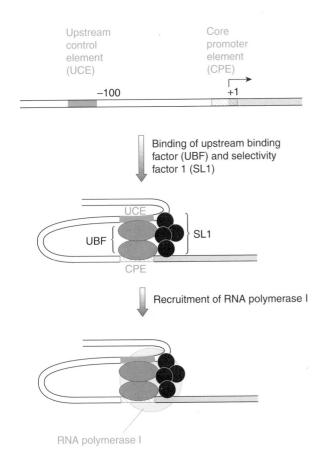

Figure 8.2: Initiation of transcription by RNA polymerase I.

One possible model envisages initial binding of the two identical subunits of the upstream binding factor to the upstream control element and the core promoter element, and forcing these two sequences to come into close proximity, enabling their subsequent binding by the selectivity factor 1 (SL1) which consists of four subunits. The stabilized structure permits subsequent binding of other factors (not shown) and subsequently RNA polymerase I.

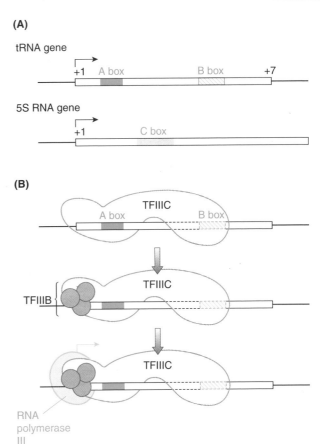

Figure 8.3: tRNA and 5S rRNA genes have promoters located within the coding sequence.

(A) Positions of promoter elements in tRNA and 5S rRNA genes. The promoter elements A and B in the tRNA genes are located in the sequences specifying the D loop and the TψCG loops respectively (see tRNA structure in *Figure 1.7B*). **(B)** Initiation of transcription of a tRNA gene. Binding of the TFIIIC transcription factor to the promoter elements permits subsequent binding of the trimeric TFIIIB factor to the sequence immediately upstream of the transcription start site. In response to binding of the TFIIIB factor, RNA polymerase III binds and initiates transcription. In the case of the 5S rRNA genes, a similar mechanism occurs but in this case an additional transcription factor TFIIIA is required to bind to the C box, and the bound TFIIIA factor permits subsequent binding of TFIIIB followed by recruitment of TFIIIC and RNA polymerase III as in the case of tRNA genes.

of snRNA gene. Like RNA polymerases I and III, RNA polymerase II is dependent on auxiliary general transcription factors (the usual nomenclature has a common prefix TF to denote transcription factor followed by a Roman numeral to denote the associated RNA polymerase). In the case of RNA polymerase II, there are a variety of auxiliary transcription factors such as TFIIA, TFIIB, TFIID, TFIIE, TFIIF, TFIIH, etc, which can be complex in structure (Nikolov and Burley, 1997). For example, the TATA box-binding protein (TBP), is only one of the multiple protein subunits that make up TFIID and the associated proteins are known as TBP-associated factors, or TAF proteins. The complex of polymerase and general transcription factors is known as the basal transcription apparatus; it constitutes all that is required to initiate transcription. Genes are constitutively expressed at a minimum rate determined by the core promoter (see below) unless the rate of transcription is increased or switched off by additional positive or negative regulatory elements (which may be located some distance away *or by intrinsic components of the promoter itself*).

Some of the genes which encode polypeptides are housekeeping genes, but unlike the products of genes transcribed by RNA polymerases I and III, a large percentage of genes transcribed by RNA polymerase II show tissue-restricted or tissue-specific expression patterns. Since the DNA in different nucleated cells of an individual is essentially identical, the identity of a cell, whether it be a hepatocyte or a T lymphocyte for instance, is defined by the proteins made by the cell. In addition to general ubiquitous transcription factors, therefore, tissue-specific or tissue-restricted transcription factors regulate the expression of many genes which encode polypeptides, by recognizing and binding specific *cis*-acting sequence elements.

Partly because of the large size of mammalian nuclear genomes and also because of the general need for more sophisticated control systems imposed by having very large numbers of interacting genes, control elements in eukaryotic cells are quite elaborate. Often, regulation of expression of individual human genes is controlled by several sets of *cis*-acting regulatory elements. While the individual regulatory elements may be composed of multiple short sequence elements (typically 4–8 nucleotides long) distributed over a few hundred base pairs, the different classes of regulatory element which modulate the expression of a single gene may be located at considerable distances. A variety of different types of *cis*-acting elements can be recognized, including promoters, enhancers, silencers, boundary elements (insulators) and response elements (see *Box 8.2*).

Box 8.2

Classes of *cis*-acting sequence elements involved in regulating transcription of polypeptide-encoding genes

Promoters are combinations of short sequence elements usually located in the immediate upstream region of the gene, often within 200 bp of the transcription start site, and serve to initiate transcription. They can be subdivided into different components.

■ **The core promoter** contains components that direct the basal transcription complex to initiate transcription of the gene and which, in the absence of additional regulatory elements, permits constitutive expression of the gene but at very low (basal) levels. Core promoter elements are typically located very close to the transcription initiation site, about from nucleotide positions −45 to +40. They include elements such as: (i) the TATA box (consensus *TATA*(A/T)A(A/T); located at position ~ −25, surrounded by GC-rich sequences and recognized by the *TATA-binding protein* subunit of TFIID); (ii) the BRE sequence (TFII<u>B</u> <u>R</u>ecognition <u>E</u>lement) which is located immediately upstream of the TATA element and is recognized by the TFIIB component; (iii) the Inr (initiator) sequence located at the start site of transcription (often a variant of the consensus sequence PyPyCAPuPu where Py = pyrimidine and Pu = purine); and (iv) the DPE (<u>D</u>ownstream <u>P</u>romoter <u>E</u>lement, located at about position +30 relative to transcription (see *Figure 8.4* and Burke and Kadonaga, 1997).

■ *Noncore promoter elements* can be found in the core promoter *region* but are typically located in the sequence upstream of it, usually from −50 to −200 bp relative to transcription initiation. This region is sometimes referred to as the proximal promoter region (as opposed to the *distal promoter region* which is sometimes used to describe regions still further upstream). There are typically multiple recognition sites for some sequence specific ubiquitous transcription factors. They include: (i) GC boxes (or Sp1 boxes; consensus GGGCGG, often found within 100 bp of transcriptional initiation site and bound by the Sp1 transcription factor); and (ii) CCAAT boxes (consensus sequence GG<u>CCAAT</u>CT; typically located at position −75 and recognized by the protein factors, CTF and CBF). CTF denotes <u>C</u>CAAT-binding <u>T</u>ranscription <u>F</u>actor and is also known as NF-I for <u>N</u>uclear <u>F</u>actor I. CBF denotes <u>C</u>CAAT box-<u>B</u>inding <u>F</u>actor and is also known as NF-Y for <u>N</u>uclear <u>F</u>actor Y.

Note that CCAAT and GC boxes serve to modulate the basal transcription of the core promoter and operate as *enhancer* sequences (see next section), while **silencer elements** (see below) may also be integral components of the promoter.

Enhancers are positive regulatory elements serving to increase the basal level of transcription which is initiated through the core promoter elements. Their functions, unlike those of the core promoter, are independent of both their orientation and, to some extent, their distance from the genes they regulate (Blackwood and Kadonaga, 1998). Enhancer elements can be very distantly located from the genes they regulated (see the example of the locus control regions for the globin genes – Section 8.5.2) and often contain, within a span of only 200–300 bp, elements recognized by ubiquitous transcription factors and also elements recognized by tissue-specific transcription factors, thereby providinig a basis for tissue-specific gene expression (see the example of the HS-40 α-globin enhancer in *Figure 8.6*). In addition, some enhancer elements may be integral components of promoters, as in the case of the CCAAT and GC boxes (see above).

Silencers serve to reduce transcription levels. Although less well-studied, two classes have been distinguished: *classical silencers* (also called *silencer elements*) are position-independent elements that direct an active transcriptional repression mechanism; and *negative regulatory elements* are position-dependent elements that result in a passive repression mechanism (Ogbourne and Antalis, 1998). Where studied in human genes, silencer elements have been reported in various positions: close to the promoter, some distance upstream and also within introns. However, the evidence for such sequences often relies on *in vitro* DNA binding studies and their significance *in vivo* is still uncertain.

Boundary elements (insulators) are regions of DNA, often spanning 0.5–3 kb, which function to block (or *insulate*) the spreading of the influence of agents that have a positive effect on transcription (enhancers) or a negative one (silencers, heterochromatin-like repressive effects; see Geyer, 1997).

Response elements function to modulate transcription in response to specific external stimuli. They are usually located a short distance upstream of the promoter elements (often within 1 kb of the transcription start site). A variety of response elements have been defined which permit transcriptional activation of specific genes in response to specific hormones (e.g. glucocorticoids or steroid hormones such as retinoic acid) or to intracellular second messengers such as cyclic AMP (see Section 8.2.4 and *Table 8.4*).

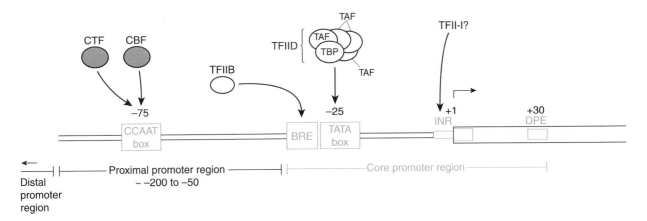

Figure 8.4: Conserved locations in complex eukaryotes for regulatory promoter elements bound by ubiquitous transcription factors.

Note that the core promoter of individual genes need not contain all elements. For example, many promoters lack a TATA box and use instead the functionally analogous initiator (INR) element (see Nikolov and Burley, 1997; Smale, 1997). GC boxes are usually found in promoters too but their locations are more variable (see *Figures 1.13* and *8.5* for some examples). Abbreviations: BRE, TFIIB recognition element; DPE, downstream promoter element; CTF, CCAAT-binding transcription factor; CBF, CCAAT box-binding factor; TBP, TATA box-binding protein; TAF, TBP-associated factors. See text.

Tissue specificity and developmental stage specificity of gene expression is often conferred by enhancer and silencer sequences and a variety of *cis*-acting sequences have been identified which are specifically recognized by **tissue-specific transcription factors**. For example, specific expression in erythroid cells is often signalled by one of two sequences: TGACTCAG (or its reverse complement CTGAGTCA) which are recognized by the erythroid-specific transcription factor NF-E2, or by the sequence

(A/T)GATA(A/G) or its reverse complement which are recognized by the GATA series of erythroid specific transcription factors (see *Figure 8.6* for an example). Some other examples of *cis*-acting sequence elements recognized by tissue-specific or tissue-restricted transcription factors are listed in *Table 8.2*.

In addition to actively promoting tissue-specific transcription, some *cis*-acting silencer elements confer tissue or developmental stage specificity by blocking

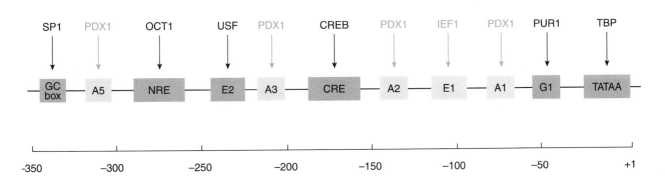

Figure 8.5: The human insulin gene promoter contains a variety of sequence elements recognized by ubiquitous and tissue-specific transcription factors.

Arrows indicate binding of transcription factors (top row) to regulatory sequence elements present upstream of the human insulin gene (bottom row). Ubiquitous or widely expressed transcription factors are shown in black; those shown in blue are specific for pancreatic beta cells. The PDX1 transcription factor (formerly called IUF1) binds to four sequence motifs of the form C(C/T)TAATG which are present in the insulin promoter (A1, A2, A3 and A5). Abbreviations: CRE, cAMP response element; NRE, negative regulatory element.

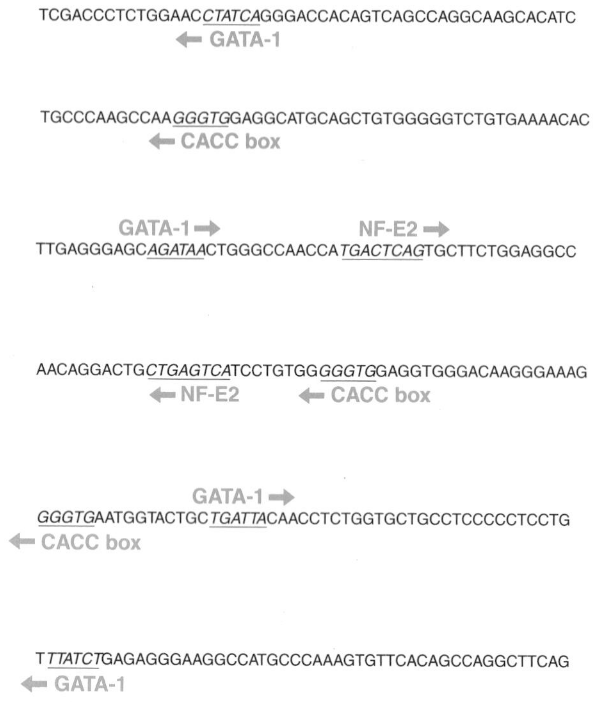

TCGACCCTCTGGAAC*CTATCA*GGGACCACAGTCAGCCAGGCAAGCACATC
⟵ GATA-1

TGCCCAAGCCAA*GGGTG*GAGGCATGCAGCTGTGGGGGTCTGTGAAAACAC
⟵ CACC box

GATA-1 ⟹   NF-E2 ⟹
TTGAGGGAGC*AGATAA*CTGGGCCAACCA*TGACTCAG*TGCTTCTGGAGGCC

AACAGGACTGC*CTGAGTCA*TCCTGTGG*GGGTG*GAGGTGGGACAAGGGAAAG
⟵ NF-E2   ⟵ CACC box

GATA-1 ⟹
*GGGTG*AATGGTACTGC*TGATTA*CAACCTCTGGTGCTGCCTCCCCCTCCTG
⟵ CACC box

T*TTATCT*GAGAGGGAAGGCCATGCCCAAAGTGTTCACAGCCAGGCTTCAG
⟵ GATA-1

Figure 8.6: The HS-40 α-globin regulatory site contains many recognition elements for erythroid-specific transcription factors.

Note that the HS-40 site appears to be a *locus control region* for the α-globin gene cluster (see Section 8.5.2).

expression in all but the desired tissue. For example, the neural restrictive silencer element (NRSE) represses expression of several genes in all tissues other than neural tissues (Schoenherr *et al.*, 1996). A transcription factor that binds to the NRSE and which is variously called the neural restrictive silencer factor (NRSF) or the RE-1 silencing transcription factor (REST) is ubiquitously expressed in non-neural tissue and neuronal precursors during early development but subsequently it is specifically *not expressed* in more mature (postmitotic) neurons.

8.2.3 Transcription factors contain conserved structural motifs that permit DNA binding

Transcription factors recognize and bind a short nucleotide sequence, usually as a result of extensive complementarity between the surface of the protein and surface features of the double helix in the region of binding. Although the individual interactions between the amino acids and nucleotides are weak (usually hydrogen bonds, ionic bonds and hydrophobic interactions), the region of DNA–protein binding is typically characterized by about 20 such contacts, which collectively ensure that the binding is strong and specific. In human and other eukaryotic transcription factors, two distinct functions can often be identified and located in different parts of the protein:

■ **An activation domain.** As the name suggests, this type of domain functions in activating transcription of the target genes once the transcription factor has bound to it. Activation domains are thought to stimulate transcription by interacting with basal transcription factors so as to assist the formation of the transcription complex on the promoter. Although not so well-studied as DNA-binding domains, some are known to be rich in aspartate and glutamate residues (*acidic activation domains*); others are rich in proline or glutamate.

■ **A DNA-binding domain.** This type of domain is necessary to permit specific binding of the transcription factor to its target genes. In contrast to activation domains, DNA-binding domains of transcription factors have been well-studied. A number of conserved structural motifs have been identified which are common to many different transcription factors with quite different specificities, including the leucine zipper, helix–loop–helix, helix–turn–helix, and zinc finger motifs which are described below. Each of the motifs uses α-helices (or occasionally β-sheets; see *Figure 1.24*) to bind to the major groove of DNA. Clearly, although the motifs in general provide the basis for DNA binding, the precise collection of sequence elements in the DNA-binding domain

Table 8.2: Examples of *cis*-acting sequences recognized by tissue-restricted and tissue-specific transcription factors

Consensus binding sequence	Transcription factor	Expression patterns
(A/T)GATA(A/G)	GATA-1, -2, etc.	Erythroid cells
TGACTCAG	NF-E2	Erythroid cells
GTTAATNATTAAC (= PE element)	HNF-1	Differentiated liver, kidney, stomach, intestine, spleen
T(G/A)TTTG(C/T)	HNF-5	Liver
GCCTGCAGGC	Ker1	Keratinocytes
(C/T)TAAAAATAA(C/T)3	MBF-1	Myocytes
(C/T)TA(A/T)AAATA(A/G)	MEF-2	Myocytes
CAACTGAC	MyoD	Myoblasts + myotubes
ATGCAAAT	OTF-2	Lymphoid cells
(C/A)A(C/A)AG	TCF-1	T cells

will provide the basis for the required sequence-specific recognition. Most transcription factors bind to DNA as homodimers, with the DNA-binding region of the protein usually distinct from the region responsible for forming dimers.

The leucine zipper motif

The **leucine zipper** is a helical stretch of amino acids rich in leucine residues (typically occurring once every seven amino acid residues, i.e. once every two turns of the helix – see *Figure 8.7*), which readily forms a dimer. Each monomer unit consists of an amphipathic *α-helix* (hydrophobic side groups of the constituent amino acids face one way; polar groups face the other way, see *Figure 1.24*). The two α-helices of the individual monomers join together over a short distance to form a coiled-coil (see Section 1.5.5) with the predominant interactions occurring between opposed hydrophobic amino acids of the individual monomers. Beyond this region the two α-helices separate, so that the overall dimer is a Y-shaped structure. The dimer is thought to grip the double helix much like a clothes peg grips a clothes line (*Figure 8.8*). In addition to forming homodimers, leucine zipper proteins can occasionally form heterodimers depending on the compatibility of the hydrophobic surfaces of the two different monomers. Such heterodimer formation provides an important combinatorial control mechanism in gene regulation.

The helix–loop–helix motif

The helix–loop–helix (HLH) motif is related to the leucine zipper and should be distinguished from the helix–turn–helix (HTH) motif described in the next section. It consists of two α-helices, one short and one long, connected by a flexible loop. Unlike the short turn in the HTH motif, the loop in the HLH motif is flexible enough to permit folding back so that the two helices can pack against each other; that is, the two helices lie in planes that are parallel to each other, in contrast to the two helices in the HTH motif (*Figure 8.7*). The HLH motif mediates both DNA binding and protein dimer formation (*Figure 8.8*) and it permits occasional heterodimer formation. In the latter case, however, heterodimers form between a full-length HLH protein and a truncated HLH protein which lacks the full length of the α-helix necessary to bind to the DNA. The resulting heterodimer is unable to bind DNA tightly. As a result, HLH dimers are thought to act as a control mechanism, by enabling inactivation of specific gene regulatory proteins.

The helix–turn–helix motif

The HTH motif is a common motif found in homeoboxes, and a number of other transcription factors. It consists of two short α-helices separated by a short amino acid sequence which induces a turn, so that the two α-helices are orientated differently (i.e. the two helices do not lie in the same plane, unlike those in the HLH motif; *Figure 8.7*). The structure is very similar to the DNA-binding motif of

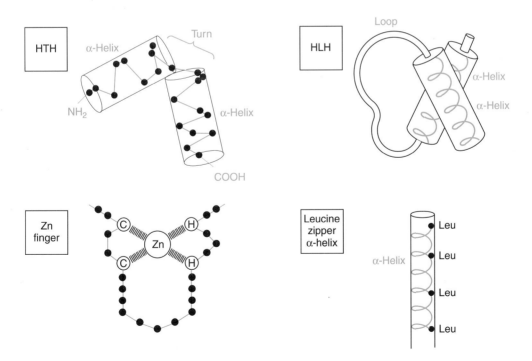

Figure 8.7: Structural motifs commonly found in transcription factors and DNA-binding proteins.

Abbreviations: HTH, helix–turn–helix; HLH, helix–loop–helix. *Note* that the leucine zipper monomer is *amphipathic* [i.e. has hydrophobic residues (leucines) consistently on one face of the helix]. Two such helices can align with their hydrophobic faces in opposition to form a coiled-coil structure.

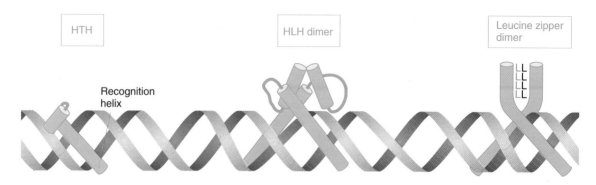

Figure 8.8: Binding of conserved structural motifs in transcription factors to the double helix.

Note that the individual monomers of the helix–loop–helix (HLH) dimer and the leucine zipper dimer are colored differently to permit distinction, but may be identical (homodimers). HLH heterodimers and leucine zipper heterodimers may provide a higher level of regulation (see text).

several bacteriophage regulatory proteins such as the λ cro protein whose binding to DNA has been intensively studied by X-ray crystallography. In the case of both the λ cro protein and eukaryotic HTH motifs, it is thought that while the HTH motif in general mediates DNA binding, the more C-terminal helix acts as a specific recognition helix because it fits into the major groove of the DNA (*Figure 8.8*), controlling the precise DNA sequence which is recognized.

The zinc finger motif

The zinc finger motif involves binding a zinc ion by four conserved amino acids to form a loop (finger), a structure which is often tandemly repeated. Although several different forms exist, common forms involve binding of a Zn^{2+} ion by two conserved cysteine residues and two conserved histidine residues, or by four conserved cysteine residues. The resulting structure may then consist of an α-helix and a β-sheet held together by coordination with the Zn^{2+} ion, or of two α-helices. In either case, the primary contact with the DNA is made by an α-helix binding to the major groove.

The so-called Cys_2/His_2 finger typically comprises about 23 amino acids with neighboring fingers separated by a stretch of about seven or eight amino acids (*Figure 8.7*).

8.2.4 A variety of mechanisms permit transcriptional regulation of gene expression in response to external stimuli

In eukaryotic cells, gene expression can be altered in a semipermanent way as cells differentiate, or in a temporary, easily reversible way in response to extracellular signals (inducible gene expression). Environmental cues such as the extracellular concentrations of certain ions and small nutrient molecules, temperature, shock, etc., can result in dramatic alteration of gene expression patterns in cells exposed to changes in these parameters. In complex multicellular animals there are also fundamental requirements for cells to communicate with each other and different modes of cell signaling are possible (*Table 8.3*). In some cases, alteration of gene expression is con-

Table 8.3: Different modes of cell signaling

Mode	Characteristics	Examples
Direct cell to cell signaling	A signal on the surface of one cell is bound by a specific receptor on another cell	Membrane-anchored growth factors and their receptors
Endocrine signaling	Hormones are secreted by specialized endocrine cells and carried through the circulation to bind to receptors in target cells at distant locations in the body	Release of glucocorticoid hormones by the adrenal glands
Paracrine signaling	A molecule released from one cell acts locally to affect nearby target cells	Neurotransmitters and receptors; nitric oxide in the immune system, nervous system etc.
Autocrine signaling	A cell produces a signaling molecule to which it also responds	T lymphocytes can respond to antigenic stimulation by synthesizing factors that drive their own proliferation

Table 8.4: Examples of response elements in inducible gene expression

Consensus response element (RE)	Response to:	Protein factor which recognizes RE
(T/G)(T/A)CGTCA	cAMP	CREB (also called ATF)
CC(A/T)(A/T)(A/T)(A/T)(A/T)(A/T)GG	Serum growth factor	Serum response factor
TTNCNNNAAA	Interferon-gamma	Stat-1
TGCGCCCGCC	Heavy metals	Mep-1
TGAGTCAG	Phorbol esters	AP1
CTNGAATNTTCTAGA	Heat shock	HSP70, etc.

Note: see also response elements for steroid receptors in *Figure 8.9.*

ducted at the translational level which can offer certain advantages (Section 8.2.5). In other cases, gene expression is altered by modulating transcription.

Transcriptional regulation in response to cell signaling can take different forms, but the endpoint is always the same: a previously inactive transcription factor is specifically activated by the signaling pathway and then subsequently binds to specific regulatory sequences located in the promoters of target genes, thereby activating their transcription. In the case of transcription regulated by signaling molecules or their intermediaries, such regulatory sequences are often referred to as **response elements** (see *Table 8.4*).

Ligand-inducible transcription factors

Small hydrophobic hormones and morphogens such as steroid hormones, thyroxine and retinoic acid are able to diffuse through the plasma membrane of the target cell and bind intracellular receptors in the cytoplasm or nucleus. These receptors (often called **hormone nuclear receptors**) are inducible transcription factors. Following binding of the homologous ligand, the receptor protein associates with a specific DNA response element located in the promoter regions of perhaps 50–100 target genes and activates their transcription.

Although thyroxine and retinoic acid are structurally and biosynthetically unrelated to the steroid hormones, their receptors belong to a common nuclear receptor superfamily. Two conserved domains characterize the family: a centrally located **DNA binding domain** of about 68 amino acids, and a ~240 amino acid **ligand-binding domain** located close to the C terminus (*Figure 8.9*). The DNA binding domain contains zinc fingers and binds as a dimer with each monomer recognizing one of two hexanucleotides in the response element. The two hexanucleotides are either inverted repeats or direct repeats which are typically separated by three or five nucleotides (*Figure 8.9*). In the absence of the ligand, the

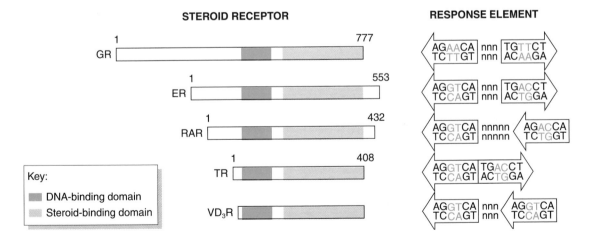

Figure 8.9: Steroid receptors and the respective response elements.

(A) Structure of members of the nuclear receptor superfamily. Numbers refer to protein size in amino acids. Abbreviations: ER, estrogen receptor; GR, gluocorticoid receptor; PR, progesterone receptor; RAR, retinoic acid receptor; TR, thyroxine receptor; VDR, vitamin D receptor. **(B)** Response elements. Note that the response elements are often perfect inverted hexanucleotide repeats, but that the response elements for retinoic acid and vitamin D₃ are imperfect direct hexanucleotide repeats. Note also that the hexanucleotides all have the general sequence AGNNCA with the central two nucleotides shown in blue conferring specificity and belonging to one of three classes: GT, AA or AC. Abbreviation: n, any nucleotide.

receptor is inactivated by direct repression of the DNA binding domain function by the ligand-binding domain, or by binding to an inhibitory protein, as in the case of the glucocorticoid receptor (*Figure 8.10*).

Activation of transcription factors by signal transduction

Unlike lipid-soluble hormones or morphogens, hydrophilic signaling molecules such as polypeptide hormones, cannot diffuse through the plasma membrane. Instead, they bind to a specific receptor on the cell surface. After binding of the ligand molecule, the receptor undergoes a conformational change and becomes activated in such a way that it passes on the signal via other molecules within the cell (signal transduction). Various classes of cell surface receptor are known but many of them have a kinase activity or can activate intracellular kinases (see *Table 8.5*). Signal transduction pathways are often characterized by complex regulatory interplay between kinases and phosphatases which can activate or repress intermediates by phosphorylation/dephosphorylation. In many cases, the phosphorylation or dephosphorylation induces an altered conformation. In the case of activation of a signaling molecule, the altered conformation often means that a signaling factor is no longer inhibited by some repressor sequence present in an inhibitory protein to which it is bound, or in a domain or sequence motif within its own structure.

In terms of transcriptional activation, two general mechanisms permit rapid transmission of signals from cell-surface receptors to the nucleus, both involving protein phosphorylation:

- protein kinases are activated and then translocated from the cytoplasm to the nucleus where they phosphorylate target transcription factors;
- inactive transcription factors stored in the cytoplasm are activated by phosphorylation and translocated into the nucleus.

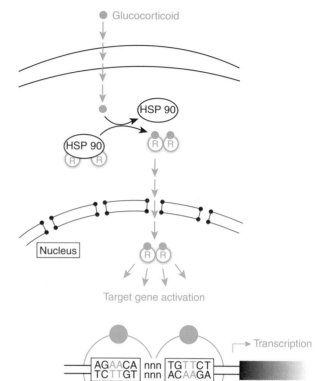

Figure 8.10: Transcriptional regulation by glucocorticoids.

The glucocorticoid receptor is normally inactivated by being bound to an inhibitor protein, Hsp90. Binding of glucocorticoids to the glucocorticoid receptor releases Hsp90, the receptor dimerizes and then activates selected genes which have a glucocorticoid response element in their promoter (see *Figure 8.9*).

Table 8.5: Major classes of cell surface receptor

Receptor class	Characteristics	Activation by
G protein-coupled	Activate heterotrimeric G-proteins (**G**TP-binding regulatory proteins). The latter consist of three subunits, α, β and γ. Upon activation, the α subunit translocates to activate a target protein (see *Figure 8.11A* for an example)	Various molecules, including peptides, hormones, e.g. epinephrine, etc.
Serine–threonine kinase	Activate intracellular proteins by phosphorylating serine or threonine residues on target proteins	Hormones, growth factors
Tyrosine kinase	Activate intracellular proteins by phosphorylating tyrosine on target proteins, e.g. PDGF receptor, insulin receptor, etc (see *Figure 8.11B*)	Hormones, growth factors
Tyrosine kinase-associated	Receptors do not phosphorylate target proteins directly, but instead rely on an associated tyrosine kinase. Examples include cytokine receptors with associated JAK (janus protein kinase) activity involved in JAK-STAT signaling	Hormones, growth factors
Ion channel-linked	Involved in rapid synaptic signaling	Neurotransmitters

The following two sections provide examples to illustrate these two mechanisms (see also Karin and Hunter, 1995).

Hormonal signaling through the cyclic AMP pathway

Cyclic AMP is an important **second messenger** (see *Table 8.6*) which acts in response to a variety of hormones and other signaling molecules. It is synthesized from ATP by a membrane bound enzyme, adenylate cyclase. Hormones which activate adenylate cyclase bind to a cell surface receptor which is of the G protein-coupled receptor class. Binding of the hormone to the receptor promotes the interaction of the receptor with a G protein which consists of three subunits, α, β and γ. Following this interaction the α subunit of the G protein is activated, causing it to dissociate and stimulate adenylate cyclase.

The increase in intracellular cAMP produced by activated adenylate cyclase can then activate the transcription of specific target sequences that contain a cAMP response element or CRE. This function of cAMP is mediated by the enzyme protein kinase A. Cyclic AMP binds to protein kinase and activates it by permitting release of the two catalytically active subunits which then enter the nucleus and phosphorylate a specific transcription factor, CREB (CRE-binding protein). Activated CREB then activates transcription of genes with the cAMP response element (*Figure 8.11A*).

Activation of NF-κB via protein kinase C signaling

NF-κB is a transcription factor which is involved in a variety of aspects of the immune response. In its inactive state, NF-κB is retained in the cytoplasm where it is complexed with an inhibitory subunit, IκB. However, the latter can be targeted for degradation following phosphorylation by protein kinase C. The consequent destruction of IκB permits NF-κB to translocate to the nucleus and activate its various target genes. Protein kinase C is activated by diacylglycerol. The latter is produced when binding of various growth factors and hormones to specific cell surface receptors triggers activation of receptor-linked phospholipase C activity. The activated enzyme converts PIP_2 (phosphatidylinositol 4,5 bisphosphate) to IP_3 (inositol 1,4,5-trisphosphate) and diacylglycerol (*Figure 8.11B*).

8.2.5 Translational control of gene expression can involve specific recognition by RNA-binding proteins of regulatory sequences within the untranslated sequences of mRNA

Different forms of translational control of gene expression are evident and an increasing number of eukaryotic and mammalian mRNA species have been shown to contain regulatory sequences in their untranslated sequences (most frequently at the 3' end; see Wickens *et al.*, 1997; Day and Tuite, 1998). Several eukaryotic and mammalian RNA-binding proteins have also been identified and shown to bind to specific regulatory sequences present in untranslated sequences, thereby providing the basis for translational control of gene expression (Siomi and Dreyfuss, 1997). A variety of different RNA-binding domains have been identified and they include elements which have previously been associated with DNA-binding properties of transcription factors such as zinc fingers and homeodomains (see Section 8.2.3 and Siomi and Dreyfuss, 1997).

Intracellular RNA localization

The interaction between *cis*-acting regulatory elements in RNA and *trans*-acting RNA-binding proteins can be envisaged to alter RNA structure in various ways: facilitating or hindering interactions with other *trans*-acting factors; altering higher order RNA structure; bringing together initially remote RNA sequences; or providing localization or targeting signals for transport of RNA molecules to specific intracellular locations. In the latter case, numerous eukaryotic and mammalian mRNAs are known to be transported as RNP particles to specific locations within cells, and transport on microtubules and actin filaments has been demonstrated in some cases to require specific molecular motors (Hazelrigg, 1998). For example, tau mRNA is localized to the proximal portions

Table 8.6: Examples of secondary messengers in cell signaling

Secondary messenger	Characteristics	Examples
Cyclic AMP (cAMP)	Produced from ATP by adenylate cyclase. Effects are usually mediated through protein kinase A	Activation of CREB transcription factor (see *Figure 8.11A*)
Cyclic GMP (cGMP)	Produced from GTP by guanylate cyclase. Best characterized role is in visual reception in the vertebrate eye	
Phospholipids/Ca^{2+}	Activated downstream of G protein-coupled receptors and protein tyrosine kinases. Hydrolysis of phosphatidylinositol 4,5-bis-phosphate (PIP_2) yields diacylglycerol and inositol 1,4,5-trisphosphate (IP_3) which activate protein kinase C and mobilize Ca from intracellular stores	See *Figure 8.11B*

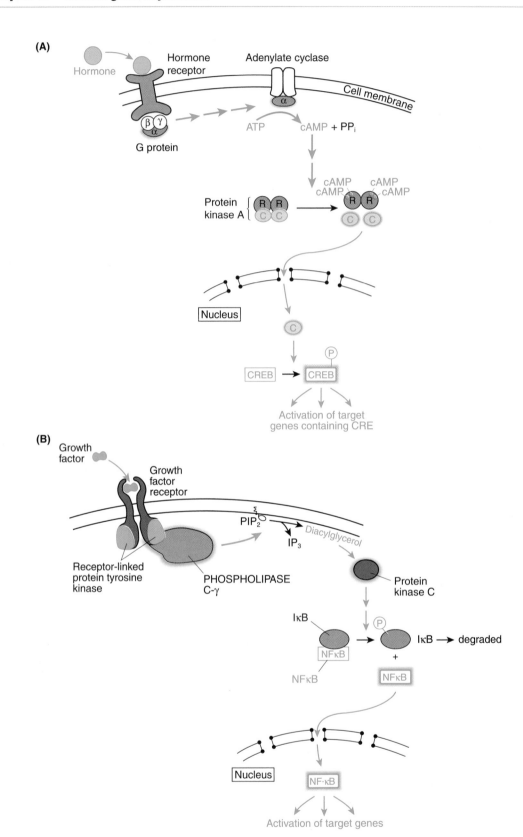

of axons rather than to dendrites, where many mRNA molecules are located in mature neurons, and myelin basic protein mRNA is transported with the aid of kinesin to the processes of oligodendrocytes.

A rationale for such intracellular mRNA localization mechanisms is that they may provide a more efficient way to localize protein products than protein targeting: as detailed below, a single mRNA can give rise to many different protein molecules, assuming that it can engage with ribosomes. Different sequential steps have been envisaged: initial translational repression, transport within the cell, localization (to the specific subcellular destination) and then localization-dependent translation. Recently, key regulatory sequences which are required for various steps in this process have been identified in the untranslated sequences, predominantly the 3' UTR, of many mRNA species (Hazelrigg, 1998). For example, two elements within the 3' UTR of myelin basic protein mRNA are required for distinct steps in its transport to the processes of oligodendrocytes: a 21 nucleotide RNA transport sequence and a longer RNA localization region (Ainger *et al.*, 1997).

Translational control of gene expression in response to external stimuli

Translational control of gene expression can permit a more rapid response to altered environmental stimuli than the alternative of activating transcription. Iron metabolism provides two useful examples. Increased iron levels stimulate the synthesis of the iron-binding protein, ferritin, without any corresponding increase in the amount of ferritin mRNA. Conversely, decreased iron levels stimulate the production of transferrin receptor (TfR) without any effect on the production of transferrin receptor mRNA. The 5'-UTR of both ferritin heavy chain mRNA and light chain mRNA contain a single iron-response element (IRE), a specific *cis*-acting regulatory sequence which forms a hairpin structure. Several such IRE sequences are also found in the 3' UTR of the transferrin receptor mRNA (see Klausner *et al.*, 1993). Regulation is exerted by binding of IREs by a specific IRE-binding protein which is activated at low iron levels (*Figure 8.12*).

Translational control of gene expression during early development

Gene expression during oocyte maturation and at the earliest embryonic stages is regulated at the level of translation, not transcription. Following fertilization of a human oocyte, no mRNA is made initially until the 4–8 cell stage when zygotic transcription is activated, that is, transcription of the genes present in the zygote. Before this time, cell functions are specified by maternal mRNA that was previously synthesized during oogenesis. While it is presently unclear to what extent the regulation of human gene expression parallels that of model organisms at this stage, extrapolation from the latter would suggest that a variety of mRNAs are stored in oocytes in an inactive form, characterized by having short oligo (A) tails. Such mRNAs were previously subject to deadenylation and the resulting short oligo (A) tail means that they cannot be translated. Subsequently, at fertilization or later in development, the stored inactive mRNA species can be activated by cytoplasmic polyadenylation, restoring the normal size poly (A) tail. Cytoplasmic polyadenylation appears to use the same type of poly (A) polymerase activity as in the standard polyadenylation of newly formed mRNA (which occurs in the nucleus). However, in addition to the AAUAAA signal, the mRNA needs to have a uridine-rich upstream *cytoplasmic polyadenylation element* (Wahle and Kuhn, 1997). Another mechanism that is used to regulate translation of some mRNAs during development is translational repression (masking) whereby RNA-binding proteins can recognize and bind specific sequences in the 3' UTRs of the mRNAs, thereby repressing translation (see Wickens *et al.*, 1997; Stebbins-Boaz and Richter, 1997).

8.3 Alternative transcription and processing of individual genes

In addition to the control that is exerted in selecting specific genes (or their transcripts) for activation or repression, control mechanisms can also select between specific alternative transcripts of a single gene. Differential promoter usage or differential RNA processing events can

Figure 8.11: Selected target genes can be actively expressed in response to extracellular stimuli by signal transduction from activated cell surface receptors.

(A) An example of activation of a protein kinase and translocation to the nucleus: hormonal signaling through the cyclic AMP–protein kinase A signal transduction pathway. Binding of hormone to a specific cell surface receptor promotes the interaction of the receptor with a G protein. The activated G protein α subunit dissociates from the receptor and stimulates the membrane-bound adenylate cyclase to synthesize cAMP. The latter phosphorylates the regulatory subunits of protein kinase A, enabling release of the catalytic subunits which migrate to the nucleus and activate the transcription factor CREB (CRE-binding protein) by phosphorylation. Activated CREB binds to cAMP response elements in the promoters of target genes.
(B) An example of activation of a cytoplasmic transcription factor (NF-κB) and translocation to the nucleus. In response to an external growth factor, a growth factor receptor-linked tyrosine kinase phosphorylates phospholipase C which can then convert phosphatidylinositol 4,5-bisphosphate (PIP_2) to diacyl glycerol and inositol 1,4,5-trisphosphate (IP_3). Diacylglycerol can activate protein kinase C which phosphorylates the inhibitory factor IκB and so targets it for degradation. As a result the transcription factor NFκB is released and migrates to the nucleus to activate its target genes.

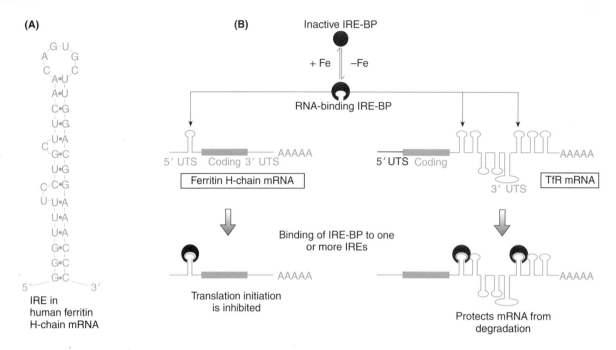

Figure 8.12: The IRE-binding protein regulates the production of ferritin heavy chain and transferrin receptor by binding to iron-response elements (IREs) in the 5'- or 3'-untranslated regions.

(**A**) Structure of the IRE in the 5'-UTR of the ferritin heavy chain. (**B**) Binding of the IRE-binding protein (IRE-BP) to ferritin and transferrin (TfR) mRNAs has contrasting effects on protein synthesis. For the sake of clarity, only some of the IRE elements in the TfR mRNA are shown to be bound by the IRE-binding protein.

result in a large number of different isoforms and these and other mechanisms have challenged the classical definition of a gene (*Box 8.3*).

8.3.1 Transcription of a single human gene can be initiated from a variety of alternative promoters and can result in a variety of tissue-specific isoforms

Several human and mammalian genes are known to have two or more alternative promoters, which can result in different isoforms with different properties (see Ayoubi and van de Ven, 1996). The isoforms can provide:

- tissue-specificity (a frequent occurrence; see the example of the human dystrophin gene below);
- developmental stage-specificity (e.g. the insulin-like growth factor II gene);
- differential subcellular localization (e.g. soluble and membrane-bound isoforms);
- differential functional capacity (as in the case of the progesterone receptor);
- sex-specific gene regulation (see the case of the *Dnmt1* methyltransferase gene in Section 8.4.2 and *Figure 8.20*).

One of the most celebrated examples of differential promoter usage in humans concerns the giant dystrophin (*DMD*) gene which comprises a total of more than 79 exons distributed over about 2.4 Mb of DNA in Xp21. At least

eight different alternative promoters can be used (Cox and Kunkel, 1997). Four of the alternative promoters are located near the conventional start site and comprise a brain cortex-specific promoter, a muscle-specific promoter located 100 kb downstream; a promoter which is used in Purkinje cells of the cerebellum and located a further 100 kb downstream, and a lymphocyte-specific promoter (see *Figure 8.13*). Usage of these promoters results in large isoforms with a molecular weight of 427 kDa (referred to as Dp427 where Dp = <u>D</u>ystrophin <u>p</u>rotein and often given a suffix to indicate tissue specificity e.g. Dp427-M to indicate the <u>m</u>uscle specific isoform). The four Dp427 isoforms differ in their extreme N-terminal amino acid sequence as a result of using four different alternatives for exon 1.

In addition to the four alternative promoters encoding the conventional large isoforms, at least four other alternative internal promoters can be used. Transcription from these promoters uses only a downstream subset of the exons, resulting in significantly smaller isoforms: a Dp260 isoform produced in retinal cells; a Dp140 isoform produced by many cells in the brain and kidney; a Dp116 isoform produced in Schwann cells and a small Dp71 isoform produced in many cell types (see *Figure 8.13*). Note that the alternative usage of promoters enforces alternative use of exons but that alternative splicing events which are independent of differential promoter usage are also very common (see Section 8.3.2). In the case of dystrophin, for example, additional isoform complexity is introduced by alternative splicing, especially at the C terminus.

Box 8.3

The classical view of a gene is no longer valid

Classically, a gene has been viewed as an entity that encodes a single RNA or polypeptide product. By contrast, the concept of several different products being encoded by a single *transcription unit* (a segment of DNA that is continuously transcribed into RNA) has long been familiar in simple genomes, such as bacterial genomes, where so-called polycistronic mRNAs are common. These arise by continuous transcription through several adjacent genes (the term *cistron* is essentially an old-fashioned word for gene). Such multigenic transcripts are then processed to generate two or more different polypeptides. In such cases, the term *transcription unit* is clearly not functionally equivalent to the term *gene* because each transcription unit corresponds to several genes. In complex genomes, such as the human genome, however, the vast majority of genes are transcribed individually and, in these cases, the terms *gene* and *transcription unit* are essentially equivalent. The immunoglobulin and T cell receptor genes, however, provide additional complexity: the individual genes are very large in germ-line DNA (see *Figure 7.7* and *Figure 8.26*) but the corresponding transcription units are much smaller and of variable size because of *cell-specific* DNA rearrangements in individual B and T lymphocytes (see *Figures 8.27* and *8.28*).

In addition, because of various other expression mechanisms listed below, human genes and transcription units often encode a variety of gene products.

Mechanism	Frequency and examples
Multigenic transcription units	Rare. Examples include 18S, 28S and 5.8S rRNA genes (see *Figure 8.1*) and mitochondrial genes (see Section 7.1.1 and *Figure 7.2*)
Use of alternative promoters	Quite common. See below and *Figure 8.13*
Alternative splicing	Very frequent. See Section 8.3.2 and *Figure 8.14*
Alternative polyadenylation	Quite common. See *Figure 8.15*
RNA editing	Extremely rare. See Section 8.3.3 and *Figure 8.16*
Post-translational cleavage	Rare. May generate functionally related polypeptides as in the case of several hormones, e.g. human insulin. See Figure *1.23*.

8.3.2 Human genes often encode more than one product as a result of alternative splicing and alternative polyadenylation events

In addition to differential use of promoters which can enforce alternative use of exons, a variety of alternative RNA processing events can also result in alternative isoforms. The primary mechanisms are alternative splicing events (that is, distinct from those induced by differential promoter usage), and alternative polyadenylation events. In many cases a combination of these mechanisms can result during the processing of a single gene. Together with the additional possibility of differential promoter usage, these mechanisms can result in very large numbers of isoforms for a single gene.

Alternative splicing
A large percentage of human genes undergo **alternative splicing** whereby different exon combinations are

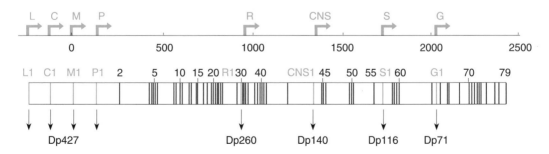

Figure 8.13: At least eight distinct promoters can be used to generate cell type-specific expression of the dystrophin gene.

The positions of the eight alternative promoters are illustrated at the top: L, lymphocyte; C, cortical; M, muscle; P, Purkinje; R, retinal; CNS, central nervous system; S, Schwann cell; G, general. The approximate positions of the exons are illustrated below. *Note* that each promoter uses its own first exon (in blue: L1, C1, M1, P1, R1, CNS1, S1 and G1) together with downstream exons (in black). All 78 downstream exons are used in the case of the full-length C-, M- and P-dystrophins to generate a product which in each case is about 427 kDa (Dp427). The other promoters are located immediately upstream of indicated exons as follows: R, exon 30; CNS, exon 45; S, exon 56; G, exon 63. *Note also* that initiation of translation for the Dp140 isoform does not occur until exon 51, although the promoter is thought to be in intron 44. In addition to the diversity of isoforms generated by usage of differential promoters, alternative splicing is known to occur, notably at the 3′ end.

Alternative splicing can alter the functional properties of a protein

The following, far from exhaustive, list is merely meant to illustrate some ways in which the biological properties of a protein can be altered as a result of alternative splicing.

■ *Tissue-specific isoforms.* Examples include isoforms encoded by the tropomyosin gene. See also the example of the calcitonin gene which is listed below (see Section 8.3.2 and *Figure 8.15*).

■ *Membrane-bound and soluble isoforms.* Protein localization can be regulated by generating soluble forms of numerous membrane receptors e.g. class I and II HLA, IgM, CD8, growth hormone receptor, IL-4, IL-5, IL-7, IL, erythropoietin, G-CSF, G-MCSF, LIF (leukaemia inhibitory factor), and the FAS apoptosis-signaling receptor.

■ *Alternative intracellular localization.* A useful example is provided by the *WT1* Wilm's tumor gene which specifies a protein with four zinc fingers at its C terminal and has many isoforms. Differential splicing can lead to inclusion or exclusion of a 17 amino acid sequence in the transregulatory domain, and also to inclusion or exclusion of a sequence of three amino acids, lysine–threonine–serine, located between the third and fourth zinc fingers (generating isoforms which are named +KTS and −KTS from the one letter amino acid code – see *Figure 8.14*). These isoforms are well-conserved in evolution and differ in several aspects, including their subnuclear distribution patterns: the +KTS isoforms are specifically localized to spliceosomal sites in the nucleus, while the −KTS isoforms are more generally distributed in the nucleoplasm (Larsson *et al.*, 1995).

■ *Altered function.* The +KTS and −KTS isoforms of the *WT1* gene product also differ in their ability to bind to specific DNA sequences in target genes. The former are thought to have a role in binding splicing factors; the latter may have a more general role in binding domains that harbor general transcription factors. Other examples include: (i) transcription factor isoforms which are transcriptional activators or repressors depending on the nature of the domains that are included or excluded from the protein product – see Lopez, 1998; (ii) apoptosis-promoting and apoptosis-inducing isoforms of various genes, such as the Ich-1 (caspase 2) gene (see Jiang *et al.*, 1998); and (iii) alternative isoforms of the calcitonin gene (*Figure 8.15*).

included in transcripts from the same gene during RNA processing. For many genes, numerous isoforms can be generated at the RNA level, but often the functional significance is poorly appreciated. In some genes, alternative splicing results in very considerable diversity in the untranslated regions. For example, in the liver alternative splicing results in at least 8 different 5′ UTR sequences for human growth hormone receptor mRNA (Pekhletsky *et al.*, 1992), but the functional significance, if any, is not understood.

Alternative splicing of coding sequence exons is also common and some of the resulting protein isoforms have been shown to be tissue specific, so that individual exons present in one isoform but not in others may be termed 'muscle-specific' , 'brain-specific' etc. The different isoforms can provide a variety of possibilities for altered functional properties but detailed knowledge of the functional significance of the different isoforms is still comparatively sparse (see *Box 8.4*).

The best understood model system for understanding the regulation of splicing is the sex determination pathway in *Drosophila* which also controls gene dosage. Alternative splicing is used in each branch of this pathway to control the expression of transcriptional regulators or chromatin-associated proteins that influence transcription, and both positive and negative control of splicing is evident (Lopez, 1998). In mammalian cells candidate splice regulators are the SR family of RNA-binding proteins (which have a distinctive C terminal domain rich in serine (S)–arginine (R) dipeptides] and some HnRNP (heterogeneous nuclear ribonucleoprotein particle) proteins. These proteins are known to promote various steps in assembly of spliceosomes and they are also known to bind to **splicing enhancer sequences**, regulatory sequences which can enhance splice site recognition (Lopez, 1998).

Alternative polyadenylation

The usage of alternative polyadenylation signals is also quite common in human mRNA, and different types of alternative polyadenylation have been identified (see Edwards-Gilbert *et al.*, 1997). In many genes, two or more polyadenylation signals are found in the 3′ UTR and the alternatively polyadenylated transcripts can show tissue specificity; in other cases, alternative polyadenylation signals may be brought into play following alternative splicing. As an example of the latter, a combination of alternative splicing and alternative polyadenylation of the calcitonin gene (*CALC*) results in tissue-specific expression of two isoforms. Calcitonin, a circulating Ca^{2+} homeostatic hormone, is produced in the thyroid; the calcitonin gene-related peptide (CGRP), which may have both neuromodulatory and trophic activities, is synthesized in the hypothalamus (*Figure 8.15*).

8.3.3 RNA editing is a rare form of post-transcriptional processing whereby base-specific changes are enzymatically introduced at the RNA level

RNA editing is a form of post-transcriptional processing which can involve enzyme-mediated insertion or deletion of nucleotides or substitution of single nucleotides *at the RNA level*. Insertion or deletion RNA editing appears to be a peculiar property of gene expression in mitochondria of kinetoplastid protozoa and slime molds. Substitution RNA editing is frequently employed in some systems, such as the mitochondria and chloroplasts of vascular

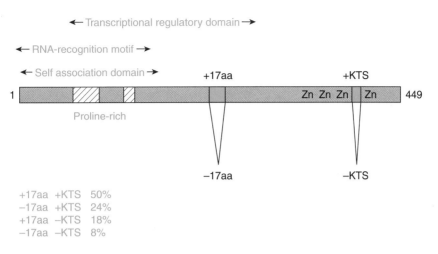

+17aa	+KTS	50%
−17aa	+KTS	24%
+17aa	−KTS	18%
−17aa	−KTS	8%

Figure 8.14: Differential splicing in the *WT1* Wilm's tumor gene.

The WT1 protein contains a transcriptional regulatory domain and a putative RNA-recognition motif at its N terminus. Four zinc fingers (Zn) are found at its C terminus, each encoded by a separate exon. Alternative splicing results in inclusion or exclusion of an exon encoding a centrally located sequence of 17 amino acids, and also a sequence specifying three amino acids (K, lysine; T, threonine; S, serine) located between the third and fourth zinc fingers. All four possible mRNA splice forms are expressed in normal WT1-expressing tissues at the approximate ratios shown.

plants where individual mRNAs may undergo multiple C → U or U → C editing events, and has also been observed in a few mammalian genes (Ashkenas, 1997). At least four different classes of RNA editing are known to occur in human cells:

■　*C ⇒ U editing*. Human *APOB* lipoprotein mRNA editing has been well-studied. In the liver the *APOB*

gene encodes a 14.1 kb mRNA transcript and a 4536 amino acid product, apoB100. However, in the intestine the same gene encodes a 7 kb mRNA which contains a premature stop codon not present in the gene and encodes a product, apoB48, which is identical in sequence to the first 2152 amino acids of apoB100. A specific cytosine deaminase, Apobec1, converts a single cytosine at nucleotide 6666 in the intestinal *APOB*

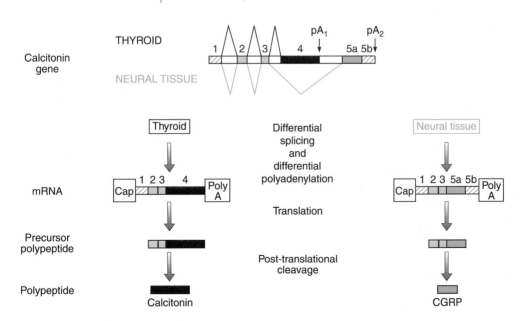

Figure 8.15: Differential RNA processing results in tissue-specific products of the calcitonin gene.

pA_1 and pA_2 represent alternative polyadenylation signals which are employed in thyroid and neural tissue respectively following alternative splicing. Exon 1 and the 3′ part of exon 5 (5b) encode 5′ and 3′ untranslated sequences respectively. Calcitonin is encoded by exon 4 sequences in the thyroid, while the calcitonin gene-related peptide (CGRP), which is synthesized in neural tissue is encoded by the 5′ part of exon 5 (5a) as a result of alternative splicing.

mRNA to uridine, thereby generating a stop codon (*Figure 8.16*).

- *A ⇒ I editing*. Genes encoding some ligand-gated ion channels including glutamate receptors and related proteins are subject to this type of mRNA editing. An adenosine is deaminated to give *inosine (I)*, a base not normally present in mRNA (the amino group at carbon 6 of adenosine is replaced by a C=O carbonyl group). Inosine base pairs preferentially with cytosine and also interacts with ribosomes during translation as if it were a G. In the case of the glutamate receptor B gene, for example, RNA editing replaces a CAG (glutamine) codon by CIG which is translated as if it were the CGG codon (arginine). This type of editing which brings about a Gln ⇒ Arg is often referred to as Q/R editing after the single letter code for the two amino acids involved.

- *Other classes of RNA editing*. Two other documented forms of editing in human mRNA are the U ⇒ C editing in mRNA from the *WT1* Wilms' tumor gene and U ⇒ A editing in α-galactosidase mRNA (Ashkenas, 1997).

8.4 Asymmetry as a means of establishing differential gene expression and DNA methylation as means of perpetuating differential expression

The concept of tissue specificity of human gene expression is long established. What is much less clear is how such patterns get laid down initially. Since the DNA content of all nucleated cells in an organism is virtually identical, genetic mechanisms cannot explain how differential gene expression first develops in cells. To explain this, CH Waddington evoked *epigenetic* mechanisms of gene control during development. In recent times, a variety of epigenetic mechanisms have been identified, including ones which can perpetuate particular states of gene expression in somatic cell lineages.

8.4.1 Selective gene expression in cells of mammalian embryos most likely develops in response to short range cell–cell signaling events

In order to explain subsequent tissue-, cell- and developmental stage-specific patterns of expression, some mechanism is required to set up an asymmetry or axis in the fertilized egg cell or in very early development. In *Drosophila*, the egg is inherently asymmetrical because of transfer of gene products from asymmetrically sited nurse cells. The embryo develops initially as a multinucleate syncytium (effectively one big cell) and regionalization depends on the response of individual nuclei to long-range gradients of regulatory molecules. In mammals, however, the egg cell is relatively small and early embryonic development creates an apparently symmetrical aggregate of individual cells. Nevertheless, development becomes asymmetric.

The generation of asymmetry in mammalian cells could derive from early positional clues. Some aspects of early development are inherently asymmetrical including

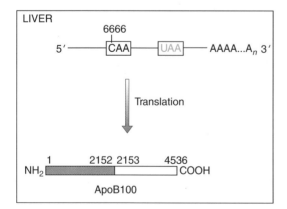

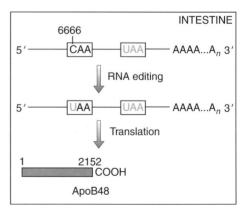

Figure 8.16: Expression of the human apolipoprotein B gene in the intestine involves tissue-specific RNA editing.

Note that codon 2153 specified by the CAA triplet at nucleotide positions 6666–6668 specifies glutamine in the ApoB100 product synthesized in liver. In the intestine, however, the CAA codon is converted by RNA editing to the stop codon **U**AA, resulting in a shorter product, ApoB48.

the point of entry of the sperm during fertilization, the attachment of the embryo to the uterine wall during implantation and the location of cells with respect to their neighbors. As the embryo develops into a ball of cells, and later on as more complex structures develop, individual cells will vary in the number of cell neighbors available. Short range intercellular signaling events (by direct cell–cell signaling or short-range intercellular signaling events) can provide a means of identifying cell position, and triggering differential gene expression. For example, if an intercellular signaling molecule has a range of, say, one cell diameter, then the cells at the outside of the blastula will receive different signals from those surrounded by neighbors on all sides, and the different positional cues may be translated into differential gene expression. As particular cell systems develop during, for example, organogenesis (mostly accomplished between the 4th and 9th embryonic weeks), particular cell type growth or differentiation factors may then induce the expression of developmental stage-specific and/or tissue-specific transcription factors.

8.4.2 Vertebrate DNA methylation is very largely confined to CpG dinucleotides and patterns of DNA methylation can be inherited when cells divide

Once differential expression patterns have been set up, epigenetic mechanisms can ensure that differential expression patterns are stably inherited when cells divide. DNA methylation is thought to play a major role in this respect, permitting the stable transmission from a diploid cell to daughter cells of chromatin states which repress gene expression. However, the precise function of DNA methylation in eukaryotes is still imperfectly understood and clearly shows species differences. Some organisms, for example, have no detectable DNA methlyation, as in the case of *Drosophila, C. elegans* and the yeast *Saccharomyces cerevisiae*. In those organisms where DNA methylation does occur, the patterns and functions of DNA methylation may differ.

Patterns of DNA methylation in vertebrates

The pattern of vertebrate DNA methylation differs from that in bacterial cells. In the latter, adenine and cytosine can both be methylated but in vertebrates methylation of DNA is restricted to cytosine residues. Only about 3% of the cytosines in human DNA are methylated, but most that are methylated are found in the CpG dinucleotide (that is, the methylated cytosines are almost always ones whose 3′ carbon atom is linked by a phosphodiester bond to the 5′ carbon atom of a guanine). In addition, a much smaller percentage of methylated cytosines occur within the sequence CpNpG.

Cytosine residues occurring in CpG dinucleotides in vertebrate DNA are targets for methylation by a specific cytosine methyltransferase. Methylation occurs at carbon atom 5 of the cytosine to generate 5-methylcytosine, which is chemically unstable and can spontaneously

deaminate to give thymine (*Figure 8.17A*). Over long periods of evolutionary time, the number of CpGs in vertebrate DNA has gradually been eroded, although regions of the normal (expected) CpG frequency are known and often mark transcriptionally active sequences (**CpG islands**, see *Box 8.5*).

Maintenance and de novo methylation during development

Unlike bacterial methylases, vertebrate cytosine methyltransferases show a strong preference for recognizing a *hemi-methylated* DNA target (i.e. one that is already methylated on one strand only). The sequence CpG shows dyad symmetry and so, following DNA replication, the newly synthesized DNA strands will receive the same CpG methylation pattern as the parental DNA (*Figure 8.17B*). As a result, the CpG methylation pattern can be stably transmitted to daughter cells. The perpetuation of a pre-existing methylation pattern is sometimes known as maintenance methylation and is carried out in mammalian cells by the product of the *Dnmt1* gene.

The pattern of 5-methlycytosine distribution in the genome of differentiated somatic cells varies according to cell type but maintenance methylation ensures that methylation patterns in individual somatic cell lineages are quite stable. During gametogenesis and in the developing embryo, however, there are dramatic changes in methylation (Razin and Kafri, 1994). The genomes of the primordial germ cells of the embryo are not methylated to any extent. After gonadal differentiation and as the germ cells begin to develop, *de novo* methylation occurs leading to substantial methylation of the DNA of mammalian sperm and egg cells (*Figure 8.19*). The sperm genome is more heavily methylated than the egg's genome, and *sex-specific differences in methylation patterns* are found, notably at imprinted loci (see Mertineit *et al.,* 1998 for references). The *Dnmt1* methyltransferase gene, in addition to being the predominant maintenance DNA methyltransferase in mammalian cells may also be the major *de novo* methyltransferase. It is highly expressed in male germ cells, mature oocytes and in the early embryo. *Dnmt1* gene expression has been shown to be subject to sex-specific regulation with oocyte- and spermatocyte-specific promoters introducing oocyte- and spermatocyte-specific exons leading to different gene products (*Figure 8.20*; Mertineit *et al.,* 1998).

The genome of the fertilized oocyte is an aggregate of the sperm and egg genomes and so it and the very early embryo are substantially methylated with methylation differences at paternal and maternal alleles of many genes. Later on, at the morula and early blastula stages in the preimplantation embryo, genome-wide demethylation occurs (*Figure 8.19*). Later still, at the pregastrulation stage, there is widespread *de novo* methylation. However, the extent of this methylation varies in different cell lineages:

- *the somatic cell lineage* is heavily methylated;
- *trophoblast-derived lineages* which give rise to the placenta, yolk sac, etc., are undermethylated;

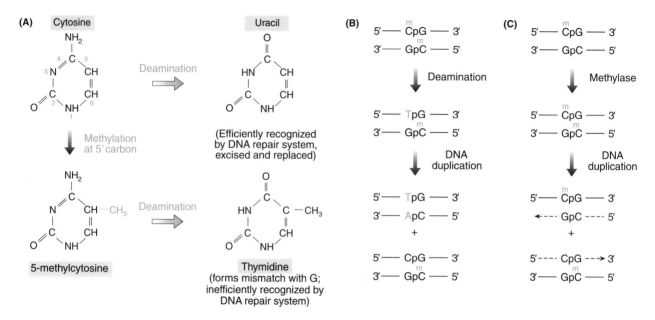

Figure 8.17: The CpG dinucleotide is underrepresented in vertebrate DNA because it is prone to methylation and deaminated 5-methylcytosine is subject to ineffective DNA repair.

(**A**) Cytosine occurring in the sequence 5′–CpG–3′ is a target for methylation at the 5′ carbon atom. The deaminated products of cytosine and its methylated derivative 5-methylcytosine are differentially recognized by DNA repair enzymes. (**B**) The methylation pattern of CpGs is perpetuated by a requirement for the specific methylase to recognize a **hemimethylated** target sequence. The sequence CpG has dyad symmetry. Following methylation of a hemimethylated target (i.e. methylated on one strand only), the two methylated strands will separate at DNA duplication and act as templates for the synthesis of two unmethylated daughter strands. The resulting daughter duplexes will now provide new hemimethylated targets for continuing the same pattern of methylation. (**C**) Deamination of 5-methylcytosine in the sequence CpG results in conversion of CpG dinucleotides to TpG and CpA dinucleotides.

CpG islands

In vertebrate DNA the sequence CpG is a signal for methylation by a specific cytosine DNA methyltransferase, which adds a methyl group to the 5′ carbon of the cytosine. The resulting 5-methylcytosine is chemically unstable and is prone to deamination, resulting in thymine (*Figure 8.17*). Over evolutionarily long periods of time the number of CpG dinucleotides in vertebrate DNA has gradually fallen because of the slow but steady conversion of CpG to TpG (and to CpA on the opposite strand). Although the overall frequency of CpG in the vertebrate genome is low (about 20% of the expected frequency), there are small stretches of DNA which are characterized by having the normal, expected frequency of CpG. Such islands of normal CpG density (**CpG islands**) are comparatively GC-rich (typically over 50% GC) and extend over hundreds of nucleotides. In the case of genes showing widespread expression, associated CpG islands are almost always found at the 5′ ends of genes, usually in the promoter region and often extending into the first exon. However, for genes which show restricted expression patterns, the associated CpG islands are quite often found some distance downstream of the transcriptional initiation site (see Jones, 1999 and *Figure 8.18*).

CpG islands can be distinguished by their susceptibility to cleavage with certain restriction nucleases. For example, the restriction nuclease *Hpa*II recognizes and cuts DNA within the sequence CCGG *unless* the central cytosine is methylated. As a result, *Hpa*II cuts frequently within islands to give small fragments (for this reason the former name for CpG islands was *HTF islands*, to signify Hpa II tiny fragments). Using this type of approach, a total of about 45 000 CpG islands have been estimated to occur in the human genome and approximately over half of all associated human genes are associated with a CpG island (Antequara and Bird, 1993). CpG island-associated genes include all housekeeping genes and genes that are widely expressed, and perhaps about 40% of genes which are expressed in a tissue-specific manner (Larsen *et al.*, 1992). While the great majority of CpGs located outside CpG islands are methylated, CpG islands are unmethylated or have very low amounts of CpG methylation. CpG islands associated with the transcription start sites of genes whose expression is restricted to certain tissues are unmethylated in these tissues but methylated in tissues where the genes are not expressed. However, CpG islands located downstream of the transcription initiation site may be methylated in tissues where the gene is expressed (see Jones, 1999 and text).

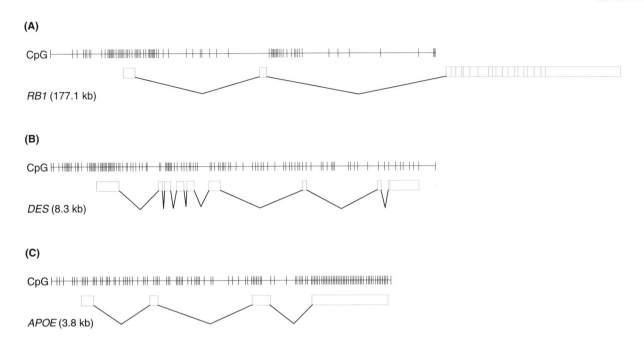

Figure 8.18: CpG island structure in three human genes.

Vertical bars represent the positions of the dinucleotide CpG in DNA sequences representing: (**A**) the human desmin *(DES)* gene; (**B**) the 5′ end of the human retinoblastoma *(RB1)* gene; and (**C**) the human apolipoprotein E *(APOE)* gene. Full gene lengths are indicated in brackets after the gene name. Boxes represent exons with splicing patterns shown by connecting thin black lines. In the case of the *RB1* gene only the first ~10 kb of sequence is shown. *Note* that CpG islands are often located at the 5′ end of a gene (as in *RBI* and *DES*), but occasionally may be found at other positions (as in the case of *APOE*).

- *early primordial germ cells* are spared; their genomic DNA remains very largely unmethylated until after gonadal differentiation and as the germ cells develop whereupon widespread *de novo* methylation occurs.

8.4.3 DNA methylation in animals has been thought to act as a form of host defense against transposons as well as a way of perpetuating patterns of transcriptional repression

Although not all eukaryotes appear to be subject to DNA methylation, its function in animal cells does appear to be critically important, and targeted muta-genesis of the cytosine methyltransferase gene in mice results in embryonic lethality. The precise function of DNA methylation in animal cells, however, remains unclear. Current views have focused in particular on two aspects of animal cells: the genome size (animals have comparatively large genomes with large numbers of genes, and also large numbers of highly repetitive DNA families belonging to the transposon class); and the mode of development (especially the variation in terms of lifespan and rate of cell turnover). Two quite contrasting views regarding the primary function of DNA methylation in animal cells have been the subject

of much controversy: the **host defense model** and the **gene regulation model**.

Host defense as a primary function for DNA methylation

Like the restriction-modification function of DNA methylation in bacteria (see *Box 4.1*), the host–defense model envisages that the primary function of DNA methylation in animal cells is to confer a form of genome protection, but in this case checking the spread of transposons (Yoder *et al.*, 1997). About one-third of the DNA sequence in the human genome can be classified as belonging to (retro)transposon families and a small fraction of these sequences in the human genome and other genomes is known to be actively transposing (see Section 7.4). Transposon families in the human and other genomes are known to be heavily methylated (about 90% of the 5-methylcytosines are thought to be located in retrotransposon families) and so DNA methylation has been viewed as a mechanism for repressing such transposition, which if left unchecked could be expected to be damaging to cells. However, recently obtained data from an invertebrate chordate, *Ciona intestinalis*, appear to be inconsistent with the genome defense model: multiple copies of an apparently active retrotransposon and a large fraction of highly repeated SINEs were predominantly

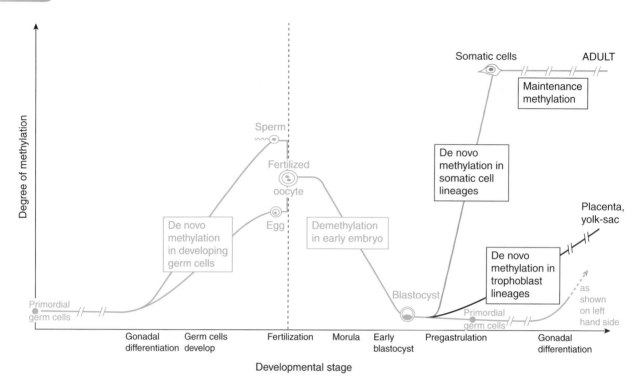

Figure 8.19: Changes in DNA methylation during mammalian development.

Developmental stages for gametogenesis and early embryo development are expanded for clarity; those for later development are contracted, as indicated by double slashes. Note the very rapid changes in DNA methylation during: (i) *gametogenesis – de novo* methylation gives rise to substantially methylated genomes in the sperm and egg (albeit with differences in both the overall level of methylation and the pattern of methylation in these genomes – see text), and in (ii) *the early embryo* where a wave of genome-wide demethylation occurs at the preimplantation stage (morula and early blastula), and is succeeded shortly afterwards by large-scale *de novo* methylation beginning at the pregastrulation stage. The latter is particularly pronounced in somatic lineages, and to a lesser extent in trophoblast lineages giving rise to placenta and yolk sac, but does not occur in the primordial germ cells (the cells of the embryo which will eventually give rise to sperm and egg cells).

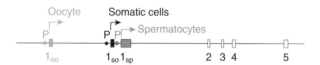

Figure 8.20: Sex-specific regulation of the *Dnmt1* methyl transferase gene.

The *Dnmt1* methyltransferase gene appears to be the predominant maintenance DNA methyltransferase in mammalian cells and may be the major *de novo* methyltransferase too. It is highly expressed in male germ cells, mature ooctes and in the early embryo. Five exons are used, but three alternatives for exon-1 are employed in different cells (Mertineit *et al.*, 1998): 1_{so} (exon 1 in somatic cells) and the two sex-specific exons, 1_{sp} (in spermatocytes) and 1_{oo} (in oocytes). The oocyte-specific exon is associated with the production of very large amounts of active Dnmt1 protein, which is truncated at the N terminus and sequestered in the cytoplasm during the later stages of growth. The spermatocyte-specific exon interferes with translation and prevents production of Dnmt1 during the crossing-over stage of male meiosis.

unmethylated, while genes, by contrast, appeared to be methylated (Simmen *et al.*, 1999).

Gene regulation as the primary function for DNA methylation

DNA methylation in vertebrates has been viewed as a mechanism for silencing transcription and may constitute a default position. DNA sequences which are transcriptionally active require to be unmethylated (at least at the promoter regions). While DNA methylation in invertebrates may serve to repress transposons and other repeated sequence families (see below), it may have acquired a special role in vertebrates as a mechanism for regulating expression of endogenous genes and reducing transcriptional noise (by silencing a large fraction of genes whose activity is not required in a cell). By reducing unnecessary gene expression, DNA methylation may have permitted the increase in gene number and in complexity that characterizes vertebrates (Bird, 1995). The counterargument is that the methylation status of the

5′ regions of tissue-specific genes cannot be correlated with expression in different tissues, and that the role of methylation in gene expression is in specialized biological functions resulting from mechanisms (e.g. imprinting) which use allele-specific gene expression (Walsh and Bestor, 1999).

DNA methylation and gene expression

The DNA of transcriptionally active and inactive chromatin differs in a number of features including the degree of compaction and the extent of its methylation (*Table 8.7*). While methylation of CpG islands downstream of promoters does not block continued transcription through these regions (Jones, 1999), there is no doubt that methylated promoter regions are correlated with transcriptional silencing. In addition, the extent of histone acetylation is an important factor. Specific histone acetyltransferases add acetyl groups to lysine residues close to the N terminus of histone proteins. The acetylated N termini then form tails that protrude from the nucleosome core. As the acetylated histones are thought to have a reduced affinity for the DNA, and possibly for each other, the chromatin may be able to adopt a more open structure that is more suited to gene expression. Deacetylation of the histones, however, promotes repression of gene expression presumably because the chromatin can become more condensed.

Recently, the processes of DNA methylation and histone deacetylation have been shown to be linked. Repression at methylated CpG sequences in promoter regions appears to be mediated by proteins which specifically bind to methylated CpG. Two of these proteins have been identified, MeCP1 and MeCP2 (methylated CpG binding proteins 1 and 2), and the latter has been shown to be essential for embryonic development and to function as a transcriptional repressor. The ability of MeCP2 to repress gene expression has been shown to involve a histone deacetylase complex (Ng and Bird, 1999). One possible

Table 8.7: Features associated with transcriptionally active and inactive chromatin

Feature	Transcriptionally active chromatin	Transcriptionally inactive chromatin
Chromatin conformation	Open, extended conformation	Highly condensed conformation; particularly apparent in heterochromatin (both facultative and constitutive; see Section 3.5)
DNA methylation	Relatively unmethylated, especially at promoter regions	Methylated, including at promoter regions
Histone acetylation	Acetylated histones	Deacetylated histones

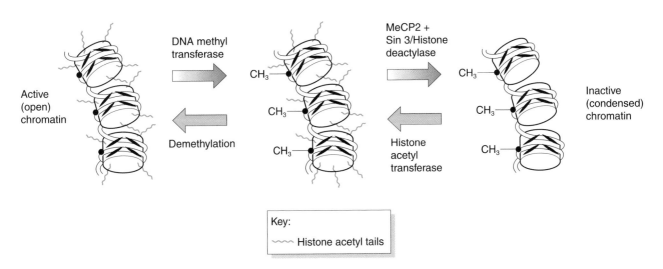

Key:
〜〜〜 Histone acetyl tails

Figure 8.21: Transcriptional repression by histone deacetylation may be mediated by DNA methylation.

CpG dinucleotides are targets for DNA methylation and, in turn, methylated CpGs are targets for specific binding by proteins such as MeCP2, which acts as a transcriptional repressor and recruits a corepressor complex consisting of the transcription factor repressor mSin3A and histone deacetylases. The latter removes acetyl groups from histones. The reverse process involves sequential histone acetylation then DNA demethylation (see Ng and Bird, 1999).

model envisages that an initial signal for transcriptional repression is the binding of MeCP2 to methylated CpG promoter sequences. The bound MeCP2 protein is then recognized by a complex consisting of a transcriptional repressor and a histone deacetylase which removes the acetyl groups from the N termini of the histones, so that the chromatin becomes more condensed (*Figure 8.21*).

8.5 Long-range control of gene expression and imprinting

8.5.1 Chromatin structure may exert long-range control over gene expression

A predominant theme in eukaryotic gene expression, distinguishing it from bacterial gene expression, has been that genes are *individually* transcribed. Promoters and related upstream elements typically control expression of a single gene with a transcription start point located within 1 kb of the control element. Some *cis*-acting elements, however, exert long-range control over a much larger chromosomal region and there is increasing evidence for coordinate regulation of gene clusters. Studies where genes are repositioned elsewhere in the genome have also suggested that chromosomes are organized into functional domains of gene expression (chromatin domains). For example, when genes are translocated to new chromosomal regions (either as a result of spontaneous chromosome breakage events, or by genetically manipulating model organisms – see Section 21.3), aberrant gene expression may often occur, even although the entire gene and the required control sequences in its immediate flanking sequences are preserved intact. Neighboring chromosome domains are envisaged to be separated by **boundary elements** (also called **insulators**) which act as barriers to the effects of distal enhancers and silencers (Geyer, 1997).

Competition for enhancers or silencers

Sometimes long-range control of gene expression appears to depend on competition between clustered genes for an enhancer. This appears to be a feature of globin gene expression as described in Section 8.5.2.

Heterochromatin-induced position effects

Studies of chromosomal rearrangements in *Drosophila* have shown that proximity to centromeres, telomeres or heterochromatic blocks may suppress gene expression, presumably by altering the structure of a large chromatin domain. Fascioscapulohumeral muscular dystrophy (FSHD; MIM 158900) is a possible example of a similar position effect in man. The gene for this autosomal dominant progressive neuromuscular disease maps close to the telomere of chromosome 4q. When Southern blots of *Eco*RI-digested DNA are hybridized to a subtelomeric probe p13H-11, a very large hybridizing fragment of over 30 kb is seen. DNA from FSHD patients consistently shows smaller bands of 14–28 kb. These patients have a

reduced copy number of a 3.2 kb repetitive sequence that is recognized by the probe, and *de novo* deletion of stretches of the repeats have been observed by FSHD patients with no previous family history.

Hopes that the 3.2 kb sequence would contain the *FSHD* gene, however, have been disappointed. Even though it contains part of a homeodomain (Hewitt *et al.*, 1994), there is no evidence that any part of it is transcribed or expressed. Most probably the *FSHD* gene is located proximally to the tandem 3.2 kb repeats, and the deletions move it closer to the 4q telomere, where it is silenced by a position effect.

Other position effects

Evidence of long-range effects controlling gene expression over large chromosomal domains has emerged from studies of disease-associated chromosome breakpoints in humans (Kleinjan and van Heyningen, 1998). Examples are aniridia (AN1; MIM 106210), which is caused by loss-of-function mutations of the *PAX6* gene on 11p13, and campomelic dysplasia (CMPD1; MIM 211970), which is caused by mutations in the *SOX9* gene on 17q24. In each case, affected patients are known who have clearly causative chromosomal breaks, but the breakpoints may be very distant (hundreds of kilobases) from the gene whose expression is affected and do not physically disrupt it. It seems likely that expression of the gene is suppressed by a long-range effect analogous to the classic position effects described above, reflecting the novel chromosomal environment created by the translocation.

Prader–Willi and Angelman syndromes (see Section 16.4.2) bring together position effects, imprinting and DNA methylation. A *cis*-acting sequence analogous to the globin locus control region has been identified which governs parent-specific methylation and gene expression of a megabase-size chromosomal region at 15q11.

X inactivation

X chromosome inactivation in mammals appears to be initiated by a single gene, *XIST*, which is uniquely expressed on the inactivated X chromosome (see Section 8.5.6). This effect is not understood but must be mediated by some sort of long-range chromatin structural change. This is so because a diffusible *XIST*-mediated agent would not be able to affect just the X chromosome on which the *XIST* gene is expressed.

8.5.2 Expression of individual genes in gene clusters may be coordinated by a common locus control region

Some human gene clusters show evidence of coordinated expression of the individual genes in the cluster. For example, individual genes in the α-globin, β-globin and the four *HOX* gene clusters are activated sequentially in a temporal sequence that corresponds exactly with their linear order on the chromosome. In the case of the globin genes, there is a clear developmental stage-specific expression: different genes can be active at the embryonic, fetal or adult stages to generate slightly different forms of hemoglobin (hemoglobin switching; *Figure 8.22*).

Recently, it has become apparent that the expression of the genes in each of the two human globin gene clusters is coordinated by a dominant control region, the **locus control region (LCR)** which is located some distance upstream of the gene cluster (see Grosveld *et al.*, 1993). Such cluster-specific LCRs are thought to organize the cluster into an active chromatin domain and to act as enhancers of globin gene transcription. The open conformation of transcriptionally active chromatin domains makes them more accessible to cleavage by the enzyme DNase I. Consistent with this relationship, the β-globin LCR has been considered to comprise short sequences at three major erythroid-specific DNase I-hypersensitive sites (HS2, HS3 and HS4) clustered over a 15 kb region located about 50–60 kb upstream of the β-globin gene, while the α-globin LCR has been identified to occur at an erythroid-specific DNase-hypersensitive site, HS-40, located 60 kb upstream of the α-globin gene (*Figure 8.23*). Each site marks the location of what is effectively an *enhancer* sequence of about 200–300 bp of DNA which contains short *cis*-acting sequence elements, including multiple sequence elements recognized by erythroid-specific transcription factors (see *Figure 8.6*). Without the respective LCRs, globin gene expression is negligible and, in the case of the β-globin LCR, it appears that the HS2, HS3 and HS4 elements interact with each other to form a larger complex that interacts with the individual globin genes.

Other DNase I-hypersensitive sites are located at the promoters of the globin genes, but show developmental stage specificity. For example, in fetal liver, the promoters of the two γ genes, the β and δ genes, are marked by DNase I-hypersensitive sites but, in adult bone marrow,

the two γ genes are no longer transcriptionally active and their promoters no longer reveal DNase I-hypersensitive sites. Developmental stage-specific switching in globin gene expression is then thought to be accomplished by competition between the globin genes for interaction with their respective LCR and stage-specific activation of gene-specific silencer elements. For example, transcription of the ε-globin gene (*HBE1*) is preferentially stimulated by the neighboring LCR at the embryonic stage. In the fetus, however, ε-globin expression is suppressed following activation of a silencer and γ-globin expression becomes dominant (*Figure 8.23*).

In addition to the human globin LCRs a number of different additional LCRs have been identified (see Kioussis and Festenstein, 1997). However, the role of the human β-globin LCR (which has relied on analysis of human transgenes) has been challenged by gene targeting studies in mice which show that the β-globin LCR has a contributory function rather than a dominant one, and that it is not required for initiation of DNase sensitivity and expression of the genes in the mouse β-globin cluster. These contradictory findings may possibly mean that there is a functional difference between the human and murine LCRs (see Grosveld, 1999 for a review).

8.5.3 Some human genes show selective expression of only one of the two parental alleles

X-linked genes in females and all autosomal genes are biallelic because both father and mother normally contribute one allele each. In males possessing one X chromosome and one Y chromosome, the great majority of sex-linked genes are **monoallelic**: most of the many genes on the X do not have a functional homolog on the Y chromosome; and some of the few genes on the Y chromosome are known to be Y-specific, for example *SRY*, the major male sex-determining locus. A few genes on the Y chromosome do have functional homologs on the X chromosome and so are biallelic. In some cases of X–Y homologous loci, both homologs are normally functional (see Section 14.3.1 and *Figure 14.9*).

We are accustomed to assuming that both the paternal and maternal alleles of biallelic genes are expressed, unless one or both copies have sustained mutations which affect expression. Clearly the expression can be tissue-specific so that in some cells both parental alleles are strongly expressed; in others, both gene copies are not apparently expressed. Thus, although there may be cell type-specific differences in expression, there is no discrimination between the capacity of the two parental alleles to be expressed, other than that due to genetic (mutational) differences between them. However, in humans and other mammals, several biallelic genes are known where the expression of one parental allele, either the paternal or the maternal allele but not both, is *normally* repressed in some cells (**allelic exclusion**). In such cells the relevant gene is said to exhibit functional hemizygosity: only one half of the maximum gene product is

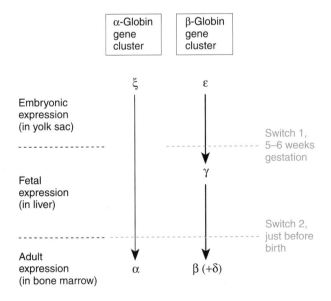

Figure 8.22: Human hemoglobin switching occurs at two distinct developmental stages.

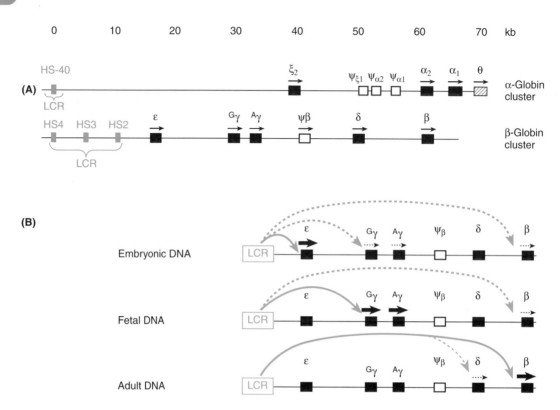

Figure 8.23: Gene expression in the α- and β-globin gene clusters is controlled by common locus control regions.

(**A**) Organization of the human α- and β-globin gene clusters. The locus control regions (LCRs) consist of one or more erythroid-specific DNase I-hypersensitive sites (HS-40, etc.) located upstream of the cluster. Arrows mark the direction of transcription of expressed genes. The functional status of the θ-globin gene is uncertain: it is expressed, but may be an expressed pseudogene (see *Box 7.3*). (**B**) Regulation of gene expression by the β-globin LCR. The strong blue arrows indicate a powerful enhancer effect by the LCR on the indicated genes, resulting in a high expression level; dotted blue arrows indicate correspondingly weak effects.

Box 8.6

Mechanisms resulting in monoallelic expression from biallelic genes in human (mammalian) cells

Mechanism of monoallelic expression	Relevant genes and cellular location
Allelic exclusion according to parent of origin	
Genomic imprinting	A small number of genes. Cellular locations depend on where an individual gene is expressed but note that some imprinted genes show monoallelic expression in some cell types but biallelic expression in others (*Table 8.8*)
Allelic exclusion but independent of parent of origin	
Allelic exclusion due to X-chromosome inactivation	Confined to certain X-linked genes in females only; expression of the allele from the noninactivated X chromosome only in all cells in which genes are expressed
Allelic exclusion following programed DNA rearrangement	Immunoglobin gene expression in B lymphocytes; T-cell receptor gene expression in T lymphocytes
Allelic exclusion by unknown mechanism	Olfatory receptor genes in neurons; NK cell receptor genes; certain interleukin genes (IL2, IL4); XIST (in cells of the early female embryo); PAX5 (in mature B cells and early progenitors)

normally obtained *even although the sequences of both parental alleles are perfectly consistent with normal gene expression and may even be identical.* In some cases the allelic exclusion may be a property of select cells or tissues while in other cells of the same individual both alleles may be expressed normally.

Although initially considered a rarity, monoallelic expression of biallelic genes has been demonstrated for a growing number of human genes. A variety of different expression mechanisms can be involved and two broad classes of mechanism are involved (see Chess, 1998; Ohlsson *et al.*, 1998):

- *Allelic exclusion according to parent of origin (imprinting).* In some cases, the choice of which of the two inherited copies is expressed is not random. This means that for some genes the allele whose expression is repressed is always the paternally inherited allele; in others it is always the maternally inherited allele (see Section 8.5.4).

- *Allelic exclusion independent of parent of origin.* Here the decision as to which of the two alleles is repressed is initially made randomly, but afterwards that pattern of allelic exclusion is transmitted stably

to daughter cells following cell division. A variety of different mechanisms may be involved (see *Box 8.6*). In some cases, complex gene regulation may be required. For example, olfactory receptor genes are found in large clusters or arrays in mammalian genes. In an individual olfactory neuron only one allelic array of olfactory receptor genes is active (see Chess, 1998). A unique form of control is the programed DNA rearrangements which are required for individual cell-specific expression of the immunoglobulin genes in B cells and the T cell receptor genes in T lymphocytes. Because of the complexity of the latter mechanisms they are discussed separately in Section 8.6.

8.5.4 Genomic imprinting involves differences in the expression of alleles according to parent of origin

Various observations in mammals have suggested that the maternal and paternal genomes in an individual are not equivalent (*Box 8.7*). In addition to genetic differences between the DNA of the sperm and oocyte genomes,

---

Box 8.7

The nonequivalence of the maternal and paternal genomes

In addition to the obvious X/Y sex chromosome difference, nonequivalence between paternally and maternally inherited autosomes and X chromosomes is indicated from the observations listed below.

Experimentally induced uniparental diploidy in mice
The male pronucleus of a fertilized mouse oocyte can be removed and replaced by a second female pronucleus to generate a gynogenote (sometimes called a parthenogenote; all 46 chromosomes are of maternal origin). If, instead, the female pronucleus were to be replaced by a second male pronucleus an androgenote is formed. Despite having normal diploid chromosomes, such embryos fail to develop and die prior to mid-gestation. Gynogenotes show severe deficiencies in extraembryonic structures but a relatively normal embryo; in contrast, in androgenotes the embryo is more severely affected than the extraembryonic structures (see Bestor, 1998 for references).

Naturally occurring uniparental diploidy in humans (see Section 2.6.4)
Human uniparental conceptuses are not uncommon. Androgenetic conceptuses develop as hydatidiform moles which consist of masses of hydropic chorionic villi and other placental structures, but lack embryonic tissues. Gynogenetic conceptuses give rise to dermoid cysts which develop into ovarian teratomas, consisting of a mass of well-differentiated but highly disorganized adult tissues, often including bone, tooth, cartilage, skin, and other tissues but usually lacking any extraembryonic structures.

Triploid abortuses may be considered to represent a combination of a diploid genome inherited from one parent and a normal haploid genome from the other. The phenotype is different depending on which parent contributes the diploid genome.

Uniparental disomy (Section 2.6.4)
Some conceptuses have a normal 46,XX or 46,XY karyotype but may have inherited two copies of the same chromosome from just one of the two parents. This may result in abnormal phenotypes which are different according to parental origin of the relevant chromosome. For example, 46,XX or 46,XY individuals who inherit both copies of chromosome 15 from their father develop Angelman's syndrome; if both copies of chromosome 15 are maternally inherited, Prader–Willi syndrome results (see *Box 16.6*).

Subchromosomal mutations causing differential abnormal phenotypes according to parent of origin

- Deletion of certain chromosomal regions produces a different phenotype when on the maternal or paternal chromosome. The best example is deletion of 15q12, which on the paternal chromosome produces Prader–Willi syndrome and on the maternal chromosome produces Angelman syndrome (see *Box 16.6*).

- Certain human characters are autosomal dominant but manifest only when inherited from one parent. In some families glomus tumors are inherited as an autosomal dominant character, but expressed only in people who inherit the gene from their father. Beckwith–Wiedemann syndrome (MIM 130650) is sometimes dominant but expressed only by people who inherit it from their mother. Example pedigrees are shown in *Figure 3.5*.

- Allele loss in many cancers (Chapter 18) preferentially involves the paternal allele.

there are also epigenetic differences. A major difference is in both the total amount of DNA methylation (the sperm genome is more extensively methylated than the oocyte genome) and the pattern of DNA methylation in specific DNA sequence classes. For example, Line 1 sequences are highly methylated in sperm cells but only partially methylated in the oocyte (Razin and Kafri, 1994; Yoder *et al.*, 1997). At some individual gene loci, too, there are major differences between the extent of methylation of paternal and maternal alleles. For example, the paternal allele of the *H19* gene is heavily methylated; the maternal allele is undermethylated.

As suggested by the observations in *Box 8.7*, differences between the paternal and maternal genomes lead to differences in expression between paternal and maternal alleles. Genomic imprinting (also called gametic or parental imprinting) in mammals describes the situation where there is nonequivalence in expression of alleles at certain gene loci, dependent on the parent of origin (Reik and Walter, 1998; Brannan and Bartolomei, 1999; Tilghman, 1999). In all (or at least some) of the tissues where the gene is expressed, the expression of either the paternally inherited allele or the maternally inherited allele, is consistently repressed, resulting in monoallelic expression. The same pattern of monoallelic expression can be faithfully transmitted to daughter cells following cell division. However, as the nucleotide sequence of the allele whose expression is repressed may be perfectly consistent with gene expression (and may even be identical to that of the expressed allele), this is an epigenetic phenomenon, not a genetic one.

Prevalence and evolution of imprinting

Most human genes are not subject to imprinting, otherwise we would not see so many simple mendelian characters. Systematic surveys have been made to identify imprinted chromosomal regions in the mouse. Unlike in humans, all mouse chromosomes are acrocentric and Robertsonian translocations can permit crosses to be set up which produce offspring having both copies of one particular chromosome derived from a single parent (**uniparental disomy, UPD**, see Section 2.6.4). These reveal that UPD for some chromosomes has no phenotypic effect; for others it produces abnormal phenotypes. The abnormal phenotypes are sometimes complementary for different parental origins, e.g. overgrowth is often seen in maternal UPD and growth retardation in paternal UPD. For some chromosomes, UPD is lethal.

Further dissection at the chromosomal and genetic levels shows that imprinting is a property of a limited number of individual genes or small chromosomal regions. Currently, a total of over 30 genes are known to be imprinted in humans and mice (electronic reference 1), but the list can be expected to grow. Thus far, two major clusters of imprinted genes are known in the human genome: a 1 Mb region at 11p15 (encompassing the Beckwith–Wiedemann region) contains at least seven imprinted genes which may be arranged in two clusters (Lee *et al.*, 1999); a 2.3 Mb cluster at 15q11–q13 region (encompassing the Prader–Willi and Angelman syndrome loci) also contains at least seven imprinted genes (Schweizer *et al.*, 1999; see *Figure 8.24*). The imprinted gene clusters contain examples of neighboring genes with different parental imprints

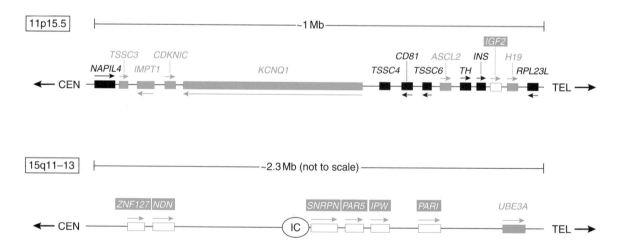

Figure 8.24: Imprinted gene clusters in 11p15.5 and 15q11–q13.

Genes not known to be imprinted are shown in black; imprinted genes are in blue. Genes shown as solid blue boxes, e.g. *KCNQ1*, *UBE3A*, show *preferential repression of paternal alleles* (so that in some tissues only the maternal allele is expressed). Genes shown in open blue boxes, e.g. *IGF2*, *ZNF127*, have the opposite pattern: *preferential repression of maternal alleles*. Arrows indicate direction of transcription. IC denotes *imprinting center* (see text). The 15q11–q13 region has been less well studied and other genes are likely to be found there. Note that some of the genes have other names in the literature, e.g. *KCNQ1* (*KvLQT1*), *CDKN1C* (*p57* or *KIP2*), *CD81* (*TAPA1*). Data from Lee *et al.* (1999) and Schweizer *et al.* (1999).

e.g. the *H19* gene is expressed only from the maternal chromosome 11 whilst the adjacent *IGF2* gene is expressed only from the paternal chromosome.

The great majority of known imprinted genes are autosomal. However, the *XIST* gene which has a major role in establishing X chromosome inactivation (see next section) may be considered an example of an imprinted X-linked gene since expression of the maternally inherited allele is preferentially repressed in trophoblast. An imprinted X-linked gene which affects cognitive function has also been suggested from differential behavior patterns in Turner syndrome. Girls with Turner syndrome lack a Y chromosome but have only one X chromosome. If the X chromosome is inherited from the mother, socially disruptive behavior is common, but if inherited from the father, the girl shows behavior closer to normal for that of a girl (Skuse *et al.*, 1997).

Imprinting is known to occur in seed plants, some insects and mammals. No major imprinting effect, as judged by phenotype, has been observed in some model organisms such as *Drosophila, C. elegans* and the zebrafish, although the potential for imprinting may exist in *Drosophila*. Mammals are unusual in the way in which embryos are totally dependent on flow of nutrients from the maternal placenta. As many imprinted genes are involved in regulating fetal growth, one explanation envisages conflict between the parental genomes: the paternal genome propagates itself best by creating an embryo which aggressively removes nutrients from the mother; the maternal genome suppresses this to protect the mother and spare some resources for future offspring. As seen in cases of uniparental diploidy (see *Box 8.7*), paternal genes are preferentially expressed in the trophoblast and extraembryonic membranes, while maternal genes are preferentially expressed in the embryo.

8.5.5 The mechanism of genomic imprinting is unclear but a key component appears to be DNA methylation

To confirm imprinting of a gene, it is necessary to identify an individual who is heterozygous for a sequence variant present in the mature mRNA. mRNA from different tissues can then be checked for monoallelic or biallelic expression, and the origin of each allele determined by typing the parents. For some genes, this type of analysis has shown that imprinting is confined to only certain tissues or to certain stages of development (*Table 8.8*). Thus imprinting allows an extra level of control of gene expression, but it is not possible to compress its functioning into a simple uniform story.

The above observations suggest that some mechanism must be able to distinguish between maternally and paternally inherited alleles: as chromosomes pass through the male and female germlines they must acquire some imprint to signal a difference between paternal and maternal alleles in the developing organism. A key component, at least in maintaining the imprinted status, is allele-specific DNA methylation (Brannan and Bartolomei, 1999; Tilghman, 1999). The imprinting of several imprinted genes has been shown to be disrupted in mutant mice that are deficient in the *Dnmt1* cytosine methyltransferase gene and all imprinted genes are characterized by CG-rich regions of differential methylation.

Intriguingly, *Dnmt1* is known to have sex-specific exons (Section 8.4.2 and *Figure 8.20*). In oocytes this results in an oocyte-specific amino-terminal truncated protein product which conceivably could specifically methylate the maternal alleles of genes such as the insulin-like growth factor II receptor. The spermatocyte-specific exon of *Dnmt1* interferes with translation of *Dnmt1* mRNA, and it is less clear how paternal-specific patterns of methylation could be acquired. During development, the imprint would be expected to be stably inherited at least for many rounds of DNA duplication (but see below). Clearly, there must also be a mechanism for erasing the imprints during transmission through the germline, as required when, for example, a man passes on an allele which he has inherited from his mother (*Figure 8.25*). Again, however, one can envisage the demethylation that occurs during the early embryo as one way of achieving this, leaving the primordial germ cells essentially unmethylated (see *Figure 8.19*).

The timing of imprinting has become more clear recently. In the female germline, the maternal imprint, including the maternal pattern of methylation, is likely to be established during oocyte maturation which is consistent with the finding that the Dmnt1 protein is not detectable in nongrowing oocytes but is produced abundantly in growing oocytes. In the male germline, the functional paternal imprint is likely to be established prior to

Table 8.8: Examples of tissue and developmental stage regulation of imprinted genes in mammals

Gene and location	Repressed allele	Differences in expression patterns
IGF2 (insulin-like growth factor type 2)	Maternal	Imprinted in many tissues but biallelic expression in brain, adult liver and chondrocytes etc.
PEG1/MEST	Maternal	Imprinted in fetal tissue but biallelically expressed in adult blood
UBE3A (ubiquitin protein ligase 3)	Paternal	Imprinted exclusively in brain; biallelically expressed in other tissues
KCNQ1 (potassium channel involved)	Paternal	Imprinted in several tissues but biallelically expressed in heart
WT1 (Wilms' tumor gene)	Paternal	Frequently imprinted in cells of placenta and brain but biallelic expression in kidney

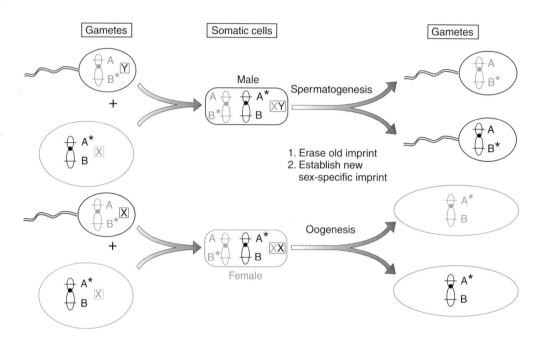

Figure 8.25: Genomic (gametic) imprinting requires erasure of the imprint in the germline.

The diagram illustrates the fate of a chromosome carrying two genes, A and B, which are subject to imprinting: A is imprinted in the female germline, B is imprinted in the male germline, as indicated by asterisks. As a result, in diploid somatic cells A is imprinted when present on a maternally inherited chromosome and B is imprinted when present on a paternally inherited chromosome. An individual chromosome may pass through the male and female germlines in successive generations: a man may transmit a chromosome inherited from his mother and a woman can transmit a chromosome inherited from her father, as indicated by the gametes in the left panel. As a result, there must be a mechanism whereby the old imprint is erased from the germline prior to establishing a new sex-specific imprint.

meiosis, possibly in the postmitotic primary spermatocyte (Brannan and Bartolomei, 1999).

Imprinted genes frequently reside in clusters with genes expressed on opposite chromosomes often located next to each other, and often containing genes which appear to encode a mature RNA (see *Figure 8.24*). Adjacent genes appear to be jointly regulated. In the case of the Prader–Willi/Angelman syndrome cluster on 15q11–q13, for example, a single region adjacent to the *SNRPN* gene, termed the imprinting center, is the dominant regulatory sequence and appears to act over comparatively large distances (see *Figure 8.24*, Brannan and Bartolomei, 1999; Tilghman, 1999). However, different mechanisms may be found in different imprinted clusters, or even within a single cluster. For example, the mouse *H19*, *Igf2* and *Ins2* genes are jointly regulated, sharing two endodermal-specific enhancers that are located 3′ to *H19*. But other genes in the cluster are not subject to this control, suggesting that multiple control mechanisms may occur within an imprinted gene cluster. Different imprinting mechanisms have been considered in relation to the *H19/Igf2* regulation, including enhancer competition, but to explain a variety of contradictory findings, an imprinting center adjacent to *H19* has been considered to function as a chromatin boundary (insulator) element (Tilghman, 1999).

8.5.6 X chromosome inactivation in mammals involves very long-range cis-acting repression of gene expression

Nature of X chromosome inactivation

X chromosome inactivation is a process that occurs in all mammals, resulting in selective inactivation of alleles on one of the two X chromosomes in females (Migeon, 1994; Lyon, 1999). It provides a mechanism of dosage compensation which overcomes sex differences in the expected ratio of autosomal gene dosage to X chromosome gene dosage (which is 2:1 in males but 1:1 in females). Males with a single X chromosome are constitutionally hemizygous for X chromosome genes, but females become functionally hemizygous by inactivating one of the parental X chromosome alleles (see also Section 2.2.3). Not all genes on the X chromosome are subject to inactivation; genes which escape X-inactivation include ones where there is a functional homolog on the Y chromosome, and some genes where gene dosage does not seem to be important (see *Figure 14.12* for examples of genes which escape X-inactivation).

In rare individuals with an abnormal number of X chromosomes (45,X; 47,XXX; 47, XXY, etc), a single X chromosome remains active no matter how many are present. By contrast, in triploid individuals either one or two X chromosomes remain active and in tetraploids two X chromosomes remain active. Thus, there must be some kind of *counting mechanism* to ensure that one X chromosome remains active for every two sets of autosomes.

In mammals, both X chromosomes are active in the early female embryo. X-inactivation occurs at an early stage in development, being initiated at the late blastula stage in mice, and most likely also in humans. In each cell that will give rise to the female fetus, one of the two parental X chromosomes is *randomly* inactivated (*note* that trophoblast cells are an exception; the paternal X chromosome is preferentially inactivated, which is a classical example of tissue-restricted imprinting). After the paternal or maternal X chromosome is inactivated in a cell, the same X chromosome usually remains inactive in all progeny cells, that is the X chromosome inactivation pattern is clonally inherited (see *Figure 2.6*). This means that female mammals are mosaics, comprising mixtures of cell lines in which the paternal X is inactivated and cell lines where the maternally inherited X is inactivated. In addition to X chromosome inactivation in female somatic cells, the X chromosome is known to be inactivated transiently during gametogenesis in both males and females.

Mechanism of X chromosome inactivation

The process of X chromosome inactivation is complex, and distinct molecular mechanisms are involved in initiation of inactivation and maintenance of the inactivation. The X-inactivation center (Xic), which in humans is located at Xq13, controls the initiation and propagation of X-inactivation. At this centre, the *XIST* gene (called *Xist* in rodents) encodes a mature 15 kb RNA product which is uniquely encoded by the inactive X chromosome. *XIST/Xist* is therefore another example of a gene that is subject to monoallelic expression. In the cells of the early embryo, the decision regarding which X chromosome to inactivate is made randomly, and so the allelic exclusion which *XIST/Xist* shows in these cells is independent of parent of origin.

XIST/Xist is essential for Xic function in initating X chromosome inactivation but is not required for maintaining X chromosome inactivation. Somehow *cis*-limited spreading of this RNA product acts so as to coat the inactivated X chromosome over very long distances. In rodents, coating of the *Xist* RNA gives the inactivated X chromosome a banded pattern suggesting a preferential association with gene-rich Giemsa-minus regions. However, the mechanism of ensuring inactivation of genes on the inactive X but not on the active X is unknown (see Duthie *et al.*, 1999 for possible models).

Although the *Xist* gene is essential for Xic function, *Xist* alone is not sufficient. The X-controlling element (Xce) affects the choice of which X chromosome remains active, and is distinct from *Xist*, being located 3' to it. In addition, deletion of a 65-kb region 3' to *Xist* produces an effect which suggests that elements involved in the counting mechanism lie distal and 3' to *Xist*. Recently, another gene has been identified as being transcribed from the opposite strand to that which is used for transcribing the

Xist gene. Because the transcription unit of the new gene completely overlaps the *Xist* gene, and is in the reverse orientation, it has been named *Tsix* (Lee *et al.*, 1999). This has given rise to the idea that *Xist* may be regulated by the *Tsix* gene (see Heard *et al.*, 1999 for possible models of *Tsix* regulation).

8.6 The unique organization and expression of Ig and TCR genes

The organization and expression of immunoglobulin (Ig) and T-cell recepter (TCR) genes is in many ways quite different from that of other genes. This is so because of the need for each individual to produce a huge variety of different Igs and TCRs. An individual B or T lymphocyte is *monospecific* and produces a single type of Ig or TCR; it is the population of different B and T cells in any one individual that enables the synthesis of so many different types of these molecules. B and T lymphocytes need to be extremely diverse because they represent the cells that provide antibody responses or cell-mediated responses to foreign antigen: by providing a large repertoire of Igs and TCRs, the possibilities for being able to recognize and bind very many different types of foreign antigen are greatly increased.

8.6.1 Ig and TCR genes exhibit a unique organization: multiple gene segments can encode each of several different regions of the polypeptide

Polypeptide structure

An Ig molecule consists of four polypeptide chains, two identical heavy chains and two identical light chains (see *Figure 7.10*). The light chains fall into two classes: kappa (κ) and lambda (λ) light chains, which are functionally equivalent. At the N-terminal segments of each type of chain are the so-called variable (V) regions, which need to bind foreign antigen; the remaining C-terminal segments are constant (C) regions. In the case of the heavy chains, there are different alternatives for the constant region which specify the tissues in which the Ig will be expressed and dictate the immunoglobulin class (*Table 8.9*). Similarly, TCRs, which provide cell-mediated immune responses to foreign antigens, consist of two types of chain. Each such chain has Ig-like variable regions which bind foreign antigen, and constant regions which anchor the molecule to the cell surface (see *Figure 7.10*). The most frequently occurring TCRs have a β and a γ chain; a minor population consists of an α chain and a δ chain.

Gene structure

The genes which encode the different types of chain in Igs and TCRs are located on different chromosomes and are

Table 8.9: Ig classes and subclasses

Class (and subclass)	Type of heavy chain	Location
IgA (IgA1,IgA2)	α (α_1, α_2)	Predominant Ig in seromucous secretions, e.g. saliva, milk, etc.
IgD	δ	Low in serum but present in large quantities on surface of many circulating B cells
IgE	ε	Especially on surface membrane of basophils and mast cells
IgG (IgG1, IgG2, IgG3, IgG4)	γ (γ_1, γ_2, γ_3, γ_4)	Major serum Ig
IgM	μ	Predominant 'early' antibody

organized as clusters of numerous gene segments (*Table 8.10*). Each such cluster is unusual in that the coding sequences for *specific segments* of each chain are often present in numerous different copies that are sequentially repeated. For example, although the constant region of human κ light chain Ig is encoded by a single C_κ sequence, the variable regions are encoded by a combination of a V_κ segment (which encodes most of the variable region) and a short J_κ segment (joining segment; encodes a small part at the C-terminal end of the variable region) which are selected from a total of about 76 alternative V_κ segments and five alternative J_κ segments. Although the λ light chain is similarly encoded by V_λ, J_λ and C_λ segments, the heavy chain Ig locus shows some differences. The variable region is encoded by a combination of a V_H gene segment, a J_H gene segment and also a D_H gene segment (encoding a diversity segment), each selected from many repeated gene segments. Additionally, there are a variety of different C_H sequences which specify the *class* of the Ig (see above). In total this cluster comprises about 140 gene segments, of which about one-third are known to be incapable of expression, and spans about 1200 kb (*Figure 8.26*).

As each Ig gene cluster or TCR gene cluster in an individual B or T lymphocyte only ever gives rise to at most one Ig or TCR polypeptide, an entire cluster can functionally be regarded as a single, albeit unusual, type of gene. However, the individual gene segments cannot be regarded as the functional equivalent of *classical exons*. This is so because individual gene segments in these clusters are sometimes composed of coding DNA and noncoding DNA and may consist of several exons. For example, each of the human C_H sequences is itself composed of three or four classical exons separated by

introns: after transcription into RNA, the intronic sequences are discarded, and only the exonic sequences are retained in the mRNA.

8.6.2 Programmed DNA rearrangements at the Ig and TCR loci occur during the maturation of B and T lymphocytes, respectively

The unique arrangement of gene segments in the Ig and TCR gene clusters reflects the very unusual way in which somatic recombinations are required in B and T lymphocytes before functional Ig and TCR genes can be assembled and then expressed (see below). Such somatic recombinations result in bringing together different combinations of the different gene segments in different individual lymphocytes. Consequently, they can be regarded as both *tissue-specific* (confined to B and T lymphocytes) and *cell-specific* events which involve *alternative DNA splicing* (as opposed to alternative RNA splicing which brings about different combinations of exons *at the RNA level* – Section 8.3.2). As a result, the original germline gene organization is altered: gene segments that were distant in the germline are spliced together at the DNA level. Because the choice of which of the many repeated gene segments are recombined to give a functional V–J or V–D–J unit is *cell specific*, individual B and T cells produce different Igs and TCRs. This means that, in a sense, every individual is a mosaic with respect to the organization of the Ig and TCR genes in B and T lymphocytes; even identical twins will diverge genetically.

The rearrangements which lead to the production of functional light chains and heavy chains of Igs are slightly different.

- *Making a light chain*. In order to generate a functional κ light chain Ig, for example, a somatic recombination event brings together a specific combination of one of the V_κ gene segments and one of the J_κ gene segments (V–J joining). Thereafter, splicing to the single C_κ sequence occurs *at the RNA level* (*Figure 8.27A*).
- *Making a heavy chain*. Two successive somatic recombinations are required, resulting first in D_H–J_H joining, and then V_H–D_H–J_H joining. Subsequently the resulting V_H–D_H–J_H coding sequence is spliced *at the RNA level* to the nearest C_H sequence, initially C_μ (*Figure 8.27B*).

Table 8.10: Functional human Ig and TCR loci

Locus	Location	Number of gene segments			
		V	D	J	C
IGH	14q32.3	86	30	9	11
IGK	2p12	76	0	5	1
IGL	22q11	52	0	7	7
TCRA	14q11–12	60	0	75	1
TCRB	7q32–33	70–100	2	13	2
TCRG	7p15	8	0	5	2
TCRD	14q11–12	6	3	3	1

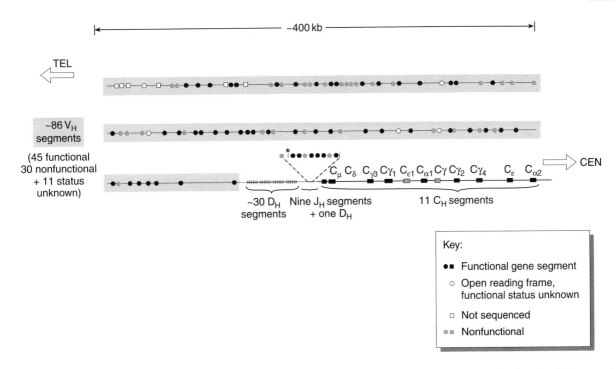

Figure 8.26: The Ig heavy chain locus on 14q32 contains about 86 variable (V) region sequences, 30 diversity (D) segments, nine joining (J) segments and 11 constant region (C) sequences.

The entire locus spans about 1200 kb of 14q32.3 and, for clarification, is shown as three segments of 400 kb from the telomeric end (top) to the centromeric end (bottom). Although the D_H segments are mostly located in a few clusters separating the V_H and J_H segments, at least one such segment is located in the J_H segment region. Segments indicated by open circles have the required open reading frames but have not been observed in productive rearrangements and so their functional status is unknown. Segments which are known to be nonfunctional are indicated in blue, and account for approximately one third of all the segments. *Note* that although this is the only functional human heavy chain locus, small clusters of V_H and D segments are also located on 15q11.2 and 16p11.2. Adapted from data in Cook *et al.* (1994) *Nature Genet.*, **7**, pp. 162–168, with permission from Nature America Inc.

Because there are three types of functional Ig gene loci in human cells (heavy chain, κ light chain and λ light chain), and because these occur on both maternal and paternal homologs, there are six chromosomal segments in which DNA rearrangments can result in production of an Ig chain. However, an individual B cell is *monospecific*: it produces only one type of Ig molecule with a single type of heavy chain and a single type of light chain. This is so for two reasons:

■ *Allelic exclusion.* A light chain or a heavy chain can be synthesized from a maternal chromosome or a paternal chromosome in any one B cell, but not from both parental homologs. As a result, there is monoallelic expression at the heavy chain gene locus in B cells. This phenomenon also applies to TCR gene clusters.

■ *Light chain exclusion.* A light chain synthesized in a single B cell may be a κ chain, or a λ chain, but never both. As a result of this requirement, plus that of allelic exclusion, there is monoallelic expression at one of the two functional light chain gene clusters and *no expression* at the other. The decision to choose which of the two heavy chain alleles and which of the four possible

segments to make a light chain appears to be random. Most likely, in each B-cell precursor, productive DNA rearrangements are attempted at all six Ig alleles but the chances of productive arrangements in more than one light chain cluster or more than one heavy chain allele may not be high. Additionally, however, there appears to be some kind of negative feedback regulation: a functional rearrangement at one of the heavy chain alleles suppresses rearrangements occurring in the other allele, and a functional rearrangement at any one of the four regions capable of encoding a light chain suppresses rearrangements occurring in the other three.

8.6.3 V–J and V–D–J joining is often achieved by intrachromatid deletions, and also by megabase inversions in the former case

The genetic mechanism leading to V–J and V–D–J joining often involves large-scale deletions which are thought to occur by an intrachromatid recombination event, similar to those used in V–D–J–C joining (see next section). In

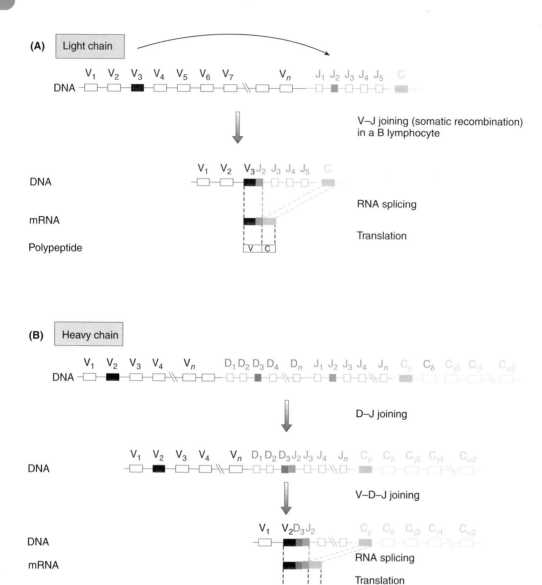

Figure 8.27: Igs are synthesized following somatic recombination of V and J, or V, D and J segments and subsequent RNA splicing to C sequences.

(**A**) Light chain synthesis. Somatic recombination (*DNA splicing*) results in joining of a specific variable (V) segment to a specific joining (J) segment; the example shows a V_3–J_2 joining which is only one of many possibilities. The VJ unit is then spliced to the constant region (C) sequence by *RNA splicing*. (**B**) Heavy chain synthesis. Two sequential somatic recombinations produce first D–J joining, then a VDJ unit. Subsequent RNA splicing results in splicing of the VDJ sequence to the C_μ sequence. As the B cell matures, however, subsequent somatic recombinations result in joining of the previously selected VDJ unit to different C genes (*heavy chain switch*, see text and *Figure 8.29*).

addition, V–J joining often occurs as a result of megabase inversions. The human κ light chain gene locus spans about 1840 kb on 2p12 and includes about 76 V_κ segments, mostly comprising pairs of duplicated V gene segments, organized as two clusters: a *proximal cluster* located adjacent to the J_κ segments and to the C_κ segment, and a *distal cluster*. This occurs as a result of an inverted repeat structure: V gene segments in the proximal V_κ cluster usually have a corresponding duplicate in a distal V_κ

cluster which is separated from the proximal cluster by about 800 kb and in the opposite orientation (*Figure 8.28*). Depending on which V segment cluster is involved, V–J joining occurs by two possible routes:

- V segments in the distal cluster are joined to J segments by inversions.
- V segments in the proximal cluster are joined to J segments by deletions (*Figure 8.28*).

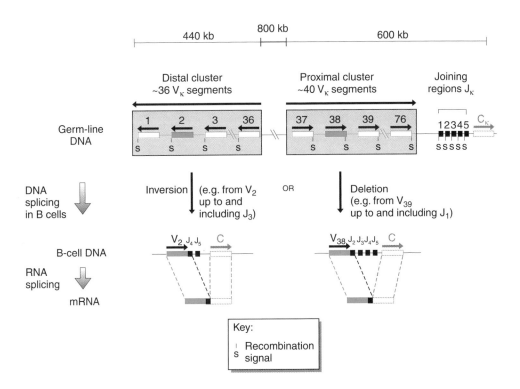

Figure 8.28: Inversion or deletion results in V–J splicing to produce functional Ig κ light chain genes.

The human κ light chain gene cluster contains about 76 V$_κ$ segments arranged in two large clusters, in opposite orientations. V segments in the distal cluster have the opposite orientation to the J$_κ$ segments and the single C$_κ$ sequence. As a result, the DNA rearrangements used to splice distal V$_κ$ segments to a J$_κ$ segment are megabase inversions (Weichhold *et al.*, 1990). Those in the proximal cluster can undergo V–J joining by a somatic recombination resulting in a deletion of the intervening chromosomal segment, most likely through an intrachromatid recombination event such as that used in class switching (see *Figure 8.29*).

Note that the joining process is imprecise, and so can also introduce a measure of variability in the sequence at the junctions of joined segments.

8.6.4 Class switching of heavy chains involves differential joining of a single VDJ unit to alternative DNA segments encoding constant regions

Although a B cell produces only one type of Ig molecule, the heavy chain class (see *Table 8.9*) can change during the cell lineage (class switching or isotype switching). Such switching involves differential joining of the same VDJ unit that was brought together by two successive somatic recombinations (see *Figure 8.27B*) to different segments encoding alternative constant regions. The initial joining of a VDJ sequence to constant region segments is accomplished *at the RNA level*. However, subsequently, class switching involves joining the same VDJ unit *at the DNA level* to alternative constant regions by yet more somatic recombination events (**V–D–J–C joining**). Class switching involves the following progression:

■ *initial synthesis of IgM only by immature B cells.* This occurs because the VDJ unit is spliced at the RNA level to a C$_μ$ sequence (*Figure 8.27B*).
■ *Later synthesis of both IgM and IgD by immature B cells.* The partial switch to making IgD occurs because the VDJ unit can be spliced at the RNA level to a C$_δ$ sequence, as a result of alternative RNA splicing (*Figure 8.29*).
■ *Synthesis of IgG, IgE or IgA by mature B cells.* Class switching events involve splicing the same VDJ unit to a C$_γ$, C$_ε$ or C$_α$ sequence, respectively, at the DNA level as a result of a somatic recombination event (VDJ–C joining). The mechanism involves deletion of the intervening sequence by intrachromatid recombination (*Figure 8.29*).

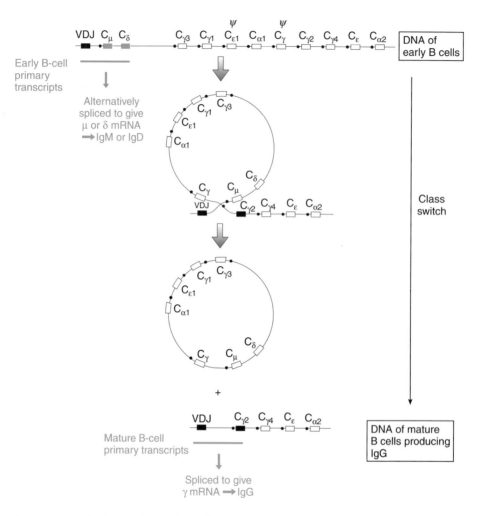

Figure 8.29 Ig heavy chain class switching is mediated by intrachromatid recombination.

Note that joining of the same VDJ unit to a C_μ or a C_δ sequence occurs at the level of RNA splicing to generate heavy chains for IgM and IgD respectively. In contrast, class switching to generate IgA, IgE or IgG involves joining of the same VDJ unit at the DNA level to, respectively, a C_α, C_ε or, as illustrated in the figure, a C_γ sequence.

Further reading

Day DA, Tuite MF (1998) Post-transcriptional gene regulatory mechanisms in eukaryotes: an overview. *J. Endocrinol.*, **157**, 361–371.

Gray NK, Wickens M (1998) Control of translation initiation in animals. *Annu. Rev. Cell Dev. Biol.*, **14**, 399–458.

Latchman D (1998) *Gene Regulation. A Eukaryotic Perspective.* Stanley Thornes, Cheltenham.

Roitt I, Brostoff J, Male D (1985) *Immunology.* Gower Medical Publishing, London.

Russo E, Martienssen RA, Riggs AD (1996) *Epigenetics and Mechanisms of Gene Regulation.* Cold Spring Harbor Laboratory Press, Cold Spring Harbor, NY.

Travers A (1993) *DNA–Protein Interactions.* Chapman & Hall, London.

van Driel R, Otte AP (1997) *Nuclear Organization, Chromatin Structure and Gene Expression.* Oxford University Press, Oxford.

Electronic references (e-Refs)

1. http://www.mgu.har.mrc.ac.uk/imprinting

References

Ainger K, Avossa D, Diana AS, Barry C, Barbarese E, Carson JH (1997) Transport and localization elements in myelin basic protein mRNA. *J. Cell Biol.*, **138**, 1077–1087.

Antequara F, Bird A (1993) Number of CpG islands and genes in human and mouse. *Proc. Natl Acad. Sci. USA*, **90**, 11995–11999.

Ashkenas J (1997) Gene regulation by mRNA editing. *Am. J. Hum. Genet.*, **60**, 2378–2383.

Ayoubi TA, Van De Ven WJ (1996) Regulation of gene expression by alternative promoters. *FASEB J.*, **10**, 453–460.

Bestor TH (1998) Cytosine methylation and the unequal developmental potentials of the oocyte and sperm genomes. *Am. J. Hum. Genet.*, **62**, 1269–1273.

Bird A (1995) Gene number, noise reduction and biological complexity. *Trends Genet.*, **11**, 94–100.

Blackwood EM, Kadonaga JT (1998) Going the distance: a current view of enhancer action. *Science*, **281**, 60–63.

Brannan CI, Bartolomei MS (1999) Mechanisms of genomic imprinting. *Curr. Opin. Genet. Dev.*, **9**, 164–170.

Burke TW, Kadonaga JT (1997) The downstream core promoter element, DPE, is conserved from Drosophila to humans and is recognized by TAFII60 of *Drosophila*. *Genes Dev.*, **11**, 3020–3031.

Chess A (1998) Expansion of the allelic exclusion principle. *Science*, **279**, 2067–2068.

Cook GP, Tomlinson, IM, Walter G, Riethman H, Carter NP, Buluwela L, Winter G, Rabbitts TH (1994) A map of the immunoglobulin V_H locus completed by analysis of the telomeric region of chromosome 14q. *Nature Genet.*, **7**, 162–168.

Cox GF, Kunkel LM (1997) Dystrophies and heart disease. *Curr. Opin. Cardiol.*, **12**, 329–343.

Day DA, Tuite MF (1998) Post-transcriptional gene regulatory mechanisms in eukaryotes. *J. Endocrinol.*, **157**, 361–371.

Duthie SM, Nesterova TB, Formstone EJ, Keohane AM, Turner BM, Zakian SM, Brockdorff N (1999) *XIST* RNA exhibits a banded localization on the inactive X chromosome and is excluded from autosomal material in *cis*. *Hum. Molec. Genet.*, **8**, 195–204.

Edwards-Gilbert G, Veraldi KL, Milcarek C (1997) Alternative poly (A) site selection in complex transcription units: means to an end? *Nucleic Acids Res.*, **13**, 2547–2561.

Geyer PK (1997) The role of insulator elements in defining domains of gene expression. *Curr. Opin. Genet. Dev.*, **7**, 242.

Grosveld F, Dillon N, Higgs D (1993) The regulation of human globin gene expression. *Baillière's Clin. Haematol.*, **6**, 31–55.

Grosveld F (1999) Activation by locus control regions? *Curr. Opin. Genet. Dev.*, **9**, 152–157.

Hazelrigg T (1998) The destinies and destinations of RNAs. *Cell*, **95**, 451–460.

Heard E, Lovell-Badge R, Avner P (1999) Anti-*Xist*entialism. *Nature Genet.*, **21**, 343–344.

Hewitt JE, Lyler R, Clark LN et al. (1994) Analysis of the tandem repeat locus D4Z4 associated with fascioscapulohumeral muscular dystrophy. *Hum. Molec. Genet.*, **3**, 1287–1295.

Jiang Z-H, Zhang W-J, Rao Y, Wu JY (1998) Regulation of Ich-1 pre-mRNA alternative splicing and apoptosis by mammalian splicing factors. *Proc. Natl Acad. Sci. USA*, **95**, 9155–9160.

Jones PA (1999) The DNA methylation paradox. *Trends Genet.*, **15**, 34–37.

Karin M, Hunter T (1995) Transcriptional control by protein phosphorylation: signal transmission from the cell surface to the nucleus. *Curr. Biol.*, **5**, 747–757.

Kioussis D, Festenstein R (1997) Locus control regions: overcoming heterochromatin-induced gene inactivation in mammals. *Curr. Opin. Genet. Dev.*, **7**, 614–619.

Klausner RD, Rouault TA, Harford JB (1993) Regulating the fate of mRNA: the control of cellular iron metabolism. *Cell*, **72**, 19–28.

Kleinjan D-J, van Heyningen V (1998) Position effect in human genetic disease. *Hum. Mol. Genet.*, **7**, 1611–1618.

Larsen F, Gundersen G, Lopez R, Prydz H (1992) CpG islands as gene markers in the human genome. *Genomics*, **13**, 1095–1107.

Larsson SH, Charlieu JP, Miyagawak K et al. (1995) Subnuclear localization of WT1 in splicing or transcription factor domains is regulated by alternative splicing. *Cell*, **81**, 391–401.

Lee JT, Davidow LS, Warshawsky D (1999) *Tsix*, a gene antisense to *Xist* at the X-inactivation centre. *Nature Genet.*, **21**, 400–404.

Lee MP, Brandenburg S, Landes GM, Adams M, Miller G, Feinberg AP (1999) Two novel genes in the center of the 11p15 imprinted domain escape genomic imprinting. *Hum. Mol. Genet.*, **8**, 683–690.

Lopez AJ (1998) Alternative splicing of pre-mRNA: developmental consequences and mechanisms of regulation. *Annu. Rev. Genet.*, **32**, 279–305.

Lyon MF (1999) X-chromosome inactivation. *Curr. Biol.*, **9**, R235–R237.

Mertineit C, Yoder JA, Taketo T, Laird DW, Trasier JM, Bestor TH (1998) Sex-specific exons control DNA methyltransferase in mammalian germ cells. *Development*, **125**, 889–897.

Migeon BR (1994) X-chromosome inactivation: molecular mechanisms and genetic consequences. *Trends Genet.*, **10**, 230–235.

Ng H-H, Bird A (1999) DNA methylation and chromatin modification. *Curr. Opin. Genet. Dev.*, **9**, 158–163.

Nikolov DB, Burley SK (1997) RNA polymerase II transcription initiation: a structural view. *Proc. Natl Acad. Sci. USA,* **94,** 15–22.

Ogbourne S, Antalis TM (1998) Transcriptional control and the role of silencers in transcriptional regulation in eukaryotes. *Biochem J.,* **331,** 1–14.

Ohlsson R, Tycko B, Sapienza C (1998) Monoallelic expression: 'there can only be one'. *Trends Genet.,* **14,** 435–438.

Pekhletsky RI, Chernov BK, Rubtsov PM (1992) Variants of the 5′-untranslated sequence of human growth hormone receptor mRNA. *Mol. Cell Endrocinol.,* **90,** 103–109.

Razin A, Kafri T (1994) DNA methylation from embryo to adult. *Prog. Nucl. Acid Res.,* **48,** 53–81.

Reik W, Walter J (1998) Imprinting mechanisms in mammals. *Curr. Opin. Genet. Dev.,* **8,** 154–164.

Schoenherr CJ, Paquette AJ, Anderson DJ (1996) *Proc. Natl Acad. Sci. USA,* **93,** 9881–9886.

Schweizer J, Zynger D, Francke U (1999) *In vivo* nuclease hypersensitivity studies reveal multiple sites of parental origin-dependent chromatin conformation in the 150 kb *SNRPN* transcription unit. *Hum. Molec. Genet.,* **8,** 555–566.

Simmen MW, Leitgeb S, Charlton J, Jones SJM, Harris B, Clark VH, Bird AP (1999) Nonmethylated transposable elements and methylated genes in a chordate genome. *Science,* **283,** 1164–1167.

Siomi H, Dreyfuss G (1997) RNA-binding proteins as regulators of gene expression. *Curr. Opin. Genet. Dev.,* **7,** 345–353.

Skuse DH, James RS, Bishop DV *et al.* (1997) Evidence from Turner's syndrome of an imprinted X-linked locus affecting cognitive function. *Nature,* **387,** 705–708.

Smale ST (1997) Transcription initiation from TATA-less promoters within eukaryotic protein-coding genes. *Biochim. Biophys. Acta,* **1351,** 73–88.

Stebbins-Boaz B, Richter JD (1997) Translational control during early development. *Crit. Rev. Eukaryotic Gene Expression,* **7,** 73–94.

Steinmetz EJ (1997) Pre-mRNA processing and the CTD of RNA polymerase II: the tail that wags the dog? *Cell,* **89,** 491–494.

Tilghman SM (1999) The sins of the fathers and mothers: genomic imprinting in mammalian development. *Cell,* **96,** 185–193.

Wahle E, Kuhn (1997) The mechanism of 3′ cleavage and polyadenylation of 3′ eukaryotic pre-mRNA. *Prog. Nucl. Acid Res. Mol. Biol.,* **57,** 41–71.

Walsh CP, Bestor TH (1999) Cytosine methylation and mammalian development. *Genes Dev.,* **13,** 26–34.

Weichhold GM, Klobeck H-G, Ohnheiser R, Combriato G, Zachau HG (1990). Megabase inversions in the human genome as physiological events. *Nature,* **347,** 90–92.

Wickens M, Anderson P, Jackson RJ (1997) Life and death in the cytoplasm: messages from the 3′ end. *Curr. Opin. Genet. Dev.,* **7,** 220–232.

Yoder JA, Walsh CP, Bestor TH (1997) Cytosine methylation and the ecology of intragenomic parasites. *Trends Genet.,* **13,** 335–340.

Instability of the human genome: mutation and DNA repair

9

As in other genomes, the DNA of the human genome is not a static entity. Instead, it is subject to a variety of different types of heritable change (mutation). Large-scale chromosome abnormalities involve loss or gain of chromosomes or breakage and rejoining of chromatids (see Section 2.6). Smaller scale mutations can be grouped into different mutation classes and can also be categorized on the basis of whether they involve a single DNA sequence (simple mutations – Section 9.2) or whether they involve

exchanges between two allelic or nonallelic sequences (Section 9.3). Three classes of small-scale mutation can be distinguished (see also *Table 9.1*):

- **Base substitutions** – involve replacement of usually a single base; in rare cases several clustered bases may be replaced simultaneously as a result of a form of **gene conversion**.
- **Deletions** – one or more nucleotides are eliminated from a sequence.
- **Insertions** – one or more nucleotides are inserted into a sequence. In rare cases this involves transposition from another locus. **Copy** or **duplicative**

Table 9.1: Incidence of mutation classes in the human genome

Mutation class	Type of mutation	Incidence
Base substitutions	All types	Comparatively common type of mutation in coding DNA but also common in noncoding DNA
	Transitions and transversions	Unexpectedly, transitions are commoner than transversions, especially in mitochondrial DNA
	Synonymous and nonsynonymous substitutions	Synonymous substitutions are considerably more common than nonsynonymous substitutions in coding DNA; conservative substitutions are more common than nonconservative
	Gene conversion-like events (multiple base substitution)	Rare except at certain tandemly repeated loci or clustered repeats
Insertions	Of one or a few nucleotides	Very common in noncoding DNA but rare in coding DNA where they produce frameshifts
	Triplet repeat expansions	Rare but can contribute to several disorders, especially neurological disorders (see *Box 16.7*)
	Other large insertions	Rare; can occasionally get large-scale tandem duplications, and also insertions of transposable elements (Section 9.5.6)
Deletions	Of one or a few nucleotides	Very common in noncoding DNA but rare in coding DNA where they produce frameshifts
	Larger deletions	Rare, but often occur at regions containing tandem repeats (Section 9.5.3) or between interspersed repeats (see Section 9.5.4 and *Figure 9.9*)
Chromosomal abnormalities	Numerical and structural	Rare as constitutional mutations, but can often be pathogenic (see Section 2.6). Much more common as somatic mutations and often found in tumor cells

transposition involves a sequence from one locus being replicated and the copy inserted into another locus. Noncopy transposition involves simple transposition of a DNA sequence from one locus to another. In human and mammalian genomes, noncopy transposition is very rare: the great majority of DNA transposition occurs via an RNA intermediate so that the insertion is of a sequence copied from another locus.

New mutations arise in single individuals, in somatic cells or in the germline. If a germline mutation does not seriously impair an individual's ability to have offspring who can transmit the mutation, it can spread to other members of a (sexual) population. Allelic sequence variation is traditionally described as a DNA polymorphism if more than one variant (allele) at a locus occurs in a human population with a frequency greater than 0.01 (a frequency high enough such that an origin as a result of chance recurrence is highly unlikely). The **mean heterozygosity** for human genomic DNA is thought to be of the order of 0.001–0.004 (i.e. approximately 1:250 to 1:1000 bases are different between allelic sequences; Cooper *et al.*, 1985; Nickerson *et al.*, 1998; Taillon-Miller *et al.*, 1998). Certain genes, notably some HLA genes, are exceptionally polymorphic and alleles can show very substantial sequence divergence (see *Figure 14.27*). Because mutation rates are comparatively low the vast majority of the differences between allelic sequences within an individual are inherited, rather than resulting from *de novo* mutations.

Mutations are the raw fuel that drives evolution, but they can also be pathogenic (Sections 9.4 and 9.5). They can be the direct cause of a phenotypic abnormality or they can result in increased susceptibility to disease. The usually low level of mutation may therefore be viewed as a balance between permitting occasional evolutionary novelty at the expense of causing disease or death in a proportion of the members of a species. Normally, most mutations arise as copying errors during DNA replication because DNA polymerases, like all enzymes, are error-prone. The error rate of a DNA polymerase (that is, the frequency of incorporating a wrong base) is significantly reduced by having a subunit of the polymerase which has a **proofreading** function. Even then, however, the size of the human genome makes huge demands on the fidelity of any DNA polymerase: a sequence of 3 billion nucleotides needs to be replicated accurately every single time a human cell divides.

DNA is also subject to significant spontaneous chemical attack in the cell. For example, every day approximately 5000 adenines or guanines are lost from the DNA of each nucleated human cell by depurination (the N-glycosidic bond linking the purine residue to the carbon 1′ of the deoxyribose is hydrolyzed and the purine is replaced by a hydroxyl group at carbon 1′). DNA is also damaged by exposure to natural ionizing radiation and to reactive metabolites. In order to minimize the mutation rate, therefore, it is necessary to have effective DNA repair systems which identify and correct many abnormalities in the DNA sequence (Section 9.6). In addition, errors that arise in the mRNA sequence during gene expression are subject to RNA surveillance mechanisms which ensure removal of mRNAs which have inappropriate termination codons (Section 9.4.6).

9.2 Simple mutations

9.2.1 Mutations due to errors in DNA replication and repair are frequent

Mutations can be induced in our DNA by exposure to a variety of mutagens occurring in our external environment or to mutagens generated in the intracellular environment. In the case of radiation-induced mutation, for example, Dubrova *et al.* (1996) reported that the normal germline mutation rate for hypervariable minisatellite loci was doubled as a consequence of heavy exposure to the radioactive fallout from the Chernobyl accident. However, under normal circumstances by far the greatest source of mutations is from *endogenous* mutation, notably spontaneous errors in DNA replication and repair. During an average human lifetime there are an estimated 10^{17} cell divisions: about 2×10^{14} divisions are required to generate the approximately 10^{14} cells in the adult, and additional mitoses are required to permit cell renewal in the case of certain cell types, notably epithelial cells (see Cairns, 1975). As each cell division requires the incorporation of 6×10^9 new nucleotides, error-free DNA replication in an average lifetime would require a DNA replication–repair process with an accuracy great enough so that the correct nucleotide was inserted on the growing DNA strands on each of about 6×10^{26} occasions.

Such a level of DNA replication fidelity is impossible to sustain; indeed, the observed fidelity of replication of DNA polymerases is very much less than this and uncorrected replication errors occur with a frequency of about 10^{-9}–10^{-11} per incorporated nucleotide (see Cooper *et al.*, 1995). As the coding DNA of an average human gene is about 1.7 kb, coding DNA mutations will occur spontaneously with an average frequency of about 1.7×10^{-6}–1.7×10^{-8} per gene per cell division. Thus, during the approximately 10^{16} mitoses undergone in an average human lifetime, each gene will be a locus for about 10^8–10^{10} mutations (but for any one gene, only a tiny minority of cells will carry a mutation). In many cases, a deleterious gene mutation in a somatic cell will be inconsequential: the mutation may cause lethality for that single cell, but will not have consequences for other cells. However, in some cases, the mutation may lead to an inappropriate continuation of cell division, causing cancer (see Chapter 18).

9.2.2 The frequency of individual base substitutions is nonrandom

Base substitutions are among the most common mutations and can be grouped into two classes:

- **Transitions** are substitutions of a pyrimidine (C or T) by a pyrimidine, or of a purine (A or G) by a purine.
- **Transversions** are substitutions of a pyrimidine by a purine or of a purine by a pyrimidine.

When one base is substituted by another, there are always two possible choices for transversion, but only one choice for a transition. For example, the base adenine can undergo two possible transversions (to cytosine or to thymine) but only one transition (to guanine; see *Figure 9.1*). One might, therefore, expect transversions to be twice as frequent as transitions. Because the substitution of alleles in a population takes thousands or even millions of years to complete, nucleotide substitutions cannot be observed directly. Instead, they are always inferred from pairwise comparisons of DNA molecules that share a common origin, such as orthologs in different species. When this is done, the transition rate in mammalian genomes is found to be unexpectedly higher than transversion rates. For example, Collins and Jukes (1994) compared 337 pairs of human and rodent orthologs and found that the transition rate consistently exceeded the transversion rate. The ratio was 1.4 to 1 for substitutions which did not lead to an altered amino acid, and more than 2 to 1 for those that did result in an amino acid change.

Transitions may be favored over transversions in coding DNA because they usually result in a more conserved polypeptide sequence (see below). In both coding and noncoding DNA the excess of transitions over transversions is at least partly due to the comparatively high frequency of $C \Rightarrow T$ transitions, resulting from instability of cytosine residues occurring in the CpG dinucleotide. In such dinucleotides the cytosine is often methylated at the 5′ C atom and 5–methylcytosines are susceptible to spontaneous deamination to give thymine (Section 8.4.2). Presumably as a result of this, the CpG dinucleotide is a hotspot for mutation in vertebrate genomes: its mutation rate is about 8.5 times higher than that of the average dinucleotide (see Cooper *et al.*, 1995). Other factors favoring transitions over transversions are likely to include differential repair of mispaired bases by the sequence-dependent proofreading activities of the relevant DNA polymerases.

9.2.3 The frequency and spectrum of mutations in coding DNA differs from that in noncoding DNA

Many mutations are generated essentially randomly in the DNA of individuals. As a result, coding DNA and noncoding DNA are about equally susceptible to mutation. Clearly, however, the major consequences of mutation are largely restricted to the approximately 3% of the DNA in the human genome which is coding DNA. Mutations which occur in this component of the genome are of two types:

- **Silent (synonymous) mutations** do not change the sequence of the gene product.
- **Nonsynonymous mutations** result in an altered sequence in a polypeptide or functional RNA: one or more components of the sequence are altered or eliminated, or an additional sequence is inserted into the product.

Silent mutations are thought to be effectively *neutral* mutations (conferring no advantage or disadvantage to the organism in whose genome they arise). In contrast, nonsynonymous mutations can be grouped into three classes, depending on their effect: those having a deleterious effect; those with no effect; and those with a beneficial effect (e.g. improved gene function or gene–gene interaction). Most new nonsynonymous mutations are likely to have a deleterious effect on gene expression and so can result in disease or lethality. However, the frequency of such mutation in the population is very much reduced because of natural selection (see *Box 9.1*). As a result, the overall mutation rate in coding DNA is much less than that in noncoding DNA. Consequently, the coding DNA component of a specific gene and the derived amino acid sequence show a relatively high degree of evolutionary conservation, as do important regulatory sequences such as the multiple elements of promoters and enhancers, and intronic sequences immediately flanking exons.

Selection pressure (the constraints imposed by natural selection) reduces both the overall frequency of surviving mutations in coding DNA and the spectrum of mutations seen. For example, deletions/insertions of one or several nucleotides are frequent in noncoding DNA but are conspicuously absent from coding DNA. This is so because often such mutations will cause a shift in the translational reading frame (**frameshift mutation**), introducing a premature termination codon and causing loss of gene expression. Even if insertions/deletions do not cause a frameshift mutation, they can often affect gene function, for example, as a result of removing a key coding sequence. Instead, coding DNA is marked by a comparatively high frequency of nonrandom base

Figure 9.1: Transversions are theoretically expected to be twice as frequent as transitions.

Blue arrows, transversions; black arrows, transitions.

Box 9.1

Mechanisms which affect the population frequency of alleles

Individuals within a population differ from each other. Much of the basis of such differences is due to inherited genetic variation. The frequency of any mutant allele in a population is dependent on a number of factors, including natural selection, random genetic drift and sequence exchanges between nonallelic sequences.

Natural selection
Natural selection is the process whereby some of the inherited genetic variation will result in differences between individuals regarding their ability to survive and reproduce successfully. The differential reproduction is due to differences between individuals in their capacity to engage in reproduction (affected by parameters such as mortality, health and mating success) and to produce healthy offspring (differences in fertility, fecundity and viability of the offspring). The *fitness* of an organism is a measure of the individual's ability to survive and to reproduce successfully. In the simplest models, the fitness of an individual is considered to be determined solely by its genetic make-up, and all loci are imagined to contribute independently to the fitness of an individual, so that each locus can be treated separately. As a result, the term fitness can also be applied to a genotype.

The great majority of new nonsynonymous mutations in coding DNA reduce the fitness of their carriers. They are therefore selected against and removed from the population (negative or purifying selection). Occasionally, a new mutation may be as fit as the best allele in the population; such a mutation is selectively *neutral*. Very rarely, a new mutation confers a selective advantage and increases the fitness of its carrier. Such a mutation will be subjected to positive or advantageous selection, which would be expected to foster its spread through a population. If we consider a locus with two alleles that have different fitnesses, the heterozygote may have a fitness intermediate between the two types of homozygote. The mode of selection in this case is *codominant* and the selection will be directional, resulting in an increase of the advantageous allele. In some cases, however, a new mutation may not be advantageous in homozygotes, but only in heterozygotes (heterozygote advantage). This situation, in which the heterozygote has a higher fitness than both the mutant homozygote *and* the normal homozygote, is a form of balancing selection known as overdominant selection (see the example of cystic fibrosis in Section 3.3.2).

Random genetic drift
Changes in allele frequency can also occur by chance (random genetic drift). Even if all the individuals in a population had exactly the same fitness so that natural selection could not operate, allele frequencies would change because of random sampling of gametes. Only a tiny fraction of the available gametes in any generation is ever passed on to the next generation. If the number of gametes contributing to the next population is not large, certain alleles may not be transmitted to the next generation at the expected frequency, simply because of sampling variation. Because of the randomness in sampling, allele frequencies will fluctuate between generations. In the absence of new mutation and other factors affecting allele frequency, such as selection, alleles subject to random genetic drift will eventually reach **fixation** (the point at which the allele frequency is 0 or 100%). Genetic drift causes rapid changes in small populations but has little effect in large ones.

Interlocus sequence exchange
Individual genes in some gene families encode essentially the same product, but there may be sequence exchanges occurring between the different gene copies. For example, human 5.8S rRNA, 18S rRNA and 28S rRNA are encoded by numerous genes which are organized in tandemly repeated arrays (see *Figure 8.1*) and are particularly prone to sequence exchanges between the different repeats. Simply as a result of the sequence exchanges between different repeats, one type of repeat can increase in population frequency (see *Figure 9.8* for an illustration of the general principle). In such cases, where multiple loci produce essentially identical products, all the genes can effectively be considered as the equivalent of alleles, although not in the mendelian sense (which normally allows a maximum of two alleles in a diploid cell). The frequency of a specific repeat ('allele') can therefore be determined in part by the frequency with which it engages in sequence exchanges.

substitution occurring at locations which lead to minimal effects on gene expression (see next section).

9.2.4 The location of base substitutions in coding DNA is nonrandom

Nucleotide substitutions occurring in noncoding DNA usually have no net effect on gene expression. Exceptions include some changes in promoter elements or some other DNA sequence that regulates gene expression, and in important intronic sequence positions, such as at splice junctions or the splice branch site (see *Figure 1.15*). Substitutions occurring in coding DNA sequences which specify polypeptides show a very nonrandom pattern of substitutions because of the need to conserve polypeptide sequence and biological function. In principle, base substitutions can be grouped into three classes, depending on their effect on coding potential (see *Box 9.2*).

The different classes of base substitution listed in the box show differential tendencies to be located at the first, second or third base positions of codons. Because of the design of the genetic code, different degrees of degeneracy characterize different sites. Base positions in codons can be grouped into three classes:

- **Nondegenerate sites** are base positions where all three possible substitutions are nonsynonymous. They include the first base position of all but eight codons, the second base position of all codons and

Classes of single base substitution in polypeptide-encoding DNA

On very rare occasions, a single nucleotide substitution within polypeptide-encoding DNA causes defective gene expression by activating a cryptic splice site within an exon (see *Figure 9.12*). Aside from this unusual mechanism, single base substitutions can be classified into synonymous substitutions, missense mutations and nonsense mutations.

Synonymous ('silent') substitutions

The substitution results in a new codon specifying the same amino acid. They are the most frequently observed in coding DNA, because they are almost always neutral mutations and not subject to selection pressure. Such substitutions often occur at the third base position of a codon: *third base wobble* means that the altered codon often specifies the same amino acid. However, base substitution at the first base position can occasionally give rise to a synonymous substitution, as in the case of some leucine and arginine codons (e.g. CUA ⇔ UUA, CUG ⇔ UUG, AGA ⇔ CGA and AGG ⇔ CGG).

Nonsense mutations

These represent a form of nonsynonymous substitution where a codon specifying an amino acid is replaced by a stop codon. Because such mutations are almost always associated with a dramatic reduction in gene function, selection pressure ensures that they are normally rare. The average human polypeptide is specified by about 550 codons, a size which would be expected to harbor about 30 termination codons, if no such functional constraints applied.

Missense mutations

These are nonsynonymous substitutions where the altered codon specifies a different amino acid. Missense mutations can be classified into two subgroups:

- A conservative substitution results in replacement of an amino acid by another that is chemically similar to it. Often, the effect of such substitutions on protein function is minimal because the side chain of the new amino acid may be functionally similar to that of the amino acid it replaces. To minimize the effect of nucleotide substitution, the genetic code appears to have evolved so that codons specifying related amino acids are themselves related. For example, the Asp (GAC, GAT) and Glu (GAA, GAG) codon pairs ensure that third base wobble in a GAX codon (where X is any nucleotide) has a minimal effect. However, some first codon position changes can also be conservative, e.g. CUX (Leu) ⇔ GUX (Val).

- A nonconservative substitution is a mutation that results in replacement of one amino acid by another which has a dissimilar side chain. Sometimes a charge difference is introduced; other changes may involve replacement of polar side chains by nonpolar ones and vice versa. Base substitutions at the first and second codon positions can often result in nonconservative substitutions, e.g. CGX (Arg) ⇒ GGX (Gly), CCX (Pro), CUX (Leu) or CAX (Gln/His), etc (where X = any nucleotide).

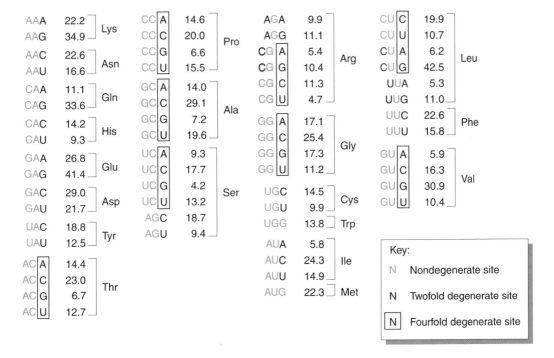

Figure 9.2: Codon frequencies in human genes and locations of nondegenerate, two- and fourfold degenerate sites.

Observed codon frequencies were derived from an analysis of 1490 human genes by Wada *et al.* (1990). *Note* that although eight of the 61 first base positions are twofold degenerate, about 96% of all possible substitutions at the first base position are nonsynonymous. 100% of base substitutions at the second base position are nonsynonymous, but only about 33% of those at the third base position.

the third base position of two codons, AUG and UGG (*Figure 9.2*). Taking into account the observed codon frequencies in human genes, they comprise about 65% of the base positions in human codons. The base substitution rate at nondegenerate sites is very low, consistent with a strong conservative selection pressure to avoid amino acid changes (*Figure 9.3*).

- **Fourfold degenerate sites** are base positions in which all three possible substitutions are synonymous and are found at the third base position of several codons (*Figure 9.2*). They comprise about 16% of the base positions in human codons. The substitution rate at fourfold sites is very similar to that within introns and pseudogenes, consistent with the assumption that synonymous substitutions are selectively neutral (*Figure 9.3*).

- **Twofold degenerate sites** are base positions in which one of the three possible substitutions is synonymous. They are often found at the third base positions of codons, but also at the first base position in eight codons (*Figure 9.2*). They· comprise about 19% of the base positions in human codons. As expected, the substitution rate for twofold degenerate sites is intermediate (*Figure 9.3*): only one out of the three possible substitutions, a transition, maintains the same amino acid. The other two possible substitutions are transversions which, because of the way in which the genetic code has evolved, are often conservative substitutions. For example, at the third base position of the glutamate codon GAA, a transition A ⇒ G is silent, while the two transversions (A ⇒ C; A ⇒ T) result in replacement by a closely similar amino acid, aspartate.

The design of the genetic code and the degree to which one amino acid is functionally similar to another affect the relative mutabilities of individual amino acids. Certain amino acids may play key roles which cannot be substituted easily by others. For example, cysteine is often involved in disulfide bonding which can play a crucially important role in establishing the conformation of a polypeptide (see *Figure 1.25*). As no other amino acid has a side chain with a sulfhydryl group, there is strong selection pressure to conserve cysteine residues at many locations, and cysteine is among the least mutable of the amino acids (Collins and Jukes, 1994). In contrast, certain other amino acids such as serine and threonine have very similar side chains, and substitutions at both the first base position of codons (<u>A</u>CX ⇒ <u>U</u>CX; where X = any nucleotide) and second base positions (A<u>C</u>Py ⇒ A<u>G</u>Py; where Py = pyrimidine) can result in serine ⇒ threonine substitutions. Presumably as a result, serine and threonine are among the most mutable of the amino acids (Collins and Jukes, 1994).

9.2.5 Protein-coding genes show enormous variation in the rate of nonsynonymous substitutions

The rate and type of substitution varies between different genes. At one extreme are proteins whose sequences are extremely highly conserved, such as ubiquitin, histones H3 and H4, calmodulin, ribosomal proteins, etc. For example, the ubiquitin proteins of humans, mouse and *Drosophila* show 100% sequence identity, and comparison with the yeast ubiquitin reveals 96.1% sequence identity. These genes are not especially protected from mutation, because the rate of synonymous codon substitution is typical of that for many protein-encoding genes. Instead, what distinguishes them is the extremely low rate of nonsynonymous codon substitution compared with other genes (see *Table 9.2* for some examples). Presumably, ubiquitin and the other highly conserved pro-

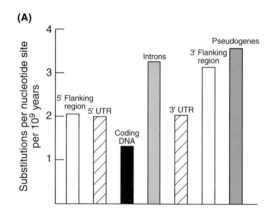 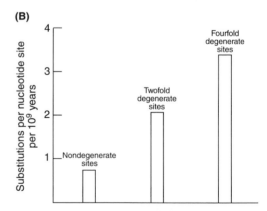

Figure 9.3: The rate of nucleotide substitution varies in different gene components and gene-associated sequences.

On the basis of the above substitution rates and the observation that an average mammalian coding DNA sequence comprises 400 codons, the coding DNA of an average human gene would be expected to undergo about one or two substitutions every million years. UTR, untranslated region. Redrawn from Li and Graur (1991) *Fundamentals of Molecular Evolution*, with permission from Sinauer Associates, Inc.

Table 9.2: Rates of synonymous and nonsynonymous substitutions in mammalian protein-coding genes

Gene	No. of codons	Nonsynonymous	Synonymous
Histone H3	135	0.00	6.38
Histone H4	101	0.00	6.12
Actin α	376	0.01	3.68
Aldolase A	363	0.07	3.59
HPRT	217	0.13	2.13
Insulin	51	0.13	4.02
α-Globin	141	0.55	5.14
β-Globin	144	0.80	3.05
Albumin	590	0.91	6.63
Ig V_H	100	1.07	5.66
Growth hormone	189	1.23	4.95
Ig κ	106	1.87	5.66
Interferon-$β_1$	159	2.21	5.88
Interferon-γ	136	2.79	8.59

Data from human–rodent comparisons abstracted from Table 1 in Chapter 4 of Li and Graur (1991) with permission from Sinauer Associates, Inc.

teins play such crucial roles that they are under huge selection pressure to conserve the sequence. At the other extreme, the fibrinopeptides are proteins which are evolving extremely rapidly and do not appear to be subject to any selective constraint. These proteins (only 20 amino acids long) are thought to be functionless – they are fragments which are generated as part of the protein *fibrinogen* and discarded when the protein is activated to form *fibrin* during blood clotting. Another extremely rapidly evolving sequence is the major sex-determining locus, *SRY*. This gene encodes a protein which contains a central 'high mobility group' domain (HMG box) of about 78 amino acids. The HMG box is central to *SRY* function and is well conserved, but the flanking N- and C-terminal segments are evolving extremely rapidly, which may indicate that the majority of the *SRY* coding sequence is not functionally significant (Whitfield *et al.,* 1993). In between the two extremes in the rate of nonsynonymous substitution are the vast majority of polypeptide-encoding genes (see *Table 9.2*).

9.2.6 The molecular clock can vary from gene to gene, and is different in different lineages

Synonymous substitutions have been considered to be effectively neutral from the point of view of selective constraints. As a result, the concept of a constant molecular clock (whereby a given gene or gene product undergoes a constant rate of molecular evolution) was suggested over 30 years ago. Since then, however, abundant evidence has been accumulated which does not support the concept (Ayala, 1999).

When substitution rates are compared for different genes, even closely related members of a gene family, there are considerable differences. For example, the genes listed in *Table 9.2* show considerable differences not only in their rates of nonsynonymous codon substi-

tutions, but also in the rate of synonymous codon substitutions. Such differences may be governed by a number of factors:

■ *Timing of DNA replication.* The DNA of different genomic components is replicated at different times. Actively transcribing genes are replicated early; transcriptionally inactive DNA such as the inactivated X chromosome is replicated late. Early replicating and late replicating DNA may be subject to different intracellular concentrations of free nucleotides and of the various enzymes involved in replication and DNA repair, causing possible differences in mutation rates.

■ *Differences in GC content.* This parameter is not independent of the previous one because the early replicating DNA is relatively GC-rich. The present evidence suggests that any relationship between the GC content of a mammalian gene and its mutation rate is not a simple one (Sharp and Matassi, 1994).

■ *Different genomes.* The mitochondrial DNA in mammals and many other animals is thought to be evolving at a much higher rate than nuclear DNA (see Section 9.4.3).

For a given gene, the molecular clock varies very considerably depending on the species lineage, and the clock runs at different rates for closely related members of a gene family (e.g. Gibbs *et al.*, 1998). In order to estimate the relative rates of nucleotide substitutions in two lineages leading to present-day species A and B, a relative rate test is used. This involves using a third reference species C which is known to have branched off earlier in evolution, before the A–B split. Pairwise comparisons of orthologs in A and C, and in B and C are then used to calculate the K value, the number of synonymous substitutions per 100 sites. The K_{AC} and K_{BC} values then provide a measure of the relative rates of mutation in the lineages leading to species A and to

Table 9.3: Rates of synonymous substitution per site per year in primates and rodents

Species pair	Number of sites	Percentage divergence	Substitution rate ($\times 10^9$)[a]
Human/chimpanzee	921	1.9	1.3 (0.9–1.9)
Human/Old World monkeys	998	11.0	2.2 (1.8–2.8)
Mouse/rat	3886	23.7	7.9 (3.9–11.8)

[a]Values represent the likely mean, and in parentheses the lower and upper ranges, according to the likely mean and lower and upper ranges for the estimated times of species divergence. The latter were estimated as follows: human/chimpanzee, 7 (5–10) million years; human/Old World monkey, 25 (20–30) million years; mouse/rat, 15 (10–30) million years but see Kumar and Hedges (1998) for more recent calculations. Reproduced from Li and Graur (1991) *Fundamentals of Molecular Evolution*, with permission from Sinauer Associates, Inc.

species B. For example, when a variety of orthologs in mouse (species A) and rat (species B) are referenced against orthologs in humans (species C), the overall K_{AC} and K_{BC} values are nearly identical (Li and Graur, 1991, p. 82). This suggests that the base substitution rates in the lineages leading to present-day mouse and rat have been nearly equal. However, similar analyses suggest that the substitution rate appears to be lower in lineages leading to the primates and lower still in the lineage leading to modern day humans (*Table 9.3*).

The data in *Table 9.3* may suggest that molecular evolution has effectively slowed down for organisms which have long generation times. With hindsight, perhaps this is not so surprising – most mutations arise when DNA is being replicated in gametogenesis (especially in males; see next section). Rodents and monkeys have comparatively shorter generation times than humans, and so will go through more generations per unit time. In addition, it has been suggested that longer-lived animals have a greater ability to repair their DNA than do short-lived species, thereby resulting in lower mutation rates (Britten, 1986).

9.2.7 Higher mutation rates in males are likely to be related to the greater number of germ cell divisions

Since Haldane first observed that most mutations resulting in hemophilia were generated in the male germline, it has been assumed that, at least in humans, mutations are preferentially paternally inherited. Two major approaches have been taken to estimate the relative mutation rates in the male and female germlines:

■ *Molecular evolutionary methods.* The starting point for such methods typically investigates known homologous genes on the X and Y chromosomes located outside the pseudoautosomal region. The sequences are compared with orthologs in another species to estimate the rate of synonymous mutations for the X chromosome gene (K_{SX}) and the Y chromosome gene (K_{SY}). Unlike autosomes, the sex chromosomes spend different amounts of time in the two sexes. The great majority of Y chromosome sequences (those outside the *pseudoautosomal region* – see Section 14.3.1) spend all their time in males. By contrast, as females have two X chromosomes whereas males have one X chromosome, X chromosome sequences spend on average 2/3 of their time in females and 1/3 of their time in males. If we use the symbol α to represent the ratio of the male mutation rate to the female mutation rate this is equivalent to setting the male mutation rate at α relative to a female mutation rate of 1. For most Y chromosome sequences, the mutation rate is therefore α. For X chromosome sequences it is 2/3 times the mutation rate in female cells (1) plus 1/3 times the mutation rate in males (α), giving a total of 2/3 + α/3. Therefore the observed K_{SX} / K_{SY} ratio = (2/3 + α/3)/α and so can be used to estimate α. This results in estimates of 3–10 for α (see *Table 9.4* and Shimmin *et al.*, 1993 for the example of the *ZFX* and *ZFY* genes).

Table 9.4: Examples of sex differences in mutation rate (see Hurst and Ellegren, 1998 and Shimmin *et al.*, 1993)

Gene	Mutation type	α (mutations originating in males/mutations originating in females)
Neurofibromatosis type I (*NFI*)	Point mutations (but excluding large deletions)	4.5
von Hippel–Lindau (*VHL*)	Point mutations	1.3
Retinoblastoma (*RBI*)	Point mutations (but excluding large deletions)	8.5
ZFX/ZFY	Mutations in last intron	6

■ *Direct observation of disease-causing mutations.* Clearly, the mutations assessed in this case are a special subset. The approach involves analysing samples from a patient with a *de novo* mutation and the two parents (the parent who passed on the faulty chromosome is typically identified by typing for markers closely flanking the disease gene). In most cases one would expect the mutation to have been transmitted through the germline (although if only a blood sample is typed, it is possible that some mutations could be postzygotic). In the case of point mutation analyses for disorders where imprinting is not suspected, estimates for α again support a higher mutation rate in males (*Table 9.4*).

The comparatively high male mutation rate may be due to different factors (Hurst and Ellegren, 1998), but a major contributory factor is thought to be the large sex difference in the number of human germ cell divisions. In females, the number of cell divisions from zygote to fertilized oocyte is constant because all of the oocytes have been formed by the fifth month of development and only two further cell divisions are required to produce the zygote (*Figure 9.4A*). The estimated number of successive female cell divisions from zygote to mature egg has been variously estimated as 24 (Vogel and Motulsky, 1996) and 31 (Li, 1997, p. 229) and is broadly similar to estimates of 30–31 male cell divisions required from zygote to stem spermatogonia at puberty. Five subsequent cell divisions are required for spermatogenesis but thereafter the spermatogenesis cycle occurs approximately every 16 days or 23 cycles per year (*Figure 9.4B*). This means that in males, the number of cell divisions required to produce sperm is age-dependent. If an average age of 13 is taken for onset of puberty and an average of 25 for male reproductive age, the total number of cell divisions is about $30 + 5 + [23 \times (25 - 13)]$, or about 310 divisions (*Figure 9.4*). Given that errors in DNA replication/repair provide the great majority of mutations, one might then expect that the male mutation rate would be substantially greater than that of the female.

9.3 Genetic mechanisms which result in sequence exchanges between repeats

In addition to very frequent simple mutations, there are several mutation classes which involve sequence exchange between allelic or nonallelic sequences, often involving repeated sequences. For example, tandemly repetitive DNA is prone to deletion/insertion polymorphism whereby different alleles vary in the number of integral copies of the tandem repeat. Such **variable number of tandem repeat (VNTR) polymorphisms** can occur in the case of repeated units that are very short (microsatellites); intermediate (minisatellites) or large.

Different genetic mechanisms can account for VNTR polymorphism depending on the size of the repeating unit (see the following two sections). In addition, interspersed repeats can also predispose to deletions/duplications by a variety of different genetic mechanisms. These are discussed particularly in the context of disease mutations and are therefore presented in Section 9.4.

9.3.1 Slipped strand mispairing can cause VNTR polymorphism at short tandem repeats (microsatellites)

There is considerable variation in the germline mutation rates at microsatellite loci, ranging from an undetectable level up to about 8×10^{-3} (Mahtani and Willard, 1993; Weber and Wong, 1993). Novel length alleles at (CA)/(TG) microsatellites and at tetranucleotide marker loci are known to be formed without exchange of flanking markers. This means that they are not generated by unequal crossover (see below). Instead, as new mutant alleles have been observed to differ by a single repeat unit from the originating parental allele (Mahtani and Willard, 1993), the most likely mechanism to explain length variation is a form of exchange of sequence information which commences by slipped strand mispairing. This occurs when the normal pairing between the two complementary strands of a double helix is altered by staggering of the repeats on the two strands, leading to incorrect pairing of repeats. Although slipped strand mispairing can be envisaged to occur in nonreplicating DNA, replicating DNA may offer more opportunity for slippage and hence the mechanism is often also called **replication slippage** or polymerase slippage (see *Figure 9.5*). In addition to mispairing between tandem repeats, slippage replication has been envisaged to generate large deletions and duplications by mispairing between noncontiguous repeats and has been suggested to be a major mechanism for DNA sequence and genome evolution (Levinson and Gutman, 1987; see also Dover, 1995). The pathogenic potential of short tandem repeats is considerable (Sections 9.5.1 and 9.5.2).

9.3.2 Large units of tandemly repeated DNA are prone to insertion/deletion as a result of unequal crossover or unequal sister chromatid exchanges

Homologous recombination describes recombination (crossover) occurring at meiosis or, rarely, mitosis between identical or very similar DNA sequences. It usually involves breakage of *nonsister chromatids* of a pair of homologs and rejoining of the fragments to generate new recombinant strands. **Sister chromatid exchange** is an analogous type of sequence exchange involving breakage of individual *sister chromatids* and rejoining fragments that initially were on different chromatids of the same chromosome. Both homologous recombination and sister chromatid exchange normally involve *equal* exchanges – cleavage and rejoining of the chromatids

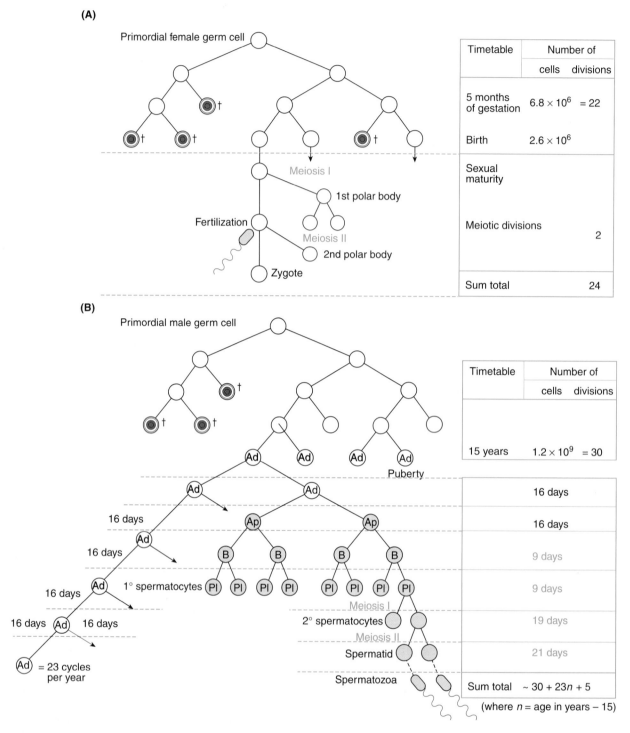

Figure 9.4: The number of cell divisions that are required to produce a human sperm cell is much greater than the number required to produce an egg cell.

(A) Human oogenesis occurs only during fetal life and ceases by the time of birth. The total population of germ cells in the female embryo rises to an estimated maximum of 6.8×10^6 during the fifth month. The expected number of binary divisions required to generate this number would be about 22 ($2^{22} = \sim 4 \times 10^6$). Note, however, that Li (1997) estimates about 29 divisions. At sexual maturity, two subsequent meiotic divisions are required to produce an egg cell. †, Cell atrophy. (B) Human spermatogenesis continues through adult life. From the early embryonic stage up to the age of puberty, the seminiferous tubules continue to become populated by so-called *Ad spermatogonia* (dark-staining A-type spermatogonia) and spermatogenesis is fully established by puberty. The number of Ad spermatogonia is estimated to be about 6×10^8 per testis, i.e. a total of about 1.2×10^9, a value which can be reached by about 30 successive cell divisions. The Ad spermatogonia then

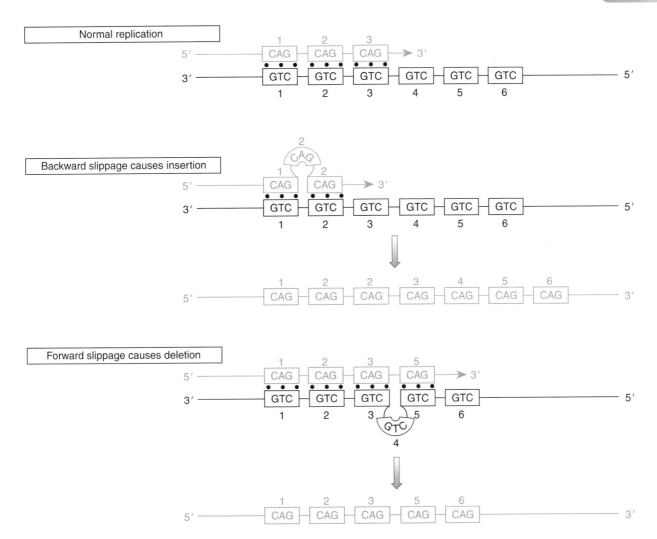

Figure 9.5: Slipped strand mispairing during DNA replication can cause insertions or deletions.

Short tandem repeats are thought to be particularly prone to slipped strand mispairing, i.e. mispairing of the complementary DNA strands of a single DNA double helix. The examples show how slipped strand mispairing can occur during replication, with the lower strand representing a parental DNA strand and the upper blue strand representing the newly synthesized complementary strand. In such cases, slippage involves a region of nonpairing (shown as a bubble) containing one or more repeats of the newly synthesized strand (backward slippage) or of the parental strand (forward slippage), causing, respectively, an insertion or a deletion on the newly synthesized strand. *Note* that it is conceivable that slipped strand mispairing can also cause insertions/deletions in nonreplicating DNA. In such cases, two regions of nonpairing are required, one containing repeats from one DNA strand and the other containing repeats from the complementary strand (Levinson and Gutman, 1987).

occurs at the same position on each chromatid. As a result, the exchanges occur *between allelic sequences* and at corresponding positions within alleles. In the case of intragenic equal crossover between two alleles, a new allele can result which is a **fusion gene** (or **hybrid gene**), comprising a terminal fragment from one allele and the remaining sequence of the second allele (*Figure 9.6*). However, equal sister chromatid exchanges cannot normally produce genetic variation because sister chromatids have identical DNA sequences.

undergo a series of cell divisions, lasting 16 days. Of the two products of cell division, one prepares for the next division into two Ad cells. Because each Ad division cycle is 16 days long, following puberty a total of 23 cycles occur each year (365/16) – see bottom left. The other product of an Ad cell division divides to give two Ap (pale-staining A type spermatogonia) cells which are precursors of sperm cells (blue cells on bottom right). The Ap cells give rise to B spermatogonia and then spermatocytes, which finally undergo two meiotic divisions to generate sperm cells. Modified from Vogel and Motulsky (1996) *Human Genetics. Problems and Approaches*, 3rd edn, with permission from Springer Verlag. © 1996, Springer-Verlag.

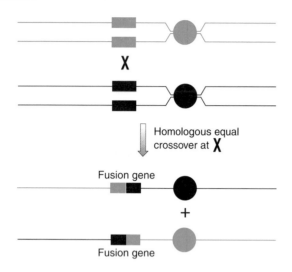

Figure 9.6: Homologous equal crossover can result in fusion genes.

The example shows how intragenic equal crossover occurring between alleles on nonsister chromatids can generate novel fusion genes composed of adjacent segments from the two alleles. *Note* that similar exchanges between genes on sister chromatids do not result in genetic novelty because the gene sequences on the interacting sister chromatids would be expected to be identical.

Unequal crossover is a form of recombination in which the crossover takes place *between nonallelic sequences* on nonsister chromatids of a pair of homologs (*Figure 9.7*). Often the sequences at which crossover takes place show very considerable sequence homology which presumably stabilizes mispairing of the chromosomes. Because crossover occurs between mispaired nonsister chromatids, the exchange results in a deletion on one of the participating chromatids and an insertion on the other. The analogous exchange between sister chromatids is called **unequal sister chromatid exchange** (see *Figure 9.7*). Both mechanisms occur predominantly at locations where the tandemly repeated units are moderate to large in size. In such cases, the very high degree of sequence homology between the different repeats can facilitate pairing of nonallelic repeats on nonsister chromatids or sister chromatids. If chromosome breakage and rejoining occurs while the chromatids are mispaired in this way, an insertion or deletion of an integral number of repeat units will result. Note that such exchanges are *reciprocal*; both participating chromatids are modified, in one case resulting in an insertion, and in the other case in a complementary deletion.

Unequal sister chromatid exchange is thought to be a major mechanism underlying VNTR polymorphism in the rDNA clusters. Unequal crossover is also expected to occur comparatively frequently in complex satellite DNA repeats and at tandemly repeated gene loci. In the latter case, unequal crossover is known to generate pathogenic deletions at some loci (see Section 9.5.3). Such exchanges can also lead to **concerted evolution** by caus-

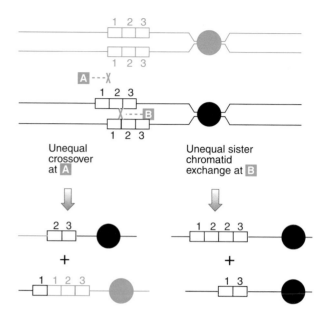

Figure 9.7: Unequal crossover and unequal sister chromatid exchange cause insertions and deletions.

The examples illustrate unequal pairing of chromatids within a tandemly repeated array. Unequal crossover involves unequal pairing of nonsister chromatids followed by chromatid breakage and rejoining. Unequal sister chromatid exchange involves unequal pairing of sister chromatids followed by chromatid breakage and rejoining. For the sake of simplicity, the breakages of the chromatids are shown to occur between repeats, but of course breaks can occur within repeats. *Note* that both types of exchange are *reciprocal* – one of the participating chromatids loses some DNA, while the other gains some.

ing a particular variant to spread through an array of tandem repeats, resulting in *homogenization* of the repeat units (see *Figure 9.8*).

Occasionally, unequal crossover and unequal sister chromatid exchanges can occur at regions where there is little homology. This is likely to be the case when such mechanisms first generate a tandemly duplicated locus following mispairing of nonallelic repeats such as two *Alu* repeats or even smaller elements (*Figure 9.9*).

9.3.3 Gene conversion events may be relatively frequent in tandemly repetitive DNA

Gene conversion describes a *nonreciprocal* transfer of sequence information between a pair of nonallelic DNA sequences (*interlocus gene conversion*) or allelic sequences (*interallelic gene conversion*). One of the pair of interacting sequences, the *donor*, remains unchanged. The other DNA sequence, the *acceptor*, is changed by having some or all of its sequence replaced by a sequence copied from the donor sequence (*Figure 9.10*). The sequence exchange is therefore a directional one; the acceptor

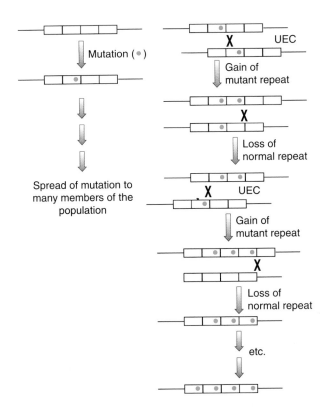

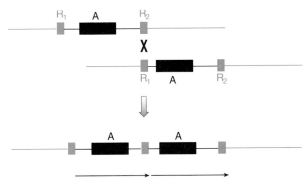

Figure 9.9: Tandem gene duplication can result from unequal crossover or unequal sister chromatid exchange, facilitated by short interspersed repeats.

The double arrow indicates the extent of the tandem gene duplication of a segment containing gene A and flanking sequences. Original mispairing of chromatids could be facilitated by a high degree of sequence homology between nonallelic short repeats (R_1, R_2). *Note* that the same mechanism can result in large-scale deletions.

Figure 9.8: Unequal crossover in a tandem repeat array can result in sequence homogenization.

Note that the initial spread of the novel sequence variant to the same position in the chromosomes of other members of a sexual population can result by random genetic drift (see *Box 9.1*). Once the mutation has achieved a reasonable population frequency (left panel) it can spread to other positions within the array (right panel). This can occur by successive gain of mutant repeats as a result of unequal crossover (or unequal sister chromatid exchanges) and occasional loss of normal repeats. Eventually the mutant repeat can replace the original repeat sequence at all positions within the array, leading to sequence homogenization for the mutant repeat. Such sequence homogenization is thought to result in species-specific concerted evolution for repetitive DNA sequences (see Section 14.4.2). UEC, unequal crossover.

sequence is modified by the donor sequence, but not the other way round.

One possible mechanism for gene conversion envisages formation of a heteroduplex between a DNA strand from the donor gene and a complementary strand from the acceptor gene. Following heteroduplex formation, conversion of an acceptor gene segment may occur by **mismatch repair** – DNA repair enzymes recognize that the two strands of the heteroduplex are not perfectly matched and 'correct' the DNA sequence of the acceptor strand to make it perfectly complementary in the converted region to the sequence of the donor gene strand (see *Figure 9.10C*).

Gene conversion has been well-described in fungi where all four products of meiosis can be recovered and studied (tetrad analysis). In humans and mammals it is not possible to do this and so gene conversion cannot be demonstrated unambiguously in higher organisms (it can never be distinguished from double crossover events, for example, although double crossovers occurring in very close proximity would normally be expected to be extremely unlikely). Despite the difficulty in identifying gene conversion in complex organisms, there are numerous instances in mammalian genomes where an allele at one locus shows a pattern of mutations which strongly resembles those found in alleles at another locus of the same species. Such evidence suggests gene conversion-like exchanges between loci.

Although simple comparisons of two sequences may be suggestive, the evidence for gene conversion is most compelling when a new mutant allele can be compared directly with its progenitor sequence. Certain highly mutable loci lend themselves to this type of analysis. In particular, some hypervariable minisatellite loci have high germline mutation rates (often 1% or more per gamete) and individual repeats often show nucleotide differences so that repeat subclasses can be recognized. Germline mutations can be studied by detecting and characterizing mutant mini-satellite alleles in individual gametes. To do this, PCR analysis has been conducted on multiple dilute aliquots of DNA isolated from the sperm of an individual (small pool PCR), where each aliquot is calibrated to contain a few, perhaps 100, input molecules (Jeffreys *et al.*, 1994). The PCR products recovered from individual pools can then be typed to identify any new mutations that result in a novel allele whose length is sufficiently different as to be distinguishable from the

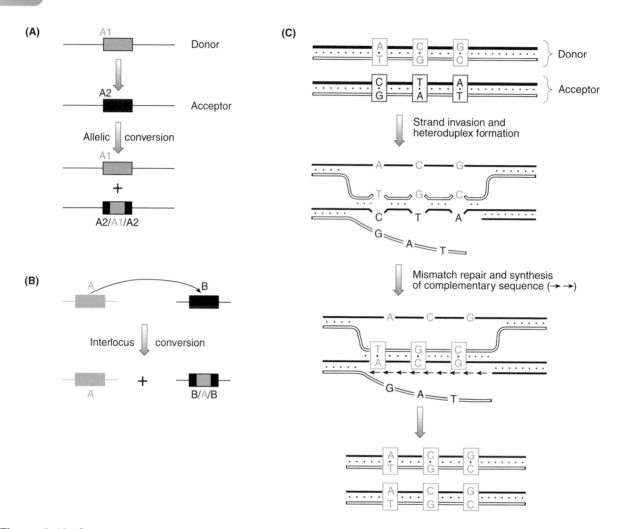

Figure 9.10: Gene conversion involves a nonreciprocal sequence exchange between allelic or nonallelic genes.

(**A**) Interallelic gene conversion. Note the nonreciprocal nature of the sequence exchange – the donor sequence is not altered but the acceptor sequence is altered by incorporating sequence copied from the donor sequence. (**B**) Interlocus gene conversion. This is facilitated by a high degree of sequence homology between nonallelic sequences, as in the case of tandem repeats. (**C**) Mismatch repair of a heteroduplex. This is one of several possible models to explain gene conversion. The model envisages invasion by one strand of the donor sequence (−) to form a heteroduplex with the complementary (+) strand of the acceptor sequence, thereby displacing the other strand of the acceptor. Mismatch repair enzymes recognize the mispaired bases in the heteroduplex and 'correct' the mismatches so that the (+) acceptor sequence is 'converted' to be perfectly complementary in sequence to the (−) donor strand. Subsequent replication of the (−) acceptor strand and sealing of nicks results in completion of the conversion.

progenitor allele. Analyses of the patterns of germline mutation at three such loci have failed to identify exchanges of flanking markers and have shown that most mutations occurring at these loci are polar, involving the preferential gain of a few repeats at one end of a tandem repeat array. There is a bias towards gain of repeats and evidence was obtained for nonreciprocal sequence exchange between alleles, suggesting interallelic gene conversion (Jeffreys *et al.*, 1994). Evidence for interlocus gene conversion has also been obtained in human genes, notably the steroid 21-hydroxylase gene (see Section 9.5.3).

9.4 Pathogenic mutations

9.4.1 There is a high deleterious mutation rate in hominids

Neutral mutations (those which are neither detrimental nor advantageous for the organism carrying them) accumulate throughout the generations at a rate equal to the mutation rate. To get an estimate of the (total) mutation rate is therefore simple: all one needs to do is to measure the rate of change of some presumed neutral sequence (e.g. intronic, pseudogene, etc). The deleterious mutation rate, by contrast, has been notoriously difficult to measure and no convincing estimate existed for any vertebrate until a study reported by Eyre-Walker and Keightley (1999). They investigated amino acid changes in 46 proteins occurring in the human ancestral line after its divergence from the chimpanzee. If all non-synonymous substitutions were neutral, 231 new substitutions would have been expected in their sample of genes (given an average neutral mutation rate of 0.0056 nonsynonymous substitutions per nucleotide and a total of 41 471 nucleotides investigated). Instead, only 143 nonsynonymous substitutions were observed and 88 such substitutions were inferred to have been removed by natural selection because they had been deleterious. On the assumption of 60 000 genes, and 240 000 generations since human–chimpanzee divergence, they estimated a deleterious rate of 1.6 mutations per person per generation out of a total of 4.2 mutations per person generation.

The estimated deleterious mutation rates in chimpanzees and gorillas were very similar but those for rodent-specific lineages are about one order of magnitude less, possibly because of much smaller numbers of germ cell divisions in rodents. The very high deleterious rate in hominids may even be an underestimate. If the total gene number were 80 000 and an average coding sequence was 1800 nucleotides, the estimated deleterious mutation rate would be 2.5 mutations per person per generation and, on other grounds too, a more likely rate has been considered to be three deleterious mutations per person per generation (Crow, 1999).

So with three genetic deaths per person why are we not extinct? The data would suggest that harmful mutations need to be weeded out in clusters at a time. One way to achieve this would be if natural selection operated such that individuals with the most mutations are preferentially eliminated (e.g. harmful mutations interact). This could only happen in a sexual species where mutations are shuffled each generation by genetic recombination, and so the existence of such a high deleterious mutation rate has been taken as further vindication that sex (meiotic recombination) is an efficient way to eliminate harmful mutations.

9.4.2 Pathogenic mutations are preferentially located at certain types of intragenic DNA sequence

Pathogenic mutations can occur at three types of DNA sequence at a gene locus.

- *The coding sequence of the gene.* This is where the great majority of recorded pathogenic mutations have been identified. Those due to nucleotide substitution are, in the vast majority of cases, nonsynonymous substitutions and mostly occur at first and second base positions of codons. However, very rarely, a synonymous codon substitution is not neutral as expected, but may cause disease by activating a cryptic splice site (Section 9.4.5). Because of its relatively high mutability, the CpG dinucleotide is often located at hotspots for pathogenic mutation in coding DNA (Cooper and Youssoufian, 1988). Other hotspots include tandem repeats within coding DNA (see below).

- *Intragenic noncoding sequences.* This is restricted to sequences which are necessary for correct expression of the gene, such as important intronic elements, notably the highly conserved GT and AG dinucleotides at the ends of introns, but also conserved elements of the untranslated sequences. Often such mutations represent a small component (~10–15%) of the total pathogenic mutations at a gene locus (Cooper *et al.*, 1995). However, in some disorders pathogenic splicing mutations may be common. In the case of the collagen disorder osteogenesis imperfecta they constitute a very common pathological mutation which is second in frequency only to substitutions leading to replacement of the highly conserved, structurally important glycine residues. The collagen genes have small exons and a comparatively large number of introns (often more than 50 and as many as 106 in the case of the *COL7A1* gene), making them exceptional targets for splicing mutations. Occasional pathogenic mutations have been recorded in the 5′ UTR (such as in the case of hemophilia B Leyden) and appear to exert their effect at the transcriptional level. Several examples are also known of pathogenic mutations in the 3′ UTR (see Cooper *et al.*, 1995).

- *Regulatory sequences outside exons.* Most mutations located in regulatory sequences have been identified in conserved elements located just upstream of the first exon, notably promoter elements. In addition, other more distantly located regulatory elements may be sites of pathological mutation. For example, deletions which eliminate the β-globin LCR (see *Figure 8.23*) but leave the β-globin gene and its promoter intact result in almost complete abolition of β-globin gene expression and contribute to β-thalassemia. Clearly, in some cases a gene may be regulated by the product of a distantly related gene. For example, in the case of rare variants of α-thalassemia with mental retardation, the α-globin gene and its promoter may show no evidence of pathological mutation and the disease maps to an X-linked gene which encodes a transcription factor, one of whose target sequences is presumably the α-globin gene (Gibbons *et al.*, 1995).

9.4.3 The mitochondrial genome is a hotspot for pathogenic mutations

Because of the very large size of the human nuclear genome, most mutations occur in nuclear DNA sequences. By comparison, the mitochondrial genome is a small target for mutation (about 1/200 000 of the size of the nuclear genome). Unlike nuclear genes, mitochondrial genes are

Box 9.3

How are new mitochondrial mutations fixed (i.e. achieve a frequency of 100% in a population)?

We inherit 23 nuclear DNA molecules (chromosomes) from each parent but perhaps as many as 100 000 mtDNA molecules in the maternal oocyte (Piko and Taylor, 1987). In normal individuals, ~99.9% of the mtDNA molecules are identical (**homoplasmy**). However, if a new mutation arises and spreads in the mtDNA population, there will be two significantly frequent mtDNA genotypes (**heteroplasmy**). A new mutation will arise on a single molecule and so, to be fixed in a population, the mutant molecule has to proliferate so that it replaces virtually all other mtDNA molecules in a whole population, thereby restoring the homoplasmy state.

The fixation process is not facilitated by a lack of recombination between mtDNA molecules in mammalian and most animal cells (although a mitochondrial recombinase has been reported its role is unknown, and there is little evidence for homologous recombination in mtDNA; see Howell, 1997;

Lightowlers *et al.*, 1997). Intuitively, therefore, one might expect that the time taken for fixation of a human mitochondrial mutation would be very much longer than that for fixation of a mutation in the nuclear genome. Paradoxically, the fixation rates for many mtDNA mutations in animal cells (including polymorphisms expected to be selectively neutral) is about 10 times that seen for mutations in the nuclear DNA (see text).

One explanation is that there is a *mtDNA bottleneck*: only a small proportion of the mtDNA molecules are selected for further amplification. A form of bottleneck occurs post-fertilization. The original fertilized oocyte has 100 000 mtDNA molecules, but following fertilization there is a substantial reduction in the number of mtDNA copies per cell. The primordial germ cells of the developing embryo only possess about 100–500 mtDNA copies and a similar number occur in the oogonium. The primary oocytes are thought to have a slightly larger number of mtDNA copies than the primordial germ cells but during maturation of the oocyte there is thought to be a >50-fold increase in mtDNA copy number.

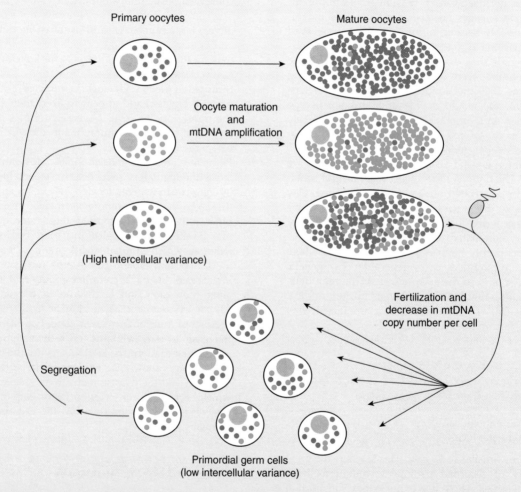

Segregation of mitochondrial genotypes in the maternal germ line.

Two subpopulations of mtDNA are shown (dark grey and blue). Primordial germ cells do not differ substantially in their levels of heteroplasmy, but a dramatic segregation of mtDNA genotypes occurs during the development of the primary oocytes. Reproduced from Lightowlers *et al.* (1997) *Trends Genet.*, **13**, pp. 450–455, with permission from Elsevier Science.

In addition to the above, conventional bottleneck theory envisages rapid segregation of mtDNA genotypes occurring during oogenesis (see Lightowlers et al., 1997; Poulton et al., 1998). The bottleneck idea was initially developed from observations of polymorphisms in the mtDNA of Holstein cows. Some point mutation differences between mothers and their offspring showed that complete switching of the mtDNA type can occur in a very few generations, even a single generation (Hauswirth and Laipis, 1982; Koehler et al., 1991). This suggests that there is first a restriction in the numbers of mtDNA to be transmitted, possibly down to even a single molecule, followed by amplification of the selected mtDNA molecule(s). However, other mutations do not show such rapid switching, and the segregation of some heteroplasmic human polymorphisms have been reported to be slow. This may mean that when genotype switching occurs it could often be incomplete, resulting in heteroplasmic individuals. For neutral polymorphisms, homoplasmy could be rapidly restored by random segregation following the bottleneck, but the maintenance of heteroplasmy for disease-associated mutations (see Section 16.6.4) may be positively selected.

Jenuth et al. (1996) analyzed neutral polymorphisms in transgenic heteroplasmic mice and observed substantial segregation. Surprisingly this occurred during the development of the primary oocyte *before* the ~ > 50-fold increase in mtDNA copy number that occurs during oocyte maturation. Therefore, the intercellular variation in heteroplasmy is absent in the primordial germ cells but is evident when the cells develop to primary oocytes (see figure). According to these authors, the level of variation in heteroplasmy seen in the primary oocytes could be explained entirely by random drift (assuming an original load of 200 mtDNA molecules in each primordial germ cell and 15 cell divisions from primordial germ cell to primary oocyte).

Recently, a study of seven oocytes donated by a woman heteroplasmic for the mtDNA nt 8993 (T ⇒ G) mutation revealed that one oocyte showed no evidence of the mutation, while in the other six more than 95% of the mtDNA copies had the mutant genotype (Blok et al., 1997). These data are also consistent with the idea of a major bottleneck occurring during oogenesis, but the rapidity of the genotype switching cannot easily be explained by the random drift model proposed by Jenuth et al. (1996). This field of research is a dynamic one and interested readers are advised to consult recent reviews.

present in numerous copies (there are thousands of copies of the mtDNA molecule in each human somatic cell; some cells, such as brain and muscle cells, have particularly high oxidative phosphorylation requirements and so more mitochondria). The mtDNA is inherited from the maternal oocyte, which is an exceptional cell with many more mtDNA molecules than somatic cells. Given that a mutation in mitochondrial DNA must arise on a single mtDNA molecule, one might intuitively expect that the chances of a single mtDNA mutation becoming *fixed* would be very low and the mutation rate correspondingly low. On these grounds, one could anticipate that the proportion of clinical disease due to pathogenic mutation in the mitochondrial genome should be extremely low. Instead, the frequency of 'mitochondrial disorders' is rather high (Section 16.6.4) and the mitochondrial genome can be considered to be a mutation hotspot. Different factors can explain this apparent paradox:

■ *Differential target size for pathogenic mutation.* Pathogenesis is associated with mutations in coding DNA and the mitochondrial genome has a much higher percentage of coding DNA (93%) than found in the nuclear genome (3%). When this is taken into consideration, however, there is still a large imbalance: about 100 Mb of coding DNA in the nuclear genome but only 15.4 kb of coding sequence in the mitochondrial genome, giving a target ratio of 6000:1 in favor of the nuclear genome.

■ *High mutation rate in mtDNA.* The mitochondrial genome is much more prone to nucleotide change than the nuclear genome. Even although about 100 000 copies of the mitochondrial genome are maternally inherited in the fertilized oocyte there are mechanisms which permit rapid fixation of mutations in mitochondrial DNA (*Box 9.3*). The combination of mtDNA instability and a high fixation rate means that the mutation rate in mitochondrial DNA is very high. Mutations have been reported to be fixed in the mitochondrial genomes of animal cells at a rate which is about 10 times greater than occurring in equivalent sequences in the nuclear genome (Brown et al., 1979). This means that the small recombination-deficient animal mtDNA molecules appear to be evolving remarkably rapidly, corresponding to about 2–4% sequence divergence per million years. In contrast, plant mtDNA molecules are comparatively large (150 kb–2.5 Mb), have introns, engage in recombination and are evolving comparatively slowly.

The high instability of mtDNA has been postulated to result from several factors. The high rate of production of reactive oxygen intermediates by the respiratory chain is thought to cause substantial oxidative damage to mtDNA which, unlike nuclear DNA, is not protected by histones. The mtDNA also has to undergo many more rounds of replication than chromosomal DNA. Although several well-characterized mtDNA repair systems are now known, some frequent mutations cannot be repaired, including thymidine dimers (Section 9.6).

9.4.4 Many different factors govern the expression of pathogenic mutations

The degree to which a pathogenic mutation results in an aberrant phenotype depends on several factors:

■ *The mutation class and the way in which the expression of the mutant gene is altered*. This may depend on the location of the mutation within the gene (*Table 9.5*). Many pathogenic mutations result in abolition or substantial reduction of gene expression, but some

Table 9.5: Effect of location and class of mutation on gene function

Location and nature of mutation	Effect on gene function	Comments
Extragenic mutation	Normally none	Rare mutations may result in inactivation of distant regulatory elements required for normal gene expression (see *Figure 8.23*)
Multigene deletion	Abolition	Associated with contiguous gene syndromes (see *Figure 16.9*)
Whole gene deletion	Abolition	
Whole gene duplication	Can have effect due to altered gene dosage	Large duplications including the peripheral myelin protein 22 gene can cause Charcot–Marie–Tooth syndrome (see *Figure 16.7*)
Whole exon deletion	Abolition or modification	May cause shift in reading frame; protein often unstable
Within exon	Abolition	If loss/change of key amino acids, shift of the reading frame or introduction of premature stop codon
	Modification	If nonconservative substitutions, small in-frame insertions or other mutations at some locations
	None	If conservative/silent substitutions or mutation at nonessential sites
Whole intron deletion	None	
Splice site mutation	Abolition or modulation of expression	Conserved GT and AG signals are critically important for normal gene expression. Mutations may induce exon skipping or intron retention
Promoter mutation	Abolition or modulation of expression	Deletion, insertion or substitution of nucleotides within promoter may alter expression. Complete deletion abolishes function
Mutation of termination codon	Modification	Additional amino acids are included at the end of the protein until another stop codon is reached
Mutation of poly(A) signal	Abolition or modulation of expression	Deletion, insertion or substitution of nucleotides within poly(A) site may alter expression. Complete deletion abolishes function
Elsewhere in introns/UTS	Usually none	

lead to inappropriate expression. For example, overexpression of a gene product may cause an abnormal phenotype where gene dosage needs to be carefully regulated, and ectopic expression, that is expression in tissues where the gene is not normally expressed, may also be harmful.

■ *The degree to which aspects of the aberrant phenotype are expressed in the heterozygote.* The presence of a single normal allele may be sufficient to maintain a clinically normal phenotype (as in recessively inherited disorders), or a milder phenotype when compared with that of mutant homozygotes, as in dominantly inherited disorders where the mutation is a simple *loss of function* mutation.

■ *The degree to which expression of a mutant phenotype is influenced by other gene products.* The same mutant allele can have different phenotypic effects on different genetic backgrounds, depending on particular alleles at other gene loci (*modifier genes*).

■ *The proportion and nature of cells in which the mutant gene is present.* Generally, mutations which are present in all the cells of an individual (inherited mutations) or in many of them (somatic mutations acquired very early in development) are likely to have a more profound effect than those present in a few cells (somatic mutations which arise at much later stages) or in cell types where the relevant gene is not expressed. Cancers, however, arise from unregulated division of cells produced from a single original mutant cell.

■ *The parental origin of the mutation.* This is only known to be important in the case of the few genes which are *imprinted* (see Section 8.5.4 and *Box 16.6*).

9.4.5 Most splicing mutations alter a conserved sequence required for normal splicing, but some occur in sequences not normally required for splicing

Many genes naturally undergo alternative forms of RNA splicing. In addition, mutations can sometimes produce an aberrant form of RNA splicing which is pathogenic. Sometimes this results in the sequences of whole exons being excluded from the mature RNA (**exon skipping**) or retention of whole introns. On other occasions, the abnormal splicing pattern may exclude part of a normal exon or result in new exonic sequences. Point mutations which alter a conserved sequence that is normally required for RNA splicing are comparatively common. Occasionally, however, aberrant splicing of a gene can be induced by mutation of other sequence elements which resemble splice donor or splice acceptor sequences but which are not normally involved in splicing (**cryptic splice sites**).

Mutations which alter important splice site signals

Often such mutations occur at the essentially invariant GT and AG dinucleotides located respectively at the start of an intron (*splice donor*) or at its end (*splice acceptor*). Flanking these important signals, however, are other conserved sequence elements (see *Figure 1.15*) which, if mutated, can also cause aberrant splicing. Mutations which alter such sequences can have different consequences:

■ *Failure of splicing causing intron retention.* This can

occasionally result, for example, when an intron is small and the neighboring sequence lacks alternative legitimate splice sites or *cryptic splice sites* (sequences which resemble the consensus splice site sequences but which are not normally used by the splicing apparatus; see *Figure 9.11A*). The introduction of intronic sequence into the coding sequence of a mature mRNA will, at the very least, introduce additional amino acids and may cause a frameshift.

■ *Exon skipping.* The splicing apparatus uses an *alternative* legitimate splice site. Mutation of a splice donor sequence results in skipping of the upstream exon while mutation of the splice acceptor sequence results in skipping of the downstream exon (*Figure 9.11A*). Often, the exclusion of an exon has a profound effect on gene expression: it may result in a frameshift, an unstable RNA transcript, or a nonfunctional polypeptide because of a loss of a critical group of amino acids.

In addition to the above, the *branch site* used in splicing (see *Figure 1.15*) may be mutated leading to defective splicing.

Mutations of sequences which are not normally important for RNA splicing

Cryptic splice sites coincidentally resemble the sequences of authentic splice sites but are not normally used in splicing, unless a mutation alters the sequence so that the splicing apparatus now recognizes it as a normal splice site. Because individual splice donor and splice acceptor sequences often show some variation from the consensus sequences shown in *Figure 1.15*, cryptic splice sites may not be difficult to find (e.g. the β-globin gene has quite a variety of cryptic splice sites; see Cooper *et al.*, 1995). The use of an intronic cryptic splice site will introduce new amino acids, while using an exonic cryptic splice site will result in a deletion of coding DNA (*Figure 9.11B*).

See *Figures 9.12* and *9.13* respectively for worked examples of activation of a cryptic splice donor within an exon, and a cryptic splice acceptor within an intron. The former is a cautionary reminder that apparently silent mutations may yet be pathogenic. Note that in some cases mutations which occur within exons but not at cryptic splice sites can also induce skipping of that exon (see next section).

9.4.6 Mutations that introduce a premature termination codon usually result in unstable mRNA but other outcomes are possible

Several different classes of mutation can introduce a premature termination codon (chain terminating mutations). Nonsense mutations produce a premature termination codon simply by substituting a normal codon with a stop codon. Frameshifting insertions and deletions usually also introduce a premature termination codon not too far downstream of the mutation site. This hap-

pens because there is no selection pressure to avoid stop codons in the other translational reading frames and so given established nucleotide frequencies, at least one stop codon is usually encountered within a stretch of 100 nucleotides downstream of the mutation site. A variety of splice site mutations too can introduce a premature termination codon e.g. by skipping of a single exon containing a number of nucleotides that cannot be divided by 3. There are several possible consequences for gene expression for chain-terminating mutations:

■ *Unstable mRNA.* This is by far the most frequent consequence. A mRNA carrying a premature codon is usually rapidly degraded *in vivo* by a form of RNA surveillance known as nonsense-mediated mRNA decay (Hentze and Kulozik, 1999; Culbertson, 1999). This can avoid the potentially lethal consequences of producing a truncated polypeptide which could interfere with vital cell functions.

■ *Truncated polypeptide.* The production of a polypeptide truncated at the C terminus is a very rare outcome *in vivo* (the well-known protein truncation test which assays for mutations introducing a premature termination codon is carried out using an *in vitro* transcription–translation system). Nevertheless, some truncated polypeptides are produced *in vivo* (see, for example, Lehrman *et al.*, 1987). The effect on gene expression may be difficult to predict and will depend among other things on the extent of the truncation, the stability of the polypeptide product and its ability to interfere with expression of normal alleles.

■ *Exon skipping.* Some nonsense mutations appear to induce skipping of constitutive exons *in vivo*. For example, a nonsense mutation in the middle of exon 51 of the *FBN1* fibrillin gene (corresponding to the C terminus of the protein) causes that exon to be skipped (Dietz *et al.*, 1993). As a result of exon skipping the abnormally spliced mRNA uses the normal stop codon and escapes nonsense-mediated mRNA decay unlike any full length mRNA which may be produced from the pre-mRNA. The abnormally spliced *FBN1* mRNA accumulates and is translated to give a dominant negative protein lacking C-terminal sequences.

9.5 The pathogenic potential of repeated sequences

The human genome, like other mammalian genomes, has a very high proportion of DNA sequences that are repeated. Tandem repeats in coding DNA include very short nucleotide repeats, moderately sized repeats and very large repeats that can include whole genes. Depending on the degree of sequence homology between the repeats, tandem repeats are liable to a variety of different genetic mechanisms causing sequence exchange between the repeats (*Table 9.6*). Often such sequence exchanges

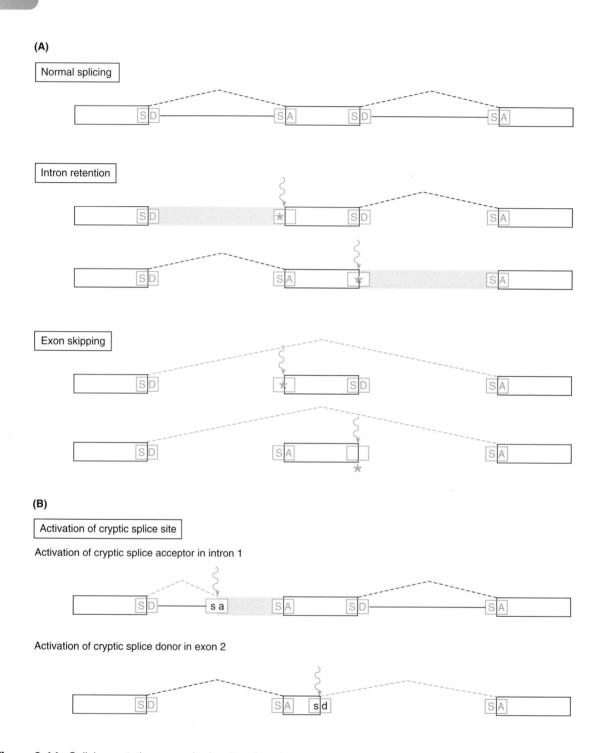

Figure 9.11: Splicing mutations can arise by alteration of conserved splice donor and splice acceptor sequences or by activation of cryptic splice sites.

(**A**) Mutations at conserved splice donor (SD) or splice acceptor (SA) sequences (see *Figure 1.15* for consensus sequences) result in (**A**) *intron retention* where there is failure of splicing and an intervening intron sequence is not excised; or in *exon skipping* where the spliceosome brings together the splice donor and splice acceptor sites of non-neighbouring exons.
(**B**) Sequences that are very similar to the consensus splice donor or splice acceptor sequences may coincidentally exist in introns and exons (sd and sa). These sequences are not normally used in splicing and so are known as *cryptic splice sites.* A mutation can activate a cryptic splice site by making the sequence more like the consensus splice donor or acceptor sequence and the cryptic splice site can now be recognized and used by the spliceosome (*activation* of the cryptic splice site). See *Figures 9.12* and *9.13* for examples of activation of an exonic and an intronic cryptic splice site, respectively.

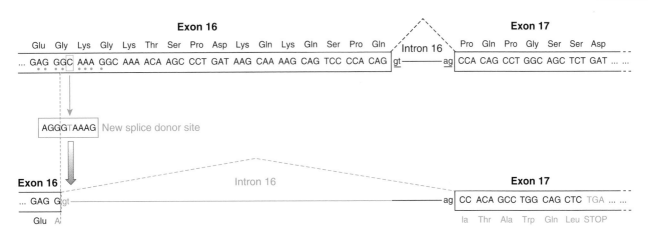

Figure 9.12: When a silent mutation is not silent.

This example shows a mutation that was identified in a LGMD2A limb girdle muscular dystrophy patient. The mutation was found in the calpain 3 gene, a known locus for this form of muscular dystrophy, but occurred at the third base position of a codon and appeared to be a silent mutation. It would lead to replacement of one glycine codon (GGC) by another glycine codon (GGT). However, the mutation is believed nevertheless to be pathogenic. The substitution results in activation of a cryptic splice donor sequence within exon 16 resulting in aberrant splicing with the loss of coding sequence from exon 16 and the introduction of a frameshift (see Richard and Beckmann, 1995).

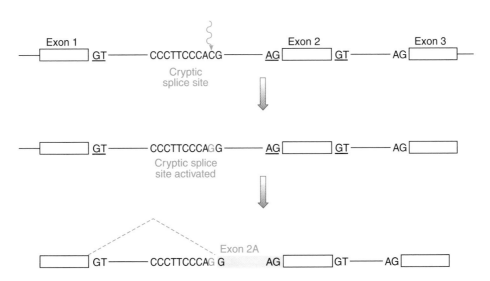

Figure 9.13: Mutations can cause abnormal RNA splicing by activation of cryptic splice sites.

This figure illustrates activation of a cryptic splice acceptor sequence located within an intron (compare *Figure 9.12* which illustrates activation of a cryptic splice donor site within an exon). A mutation can result in the alteration of a sequence which is not important for RNA splicing so as to create a new, alternative splice site. In the example illustrated, the mutation is envisaged to change a single nucleotide in intron 1. The nucleotide happens to occur within a *cryptic splice site* sequence that is closely related to the splice acceptor consensus sequence but, unlike the splice acceptor sites in *Figure 9.11*, shows a difference with respect to the conserved AG dinucleotide (see *Figure 1.15*). The mutation overcomes this difference and so can activate the cryptic splice site so that it competes with the natural splice acceptor site. If it is used by the splicing apparatus, a novel exon, exon 2A, results, which contains additional sequence which may or may not result in a frameshift.

Table 9.6: Repeated DNA sequences often contribute to pathogenesis

Type of repeated DNA	Type of mutation	Mechanism and examples
Tandem repeats		
Very short repeats within genes	Deletion	Slipped strand mispairing (see *Figure 9.5*). Examples in *Figure 9.14*
	Frameshifting insertion	Slipped strand mispairing
	Triplet repeat expansion	Initially by slipped strand mispairing?; subsequently large-scale expansion by unknown mechanism
Moderate sized intragenic repeats	Intragenic deletion	UEC/UESCE[a] (see *Figure 9.7*)
Large tandem repeats containing whole genes	Partial or total gene deletion	UEC/UESCE[a] (*Figure 9.7*). Examples in *Figure 9.17*
	Alteration of gene sequence	Gene conversion (*Figure 9.10*). Examples in *Figures 9.17* and *9.18*
	Duplication causing gene dosage-related aberrant expression	UESCE[a] – 1.5 Mb duplication in Charcot–Marie–Tooth 1A (see *Figure 15.6*)
Interspersed repeats		
Short direct repeats	Deletion	Slipped strand mispairing or intrachromatid recombination?
Interspersed repeat elements (e.g. *Alu* repeats)	Deletion	UEC/UESCE[a]
	Duplication	UEC/UESCE[a]
Inverted repeats	Inversion	Intrachromatid exchange, e.g. Factor VIII (see *Figure 9.20*)
Active transposable elements	Intragenic insertion by retrotransposons	Retrotransposition (*Figures 7.13 and 7.17*). Examples, see Section 9.5.6

[a]UEC, unequal crossover; UESCE, unequal sister chromatid exchange.

result in changes in the number of tandem repeats. A reduction in repeat number can often result in a pathogenic deletion, but expansion by sequence duplication can be pathogenic too (Mazzarella and Schlessinger, 1998). Certain chromosomal regions, notably the subtelomeric and pericentromeric regions, harbor large tracts of duplicated DNA and instability of such regions can predispose to disease (Eichler, 1998). Interspersed repeats can also cause pathogenic mutations by a different variety of mechanisms (see *Table 9.6*).

9.5.1 Slipped strand mispairing of short tandem repeats predisposes to pathogenic deletions and frameshifting insertions

Insertions and deletions in coding DNA are rare because they usually introduce a translational frameshift. However, occasionally, a series of tandem repeats of a small number of nucleotides occurs by chance in the coding sequence for a polypeptide. Such repeats, like classical microsatellite loci, are comparatively prone to mutation by slipped strand mispairing. As a result, the copy number of tandem repeats is liable to fluctuate, introducing a deletion or an insertion of one or more repeat units. If the mutation occurs in polypeptide-encoding DNA, a resulting deletion

will often have a profound effect on gene expression. Frameshifting deletions will normally result in abolition of gene expression. Even if the deletion does not produce a frameshift, deletions of one or more amino acids can still be pathogenic (*Figure 9.14*). Small frameshifting insertions will also be expected to lead to loss of gene expression and often the insertion is a tandem repeat of sequences flanking it. However, nonframeshifting insertions would often not be expected to be pathogenic, unless the insertion occurs in a critically important region, destabilizing an essential structure or impeding gene function in some way. Note that large triplet repeat expansions can lead to disease by mechanisms that are not understood at present (see next section).

9.5.2 Rapid large-scale expansion of intragenic triplet repeats can cause a variety of diseases but the mutational mechanism is not well understood

Sometimes microsatellites within or in the immediate vicinity of a gene can expand to considerable lengths and affect gene expression, causing disease. In some cases, a modestly expanded repeat which causes disease may be perfectly stable and be propagated without change in size through several generations. For example, triplet repeat

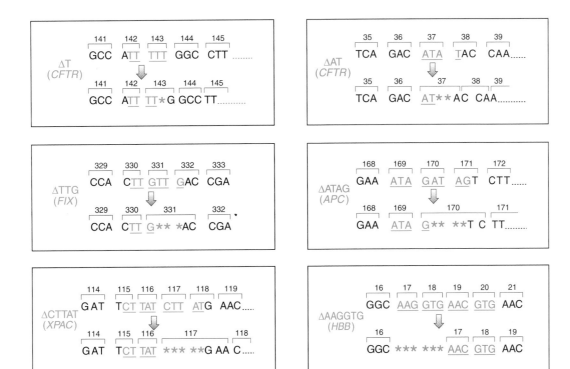

Figure 9.14: Short tandem repeats are deletion/insertion hotspots.

The six deletions illustrated are examples of pathogenic deletions occurring at tandemly repeated units of from 1 to 6 bp and have probably arisen as a result of replication slippage (*Figure 9.5*). The deletions of 3 and 6 bp do not cause frameshifts, and pathogenesis is thought to be due to removal of one or two amino acids that are critically important for polypeptide function. *Note* that in the case of the 6-bp deletion the original tandem repeat is not a perfect one. Genes (and associated diseases) are: *CFTR*, cystic fibrosis transmembrane regulator; *FIX*, factor IX (hemophilia B); *APC*, adenomatous polyposis coli; *XPAC*, xeroderma pigmentosa complementation group C; *HBB*, β-globin (β-thalassemia). Original references are listed in Appendix 3 of Cooper and Krawczak (1993). Though not illustrated here, small insertions are often tandem repeats of sequences flanking them (see Table 8.1 of Cooper and Krawczak, 1993).

expansion leading to long polyalanine tracts in the *HOXD13* gene cause a form of synpolydactyly, probably as a result of unequal crossover (Warren, 1997), but the expanded repeat is stable (Akarsu *et al.*, 1996). In other cases, however, the expanded triplet repeat is *unstable* and the discovery that human disease can be caused by large-scale expansion of highly unstable trinucleotide repeats was quite unexpected. Studies in other organisms had not revealed precedents for such a phenomenon, but the list of human examples is now considerable (see *Box 16.7*). In addition to unstable triplet repeat expansion, the majority

of disease alleles at the cystatin B gene which cause progressive myoclonus epilepsy involve expansions of a 12 nucleotide repeat (CCCCGCCCCGCG; Lalioti *et al.*, 1998). The pathological mechanisms by which unstable expanded repeats cause disease are discussed in Chapter 16. Here we are concerned with the nature and mechanism of the DNA instability (see also Djian, 1998; Sinden, 1999).

Tandem trinucleotide repeats are not infrequent in the human genome. Although there are 64 possible trinucleotide sequences, when allowance is made for cyclic permutations $(CAG)_n = (AGC)_n = (GCA)_n$ and reading from either strand [5'$(CAG)_n$ on one strand = 5'$(CTG)_n$ on the other], there are only 10 different trinucleotide repeats (*Figure 9.15*). Most of these are known as usefully polymorphic microsatellite markers but, in addition, certain repeats show anomalous behavior which can cause abnormal gene expression. In each case, repeats below a certain length are stable in mitosis and meiosis while, above a certain threshold length, the repeats become extremely unstable. These unstable repeats are virtually never transmitted unchanged from parent to child. Both expansions and contractions can occur, but there is a bias towards expansion. The average size change often

AAC/GTT	AGG/CCT
AAG/CTT	ATC/GAT
AAT/ATT	
ACC/GGT	CAG/CTG
ACG/CGT	CCG/CGG
ACT/AGT	

Figure 9.15: The ten possible trinucleotide repeats.

Both DNA strands are shown. All other trinucleotide repeats are cyclic permutations of one or another of these (see text).

depends on the sex of the transmitting parent, as well as the length of the repeat. Genes containing unstable expanding trinucleotide repeats fall into two major classes (see *Box 16.7*):

■ *Genes which show modest expansions of (CAG)ₙ repeats within the coding sequence.* Typically, the stable and nonpathological alleles have 10–30 repeats, while unstable pathological alleles have modest expansions, often in the range of 40–200 repeats. Transcription and translation of the gene are not affected by the expansion. The resulting protein product shows a gain of function: its long polyglutamine tract causes it to aggregate within certain cells and kill them.

■ *Genes which show very large expansions of a noncoding repeat.* For some genes, various types of triplet repeat (e.g. CGG, CCG, CTG, GAA) found in the promoter, the untranslated regions or intronic sequences can undergo very large expansions in such a way as to inhibit gene expression, causing loss of function. Typically, the stable and nonpathological alleles have 5–50 repeats, while unstable pathological alleles have several hundreds or thousands of copies (see *Box 16.7*).

Intergenerational changes are normally reported as parent–child comparisons of blood lymphocyte DNA. There is little information about when in gametogenesis, fertilization or embryogenesis the changes arise. Limited studies of sperm show that highly expanded *DM* and *FRAXA* (fragile-X syndrome) repeats are not transmitted by affected males, although modest expansions can be. The largest expansions in Huntington disease (which, however, are small compared with large *FRAXA* or *DM*

expansions) are seen in sperm, consistent with the observation that the severest cases inherit the disease from their father. At least for *FRAXA, DM* and Kennedy disease, the expanded repeats are mitotically unstable, so that a blood sample shows a smear of heterogeneous expanded repeats sizes. However, *in vitro*, even large repeats are stable. Thus, whatever the mechanism, it is not operative in all cells.

The basis of the unstable expansions is very largely unknown and this type of mutagenic event has not been identified thus far in genetically tractable organisms such as *E. coli*, yeast or *Drosophila*. There is also evidence that the unstable expansion mechanism may not have a parallel in some other mammals such as mice. Human transgenes containing long trinucleotide repeats show virtually no instability after being propagated through several generations in transgenic mice whereas the same sequences may show a 100% probablity of expansion when transmitted in the human germline (Djian, 1998). Investigations have also suggested that arrays of triplet repeats may be able to form alternative DNA structures, such as DNA hairpins, triplex DNA and quadruplex DNA (Sinden, 1999) but their significance if formed *in vivo* is unknown. The repeats have also been envisaged as possible protein-binding sites and protein-binding at the RNA level has also been envisaged to contribute to pathogenesis in some cases, notably in myotonic dystrophy (Philips *et al.*, 1998).

Slipped strand mispairing (see *Figure 9.5*) is likely to be a component of the expansion mechanism, given the observation that interrupted repeats appear to be stable and only homogeneous repeats are unstable. For example, in spinocerebellar ataxia type 1, 123/126 normal

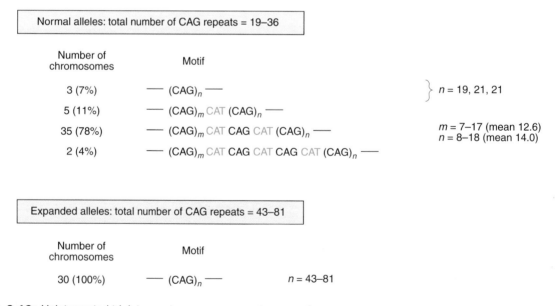

Figure 9.16: Uninterrupted triplet repeats are more prone to expansion.

Analysis of the *SCA1* spinocerebellar ataxia gene by Chung *et al.* (1993) showed that all the presumed stable alleles from normal subjects had interrupted repeats except the three with the shortest runs. However, all of the unstable expanded alleles found on disease chromosomes had uninterrupted repeats.

sized CAG repeats were interrupted by one or two CAT triplets, while 30/30 expanded alleles contained no interruption (Chung *et al.*, 1993; see *Figure 9.16*). One problem with all these mispairing mechanisms is that they should result in contractions as well as expansions and this is not seen. Instead, after a certain threshold size, there appears to be a clear bias towards continued expansion of the size of the repeat unit array. Because understanding of trinucleotide repeats is progressing very rapidly at the time of writing, the reader is advised to consult a recent review for more information.

9.5.3 Tandemly repeated and clustered gene families may be prone to pathogenic unequal crossover and gene conversion-like events

Many human and mammalian gene clusters contain nonfunctional pseudogenes which may be closely related to functional gene members. Interlocus sequence exchanges between pseudogenes and functional genes can result in disease by removing or altering some or all of the sequence of a functional gene. For example, unequal crossover (or unequal sister chromatid exchange) between a functional gene and a related pseudogene can result in deletion of the functional gene or the formation of fusion genes containing a segment derived from the pseudogene. Alternatively, the pseudogene can act as a donor sequence in gene conversion events and introduce deleterious mutations into the functional gene.

The classical example of pathogenesis due to gene–pseudogene exchanges is steroid 21-hydroxylase deficiency, where over 95% of pathogenic mutations arise as a result of sequence exchanges between the functional 21-hydroxylase gene, *CYP21B*, and a very closely related pseudogene, *CYP21A*. The two genes occur on tandemly repeated DNA segments approximately 30 kb long which also contain other duplicated genes, notably the

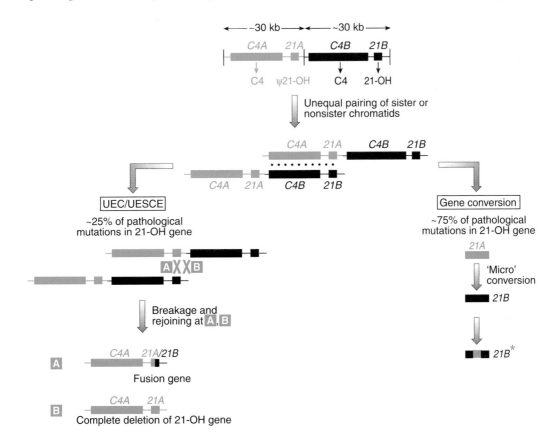

Figure 9.17: Almost all 21-hydroxylase gene mutations are due to sequence exchange with a closely related pseudogene.

The duplicated complement *C4* genes and steroid 21-hydroxylase genes are located on tandem 30 kb repeats which show about 97% sequence identity. Both the *C4A* and *C4B* genes are expressed to give complement C4 products; the *CYP21B* gene (*21B*) encodes a 21-hydroxylase product, but the *CYP21A* (*21A*) gene is a pseudogene. About 25% of pathological mutations at the 21-hydroxylase locus involve a 30 kb deletion resulting from unequal crossover (UEC) or unequal sister chromatid exchange (UESCE). The remaining mutations are point mutations where small-scale gene conversion of the *CYP21B* gene occurs – a small segment of the *CYP21A* gene containing deleterious mutations is copied and inserted into the *CYP21B* gene replacing a short segment of the original sequence (see *Figure 9.10C* for one possible mechanism). Possibly gene conversion events are, like UEC and UESCE, primed by unequal pairing of the tandem repeats on sister or nonsister chromatids.

complement C4 genes, *C4A* and *C4B* (*Figure 9.17*). Large pathogenic deletions uniformly result in removal of about 30 kb of DNA, corresponding to one repeat unit length, and analysis of *de novo* 21-hydroxylase deficiency mutations has provided strong evidence for pathogenic deletions arising as a result of meiotic unequal crossover (Sinnott *et al.*, 1990).

Virtually all of the 75% of pathogenic point mutations are copied from deleterious mutations in the pseudogene, suggesting a gene conversion mechanism (*Figures 9.17* and *9.18*). Analysis of one such mutation which arose *de novo* suggests that the conversion tract is a maximum of 390 bp (Collier *et al.*, 1993). Gene conversion events are also found in the duplicated C4 genes, both of which are normally expressed. A likely priming event for conversions in the *CYP21–C4* gene cluster is unequal pairing of chromatids so that a *CYP21A–C4A* unit pairs with a *CYP21B–C4B* unit (*Figure 9.17*).

9.5.4 Interspersed repeats often predispose to large deletions and duplications

Short direct repeats
In several cases, the endpoints of deletions are marked by very short direct repeats. For example, the breakpoints in numerous pathogenic deletions in mtDNA occur at perfect or almost perfect short direct repeats. Of these, the most common is a deletion of 4977 bp which has been found in multiple patients with Kearns–Sayre syndrome, an encephalomyopathy characterized by external ophthalmoplegia, ptosis, ataxia and cataract. The deletion results in elimination of the intervening sequence between two perfect 13 bp repeats and loss of the sequence of one of the repeats (*Figure 9.19*). The mitochondrial genome is recom-

bination-deficient and Shoffner *et al.* (1989) have postulated that such deletions arise by a replication slippage mechanism, similar to that occurring at short tandem repeats (see *Figure 9.5*). Partial duplications of the mitochondrial genome are also distinctive features of certain diseases, notably Kearns–Sayre syndrome. The ends of the duplicated sequences, like those of the common deletions, are often marked by short direct repeats, and the mechanisms of duplication and deletion appear to be closely related (Poulton and Holt, 1994).

The *Alu* repeat as a recombination hotspot
Some large-scale deletions and insertions may be generated by pairing of nonallelic interspersed repeats, followed by breakage and rejoining of chromatid fragments. For example, the *Alu* repeat occurs approximately once every 4 kb and mispairing between such repeats has been suggested to be a frequent cause of deletions and duplications. Some large genes have many internal *Alu* sequences in their introns or untranslated sequences, making them liable to frequent internal deletions and duplications. For example, the 45-kb low density lipoprotein receptor gene has a relatively high density of *Alu* repeats (approximately one every 1.6 kb). A very high frequency of pathogenic deletions in this gene are likely to involve an *Alu* repeat, usually at both endpoints, and occasional pathogenic intragenic duplications also involve *Alu* repeats (Hobbs *et al.*, 1990). Such observations have suggested a general role for *Alu* sequences in promoting recombination and recombination-like events. Initial gene duplications in the evolution of clustered multigene families may often have involved an unequal crossover event between *Alu* repeats or other dispersed repetitive

Location of mutation	Normal 21-OH gene sequence (*CYP21B*)	21-OH pseudogene sequence (*CYP21A*)	Mutant 21-OH gene sequence
Intron 2	CCCAСCTCC	CCCAGCTCC	CCCAGCTCC
Exon 3 (codons 110–112)	**GGA GAC TAC** TC **Gly Asp Tyr** Ser	G(.........)TC	G(.........)TC Val
Exon 4 (codon 172)	ATC **ATC** TGT Ile **Ile** Cys	ATC AAC TGT	ATC AAC TGT Ile Asn Cys
Exon 6 (codons 235–238)	ATC **GTG** GAG **ATG** Ile **Val** Glu **Met**	AAC GAG GAG AAG	AAC GAG GAG AAG Asn Glu Glu Lys
Exon 7 (codon 281)	CAC **GTG** CAC His **Val** His	CAC TTG CAC	CAC TTG CAC His Leu His
Exon 8 (codon 318)	CAC **CAG** GAG His **Gln** Glu	CTG TAG GAG	CTG TAG GAG Leu STOP
Exon 8 (codon 356)	CTG **CGG** CCC Leu **Arg** Pro	CTG TGG CCC	CTG TGG CCC Leu Trp Pro

Figure 9.18: Pathogenic point mutations in the steroid 21-hydroxylase gene originate by copying sequences from the 21-hydroxylase pseudogene.

The copying is thought to involve a gene conversion-like mechanism (see *Figures 9.10C* and *9.17*).

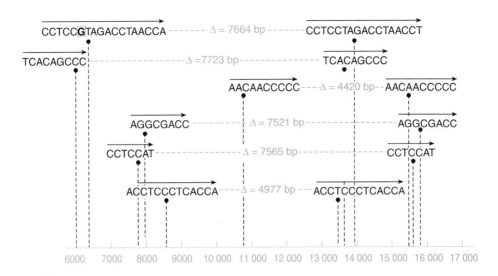

Figure 9.19: Short direct repeats mark the endpoints of many pathogenic deletions in the mitochondrial genome.

Scale represents nucleotide position in the mitochondrial genome. Position 1 occurs within the D loop region and numbering increases in a clockwise direction for the illustration in *Figure 7.2. Note* that as recombination does not occur within the mitochondrial genome, one likely mechanism to explain the deletions is slipped strand mispairing (see text).

elements. It should be noted, however, that some *Alu*-rich genes do not appear to be loci for frequent *Alu*-mediated recombination.

9.5.5 Pathogenic inversions can be produced by intrachromatid recombination between inverted repeats

Occasionally, clustered inverted repeats with a high degree of sequence identity may be located within or close to a gene. The high degree of sequence similarity between inverted repeats may predispose to pairing of the repeats by a mechanism that involves a chromatid bending back upon itself. Subsequent chromatid breakage at the mispaired repeats and rejoining can then result in an inversion, in much the same way as the natural mechanism used for the production of some immunoglobulin κ light chains (see *Figure 8.28*).

The classic example of pathogenic inversions is a mutation which accounts for more than 40% of cases of severe hemophilia A. Intron 22 of the factor VIII gene, *F8*, contains a CpG island from which two internal genes are transcribed: *F8A* in the opposite direction to the host gene *F8*, and *F8B* in the same direction as *F8* (see *Figure 9.20*). *F8A* belongs to a gene family with two other closely related members located several hundred kilobases upstream of *F8* gene and transcribed in the opposite direction to *F8A*. As a result, the region between the *F8A* gene and the other two members is susceptible to inversions – the *F8A* gene can pair with either of the other two members on the same chromatid, and subsequent chromatid breakage and rejoining in the region of the paired repeats results in an

inversion which disrupts the factor VIII gene (Lakich *et al.*, 1993, see *Figure 9.20*).

9.5.6 DNA sequence transposition is not uncommon and can cause disease

As described in Sections 7.4.4 to 7.4.6, a proportion of moderately and highly repeated interspersed elements are capable of transposition via an RNA intermediate. Defective gene expression due to DNA transposition is comparatively rare and represents only a small component of molecular pathology. However, several examples have been recorded of genetic deficiency due to insertional inactivation by retrotransposons. For example, in one study, hemophilia A was found to arise in two out of 140 unrelated patients as a result of a *de novo* insertion of a LINE-1 (*Kpn*) repeat into an exon of the factor VIII gene (Kazazian *et al.*, 1988). Other instances are known of insertional inactivation by an actively transposing *Alu* element, as in a case of neurofibromatosis type 1 (Wallace *et al.*, 1991). Additionally, a number of other examples have been recorded of pathogenesis due to intragenic insertion of undefined DNA sequences.

9.6 DNA repair

DNA in cells suffers a wide range of damage:

- purine bases are lost by spontaneous fission of the base–sugar link;
- cytosines, and occasionally adenines, spontaneously deaminate to produce uracil and hypoxanthine respectively;

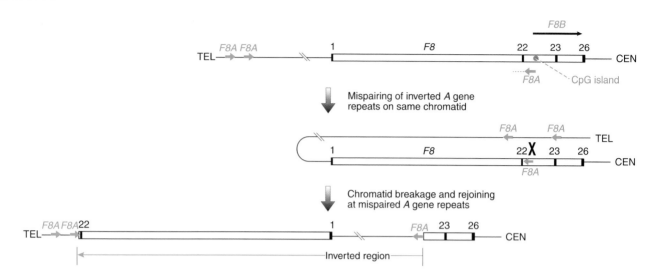

Figure 9.20: Inversions disrupting the factor VIII gene result from intrachromatid recombination between inverted repeats.

For the sake of clarity only exons 22, 23 and the first and last exons (1 and 26) of the factor VIII gene (*F8*; open box) are shown. Intron 22 of this gene contains a CpG island from which the *F8B* gene internal is transcribed in the same direction as the factor VIII gene, and expression involves splicing of a novel exon within intron 22 on to exons 23–26 of the factor VIII gene. The internal *F8A* gene is also transcribed from the intron 22 CpG island but in the opposite direction. Two other sequences closely related to *F8A* are found about 500 kb upstream of the factor VIII gene and are transcribed in the same direction as the factor VIII gene and the opposite direction to that of the *F8A* gene. The high degree of sequence identity between the three members of the *F8A* gene family means that pairing of the *F8A* gene with one of the other two members on the same chromatid can occur by looping back of the chromatid. Subsequent chromatid breakage and rejoining can result in an inversion of the region between the *F8A* gene and the other paired family member, resulting in disruption of the factor VIII gene (see Lakich *et al.*, 1993).

- many chemicals, for example alkylating agents, form adducts with DNA bases;
- reactive oxygen species in the cell attack purine and pyrimidine rings
- ultraviolet light causes adjacent thymines to form a stable chemical dimer;
- mistakes in DNA replication result in incorporation of a mismatched base;
- ionizing radiation causes single- or double-strand breaks;
- mistakes in replication or recombination leave strand breaks in DNA.

All these lesions must be repaired if the cell is to survive. The importance of effective DNA repair systems is highlighted by the severe diseases affecting people with deficient repair systems (see below).

9.6.1 DNA repair usually involves cutting out and resynthesizing a whole area of DNA surrounding the damage

To cope with all these forms of damage, cells must be capable of several different types of DNA repair (for reviews, see the October 1995 issue of *Trends in Biochemical Sciences*). DNA repair seldom involves simply undoing the change that caused the damage. Almost always a stretch of DNA containing the damaged nucleotide(s) is excised and the gap filled by resynthesis. There are at least five main types of DNA repair in human cells:

- *Direct repair* reverses the DNA damage. A specific enzyme is able to dealkylate O^6-alkyl guanine directly. In bacteria thymine dimers can be removed in a photoreactivation reaction that depends on visible light and an enzyme, photolyase. Mammals possess enzymes related to photolyase, but use them for a quite different purpose, to control their circadian clock (Van der Horst *et al.*, 1999).
- *Base excision repair (BER)* uses glycosidase enzymes to remove abnormal bases. An endonuclease, AP endonuclease, cuts the sugar–phosphate backbone at the position of the missing base. A few nucleotides of the DNA strand are stripped back by exonucleases, the gap is filled by resynthesis, and the remaining nick is sealed by DNA ligase III. The same process is used to repair spontaneous depurination. Interestingly, no human diseases caused by defective BER are known. Maybe any such defect would be lethal, since BER corrects much the commonest type of DNA damage.
- *Nucleotide excision repair (NER)* removes thymine dimers and large chemical adducts. *Figure 9.21* illustrates the process. Defects in NER cause the autosomal recessive disease xeroderma pigmentosum (Lambert *et al.*, 1998). Seven complementation

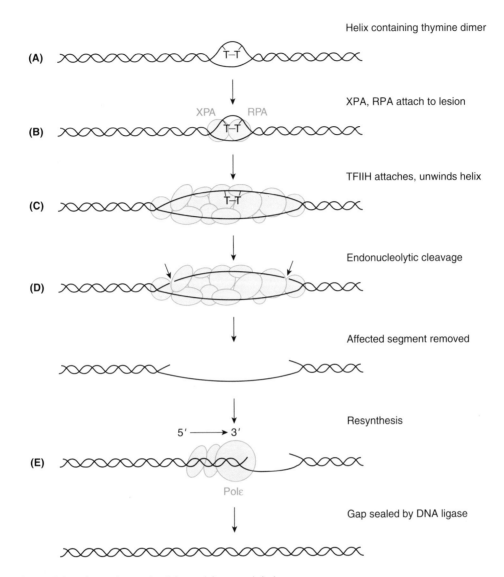

Figure 9.21: A possible scheme for nucleotide excision repair in humans.

(**A**) XPA protein recognizes damaged DNA and binds to it, directly or by binding to RPA, a single-strand binding protein. (**B**) The DNA–XPA–RPA complex recruits the TFIIH transcription factor. TFIIH is a multiprotein complex that includes the XPB and XPD proteins. These are helicases of opposite polarity, and they open up a single-stranded bubble in the DNA, about 30 nucleotides long. (**C**) Two cuts are made in the sugar–phosphate backbone of the damaged strand. XPF + ERCC1 cut at the 5′ end, and XPG cuts the 3′ end. (**D**) DNA polymerase ε together with replication factor C and the DPE2 subunit synthesize DNA to fill the gap. (**E**) DNA ligase seals the gap. Over 30 proteins are involved in mammalian nucleotide excision repair, and this simplified scheme does not include the likely requirement to remodel chromatin structure as part of the process (after Lehmann, 1995).

groups, XPA–XPG, have been defined by cell fusion studies. XP patients are exceedingly sensitive to UV light. Sun-exposed skin develops thousands of freckles, many of which progress to skin cancer.

■ *Post-replication repair* is required to correct double-strand breaks (Haber, 1999). The usual mechanism is a gene conversion-like process (recombinational repair), where a single strand from the homologous chromosome invades the damaged DNA helix. Alternatively, broken ends are rejoined regardless of their sequence, a desperate measure that is likely to cause mutations. The eukaryotic machinery for recombination repair is less well defined than the excision repair systems. Human genes involved in this pathway include *NBS* (mutated in Nijmegen breakage syndrome, MIM 251260, Section 12.4.1), *BLM* (mutated in Bloom syndrome, MIM 210900) and the *BRCA2* and maybe *BRCA1* breast cancer susceptibility genes (Section 19.5.3; Zhang *et al.*, 1998).

■ *Mismatch repair* corrects mismatched base pairs caused by mistakes in DNA replication. Cells deficient in mismatch repair have mutation rates

100–1000 times higher than normal, with a particular tendency to replication slippage in homopolymeric runs (*Figure 9.5*). In humans the mechanism involves at least five proteins and defects cause hereditary nonpolyposis colon cancer (Section 18.7.1; *Figure 18.17*).

All these systems, except for direct repair, require exo- and endonucleases, helicases, polymerases and ligases, usually acting in multiprotein complexes that have some components in common. Sorting out the individual pathways has been greatly aided by the very strong conservation of repair mechanisms across the whole spectrum of life. Not only the reaction mechanisms but also the protein structures and even gene sequences are often conserved from *E. coli* to man. A downside of the conservation is a confusing gene nomenclature, referring sometimes to human diseases (*XPA* etc.), sometimes to yeast mutants (*RAD* genes) and sometimes to mammalian cell complementation systems (ERCC = excision repair cross-complementing) – for example *XPD*, *ERCC2* and *RAD3* are the same gene in man, mouse and yeast. Generally eukaryotes have multiple proteins corresponding to each single protein in *E. coli*, so that, for example, nucleotide excision repair requires six proteins in *E. coli* but at least 30 in mammals.

9.6.2 DNA repair systems share components and processes with the transcription and recombination machinery

As well as sharing components with each other, many repair systems share components with the machinery for DNA replication, transcription and recombination. DNA polymerases and ligase are required for both DNA replication and resynthesis after excision of a defect. The recombination machinery is involved in double-strand break repair. The link with transcription is particularly intriguing (Lehmann, 1995). The general transcription factor TFIIH is a multiprotein complex that includes the XPB and XPD proteins. TFIIH exists in two forms. One form is concerned with general transcription and the other with repair, probably specifically repair of transcriptionally active DNA. This system is deficient in two rare diseases, Cockayne syndrome (MIM 216400) and trichothiodystrophy (MIM 601675). Clinically and in cell biology, CS and TTD both overlap XP, and in some cases the same genes are responsible, but CS and TTD patients have developmental defects that presumably reflect defective transcription, and they do not have the cancer susceptibility of XP patients.

9.6.3 Hypersensitivity to agents that damage DNA is often the result of an impaired cellular response to DNA damage, rather than defective DNA repair

Many human diseases that involve hypersensitivity to DNA-damaging agents, or a high level of cellular DNA damage, are not caused by defects in the DNA repair systems themselves, but by a defective cellular response to DNA damage. Normal cells react to DNA damage by stalling progress through the cell cycle at a checkpoint until the damage has been repaired, or triggering apoptosis if the damage is irreparable. Part of the machinery for doing this involves the ATM protein. The role of ATM is described in Section 18.7.3. Briefly, it senses DNA damage and relays the signal to the p53 protein, the 'guardian of the genome'. People with no functional ATM have ataxia telangiectasia (MIM 208900; Lambert *et al.*, 1998). Their cells are hypersensitive to radiation, and they have chromosomal instability and a high risk of malignancy, but the DNA repair machinery itself is intact. Fanconi anemia (MIM 227650) is another heterogeneous group of diseases (at least five complementation groups) marked by defective responses to DNA damage, without specific defects in DNA repair.

Further reading

Cooper DN, Krawczak M (1993) *Human Gene Mutation*. BIOS Scientific Publishers, Oxford.
Li W-H (1997) *Molecular Evolution*. Sinauer Associates, Inc., Sunderland, MA.

References

Akarsu AN, Stoilov I, Yilmaz E, Sayli BS, Sarfarazi M (1996) Genomic structure of *HOXD13* gene: a nine polyalanine duplication causes synpolydactyly in two unrelated families. *Hum. Mol. Genet.*, **5**, 945–952.
Ayala FJ (1999) Molecular clock mirages. *Bioessays*, **21**, 71–75.
Blok RB, Gook DA, Thorburn DR, Dahl H-H M (1997) Skewed segregation of the mtDNA nt 8993 (T $\Rightarrow$ G mutation in human oocytes. *Am. J. Hum. Genet.*, **60**, 1495–1501.
Britten RJ (1986) Rate of DNA sequence evolution between different taxonomic groups. *Science*, **231**, 1393–1398.
Brown WM, George Jr M, Wilson AC (1979) Rapid evolution of animal mitochondrial DNA. *Proc. Natl Acad. Sci. USA*, **76**, 1967–1971.

Cairns J (1975) Mutation selection and the natural history of cancer. *Nature,* **255**, 197–200.

Chung MY, Ranum LPW, Duvick IA, Servadio A, Zoghbi HY, Orr HT (1993) Evidence for a mechanism predisposing to intergenerational CAG repeat instability in spinocerebellar ataxia type 1. *Nature Genet.,* **5**, 254–258.

Collier PS, Tassabehji M, Sinnott PJ, Strachan T (1993) A *de novo* pathological point mutation at the 21-hydroxylase locus: implications for gene conversion in the human genome. *Nature Genet.,* **3**, 260–265 and *Nature Genet.,* **4**, 101.

Collins DW, Jukes TH (1994) Rates of transition and transversion in coding sequences since the human–rodent divergence. *Genomics,* **20**, 386–396.

Cooper DN, Krawczak M (1993) *Human Gene Mutation.* BIOS Scientific Publishers, Oxford.

Cooper DN, Youssoufian H (1988) The CpG dinucleotide and human genetic disease. *Hum. Genet.,* **78**, 151–155.

Cooper DN, Smith BA, Cooke HJ, Niemann S, Schmidtke J (1985) An estimate of unique DNA sequence heterozygosity in the human genome. *Hum. Genet.,* **69** (3), 201–205.

Cooper DN, Krawczak M, Antonarakis SE (1995) The nature and mechanisms of human gene mutation. In: *Metabolic and Molecular Bases of Inherited Disease,* 7th edn. (C Scriver, AL Beaudet, WS Sly, D Valle eds), pp. 259–291. McGraw-Hill, New York.

Crow JF (1999) The odds of losing at genetic roulette. *Nature,* **397**, 293–294.

Culbertson MR (1999) RNA surveillance. *Trends Genet.,* **15**, 74–80.

Dietz HC, Valle D, Francomano C, Kendzior Jr RJ, Pyeritz RE, Cutting GR (1993) The skipping of constitutive exons *in vivo* induced by nonsense mutations. *Science,* **259**, 680–683.

Djian P (1998) Evolution of simple repeats in DNA and their relation to human disease. *Cell,* **94**, 155–160.

Dover GA (1995) Slippery DNA runs on and on and on . . . *Nature Genet.,* **10**, 254–256.

Dubrova YE, Nesterov VN, Krouchinsky NG, Ostapenko VN, Neumann R, Neil DL, Jeffreys AJ (1996) Human minisatellite mutation rate after the Chernobyl accident. *Nature,* **380**, 683–686.

Eichler EE (1998) Masquerading repeats: paralogous pitfalls of the human genome. *Genome Res.,* **8**, 758–762.

Eyre-Walker A, Keightley PD (1999) High genomic deleterious mutation rates in hominids. *Nature,* **397**, 344–347.

Gibbons RJ, Picketts DJ, Villard L, Higgs DR (1995) Mutations in a putative global transcriptional regulator cause X-linked mental retardation with α-thalassemia (ATR-X syndrome). *Cell,* **80**, 837–845.

Gibbs PEM, Witke WF, Dugaiczyk A (1998) The molecular clock runs at different rates among closely related members of a gene family. *J. Mol. Evol.,* **46**, 552–561.

Haber JA (1999) Gatekeepers of recombination. *Nature,* **398**, 665–667.

Hauswirth W, Laipis P (1982) Mitochondrial DNA polymorphism in a maternal lineage of Holstein cows. *Proc. Natl Acad. Sci. USA.,* **79**, 4686–4690.

Hentze MW, Kulozik AE (1999) A perfect message: RNA surveillance and nonsense-mediated decay. *Cell,* **96**, 307–310.

Hobbs HH, Russell DW, Brown MS, Golding JL (1990) The LDL receptor locus in familial hyper-cholesterolaemia: mutational analysis of a membrane protein. *Annu. Rev. Genet.,* **24**, 133–170.

Howell N (1997) mtDNA recombination: what do in vitro data mean? *Am. J. Hum. Genet.,* **61**, 19–22.

Hurst LD, Ellegren H (1998) Sex biases in the mutation rate. *Trends Genet.,* **14**, 446–452.

Jeffreys A, Tamaki K, MacLeod A, Monckton DG, Neil DL, Armour JAL (1994) Complex gene conversion events in germline mutation at human microsatellites. *Nature Genet.,* **6**, 136–145.

Jenuth JP, Peterson AC, Fu K, Shoubridge EA (1996) Random genetic drift in the female germline explains the rapid segregation of mammalian mitochondrial DNA. *Nature Genet.,* **14**, 146–150.

Kazazian HH, Wong C, Youssoufian H, Scott AF, Phillips DG, Antonarakis SE (1988) Haemophilia A resulting from de novo insertion of L1 sequences represents a novel mechanism for mutation in man. *Nature,* **332**, 164–166.

Koehler CM, Lindberg GL, Brown DR, Beitz DC, Freeman AE, Mayfield JE, Myers AM (1991) Replacement of bovine mitochondrial DNA by a sequence variant within a single generation. *Genetics,* **129**, 247–255.

Kumar S, Hedges SB (1998) A molecular timescale for vertebrate evolution. *Nature Genet.,* **5**, 917–920.

Lakich D, Kazazian Jr HH, Antonarakis SE, Gitschier J (1993) Inversions disrupting the factor VIII gene are a common cause of severe haemophilia A. *Nature Genet.,* **5**, 236–241.

Lalioti MD, Scott HS, Buresi C, Rossier C, Bottani A, Morris MA, Malafosse A, Antonarakis SE (1997) Dodecamer repeat expansion in cystatin B gene in progressive myoclonus epilepsy. *Nature,* **386**, 847–851.

Lambert WC, Kuo H-R, Lambert MW (1998) Xeroderma pigmentosum and related disorders. In: Jameson JL (ed.) *Principles of Molecular Medicine.* Humana Press, NJ.

Lehmann AR (1995) Nucleotide excision repair and the link with transcription. *Trends Biochem. Sci.,* **20**, 402–405.

Lehrman MA, Schneider WJ, Brown MS, Davis CG, Elhammer A, Russell DW, Goldstein JL (1987) The Lebanese allele of the low density lipoprotein receptor locus: nonsense mutation produces truncated receptor that is retained in the endoplasmic reticulum. *J. Biol. Chem.,* **262**, 401–410.

Levinson G, Gutman GA (1987) Slipped strand mispairing: a major mechanism for DNA sequence evolution. *Mol. Biol. Evol.,* **4**, 203–221.

Li W-H (1997) *Molecular Evolution,* p. 229. Sinauer Associates, Inc., Sunderland, MA.

Li W-H, Graur D (1991) *Fundamentals of Molecular Evolution.* Sinauer Associates, Inc., Sunderland, MA.

Lightowlers RN, Chinnery PF, Turnbull DM, Howell N (1997) Mammalian mitochondrial genetics: heredity, heteroplasmy and disease. *Trends Genet.,* **13**, 450–455.

Mahtani MM, Willard HF (1993) A polymorphic X-linked tetranucleotide repeat locus displaying a high rate of new mutation: implications for mechanisms of mutation at short tandem repeat loci. *Hum. Molec. Genet.,* **2**, 431–437.

Mazzarella R, Schlessinger D (1998) Pathological consequences of sequence duplications in the human genome. *Genome Res.,* **8**, 1007–1021.

Nickerson DA, Taylor SL, Weiss KM *et al.* (1998) DNA sequence diversity in a 9.7 kb region of the human lipoprotein lipase gene. *Nature Genet.,* **19**, 233–240.

Philips AV, Timchenko LT, Cooper TA (1998) Disruption of splicing regulated by a CUG-binding protein in myotonic dystrophy. *Science,* **280**, 737–741.

Piko L, Taylor KD (1987) Amounts of mitochondrial DNA and abundance of some mitochondrial gene transcripts in early mouse embryos. *Dev. Biol.,* **123**, 364–374.

Poulton J, Holt IJ (1994) Mitochondrial DNA: does more lead to less? *Nature Genet.,* **8**, 313–315.

Poulton J, Macaulay V, Marchington DR (1998) Is the bottleneck cracked? *Am. J. Hum. Genet.,* **62**, 752–757.

Richard I, Beckmann JS (1995) How neutral are synonymous codon mutations? *Nature Genet.,* **10**, 259.

Sharp P, Matassi G (1994) Codon usage and genome evolution. *Curr. Opin. Genet. Dev.,* **4**, 851–860.

Shimmin LC, Chang BH-J, Li W-H (1993) Male driven evolution of DNA sequences. *Nature,* **362**, 745–747.

Shoffner JM, Lott MT, Voljavec AS, Soueidan SA, Costigan DA, Wallace DC (1989) Spontaneous Kearns–Sayre/chronic external ophthalmoplegia plus syndrome associated with a mitochondrial DNA deletion: a slip-replication model and metabolic therapy. *Proc. Natl Acad. Sci. USA.,* **86**, 7952–7956

Sinden RR (1999) Biological implications of the DNA structures associated with disease-causing triplet repeats. *Am. J. Hum. Genet.,* **64**, 346–353.

Sinnott PJ, Collier S, Costigan C, Dyer PA, Harris R, Strachan T (1990) Genesis by meiotic unequal crossover of a *de novo* deletion that contributes to 21-hydroxylase deficiency. *Proc. Natl Acad. Sci. USA,* **87**, 2107–2111.

Taillon-Miller P, Gu Z, Li Q, Hillier L, Kwok P-Y (1998) Overlapping genomic sequences: a treasure trove of single-nucleotide polymorphisms. *Genome Res.,* **8**, 748–754.

Van der Horst GTJ, Muijtjens M, Kobayashi K *et al.* (1999) Mammalian Cry1 and Cry2 are essential for maintenance of circadian rhythms. *Nature,* **398**, 627–630.

Vogel F, Motulsky AG (1996) *Human Genetics. Problems and Approaches,* 3rd edn. Springer-Verlag, Berlin.

Wada K, Aota S, Tsuchiya R, Ishibashi F, Gojobori T, Ikemura T (1990) Codon usage tabulated from the GenBank genetic sequence data. *Nucl. Acid Res.,* **18** (suppl.), 2367–2400.

Wallace MR, Andersen LB, Saulino AM, Gregory PE, Glover TW, Collins FS (1991) A *de novo Alu* insertion results in neurofibromatosis type I. *Nature,* **353**, 864–868.

Warren ST (1997) Polyalanine expansion in synpolydactyly might result from unequal crossing-over of *HOXD13. Science,* **275**, 408–409.

Weber JL, Wong C (1993) Mutation of human short tandem repeats. *Hum. Molec. Genet.,* **2**, 1123–1128.

Whitfield LS, Lovell-Badge R, Goodfellow PN (1993) Rapid sequence evolution of the mammalian sex-determining gene *SRY. Nature,* 364, 713–717.

Zhang H, Tombline G, Weber BL (1998) BRCA1, BRCA2 and DNA damage response: collision or collusion? *Cell,* **92**, 433–436.

Physical and transcript mapping

10

A wide variety of physical mapping strategies have been used to analyze the DNA of complex eukaryotic genomes. In the following sections, an arbitrary distinction is made between two classes of physical mapping:

- **low resolution physical mapping** – the smallest map unit that can be resolved is typically one to several megabases of DNA;
- **high resolution physical mapping** – the resolution is typically very high, from hundreds of kilobases to a single nucleotide.

Since mammalian DNA has only a very small percentage of coding DNA (~3% in the case of the human genome), a variety of transcript mapping methods have been developed for selectively identifying and studying transcribed sequences, the most important and, arguably, most interesting component of the genome.

10.1 Low resolution physical mapping

10.1.1 Somatic cell hybrid panels can permit chromosomal localization of any human DNA sequence

Under certain experimental conditions, cultured cells from different species can be induced to fuse together, thereby generating **somatic cell hybrids**. In human genetic mapping, hybrid cells are typically constructed by fusing human cells and mouse or hamster cells. The initial fusion products are described as heterokaryons because the cells contain both a human and a rodent nucleus. Eventually, heterokaryons proceed to mitosis, and the two nuclear envelopes dissolve. Thereafter, the human and rodent chromosomes are brought together in a single nucleus. The hybrid cells are unstable. For reasons that remain unknown, most human chromosomes fail to replicate in subsequent rounds of cell division and are lost. This gives rise, eventually, to a variety of more-or-less stable hybrid cell lines, each with the full set of

rodent chromosomes plus a few human chromosomes (*Figure 10.1*). The loss of the human chromosomes occurs essentially at random, but can be controlled by selection (see *Box 10.1*).

The human chromosomes in somatic cell hybrids can conveniently be identified by PCR screening with sets of chromosome-specific primers (Abbott and Povey, 1991). By collecting hybrid cell lines with different human chromosome contents it is possible to generate a **hybrid cell panel** that can be used to map any human DNA sequence to a specific chromosome. To do this, each of the hybrid cell lines is tested for the presence of the human sequence of interest. A PCR assay can be used with primers specific for that sequence (this is a form of **STS content mapping**; see *Box 10.3*) or the relevant DNA sequence can be labeled and used as a hybridization probe. Both of these methods require that the human sequence should be distinguishable from any homologous rodent sequence. Somatic cell hybrids could also be used to establish which chromosome encodes a human enzyme or other gene product by biochemical typing of the hybrids. Whichever human chromosome contains the sequence of interest, it will be present in all the cell lines that are positive for the test and absent from all cell lines that are negative for the test (*Table 10.1*).

Monochromosomal hybrids generated by microcell fusion allow direct and rapid chromosomal localization of human DNA clones

Traditional somatic cell hybrids generally contain several human chromosomes. Microcell fusion (Fournier and Ruddle, 1977) can be used to generate monochromosomal hybrids that contain only a single human chromosome. The donor human cells are subjected to prolonged mitotic arrest by continued exposure to an inhibitor of mitotic spindle formation, such as colcemid. This causes the chromosome content of a cell to become partitioned into discrete subnuclear packets (micronuclei). Centrifugation in the presence of cytochalasin B (a mitotic spindle inhibitor), results in the formation of microcells. Microcells consist of a single micronucleus with a thin rim of

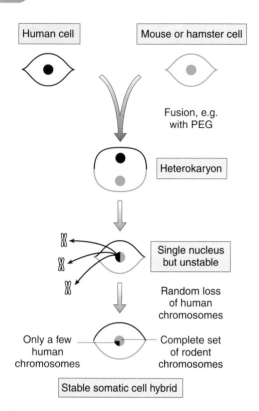

Figure 10.1: Fusion of cells from different species can result in stable somatic cell hybrids.

The example shows how stable human–rodent somatic cell hybrids can be generated following initial fusion using polyethylene glycol (PEG). For reasons that are not understood, human chromosomes are selectively lost from the initial fusion products. The loss occurs essentially at random so that eventually the stable products of a single fusion experiment will include a variety of cells with different complements of human chromosomes. They can be cloned to establish individual cell lines with a specific complement of human chromosomes. The identity of the human chromosomes can be established by PCR-based typing for chromosome-specific markers (see text).

cytoplasm, surrounded by an intact plasma membrane. In much the same way as for normal donor cells, the microcells can be fused with recipient rodent cells to generate hybrids (microcell-mediated chromosome transfer). Some microcell hybrids contain a few donor chromosomes, but the simplest contain a single donor chromosome (see, for example, Warburton *et al.*, 1990).

Monochromosomal human–rodent hybrid cell lines have been established for each of the human chromosomes (Cuthbert *et al.*, 1995). If the sequence of a particular human DNA is known, a PCR assay can be established and used to type a complete monochromosomal hybrid panel, thereby establishing the chromosomal location quickly. Alternatively, if no DNA sequence is available, a human fragment of interest can still be labeled and used as a hybridization probe against a Southern blot of

Box 10.1

Selecting for the chromosome contents of hybrids

Hybrids can be selected for retention of a given human chromosome or chromosome fragment if it corrects an otherwise lethal abnormality in the rodent cell. Frequently used systems include :

- **HAT selection.** Somatic cell hybrids can be forced to retain human chromosome 17 by using thymidine kinase deficient (TK⁻) rodent cells and growing the hybrids in *HAT* (hypoxanthine-aminopterin-thymidine) medium. TK⁻ cells are killed in HAT medium, but are rescued by the human TK gene on chromosome 17.
- **G418 selection.** Hybrids can be selected for the presence of a particular human chromosome segment if it has been tagged by incorporation of a neomycin resistance (*neo*^{R}) gene. The neomycin analog G418 kills nonresistant cells. Neo^R is a typical example of a dominant selectable marker.

genomic DNA samples from a monochromosomal hybrid panel.

10.1.2 Subchromosomal mapping is possible using hybrid cells containing defined portions of a human chromosome

Conventional somatic cell hybrids are a relatively crude tool for physical mapping. More refined mapping is possible using hybrids that contain only part of a particular human chromosome. Translocation hybrids and deletion hybrids are made using donor human cells that have a chromosomal translocation or deletion. To be useful, the hybrids must lack the normal homolog of the chromosome of interest. Such hybrids can be used for subchromosomal mapping of a human sequence-tagged site or biochemical marker (*Figure 10.2*). They are especially useful for defining the sequences removed by microdeletions, by segregating the deletion-carrying chromosome away from its normal homolog.

10.1.3 Human DNA clones can be physically mapped to any desired resolution by STS mapping in panels of hybrid cells that contain defined fragments of the human genome

A more general approach to producing tools for physical mapping involves artificially breaking human chromosomes and transferring the fragments into rodent cells.

Chromosome-mediated gene transfer

One of the first techniques to use this approach was **chromosome-mediated gene transfer (CMGT)**. Fragments of purified mitotic chromosomes from a donor, such as a

Table 10.1: Mapping of a gene for microfibril-associated glycoprotein (MAGP) to human chromosome 1 using a panel of 16 somatic cell hybrids

MAGP/chromosome	Human chromosome																						
	1	2	3	4	5	6	7	8	9	10	11	12	13	14	15	16	17	18	19	20	21	22	X
Concordant hybrids																							
+/+	7	3	4	3	2	5	0	6	4	1	2	5	2	6	4	6	2	6	6	3	6	7	2
−/−	9	8	3	6	6	6	7	6	4	9	4	6	3	3	4	6	9	5	5	4	6	5	3
Discordant hybrids																							
+/−	0	3	2	2	5	1	5	1	4	6	2	2	5	1	3	1	5	1	0	4	0	0	0
−/+	0	2	7	3	3	3	2	4	3	1	6	4	6	5	6	3	1	5	5	6	4	4	2
Total discordant hybrids	0	5	9	5	8	4	7	5	7	7	8	6	11	6	9	4	6	6	5	10	4	4	2
Total informative hybrids[a]	16	16	16	14	16	15	14	17	15	17	14	17	16	15	17	16	17	17	16	17	16	16	7
Percentage discordant hybrids	0	31	56	36	50	27	50	29	47	41	57	35	69	40	53	25	35	35	31	59	25	25	29

[a] Chromosomes with rearrangements or present at a frequency of 0.1 or less were excluded.

The assignment to human chromosome 1 is indicated by complete concordance of the informative hybrids: all seven hybrids which possessed human chromosome 1 tested positive for MAGP and all nine hybrids which lacked human chromosome 1 tested negative for MAGP. There was discordance between the hybrids for all of the other human chromosomes. *Note* that the chromosome complement of individual hybrids, although generally stable, may occasionally undergo changes, and apparent discordance between hybrids may sometimes be found for a chromosome that contains the locus of interest. Reproduced from Faraco *et al.* (1995) *Genomics*, **25**, pp. 630–637, with permission from Academic Press Inc.

human fibroblast, are coprecipitated with calcium phosphate on to the surface of a recipient rodent cell line in monolayer culture (see Porteous, 1987). Human chromosome fragments enter the recipient cells, such as mouse fibroblasts, and integrate into the chromosomes, resulting in stable transformation. As a result, hybrids can be established that retain segments of human DNA (**transgenomes**) of a size that is useful for mapping (usually in the range of 1–50 Mb). However, the transgenomes are prone to frequent rearrangements, so CMGT is more suited to functional assays of complex loci than as a mapping tool.

Irradiation fusion gene transfer

The most valuable hybrids for gene mapping are **radiation hybrids** (see Walter *et al.*, 1994). Donor cells are subjected to a lethal dose of radiation which fragments their chromosomes. The average size of a fragment is a function of the dose of radiation. After irradiation the donor cells are fused with recipient cells of a different species. A selection system is used to pick out recipient cells that have taken up some of the donor chromosome fragments (see *Box 10.1*). These cells are useful for mapping insofar as they have taken up a random set of other chromosome fragments from the donor, as well as the selected fragment. Stably incorporated donor fragments are either integrated into rodent chromosomes or are assembled into novel human minichromosomes formed around fragments containing a functional centromere.

Although this procedure was first proposed by Goss and Harris in 1975, it was not used seriously until 1990, when hybrids were constructed using irradiated monochromosomal hybrid cells as donors (Cox *et al.*, 1990).

When a set of DNA markers from the human chromosome is assayed in a panel of such radiation hybrids, the patterns of cross-reactivity can be used to construct a map (*Figure 10.3*). The principle is very similar to meiotic linkage analysis as described in Chapter 11: the nearer together two DNA sequences are on a chromosome, the lower the probability that they will be separated by the chance occurrence of a breakpoint between them. The frequency of breakage between two markers can be defined by a value θ, analogous to the recombination frequency in meiotic mapping. θ varies from 0 (the two markers are never separated) to 1.0 (the two markers are always broken apart). As in meiotic mapping, θ underestimates the distance between markers that are far apart on the same chromosome, in this case because a cell can take up two markers on separate fragments. A more accurate estimate is provided by a **mapping function**, $D = -\ln (1 - \theta)$, which is analogous to the Haldane mapping function used in meiotic linkage analysis (Section 11.1.3). D is measured in **centiRays (cR)**. D is dependent on the dosage of radiation, so it is referenced against the number of rads. For example, a distance of 1 cR_{8000} between two markers represents a 1% frequency of breakage between them after exposure to 8000 rad of X-rays.

Radiation hybrids derived from monochromosomal hybrid donor cells have been superseded by whole-genome radiation hybrids where the donor is an irradiated normal human diploid cell. The first such panel consisted of 199 hybrids made by fusing an irradiated 46,XY human fibroblast cell line to TK− hamster cells (Walter *et al.*, 1994). Gyapay *et al.* (1996) used 404 microsatellite markers of known location to show that this hybrid panel could generate accurate maps, and then

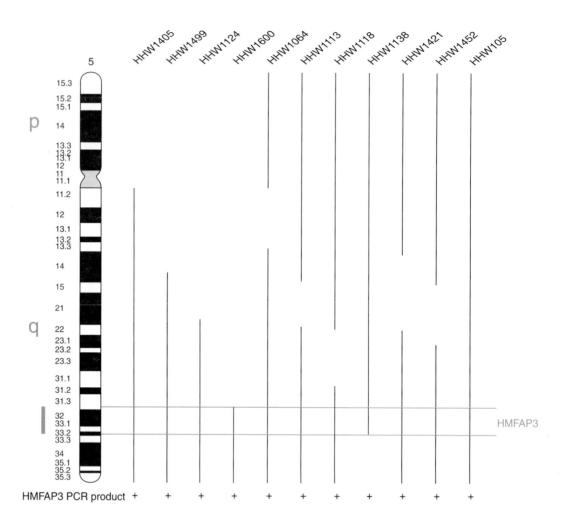

Figure 10.2: Subchromosomal localization can be achieved by mapping against a panel of hybrid cells containing translocation or deletion chromosomes.

The figure illustrates PCR-based mapping of the human microfibrillar protein MFAP3 using a panel of 5q translocation and deletion hybrids. Vertical black bars to the right indicate the extent of human chromosome 5 sequences which are retained in the hybrids. Hybrids HHW1405, 1499, 1124 and 1600 contain translocation chromosomes with 5q breakpoints and retention of the segment distal to the breakpoint. By contrast, translocation hybrid HHW1138 retains material proximal to the 5q breakpoint. Hybrids HHW1064, 1113, 1118, 1421 and 1452 have different interstitial deletions of 5q. The solid blue vertical bar to the left indicates the inferred subchromosomal location as defined by breakpoints in hybrids HHW1600 and HHW1138 (blue horizontal lines near bottom). Reproduced from Abrams *et al.* (1995) *Genomics*, **26**, pp. 47–54, with permission from Academic Press, Inc.

used it to map 374 unmapped ESTs. A subset of 93 of the hybrids has been made widely available as the Genebridge 4 panel. The 93 hybrids average 32% retention of any particular human sequence, with an average fragment size of 25 Mb. Laboratories can map any unknown STS by scoring the 93 Genebridge hybrids and comparing the pattern with patterns of previously mapped markers held on a central server (*Figure 10.4*).

This has turned into an extremely powerful and convenient tool for physically mapping any STS or EST. A second human–hamster panel, Stanford G3, was made using a higher dose of radiation, so that the average human fragment size is smaller. The 83 hybrids in G3 average 16% retention of the human genome, with

an average fragment size of 2.4 Mb. Thus G3 can be used for finer mapping. The impressive results of large-scale use of these panels can be accessed at http://www.ncbi.nlm.nih.gov/genemap98/ (Deloukas *et al.*, 1998).

10.1.4 Chromosomal in situ hybridization has been revolutionized by fluorescence in situ hybridization techniques

Chromosomal *in situ* hybridization typically uses an air-dried microscope slide preparation of metaphase chromosomes, in which the chromosomal DNA has been denatured by exposure to formamide before hybridization

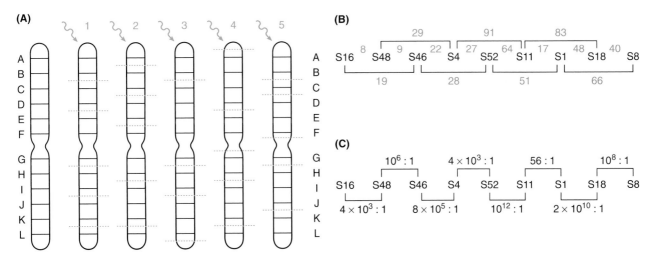

Figure 10.3: Constructing radiation hybrid maps.

(A) Breakpoints occur randomly. Five possible examples of breakpoints (dashed blue lines) on the same type of chromosome are shown. Markers close together will tend to occur on the same fragment, e.g. A and B in all cases other than example 2. Thus, if a radiation hybrid contains marker A it will frequently also contain marker B, but rarely a distant marker such as L. **(B)** Ordering of markers on human 21q. The order of markers *D21S16–D21S8* as inferred by Cox *et al.* (1990) from radiation hybrid mapping is shown. Figures on the top panel refer to distances between markers in centiRays$_{8000}$. For example, the S16–S48 interval is 8 cR$_{8000}$: at a radiation dose of 8000 rad, there is 8% frequency of breakage between them, and so a 92% chance they will occur together on one fragment. **(C)** Odds ratios refer to the likelihood of the indicated order for pairs of markers compared with that with the markers inverted. For example, the calculated likelihood for the order S16–S48–S46–S4 is 10^6 times greater than for the order S16–S46–S48–S4.

with a suitable probe (see Section 5.3.4). Isotopes such as ^{32}P are not suitable for probe labeling because their high energy causes scattering of the signal. Instead, isotopes with weaker emission of beta radiation have been used. Owing to the high background noise when using such isotopes, however, the genuine signal is not easily identified and complex statistical analysis is required to differentiate genuine signal from background noise.

More recently, the sensitivity and resolution of *in situ* hybridization has been increased significantly by the development of **fluorescence *in situ* hybridization** (**FISH**) (Trask, 1991; van Ommen *et al.*, 1995). In this technique, the DNA probe is either labeled directly by incorporation of a fluorescent-labeled nucleotide precursor, or indirectly by incorporation of a nucleotide containing a **reporter molecule** (such as biotin or digoxigenin) which after incorporation into the DNA is then bound by fluorescently labeled affinity molecule (see Section 5.1.2). To increase the intensity of the hybridization signal, large DNA probes are preferred, usually cosmid clones containing around 40 kb of insert. Because such large sequences will contain many interspersed repetitive DNA sequences, it is necessary to use chromosome *in situ* suppression hybridization (Lichter *et al.*, 1990). Essentially, this is a form of competition hybridization (see *Box 5.3*): before the main hybridization, the probe is mixed with a large excess of unlabeled total genomic DNA and denatured, thereby saturating the repetitive elements in the probe, so that they no longer mask the signal generated by the unique sequences.

FISH has the advantage of providing rapid results which can be scored conveniently by eye using a fluorescence microscope (see *Figure 5.5*). In metaphase spreads, positive signals show as double spots, corresponding to probe hybridized to both sister chromatids (*Figure 10.5*). Using sophisticated image processing equipment and reporter-binding molecules carrying different fluorophores, it is possible to map and order several DNA clones simultaneously. The maximum resolution of conventional FISH on metaphase chromosomes is several megabases. The use of the more extended prometaphase chromosomes can permit 1 Mb resolution but, because of problems with chromatin folding, two differentially labeled probe signals may appear to be side-by-side, unless they are separated by distances greater than 2 Mb. Recently, however, new variations have been developed, permitting very high resolution (see Section 10.2.1). There are numerous applications for chromosome FISH but a special one is chromosome painting which has permitted a form of molecular karyotyping (*Box 10.2*).

10.1.5 Flow cytometry permits sorting of individual chromosomes and the construction of chromosome-specific DNA libraries

Human chromosomes show considerable variation in size and DNA content (see *Table 7.2*). In addition, the base composition can vary considerably: chromosomes that are known to be gene-rich or gene-poor (see *Figure 7.4*)

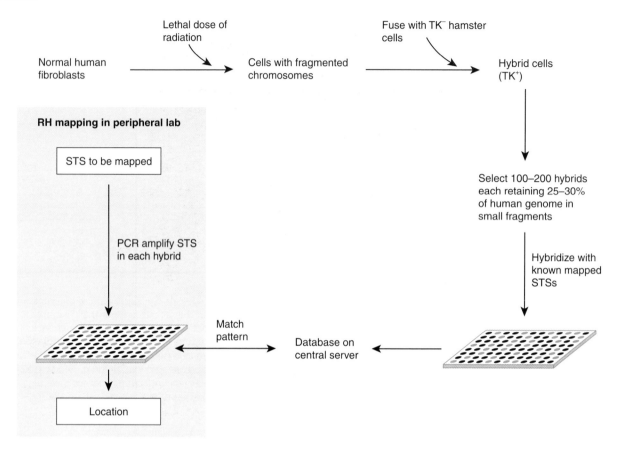

Figure 10.4: Use of the Genebridge 4 radiation hybrid panel for physical mapping.

The panel consists of 93 hamster cells each containing on average 32% of the human genome in the form of many small fragments (average size 25 Mb). A central database stores data on the presence or absence in each hybrid of hundreds of markers from known chromosomal locations. A new STS is mapped by attempting to PCR-amplify it from each of the 93 hybrids, and comparing the pattern of positives and negatives with patterns in the database.

have a comparatively high % (G + C) or % (A + T) respectively. For example, chromosomes 21 and 22 are similar in size and in DNA content, but chromosome 21 is gene-poor and has a relatively high % (A + T), whereas chromosome 22 is gene-rich and has a relatively high % (G + C). As a result of the size differences and differences in base composition, human chromosomes can be separated by **flow sorting** (also called **flow cytometry**; see Bartholdi *et al.*, 1987). In this technique, chromosome preparations are stained with a DNA-binding dye which can fluoresce in a laser beam. The amount of fluorescence exhibited by a given chromosome is proportional to the amount of dye bound, which in turn is largely proportional to the amount of DNA, and hence the size of the chromosome. Additionally, certain dyes, such as chromomycin A_3, show preferential binding to GC-rich sequences, while others, such as Hoechst 33258, bind preferentially to AT-rich regions.

Because individual chromosomes can bind different amounts of fluorescent dyes, they can be fractionated in a flow cytometer. A stream of droplets containing stained chromosomes is passed through a finely focused laser

beam at a rate of about 1000–2000 chromosomes per second, and the fluorescence of individual chromosomes present in individual droplets is monitored by a photo-multiplier. The intensity of fluorescence of the different chromosomes can then be recorded. By using sort windows to define a range of fluorescence intensities corresponding to a particular type of chromosome, droplets containing the desired chromosome can be given an electric charge and deflected from the main stream on to a separate collecting grid (*Figure 10.7A*).

The resulting flow karyogram (see Figure 10.7B) may show good separation of some human chromosomes; others such as chromosomes 9–12 are not well separated. However, monochromosomal somatic cell hybrids (Section 10.1.1) can be used instead, and often result in good separation of the human chromosome. As a result, panels of chromosome-specific DNA representing each of the 24 human chromosomes have been obtained. These can be used to obtain a chromosomal localization for any human DNA fragment: the fragment is labeled and hybridized to dot-blots of DNA from the 24 chromosomes ('flow-blots'), or a PCR assay can be used, if

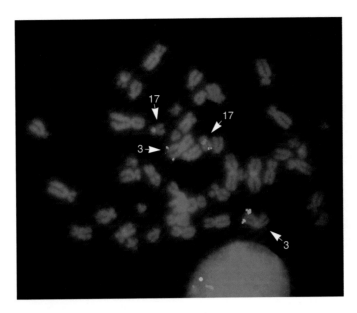

Figure 10.5: Chromosome FISH (fluorescence *in situ* hybridization).

Green signals indicate positive hybridization of a YAC from human 3q26.3 to metaphase chromosomes from a patient with Cornelia de Lange syndrome with a *de novo* balanced translocation (breakpoints at 3q26.3 and 17q23.1). Red signals indicate simultaneous hybridization with a chromosome 17 centromere probe. The YAC spans the 3q26.3 breakpoint (green signals on one normal chromosome 3 plus the two translocation chromosomes). One translocation chromosome is small and carries a chromosome 17 centromere (red signal); the other has a chromosome 3 centromere and is about the same size as the normal chromosome 3. Figure kindly provided by Melanie Smith, University of Newcastle upon Tyne and sponsored by Applied Imaging International Ltd.

Chromosome painting

A special application of FISH has been the use of DNA probes where the starting DNA is composed of a large collection of different DNA fragments from a single type of chromosome. Such probes can be prepared by combining all human DNA inserts in a chromosome-specific DNA library (see Section 10.1.5). The resulting hybridization signal represents the combined contributions of many loci spanning a whole chromosome and causes whole chromosomes to fluoresce (chromosome painting; Ried *et al.*, 1998).

Chromosome painting was initially limited by the relatively small number of differently colored fluorescent dyes (fluorophores) one could use to distinguish different chromosomes. To increase the number of different targets that can be detected beyond the number of available differently colored fluorophores, two approaches have been used: (i) combinatorial labeling involves labeling individual probes with more than one type of fluorophore; and (ii) ratio labeling uses a combination of different fluorophores but also in different ratios (Lichter, 1997). The mixed colors are not detected by standard fluorescence microscopy using appropriate filters (see *Figure 5.5*). Instead, automated digital image analysis is preferred by which various combinations of fluorophores are assigned artificial pseudocolors. The long awaited goal of simultaneously visualizing all 24 different human chromosomes was achieved using two approaches:

- **Multiplex FISH (M-FISH).** This approach reported by Speicher *et al.* (1996) uses digital images acquired separately for each of five different fluorophores using a CCD (charge coupled device) camera. The images are analyzed by a software package which generates a composite image in which each chromosome is given a different pseudocolor depending on the fluorophore composition (see *Figure 18.6*).
- **Spectral karyotyping (SKY).** Schrock *et al.* (1996) reported this approach, in which CCD imaging is combined with Fourier spectroscopy. The spectrum of fluorescent wavelengths for each pixel (picture element) is assessed using an interferometer and a dedicated computer program assigns a specific pseudocolor depending on the particular fluorescence spectrum identified.

Chromosome painting has found increasing applications in defining *de novo* rearrangements and marker chromosomes (see *Box 2.6*) in clinical and cancer cytogenetics (see *Figure 10.6* for an example). It is particularly helpful in cancer cytogenetics for two reasons. Chromosome preparations from tumors are often of poor quality, but information can often be obtained with the aid of chromosome painting. Additionally, complex chromosome rearrangements are particularly frequent in tumor samples and chromosome paints can be used in combination with standard probes to help recognize particular chromosome segments (see *Figure 18.6*).

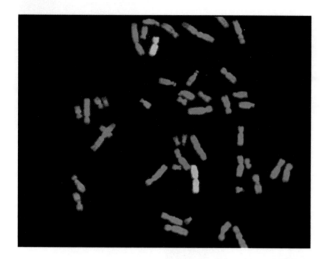

Figure 10.6: Chromosome painting can be used to define chromosome rearrangements.

In this case an abnormal chromosome 8 with a large short arm, identified by karyotyping from an amniocentesis sample, was investigated using a whole chromosome 8 paint (pink). The normal chromosome 8 is at the top and the larger chromosome 8 towards the bottom. The data indicate a subchromosomal duplication of material on 8p. Figure kindly provided by Ian Cross, Institute for Human Genetics, University of Newcastle upon Tyne and sponsored by Applied Imaging International Ltd.

appropriate. Additionally, purified DNA from a specific type of chromosome can be amplified by cell-based or PCR-based DNA cloning to generate chromosome-specific DNA libraries (Davies et al., 1981). Chromosome libraries constructed using cosmid vectors have been particularly useful because YACs from a specific chromosome can be related quickly to cosmid clones from the same region. In addition, it has been possible in some cases to construct human chromosome-specific YAC libraries (e.g. McCormick et al., 1993).

10.2 High resolution physical mapping: chromatin and DNA fiber FISH and restriction mapping

The physical mapping methods described above typically have a lower resolution limit of one to several megabases; they are complemented by molecular mapping methods which can map DNA in the range 1 bp to several megabases. DNA sequencing provides the ultimate physical map by determining the linear order of single nucleotides, but mapping large DNA regions by sequencing is technically arduous. Two major techniques provide map resolution of less than 1 Mb: restriction mapping and

high resolution FISH on naturally extended chromosomes or artificially extended chromatin or DNA fibers.

10.2.1 Very high resolution FISH mapping can be achieved by hybridizing probes to extended chromosomes or artificially extended chromatin or DNA fibers

The linear length of DNA in an average sized human chromosome is about 5 cm, but it is compacted by various hierarchies of folding in metaphase chromosomes to a few microns (see *Figure 2.7*). As a result, standard *in situ* hybridization against the highly condensed chromatin of metaphase and prometaphase chromosomes does not have a high mapping resolution, with typical upper limits of about one to several megabases of DNA. To obtain higher resolution, DNA probes are hybridized to the naturally extended interphase chromosomes or to artificially stretched chromatin or DNA fibers prepared by a variety of different methods.

In situ *hybridization using conventional interphase chromosomes*

The chromosomes of interphase nuclei are much more extended than metaphase or prometaphase chromosomes. As a result, FISH analyses of the chromosomes of interphase nuclei can permit a high mapping resolution, and can help to determine the physical mapping order of some syntenic DNA clones. Because the extended chromosomes may loop back at certain regions, an accurate linear order for syntenic DNA clones requires a statistical analysis on the mapping results from many individual interphase nuclei (see *Figure 10.8*). Even so, the method is best suited to ordering sequences which are separated by intervals within the 50–500 kb range.

FISH mapping on artificially extended chromatin or DNA fibers

High resolution FISH can also be conducted using procedures which cause the DNA of chromosomes on a microscope slide to be extended prior to hybridization (extended chromatin fiber FISH; see Houseal and Klinger, 1994, for references). One such method, DIRVISH (direct visual hybridization), involves lysing cells with detergent at one end of a glass slide, tipping the slide, and allowing the DNA in solution to stream down the slide. Such preparations permit extremely high mapping resolutions: from over 700 kb to under 5 kb (see *Figure 10.9*). More recently, the principle of artificially stretching chromatin fibers has been extended to protein-free DNA: the target DNA is prepared from cells embedded in pulsed field gel electrophoresis blocks (see Section 10.2.2). This DNA fiber FISH method used unfixed linearized DNA fibers on a microscope slide and has a resolution of from 500 kb to a few kilobases (see Heiskanen *et al.*, 1996 for a review on high resolution FISH mapping).

(A)   Metaphase chromosomes stained with fluorescent dye

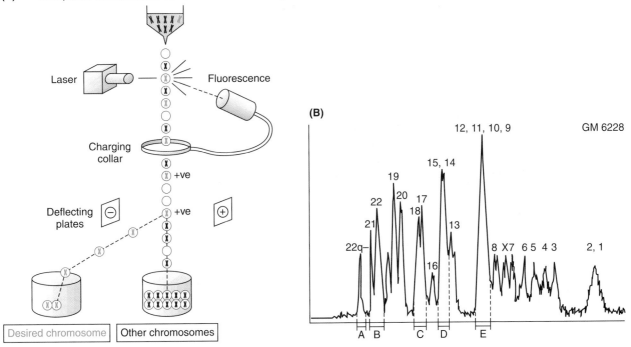

Figure 10.7: Fractionating chromosomes in a flow cytometer.

(**A**) Principle of chromosome fractionation. The stream of fluorescent dye-labeled chromosomes consists of very fine droplets, with each droplet containing at most one chromosome. As they pass through the laser beam they fluoresce and a photomultiplier tube records the intensity. Droplets containing a desired chromosome (as measured by the fluorescence intensity) can be given an electric charge using a charging collar, e.g. a positive (+ve) charge. After passing between electrically charged deflecting plates, the charged droplets can be deflected from the uncharged droplets, allowing sorting of a desired chromosome type (shown in blue). (**B**) A flow karyogram. The example is from the cell line GM6228 which has an unbalanced constitutional translocation t(11;22) (q23;q11). A–E: sorting windows used to collect specific chromosomes or sets of chromosomes. Reproduced from Cotter *et al.* (1989) *Genomics*, **5**, 470–474, with permission from Academic Press Inc.

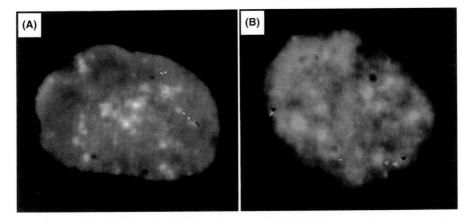

Figure 10.8: Determining the map order of syntenic DNA clones by three-color interphase FISH.

The examples illustrate YAC clones from 3p14 as follows. (**A**) YACs 74B2 (red, R), 468B10 (orange, O) and 403B2 (green, G). (**B**) YACs D20F4 (red, R), 168A8 (orange, O) and 258B7 (green, G). In example **A**, the orange color appears between green and red on both chromosomes 3. The order 74B2–468B10–403B2 is presumed to be correct because the order ROG was observed on 33/37 occasions. In example B the observed RGO order was seen in 30/34 occasions, consistent with a correct order of D20F4–258B7–168A8. Alternative map orders are thought to be due to fold-back of the extended chromosomes. Reproduced from Wilke *et al.* (1994) *Genomics*, **22**, pp. 319–326, with permission from Academic Press Inc.

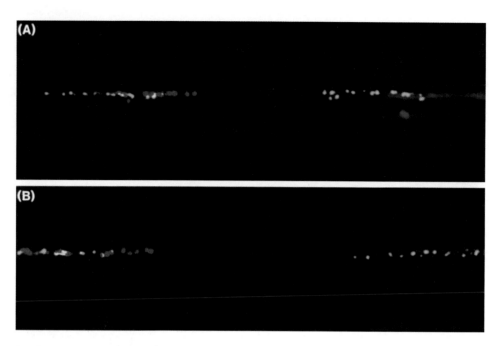

Figure 10.9: Extended chromatin fiber (ECF) FISH.

The example illustrates two-color FISH on artificially extended chromatin fibers. Signals are generated from three YACs which map to the 5q34-q35 region. YACs 786F6 and 935F5 known to map to opposite sides of a tumor translocation breakpoint were visualized using a green FITC (fluorescein isothiocyanate) labeling system. YAC 746B2, which maps to the same side of the tumor breakpoint as 786F6, was visualized using a red (rhodamine) labeling system. *Note* the partial overlap between 786F6 and 746B2, and the gap between them and 953F5, suggesting the order: 746B2/786F6–953F5. Reproduced from Haaf and Ward (1994) *Hum. Mol. Genet.*, **3**, 629–633, by permission of Oxford University Press.

10.2.2 Long-range restriction mapping requires enzymes that cut DNA infrequently and fractionation of large restriction fragments by pulsed-field gel electrophoresis

Restriction mapping permits molecular mapping with resolutions which depend on the frequency of the recognition site. Most restriction enzymes which recognize a 4 or 6 bp sequence typically cut vertebrate DNA once every few hundred or few thousand base pairs. The recognition sequences for **rare-cutter** restriction nucleases are typically 6–8 bp long and contain one or more CpG dinucleotides which are rare in vertebrate DNA (see *Box 8.5*). As a result, they generate fragments that are typically several hundred kilobases in size (see *Table 4.1* for some examples). Small restriction fragments can be size fractionated conveniently by agarose gel electrophoresis. However, the ability of conventional agarose gel electrophoresis to fractionate large DNA fragments is very limited. The method relies on a *sieving* effect: DNA molecules pass through pores in the agarose gel and small molecules are able to migrate more quickly through the uniform size pores (see *Figure 10.10*). Above a certain size of DNA fragment, however, the sieving effect is no longer effective and the resolution of DNA fragments above 40 kb is extremely limited.

Large-scale restriction mapping of human DNA can be done directly on the large inserts of YACs or by means of Southern blot hybridization of genomic DNA. Conventionally prepared genomic DNA is not a suitable target for large-scale restriction mapping, because the procedures involved in lysing the cells and purifying the DNA result in shear forces causing considerable fragmentation of the DNA. Instead, the DNA is isolated in such a way as to minimize artificial breakage of the large molecules prior to digesting with restriction endonucleases. To prepare high molecular weight DNA, samples of cells, for example white blood cells, are mixed with molten agarose and then transferred into wells in a block-former and allowed to cool. As a result, the cells become entrapped in solid agarose blocks (*Figure 10.11*). The agarose blocks are removed and incubated with hydrolytic enzymes which diffuse through the small pores in the agarose and digest cellular components, but leave the high molecular weight chromosomal DNA virtually intact. Individual blocks containing purified high molecular weight DNA can then be incubated in a buffer containing a rare-cutter restriction endonuclease.

In order to separate large restriction fragments, pulsed-field gel electrophoresis (PFGE), a modified form of agarose gel electrophoresis, is used. Agarose blocks containing the large DNA fragments to be separated are placed in wells at one end of an agarose gel and the DNA migrates in the electric field. However, during a PFGE run, the relative orientation of the gel and the electric field

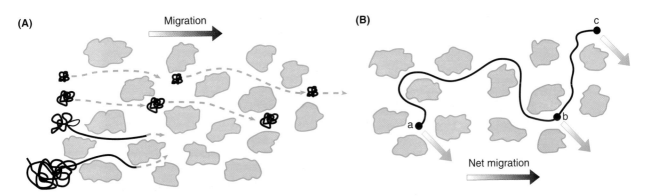

Figure 10.10: Migration of DNA fragments during conventional agarose gel electrophoresis and pulsed-field gel electrophoresis.

(**A**) Conventional agarose gel electrophoresis. Small DNA fragments can pass through the pores of the gel by a sieving effect: very small fragments migrate more rapidly than larger fragments. Above a certain size threshold, DNA fragments are so large that compact forms are too large to pass through the pores: the DNA fragment needs to adopt an extended conformation with a leading end migrating into the gel, and separation according to size is poor. (**B**) Pulsed-field gel electrophoresis. Extended DNA fragments need to adopt a new conformation in response to an altered electric field. In the example, alternating electric fields are imagined to operate in two directions (top left ⇒ bottom right; bottom left ⇒ top right), resulting in net migration from left to right (see *Figure 10.11*). In response to the new electric field shown by blue arrows, the DNA strand needs to reorientate: the end marked c, which was previously the leading end (when the electric field was in the bottom left ⇒ top right orientation), may become the leading end again, possibly after initial migration of a loop at position b. The time taken to reorientate and present the best conformation for advancing in the new direction is strictly size dependent, resulting in good separation of very large fragments.

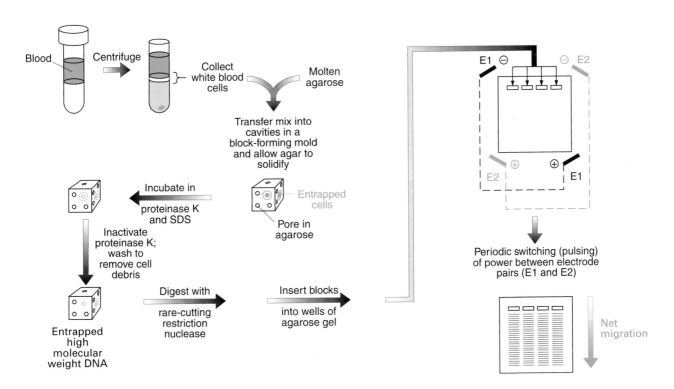

Figure 10.11: Fractionation of high molecular weight DNA from blood cells by pulsed field gel electrophoresis.

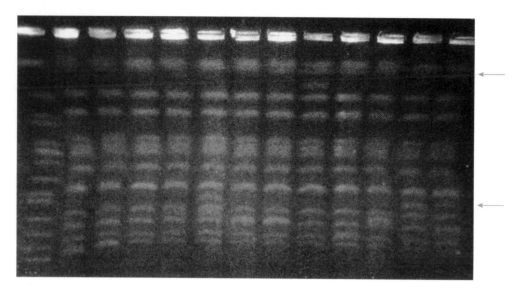

Figure 10.12: Pulsed field gel electrophoresis can be used to size-fractionate whole (small) chromosomes.

The lane on the extreme left is a size marker. Bands represent separated chromosomes from the *S. cerevisiae* strain YNN 295. Other lanes show a background of normal yeast chromosomes from the *S. cerevisiae* strain AB1380, plus occasional additional bands (shown by arrows) representing YACs with inserts from human chromosome 10. Note that sometimes YACs may not be evident on such gels because they comigrate with one of the normal chromosomes from the host yeast cell. Figure kindly provided by Dr Mike Jackson, University of Newcastle upon Tyne, UK.

is periodically altered, typically by setting a switch to deliver brief pulses of power, alternatively activating two differently oriented fields (*Figure 10.11*). Variants of the technique use a single electric field but with periodic reversals of the polarity (*field inversion gel electrophoresis*), or periodic rotation of the gel or electrodes. In common to each variation is the principle of a discontinuous electric field so that the DNA molecules are intermittently forced to change their conformation and direction of migration during their passage through the gel. The time taken for a DNA molecule to alter its conformation and reorient itself in the direction of the new electric field is strictly size dependent; as a result, DNA fragments up to several megabases in size can be fractionated efficiently (see *Figure 10.12*).

10.3 Assembly of clone contigs

10.3.1 A primary goal of physical mapping is to assemble comprehensive series of DNA clones with overlapping inserts (clone contigs)

The construction of the ultimate physical map (the complete nucleotide sequence) requires considerable time and effort in the case of a very large DNA molecule such as that found in a chromosome. In order to provide a framework for this to be done efficiently, a series of cloned DNA fragments need to be assembled which collectively provide full representation of the sequence of interest. To ensure that there is complete representation,

and no gaps, the series of clones should contain overlapping inserts forming a comprehensive clone contig (*Figure 10.13A*). In principle, contig assembly is facilitated by the way in which genomic DNA libraries are constructed: as part of the strategy for maximizing the representation of a library, the genomic DNA is deliberately subjected to **partial digestion** with a restriction endonuclease (by reducing the time of incubation and by using low concentrations of enzyme). As a result, individual genomic DNA clones usually contain DNA sequences that partially overlap with the insert DNA of at least some other clones in the library (see *Figure 10.13B*). The cloning step means that the individual DNA fragments are sorted into different cells and so the original positional information of the fragments (how they were related to each other on the original chromosomes) is lost. However, such information can be retrieved by a variety of methods which can identify clones with overlapping inserts.

10.3.2 Chromosome walking means establishing clone contigs from fixed starting points

One widely used technique for identifying clones with overlapping inserts is to use a specific DNA probe from one clone to screen a DNA library. The positively hybridizing clones should contain a DNA sequence that is closely related to the probe, including clones which contain sequences which partly overlap that found in the probe. This has often involved the preparation of a so-called end probe from the starting DNA clone: a fragment located at one end of the insert DNA and preferably pre-

(A)

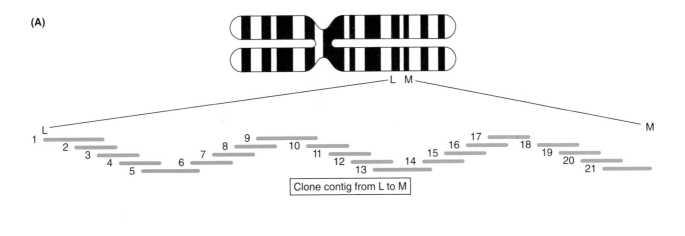

Clone contig from L to M

(B)

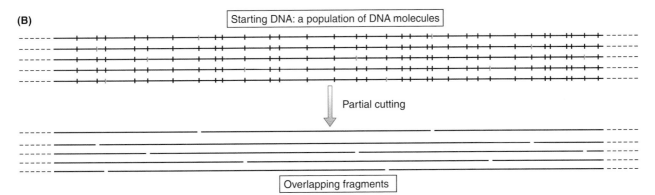

Starting DNA: a population of DNA molecules

Partial cutting

Overlapping fragments

Figure 10.13: A clone contig consists of a linear series of DNA clones with overlapping inserts which originated by partial cutting of genomic DNA during DNA library construction.

(A) Clone contig. The contig consists of 21 clones arranged as a series of clones with partially overlapping inserts. All the genomic DNA between chromosomal sites L and M is represented in the contig. **(B)** Generating overlapping DNA fragments by partial restriction cutting. During the construction of a genomic DNA library, the genomic DNA is *partially digested* with a restriction endonuclease. Of the available restriction sites in the chromosomal DNA (vertical bars in top panel), the chosen conditions will ensure that only a very small minority (blue vertical bars) will be cleaved. Because the choice of which site is cut is essentially random, and because the starting DNA will contain numerous identical sequences, a series of overlapping fragments will be produced.

sent as single copy DNA, is purified and labeled. Positively hybridizing clones can then be purified and new, distal end probes can be prepared for further rounds of hybridization screening of the DNA library. In the case of genomic DNA libraries, this permits the assembly of a clone contig by bidirectional **chromosome walking** from a fixed starting point (*Figure 10.14*). In the case of mammalian genomes, chromosome walking has frequently involved screening cosmid libraries (cosmid walking). Modern cosmid walking has been simplified enormously by avoiding the need to prepare end probes. Instead, the entire insert can be labeled and used as a probe under conditions which suppress the hybridization signal from DNA sequences in the probe which are highly repetitive in the human genome (see *Box 5.3*). Nevertheless, the signal from some other repetitive sequences may be difficult to suppress, causing occasional difficulties with this approach.

10.3.3 Chromosome walking with YACs often involves PCR-based library screening

Chromosome walking by whole clone–clone hybridization is not practically feasible with mammalian YACs: the large amount of repetitive DNA in the inserts means that blocking of the repetitive DNA signal during hybridization (see *Box 5.3*) is technically difficult. Instead, techniques are used to recover short end fragments from individual YACs. The YAC DNA is cleaved with a restriction enzyme which is known to cleave the YAC vector sequence and which cuts frequently in human genomic DNA. Among the cleavage products there will be end fragments containing both the unknown terminal sequence from the insert DNA and the adjacent known vector sequence. Such sequences can then be amplified using various PCR-based methods whereby a primer for a characterized sequence (in this case the sequence of the

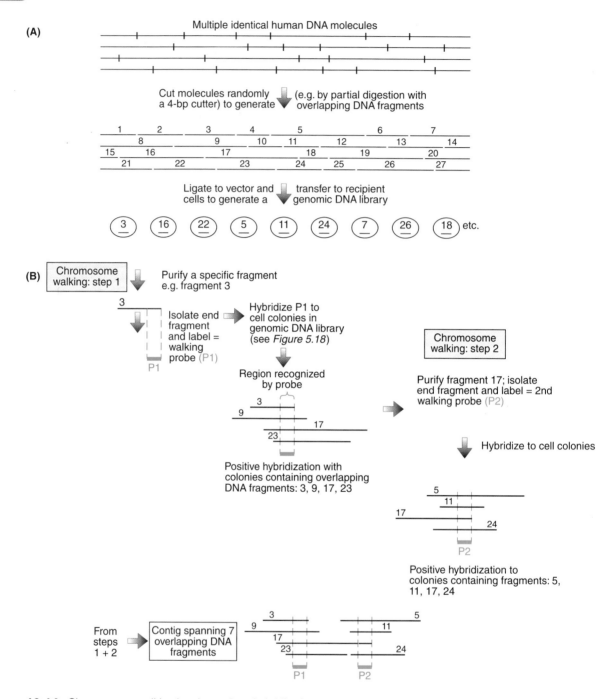

Figure 10.14: Chromosome walking by clone–clone hybridization.

The starting DNA is a population of DNA molecules from many cells. Random cutting by partial restriction endonuclease digestion of identical DNA molecules generates a series of overlapping fragments. For example, fragment 3 from the first molecule shows partial overlaps with fragments 9, 16, 17, 22 and 23 from the other three molecules. After cell-based cloning to make a library, the different fragments are sorted into different host cells. In this example, the starting point is cloned fragment 3 and a monodirectional chromosome walk (towards the right) is shown by using end fragments as hybridization probes for screening a genomic DNA library. A bidirectional walk starting from fragment 3 would involve isolating a second end fragment from the other end of the insert. In the example shown this would be expected to hybridize to clones containing fragments 9, 16 and 22, and subsequent isolation of a second walking probe from, say, the distal end of fragment 22. *Note* that, for cosmid clones, it is no longer necessary to isolate end probes: walking can be done by using whole cosmids as hybridization probes under conditions that suppress signals from repetitive DNA elements (whole clone–clone hybridization).

YAC vector adjacent to the cloning site) is used to permit access to an adjacent uncharacterized sequence (in this case the sequences at the ends of the insert). Frequently used methods include the following:

- **Inverse-PCR**. This is a general method for amplifying DNA flanking a previously characterized region (Ochman *et al.*, 1988). It utilizes primers derived from the extremities of a characterized region which are divergent: the 5′ → 3′ direction of the two primers point away from each other instead of towards each other. Normally, PCR amplification would be impossible with such primers. However, inverse PCR involves first cutting the DNA so that the desired region is present on a small restriction fragment and then encouraging the DNA to circularize (by conducting the ligations at very low DNA concentrations, the two ends of a single molecule are more likely to come into contact than those on different molecules). Amplification of the uncharacterized

flanking regions is then possible using standard conditions (see *Figure 10.15*).

- **Vectorette PCR** (also called **bubble-linker PCR**) is a type of anchored PCR (Section 6.2.5) which uses a specially designed double-stranded oligonucleotide linker ligated to the target DNA (Riley *et al.*, 1990). The double-stranded linker does not contain perfectly complementary oligonucleotides (as is normal, except for overhangs designed to help ligation to target DNA). Instead, the two oligonucleotides of the linker are designed to be complementary at the ends, but not in the middle creating a bubble-like formation. Because the two sequences in the region of the bubble are noncomplementary, it is possible to design a linker strand-specific primer which, with a convergent target DNA-specific primer, permits amplification of the uncharacterized sequence flanking one side of the characterized DNA (see *Figure 10.16*).

Once retrieved, YAC insert end fragments can be used as hybridization probes to screen colony filters from a YAC library. A widely used alternative, however, is to sequence an end fragment and then design oligonucleotide primers to permit a specific PCR assay for this sequence. The resulting PCR assay can be used to screen YAC libraries which have been distributed as hierarchical sets of agarose plugs each containing pools of YAC clones (see *Figure 10.17*). Positive results will identify YACs with overlapping inserts from which new end fragments can be rescued to continue the YAC walk. As, however, YAC clones often have **chimeric inserts** (composed of two or more fragments that are derived from noncontiguous portions of the genome), the chromosomal location of any new YACs identified in a YAC walk ought to be verified by FISH analyses.

10.3.4 Clone contigs can be assembled rapidly over large fractions of a genome by random clone fingerprinting

Chromosome walking is a highly directional and location-restricted procedure for generating clone contigs: a great deal of effort is invested in generating a contig from a fixed starting point on a specific chromosome, often as a prelude to positional cloning (see Section 15.3). In order to build clone contigs spanning large amounts of a chromosome, or even a whole genome, a more general, labor-efficient mapping approach is needed. To meet this need, a variety of **clone fingerprinting** techniques have been used to type clones at random and then integrate the information in order to identify clone contigs over large regions of a genome. For example, this approach has been used to great effect in assembling YAC contigs over significant amounts of the human genome (see Section 13.2.2). Clone fingerprinting techniques should be simple and rapid, and the methods described below have been widely used.

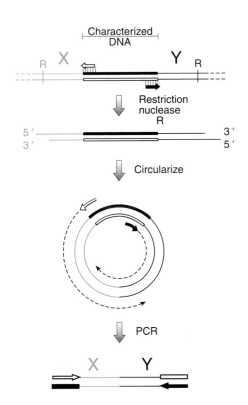

Figure 10.15: Inverse PCR permits cloning of flanking DNA sequences from a circularized template.

The two primer sequences are designed to bind to the characterized DNA sequence but with their 3′ ends facing away from each other, instead of towards each other. After the characterized sequence and its two flanking regions have been recovered on a short restriction fragment, the two flanking regions can be brought next to each other by forcing the fragment to circularize (by ligation at very low DNA concentrations). The PCR primers can now permit amplification of the flanking regions.

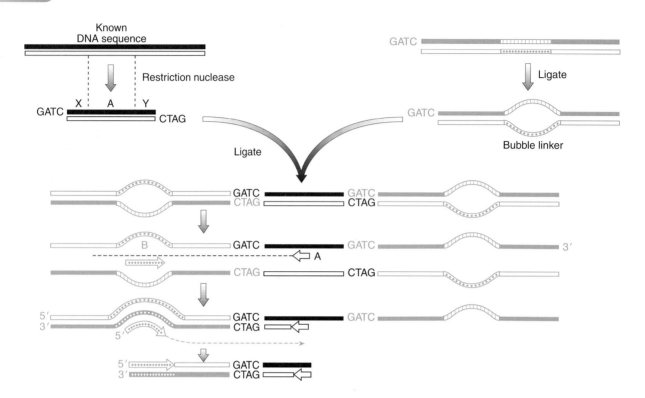

Figure 10.16: Vectorette (bubble linker) PCR permits amplification of uncharacterized sequences flanking a known DNA sequence.

Note that the vectorettes contain complementary sequences at their ends but unrelated sequences within the bubble. A restriction fragment containing a known sequence A flanked by uncharacterized regions X and Y is ligated to a vectorette linker containing a suitable overhang at one end. PCR amplification of the uncharacterized sequence X is then possible using a primer specific for the known DNA sequence (A) and a primer specific for one of the strands of the bubble in the vectorette linker (B). The vectorette linker primer B cannot prime DNA synthesis initially as there is no sequence to which it can bind: it is identical, not complementary in sequence to one strand of the bubble, and unrelated in sequence to the other. However, primer A initiates synthesis of a complementary DNA strand which will contain a sequence *complementary to one of the unique sequences within the bubble linker*. As a result, primer B can bind to this newly synthesized strand and initiate new strand synthesis to start a PCR reaction. The flanking sequence Y can similarly be isolated in another reaction using a suitable A-specific primer and a bubble-specific primer derived from the strand opposite to that used for making primer B.

Repetitive DNA fingerprinting

Mammalian DNA has a very high component of highly repeated, interspersed DNA sequences such as the human *Alu* and LINE-1 (*Kpn*) repeats (occurring on average about once every 3 kb and 10 kb respectively). Because the spacing between such interspersed repetitive DNA elements will vary in different genomic regions, clones which contain overlapping DNA sequences may be identified by similar patterns following hybridization of repetitive DNA probes to Southern blots of the clone DNA digested with a suitable restriction endonuclease. For example, a whole genome clone fingerprinting approach has been applied to mapping the human genome largely on the basis of repetitive DNA finger-printing of YACs (Bellanne-Chantelot *et al.*, 1992).

Sequence-tagged site content mapping

If a small amount of DNA sequence (e.g. a few hundred nucleotides) is available for a specific DNA clone, a PCR assay can be developed which is specific for that sequence. The DNA sequence is examined to identify short regions from which oligonucleotide primers can be designed for use in a standard PCR reaction. The site on the original genomic DNA from which the sequence is derived can then be thought of as being tagged by the ability to assay for that sequence. A site for which a specific PCR assay is available is therefore described as a **sequence-tagged site** (**STS**) (*Box 10.3*). If numerous individual STS sequences have been positioned on a chromosome or subchromosomal area, and YACs have been mapped to the same general area (e.g. by FISH), the ability to type rapidly for STSs allows a rapid method for identifying YACs with overlapping sequences (**STS content mapping**; see *Figure 10.18*).

Hybridization using Alu-PCR products

Alu-PCR is an exceptional type of PCR reaction in which only a single primer is used, corresponding to a sequence

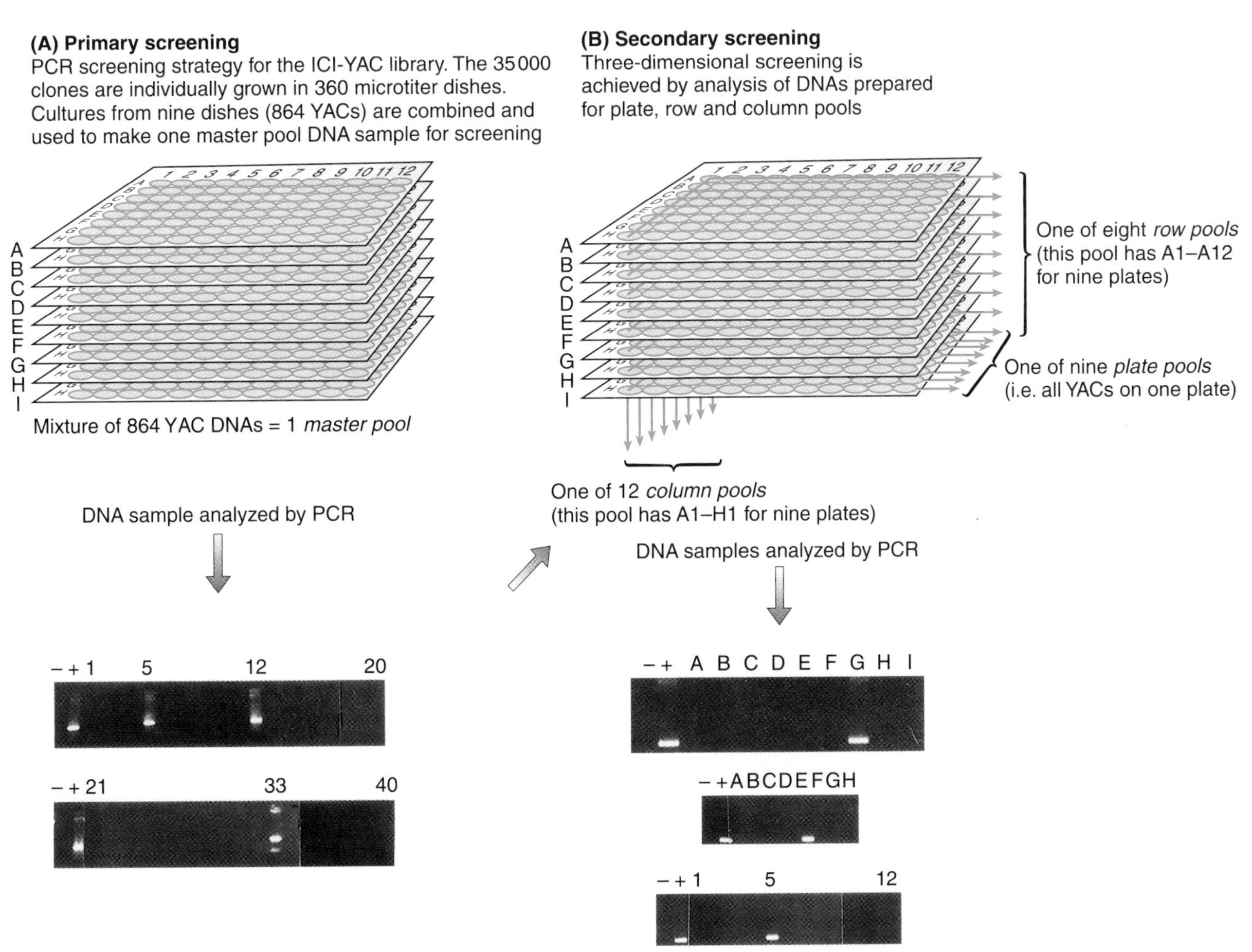

(A) Primary screening

PCR screening strategy for the ICI-YAC library. The 35 000 clones are individually grown in 360 microtiter dishes. Cultures from nine dishes (864 YACs) are combined and used to make one master pool DNA sample for screening

(B) Secondary screening

Three-dimensional screening is achieved by analysis of DNAs prepared for plate, row and column pools

Figure 10.17: Screening of YAC libraries by PCR.

A YAC library can be screened for the presence of clones containing a specific sequence, provided there is a specific PCR assay for that sequence. Amongst other applications, this provides a convenient method for chromosome walking using YACs: a PCR assay for a sequence at the end of one YAC can be used to identify other YACs with overlapping sequences. The example illustrates screening of the human ICI YAC library generated by Anand *et al*. (1990). Approximately 35 000 individual clones were individually deposited into the 96 wells of 360 microtiter dishes. To facilitate screening, a total of 40 *master pools* were generated by combining all 864 clones in sets of 9 microtiter dishes (plates A–I). Modified from Jones *et al*. (1994) *Genomics*, **24**, pp. 266–275, with permission from Academic Press, Inc.

which is chosen to be close to the end of *Alu* repeat consensus sequence (Nelson *et al.*, 1989). Since the *Alu* repeat is present at very high frequency in human DNA, two such repeats will often be found in close proximity and sometimes *in opposite orientations*. A single *Alu*-end primer can bind to each of two closely located, oppositely orientated *Alu* repeats and permit amplification of the sequence between them (see *Figure 10.19*).

If the starting human DNA is a complex genomic clone such as a BAC or YAC clone, the *Alu*-PCR reaction will generate a diagnostic series of bands of specific sizes which can be compared with equivalent products from other library clones to check for overlapping inserts. In addition, since *Alu*-PCR products often consist of single-copy DNA (except for the incorporated primers), individual or pooled products from a single starting clone can

be used as hybridization probes for screening other YACs or collections of inter *Alu*-PCR products derived from them. A similar type of amplification is possible for any class of highly repetitive interspersed repeat provided the repeat is suitably frequent so that neighboring elements are often found within a few kilobases, and there is sufficient sequence conservation to design a useful consensus primer.

Restriction site-based fingerprinting

Long-established clone fingerprinting methods have involved restriction nuclease treatment of clones and separation of radioisotope-labeled restriction fragments on gels. Modern methods of clone fingerprinting using restriction nucleases seek to achieve a high degree of automation using fluorescence labeling and detection

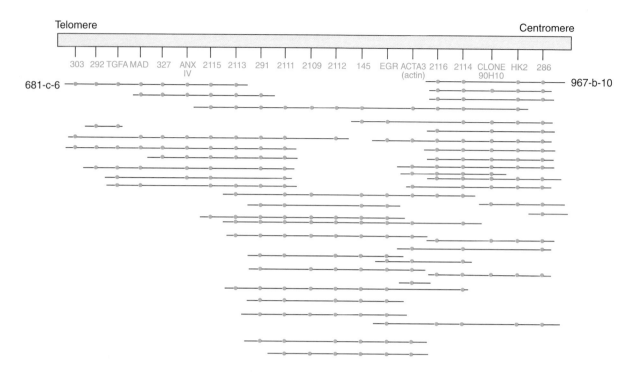

Figure 10.18: Assembling YAC contigs by STS content mapping.

The YAC contig shown was assembled for a region of human 2p12 which contains the limb girdle muscular dystrophy locus *LGMD2B*. Vertical lines indicate the positions of indicated STSs. STSs indicated by simple numbers represent microsatellite markers; others are known genes or an expressed sequence (clone 90H10). Horizontal lines indicate individual YACs, with filled circles denoting the presence of specific STSs. For the sake of clarity, YACs other than the most distal YAC (681–c-6) and the most proximal YAC (967–b-10) are not labeled. Data provided by Dr Rumaisa Bashir, University of Newcastle upon Tyne.

systems (fluorescent fingerprinting). As an example of this approach, the method of Gregory *et al.* (1997) involves digesting clones with *Hin*dIII and *Sau*3AI, specific end-labeling of the *Hin*dIII ends with fluorescently labeled dideoxyATP, and size fractionation on a model 377 automated DNA sequencer (see *Figure 10.20*).

10.3.5 Clone gridding and increasing automation have facilitated large-scale physical mapping efforts

Using current technologies, physical mapping of complex genomes is arduous and continues to be hampered by various practical difficulties:

■ **Chimeric DNA clones.** Cloning artifacts can mean that cells contain two noncontiguous pieces of DNA from the desired genome. This may occur as a result of *co-ligation* of two different restriction fragments (*Figure 4.2A*), internal deletion and *cotransformation* (*Figure 4.2B*). Yeast cells that have been transformed by YACs and used to propagate have often been much more recombination proficient than typical *E. coli* hosts, and so YACs are often chimeric.

■ **Repetitive DNA.** Arranging clones in a physical order means relying on some clone-specific features.

DNA sequences that are repeated can make the task of ordering clones a difficult one, and may cause problems in sequencing DNA.

■ **Unclonable sequences.** Some DNA sequences, including classes of repetitive DNA that are not well represented in bacterial and yeast cells, may not easily be propagated in such cells. A variety of different genomic libraries may need to be screened (including those constructed with vectors containing a low copy number origin of replication such as BAC and P1 libraries).

A major requirement for efficient physical mapping of complex genomes is the need for gridded DNA libraries. Following transformation of the host cells, individual bacterial or yeast colonies are picked into individual wells of microtiter dishes and stored in numerous multi-well dishes. As a result, each cell clone can be assigned an identifying grid coordinate, comprising dish number, row number and column number. When libraries need to be screened by hybridization with a labeled DNA probe, it is again most efficient to have previously prepared high density gridded colony filters in which the colonies have been spotted on to the appropriate membrane in a gridded fashion with a uniform number of rows and columns (see *Figure 5.19*). This simplifies identification

The importance of sequence tagged sites (STSs)

Sequence tagged sites are important mapping tools simply because the presence of that sequence can very conveniently be assayed by PCR. Most STSs are nonpolymorphic and can be assigned to a specific chromosomal location by typing a panel of monochromosomal hybrids, and to subchromosomal localizations by typing panels of suitable hybrids (Sections 10.1.2 and 10.1.3). Some STSs may be known to derive from larger clones (e.g. gene clones) that have been physically mapped to a particular subchromosomal localization. In addition, microsatellite markers are a useful subset of STSs which are highly polymorphic and can be assigned a subchromosomal localization as a result of linkage analyses (see Section 11.4). An example of how an STS is developed from a DNA sequence is shown below.

Rough sequence: 200 nucleotides (see below)

Primers: chosen from underlined sequences, both 16 nucleotides long.
A (forward primer in bold, identical to the sense sequence from +50 to +66).
B (reverse primer, in bold, corresponding to antisense strand for the sequence from + 175 to +190).

Note that the sequences from +1 to +50 and from +191 to +200 are extremely GC-rich with multiple runs of G and of C, which are not optimal sequences for designing PCR primers. Hence the decision to use internal primers.

STS: 141 nucleotides defined by primer ends: +50 to +190

```
1                                        40
5′ CCCAGCGGGC CCGCGGCGCA GGGGCCCGGC GGGGCCCTGG


41              5′ — PRIMER A → 3′            80
   GGCCGCCCGG CAGTGAGCAT CAGATACAGA ACCTAGACGA


81                                       120
   ACCTAGGACC AGTACCTACA AGGTACTCTA GATGATCTAT


121                                      160
   ACTGAGGATC CTATTCAGAT CCTAGGTACC ACACTGATTA


161                                      200
   AGGATACTAG CTATACGGAC ATGGCATTAC ACCCCCGGGG   3′
                      |||||||||||||||||
           ← 3′ TGCCTG TACCGTAATG  5′
                      PRIMER B
```

In a complex genome such as the human genome, the chances of a 16 nucleotide primer binding by chance to a related but different sequence to the intended target is not insignificant. However, the chances of both primers binding to unintended related sequences which just happen to be *both in close proximity and also in a suitable orientation* is normally very low. The specificity of the reaction can be assayed simply by size-fractionating the amplification products on an agarose gel. If there is a single strong PCR product of the expected approximate size (141 bp in the above example), there is an excellent chance that the assay is specific for the intended target sequence.

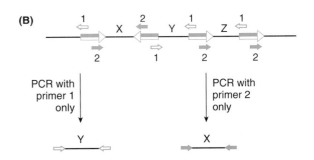

Figure 10.19: *Alu*-PCR permits amplification of DNA sequences located between two closely positioned but oppositely orientated *Alu* repeats.

(**A**) Primers can be designed from the *Alu* consensus sequence to hybridize to sequences at either terminus of the repeat element, with the 5′ → 3′ direction pointing away from the repeat element. (**B**) A single such primer can be used in a PCR reaction to amplify sequences located between two *Alu* repeats which are in opposite orientation, and arranged such that the two bound primers are *convergent* (e.g. PCR with primer 2 alone can amplify sequence X but not Y or Z; primer 1 alone can amplify sequence Y but not X or Z). A combination of primers 1 and 2 should, in theory, permit amplification of sequences between repeats in the same orientation. In practice, PCR amplification may be difficult or impossible if the distance between the repeats is too great (see Section 6.1.3).

of hybridization-positive colonies. More importantly, it facilitates integration of the results from numerous independent research groups. This is possible because of increasing automation. Robots have been designed to pick and spot colonies, and numerous copies of reference libraries can be distributed, either as cell cultures or, more conveniently, as sets of colony filters and sent to researchers throughout the world. Hybridization data can then be sent back from individual laboratories to the distributing centers and the fixed geometry of the library coordinates simplifies the task of integrating the information received.

The Human Genome Project (see Chapter 13) is one of several projects to benefit from such technologies, and sets of hybridization filters containing gridded colonies for total genome YAC libraries, chromosome-specific cosmid libraries and other libraries have been distributed extensively. In addition, pools of yeast colonies have been distributed in agarose plugs to permit PCR typing for a sequence of interest. Again, this is done in a hierarchical coordinate manner to simplify the screening of libraries (see *Figure 10.17*).

Figure 10.20: Fluorescence fingerprint.

The figure shows a portion of a fluorescent fingerprint of a variety of cosmids from the Cornelia de Lange syndrome region at 3q26.3 region. Each lane shows a marker (a *Bsa*JI digest of lambda DNA) labeled with ROX-ddCTP (red) plus up to three individual cosmid clones labelled with TET-ddATP (blue), HEX-ddATP (green) or NED-ddATP (yellow) fluorophores (see Gregory *et al.*, 1997). ABI Prism technology was used. Data provided by Melanie Smith, University of Newcastle upon Tyne and Dr Simon Gregory, Sanger Centre. Figure kindly sponsored by PE Biosystems, a PE Corporation Business.

10.4 Constructing transcript maps and identifying genes in cloned DNA

10.4.1 A wide variety of different methods can be used to identify genes in cloned DNA

A primary goal for physical mapping is identifying the locations of genes within a clone contig that has been localized to a specific chromosomal region. In principle, two major features permit the DNA of genes to be distinguished from DNA that does not have a coding function.

- **Expression**. All active genes are capable of making an RNA product, which, in the vast majority of cases, is mRNA. Mammalian genes usually contain exons and so the initial RNA transcript usually needs to undergo splicing.
- **Sequence conservation**. Because genes execute important cellular functions, mutations which alter the sequence of the product will often be disadvantageous and are rapidly eliminated by natural selection (see *Box 9.1*). The sequence of coding DNA and

important regulatory sequences is therefore more strongly conserved in evolution than that of noncoding DNA. In addition, premature termination codons in coding DNA are selected against. This means that genes often contain comparatively long **open reading frames (ORFs)**; in noncoding DNA, the DNA triplets corresponding to termination codons are not selected against and ORFs are usually comparatively short. Some exons, however, may be quite small but can often be detected using computer programs to analyze the relevant DNA sequence (see Section 10.4.5).

In addition, vertebrate genes are often associated with **CpG islands** (see *Box 8.5*). These features have permitted a variety of different methods for identifying genes in cloned vertebrate DNA (Monaco, 1994) of which the most commonly used are described in *Box 10.4*.

Genes can be identified easily by hybridizing DNA clones against northern blots, cDNA libraries, zoo blots and Southern blots of genomic DNA digested with rare-cutter restriction endonucleases. Of the various methods of identifying genes, the simpler methods are normally used as a first line of attack. In each case, the method involves using the genomic DNA clone as a hybridization probe. Inserts in cosmid clones and subclones are often useful hybridization probes. Larger clones such as YACs occasionally can be used as hybridization probes, but are generally less amenable to such approaches.

Hybridization to RNA/cDNA

A candidate DNA clone can be hybridized against a northern blot containing a panel of mRNA or total RNA samples isolated from a variety of different tissues (brain, heart, lung, liver, kidney, etc.). Positive hybridization may indicate the presence of a gene within the cloned fragment and may suggest a suitable cDNA library for screening. This approach has been facilitated enormously by using whole cosmid clones as hybridization probes in competition hybridization (see *Box 5.3*). They may fail, however, for two reasons. First, significant expression of the gene may be restricted to a cell population or developmental stage that is not represented in the northern blot panel or those cDNA libraries which are selected for screening. Additionally, there may be a problem with the intensity of the hybridization signal. For example, the proportion of exon sequence in the probe may be very low (the average exon size is only about 200 bp and a 40 kb insert in a cosmid may by chance contain only one exon). If this is the case and if the gene is not strongly expressed in the relevant tissue (so that it is not well represented in the RNA samples or cDNA libraries which are being screened), detection of positive hybridization signals may be difficult. Occasional transcribed repeats may also be problematic.

Zoo blot hybridization

Coding DNA sequences are subject to considerable selection pressure to conserve biologically important sequences. By contrast, noncoding DNA sequences accu-

Box 10.4

Commonly used methods for identifying genes in cloned DNA

Method	Comments
Zoo blotting	A DNA clone is hybridized at reduced hybridization stringency against a Southern blot of genomic DNA samples from a variety of animal species, a **zoo blot**. Depends on coding DNA being more strongly conserved in evolution than non-coding DNA (*Figure 10.21*).
CpG island identification	Many vertebrate genes have associated **CpG islands**, hypomethylated GC-rich sequences usually having multiple rare-cutter restriction sites (Cross and Bird, 1995).
	■ **Identification by restriction mapping**. DNA clones are usually hybridized against Southern blots of genomic DNA cut with *Sac*II, *Eag*I or *Bss*HII to identify clustering of rare-cutter sites (*Figure 10.22*).
	■ **Island-rescue PCR**. This is a way of isolating CpG island sequences from YACs by amplifying sequences between islands and neighbouring *Alu* repeats.
Hybridization	A genomic DNA clone can be hybridized against a Northern to mRNA/cDNA blot of mRNA from a panel of culture cell lines, or against appropriate cDNA libraries.
Exon trapping	This is essentially an artificial RNA splicing assay (see *Figure 10.23*). It relies on the observation that the vast majority of mammalian genes contain multiple exons which need to be spliced together at the RNA level.
cDNA selection or capture	These techniques involve repeated purification of a subset of genomic DNA clones which hybridize to a given cDNA population (see *Figure 10.24*).
Computer analysis of DNA sequence	■ **Homology searches**. Any DNA sequence obtained from a genomic clone can be compared against all other sequences in sequence data-bases. Significant homology to known coding DNA or gene-associated sequences may indicate a gene (see Section 20.1.4)
	■ **Gene searching algorithms**. A variety of computer programs have been developed to search sequences for exons and other gene-associated motifs (see *Figure 10.25* and Section 20.1.4).

mulate mutations comparatively rapidly and are not well conserved between species. A **zoo blot** is a Southern blot of genomic DNA samples from a wide variety of different species. A genomic DNA clone which shows positive hybridization signals against the DNA of a variety of different species would be expected, therefore, to contain coding DNA sequences that have been strongly conserved during evolution. Some mammalian genes, often with a crucially important function in development, are so highly conserved that they will show significant hybridization signals with evolutionarily distant species such as yeast, *Drosophila* and *Caenorhabditis elegans*. Others may only show significant hybridization signals to mammals (*Figure 10.21*).

CpG island identification

CpG islands (see *Box 8.5*) are short (~1 kb) hypomethylated GC-rich sequences which are often found at the 5′ ends of vertebrate genes. In the human genome, an estimated 56% of genes are associated with such sequences (Antequera and Bird, 1993). They include all examples of housekeeping genes and genes that are widely expressed, and a significant portion, perhaps about 40% of genes which show tissue-specific or restricted expression patterns (see Larsen *et al.*, 1992).

The % (G + C) of CpG island sequences generally exceeds 60% and they have a high concentration of the CpG dinucleotide (typically the CpG frequncy is 10–20

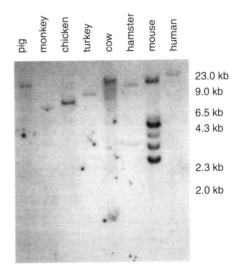

Figure 10.21: Zoo blot hybridization is an assay for DNA sequences that are highly conserved between species.

The example shows hybridization using a cDNA clone from the *NF2* (neurofibromatosis type 2) gene against a Southern blot of genomic DNA samples from the indicated species. Reproduced from Claudio *et al.* (1994) *Hum. Mol. Genet.*, **3**, pp. 185–190, by permission of Oxford University Press.

times greater than that in the bulk vertebrate DNA). As a result, CpG islands often have restriction sites for a variety of rare-cutter restriction nucleases which cleave at GC-rich sequences containing one or two CpG dinucleotides. For example, each of the enzymes *Sac*II (CCGCGG), *Eag*I (CGGCCG) and *Bss*HII (GCGCGC) is expected to cut, on average, about 1.2 times within an island (see Cross and Bird, 1995), but very rarely outside islands. Close clustering of such restriction sites in genomic DNA is often indicative of a CpG island, and can be identified by using genomic DNA clones as hybridization probes against southern blots of suitably digested genomic DNA samples (*Figure 10.22*). Note, however, that the method is not applicable to the substantial number of genes which have no associated CpG islands.

10.4.2 Exon trapping identifies expressed sequences by using an artificial RNA splicing assay

The gene identification methods described above have their limitations: restricted expression patterns of some genes may make them difficult to identify; genomic DNA clones may give very weak hybridization signals if the percentage of exon sequence is very low; and many genes which are expressed in a tissue-specific or restricted manner do not have associated CpG islands. An alternative is to identify a gene by virtue of the ability of its exons to engage in an artificial RNA splicing assay. RNA splicing involves fusion of exonic sequences at the RNA level and excision of intronic sequences. Spliceosomes are able to accomplish this *in vivo* by recognizing certain sequences at exon/intron boundaries: a

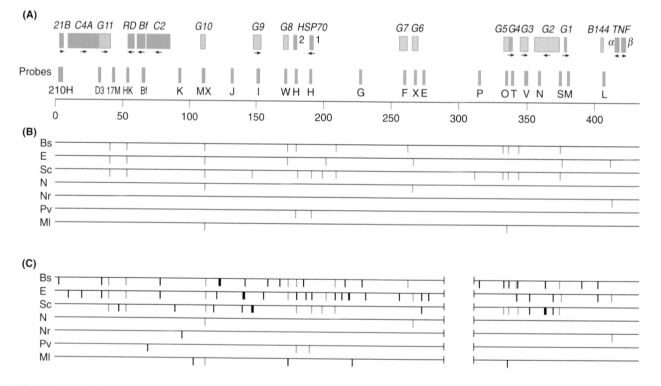

Figure 10.22: Identification of CpG island-associated genes by restriction mapping.

The figure shows restriction mapping in the class III region of the human leukocyte antigen (HLA) complex using the enzymes *Bss*HII (Bs), *Eag*I (E), *Sac*II (Sc), *Not*I (N), *Nru*I (Nr), *Pvu*I (Pv) and *Mlu*I (Ml). The restriction mapping was done by two methods. In one method, probes derived from the indicated regions (**A**) were hybridized to Southern blots of digested human genomic DNA to give the maps shown in (**B**). The second method involved direct restriction mapping on isolated cosmid clones to give the maps shown in (**C**). CpG islands were indicated by close clustering of rare-cutter sites (especially those for *Bss*HII, *Eag*I and *Sac*II) and were found in the genomic DNA associated with several genes, e.g. G11, RD, G10, etc. *Note* that many of the rare-cutter enzymes do not cleave at methylated sequences. As a result, there are numerous restriction sites in the cosmids because propagation of the human sequences in *E. coli* leads to removal of methyl groups originally present in genomic DNA. As CpG islands are hypomethylated in genomic DNA, however, they are susceptible to cleavage. Redrawn from Sargent *et al.* (1986) *EMBO J.*, **8**, pp. 2305–2312, by permission of Oxford University Press.

splice donor sequence at the junction between an exon and its downstream (3′) intron, and a *splice acceptor* sequence at the junction between an exon and its upstream (5′) intron (see *Figure 1.15*). A cosmid or other suitable genomic DNA clone containing an internal exon flanked by intronic sequences will therefore contain functional splice donor and acceptor sequences.

Exons can be identified in cloned genomic DNA by subcloning the DNA into a suitable expression vector and transfecting into an appropriate eukaryotic cell line in which the insert DNA is transcribed into RNA and the RNA transcript undergoes RNA splicing. Such techniques are known as **exon trapping** (often called exon amplification if a PCR reaction is employed to recover the exons from a cDNA copy of the spliced RNA). For example, in the method of Church *et al.* (1994), the DNA is subcloned into a plasmid expression vector pSPL3 (*Figure 10.23*) which contains an artificial minigene that can be expressed in a suitable host cell. The minigene consists of:

- a segment of the simian virus 40 (SV40) genome which contains an origin of replication plus a powerful promoter sequence;
- two splicing-competent exons separated by an intron which contains a multiple cloning site;
- an SV40 polyadenylation site.

The recombinant DNA is transfected into a strain of monkey cells, known as COS cells. COS cells were derived from monkey C̲V-1 cells by artificial manipulation, leading to integration of a segment of the SV40 genome containing a defective o̲rigin of S̲V40 replication. The integrated SV40 segment in COS cells allows any circular DNA which contains a functional SV40 origin of replication to replicate independently of the cellular DNA. Transcription from the SV40 promoter results in an RNA transcript which normally splices to include the two exons of the minigene. If the DNA cloned into the intervening intron contains a functional exon, however, the foreign exons can be spliced to the exons present in the vector's minigene. After making a cDNA copy using reverse transcriptase, PCR reactions using primers specific for vector exon sequences should distinguish between normal splicing and splicing involving exons in the insert DNA (*Figure 10.23*).

10.4.3 Genes in complex DNA clones, such as YACs, can be identified by forming heteroduplexes with cDNAs and by PCR-based methods of amplifying CpG island sequences

The techniques described in the previous sections have been popular when dealing with inserts from comparatively small genomic clones, such as cosmid clones or subclones. In some cases, however, attempts have been made to identify genes within large genomic clones, notably YAC clones. Two popular methods are cDNA selection and island-rescue PCR.

cDNA selection/capture

Direct cDNA selection techniques (also called direct selection and **cDNA selection**) involve forming genomic DNA/cDNA heteroduplexes by hybridizing a complex cloned DNA, such as the insert of a YAC, to a complex mixture of cDNAs, such as the inserts from all cDNA clones in a cDNA library (Lovett, 1994). The principle underlying the technique is that cognate cDNAs corresponding to genes found within the YAC will bind preferentially to the YAC DNA; several rounds of hybridization should lead to a huge enrichment of the desired cDNA sequences, enabling the identification of the corresponding genes. Considerable blocking of repetitive DNA sequences is required. Early approaches used immobilized YACs, but more modern approaches have used a solution hybridization reaction and biotin–streptavidin capture methods (*Figure 10.24*). Like all expression-based systems, the method depends on appropriate levels of gene expression (the cognate cDNAs should not be too rare in the starting population). Additionally, genes containing very short exons may be missed because the heteroduplexes formed with cognate cDNAs may not be sufficiently stable. Another problem is that cDNAs may bind to pseudogenes which show a high degree of homology to the cognate functional genes.

Island rescue PCR

Island rescue PCR (IRP) can permit selective amplification of CpG island sequences from human YACs. The method depends on the high copy number of *Alu* repeats (so that there is a high chance of an *Alu* repeat in the vicinity of a CpG island) and the frequent occurrence of restriction sites for rare-cutters such as *Bss*HII in CpG islands. The YAC DNA is cut with a suitable rare-cutter restriction nuclease and fragments are ligated to a vectorette linker (see *Figure 10.16*) with a suitably complementary overhang. CpG island-*Alu* PCR is then possible using an *Alu*-specific primer and a vectorette-specific primer (Valdes *et al.*, 1994).

10.4.4 Transcript maps can be obtained by physically mapping randomly generated expressed sequence tags

The expressed component of complex genomes, such as mammalian genomes, may constitute only a few percent of the total genome. As described in the preceding sections, one way of building transcript maps is to identify genes in defined clone contigs. An alternative, and more general, method is to obtain partial sequences of numerous randomly selected cDNA clones and then place these on physical maps. Even a short sequence of, say, 200 bp from a cDNA clone permits sequence-specific primers to be designed so that a PCR assay can be developed that is specific for that sequence. This effectively means that an STS is available for an expressed sequence, a so-called expressed sequence tag (EST).

An individual EST can be mapped by a PCR assay to: a chromosome (using monochromosomal hybrids, Section

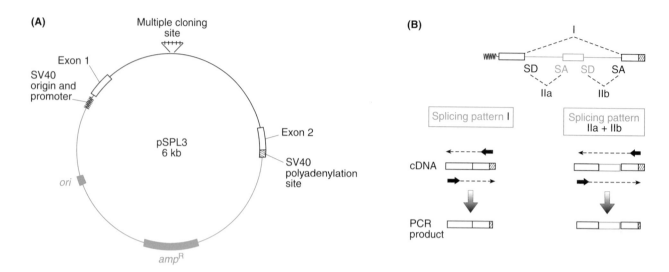

(A)

Multiple cloning site

Exon 1

SV40 origin and promoter

pSPL3 6 kb

ori

Exon 2

SV40 polyadenylation site

amp^R

(B)

I

SD   SA   SD   SA

IIa   IIb

Splicing pattern I

Splicing pattern IIa + IIb

cDNA

PCR product

1. Clone genomic DNA fragment into multiple cloning site
2. Transfect into COS cells
3. Expression from SV40 promoter → RNA product
4. Isolate RNA and use as template for making cDNA
5. Amplify in PCR reaction using primers specific for exons in vector

Figure 10.23: Exon trapping using the pSPL3 vector.

(A) The pSPL3 plasmid vector. This shuttle vector can be propagated in *E. coli* (using the *ori* origin of replication and selection for ampicillin resistance) and also in monkey COS cells (using the functional SV40 origin of replication) (Church *et al.*, 1994). The pSPL3 vector contains a minigene (in black): transcription occurs from the SV40 promoter and the RNA undergoes splicing under control of the host cell's RNA splicing machinery, resulting in fusion of the two vector exon sequences.
(B) Splicing patterns. The normal splicing pattern which is seen when only the vector exons are present is indicated by splicing pattern I. If a genomic DNA fragment cloned into pSPL3 contains an exon with functional splice donor (SD) and splice acceptor (SA) sequences, a different splicing pattern (IIa + IIb) may occur. The two splicing patterns can be distinguished at the cDNA level by using various vector-specific PCR primers and size-fractionation on gels can lead to recovery of the amplified exon from genomic DNA.

10.1.1); a subchromosomal location (using radiation, translocation and deletion hybrids, Sections 10.1.2 and 10.1.3); or to defined genomic clones (e.g. using panels of YACs from specific YAC contigs). The genome projects have placed considerable reliance on mapping ESTs using radiation hybrid panels in order to prepare first generation gene maps (see Section 13.2.3).

10.4.5 Genes can be identified in cloned DNA by computer analysis of the DNA sequence

Once a DNA clone has been sequenced computer analyses can be used to determine whether the sequence is likely to represent part of a gene (see Fickett, 1996; Borsani *et al.*, 1998). Two major types of software are used. One type is designed to search sequence databases for coding sequences which are similar to a nucleotide sequence obtained from a genomic clone (homology searches). Another type of software interrogates the DNA sequence from a genomic clone and looks for sequences that show similarity to gene-specific features, notably exons (exon-prediction programs). Modern nucleotide sequence analysis is typically performed using integrated software packages which include a large range of com-

puter programs for both homology searching and for identifying exons and gene-associated motifs.

Homology searches against sequence databases

The nucleotide sequence and the inferred amino acid sequences for all six possible translational reading frames (three frames for each of the two DNA strands) are compared against all available DNA and protein sequences which have been recorded in electronic sequence databases. The largest nucleotide sequence databases are the EMBL, GenBank and DNA Database of Japan databases and general protein sequence databases such as Swiss-Prot and PIR (see *Table 13.1*). The most popular algorithms that are used in sequence searching are *BLAST* (basic local alignment sequence tool; Altschul *et al.*, 1990) and *FASTA* (Pearson, 1990; see Section 20.1.4 for fuller details). Any significant matching between the test sequence and the sequence of a known gene, cDNA or protein, whether of human or non-human origin, indicates a gene-associated sequence (either a functional gene, pseudogene or gene fragment (see *Figure 20.4* for a typical positive output).

In addition to the above general sequence homology search programmes, a variety of computer programs permit searching for gene-associated motifs (see Section 20.1.4).

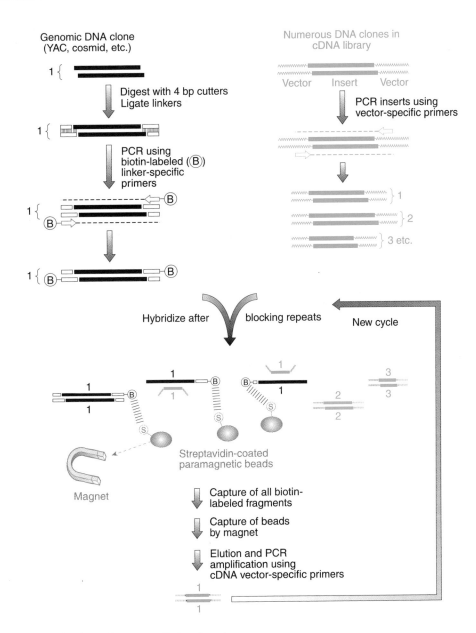

Figure 10.24: cDNA selection using magnetic bead capture.

The method relies on heteroduplex formation between single strands of a single genomic DNA clone (in black, numbered 1) and of a complex cDNA population, such as the inserts of a cDNA library (in blue, numbered 1, 2, 3, etc.). The genomic DNA strands are labeled with a biotin group (attached to PCR primers which are incorporated during amplification). The hybridization reaction will favor heteroduplex formation involving those cDNA clones *cognate* with the genomic DNA clone. In this example, genomic DNA clone 1 and cDNA clone 1 are envisaged to be cognate, i.e. contain common sequences, allowing opposite sense strands to bond together, giving a heteroduplex. Hybridization products with a biotin group (including genomic DNA–cDNA heteroduplexes) will bind to streptavidin-coated paramagnetic beads and can be removed from other reaction components by a magnet. The separated beads can then be treated to elute the biotin-containing molecules and, by using PCR primers specific for the vector sequences flanking the cDNA, the bound cDNA can be amplified. This population is submitted to further hybridization cycles to enrich for the desired cDNA.

Exon prediction programs

These programs are becoming increasingly important as large-scale DNA sequencing projects gather momentum. They are designed to scan a DNA sequence in order to identify the locations of likely exons by screening for con-served sequences found at exon/intron junctions and the splice branch site (see *Figure 1.15*), and the presence of comparatively long ORFs, etc. As yet, however, even the best such programs, such as the *GENSCAN* and *GRAIL2* software (see Burge and Karlin, 1997) have been only

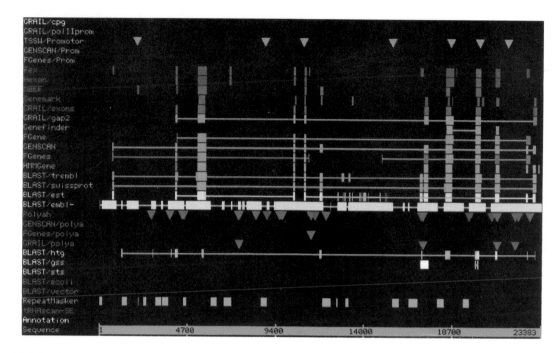

Figure 10.25: Gene finding by computer-based analysis of genomic sequences.

This example shows NIX analysis of PAC sequence from a region of chromosome 12q24.1 encompassing the Darier's disease locus. The nucleotide size of the region is illustrated by the bar at the bottom. Analyses include the use of programs to scan for gene-associated motifs such as promoter sequences (green inverted triangles at top), polyadenylation sites (ochre-colored inverted triangles), and various exon prediction programs (GRAIL, GENSCAN, etc.). Significant homologies to other sequences at the nucleotide level and at the protein level are indicated by the boxes for the various BLAST programs. Data provided by Dr Victor Ruiz-Perez and Simon Carter, University of Newcastle upon Tyne, UK.

moderately successful in identifying exons when tested against genes whose exon organizations had previously been established. When the relevant gene is GC-rich, however, exon prediction can be quite accurate. An example of a successful application is provided by analysis of the adult polycystic kidney disease gene *PKD1*: the GRAIL2 program was able to predict a total of 46 exons in 54 kb of genomic DNA sequence upstream of the poly(A) site (American PKDI Consortium, 1995), a figure which was subsequently confirmed using more conventional methods.

Comprehensive integrated gene-finding software packages

Recently, a variety of software packages have been produced which use general sequence homology-based database searching programs together with programs designed to identify gene-associated motifs and exons and produce the output in a graphical format. Two popular packages are the NIX (nucleotide identification) and Genotator packages which are maintained respectively at the UK Human Genome Mapping Project Resource Centre and the US Lawrence Berkeley National Laboratory (*Figure 10.25* and Borsani *et al.*, 1998).

Further Reading

Anand R (ed.) (1992) *Techniques for the Analysis of Complex Genomes.* Academic Press, London.
Dracopoli NC et al. (eds) (1995) *Current Protocols in Human Genetics.* John Wiley & Sons, Chichester.

References

Abbott C, Povey S (1991) Development of human chromosome-specific PCR primers for characterization of somatic cell hybrids. *Genomics*, **9**, 73–77.
Abrams WR, Ma RI, Kucich V, Bashir MM, Decker S, Tsipouras P, McPherson JD, Wasmuth JJ, Rosenbloom J (1995) Molecular cloning of the microfibrillar protein MFAP3 and assignment of the gene to human chromosome 5q32–q33.2. *Genomics*, **26**, 47–54.

Altschul SF, Gish W, Myers EW, Lipman DJ (1990) Basic local alignment search tool. *J. Mol. Biol.,* **215**, 403–410.

Anand R, Riley JH, Butler R, Smith JC, Markham AF (1990) A 3.5 genome equivalent multi access YAC library: Construction, characterisation, screening and storage. *Nucleic Acids Res.,* **18**, 1951–1956.

Antequera F, Bird A (1993) Number of CpG islands and genes in human and mouse. *Proc. Natl Acad. Sci. USA,* **90**, 11995–11999.

Bartholdi M, Meyne J, Albright K *et al.* (1987) Chromosome sorting by flow cytometry. *Meth. Enzymol.,* **151**, 252–267.

Bellanne-Chantelot C, Lacroix B, Ougen P *et al.* (1992) Mapping the whole human genome by fingerprinting yeast artificial chromosomes. *Cell,* **70**, 1059–1068.

Borsani G, Ballabio A, Banfi S (1998) A practical guide to orient yourself in the labyrinth of genome databases. *Hum. Mol. Genet.,* **7**, 1641–1648.

Burge C, Karlin S (1997) Prediction of complete gene structures in human genomic DNA. *J. Mol. Biol.,* **268**, 78–94.

Burn TC, Connors TD, Dackowski WR *et al.* (1995) Analysis of the genomic sequence for the autosomal dominant polycystic kidney disease (PKD1) gene predicts the presence of a leucine-rich repeat. The American PKD1 Consortium (APKD1 Consortium). *Hum. Mol. Genet.,* **4**, 575–582.

Church DM, Stotler CJ, Rutter JL, Murrell JR, Trofatter JA, Buckler AJ (1994) Isolation of genes from complex sources of mammalian genomic DNA using exon amplification. *Nature Genet.,* **6**, 98–105.

Claudio JO, Marineau C, Rouleau GA (1994) The mouse homologue of the neurofibromatosis type 2 gene is highly conserved. *Hum. Mol. Genet.,* **3**, 185–190

Cotter F, Nasipuri S, Lam G, Young BD (1989) Gene mapping by enzymatic amplification from flow-sorted chromosomes. *Genomics,* **5**, 470–474.

Cox DR, Burmeister M, Proce ER, Kim S, Myers RM (1990) Radiation hybrid mapping: a somatic cell genetic method for constructing high resolution maps of mammalian chromosomes. *Science,* **250**, 245–250.

Cross SH, Bird AP (1995) CpG islands and genes. *Curr. Opin. Genet. Dev.,* **5**, 309–314.

Cuthbert AP, Trott DA, Ekong RM, Jezzard S, England NL, Themis M, Todd CM, Newbold RF (1995) Construction and characterization of a highly stable human:rodent monochromosomal hybrid panel for genetic complementation and genome mapping studies. *Cytogenet. Cell Genet.,* **71**, 68–76.

Davies K, Young BD, Elles RG, Hill ME, Williamson R (1981) Cloning of a representative genomic library of the human X chromosome after sorting by flow cytometry. *Nature,* **293**, 374–376.

Deloukas P, Schuler GD, Gyapay G *et al.* (1998) A physical map of 30,000 human genes. *Science,* **282**, 744–746.

Faraco J, Bashir M, Rosenbloom J, Francke U (1995) Characterization of the human gene for microfibril-associated glycoprotein (MFAP2), assignment to chromosome 1p36.1–p35 and linkage to D1S170. *Genomics,* **25**, 630–637.

Fickett JW (1996) Finding genes by computer: the state of the art. *Trends Genet.,* **12**, 316–320.

Fournier REK, Ruddle FH (1977) Microcell-mediated transfer of murine chromosomes into mouse, Chinese hamster and human somatic cells. *Proc. Natl Acad. Sci. USA,* **74**, 319–323.

Goss SJ, Harris H (1975) New method for mapping genes in human chromosomes. *Nature,* **255**, 680–684.

Gregory SG, Howell GR, Bentley DR (1997) Genome mapping by fluorescent fingerprinting. *Genome Res.,* **7**, 1162–1168.

Gyapay G, Schmitt K, Fizames C *et al.* (1996*)*. A radiation hybrid map of the human genome. *Hum. Mol. Genet.,* **5**, 339–346.

Haaf T, Ward D (1994) High resolution ordering of YAC contigs using extended chromatin and chromosomes. *Hum. Mol. Genet.,* **3**, 629–633.

Heiskanen M, Peltonen L, Palotie A (1996) Visual mapping by high resolution FISH. *Trends Genet.,* **12**, 379–384.

Houseal TW, Klinger KW (1994) Commentary: what's in a spot? *Hum. Mol. Genet.,* **3**, 1215–1216.

Jones MH, Khwaja OS, Briggs H *et al.* (1994) A set of ninety-seven overlapping yeast artificial chromosome clones spanning the human Y chromosome euchromatin. *Genomics,* **24**, 266–275.

Larsen F, Gundersen G, Lopez R, Prydz H (1992) CpG islands as gene markers in the human genome. *Genomics,* **13** (4), 1095–1107.

Lichter P (1997) Multicolor FISHing: what's the catch? *Trends Genet.,* **13**, 475–479.

Lichter P, Tang CJ, Call K *et al.* (1990) High resolution mapping of human chromosome 11 by *in situ* hybridization with cosmid clones.

Lovett M (1994) Fishing for complements: finding genes by direct selection. *Trends Genet.,* **10**, 352–357.

McCormick MK, Campbell E, Deaven L, Moyzis R (1993) Low-frequency chimeric yeast artificial chromosome libraries from flow-sorted human chromosomes 16 and 21. *Proc. Natl Acad. Sci. USA,* **90**, 1063–1067.

Monaco AP (1994) Isolation of genes from cloned DNA. *Curr. Opin. Genet. Dev.,* **4**, 360–365.

Nelson DL, Ledbetter SA, Corbo L, Victoria MF, Ramirez-Solis R, Webster TD, Ledbetter DH, Caskey CT (1989) Alu polymerase chain reaction: a method for rapid isolation of human-specific sequences from complex DNA sources. *Proc. Natl Acad. Sci. USA*, **86**, 6686–6690.

Ochman H, Gerber AS, Hartl DL (1988) Genetic applications of an inverse polymerase chain reaction. *Genetics*, **120**, 621–623.

Pearson WR (1990) Rapid and sensitive sequence comparison with FASTP and FASTA. In *Methods in Enzymology* (RF Doolittle, ed.), Vol. 183, pp. 63–98. Academic Press, San Diego, CA.

Porteous DJ (1987) Chromosome-mediated gene transfer: a functional assay for complex loci and an aid to human genome mapping. *Trends Genet.*, **3**, 177–182.

Ried T, Schrock E, Ning Y, Wienberg J (1998) Chromosome painting: a useful art. *Hum. Mol. Genet.*, **7**, 1619–1626.

Riley J, Butler R, Ogilvie D et al. (1990) A novel, rapid method for the isolation of terminal sequences from yeast artificial chromosome (YAC) clones. *Nucleic Acids Res.*, **18**, 2887–2890.

Sargent CA, Dunham I, Campbell RD (1986) Identification of multiple HTF-island associated genes in the human major histocompatibility complex class III region. *EMBO J.*, **8**, 2305–2312.

Schrock S, DuManoir S, Veldman T et al. (1996) Multicolor spectral karyotyping of human chromosomes. *Science*, **273**, 494–497.

Speicher MR, Ballard GS, Ward DC (1996) Karyotyping human chromosomes by combinatorial multi-fluor FISH. *Nature Genet.*, **12**, 368–375.

Trask BJ (1991) Fluorescence *in situ* hybridization: applications in cytogenetics and gene mapping. *Trends Genet.*, **7**, 149–154.

Valdes JM, Tagle DA, Collins FS (1994) Island rescue PCR: a rapid and efficient method for isolating transcribed sequences from yeast artificial chromosomes. *Proc. Natl Acad. Sci. USA*, **91**, 5377–5381.

van Ommen G-JB, Breuning MH, Raap AK (1995) FISH in genome research and molecular diagnostics. *Curr. Opin. Genet. Dev.*, **5**, 304–308.

Walter MA, Spillett DJ, Thomas P, Weiseenbach J, Goodfellow PN (1994) A method for constructing radiation hybrid maps of whole genomes. *Nature Genet.*, **7**, 22–28.

Warburton D, Gersen D, Yu M-T, Jackson C, Handelin B, Houseman D (1990) Monochromosomal rodent–human hybrids from microcell fusion of human lymphoblastoid cells containing an inserted dominant selectable marker. *Genomics*, **6**, 358–366.

Wilke CM, Guo SW, Hall BK et al. (1994) Multicolor FISH mapping of YAC clones in 3p14 and identification of a YAC spanning both *FRA3B* and the t(3;8) associated with hereditary renal cell carcinoma. *Genomics*, **22**, 319–326.

Genetic mapping of mendelian characters

11

11.1 Recombinants and nonrecombinants

In principle, genetic mapping in humans is exactly the same as genetic mapping in any other sexually reproducing diploid organism. The aim is to discover how often two loci are separated by meiotic recombination. Consider a person who is heterozygous at two loci, and so types as A_1A_2 B_1B_2. Suppose the alleles A_1 and B_1 in this person came from one parent, and A_2 and B_2 from the other. Any of that person's children who inherit one of these parental combinations (A_1B_1 or A_2B_2) is nonrecombinant, whereas children who inherit A_1B_2 or A_2B_1 are **recombinant** (*Figure 11.1*). The proportion of children who are recombinant is the **recombination fraction** between the two loci A and B.

11.1.1 The recombination fraction is a measure of genetic distance

If two loci are on different chromosomes, they will segregate independently. Considering spermatogenesis in individual II_1 in *Figure 11.1*, at the end of meiosis I, whichever sperm receives allele A_1, there is a 50% chance that it will receive allele B_1 and a 50% chance it will receive B_2. Thus, on average, 50% of the children will be recombinant and 50% nonrecombinant. The recombination fraction is 0.5. If the loci are **syntenic**, that is if they lie on the same chromosome, then they might be expected always to segregate together, with no recombinants. However, this simple expectation ignores meiotic crossovers. During prophase of meiosis I, pairs of homologous chromosomes synapse and exchange segments (*Figure 2.14*). Only two of the four chromatids are

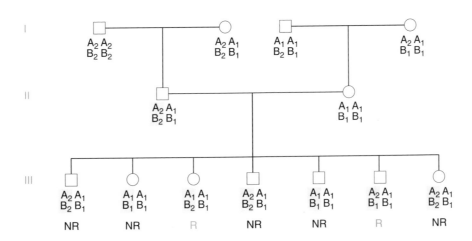

Figure 11.1: Recombinants and nonrecombinants.

Alleles at two loci (locus A, alleles A_1 and A_2; locus B, alleles B_1 and B_2) are segregating in this family. Where this can be deduced, the combination of alleles a person received from his or her father is boxed. Persons in generation III who received either A_1B_1 or A_2B_2 from their father are the product of nonrecombinant sperm; persons who received A_1B_2 or A_2B_1 are recombinant. The information shown does not enable us to classify any of the individuals in generations I and II as recombinant or nonrecombinant, nor does it identify recombinants arising from oogenesis in individual II_2.

involved in any particular crossover. A crossover, if it occurs between the positions of the two loci, will create two recombinant chromatids carrying A_1B_2 and A_2B_1, and leave the two noninvolved chromatids nonrecombinant. Thus one crossover generates 50% recombinants between loci flanking it.

Recombination will rarely separate loci that lie very close together on a chromosome, because only a crossover located precisely in the small space between the two loci will create recombinants. Therefore sets of alleles on the same small chromosomal segment tend to be transmitted as a block through a pedigree. Such a block of alleles is known as a **haplotype**. Haplotypes mark recognizable chromosomal segments which can be tracked through pedigrees and through populations. When not broken up by recombination, haplotypes can be treated for mapping purposes as alleles at a single highly polymorphic locus.

The further apart two loci are on a chromosome, the more likely it is that a crossover will separate them. Thus the recombination fraction is a measure of the distance between two loci. Recombination fractions define **genetic distance**, which is not the same as **physical distance**. Two loci that show 1% recombination are defined as being 1 **centimorgan (cM)** apart on a genetic map.

11.1.2 Recombination fractions do not exceed 0.5 however great the physical distance

A single recombination event produces two recombinant and two nonrecombinant chromatids. When loci are well separated there may be more than one crossover between them. Double crossovers can involve two, three or four chromatids, but *Figure 11.2* shows that the overall effect, averaged over all double crossovers, is to give 50% recombinants. Loci very far apart on the same chromosome might be separated by three, four or more crossovers. Again, the overall effect is to give 50% recombinants. Recombination fractions never exceed 0.5, however far apart the loci are.

11.1.3 Mapping functions define the relationship between recombination fraction and genetic distance

Because recombination fractions never exceed 0.5, they are not simply additive across a genetic map. If a series of loci, A, B, C, ... are located at 5 cM intervals on a map, locus M may be 60 cM from locus A, but the recombina-

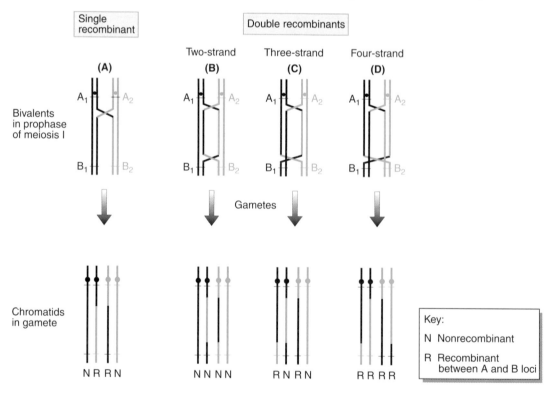

Figure 11.2: Single and double recombinants.

Each crossover involves two of the four chromatids of the two synapsed homologous chromosomes. The black chromosome carries alleles A_1 and B_1 at two loci, while the blue chromosome carries alleles A_2 and B_2. Gametes in which the chromatid is the same color at the two loci are nonrecombinant for these loci, those where the chromatids are different colors are recombinant. **(A)** A single crossover generates two recombinant and two nonrecombinant chromatids. **(B)** A two-strand double crossover leaves flanking markers nonrecombinant on all four chromatids. **(C)** A three-strand double crossover leaves flanking markers recombinant on two of the four strands. **(D)** A four-strand double crossover generates 100% recombinants. The three types of double crossover occur in random proportions, so the average effect of a double crossover is to give 50% recombinants.

tion fraction between A and M will not be 60%. The mathematical relationship between recombination fraction and genetic map distance is described by the **mapping function**. If crossovers occurred at random along a bivalent and had no influence on one another, the appropriate mapping function would be Haldane's function:

$$w = -\tfrac{1}{2} \ln(1 - 2\theta)$$
$$\text{or} \qquad \theta = \tfrac{1}{2} [1 - \exp(-2w)]$$

where w is the map distance and θ the recombination fraction; as usual ln means logarithm to the base e, and exp means 'e to the power of'. However, we know that crossovers do not occur at random. The presence of one chiasma inhibits formation of a second chiasma nearby. This phenomenon is called **interference**. A variety of mapping functions exist that allow for varying degrees of interference. A widely used function for human mapping is Kosambi's function:

$$w = \tfrac{1}{4} \ln [(1 + 2\theta) / (1 - 2\theta)]$$
$$\text{or} \qquad \theta = \tfrac{1}{2} [\exp(4w) - 1] / [\exp(4w) + 1]$$

A mapping function is needed in multipoint mapping (Section 11.4) to convert the raw data on the recombination fraction into a genetic map. The interested reader should consult Ott's book (see Further reading) and Broman and Weber (1998) for a fuller discussion of mapping functions.

11.1.4 The relation between physical and genetic distances is not constant across the genome

Chiasma counts in human male meiosis show an average of 49 crossovers per cell (Morton *et al.*, 1982). Since each crossover gives 50% recombinants, the chiasma count implies a total male genetic map length of 2450 cM. The current version of the Location Database (Collins *et al.*, 1996) suggests a total male map length of 2851 cM. Chiasmata are more frequent in female meiosis (exemplifying Haldane's rule that the heterogametic sex has the lower chiasma count), and the total female map length in the Location Database is 4296 cM (excluding the X). Thus over the 3000 Mb autosomal genome, 1 male cM averages 1.05 Mb and 1 female cM averages 0.70 Mb; the sex-averaged figure is 1 cM = 0.88 Mb.

The approximation 1 cM = 1 Mb is a useful rule of thumb, but the actual correspondence varies widely for different chromosomal regions. In general, there is more recombination towards the telomeres of chromosomes in males, while centromeric regions have recombinants in females but not in males (see *Figure 11.3* and Broman *et al.*, 1998). The most extreme deviation is shown by the pseudoautosomal region at the tip of the short arms of the X and Y chromosomes (see *Figure 14.7*). Males have an *obligatory* crossover within this 2.6 Mb region, so that it is 50 cM long. Thus, for this region in males 1 Mb = 19 cM, whereas in females 1 Mb = 2.7 cM. Uniquely, the Y chromosome, outside the pseudoautosomal region, has no genetic map because it is not subject to synapsis and crossing over in normal meiosis. The X chromosome of course undergoes normal recombination in females, and can be genetically mapped in female meioses.

11.2 Genetic markers

11.2.1 Mapping human disease genes requires genetic markers

Since most human geneticists are interested in diseases, we would like a map to show the order and distance apart of all disease genes. Scoring the recombination fraction between pairs of diseases would be the obvious way to construct such a map, but disease–disease mapping is not possible in humans. Defining recombinants, as we have seen (*Figure 11.1*) requires double heterozygotes. People heterozygous for two different diseases are extremely rare. Even if they can be found, they will probably have no children, or be unsuitable for genetic analysis in some other way. For this reason human genetic mapping depends on **markers**. Any mendelian character can in principle be used as a genetic marker. It helps if the character can be scored easily and cheaply using readily available material (blood cells rather than a brain biopsy), but the crucial thing is that it should be sufficiently polymorphic that a randomly selected person has a good chance of being heterozygous. *Box 11.1* summarizes the development of human genetic markers, from blood groups and polymorphisms of serum proteins through to the present generation of DNA microsatellites and single nucleotide polymorphisms.

Gene mappers could not set out to map a disease with a reasonable hope of success until markers were available that were spaced throughout the genome. Disease-marker mapping, if it is not to be a purely blind exercise, requires framework maps of markers. These are generated by marker–marker mapping. Although in theory linkage can be detected between loci 40 cM apart, the amount of data required to do this is prohibitive. Ten meioses are sufficient to give evidence of linkage if there are no recombinants, but 85 meioses would be needed to give equally strong evidence of linkage if the recombination fraction was 0.3 (see *Box 11.3* for a guide to these calculations). Obtaining enough family material to test much more than 30 meioses can be seriously difficult for a rare disease. Thus mapping requires markers spaced at intervals no greater than about 20 cM across the genome. Given the genome lengths calculated above, this means that we need a minimum of 150 markers. Allowing for imperfect informativeness (see below), we need at least 300. In fact much denser maps, down to 1 cM or less average spacing of markers, are needed to guide progress from initial mapping of a disease through to cloning the gene. A major achievement of the Human Genome Project has been to generate upwards of 10 000 highly polymorphic markers and place them on framework maps (Collins *et al.*, 1996; Broman *et al.*, 1998).

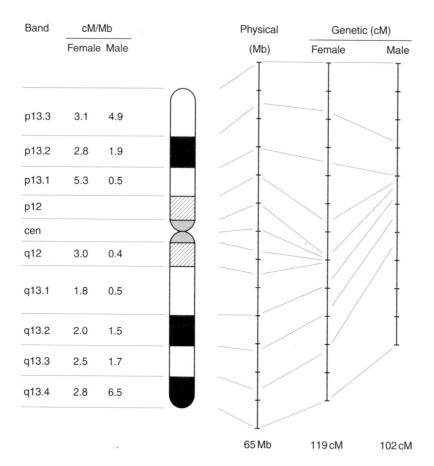

Band	cM/Mb		Physical	Genetic (cM)	
	Female	Male	(Mb)	Female	Male
p13.3	3.1	4.9			
p13.2	2.8	1.9			
p13.1	5.3	0.5			
p12					
cen					
q12	3.0	0.4			
q13.1	1.8	0.5			
q13.2	2.0	1.5			
q13.3	2.5	1.7			
q13.4	2.8	6.5			

65 Mb   119 cM   102 cM

Figure 11.3: Relation of physical and genetic maps of chromosome 19.

180 markers from chromosome 19 were mapped genetically and physically. The physical map of the 65 Mb chromosome is compared with genetic maps separately computed for male and female meioses. Note the uneven distribution of recombinants along the chromosome, with more recombination towards the telomeres, and the varying male:female recombination ratio. The female map is about 10% longer than the male map; for most chromosomes the difference is more marked. Data from Mohrenweiser *et al.* (1998).

11.2.2 The heterozygosity or polymorphism information content measure how informative a marker is

For linkage analysis we need **informative meioses** (see Box 11.2). The examples in the box show that a meiosis is not informative with a given marker if the parent is homozygous for the marker, and also in half of the cases where both parents have the same heterozygous genotype. For most purposes the mean **heterozygosity** of a marker (the chance that a randomly selected person will be heterozygous) is used as the measure of informativeness. If there are marker alleles A_1, A_2, A_3 ... with gene frequencies p_1, p_2, p_3 ..., then the proportion of people who are heterozygous is $1 - (p_1^2 + p_2^2 + p_3^2 + ...)$ (Section 3.1). A more sophisticated measure, the polymorphism information content (PIC) allows for couples who are both heterozygous A_1A_2. Half their children will also be A_1A_2 and therefore uninformative. The PIC of a marker is given by:

$$\text{PIC} = 1 - \sum_{i=1}^{n} p_i^2 - \sum_{i=1}^{n} \sum_{j=i+1}^{n} 2p_i^2 p_j^2$$

where p_i is the frequency of the ith allele. The third term takes out half the matings of similar heterozygotes. For X-linked markers the PIC and heterozygosity are the same; for autosomal markers the heterozygosity somewhat overstates the informativeness, especially for 2–allele markers. For an autosomal marker with two alleles of equal frequency the heterozygosity is 0.5 but the PIC is only 0.375.

11.2.3 DNA polymorphisms are the basis of all current genetic markers

In the early 1980s, DNA polymorphisms provided, for the first time, a set of markers that were sufficiently numerous and spaced across the entire genome. DNA markers have the additional advantage that they can all be typed by the same technique. Moreover their chromosomal

Box 11.1

The development of human genetic markers

Type of marker	No. of loci	Features
Blood groups 1910–1960	~20	May need fresh blood, rare antisera Genotype cannot always be inferred from phenotype because of dominance No easy physical localization
Electrophoretic mobility variants of serum proteins 1960–1975	~30	May need fresh serum, specialized assays No easy physical localization Often limited polymorphism
HLA tissue types 1970–	1 (haplotype)	One linked set Highly informative Can only test for linkage to 6p21.3
DNA RFLPs 1975–	$>10^5$ (potentially)	Two allele markers, maximum heterozygosity 0.5 Initially required Southern blotting, now PCR Easy physical localization
DNA VNTRs (minisatellites) 1985–	$>10^4$ (potentially)	Many alleles, highly informative Type by Southern blotting Easy physical localization Tend to cluster near ends of chromosomes
DNA VNTRs (microsatellites) (di-, tri- and tetranucleotide repeats) 1989–	$>10^5$ (potentially)	Many alleles, highly informative Can type by automated multiplex PCR Easy physical localization Distributed throughout genome
DNA SNPs (single nucleotide polymorphisms) 1998–	$>10^6$ (potentially)	Less informative than microsatellites Can be typed on a very large scale by automated equipment without gel electrophoresis

VNTR, variable number of tandem repeats

location can be determined using FISH or radiation hybrid mapping (Sections 10.1 and 10.2), allowing DNA-based genetic maps to be cross-referenced to physical maps. This avoids the frustrating situation that arose when the long-sought cystic fibrosis gene (*CFTR*) was first mapped. Linkage was established to a protein polymorphism of the enzyme paraoxonase, but the chromosomal location of the paraoxonase gene was not known. The development of DNA markers allowed human gene mapping to start in earnest.

Restriction fragment length polymorphisms (RFLPs)

The first generation of DNA markers were restriction fragment length polymorphisms (RFLPs). RFLPs were initially typed by preparing Southern blots from restriction digests of the test DNA, and hybridizing with radio-labeled probes (see *Figure 5.12*). This technology required plenty of time, money and DNA, and made a whole genome search a heroic undertaking. Nowadays this is less of a problem because RFLPs can usually be typed by PCR. A sequence including the variable restriction site is amplified, the product is incubated with the appropriate restriction enzyme and then run out on a gel to see if it

has been cut (see *Figure 6.6*). A more fundamental limitation is their limited informativeness. RFLPs have only two alleles: the site is present or it is absent. The maximum heterozygosity is 0.5. Disease mapping using RFLPs is frustrating because all too often a key meiosis in a family turns out to be uninformative.

Minisatellites

Minisatellite VNTR (variable number tandem repeat) markers were a great improvement. The VNTRs have many alleles and high heterozygosity. Most meioses are informative. However, the technical problems of Southern blotting and radioactive probes were still an obstacle to easy mapping, and VNTRs are not evenly spread across the genome.

Microsatellites

The advent of PCR finally made mapping relatively quick and easy. Minisatellites are too long to amplify well, and so the standard tools for PCR linkage analysis are **microsatellites**. These are mostly $(CA)_n$ repeats. Tri- and tetranucleotide repeats are gradually replacing dinucleotide repeats as the markers of choice because they give cleaner results – dinucleotide repeat sequences are

Informative and uninformative meioses

A meiosis is informative for linkage when we can identify whether or not the gamete is recombinant. Consider the male meiosis which produced the paternal contribution to the child in the four pedigrees below. We assume that the father has a dominant condition that he inherited along with marker allele A_1.

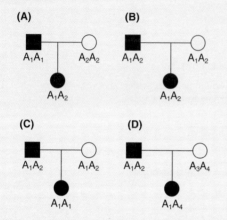

(A)   (B)

(C)   (D)

(A) This meiosis is uninformative: the marker alleles in the homozygous father cannot be distinguished.

(B) This meiosis is uninformative: the child could have inherited A_1 from father and A_2 from mother, or vice versa.

(C) This meiosis is informative: the child inherited A_1 from the father.

(D) This meiosis is informative: the child inherited A_1 from the father.

peculiarly prone to replication slippage during PCR amplification. Each allele gives a little ladder of 'stutter bands' on a gel, making it hard to read (see *Figure 6.8*). Much effort has been devoted to producing compatible sets of microsatellite markers that can be amplified together in a multiplex PCR reaction and give nonoverlapping allele sizes, so that they can be run in the same gel lane. With fluorescent labeling in several colors, it is possible to score perhaps ten markers on a sample in a single lane of an automated gel.

Single nucleotide polymorphisms (SNPs)

After 10 years of developing more and more polymorphic markers, it may seem perverse that the newest generation of markers are 2-allele single nucleotide polymorphisms. They include the classic RFLPs, but also polymorphisms that do not happen to create or abolish a restriction site. The advantage of SNPs is that they can be scored on solid-state arrays without recourse to gel electrophoresis (Wang *et al.*, 1998). The gain in throughput more than offsets the lower informativeness of SNPs. Typically the test DNA is PCR amplified in a very large multiplex and hybridized to an array comprising a series of anchored

oligonucleotide primers, each terminating with a polymorphic nucleotide. A single primer extension step is carried out on the array, using a mixture of four fluorescently-labeled dideoxynucleotides. Label adds to primers that perfectly match the test DNA, but not to those with a 3' mismatch (see *Figure 17.10*). Reading the cells of the array for presence or absence of fluorescence allows the types for every SNP on the array to be read off. Although the technology is still being put together, it is hoped that an array of a few thousand primers can be used to genotype markers spaced closely across the whole genome in a single chip hybridization.

11.3 Two-point mapping

11.3.1 Scoring recombinants in human pedigrees is not always simple

Having collected families where a mendelian disease is segregating, and typed them with an informative marker, how do we know when we have found linkage? There are two aspects to this question:

(i) How can we work out the recombination fraction?
(ii) What statistical test should we use to see if the recombination fraction is significantly different from 0.5, the value expected on the null hypothesis of no linkage?

In some families the first question can be answered very simply by counting recombinants and nonrecombinants. The family shown in *Figure 11.1* is one example. There are two recombinants in seven meioses and the recombination fraction is 0.28. *Figure 11.4A* shows another example. The double heterozygote who is informative for linkage (individual II_1 in both *Figure 11.1* and *Figure 11.4A*), is phase-known: we know which alleles were inherited from which parent, and so we can unambiguously score each meiosis as recombinant or nonrecombinant. In *Figure 11.4B*, individual II_1 is again doubly heterozygous, but this time phase-unknown. Among her children, either there are five nonrecombinants and one recombinant, or else there are five recombinants and one nonrecombinant. We can no longer identify recombinants unambiguously, even if the first alternative seems much more likely than the second. *Figure 11.4C* adds yet more complications, yet if this is a family with a rare disease no researcher would be willing to discard it. Some method is needed to extract the linkage information from a collection of such imperfect families.

11.3.2 Computerized lod score analysis is the best way to analyze complex pedigrees for linkage between mendelian characters

In the pedigree shown in *Figure 11.4B* it is not possible to identify recombinants unambiguously and count them. It is possible, however, to calculate the overall likelihood of

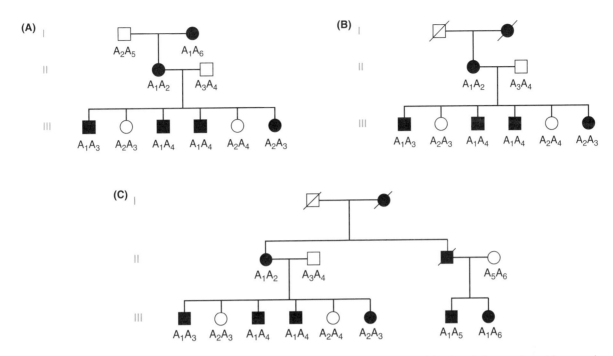

Figure 11.4: Recognizing recombinants: three versions of a family with an autosomal dominant disease, typed for a marker A.

(**A**) All meioses are phase-known. We can identify III$_1$–III$_5$ unambiguously as nonrecombinant and III$_6$ as recombinant.
(**B**) The same family, but phase-unknown. The mother, II$_1$, could have inherited either marker allele A$_1$ or A$_2$ with the disease; thus her phase is unknown. *Either* III$_1$–III$_5$ are nonrecombinant and III$_6$ is recombinant; *or* III$_1$–III$_5$ are recombinant and III$_6$ is nonrecombinant. (**C**) The same family after further tracing of relatives. III$_7$ and III$_8$ have also inherited marker allele A$_1$ along with the disease from their father, but we cannot be sure whether their father's allele A$_1$ is identical by descent to the allele A$_1$ in his sister II$_1$. Maybe there are two copies of allele A$_1$ among the four grandparental marker alleles. The likelihood of this depends on the gene frequency of allele A$_1$. Thus although this pedigree contains linkage information, extracting it is problematic.

the pedigree, on the alternative assumptions that the loci are linked (recombination fraction = θ) or not linked (recombination fraction = 0.5). The ratio of these two likelihoods gives the odds of linkage, and the logarithm of the odds is the **lod score**. Morton (1955) demonstrated that lod scores represent the most efficient statistic for evaluating pedigrees for linkage, and derived formulae to give the lod score (as a function of θ) for various standard pedigree structures. *Box 11.3* shows how this is done for simple structures. Being a function of the recombination fraction, lod scores are calculated for a range of θ values. In a set of families, the overall probability of linkage is the product of the probabilities in each individual family, therefore lod scores (being logarithms) can be added up across families.

Calculating the full lod score for the family in *Figure 11.4C* is difficult. To calculate the likelihood that III$_7$ and III$_8$ are recombinant or nonrecombinant, we must take likelihoods calculated for each possible genotype of I$_1$, I$_2$ and II$_3$, weighted by the probability of that genotype. For I$_1$ and I$_2$, the genotype probabilities depend on both the gene frequencies and the observed genotypes of II$_1$, III$_7$ and III$_8$. Genotype probabilities for II$_3$ are then calculated by simple mendelian rules. Human linkage analysis, except in the very simplest cases, is entirely dependent

on computer programs that implement algorithms for handling these branching trees of genotype probabilities, given the pedigree data and a table of gene frequencies.

11.3.3 Lod scores of +3 and −2 are the criteria for linkage and exclusion (for a single test)

The result of linkage analysis is a table of lod scores at various recombination fractions, like the two tables in *Box 11.3*. Positive lods give evidence in favor of linkage and negative lods give evidence against linkage. Note that only recombination fractions between 0 and 0.5 are meaningful, and that all lod scores are zero at $\theta = 0.5$ (because they are then measuring the ratio of two identical probabilities, and $\log_{10}(1) = 0$). The results can be plotted to give curves like those in *Figure 11.5*.

Returning to the two questions posed at the start of this section, we now see that the most likely recombination fraction is the one at which the lod score is highest. If there are no recombinants, the lod score will be maximum at $\theta = 0$. If there are recombinants, Z will peak at the most likely recombination fraction ($0.167 = 1/6$ for the family in *Figure 11.4A*, but harder to predict for *Figure 11.4B*).

Box 11.3

Calculation of lod scores for the families in *Figure 11.4*

■ Given that the loci are truly linked, with recombination fraction θ, the likelihood of a meiosis being non-recombinant is $1 - \theta$ and the likelihood of it being recombinant is θ.

■ If the loci are in fact unlinked, the likelihood of a meiosis being either recombinant or nonrecombinant is 1/2.

Family A

There are five recombinants and one nonrecombinant.
The overall likelihood, given linkage, is $(1 - \theta)^5.\theta$
The likelihood given no linkage is $(1/2)^6$
The likelihood ratio is $(1 - \theta)^5.\theta / (1/2)^6$
The lod score, Z, is the logarithm of the likelihood ratio.

θ	0	0.1	0.2	0.3	0.4	0.5
Z	$-$ infinity	0.577	0.623	0.509	0.299	0

Family B

II$_1$ is phase-unknown.
If she inherited A_1 with the disease, there are five non-recombinants and one recombinant.
If she inherited A_2 with the disease, there are five recombinants and one nonrecombinant.
The overall likelihood is $^1\!/_2\,[(1 - \theta)^5.\theta / (1/2)^6] + ^1\!/_2\,[(1 - \theta).\theta^5 / (1/2)^6]$. This allows for either possible phase, with equal prior probability.
The lod score, Z, is the logarithm of the likelihood ratio.

θ	0	0.1	0.2	0.3	0.4	0.5
Z	$-$ infinity	0.276	0.323	0.222	0.076	0

Family C

At this point nonmasochists turn to the computer.

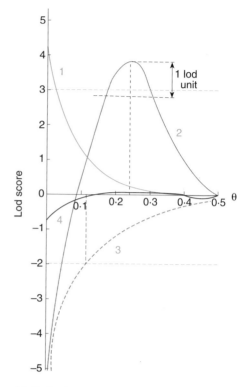

Figure 11.5: Lod score curves.

Graphs of lod score against recombination fraction from a hypothetical set of linkage experiments. Curve 1: evidence of linkage ($Z > 3$) with no recombinants. Curve 2: evidence of linkage ($Z > 3$) with the most likely recombination fraction being 0.23. Curve 3: linkage excluded ($Z < -2$) for recombination fractions below 0.12; inconclusive for larger recombination fractions. Curve 4: inconclusive at all recombination fractions.

The second question concerned the threshold of significance. Here the answer is at first sight surprising. $Z = 3.0$ is the threshold for accepting linkage, with a 5% chance of error. Linkage can be rejected if $Z < -2.0$. Values of Z between -2 and $+3$ are inconclusive. For most statistics $p < 0.05$ is used as the threshold of significance, but $Z = 3.0$ corresponds to 1000 : 1 odds ($\log_{10}(1000) = 3.0$). The reason why such a stringent threshold is chosen lies in the inherent improbability that two loci, chosen at random, should be linked. With 22 pairs of autosomes to choose from, it is not likely they would be located on the same chromosome (syntenic) and, even if they were, loci well separated on a chromosome are unlinked. Common sense tells us that if something is inherently improbable, we require strong evidence to convince us that it is true. This common sense can be quantified in a Bayesian calculation (see *Box 11.4*), which shows that 1000 : 1 odds in fact corresponds precisely to the conventional $p = 0.05$ threshold of significance. The same logic suggests a threshold lod of 2.3 for establishing linkage between an X-linked character and an X-chromosome marker (prior probability of linkage $\cong 1/10$).

Confidence intervals are hard to deduce analytically, but a widely accepted support interval extends to recombination fractions at which the lod score is 1 unit below the peak value (the lod-1 rule). Thus, curve 2 in *Figure 11.5* gives acceptable evidence of linkage ($Z > 3$) with the most likely recombination fraction 0.23 and support interval 0.17–0.32. The curve will be more sharply peaked the greater the amount of data, but in general peaks are quite broad. It is important to remember that distances on human genetic maps are often very imprecise estimates.

Negative lod scores exclude linkage for the region where $Z < -2$. Curve 3 on *Figure 11.5* excludes the disease from 12 cM either side of the marker. While gene mappers hope for a positive lod score, exclusions are not without value. They tell us where the disease is not (**exclusion mapping**). This can exclude a possible candidate gene, and if enough of the genome is excluded, only a few possible locations may remain.

Box 11.4

Bayesian calculation of linkage threshold

The likelihood that two loci should be linked (the prior probability of linkage) has been argued over, but estimates of about one in 50 are widely accepted.

Hypothesis	Loci are linked (recombination fraction = θ)	Loci are not linked (recombination fraction = 0.5)
Prior probability	1/50	49/50
Conditional probability: 1000 : 1 odds of linkage (lod score $Z(\theta) = 3.0$)	1000	1
Joint probability (prior × conditional)	20	~1

Because of the low prior probability that two randomly chosen loci should be linked, evidence giving 1000 : 1 odds in favor of linkage is required in order to give overall 20 : 1 odds in favor of linkage. This corresponds to the conventional $p = 0.05$ threshold of statistical significance. The calculation is an example of the use of Bayes' formula to combine probabilities (see *Box 17.1* and *Figure 17.14*). See text for description of the lod score.

11.3.4 For whole genome searches a genome-wide threshold of significance must be used

In disease studies, families are typed for marker after marker until positive lods are obtained. The appropriate threshold for significance is a lod score such that there is only a 0.05 chance of a false positive result occurring *anywhere during a search of the whole genome*. As shown in *Box 11.4*, a lod score of 3.0 corresponds to a significance of 0.05 at a single point. But if 50 markers have been used, the chance of a spurious positive result is greater than if only one marker is used. A stringent procedure would multiply the p value by 50 before testing its significance. The threshold lod score for a study using n markers would be $3 + \log(n)$, that is a lod score of 4 for 10 markers, 5 for 100, etc. However, this is over-stringent. Linkage data are not independent. If a character is mendelian, then it is determined at a single chromosomal location. If it does not map to one location, then the prior probability that it maps to another location is raised. The threshold for a genome-wide significance level of 0.05 has been much argued over, but a widely accepted answer for mendelian characters is 3.3 (Lander and Schork, 1994). For non-mendelian characters see Section 12.5. In practice, lod scores below 5, whether with one marker or many, should be regarded as provisional.

11.4 Multipoint mapping is more efficient than two-point mapping

11.4.1 Multipoint linkage can locate a disease locus on a framework of markers

Linkage analysis can be more efficient if data for more than two loci are analyzed simultaneously. Multilocus analysis is particularly useful for establishing the chromosomal order of a set of linked loci. Experimental geneticists have long used three-point crosses for this purpose. The rarest recombinant class is that which requires a double recombination. In *Table 11.1*, the gene order A–C–B is immediately apparent. This procedure is more efficient than estimating the recombination fractions for intervals A–B, A–C and B–C separately in a series of two-point crosses. Ideally, in any linkage analysis the whole genome would be screened for linkage, and the full dataset would be used to calculate the likelihood at each location across the genome.

A second advantage of multilocus mapping in humans is that it helps overcome problems caused by the limited informativeness of markers. Some meioses in a family might be informative with marker A, and others uninformative for A but informative with the nearby marker B. Only simultaneous linkage analysis of the disease with markers A and B extracts the full information. This is less important for mapping using highly informative microsatellite markers rather than two-allele RFLPs, but it will resurface if SNPs (*Box 11.1*) become the main mapping tool.

Table 11.1 Gene ordering by three-point crosses

Class of offspring	Position of recombination (×)	Number
ABC/abc abc/abc	Nonrecombinant	853
ABc/abc abC/abc	(A, B)–×–C	5
Abc/abc aBC/abc	A–×–(B, C)	47
AbC/abc aBc/abc	B–×–(A, C)	95

A cross has been set up between mice heterozygous at three linked loci (ABC/abc) and triple homozygotes abc/abc. The offspring are classified as shown. The rarest class of offspring will be those whose production requires two crossovers. Of the 1000 animals, 142 (95 + 47) are recombinant between A and B, 52 (47 + 5) between A and C, and 100 (95 + 5) between B and C. Only five animals are recombinant between A and C, but not between A and B, so these must have double crossovers, A–×–C–×–B. Therefore the map order is A–C–B. The genetic distances are approximately A–(5 cM)–C–(10 cM)–B.

11.4.2 Multipoint mapping by computer

For disease–marker mapping the starting point is usually a two-point lod score showing that the disease maps near one particular marker, plus a map of the framework of markers. The marker map is taken as given, and the aim is to locate the disease gene in one of the intervals of the framework. Programs such as LINKMAP (part of the Linkage package) or GENEHUNTER can notch the disease locus across the marker framework, calculating the overall likelihood of the pedigree data at each position. The result (*Figure 11.6*) is a curve of likelihood against map location. The y-axis is usually a lod score, the log likelihood ratio for this location versus a location off the end of the map. Occasionally, for reasons based on statistical theory, a location score is used. Location scores are twice the natural logarithm of the likelihood ratio, i.e. 4.6 × the lod score. This method is also useful for exclusion mapping: if the curve stays below a lod score of −2 across the region, then the disease locus is excluded from that region.

The apparently quantitative nature of *Figure 11.6* is largely spurious. Peak heights depend crucially on the precise distances between markers and on the mapping function (Section 11.1.3). In reality these are seldom accurately known. The distances on marker–marker maps should be regarded as only rough guides, and moreover none of the mapping functions in linkage programs even approximates to the real complexities of chiasma distribution (see *Figure 11.3*). However, unless the marker map is radically wrong, it remains true that the highest peak marks the most likely location.

11.4.3 Multipoint linkage is essential for constructing marker framework maps

Disease–marker mapping suffers from the necessity of using whatever families can be found where the disease of interest is segregating. Such families will rarely have ideal structures. All too often the number of meioses is undesirably small, and missing persons mean that some meioses are phase-unknown. Marker–marker mapping can avoid these problems. Markers can be studied in any family, so families can be chosen that have plenty of children and ideal structures for linkage, like the family in *Figure 11.1*. Construction of marker framework maps has benefited greatly from a collection of families (the CEPH families) assembled specifically for the purpose by the Centre pour l'Étude du Polymorphisme Humain (now the Centre Jean Dausset) in Paris. Immortalized cell lines from every individual ensure a permanent supply of DNA, and sample mix-ups and non-paternity have long since been ruled out by typing with many markers. The first goal of the Human Genome Project was to produce high-density framework maps of highly polymorphic markers. This phase is now complete. As an example, the current map from CHLC (Cooperative Human Linkage Center) is based on the results of scoring eight CEPH families with 8325 microsatellites, resulting in over 1 million genotypes (Broman *et al.*, 1998).

11.4.4 Integrated maps combine genetic and physical data

Ordering the loci in multipoint mapping is not a trivial problem. There are $n!/2$ possible orders for n markers, and

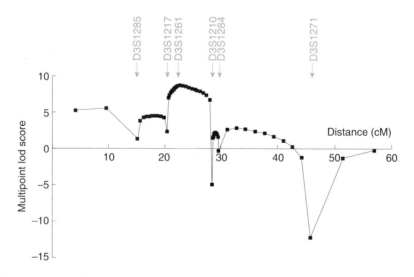

Figure 11.6: Multipoint mapping in man.

The horizontal axis is a map of markers and the vertical axis is the lod score. The LINKMAP program has moved the unmapped disease locus across the map, calculating the lod score at each position. Lod scores dip to strongly negative values near to the position of markers which show recombinants with the disease. The highest peak shows the most likely location. Odds in favor of this position are measured by the degree to which the highest peak overtops its rivals. Redrawn from Hughes *et al.* (1994) with permission from the author.

current maps have hundreds of markers per chromosome. Something more intelligent than brute force computing must be used to work out the correct order. Physical mapping information can be immensely helpful here. Markers that can be typed by PCR can be used as sequence-tagged sites (STS) and grouped into physically localized sets using radiation hybrids or YAC clones. Within a set, the number of possible orders should be small enough to test against the multipoint mapping data. As the genome is increasingly covered with clone contigs, physical distance data are becoming available for more markers. The overall goal of mapping is an integrated map, that lists features in order of chromosomal location and gives their distances on both genetic (preferably separate male and female cM) and physical scales, and relates all this to the chromosomal bands. The Location Database (Collins *et al.*, 1996) contains such integrated maps, and the latest versions can be consulted at http://cedar.genetics.soton.ac.uk/public_html/.

11.5 Standard lod score analysis is not without problems

Standard lod score analysis is a tremendously powerful method for scanning the genome in 20-Mb segments to

locate a disease gene, but it can run into difficulties. These include:

- vulnerability to errors;
- computational limits on what pedigrees can be analyzed;
- problems with locus heterogeneity;
- limits on the ultimate resolution achievable;
- the need to specify a precise genetic model, detailing the mode of inheritance, gene frequencies and penetrance of each genotype.

11.5.1 Errors in genotyping and misdiagnoses can generate spurious recombinants

With highly polymorphic markers, common errors such as misread gels, switched samples or nonpaternity will usually result in a child being given a genotype incompatible with the parents. The linkage analysis program will stall until such errors have been corrected. Errors that introduce possible but wrong genotypes are more of a problem. These include misdiagnosis of somebody's disease status. Such errors inflate the length of genetic maps by introducing spurious recombinants, because if a child has been assigned the wrong parental allele, it will appear to be a recombinant. Multilocus analysis can help, because

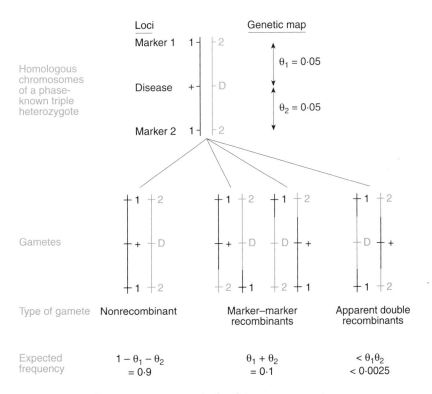

Figure 11.7: Apparent double recombinants suggest errors in the data.

Because of interference (Section 11.1.3), the probability of a true double recombinant with markers 5 cM apart is small, well below 0.05 × 0.05 = 0.0025. Apparent double recombinants usually signal an error in typing the markers, a clinical misdiagnosis, or locus heterogeneity such that the disease in this case does not map to locus D but elsewhere in the genome. Mutation in one of the genes or germinal mosaicism are rarer causes.

spurious recombinants appear as close double recombinants (*Figure 11.7*). Error-checking routines test the extent to which the map can be shortened by omitting any single test result (see Broman *et al.*, 1998). Results that significantly lengthen the map (i.e. add recombinants) are suspect.

11.5.2 Computational difficulties limit the pedigrees that can be analyzed

As we saw in Section 11.3.2, human linkage analysis depends on computer programs that implement algorithms for handling branching trees of genotype probabilities, given the pedigree data and gene frequencies. LIPED was the first generally useful program, and MLINK (part of a package called LINKAGE) used the same basic algorithm, the Elston–Stewart algorithm, but extended it to multipoint data. The Elston–Stewart algorithm can handle arbitrarily large pedigrees, but the computing time increases exponentially with increasing numbers of possible haplotypes (more alleles and/or more loci). This limits the ability of MLINK to analyse multipoint data. An alternative algorithm, the Lander–Green algorithm, can cope with any number of genotypes but the computing time increases exponentially with the size of the pedigree. This algorithm is implemented in the GENEHUNTER program (see Section 12.2.4), which is particularly good for analysing whole-genome searches of modest sized pedigrees. The general theory of linkage analysis is excellently covered in the book by Ott (Further reading), while the book by Terwilliger and Ott (Further reading) is full of practical advice indispensable to anybody undertaking human linkage analysis.

11.5.3 Locus heterogeneity is always a pitfall in human gene mapping

As we saw in Section 3.1.4, it is common for mutations in several unlinked genes to produce the same clinical phenotype. Even a dominant condition with large families can be hard to map if there is locus heterogeneity within the collection of families studied. It took years of collaborative work to show that tuberous sclerosis was caused by mutations at either of two loci, *TSC1* (MIM 191100) at 9q34 and *TSC2* (MIM 191092) at 16p13. With recessive conditions, the difficulty is multiplied by the need to combine many small families. Autozygosity mapping (Section 11.5.5) is the main solution in such cases.

GENEHUNTER or HOMOG and related programs (see Terwilliger and Ott, 1994) can compare the likelihood of the data on the alternative assumptions of locus homogeneity (all families map to the location under test) and heterogeneity (a proportion α of unlinked families), and give a maximum likelihood estimate of α.

11.5.4 The limited resolution of human genetic mapping may be overcome by typing single sperm or by using linkage disequilibrium

Once a marker is found for which all meioses are informative and nonrecombinant, linkage analysis comes to a halt. In typical collections of disease families, the target region thus identified is likely to be 1 Mb or more. This is uncomfortably large for positional cloning of an unknown disease gene. One possible way to increase the resolution of marker–marker mapping is to type sperm instead of children. Humans have far too few children for optimal linkage analysis, but men produce untold millions of sperm, and modern PCR technology allows markers to be scored on single separated sperm from a doubly heterozygous man. Yu *et al.* (1996) show examples. Apart from technical problems, one drawback is that a single sperm cannot be resampled repeatedly to confirm interesting results, in the same way as a child can. Whole genome amplification (Zhang *et al.*, 1992) partially circumvents this problem. Individual spermatozoa are subjected to whole genome amplification followed by multiplex PCR amplification of markers from an aliquot. Further aliquots can be used to check any recombinants. Unfortunately sperm typing could not be used for disease–marker mapping, unless the disease mutations were already characterized.

Linkage disequilibrium provides the best hope of narrowing down the candidate region in disease–marker mapping. Genotypes or haplotypes for markers spread across the candidate region are examined in a series of unrelated affected patients. If the patients all carry independent mutations, as may very well be the case for a dominant or X-linked disease, this exercise will reveal nothing of interest. However, if a proportion of the disease genes in apparently unrelated patients derive from a common ancestor, as often happens with recessive conditions, it may be possible to find a shared ancestral haplotype that defines a small part of the candidate region. This approach is illustrated in Section 12.4.1.

11.5.5 Autozygosity mapping can map recessive conditions efficiently in extended inbred families

Autozygosity is a term used to mean homozygosity for markers identical by descent, inherited from a recent common ancestor. People with rare recessive diseases in consanguineous families are likely to be autozygous for markers linked to the disease locus. Suppose the parents are second cousins: they would be expected to share 1/32 of all their genes because of their common ancestry, and a child would be autozygous at only 1/64 of all loci. If a child is homozygous for a particular marker allele, this could be because of autozygosity, or it could be because a second copy of the same allele has entered the family independently. The rarer the allele is in the population, the greater the likelihood that homozygosity represents autozygosity. For an infinitely rare allele, a single homozygous affected child born to second cousin parents generates a lod score of $\log_{10}(64) = 1.8$. If there are two other affected sibs who are both also homozygous for the same rare allele, the lod score is 3.0 ($\log_{10}(64 \times 4 \times 4)$); the chance that a sib would have inherited the same pair of parental haplotypes even if they are unrelated to the disease is 1 in 4).

Thus quite small inbred families can generate significant lod scores, and autozygosity mapping becomes a powerful tool for linkage analysis if families can be found with multiple affected people in two or more sibships, linked by inbreeding. Suitable families may be found in Middle Eastern countries where inbreeding is common. The method has been applied with great success to locating genes for autosomal recessive hearing loss, which otherwise presents intractable problems because of extensive locus heterogeneity (Guilford *et al.*, 1994). An example is shown in *Figure 11.8*.

The same principle can be extended to populations where the common ancestry is inferred rather than demonstrated. A bold application of this principle enabled Houwen *et al.* (1994) to map the rare recessive condition, benign recurrent intrahepatic cholestasis, using only four affected individuals (two sibs and two supposedly unrelated people) from an isolated Dutch village. The more remote the shared ancestor, the smaller is the proportion of the genome that is shared by virtue of that common ancestry, and therefore the greater the significance for linkage if autozygosity can be demonstrated. But at the same time, the remoter the common ancestor, the more chances there are for a second independent allele to enter the family from outside, and so the less likely is it that homozygosity represents autozygosity, either for the disease or for the markers. With remote common ancestry, as in the study of Houwen *et al.*, everything depends on finding people with a very rare recessive condition who are homozygous for a very rare marker allele or (more likely) haplotype. The power of Houwen's study seems almost miraculous, but it is important to remember that this methodology applies only to diseases and populations where most affected people are descended from a common ancestor who was a carrier. The wider use of allelic association is described in the next chapter (Sections 12.3 and 12.4).

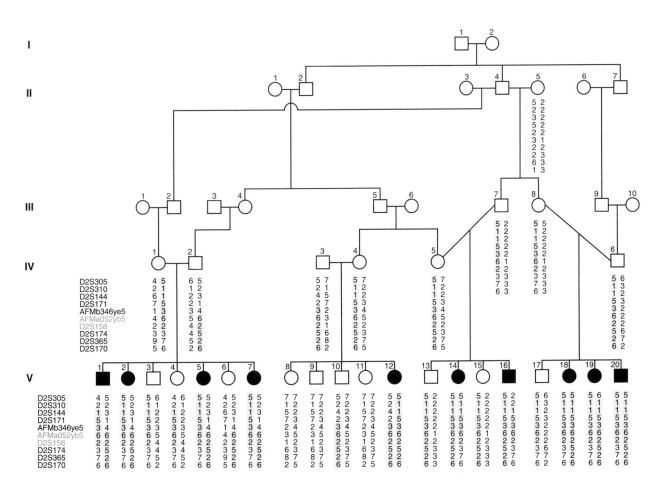

Figure 11.8 Autozygosity mapping.

A large multiply inbred family in which several members suffer from profound congenital deafness (filled symbols). A whole genome screen with 160 polymorphic microsatellite markers showed that all affected family members were homozygous for markers D2S2144 and D2S158, thus mapping the *DFNB9* locus to the 2 cM region between the flanking markers, D2S2303 and D2S174. Redrawn from Chaib *et al.* (1996) *Hum. Mol. Genet.* **5**, 155–158, with permission from Oxford University Press.

11.5.6 Characters whose inheritance is not mendelian are not suitable for mapping by the methods described in this chapter

The methods of lod score analysis described in this chapter require a precise genetic model that specifies the mode of inheritance, gene frequencies and penetrance of each genotype. For mendelian characters, penetrance is the main problem area. If no allowance is made for unaffected people being nonpenetrant gene carriers, or affected people being phenocopies, then these people may be wrongly scored as recombinant. On the other hand, if the penetrance is set too low there is a reduction in the power to detect linkage, because a less precise hypothesis is being tested. Errors in the order of markers on marker framework maps can cause problems, but these are diminishing as genetic maps are cross-checked against physical mapping data. Given sufficient meioses, the main obstacle in linkage analysis of mendelian characters is locus heterogeneity. However, for common complex diseases like diabetes or schizophrenia, the problems are far more intractable. Any genetic model is no more than a hypothesis – we have no real idea of the gene frequencies or penetrance of any susceptibility alleles, or even the mode of inheritance. This makes it near-impossible to apply the methods we have described in this chapter to such diseases. Nevertheless, identifying the genetic components of susceptibility to complex diseases is now a major part of human genetics research. The ways one can attempt to do this are the subject of the next chapter.

Further reading

Ott J (1991) *Analysis of Human Genetic Linkage,* revised edn. Johns Hopkins University Press, Baltimore, MD.

References

Broman KW, Weber JL (1998) Characterization of human crossover interference. *Am. J. Hum. Genet.,* **63** (suppl.), A1632.

Broman KW, Murray JC, Sheffield VC, White RL, Weber JL (1998) Comprehensive human genetic maps: individual and sex-specific variation in recombination. *Am. J. Hum. Genet.,* **63**, 861–869.

Chaib H, Place C, Salem N et al. (1996) A gene responsible for a sensorineural nonsyndromic recessive deafness maps to chromosome 2p22–23. *Hum. Molec. Genet.,* **5**, 155–158.

Collins A, Frezal J, Teague J, Morton NE (1996) A metric map of humans: 23,500 loci in 850 bands. *Proc. Natl Acad. Sci. USA,* **93**, 14771–14775.

Guilford P, Ben Arab S, Blanchard S, Levilliers J, Weissenbach J, Belkahia A, Petit C (1994) A non-syndromic form of neurosensory, recessive deafness maps to the pericentromeric region of chromosome 13q. *Nature Genet.,* **6**, 24–28.

Houwen RHJ, Baharloo S, Blankenship K, Raeymaekers P, Juyn J, Sandkuijl LA, Freimer NB (1994) Genome screening by searching for shared segments: mapping a gene for benign recurrent intrahepatic cholestasis. *Nature Genet.,* **8**, 380–386.

Hughes A, Newton VE, Liu XZ, Read AP (1994) A gene for Waardenburg syndrome Type 2 maps close to the human homologue of the microphthalmia gene at chromosome 3p12-p14.1. *Nature Genet.,* **7**, 509–512.

Lander ES, Schork NJ (1994) Genetic dissection of complex traits. *Science,* **265**, 2037–2048.

Mohrenweiser HW, Tsujimoto S, Gordon L, Olsen AS (1998) Regions of sex-specific hypo- and hyper-recombination identified through integration of 180 genetic markers into the metric physical map of human chromosome 19. *Genomics,* **47**, 153–162.

Morton NE (1955) Sequential tests for the detection of linkage. *Am. J. Hum. Genet.,* **7**, 277–318.

Morton NE, Lindsten J, Iselius L, Yee S (1982) Data and theory for a revised chiasma map of man. *Hum. Genet.,* **62**, 266–270.

Terwilliger J, Ott J (1994) *Handbook for Human Genetic Linkage.* Johns Hopkins University Press, Baltimore, MD.

Wang DG, Fan JB, Siao CJ et al. (1998) Large-scale identification, mapping and genotyping of single nucleotide polymorphisms in the human genome. *Science,* **280**, 1077–1082.

Yu J, Lazzeroni L, Qin J, Huang M-M, Navidi W, Ehrlich H, Arnheim N (1996) Individual variation in recombination among human males. *Am. J. Hum. Genet.,* **59**, 1186–1192.

Zhang L, Cui X, Schmitt K, Hubert R, Navidi W, Arnheim N (1992) Whole genome amplification from a single cell: implications for genetic analysis. *Proc. Natl Acad. Sci. USA,* **89**, 5847–5851.

Genetic mapping of complex characters

12

As we saw at the end of Chapter 3, now that most mendelian diseases have been mapped and most genes at least partially cloned, many researchers see unraveling the genetic determinants of nonmendelian diseases as the next frontier in human genetics. For healthcare, it is certainly an important task. The main genetic contribution to morbidity and mortality in the developed world is through the genetic component of common diseases. Identifying the genes involved may suggest new means of prevention or treatment. Pharmaceutical companies, too, have shifted much of their research into genomics in the belief that this represents the best way to identify new drug targets. However, there are problems in tackling complex diseases with the methods described in Chapter 11 that served so well for mapping mendelian characters. This chapter discusses the approaches that might be used for mapping nonmendelian characters, and the second half of Chapter 19 describes how well they have worked when tried.

12.1 Parametric linkage analysis and complex diseases

12.1.1 Standard lod score analysis is usually inappropriate for nonmendelian characters

Standard lod score analysis is called parametric because it requires a precise genetic model, detailing the mode of inheritance, gene frequencies and penetrance of each genotype. As long as a valid model is available, parametric linkage provides a wonderfully powerful method for scanning the genome in 20-Mb segments to locate a disease gene. For mendelian characters, specifying an adequate model should be no great problem. Nonmendelian conditions, however, are much less tractable.

A major problem is establishing diagnostic criteria. With mendelian syndromes it is usually fairly obvious which features of a patient form part of the syndrome and which are coincidental. Different features may have

different penetrances, but basically the components of the syndrome are those that cosegregate. No such check exists for nonmendelian conditions. Great efforts are made, especially with psychiatric diseases, to establish diagnostic categories that are valid, in the sense that two independent psychiatrists will agree whether or not a certain label applies to a given patient. But a diagnostic label can be valid without being biologically meaningful. Any mendelian pattern *must* be biologically meaningful. Without a mendelian pattern, sometimes physiology will provide an alternative reality check but, especially for psychiatric and behavioral phenotypes, the diagnostic criteria are often biologically arbitrary. Adhering to them helps make different studies comparable, but does not guarantee that the right genetic question is being asked.

Once diagnostic criteria are agreed, segregation analysis (Section 19.4) can identify the most likely mode of inheritance, gene frequencies and penetrances. However, these estimates are averages over a probably heterogeneous set of families, and over all the loci within a family, and they are rarely much use for gene mapping. In the face of all these difficulties, there are several possible ways to proceed:

■ Seek families in which the disease segregates in a near-mendelian manner.
■ Use affected pedigree members only in a parametric analysis.
■ Use a nonparametric (model-free) method of linkage analysis.

12.1.2 Near-mendelian families can be selected for parametric linkage analysis – but the results may be misleading

Both breast cancer and schizophrenia are, in most cases, nonmendelian, but rare families can be found with many affected people in a pattern consistent with autosomal dominant inheritance, albeit with reduced penetrance and, for breast cancer, sex limitation. In each case, these families have been used for a genome search using standard lod score analysis. There are two justifications for this strategy. First, the disease may be heterogeneous and include one or more mendelian conditions phenotypically indistinguishable from the nonmendelian majority. Second, the near-mendelian families may represent cases where, by chance, many determinants of the disease are already present in most people, so that the balance is tipped by the mendelian segregation of just one of the normal susceptibility factors. In the first case, identifying the mendelian subset does not necessarily cast any light on the causes of the nonmendelian disease. In the second case, the loci mapped are also susceptibility factors for the common nonmendelian disease.

The breast cancer work led to the identification of the *BRCA1* and *BRCA2* genes, as described in Chapter 19, whereas the first such attempt in schizophrenia produced a lod score of 6 that is now generally agreed to have been spurious. What was the difference? With hindsight, it is clear that whereas a subset of breast cancer patients really do have a mendelian form of the disease, the apparently mendelian schizophrenia families must have been chance aggregations of affected people within one family. Of itself, this should have simply produced negative lod scores across the whole genome in the schizophrenia families. The other problem (apart from bad luck) was multiple testing. Because the diagnostic criteria for schizophrenia are arbitrary, the researchers tried a number of different criteria, and checked which one gave the highest lod score. This is a perfectly valid procedure – any number of variables can be estimated from a given dataset – but each variable adds more degrees of freedom, and the raw p value needs correcting accordingly.

12.1.3 Using affected pedigree members only avoids the need to specify the penetrance

One solution to the problem of having to specify the penetrance in parametric linkage analysis is to use a parametric method but analyze only the affected family members. The penetrance is irrelevant for affected people, and unaffected members are scored as having an unknown disease phenotype. If the penetrance is low, unaffected people provide relatively little information. We can infer the genotype of affected people (they must have the susceptibility allele), but not of unaffected people, therefore not too much is lost by ignoring unaffected family members. This strategy is useful for testing candidate susceptibility loci for oligogenic diseases. It is often sensible to check for linkage before starting to screen a candidate gene for mutations. Since a parametric analysis is used, it is still necessary to specify a genetic model, and so there is still the danger of getting meaningless results if the model is wrong. The risk of false positives is reduced if the analysis is restricted to checking a few candidate loci. It helps if the disease is rare but distinctive, so that the risk of heterogeneity and of phenocopies is minimized.

12.2 Nonparametric linkage analysis does not require a genetic model

If the need to specify a complete genetic model is too daunting, one can use model-free or nonparametric methods of linkage analysis. These methods ignore unaffected people, and look for alleles or chromosomal segments that are shared by affected individuals. Shared segment methods can be used within nuclear families (sib pair analysis, see below), within known extended families, or in whole populations. At the population level they constitute association studies, which are considered in the following section.

12.2.1 Identity by state is not the same as identity by descent

It is important to distinguish segments **identical by descent (IBD)** from those **identical by state (IBS)**. IBS alleles look the same, and may have the same DNA sequence, but they are not derived from a *known* common ancestor. Alleles IBD are demonstrably copies of the same ancestral (usually parental) allele. If two sibs each have allele A_1 (*Figure 12.1*), the shared allele is IBS, but it may or may not be IBD. For very rare alleles, two independent origins are unlikely, so IBS generally implies IBD, but this is not true for common alleles. Multiallele microsatellites are more efficient than two-allele markers for defining IBD, and multilocus multiallele haplotypes are better still, because any one haplotype is likely to be rare. Shared segment analysis can be conducted using either IBS or IBD data, provided the appropriate analysis is used. IBD is the more powerful, but requires parental samples.

12.2.2 Affected sib pairs allow model-free analysis in nuclear families

Picking a chromosomal segment at random, pairs of sibs are expected to share 0, 1 or 2 parental haplotypes with frequency 1/4, 1/2 and 1/4, respectively. However, if both sibs are affected by a genetic disease, then they are

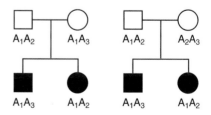

Figure 12.1: Identity by state (IBS) and identity by descent (IBD).

Both sib pairs share allele A_1. The first sib pair have two independent copies of A_1 (IBS but not IBD); the second sib pair share copies of the same paternal A_1 allele (IBD). The difference is only apparent if the parental genotypes are known.

likely to share whichever segment of chromosome carries the disease locus. On the simplest assumption, that everybody with the disease carries a mutant allele at this locus, then if the disease is dominant, they will share at least one parental haplotype, and if the disease is recessive they will share both haplotypes. This allows a simple form of linkage analysis (*Figure 12.2*). Affected sib pairs (ASP) are typed for markers, and chromosomal regions sought where the sharing is above the random $1:2:1$ ratios of sharing 2, 1 or 0 haplotypes identical by descent. If the sib pairs are tested only for identity by state, the expected sharing on the null hypothesis is a function of the gene frequencies. Multipoint analysis is preferable to single-point analysis because it more efficiently extracts the information about IBD sharing across the chromosomal region. The MAPMAKER/SIBS program of Kruglyak and Lander (1995) is widely used to analyze multipoint ASP data and produce nonparametric lod scores.

Because **sib pair analysis** is model-free, it can be performed without making any assumptions about the genetics of the disease. Thus it has been used as one of the main tools for seeking genes conferring susceptibility to common nonmendelian diseases like diabetes or schizophrenia. One drawback is that candidate regions defined by sib pair analysis are usually uncomfortably large for positional cloning. Sib pair analysis has no process analogous to the end-game of mendelian mapping, where closer and closer markers are tested until there are no more recombinants. It is not likely that a chromosomal segment can be defined that is shared by all affected sib pairs. If a susceptibility factor is neither necessary nor sufficient for disease, then not all affected sib pairs will share the chromosomal segment that contains the susceptibility locus. Moreover, sib pairs share many segments by chance, including, perhaps, segments that coincidentally lie close to a susceptibility locus. The mathematics of ASP analysis have been detailed by Sham and Zhao (1998), and examples of some systematic applications of ASP analysis to complex diseases are given in Section 19.5.

12.2.3 Nonparametric affected pedigree member analysis generalizes affected sib pair analysis

The affected pedigree member (APM) method of Weeks and Lange (1992) extends the logic of affected sib pair analysis to other relationships. In a complex pedigree with several affected people, for each pair of affected pedigree members the distribution of alleles identical by state is observed, and compared to the expectation on the null hypothesis of no linkage. APM allows multipoint data to be analysed in large pedigrees; however, because it uses IBS and not IBD data, it does not necessarily use all the linkage information that could in theory be extracted from a pedigree.

12.2.4 The GENEHUNTER program allows nonparametric lod scores to be calculated – but they must be interpreted with care

A more radical approach to nonparametric analysis of complex pedigrees is implemented in the GENEHUNTER program of Kruglyak et al. (1996). This is based on a generalization of the MAPMAKER/SIBS program for analysis of multipoint ASP data mentioned above. The basic algorithm in these programs is able to handle any number of loci (the computing time increases linearly with the number of loci), but is limited to fairly small pedigrees. Pedigrees contain founders (people whose parents are not included in the pedigree) and nonfounders (people whose parents are included). If somebody has a sib in the pedigree, then they must be nonfounders, because the only way to tell the computer that they are sibs is to include the parents. If a pedigree contains *f* founders and *n* nonfounders, the GENEHUNTER computing time increases exponentially with $(2n - f)$. Current versions fail to cope with pedigrees where $2n - f > 16$.

Provided a pedigree falls within the size limit, GENEHUNTER can include any number of loci in a multipoint analysis. It is in fact able to compute parametric lod scores,

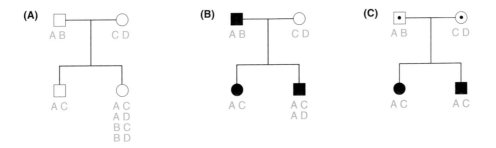

Figure 12.2: Sib pair analysis.

(**A**) By random segregation sib pairs share 0, 1 or 2 parental haplotypes $\frac{1}{4}, \frac{1}{2}$ and $\frac{1}{4}$ of the time, respectively. (**B**) Pairs of sibs who are both affected by a dominant condition share one or two parental haplotypes for the relevant chromosomal segment. (**C**) Pairs of sibs who are both affected by a recessive condition share both parental haplotypes for the relevant chromosomal segment.

if a concrete genetic model is provided. For complex characters where no model can be provided, the result is expressed as a nonparametric lod (NPL) score. These are based on calculating the extent to which affected relatives share alleles identical by descent, and comparing the result across all affected pedigree members with the null hypothesis of simple mendelian segregation (markers will segregate according to mendelian ratios unless the segregation is distorted by linkage or association). This method appears to extract the linkage information from a pedigree more efficiently than the APM method. However, the threshold of significance for a NPL is not so obvious as with the parametric lod score that would be calculated for a single pair of mendelian characters. The significance is best expressed as a genome-wide p value, as discussed in Section 12.5.2.

12.3 Association is in principle quite distinct from linkage, but where the family and the population merge, linkage and association merge

12.3.1 Linkage is a relation between loci, but association is a relation between alleles

In principle, linkage and association are totally different phenomena. Association is simply a statistical statement about the co-occurrence of alleles or phenotypes. Allele A is associated with disease D if people who have D also have A more (or maybe less) often than would be predicted from the individual frequencies of D and A in the population. For example, HLA-DR4 is found in 36% of the general UK population but 78% of people with rheumatoid arthritis. An association can have many possible causes, not all genetic (see below). Linkage, on the other hand, is a specific genetic relationship between loci (not alleles or phenotypes). Linkage does not of itself produce any association in the general population. The STR45 locus is linked to the dystrophin locus. Within a family where a dystrophin mutation is segregating, we would expect affected people to have the same allele of STR45, but over the whole population the distribution of STR45 alleles is just the same in people with and without muscular dystrophy. Thus linkage creates associations within families, but not among unrelated people. However, if two supposedly unrelated people with disease D have actually inherited it from a distant common ancestor, they may well also tend to share particular ancestral alleles at loci closely linked to D. Where the family and the population merge, linkage and association merge.

12.3.2 Population associations depend on population history

All humans are related, if we go back far enough. A simplified calculation suggests that in the UK two 'unrelated'

people would typically share common ancestors not more than 22 generations ago. If fully outbred, they would have $2^{22} = 4$ million ancestors each at that time. Twenty-two generations is about 500 years, and in 1500 the population of Britain was around 4 million (*Figure 12.3*). Therefore, if the UK population interbred freely, not more than 44 meioses would separate our two unrelated people. Suppose the two 'unrelated' people each inherit a disease susceptibility allele from their common ancestor. During the many generations and many meioses that separate them from their common ancestor, repeated recombination will have reduced the shared chromosomal segment to a very

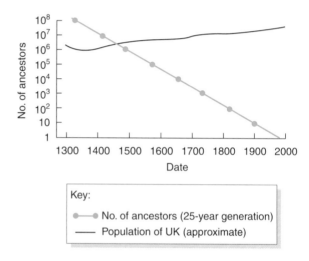

Figure 12.3: Merging into the gene pool.

A fully-outbred person has 2^n ancestors n generations ago. If the UK population were fully outbred, two 'unrelated' present-day people would have shared ancestors in 1500, if not more recently. Reprinted from Read (1989) *Medical Genetics: An Illustrated Outline*, by permission of Mosby.

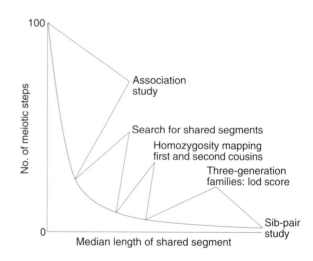

Figure 12.4: Family linkage and population association studies are two ends of a continuum.

small region (*Figure 12.4*). Only alleles at loci tightly linked to the disease susceptibility locus will still be shared.

For a locus showing recombination fraction θ with the susceptibility locus, a proportion θ of ancestral chromosomes will lose the association each generation, and a proportion $(1-\theta)$ will retain it. After n generations, a fraction $(1-\theta)^n$ of chromosomes will retain the association. Considering the 44 meioses that may separate our two patients, loci showing 1% recombination per meiosis would have a better than 50% chance of remaining in the same combination, since $(0.99)^{44} = 0.64$. Loci 3 cM apart would have $(0.97)^{44} = 0.26$ chance of remaining together. This argument is grossly simplified because it ignores population substructure and assumes the entire British population has been one freely interbreeding unit over the past 500 years. However, for what it is worth, it suggests that allelic associations reflecting sharing of ancestral chromosomes in the British population might begin to be noticeable for loci within 2–3 cM of each other.

Although this calculation is crude, it does show how the extent of allelic association depends on the history of the population concerned. There has been considerable debate about the differences to be expected when comparing a population that has expanded rapidly from a relatively recent bottleneck with one that has maintained or only gradually increased its size over many generations. The Finns would typify the former and the British the latter. Diseases, being subject to selection, are likely to be more mutationally homogeneous in a rapidly expanded population; the mutant alleles found will reflect those present in the founder population. For selectively neutral markers the position is less clear. In a study of 20 microsatellites from chromosome 18q21 in 664 British and 430 Finnish subjects, Eaves *et al.* (1998) observed significant disequilibrium for all 53 pairs of loci less than 1 cM apart in both populations, for 20/75 (UK) and 61/75 (Finland) pairs 1–3 cM apart, and for 0/62 (UK) and 20/66 (Finland) pairs more than 3 cM apart. In other words, disequilibrium is present in both populations, but extends over a longer range in Finns.

Searching for population associations is an attractive option for identifying disease susceptibility genes. Association studies are easier to conduct than linkage analysis, because no multicase families or special family structures are needed. Also, because linkage disequilibrium is a short-range phenomenon, if an association is found, it defines a small candidate region in which to search for the susceptibility gene. Finally, recent work suggests that association is more powerful than linkage for detecting weak susceptibility alleles (Section 12.5.4). However, there are several pitfalls to be avoided if a claimed association is to provide a reliable pointer to a nearby susceptibility locus.

12.3.3 Not all population associations are caused by linkage disequilibrium

Linkage disequilibrium is not the only possible reason for

an association between a disease D and allele A. Possible causes include the following:

- *Direct causation* – having allele A makes you susceptible to disease D. Possession of A is neither necessary nor sufficient for somebody to develop D, but it increases the likelihood. In this case one would expect to see the same allele A associated with the disease in any population studied (unless the causes of the disease vary from one population to another).

- *Natural selection* – people who have disease D might be more likely to survive and have children if they also have allele A.

- *Population stratification* – the population contains several genetically distinct subsets. Both the disease and allele A happen to be particularly frequent in one subset. Lander and Schork (1994) give the example of the association in the San Francisco Bay area between *HLA-A1* and ability to eat with chopsticks. *HLA-A1* is more frequent among Chinese than among Caucasians.

- *Statistical artefact* – association studies often test a range of loci, each with several alleles, for association with a disease. The raw p values need correcting for the number of questions asked (Section 12.5.1). In the past, researchers often applied inadequate corrections, and associations were reported that could not be replicated in subsequent studies.

- *Linkage disequilibrium* – close linkage can produce allelic association at the population level, provided that most disease-bearing chromosomes in the population are descended from one or a few ancestral chromosomes. If linkage disequilibrium is the cause of the association, there should be a gene near to the A locus that has mutations in people with disease D. The particular allele at the A locus that is associated with disease D may be different in different populations

Direct causation and selective advantage are unlikely if the associated allele is a variant in the noncoding DNA and not closely associated with any gene, but studies in several ethnically distinct populations are useful to help distinguish these causes of association from linkage disequilibrium. Statistical artefacts are reduced by proper correction of probabilities (see Section 12.5.1).

The choice of the control group in association studies is crucial. Many studies in the past have used published gene frequencies, often without adequate certainty that these frequencies were representative of the population from whom the patients were recruited. Alternatively, students or staff from the investigator's university may be used as a control series. Again, this is undesirable because they may well not be typical of the population from which the patients were drawn. Thus, when an association is found, it may be impossible to know whether it is caused by linkage disequilibrium with a susceptibility locus or by inadequately matched controls.

12.3.4 The transmission disequilibrium test (TDT) overcomes many of the problems of classical disease–marker association studies

Recently, a clutch of methods have been developed that largely circumvent the stratification problem. Collectively they can be called association studies with internal controls. They involve 50% more work than standard case-control studies because three people (proband and parents) are typed in each family. This seems a small price to pay for the gain in reliability. Parents must be available, which restricts the usefulness of these tests for late-onset diseases. One method, the haplotype relative risk (HRR) method, handles the data like typical case-control data, except that the control is not a real person but is made out of the two alleles that the parents did not transmit to their affected offspring.

The most popular method is the **transmission disequilibrium test (TDT**; Schaid, 1998). The TDT starts with couples who have one or more affected offspring. It is irrelevant whether either parent is affected or not. To test whether marker allele M_1 is associated with the disease, we select those parents who are heterozygous for M_1. The test simply compares the number of such parents who transmit M_1 to their affected offspring with the number who transmit their other allele (*Box 12.1*). The result is unaffected by population stratification. The TDT can be used when only one parent is available, but this may bias the result (Schaid, 1998). There has been some argument about whether the TDT is a test of linkage or association. Since it asks questions about alleles and not loci, it is fundamentally a test of association. The associated allele may itself be a susceptibility factor, or it may be in linkage disequilibrium with a susceptibility allele at a nearby locus. The TDT cannot detect linkage if there is no disequilibrium – a point to remember when considering schemes to use the TDT for whole-genome scans.

12.4 Linkage disequilibrium as a mapping tool

12.4.1 Cystic fibrosis and Nijmegen breakage syndrome illustrate the use of linkage disequilibrium to narrow down a candidate region for positional cloning of a mendelian disease locus

Association studies are not restricted to nonmendelian conditions. We have already seen how an association was used to map autosomal recessive familial benign intrahepatic cholestasis (Section 11.5.5). More commonly a population association has been used to narrow down a candidate region that was initially defined by standard parametric linkage analysis. Cystic fibrosis is a disease of northern Europeans, where the sort of large multiply inbred family used for autozygosity mapping (Section 11.5.5) is excessively rare. Thus mapping CF depended on rare unfortunate nuclear families with more than one affected child. Using these, CF was mapped to 7q32, but after all available recombinants had been used, the candidate region was still dauntingly large. The initial markers (*MET* and *D7S8*) showed little or no linkage disequilibrium, but a new set of markers from within the candidate region, XV2.c and KM19 showed strong association between the X_1, K_2 (XV2.c*1, KM19*2) haplotype and CF. Typical data are shown in *Table 12.1*. As more markers were isolated, the gradient of linkage disequilibrium helped indicate the location of the CF gene.

A more recently cloned gene, governing Nijmegen breakage syndrome, shows how ancestral haplotypes can be identified and used to define the exact position of the disease gene. Nijmegen breakage syndrome (NBS; MIM 251260) is a very rare autosomal recessive disease characterized by chromosome breakage, growth retardation, microcephaly, immunodeficiency and a predisposition to cancer. The suspected cause is a defect in DNA repair. Conventional linkage analysis in small nuclear families located the NBS locus to an 8-Mb target region between markers D8S271 and D8S270 on chromosome 8p21. There were no recombinants within the candidate region. Fifty

Table 12.1: Allelic association in cystic fibrosis

Marker alleles	CF chromosomes	Normal chromosomes
X_1, K_1	3	49
X_1, K_2	147	19
X_2, K_1	8	70
X_2, K_2	8	25

Data from typing for the RFLP markers XV2.c (alleles X_1 and X_2) and KM19 (alleles K_1 and K_2) in 114 British families with a cystic fibrosis (CF) child. Chromosomes carrying the CF disease mutation tend also to carry allele X_1 of XV2.c and allele K_2 of KM19. Data derived from Ivinson *et al.* (1989).

one individual apparently unrelated patients were typed for a series of microsatellite markers spaced across the candidate region. Among the haplotypes of the patients, 74 are apparently related to a common ancestral haplotype (*Figure 12.5*). It is particularly associated with Slav ancestry. Some patients do not have this haplotype at all, and presumably carry independent NBS mutations. Others share only part of the haplotype, showing the effect of recombination in distant ancestors. These apparently recombinant haplotypes still share the region

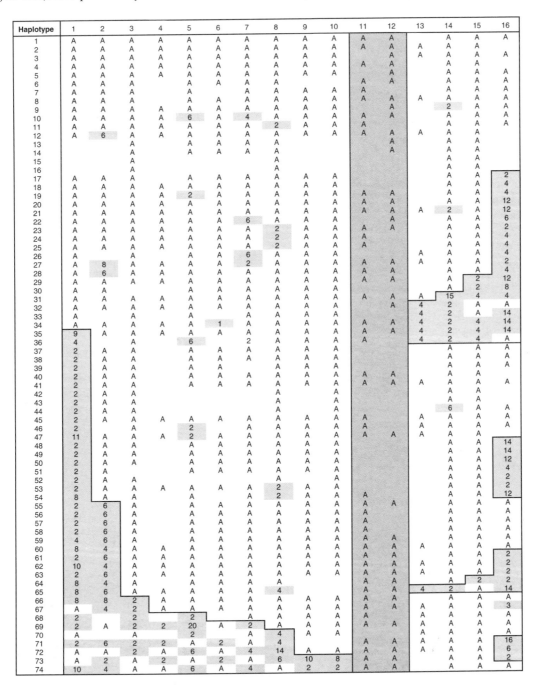

Figure 12.5: An ancestral haplotype in European patients with Nijmegen breakage syndrome.

16 markers from 8p21, shown in chromosomal order across the top of the table, were used to generate 74 NBS-associated haplotypes in unrelated patients. Alleles attributed to an ancestral haplotype are marked A. Other alleles are shaded. Where there are no data, cells are left blank. Non-ancestral alleles are marked with the number of nucleotides by which they differ from the ancestral allele. Differences of two nucleotides are often the result of a mutation of the marker, but larger differences are likely to be the result of ancestral recombinations. All 74 haplotypes have the ancestral alleles at markers 11 and 12, which therefore indicated the likely location of the NBS gene. After Varon et al, 1998.

between polymorphic markers H4CA and H5CA (markers 11 and 12 in *Figure 12.5*), thus defining the likely location of the NBS gene. Subsequently a gene encoding a novel protein was cloned from this location and shown to carry mutations in NBS patients. As predicted, patients with the common haplotype all have the same mutation (Varon *et al.*, 1998).

12.4.2 Linkage disequilibrium can be quantified, but the gradients of disequilibrium around a disease gene can be hard to understand

For positional cloning of a disease where a large number of patients are available, quantitative measures of linkage disequilibrium can be calculated for a series of markers across the target region. Hopefully the disease gene will be located at the peak of disequilibrium. The simplest measures of disequilibrium are affected by the gene frequencies. A better measure is the Yule coefficient (Krawczak and Schmidtke, 1998). For two loci A and B with alleles A_1, A_2, B_1 and B_2, this is

$$(p_{1,1} - p_{1,2}) / (p_{1,1} + p_{1,2} - 2p_{1,1}p_{1,2})$$

where $p_{1,1}$ and $p_{1,2}$ are the frequency of allele A_1 on chromosomes carrying alleles B_1 and B_2, respectively.

This approach was used with Huntington disease. The result was helpful, but not simple to interpret. In *Figure 12.6* we can see cases of a strong association with a more distant marker and a weak association with a closer marker. Even more curious, the marker D4S95, closely linked to the *HD* locus, detects RFLPs with three

enzymes, *Taq*I, *Mbo*I and *Acc*I. Results confirmed in several independent studies show a strong association with a particular *Acc*I and a particular *Mbo*I allele, but no association with either *Taq*I allele. Probably the associations reflect a complex history, with a combination of several independent mutations, recombination in one of a small founder population of disease chromosomes, and maybe an origin of some marker polymorphisms more recently than some disease mutations. Xiong and Guo (1997) give a nice overview of the problems and the literature of linkage disequilibrium mapping, before proposing a sophisticated method of analysis based on maximum likelihood estimation of multipoint data. This is one of several approaches that appear able to predict gene locations much better than the simple analysis in *Figure 12.6*.

12.5 Thresholds of significance are an important consideration in analysis of complex diseases

Whereas most mendelian loci localized by significant lod scores have been successfully cloned, the history of complex disease analysis has been marked by a succession of false dawns and irreproducible results. Innumerable HLA-disease associations have been reported, but few proved reproducible. Positive lod scores in families with schizophrenia turned into a serious embarrassment (reviewed by Byerley, 1989). More recently, there are a number of complex diseases where several different groups have undertaken independent large-scale sib pair

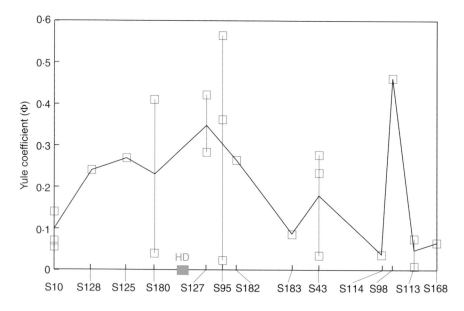

Figure 12.6: Allelic association around the locus for Huntington disease.

S10, S125, etc. are shorthand for the DNA markers D4S10, D4S125, etc., shown in their map positions relative to the *HD* locus. The total distance represented is 2500 kb. For some loci, several different RFLPs exist, which sometimes show very different allelic association, for example marker S95 (see text). Linkage disequilibrium is measured by the Yule coefficient. From Krawczak and Schmidtke (1998) *DNA Fingerprinting*, 2nd edn, BIOS Scientific Publishers.

analyses. The candidate regions defined in the different studies have seldom coincided. Risch and Botstein (1996) outline a typical history, that of manic-depressive psychosis, and similar results with multiple sclerosis are discussed in Section 19.5.5. Whatever the exact cause of these problems in the various cases, a clear common thread is the difficulty of deciding when to call the results of a linkage or association study significant.

12.5.1 Probabilities calculated from association studies must be corrected for the number of questions asked

A mendelian condition must map somewhere and so, in linkage analysis, no matter how many markers are used in finding the location, the risk of a false positive result remains manageably low (Section 11.3.4). This is not the case for association studies. There may well be no association to find, and so each test performed carries an independent risk of a false positive result. To avoid errors, a correction has to be applied. The threshold of significance is set, not at the conventional $p = 0.05$, but at $p = 0.05/n$, where n is the number of independent potential associations checked (*Table 12.2*). This is called the *Bonferroni correction*. All too few published disease association studies apply the rigorous correction factor, $n(m - 1)$ for the testing of n loci with m alleles each, and all too often associations reported in one study cannot be confirmed in a second independent sample of patients.

12.5.2 Genome-wide significance levels for analysis of complex diseases are controversial

The problems of deciding appropriate thresholds of significance are partly technical and partly philosophical. We have already noted the distinction between pointwise (or nominal) and genome-wide significance (Section 11.3.4):

■ The **pointwise** p **value** of a linkage statistic is the probability of exceeding the observed value at a specified position in the genome, assuming the null hypothesis of no linkage.
■ The **genome-wide** p **value** is the probability that the observed value will be exceeded anywhere in the genome, assuming the null hypothesis of no linkage.

For a whole-genome study, the appropriate significance threshold is a value where the probability of finding a false positive *anywhere in the genome* is 0.05. This will be more stringent than the pointwise threshold for a single test. But suppose an association study finds a significant result (pointwise $p < 0.05$) with the very first marker tested. Had the result been negative, the researchers would no doubt have gone on to test marker after marker until either they found something or else they had got negative results across the whole genome. Should they apply the genome-wide threshold, even though they did only one test?

According to Lander and Kruglyak (1995), the genome-wide false-positive rate, $\alpha_T{}^*$ is related to the pointwise false positive rate, α_T by the equation

$$\alpha_T{}^* \simeq [C + 9.2\rho GT]\alpha_T$$

T is the threshold lod score; C = 23, the number of chromosomes, and G = 33, the total length of the genome in Morgans. The parameter ρ measures the crossover rate, and takes different values depending on the relationship being studied, so that the formula cannot be simply applied to complex pedigrees. For affected sib pairs the formula suggests genome-wide lod score thresholds of 3.6 for IBD testing and 4.0 for IBS testing. Note that the formula applies strictly only to large samples and to stringent thresholds.

Because the associations underlying TDT tests operate over much shorter chromosomal distances than the linkage underlying ASP testing, and because TDT, as an association test, must be performed separately for every allele of each locus, the total number of tests needed for a

Table 12.2: p values from a hypothetical association study

	D_1	D_2	D_3	D_4	D_5	D_6	D_7	D_8	D_9	D_{10}
M_1	0.29	0.47	0.80	0.47	0.36	0.13	0.93	0.15	0.08	0.08
M_2	0.21	0.26	0.38	0.55	0.96	0.61	0.46	0.28	0.10	0.40
M_3	0.36	0.87	0.61	0.76	0.80	0.51	0.44	0.11	0.76	0.99
M_4	0.12	0.77	0.20	0.68	0.88	0.47	0.39	0.05	0.50	0.53
M_5	0.09	0.56	0.01	0.93	0.24	0.81	0.18	0.28	0.04	0.18
M_6	0.61	0.83	0.27	0.95	0.66	0.03	0.24	0.05	0.03	0.87
M_7	0.63	0.64	0.12	0.33	0.76	0.09	0.54	0.77	0.42	0.09
M_8	0.24	0.12	0.06	0.65	0.98	0.52	0.91	0.63	0.68	0.23
M_9	0.36	0.03	0.15	0.62	0.68	0.88	0.15	0.96	0.94	0.55
M_{10}	0.27	0.94	0.31	0.32	0.54	0.06	0.20	0.63	0.53	0.38

Panels of patients with diseases (D_1–D_{10}) and a panel of controls were typed for markers (M_1–M_{10}). For each possible association the p value is tabulated. In reality none of the diseases is associated with any of the markers, but five of the 100 p values are significant at the 5% level, including one at the 1% level. This is of course exactly what is expected of a series of 100 random numbers. If n questions are asked, the appropriate threshold of significance is $p = 0.05/n$ (Bonferroni correction).

genome-wide scan by TDT is huge. Risch and Merikangas (1996) considered the ultimate case of testing five diallelic polymorphisms at each of 100 000 gene loci by TDT. Applying a full Bonferroni correction for 1 million independent tests means the threshold significance for a positive result is $p = 5 \times 10^{-8}$.

Most complex disease studies avoid these theoretical approaches by basing the significance threshold on simulation. Typically, 1000 replicates of the family collection are generated by computer with random genotypes, but based on correct allele frequencies, recombination fractions, etc. A whole-genome search is conducted in each simulated dataset and the maximum lod score noted. The genome-wide threshold of significance is taken as a score that is exceeded in less than 5% of replicates.

12.5.3 Criteria for suggestive and significant linkage have been suggested for complex diseases

In response to the frequent failure to replicate claimed localizations of disease susceptibility genes, Lander and Kruglyak (1995) proposed a series of thresholds:

- **Suggestive linkage** is a lod score or *p* value that would be expected to occur once by chance in a whole genome scan.
- **Significant linkage** is a lod score or *p* value that would be expected to occur by chance 0.05 times in a whole genome scan (i.e. the conventional $p = 0.05$ threshold of significance)
- **Highly suggestive linkage** is a lod score or *p* value that would be expected to occur by chance 0.001 times in a whole genome scan.
- **Confirmed linkage** – linkage is to be regarded as confirmed when a significant linkage observed in one study is confirmed by finding a lod score or *p* value that would be expected to occur 0.01 times by chance in a specific search of the candidate region.

The pointwise *p* values for significant linkage work out at $1 - 5 \times 10^{-5}$ for different genome-wide study designs. Note that these values do *not* imply threshold lod scores of 4.3–5.0. A lod score of 5 means that the data are 10^5 times more likely on the given linkage hypothesis than on the null hypothesis; a *p* value of 10^{-5} means that the stated lod score will be exceeded only once in 10^5 times, given the null hypothesis. The two measures are not the same. The lod scores for genome-wide significant linkage are in the range 3.3–4.0, again depending on the study design. For some discussion of the Lander and Kruglyak criteria, see the correspondence section of the April 1996 issue of *Nature Genetics*.

12.5.4 For detecting alleles of modest effect, association tests are likely to be more powerful than linkage tests

An important paper by Risch and Merikangas (1996) compared the power of affected sib pair and TDT testing

to detect associations between a marker at a susceptibility locus and a complex disease. They calculated the number of ASPs or TDT trios required to obtain a given power and significance level in order to distinguish a genetic effect from the null hypothesis. *Box 12.2* illustrates their method (consult the original paper for more detail). *Table 12.3* shows typical results of applying their formulae.

Box 12.2

Sample sizes needed to find a disease susceptibility locus by a whole genome scan using either affected sib pairs (ASP) or the transmission disequilibrium test (TDT)

Risch and Merikangas (1996) calculated the sample sizes needed to distinguish a genetic effect from the null hypothesis with power $(1 - \beta)$ and significance level α. This Box summarizes their formulae and equations, but the original paper should be consulted for the derivations and for details.

A standard piece of statistics tells us that the sample size *M* required is given by

$$(Z_\alpha - \sigma Z_{1-\beta})^2/\mu^2,$$

where *Z* refers to the standard normal deviate. The mean μ and variance σ^2 are calculated as functions of the susceptibility allele frequency (*p*) and the relative risk γ conferred by one copy of the susceptibility allele. The model assumes that the relative risk for a person carrying two susceptibility alleles is γ^2; that the marker used is always informative; and that there is no recombination with the susceptibility locus.

For ASP, the expected allele sharing at the susceptibility locus is given by

$$Y = (1 + w)/(2 + w),$$

where $w = [pq(\gamma - 1)^2]/(p\gamma + q)$. $\mu = 2Y - 1$ and $\sigma^2 = 4Y(1 - Y)$. The genome-wide threshold of significance (probability of a false positive anywhere in the genome = 0.05; testing for sharing IBD) requires a lod score of 3.6, corresponding to $\alpha = 3 \times 10^{-5}$, and $Z_\alpha = 4.014$. For 80% power to detect an effect, $1 - \beta = 0.2$ and $Z_{1-\beta} = -0.84$.

For the TDT, the probability that a parent will be heterozygous for the allele in question is

$$h = pq(\gamma + 1)/(p\gamma + q)$$

The probability that such a heterozygous parent will transmit the high-risk allele to the affected child is

$$P(trA) = \gamma/(1 + \gamma)$$

$\mu = \sqrt{h}(\gamma - 1)/(\gamma + 1)$, and $\sigma^2 = 1 - [h(\gamma - 1)^2/(\gamma + 1)^2]$. As discussed above, for an ultimate genome screen involving 1 000 000 tests, $\alpha = 5 \times 10^{-8}$, $Z_\alpha = 5.33$ and, as before, $Z_{1-\beta} = -0.84$.

In *Table 12.3* the Z_α, $Z_{1-\beta}$, μ and σ^2 values are used to calculate sample sizes by substituting in the formula

$$M = (Z_\alpha - \sigma Z_{1-\beta})^2/\mu^2$$

For the TDT, the answer is halved because each parent–child trio allows two tests, one on each parent.

Table 12.3: Sample sizes for 80% power to detect significant linkage or association in a genome-wide search.

γ	p	ASP analysis		TDT analysis	
		Y	N-ASP	$P(trA)$	N-TDT
5	0.01	0.534	2530	0.830	747
	0.1	0.634	161	0.830	108
	0.5	0.591	355	0.830	83
3	0.01	0.509	33797	0.750	1960
	0.1	0.556	953	0.750	251
	0.5	0.556	953	0.750	150
2	0.1	0.518	9167	0.667	696
	0.5	0.526	4254	0.667	340
1.5	0.1	0.505	115537	0.600	2219
	0.5	0.510	30660	0.600	950
1.2	0.1	0.501	3951997	0.545	11868
	0.5	0.502	696099	0.545	4606

γ is the relative risk for individuals of genotype Aa compared to aa; p is the frequency of the A susceptibility allele. For affected sib pair (ASP) analysis, Y is the expected allele sharing and N-ASP the number of pairs required for significance, based on IBD testing ($\alpha = 3 \times 10^{-5}$). For transmission disequilibrium testing (TDT), $P(trA)$ is the probability that an Aa parent will transmit A to an affected child, and N-TDT is the number of parent–child trios required for significance. After Risch and Merikangas (1996).

Their conclusion is clear: ASP analysis would require unfeasibly large samples to detect susceptibility loci conferring a relative risk of less than about 3, whereas TDT might detect loci giving a relative risk below 2 with manageable sample sizes. Susceptibility genes conferring a relative risk below 1.5 would be very hard to find by either method. Note, however, that their result is obtained with one particular genetic model, and might not apply to all. In particular, linkage disequilibrium is not necessarily present between alleles at tightly-linked loci.

12.6 Strategies for complex disease mapping usually involve a combination of linkage and association techniques

In many ways linkage and association provide complementary data. Linkage operates over a long chromosomal range. Linkage analysis, whether parametric or nonparametric, can scan the entire genome in a few hundred tests. A typical study of 250 sib pairs with 300 markers would require 1.5–3×10^5 genotypes to be generated (depending whether or not the parents were typed). Such a study might be completed in a few months by a well-organized and well-funded laboratory using an automated fluorescence sequencer. However, as noted (Section 12.2.2), candidate regions defined by linkage are usually uncomfortably large for positional cloning.

Association tests like the TDT have the opposite characteristics. Linkage disequilibrium is seldom striking over more than a megabase, so a genome screen by TDT would involve huge numbers of tests; on the other hand, a positive result would localize the susceptibility factor rather accurately. A natural study design is therefore to start with a genome-wide screen by linkage, probably in affected sib pairs, and then, once an initial localization has been achieved, to narrow the candidate region by linkage disequilibrium mapping.

It is important to remember that linkage disequilibrium is not an inevitable result of tight linkage. Association due to disequilibrium will be seen only if a significant proportion of the disease chromosomes derive from one not too distant common ancestor. There is a balance in this. Some serious dominant or X-linked mendelian diseases show no linkage disequilibrium because natural selection ensures a rapid turnover of disease genes, and most affected people are the result of independent mutations. For susceptibility factors in common disease, the problem is more likely to lie at the opposite end of the spectrum. Susceptibility factors may be common variants that have existed in the population at high frequency for a very long time, and that are nonpathogenic except when they get into bad company. A very old variant may have reached linkage equilibrium with adjacent markers. Equally, if many different changes to a given gene each acts as a susceptibility factor (in the same way that many different changes can cause loss of function, see *Figure 16.1*), then there may be no linkage disequilibrium. Therefore even if a susceptibility factor can be roughly localized by linkage, it does not necessarily follow that it can be fine-mapped by linkage disequilibrium or a method such as TDT that relies on it.

A major problem with sib pair analysis is that it will detect only rather strong susceptibility factors. The calculations in *Table 12.3* show that for $\gamma \leq 2$ the excess of allele sharing by affected sib pairs is very small, and detecting it would require huge numbers. It appears that large-scale association testing offers the best chance of finding these weak susceptibility factors through genome screening. But this would require work on a hitherto unprecedented scale. If we needed to be within 1 cM of a susceptibility locus to detect linkage disequilibrium, we would need about 3000 markers for a genome-wide screen. Given the quirky nature of linkage disequilibrium, with its dependence on details of population history, it might be prudent to use a denser set of maybe 10 000 markers. Scoring 1000 parent–child trios with a panel of 10 000 markers would mean generating 3×10^7 genotypes – an increase of two orders of magnitude on the current best technology. This may be achievable with diallelic single nucleotide polymorphisms (Section 11.2.3) scored on high-density DNA chips. For many common diseases, such a scale of operation may be necessary before susceptibility factors can be reliably identified. As will become apparent in Chapter 19, the present generation of studies lack the power to detect weak susceptibility loci. Nevertheless, for the present, TDT and other association testing is limited to testing candidate loci or regions.

Further reading

Ott J (1991) *Analysis of Human Genetic Linkage,* revised edn. Johns Hopkins University Press, Baltimore.

Terwilliger J, Ott J (1994) *Handbook for Human Genetic Linkage.* Johns Hopkins University Press, Baltimore.

References

Byerley WF (1989) Genetic linkage revisited. *Nature,* **340**, 340–341.

Eaves IA, Merriman TR, Barber RA, *et al.* (1998) Comparison of linkage disequilibrium in populations from the UK and Finland. *Am. J. Hum. Genet.,* **63** (suppl.), A1212.

Ivinson AJ, Read AP, Harris R, Super M, Schwarz M, Clayton Smith J, Elles R (1989) Testing for cystic fibrosis using allelic association. *J. Med. Genet.,* **26**, 426–430.

Krawczak M, Schmidtke J (1998) *DNA Fingerprinting,* 2nd edn. BIOS Scientific Publishers, Oxford.

Kruglyak L, Lander ES (1995) Complete multipoint sib-pair analysis of qualitative and quantitative traits. *Am. J. Hum. Genet.,* **57**, 439–454.

Kruglyak L, Daly MJ, Reeve-Daly MP, Lander ES (1996) Parametric and nonparametric linkage analysis: a unified multipoint approach. *Am. J. Hum. Genet.,* **58**, 1347–1363.

Lander ES, Kruglyak L (1995) Genetic dissection of complex traits: guidelines for interpreting and reporting linkage results. *Nature Genet.,* **11**, 241–247.

Lander ES, Schork N (1994) Genetic dissection of complex traits. *Science,* **265**, 2037–2048.

Read A (1989) *Medical Genetics: An Illustrated Outline.* Mosby, London.

Risch N, Botstein D (1996) A manic depressive history. *Nature Genet.,* **12**, 351–353.

Risch N, Merikangas K (1996) The future of genetic studies of complex human diseases. *Science,* **273**, 1516–1517. (See also *Science,* **275**, 1327–1330, 1997, for discussion.)

Schaid DJ (1998) Transmission disequilibrium, family controls and great expectations. *Am. J. Hum. Genet.,* **63**, 935–941.

Sham S, Zhao J (1998) Linkage analysis using affected sib-pairs. In: *Guide to Human Genome Computing,* 2nd edn. (MJ Bishop ed.). Academic Press, London.

Varon R, Vissinga C, Platzer M, *et al.* (1998) Nibrin, a novel DNA double-stranded break repair protein, is mutated in Nijmegen Breakage Syndrome. *Cell,* **93**, 467–476.

Weeks DE, Lange K (1992) A multilocus extension of the affected pedigree member method of linkage analysis. *Am. J. Hum. Genet.,* **50**, 859–868.

Xiong M, Guo S-W (1997) Fine-scale genetic mapping based on linkage disequilibrium: theory and applications. *Am. J. Hum. Genet.,* **60**, 1513–1531.

Genome projects

13

The great discoveries cataloged in our history have mostly been achieved by exploring our environment, whether it be Eratsothenes calculating the Earth's circumference, Columbus sailing across the Atlantic, Galileo peering at the moons of Jupiter, or Darwin studying finches in the Galapagos. After many centuries, we have built up an approximate understanding of our external universe, but the universe within us has only very recently been the subject of serious study. The application of microscopy to the study of cells and subcellular structures provided one major route into this world, to be followed by pioneering advances in biochemistry and then molecular biology. Now, as we enter the next millennium, we are on the threshold of a truly momentous achievement that will have enormous implications for the future. For the first time, we will know our genetic endowment – the sequence of our DNA. Then our voyage into the *universe within* really will have begun. Sequencing our DNA will be just the beginning of a huge effort to understand exactly how this sequence can specify a person, and how the DNA of other organisms is related to us and to their biologies.

Because of the scale of the effort, the endeavor to sequence our DNA represents biology's first 'big project'. The Human Genome Project is a truly international effort, the primary aim of which is to deliver the complete nucleotide sequence of our DNA. The human mtDNA sequence was established in 1981 (Section 7.1.1), so the genome in question is the nuclear genome. It has been paralleled by many other genome projects which seek to sequence the DNA of a variety of model organisms. Some genome projects have already been completed; others, including the Human Genome Project, are nearing completion. Added to this are ancillary studies. Some are investigating the extent of *sequence variation* within the human genome. Others are examining ethical and legal implications of the new knowledge that will be obtained and its likely impact on society.

13.1 The history, organization, goals and value of the Human Genome Project

13.1.1 The Human Genome Project was conceived out of the need for a large-scale project to develop new mutation detection methods

A workshop held in Alta, Utah, in December 1984 was a major catalyst in the development of the Human Genome Project. Sponsored partially by the US Department of Energy (DOE), the workshop was intended to evaluate the current state of mutation detection and characterization and to project future directions for technologies to address current technical limitations. The growing roles of novel DNA technologies were discussed, notably the emerging gene cloning and sequencing technologies. Although such technologies had been in operation for about a decade, the efforts of individual laboratories to try to clone and characterize one gene at a time were considered to be wasteful of scientists' time and research resources. Because of the perceived technical obstacles, a principal conclusion was that methods were incapable of measuring mutations with sufficient sensitivity unless an enormously large, complex and expensive program was undertaken. A subsequent report on *Technologies for Detecting Heritable Mutations in Human Beings* sparked the idea for a dedicated human genome project by the DOE, and in March 1986 it sponsored an international meeting in Santa Fe, New Mexico, to assess the desirability and feasibility of ordering and sequencing DNA clones representing the entire human genome. Virtually all participants concluded that such a project was feasible and would be an oustanding achievement in biology.

After extensive discussions with the US scientific community, the DOE responded to the Santa Fe meeting by issuing a Report on the Human Genome Initiative in the spring of 1987. Three major objectives were to be implemented: the generation of refined physical maps of human chromosomes, the development of support technologies and facilities for human genome research, and expansion of communication networks and of computational and database capacities. As implementation of this program began with a small number of pilot projects, other US organizations initiated their own studies of policy and strategy. In 1988, two additional widely circulated reports from the US Office of Technology Assessment and National Research Council appeared, and the US National Institute of Health (NIH) set up an Office of Human Genome Research (later renamed the National Center for Human Genome Research) to coordinate NIH genome activities in cooperation with other US organizations. In the same year, the US congress officially gave approval to a 15-year US human genome project commencing in 1991. The required funding was estimated to be about $3 billion.

13.1.2 Organization and goals of the Human Genome Project

Organization of the Human Genome Project

The US Human Genome Project remains the major contributor to international research in this area, but several other countries quickly developed their own Human Genome Projects. Centers in the UK and France have made major contributions and large programs are also underway in some other countries, notably Germany and Japan. In order to coordinate the different national efforts, the Human Genome Organization (HUGO) was established in 1988 with a remit of facilitating exchange of research resources, encouraging public debate and advising on the implications of human genome research (McKusick, 1989). Currently, much of the effort that is going into the Human Genome Project is concentrated in a few very large genome centers, but interacting with them is a worldwide network of small laboratories, mostly attempting to map and identify disease genes (*Figure 13.1*).

Communication between the network of genome centers and interacting laboratories is very largely based on electronic communication. This has evolved because of the need to manage and store the huge amount of data that are being produced. The data are entered into large electronic databases which, at least for publically funded mapping and sequencing efforts, are freely accessible through the Internet. Analyses can then be conducted from remote computer terminals throughout the world. Depending on the source of input data, there are two types of database:

■ *Central repositories for storing globally produced mapping and sequence data* (*Table 13.1*). Universal DNA and protein sequence databases were established decades before the onset of the genome projects, but specialized *species-specific* mapping databases were established comparatively recently. For example, the Genome database (GDB) is a mapping database

which is aimed specifically at storing human data (*note* that there is a specific nomenclature for naming DNA segments and genes in different species – see Further reading and *Box 13.1* for the human nomenclature).

■ *Databases for storing locally produced data.* In order to improve their own efficiency, the big genome mapping and sequencing centers have stored data produced in their own laboratories in dedicated databases. Unlike data input, data access is freely available through the Web from publically funded genome centers.

Goals

The major rationale of the Human Genome Project is to acquire fundamental information concerning our genetic make-up which will further our basic scientific understanding of human genetics and of the role of various genes in health and disease. As a first step towards achieving this, high-resolution genetic maps were constructed and used as a *framework* for constructing high-resolution physical maps, culminating in the ultimate physical map – the complete sequence of the human genome. The first major task, high-resolution genetic maps, was achieved by 1994. By the time of writing (May 1999), large clone contigs had been assembled for much of the genome and large-scale DNA sequencing was very much underway. In short, rapid progress has been made and the goals that had been set for 1998 have been achieved or surpassed. Recently, goals for the next 5 years have been established with an anticipated completion date for human genome sequencing by 2003 (*Table 13.2*; see Collins *et al.*, 1998).

13.1.3 The medical and scientific benefits of the Human Genome Project are expected to be enormous

For many human biologists and geneticists, the Human Genome Project represents an exciting, historic mission. Since its outset, the project has been especially justified by the expected medical benefits of knowing the structure of each human gene. Inevitably, this information will provide more comprehensive prenatal and presymptomatic diagnoses of disorders in individuals judged to be at risk of carrying a disease gene. The information on gene structure will also be used to explore how individual genes function and how they are regulated. Such information will provide sorely needed explanations for biological processes in humans. It would also be expected to provide a framework for developing new therapies for diseases, in addition to simple gene therapy approaches. More importantly, as mutation screening techniques develop, an expected benefit would be to alter radically the current approach to medical care, from one of treating advanced disease to *preventing disease* based on the identification of individual risk (Cantor, 1998).

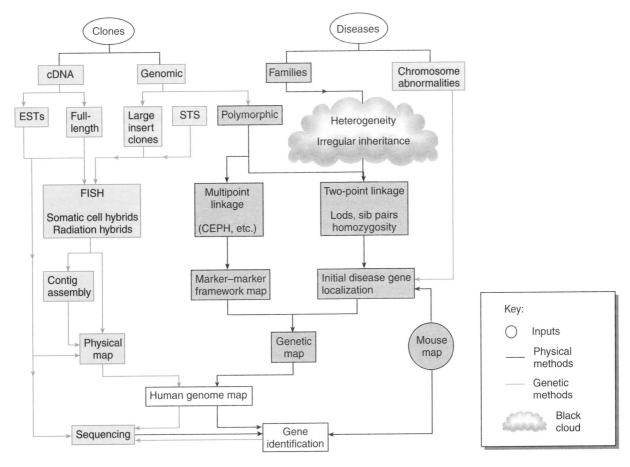

Figure 13.1: Major scientific strategies and approaches being used in the Human Genome Project.

The major scientific thrust of the Human Genome Project begins with the isolation of human genomic and cDNA clones (by cell-based cloning or PCR-based cloning). These are then used to construct high-resolution genetic and physical maps prior to obtaining the ultimate physical map, the complete nucleotide sequence of the 3300 Mb nuclear genome. Inevitably, the project interacts with research on mapping and identifying human disease genes. In addition, ancillary projects include studying genetic variation (Section 13.2.5); genome projects for model organisms (Section 13.3) and research on ethical, legal and social implications. The data produced are being channeled into mapping and sequence databases permitting rapid electronic access and data analysis. Abbreviations: EST, expressed sequence tag; STS, sequence tagged site.

Exciting though such possibilities are, there may be unexpected difficulties in understanding precisely and comprehensively how some genes function and are regulated (cautionary precedents are the lack of progress in predicting protein structure from the amino acid sequence, and the imperfect understanding of the precise ways in which the regulation of globin gene expression is coordinated, decades after the relevant sequences have been obtained). In addition, the single-gene disorders which should be the easiest targets for developing novel therapies are very rare; the most common disorders are multifactorial. Hence, although the data collected in the Human Genome Project will inevitably be of medical value, some of the most important medical applications may take some time to be developed.

13.2 Genetic and physical mapping of the human genome

13.2.1 The first comprehensive human genetic map was published only in 1987, but within 7 years very high resolution maps were achieved, mostly using microsatellite markers

Human genetic maps could not easily be constructed using classical genetic mapping

Classical genetic maps for experimental organisms such as *Drosophila* and mouse are based on *genes*. They have been available for decades, and have been refined

Table 13.1: Examples of databases relevant to the human genome project which store globally produced data

Database	Description and electronic addresses
GenBank	DNA and protein sequences. One of many databases distributed by the US National Center for Biotechnology Information (NCBI), NIH http://www.ncbi.nlm.nih.gov
EMBL	DNA sequences. Distributed by the European Bioinformatics Institute (EBI) at Cambridge, UK, together with 30 other molecular biology databases http://www.ebi.ac.uk
DDBJ	DNA database of Japan (Mishima) http://www.nig.oc.jp/home.html
PIR (protein information resource)	Protein sequences (single database distributed as a collaboration by: the US National Biomedical Research Foundation, Washington; the Martinsried Institute for Protein Sequences, Germany; the Japan International Protein Information Database, Tokyo). Accessible at various centers including the US NCBI (see above)
SWISS-PROT	Protein sequences. Maintained collaboratively by the University of Geneva and the EMBL data library. Distributed by several centers, such as the NCBI and EBI. Access via NCBI or EBI (see above)
GDB (genome database)	The major human mapping database. Permits interaction with the OMIM database (see below) http://gdbwww.gdb.org/
OMIM (on-line Mendelian	Electronic catalog of Victor McKusick's *Mendelian Inheritance in Man*, listing all known inherited human disorders. Established at the Johns Hopkins University. Access via various centers such as the US NCBI (see above) or the UK Human Genome Mapping Project Resource Centre (http://www.hgmp.mrc.ac.uk/omim/)

Table 13.2: Goals for the US Human Genome Project (1998–2003) referenced against previous goals and achievements.

Area	Goals 1993–1998	Status as of Oct. 1998	Goals 1998–2003
Genetic map	Average 2–5 cM resolution	1 cM map published Sept. 1994	Completed
Physical map	Map 30 000 STSs	52 000 STSs mapped	Completed
DNA sequence	Complete 80 Mb for all organisms by 1998	180 Mb human plus 111 Mb nonhuman	Finished a third of human sequence by end of 2001 Working draft of remainder by end of 2001 Complete human sequence by end of 2003
Sequencing technology	Evolutionary improvements and innovative technologies	90 Mb per year capacity at ~$0.50 per base Capillary array electrophoresis validated Microfabrication feasible	Integrate and automate to achieve 500 Mb per year at ⩽$0.25 per base Support innovation
Human sequence variation	Not a goal	—	100 000 mapped SNPS Develop technology
Gene identification	Develop technology	30 000 ESTs mapped	Full-length cDNAs
Functional analysis	Not a goal	—	Develop genomic-scale technologies
Model organisms	*E. coli* complete sequence Yeast: complete sequence *C. elegans:* most of sequence *Drosophila*: begin sequencing Mouse: map 10 000 STSs	Published Sept 1997 Released April 1996 80% complete 9% done 12 000 STSs mapped	— — Complete Dec 1998 Sequence by 2002 Develop extensive genomic resources Lay basis for finishing sequence by 2005 Produce working draft before 2005

Reprinted with permission from Collins *et al.*, *Science*, **282**, pp. 682–689. Copyright 1998 American Association for the Advancement of Science.

Human gene and DNA segment nomenclature

The nomenclature used is decided by the HUGO nomenclature committee. Genes and pseudogenes are allocated symbols of usually two to six characters; a final P indicates a pseudogene. For anonymous DNA sequences, the convention is to use D (= DNA) followed by 1–22, X or Y to denote the chromosomal location, then S for a unique segment, Z for a chromosome-specific repetitive DNA family or F for a multilocus DNA family, and finally a serial number. The letter E following the number for an anonymous DNA sequence indicates that the sequence is known to be expressed.

Symbol	Interpretation
CRYBA1	Gene for crystallin, beta A1 polypeptide
GAPD	Gene for glyceraldehyde-3-phosphate dehydrogenase
GAPDL7	GAPD-like gene 7, functional status unknown
GAPDP1	GAPD pseudogene 1
AK1	Gene for adenylate kinase, locus 1
AK2	Gene for adenylate kinase, locus 2
*PGK1*2*	Second allele at *PGK1* locus
B3P42	Breakpoint number 42 on chromosome 3
DYS29	Unique DNA segment number 29 on the Y chromosome
D3S2550E	Unique DNA segment number 2550 on chromosome 3, known to be expressed
D11Z3	Chromosome 11-specific repetitive DNA family number 3
DXYS6X	DNA segment found on the X chromosome, with a known homolog on the Y chromosome, and representing the 6th XY homolog pair to be classified
DXYS44Y	DNA segment found on the Y chromosome, with a known homolog on the X chromosome, 44th XY homolog pair
D12F3S1	DNA segment on chromosome 12, first member of multilocus family 3
DXF3S2	DNA segment on chromosome X, second member of multilocus family 3
FRA16A	Fragile site A on chromosome 16

continuously. They are constructed by crossing different mutants in order to determine whether the two gene loci are linked or not. For much of this period, human geneticists were envious spectators, because the idea of constructing a human genetic map was generally considered unattainable. Unlike the experimental organisms, the human genetic map was never going to be based on genes because the frequency of mating between two individuals suffering from different genetic disorders is extremely small.

The only way forward for a human genetic map was to base it on *polymorphic markers* which were not necessarily related to disease or to genes. As long as the markers showed mendelian segregation and were polymorphic

enough so that recombinants could be scored in a reasonable percentage of meioses, a human genetic map could be obtained. The problem here was that, until recently, suitably polymorphic markers were just not available. Classical human genetic markers consisted of protein polymorphisms, notably blood group and serum protein markers, which are both rare and not very informative (see *Box 11.1*). By 1981, only very partial human linkage maps had been obtained, and then only in the case of a few chromosomes.

The identification of DNA-based polymorphisms transformed human genetic mapping

Unlike classical markers, DNA-based polymorphisms were not simply confined to the 3% of the DNA that was expressed (genes): they were also available in noncoding DNA. Since the latter was not so strongly conserved in evolution, changes in the DNA were comparatively frequent. The realization that DNA polymorphisms could be abundant called for a radical revision of thinking, and the early 1980s saw serious discussion of the possibility of constructing a complete human genetic map for the first time (see Botstein *et al.*, 1980).

The first comprehensive human genetic map was based on RFLPs

The desirability of a complete linkage map of the human genome was clear. In addition to providing a framework for studying the nature of recombination in humans, it would permit rapid gene localization, assist gene cloning, and facilitate genetic diagnosis Almost inevitably, the realization that a comprehensive human genetic map was now attainable sparked serious efforts to construct one. In 1987, after a huge effort, the first such map was published based on the use of 403 polymorphic loci, including 393 RFLP markers (Donis-Keller *et al.*, 1987). Although this achievement was important, there remained some serious drawbacks with the map: the average spacing between the markers (>10 cM) was still considerable, and, more significantly, RFLP markers are not very informative and are difficult to type (see *Box 11.1*).

High-resolution human genetic maps were obtained largely through the use of microsatellite markers

Hypervariable minisatellite polymorphisms are highly polymorphic, but their applicability to genome-wide maps is limited because they are mostly restricted to chromosomal regions near the telomeres. *Microsatellite markers* (also described as *short tandem repeat polymorphisms*, or *STRPs*) have the advantage of being abundant, dispersed throughout the genome, highly informative and easy to type (see *Box 11.1*). By focusing on this type of marker, researchers at the Généthon laboratory in France were quickly able to provide a second-generation linkage map of the human genome (Weissenbach *et al.*, 1992). Subsequently, maps have been produced with ever increasing numbers of genetic markers, especially microsatellite markers, and ever increasing resolution. Within a further two years, a genetic map with 1 cM resolution had been achieved (Murray *et al.*, 1994). After this time the major effort switched to the construction of high-resolution physical maps.

13.2.2 Different physical maps of the human genome have been constructed, but clone contig maps are the primary templates for DNA sequencing

The variety of physical maps of human DNA

Like the genetic map, a physical map of the human genome will consist of 24 maps, one for each chromosome. The different genetic maps of the human genome that have been assembled so far all represent the same concept – sets of linked polymorphic markers (*linkage groups*) corresponding to different chromosomes. However, unlike this uniformity, a variety of different types of physical map are possible (*Table 13.3, Figure 13.2*). The first physical map of the human genome was obtained more than 40 years ago when cytogenetic banding techniques were used not only to distinguish the different chromosomes, but also to provide discrimination of different subchromosomal regions (see the human karyogram in *Figure 2.17*). Although the resolution is coarse (an average sized chromosome band in a 550-band preparation contains ~6 Mb of DNA), it has been very useful as a framework for ordering the locations of human DNA sequences by chromosome *in situ* hybridization techniques.

Other maps have been obtained by mapping *natural chromosome breakpoints* (using translocation and deletion hybrids; see Section 10.1.2), or by mapping *artificial chromosome breakpoints* using radiation hybrids (Section 10.1.3), but the resolution achieved can be quite limited. Such maps have, however, been useful frameworks for mapping genes (*transcript maps*; see below). Large-scale restriction maps

have also been generated, such as the *Not*I restriction map of 21q (Ichikawa *et al.*, 1993; *Figure 13.2*). However, the most important maps are clone contig maps because these are the immediate templates for DNA sequencing.

YAC clone contig maps: a first-generation physical map of the human genome

A complete clone contig map of a chromosome would comprise all the DNA without any gaps (*contig* originated as a shortened form of the word *contiguous*; Section 10.3). Because of their large inserts, *yeast artificial chromosome* (*YAC*) clones have been particularly useful in generating first-generation physical maps of human chromosomes. Different methods of identifying overlaps between clones have been used, but *STS markers* (both polymorphic and nonpolymorphic), which had previously been mapped to the chromosome of interest, have been particularly useful. Significant contig maps for individual human chromosomes were first reported in 1992 for chromosome 21 and the Y chromosome and, subsequently, a first-generation clone contig map of the human genome was reported by workers at the CEPH lab in Paris (Cohen *et al.*, 1993). An updated YAC contig map, covering perhaps 75% of the human genome and consisting of 225 contigs with an average size of 10 Mb, was subsequently published by the same group (Chumakov *et al.*, 1992). While these physical maps were recognized to be far from complete, this was an outstanding achievement and provided a good framework for the scientific community to build upon in order to produce future detailed maps of all the chromosomes. Complementing this approach, good STS-based physical maps of the human genome have been developed, such as the one constructed at the Whitehead Institute in Massachussetts (Hudson *et al.*, 1995). These have been achieved in part by mapping STSs against panels of whole-genome radiation hybrids).

Table 13.3: Different types of physical map can be used to map the human nuclear genome

Type of map	Examples/methodology	Resolution
Cytogenetic	Chromosome banding maps	An average band has several Mb of DNA
Chromosome breakpoint maps	Somatic cell hybrid panels containing human chromosome fragments derived from natural translocation or deletion chromosomes	Distance between adjacent chromosomal breakpoints on a chromosome is usually several Mb
	Monochromosomal radiation hybrid (RH) maps Whole genome RH maps	Distance between breakpoints is often many Mb Resolution can be as high as 0.5 Mb
Restriction map	Rare-cutter restriction maps, e.g. *Not*I maps	Several hundred kb
Clone contig map	Overlapping YAC clones Overlapping cosmid clones	Average YAC insert has several hundred kb of DNA Average cosmid insert is 40 kb
Sequence-tagged site (STS) map	Requires prior sequence information from ordered clones so that STSs can be ordered	A desired goal is an average spacing of 100 kb
Expressed sequence tag (EST) map	Requires cDNA sequencing then mapping cDNAs back to other physical maps	Highest possible average spacing is ~40 kb
DNA sequence map	Complete nucleotide sequence of chromosomal DNA	1 bp

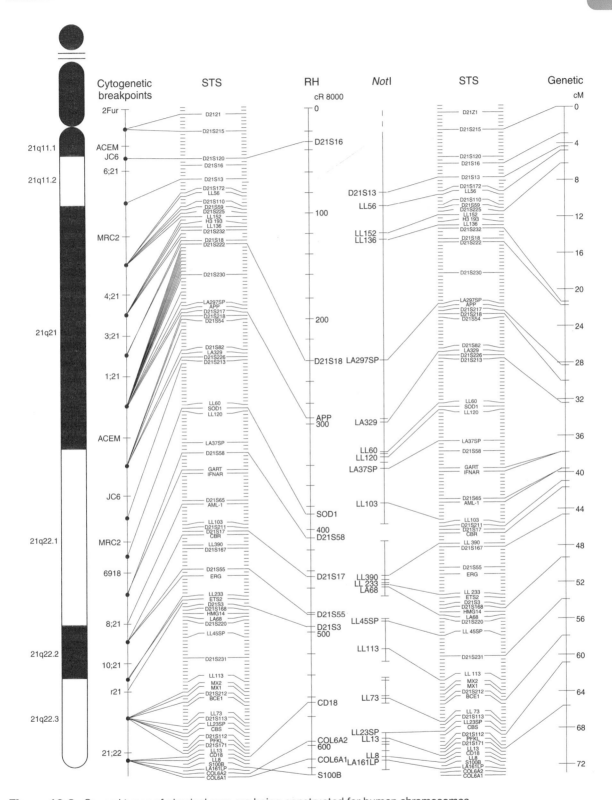

Figure 13.2: Several types of physical map are being constructed for human chromosomes.

The figure shows integration of several physical maps for the long arm of human chromosome 21. Next to the standard cytogenetic map on the left are the positions of chromosome 21 breakpoints observed largely from studying chromosome 21 translocations (6;21, 4;21, 3;21, 2;21, etc.). The STS map is shown twice to facilitate comparisons with the other maps, which also include the genetic linkage map of Chumakov *et al.* (1992), the PFGE-based *Not*I restriction map (Ichikawa *et al.,* 1993) and the radiation hybrid map (measured in centiRays after exposure to 8000 rad). Reproduced with permission from Chumakov *et al.* (1992) *Nature*, **359**, p. 385. Copyright 1992 Macmillan Magazines Limited.

BAC/PAC clone contig maps: the major templates for DNA sequencing

The utility of YAC contig maps is limited because YAC inserts are often not faithful representations of the original starting DNA; many YAC clones are *chimeric* or have internal deletions (see Section 10.3.5). As a result, second-generation clone contig maps have relied on *bacterial artificial chromosomes* (BACs) and *P1 artificial chromosomes* (PACs). Although the insert sizes of these clones (typically 70–250 kb) are much smaller than that of YACs, this disadvantage is more than outweighed by their greater stability, making them more faithful representations of the original DNA. Recently, the large genome centers have focused greatly on constructing large BAC contigs as a prelude to large-scale DNA sequencing.

13.2.3 An early priority in the Human Genome Project was the construction of gene (transcript) maps

Coding-DNA sequencing or whole-genome sequencing?

At the outset of the Human Genome Project there was much debate over whether to go for an all-out assault (indiscriminate sequencing of all 3 billion bases), or whether to focus initially just on the coding-DNA sequences. The average coding DNA of a human gene is about 1.7 kb, but human genes occur, on average, once every 40–50 kb of DNA. As a result, coding DNA accounts for a mere 3% of the human genome (unlike *Saccharomyces cerevisiae* and *Caenorhabditis elegans*, where the gene density is much higher – 1 per 2 kb and 1 per 5 kb, respectively). To obtain coding-DNA sequences, the easiest approach would be to make a range of human cDNA libraries, then sequence cDNA clones at random.

The priority of coding-DNA sequencing was dependent on two arguments: (i) coding DNA contains the information content of the genome and so is by far the most interesting and medically relevant part; and (ii) it is such a small percentage of the genome that it can be achieved very quickly and cheaply, when compared with efforts to sequence the entire genome. Supporters of whole-genome sequencing emphasized that finding all genes could be difficult (some genes may not be well-represented in available cDNA libraries if they are very restricted in expression, or expressed transiently during early development). In addition, at least some of the non-coding DNA is functionally important, e.g. in the case of regulatory elements and sequences that are important for chromosome function.

The first comprehensive human gene map was based on short sequence tags from cDNA clones

The coding sequence priority prevailed and the first reasonably comprehensive human gene maps were constructed. This involves essentially three steps:

■ *Random cDNA sequencing.* Initially this meant sequencing short (~ 300 bp) sequences at the 3' ends of cDNA clones from a variety of human cDNA libraries. These short sequences were described as **expressed sequence tags (ESTs)** because they permitted a simple and rapid PCR assay for a specific expressed sequence (gene) (Adams *et al.*, 1991). In this sense, therefore, an EST is simply the gene equivalent of an STS (a term used to describe *any type of sequence*, but often noncoding DNA, which is specific for a particular *locus*). Because the 3' UTR of almost all human genes exceeds 300 bp, the 3' ESTs typically did not contain coding sequence.

■ *Mapping ESTs to specific chromosomes.* 3' UTR sequences are not as frequently interrupted by introns as coding DNA. This means that it is usually easy to design PCR primers from an EST that will amplify the specific sequence in a *genomic DNA* sample. Because 3' UTR sequences are not very well conserved during evolution, it is also possible to screen human–rodent somatic cell hybrids for the presence of a human EST (the orthologous rodent sequences are usually so diverged that they do not amplify). By using a panel of human *monochromosomal* somatic cell hybrids (Section 10.1.1), an EST can be mapped to a specific human chromosome.

■ *Mapping ESTs to subchromosomal locations.* A huge effort has been mounted at some centers to establish integrated STS-based and EST-based maps, such as those produced by the Whitehead Institute. This has involved using PCR primers that are specific for an EST (or STS) to type YACs and other clones within clone contig maps that have been produced for the relevant chromosome and/or typing of a panel of *whole genome radiation hybrids.* Two such panels have been used in particular (Section 10.1.3): the Genebridge 4 panel (average fragment size 25 Mb), and for higher resolution, the Stanford G3 panel (average fragment size 2.4 Mb).

Using the above approaches, the number of human genes that were placed on the physical map increased exponentially (*Figure 13.3*). The latest human gene map, published in October 1998, was achieved by a radiation hybrid mapping consortium led by the Sanger Centre, UK, together with various other centers, notably Stanford Human Genome Center, the Généthon lab in Paris, the Whitehead Institute and the Wellcome Trust Centre for Human Genetics at Oxford, UK. In all, map positions for over 30 000 human genes were reported (Deloukas *et al.,* 1998; electronic reference 1), representing possibly 30–40% of the total human gene catalog.

In many cases, there is little or no coding sequence for the mapped genes and considerable effort is being devoted to sequencing large inserts of human cDNA clones in various laboratories throughout the world. Different research programs are investigating gene expression in specific tissues or in specific states. For example, the Cancer Genome Anatomy Project (electronic reference 2), a program devised at the US National Cancer Institute,

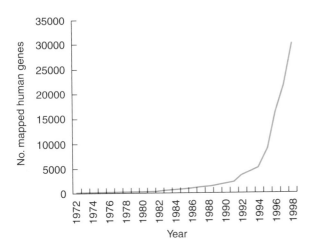

Figure 13.3: Progress in human gene mapping.

Data were reproduced from the GeneMap'99 website (electronic reference 1).

is devoted to studying expression of genes in various human tumor cells, including sequencing of large insert cDNA clones from cDNA libraries made from human tumor cells and large-scale expression profiling using microarrays (Section 20.2.2).

13.2.4 Accelerated sequencing efforts mean that the ultimate physical map, the complete nucleotide sequence of the human genome, should be delivered by the year 2003

At the outset of the Human Genome Project, DNA sequencing was expensive and not very efficient. It was anticipated, however, that technological developments would lead to considerable reductions in costs and much more efficient sequencing. The sequencing of the human genome at that time seemed an immense challenge because there was so little experience in sequencing large genomes. All that has changed, and some very large genomes have already been sequenced (*Figure 13.4*). There have been no significant changes in the basic sequencing technology; the dideoxy sequencing approach invented by Fred Sanger and his colleagues at Cambridge, UK, more than 20 years ago is still used. Instead, efficiency gains have been made through the use of automated fluorescence-based systems and *capillary gel electrophoresis*.

While the first few years of the Human Genome Project were devoted to producing high-resolution genetic and physical maps, large-scale human genome sequencing is now very much underway and 10% of the human genome had been sequenced by May 1999 (*Figure 13.5*; electronic reference 3). Funded largely by the Wellcome Trust, the greatest single contributor has been the Sanger Centre at Hixton, UK (*Figure 13.6*). By May 1999 the Sanger Centre had contributed over 100 Mb of finished human sequence (out of a global total of 300 Mb), and had also achieved a further 65 Mb of unfinished sequence (sequences which have not yet been compiled into large contigs). In order to avoid wasteful duplication of effort, the HUGO-sponsored Human Genome Sequencing Index identifies priority chromosomes or subchromosomal regions targeted by individual sequencing centers (electronic reference 4). Currently, chromosome 22 is set to be the first human chromosome to be completed (*Figure 13.7*).

Partly in response to competition from the private sector (*Box 13.2*), the UK Wellcome Trust and the US National Human Genome Research Institute have collaborated to

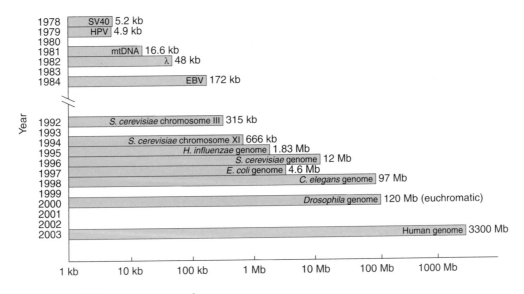

Figure 13.4: Landmarks in genome sequencing.

Viral genome abbreviations are as follows: SV40, simian virus 40; HPV, human papilloma virus; λ, lambda phage; EBV, Epstein Barr virus.

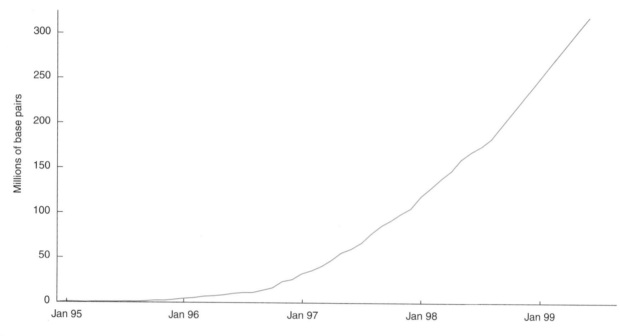

Figure 13.5: Progress in human genome sequencing.

See also the Genome Monitoring Table maintained at the European Bioinformatics Institute (EBI; electronic reference 4).

Figure 13.6: Large-scale DNA sequencing at the Sanger Centre.

The Sanger Centre at Hinxton, UK, is the largest single contributor to human genome sequencing. Data can be accessed at http://www.sanger.ac.uk.

bring forward the timescale for completion of the Human Genome Project. The aim is to produce a working draft, comprising about 90% of the human genome, by the year 2000. The Sanger Centre is expected to produce 33% of the working draft and the three major American genome sequencing centers (Washington University School of Medicine at St. Louis, Baylor College of Medicine and the Whitehead/Massachussetts Institute of Technology) are expected to achieve 60% between them. Other centers, notably in France, Germany and Japan, are also committed to sequencing specific subchromosomal regions (elec-

tronic reference 4). After completion of the working draft, the full genome sequence is expected to be achieved around 2002–2003.

13.2.5 Programs looking at human genome sequence variation aim to understand human evolution and to facilitate identification of disease genes

At the outset, the Human Genome Project was conceived as a project to obtain the nucleotide sequence of a collection of cloned human DNA fragments collectively amounting to one or a very few haploid genomes. What it did not consider was the genetic diversity of humans. Information on human genetic diversity has subsequently been considered desirable in three major contexts:

- *Human evolution.* The information should be of help in anthropological and historical research in tracking human origins, prehistoric population movements and social structure.
- *Identification of common disease genes and factors which confer susceptibility to or protection from disease.* Common diseases are multifactorial, and it can be frustratingly difficult to identify the underlying genes. Genetic differences between human populations can also make some populations more susceptible to particular diseases while others are comparatively protected.
- *Forensic anthropology.* The accuracy of DNA fingerprinting, a widely used tool in forensic science, is dependent, in part, on knowing how the DNA markers detected in fingerprinting vary from one population to the next.

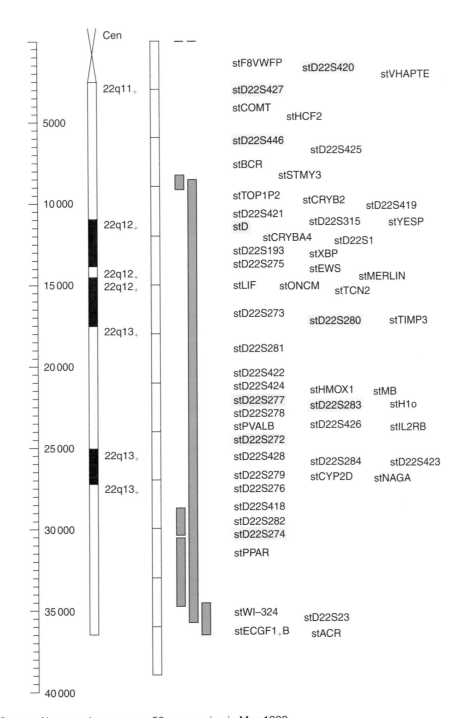

Figure 13.7: Status of human chromosome 22 sequencing in May 1999.

The figure represents the map status for chromosome 22 accessed from the Sanger Centre's website (http://www.sanger.ac.uk) during May 1999. Because of heterochromatic regions on 22p, the major effort is going into sequencing 22q. The format is a graphical display constructed using the *acedb* database system. The cytogenetic map on the left is referenced against a kilobase linear DNA map at extreme left. DNA markers are to the right and dark blue boxes indicate sequenced clone contigs.

The Human Genome Diversity Project

The idea of a global effort to study human genome sequence diversity, the Human Genome Diversity Project, was proposed by Cavalli-Sforza *et al.* (1991). The emphasis was predominantly focused on the need to collect DNA samples from a large number of ethnic groups. However, although supported by HUGO, this project has been in considerable difficulty (Harding and Sajantila, 1998). From the outset it has been dogged by a conspicuous lack of funding. As the primary aim of this project is

Box 13.2

Co-operation, competition and controversy in the genome projects

Because of their scale, the genome projects are major undertakings. In many cases, there have been laudable examples of cooperation: sharing of resources between different centers and labs, and agreed subdivision of different project tasks, etc. In other cases, tensions have been evident as a result of fierce competition between different laboratories and wasteful duplication of effort. The *E. coli* genome project involved a competitive race between American and Japanese groups; at the end the American lab deposited its sequence in GenBank one week before that of the Japanese group. By contrast, the yeast genome project was a model of international cooperation, especially the cooperation between different European centers which has subsequently extended to cooperation on functional analyses.

A powerful tension in the Human Genome Project has also emerged as a result of the different priorities of the public and private sectors. One major area has concerned gene patents. This issue first appeared in 1991 when the US NIH applied for a patent for more than 7000 fragments of brain cDNA clones whose sequences had been established as part of an *EST mapping* exercise led by Dr Craig Venter. This attempt met with widespread opposition from the scientific community, especially since nothing was known about the functions of the expressed sequences. Under pressure, the US Patents Office rejected the applications. For the first time, the question of *who owns the human genome* had been raised (Thomas *et al.*, 1996) and the idea of commercial monopoly of what is quite simply our genetic heritage appeared alarming and offensive to many. Dr Venter subsequently left NIH to set up a new commercially backed institute, the Institute of Genome Research, and by adopting a factory-style approach to EST sequencing, he quickly compiled the world's largest human gene databank. In April 1994, the drug company SmithKline Beecham invested £80 million for an exclusive stake in Venter's database and announced to scientists that they could have access to it only if they agreed to concede first rights to any patentable discovery. Again the prospect of a corporation trying to monopolize

control of a large part of the expressed human genome alarmed the scientific community. Many felt that a case could be made for patenting after identifying the function of a gene, but not before. Thousands of patents have been awarded for human DNA sequences where the sequence has been associated with some functional feature (Thomas *et al.*, 1996) but in October 1998 the US patent office awarded the first patent for an EST sequence (to Incyte Pharmaceuticals). As described by Knoppers (1999), the issue of patentability of the human genome continues to be problematic.

Private sector-funded attempts to sequence the human genome have also been controversial and have been viewed to be competing directly with publicly funded human genome sequencing. While publicly funded laboratories have been committed to making new sequence data immediately available (accessible through the Internet) private-sector laboratories have declined to make their data so readily and freely available. Recently, the timetable for publicly funded sequencing of the human genome has been dramatically brought forward, partly in response to the announcement by Celera Genomics of its intention to sequence the human genome using a rapid shotgun approach described by Venter *et al.* (1996). Unlike the physical map-based approach to DNA sequencing used in publicly funded human genome sequencing, this rapid random clone sequencing approach may have difficulty in closing gaps within clone contigs. As a test system, the *Drosophila* genome will be sequenced using this approach, but in this case there will be cooperation with publicly funded *Drosophila* genome sequencing efforts, and, unusually, the sequence data will be made available to the international scientific community.

The issue of *single nucleotide polymorphism maps* which may be very helpful in identifying genes for common disorders (see Section 13.2.5) has also been controversial. Various companies have been compiling SNP databases and a number of SNP patents have been filed. The rapid momentum of the private effort appears to have acted as a stimulus to the public effort to develop freely accessible large SNP databases. The latest 5-year plan for the US Human Genome Project envisages creating a map containing at least 100 000 human SNPs (Collins *et al.*, 1998).

---

simply to find markers for ethnic groups and to trace the origins of human migrations and ancestral lineages, the case for large-scale funding has not been persuasive. The project has also been beset by controversy. In some cases, researchers have visited isolated human populations on the margins of survival, obtained samples quickly without taking time to explain their significance and then left, with little or no subsequent communication. While proponents refer to the need to safeguard our cultural heritage, critics have witheringly used terms such as *helicopter genetics* to refer to the insensitive 'quickly-in, quickly-out' approach often used to obtain samples.

Single nucleotide polymorphism (SNP) maps
Human genetic maps based on microsatellite markers, although extremely valuable, have some limitations. In particular, although such markers are found all over the genome their density is limited to about one per 30 kb.

In addition, typing of microsatellite markers is not so amenable to automation on a very large scale. By contrast, *single nucleotide polymorphisms* (*SNPs*) are very frequent (about 1 per kb) and typing is easily automated because they have only two alleles (see Section 11.2.3). As a result, they have been envisaged to have potentially powerful applications in association studies to identify genes underlying polygenic disease (Collins *et al.*, 1997; Schafer and Hawkins, 1998). The first steps toward establishing a third-generation SNP-based genetic map have recently been described by Wang *et al.* (1998). Partly in response to initiatives from the private sector (see *Box 13.2*), the US Human Genome Project and the UK Wellcome Trust have committed funds for the construction of a map containing 100 000 SNPs by 2003. However, the utility of SNPs remains unproven and some data are emerging which have dampened the initial huge optimism (Pennisi, 1998).

13.3 Model organism and other genome projects

Mapping the human genome is not the only scientific focus of the Human Genome Project; at its outset, the value of sequencing genomes of model organisms was recognized. Such organisms include a variety of species, some of which have been particularly amenable to genetic analysis (*Box 13.3*). In part, the sequencing of smaller genomes was also considered as a pilot for large-scale sequencing of the human genome. By 1999 the genomes of about 100 organisms were being sequenced or had already been sequenced (electronic reference 5).

13.3.1 The genomes of many prokaryotic organisms have been sequenced, including well-studied experimental models and disease-associated organisms

The diversity of prokaryotic genome sequencing projects

Prokaryotic genomes are typically small (often only one or a few Mb) and are therefore amenable to comparatively rapid sequencing. By 1999 the genomes of a total of 75 different prokaryotes were being sequenced or had already been sequenced (electronic reference 5). The first to be completed (in 1995) was the 1.83 Mb genome of *Haemophilus influenzae*. This truly was a landmark because it was the first time that the complete genome of a free-living organism had been achieved (Tang *et al.*, 1997). Subsequently, a variety of other firsts were achieved: the genome of the smallest autonomous self-replicating entity (*Mycoplasma genitalium* in 1995), the first archaeal genome (*Methanococcus jannaschii* in 1996), and then the important achievement of the complete sequence of the 4.6 Mb of

E. coli achieved by an American group led by Fred Blattner and, independently, by a Japanese group led by Hirotada Mori and Takashi Horiuchi (see Pennisi, 1997).

The list of prokaryotic organisms whose genomes have been sequenced reveals different priorities. In some cases, the driving force was to understand evolutionary relationships between different organisms, as in the case of the archaeal genomes (Olsen and Woese, 1997), and in the case of *Mycoplasma genitalium* it was to understand what constitutes a *minimal genome*. In other cases, as in *E. coli* and *B. subtilis*, the priority was simply to further basic research using organisms that had been well studied in the laboratory. For many researchers, the big prize has been *E. coli*, the bacterium that has been the most intensively studied, biochemically and genetically. Surprisingly, given the huge amount of prior investigation, almost 40% of the initially identified 4288 genes have no known function and are now the subject of intensive investigation. For many other organisms, however, the primary motivation for genome sequencing has been their medical relevance.

Disease-related prokaryotic genome projects

The UK Wellcome Trust has been a major supporter of genome projects for microbial pathogens and its Beowulf Genomics program has established a large number of such programs at the Sanger Centre (electronic reference 6). Various other organizations have supported similar programs. In some cases, prokaryotes have been selected for genome sequencing because of their known associations with chronic diseases (Danesh *et al.*, 1997). They include *Helicobacter pylori* (associated with peptic ulcers) and *Chlamydia penumoniae* (associated with respiratory disease and also with coronary heart disease). Other completed projects have yielded the genomes of prokaryotes known to be causative agents of disease (*Table 13.4*), such as *Mycobacterium tuberculosis* (Cole *et al.*, 1998),

Table 13.4: Germ wars: examples of pathogenic microorganisms for which genome projects have been developed

Organism	Type and genome size	Genome size	Associated disease
Bordetella pertussis	Bacterium	3.88 Mb	Whooping cough
Borrelia burgdorferi	Spirochete bacterium	0.95 Mb	Lime disease
Chlamydia pneumoniae	Intracellular bacterium	1.0 Mb	Respiratory disease; coronary heart disease
Chlamydia trachomatis	Intracellular bacterium	1.7 Mb	Trachoma is a major cause of blindness
Clostridium difficile	Bacterium	4.4 Mb	Antibiotic-associated diarrhoea; pseudomembranous colitis
Helicobacter pylori	Bacterium	1.67 Mb	Peptic ulcers
Leishmania major	Parasitic protozoan	34 Mb	Leishmaniasis
Mycobacterium leprae	Bacterium	2.8 Mb	Leprosy
Mycobacterium tuberculosis	Bacterium	4.4 Mb	Tuberculosis
Plasmodium falciparum	Parasitic protozoan	30 Mb	Malaria
Rickettsia prowazekii	Bacterium	1.1 Mb	Typhus
Salmonella typhi	Bacterium	4.5 Mb	Typhoid fever
Treponema pallidum	Spirochete bacterium	1.1 Mb	Syphilis
Trypanosoma brucei	Parasitic protozoan	30 Mb	African trypanosomiasis (sleeping sickness)
Trypanosoma cruzi	Parasitic protozoan	30 Mb	American trypanosomiasis
Vibrio cholerae	Bacterium	2.5 Mb	Cholera
Yersinia pestis	Bacterium	4.38 Mb	Plague; schistosomiasis; filariasis

Box 13.3

Model organisms for which genome projects are considered particularly relevant to the Human Genome Project

Escherichia coli

This bacterium has been the most extensively studied, both biochemically and genetically, and was therefore an early priority for the Human Genome Project. The full genome sequence was released in 1997 but since then there have been a very large number of other prokaryotic genome projects, many of which have been completed already, including many pathogenic organisms (Section 13.3.1).

Saccharomyces cerevisiae

This budding single-celled yeast is one of the most extensively studied eukaryotes. Partly because of the high frequency of nonhomologous recombination, it has been very amenable to genetic analysis, and a large amount of information was known about its gene structure, regulation and function. Surprisingly, though, a very considerable fraction of the genes that have now been identified have no known function, but strenuous efforts are being developed to look at the expression and function of every one of the 6000 or so genes. A significant percentage of yeast genes have human homologs, and some surveys of human disease genes have suggested that about 20% have yeast homologs (Andrade *et al.*, 1998). Some essential cellular functions, such as DNA repair, have been conserved between human and yeast cells, and as a result, yeast cells have provided useful models for understanding such processes.

Caenorhabditis elegans

This 1 mm long nematode consists of just 959 somatic cells and is a form of roundworm. These worms outnumber all other complex creatures on the planet and are found almost everywhere in the temperate world, flourishing in compost. They infect a billion humans, spread diseases including river blindness and elephantiasis, and devour crops. *C. elegans* is an important model of development and it is also amenable to genetic analysis. In the latter case, in addition to the standard methods of knocking out gene function at the DNA level, transient inactivation of expression can be achieved for specific genes by *RNA interference (RNAi) technology*. In this approach, an investigator injects double-stranded RNA for the gene of interest into the oocyte. Although not clearly understood, somehow the double-stranded RNA inactivates expression of the relevant gene and the resulting mutant phenotypes can be examined for clues to gene function (Fire *et al.*, 1998).

Several features make *C. elegans* a good model organism for developmental biology studies:

■ *Expression studies.* Because *C. elegans* is transparent, the green fluorescent protein gene (GFP; see Section 20.2.6) can be linked to a *C. elegans* gene and used to find out where the gene is expressed in the worm.
■ *Lineage studies.* The transparency of *C. elegans* means that every cell can be seen and followed during development. As a result, the exact lineage of every cell in *C. elegans* is known – information which is unknown in all other multicellular organisms.

■ *Nervous system.* There is also a complete wiring diagram of the nervous system: all 302 neurons and the connections between them are known. *C. elegans* also possesses genes for most of the known molecular components of vertebrate brains. While many scientists believe that the human brain is so complex that we can never hope to fully understand it, understanding the simple nervous system of *C. elegans* will provide enormous insights, and knowing all the *C. elegans* genes will be extremely important in understanding its nervous system.

In addition to neurobiology, several other areas are being explored that can be expected to produce insight into human systems. For example, aging is readily studied: the worm develops from one cell into the fully grown form within 3 days and survives for only 2 weeks. Apoptosis is another very relevant area. Genes involved in apoptosis are often highly conserved and the pattern of apoptosis is well understood in *C. elegans* development (a total of 1090 cells develop initially, but 131 cells undergo cell suicide).

Drosophila melanogaster

The fruit fly has a short life cycle, is particularly amenable to sophisticated genetic analyses (see Perrimon, 1998, for a summary of new advances in studying gene function), and has been studied extensively over many decades. Large polytene chromosomes allow precise (tens of kb) localization of chromosome breakpoints and precise localization of DNA clones by *in situ* hybridization. A particular *Drosophila* transposable element known as the P element permits several types of experimental manipulation, including P element-mediated mutagenesis (Spradling *et al.*, 1995) and P element-mediated transgenesis (see Section 21.2.1). Unequal recombination between adjacent P element inserts can also produce precise deletions. Spatially and temporally restricted expression of transgenes is possible using the GAL4-UAS system of conditional gene expression. Large scale mutagenesis screens are possible and *RNAi technology* (see above) has recently been used to transiently inactivate specific genes. In many cases (perhaps two-thirds of the 12 000 *Drosophila* genes), loss of function does not result in a mutant phenotype, but transgene misexpression often gives clues to gene function by producing dominant/dominant negative phenotypes. One-generation screens for suppressors/enhancers of dominant mutant phenotypes can identify interacting genes. The yeast flp-frt recombinase system can be used to induce mitotic clones and so form homozygous patches permitting observation of phenotypes of lethal recessive mutations at late stages of development. Mitotic recombination can also be used in a one-generation screen to score mutant phenotypes in clones and recover lethal mutations that affect late development.

Although *Drosophila* is an invertebrate, and the number of *Drosophila* genes is perhaps only a quarter of the number of human genes, there are nevertheless some remarkable similarities between human and *Drosophila* genes. Much of the difference in gene number is due to gene duplication events resulting in larger gene families in humans, and a considerable proportion of *Drosophila* genes have human homologs, including genes that underlie many genetic disorders and cancer genes (St John and Xu, 1997). The level of homology has permitted successful electronic screening to identify human cDNAs corresponding to mutant *Drosophila* genes (Banfi *et al.*, 1996). Many of the highly conserved genes play important roles in early development, which is comparatively well understood in *Drosophila*, and many of the relevant pathways in early development and in some other crucial cellular processes are essentially conserved from *Drosophila* to mammals. As a result, *Drosophila* can be used as a model system for exploring gene function and identifying interacting partners of genes which have direct relevance to human systems.

The mouse
The model organism considered most relevant to the Human Genome Project (Meisler, 1996). It is the mammalian species with the most highly developed genetics and it is an extensively used model of mammalian development. Its small size and short generation time have allowed large-scale mutagenesis programs and extensive genetic crosses, and various features aid in mapping genes and phenotypes (see *Box 15.4*). Because of the comparatively high level of sequence conservation between human and mouse coding sequences (Section 14.6.1), almost all human genes have an easily identifiable mouse homolog. Large chromosomal segments are conserved between mouse and man (see *Figure 14.23*), and if a region of the mouse genome is mapped to high resolution, the information can be used to make predictions about the orthologous region of the human genome (and vice versa). This is particularly relevant to medical research because orthologous mouse and human mutants often show similar phenotypes, so that positional cloning of a disease gene in one species may have considerable relevance to the other species (Section 15.4.3). The ability to construct mice with pre-determined genetic modifications to the germline (by transgenic technology and gene targeting in embryonic stem cells) has been a powerful tool in studying gene expression and function and in creating mouse models of human disease. See Chapter 21 for details.

Other model organisms
Some other model organisms have also been particularly relevant to human research, although genome projects are undeveloped. They include:

- *The rat.* Rats, being considerably larger than mice, have for many years been the mammal of choice for physiological, neurological, pharmacological and biochemical analysis. They may also provide genetic model systems for complex human vascular and neurological disorders, such as hypertension and epilepsy (for various reasons, there are no mouse models for such diseases). Genetic analysis in laboratory rats, however, is much less advanced than in mice, partly because of the relatively high cost of rat breeding programs and the current inability to modify the rat germline by gene targeting. Recently, however, high-resolution genetic and physical maps have been constructed.

- *The pufferfish* (Fugu rubripes rubripes). The value has mostly been in *comparative genomics* (see Section 13.4.1). The pufferfish has an extremely compact genome with about the same number of genes as mammals compressed into a genome only about one-seventh of the size of the human or mouse genomes. Conservation of exons and important regulatory sequences aids identification of human equivalents by comparative mapping (see Clark, 1999).

- *Zebrafish.* The zebrafish is a good model of vertebrate development. The embryo is transparent, facilitating identification of developmental mutants. Genes that are important in vertebrate development are often very highly conserved, so human developmental control genes normally have easily identifiable orthologs in zebrafish. Large-scale mutagenesis screens have produced many valuable developmental mutants and some of these have been used to model human disorders (see *Box 21.2*). *RNA interference technology* (see above) is also being used to inactivate specific genes. Establishment of good genetic and physical maps will aid gene identification (see Roush, 1997).

Treponema pallidum (the causative agent of syphilis) and *Rickettsia prowazekii* (the causative agent of typhus, also of interest because it is thought to be closely related to the prokaryotic precursor of mitochondria – see Section 14.1.1). In addition to having a more complete understanding of these organisms, the new information can be expected to lead to more sensitive diagnostic tools and new targets for establishing drugs and vaccines.

13.3.2 The S. cerevisiae genome project was the first of several protist genome projects to be completed and provided for the first time the entire DNA sequence of a eukaryote

The Saccharomyces cerevisiae *genome project*
The budding yeast *S. cerevisiae* is a single-celled eukaryote (*protist*) which has been the subject of intensive genetic analyses. Its 16 chromosomes were sequenced by European and American consortia and the complete sequence was reported by Goffeau *et al.* (1996). This represented another milestone in biology: the first complete sequence for a eukaryotic cell. The data indicate that yeast genes are closely clustered, being spaced, on average, once every 2 kb. Of the 6340 genes, about 7% specify a mature RNA species. Although it has been one of the most intensely studied organisms, 60% of its genes had no experimentally determined function. However, a sizeable fraction of yeast genes have an identifiable mammalian homolog, and in only about 25% of yeast genes is there no clue whatsoever to their function (Botstein *et al.*, 1997). The successful conclusion of this project has now opened up large-scale functional analyses (see Section 13.4.1).

Other protist genome projects
The other protist genome projects include organisms that have been well studied and are amenable to biochemical and genetic analyses, including the fission yeast *Schizosaccharomyces pombe* and the molds *Aspergillus nidulans*

and *Neurospora crassa*. In addition, various organizations, including the Wellcome Trust and the World Health Organization, have been involved in supporting genome projects for various animal protists (protozoans) involved in human parasitic infections (see *Table 13.4*). In most cases the genome sizes are substantial. For example, the *Plasmodium falciparum* genome is about 30 Mb in size and contains 14 chromosomes, some of which had been sequenced at the time of writing (Gardner *et al.*, 1998).

13.3.3 The Caenorhabditis elegans *genome project was the first of several animal genome projects to be completed*

The Caenorhabditis elegans *genome project*

In addition to the numerous microbial genome projects, a variety of animal and plant genome projects are underway, of which the *C. elegans* project was the first to be completed. Because of its large genome size (nearly 100 Mb), the *C. elegans* genome project was viewed as the major pilot model for large-scale sequencing of the human genome. From the research point of view, this organism, although a simple one only about 1 mm long, was an important model of development and was also useful for modeling other processes relevant to human cells (see *Box 13.3*). A consortium of the Sanger Centre and Washington University Sequencing Center in St Louis reported the essentially finished sequence at the end of 1998 (*C. elegans* Sequencing Consortium, 1998).

The genome project identified a total of 19 099 polypeptide-encoding genes and over 1000 RNA-encoding genes, giving an average spacing of one gene every 5 kb. A surprisingly high number of the genes appear to occur as part of an operon, where individual genes are transcribed as part of a large multigenic RNA transcript. Comparison with published sequences from elsewhere reveals that about one in three *C. elegans* genes shows similarities to previously known genes, and 12 000 of the polypeptide-encoding genes are of unknown function. Now major efforts are being made to investigate specific gene function, and large-scale chemical mutagenesis programs are seeking to produce large numbers of mutant phenotypes.

Other animal genome projects

Other projects include the following:

- *The* D. melanogaster *genome project*. This was established largely as a collaboration between the University of California at Berkeley laboratory and a consortium of European laboratories (Rubin, 1998). Of the 165 Mb genome, about 125 Mb is euchromatic. Initially, the target for sequencing the 125 Mb euchromatic component was the year 2001, but collaboration with industrial partners may mean that it is finished much earlier than this (see *Box 13.2*). Transposable P element insertion is being employed to carry out insertional mutagenesis on a large scale (Spradling *et al.*, 1995).
- *The mouse genome project*. Because of various features

(see *Box 13.3*), the mouse provides the model genome which is most relevant to the Human Genome Project and is expected to have about the same number of genes. Good genetic maps exist. By May 1999 more than 10 000 genes, of which about 7 000 had been mapped, had been entered into the mouse genome database maintained at the Jackson Laboratories, Bar Harbor (electronic reference 7), and large mouse EST sequencing programs were under way. A working draft of the sequence of the 3000 Mb genome is expected by 2005.

13.4 Life in the post-genome (sequencing) era

Once the sequence of the human genome is known, what difference will it make? Certainly, there will be a huge boost to basic research as we grapple with the fundamental biological question of how our genome is interpreted to specify a person. In the so-called *post-genome era*, accurate genetic testing will become widely available, not just for genetic disorders, but also in terms of genetic susceptibility to a variety of different conditions, including infectious diseases. But there may be a downside in terms of discrimination against individuals. Improved treatments can also be expected. The much vaunted gene therapy approaches may prove technically difficult, but the new infomation will undoubtedly assist the development of novel therapies. The following two sections are selective and merely illustrate some of the implications of knowledge of our genome.

13.4.1 *Comparative and whole-genome analyses permit large-scale studies of DNA organization and evolution and of gene expression and function*

The human genome project had not reached its half-way point before serious consideration was given to what the research priorities should be in the post-genome era. Certainly, the sequencing of whole genomes will provide revolutionary approaches to biomedical research. For the first time, there are opportunities to compare whole genomes and the newly developed field of bioinformatics is set to take off (Gershon *et al.*, 1997; Smith, 1998). *Genome-wide analyses* of gene expression and function will become a major area of investigation.

Comparative genomics

Comparative genomics involves analysis of two or more genomes to identify the extent of similarity of various features, or large-scale screening of a genome to identify sequences present in another genome. The examples below are merely meant to be illustrative of some of the applications.

- *Evolutionary relationships*. One early application involved comparison of archaeal genomes with eubacterial and eukaryotic genomes to infer evolutionary relatedness (see Section 14.1.1). Once eukaryotic genomes were sequenced, they too could be compared. For example, sequencing of the *C. elegans* genome permitted an evaluation of how its genes compared with those of a simple eukaryote (*S. cerevisiae*) and a bacterium (*E. coli*) (*Figure 13.8*; *C. elegans* Sequencing Consortium, 1998).

■ *Identification of regulatory elements*. We have very limited information about regulatory elements in complex eukaryotic genomes. By referring to databases of known regulatory element sequences, computer programs can inspect new genomic sequences for the presence of regulatory elements, but the efficiency is very low. An alternative approach is to use high evolutionary conservation as a screen for regulatory elements within otherwise poorly conserved noncoding DNA. Large-scale sequencing of orthologous chromosomal regions in reasonably distantly related species should lead to identification of common regulatory elements, in addition to common exons. One example of this approach involves sequencing the genome of *Caenorhabditis briggsae* and comparing it with that of *C. elegans*. Although the genomes of these two nematodes are essentially colinear, they diverged from a common ancestor about 80 million years ago. As a result, noncoding sequences other than regulatory regions have undergone complete sequence divergence. Equivalent large-scale comparative sequencing in orthologous human and mouse chromosomal regions is being conducted for the same purpose and pufferfish comparisons have also been important (Clark, 1999).

■ *Gene identification*. Electronic screening of EST databases can identify homologs of biologically interesting genes in other species. For example, systematic screening of the *dbEST* database of ESTs has revealed many potentially interesting human homologs of *Drosophila* genes known to be loci for mutant phenotypes (Banfi *et al.*, 1996).

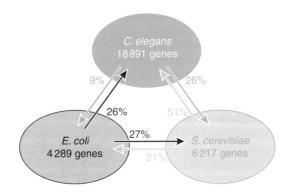

Figure 13.8: Percentage of matching proteins in three fully sequenced organisms.

The gene number given for the three organisms is the number of *polypeptide encoding genes*. As expected, the percentage of the 18 891 different *C. elegans* polypeptides which found a match in the *E. coli* proteome was low (9%), whereas 26% of the *C. elegans* polypeptides had a match within the *S. cerevisiae* proteome. Data from the *C. elegans* Sequencing Consortium (1998).

Functional genomics

Functional genomics refers to large-scale or global investigations of gene function. For example, the way in which a cell responds to a particular signal or environmental stimulus can be monitored by simultaneously analysing the expression patterns of every single gene. Already, microarrays have been devised to track the expression of virtually all 6200 yeast genes (*Figure 13.9*; Wodicka *et al.*, 1997). Strenuous efforts are now being mounted to investigate the function of each of the 6000 or so polypeptide-encoding yeast genes (Winzeler and Davis, 1997).

Similar types of approaches will also permit extensive investigation of human gene function and dissection of complex regulatory pathways. Once all human genes are known, we can know all the products. We are accustomed to large-scale DNA analyses (which bred the term *genome*); now as we explore RNA and polypeptide

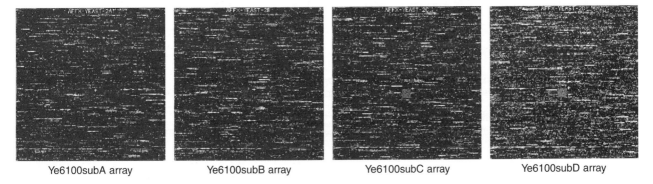

| Ye6100subA array | Ye6100subB array | Ye6100subC array | Ye6100subD array |

Figure 13.9: Genome-wide expression screening in *Saccharomyces cerevisiae*.

The mRNA levels for practically all (about 6100) yeast genes can be monitored simultaneously using the GeneChip® Ye 6100 set from Affymetrix. Each gene is analyzed with about 20 pairs of specific, unique 25 mer oligonucleotide probes. See also Section 5 and *Figure 20.6*. Image courtesy of Affymetrix, Inc. (Santa Clara, CA). GeneChip® and Affymetrix® are US registered trademarks used by Affymetrix, Inc.

expression products on a global scale, new terms are being coined: the **transcriptosome** (the total collection of RNA transcripts in a cell) and the **proteome** (the total collection of polypeptides/proteins expressed in a cell). The new science of **proteomics** is devoted to the study of global changes in protein expression and the systematic study of protein–protein interactions (Blackstock and Weir, 1999; Dove, 1999). The former can be tracked by using high-throughput expression assays (such as 2D gel electrophoresis and mass spectrometry); the latter by large-scale application of methods such as *two hybrid screening* (Section 20.4.1). For example, this approach is now being used to systematically screen the yeast proteome (Lecrenier *et al.,* 1998).

13.4.2 Without proper safeguards, the Human Genome Project could lead to discrimination against carriers of disease genes and to a resurgence of eugenics

Any major scientific advance carries with it the fear of exploitation. The Human Genome Project is no exception, and the perceived benefits of the project can also have a downside. For example, once we know all the human genes and can detect large numbers of disease-associated mutations, there will be enormous benefit in targeting prevention of disease to those individuals who can be shown to carry disease genes. However, the same information can also be used to discriminate against such individuals by insurance companies. For example,

there is the very real prospect of insurance companies insisting on genetic screening tests for the presence of genes that confer susceptibility to common disorders, such as diabetes, cardiovascular disease, cancers and various mental disorders. Perfectly healthy individuals who happen to be identified as carrying such disease-associated alleles may then be refused life or medical insurance. Clearly, such discrimination is practised on a small scale at the moment; what is alarming to many is the prospect of discrimination against a very large percentage of the individuals in our society. It is also important to preserve people's right not to know. A fundamental ethical principle in all genetic counseling and genetic testing is that genetic information should be generated only in response to an explicit request from a fully informed adult patient.

Another troublesome area is the question of biological determinism and whether comprehensive knowledge of human genes could foster a revival of *eugenics,* the application of selective breeding or other genetic techniques to 'improve' human qualities (Garver and Garver, 1994). In the past, negative eugenic movements in the US and Germany severely discriminated against individuals who were adjudged to be inferior in some way, notably by forcing them to be sterilized. The possibility also exists of a preoccupation with *genetic enhancement* to positively select for heritable qualities that are judged to be desirable (see Section 22.6). In recognition of the above problems, the US Human Genome Project has devoted considerable resources to support research into the ethical, legal and social impact of the project.

Further reading

Borsani G, Ballabio A, Banfi S (1998) A practical guide to orient yourself in the labyrinth of genome databases. *Hum. Mol. Genet.,* **7**, 1641–1648.

Database issue of Nucleic Acid Research (1999) *Nucleic Acids Res.,* **17**, 3441–3665.

Guyer MS, Collins FS (1993) The Human Genome Project and the future of medicine. *Am. J. Dis. Child.,* **147**, 1145-1152.

Wilkie T (1993) *Perilous Knowledge: the Human Genome Project and its Implications.* Faber and Faber, New York.

US Department of Health and Human Services and US Department of Energy (1990) Understanding our genetic inheritance. The US Human Genome Project: the first five years FY 1991–1995.

US National Institute of Health and Department of Energy (1993) Genetic Information and Health Insurance. US NIH–DOE Working Group on Ethical and Social Implications of Human Genome Research.

Electronic information on the Human Genome Project (and related projects)

Useful websites are found at:

- the US National Human Genome Research Institute (NGHRI) at http://www.nhgri.nih.gov/
- the US National Center for Biotechnology Information (NCBI) at http://www.ncbi.nlm.nih.gov/
- the Genome Web maintained at the UK Human Genome Mapping Project Resource Centre at http://www.hgmp.mrc.ac.uk/GenomeWeb/

Electronic references

1. GeneMap'99 at http://www.ncbi.nlm.nih.gov/genemap99/
2. The Cancer Genome Anatomy Project at http://www.ncbi.nlm.nih.gov/CGAP/
3. The Human Genome Sequencing Index at http://www.ncbi.nlm.nih.gov/HUGO/
4. The Genome Monitoring Table maintained at the European Bioinformatics Institute at http://www.ebi.ac.uk/~sterk/genome-MOT/
5. Terry Gaasterland's running list of genome projects at http://www-fp.mcs.anl.gov/~gaasterland/genomes.html
6. Beowulf Genomics at http://www.beowulf.org.uk/
7. The Mouse Genome Database at http://www.jax.org/

References

Adams MD, Kelley JM, Gocayne JD *et al.* (1991) Complementary DNA sequencing: expressed sequence tags and Human Genome Project. *Science*, **252**, 1651–1656.

Andrade MA, Sander C, Valencia A (1998) Updated catalogue of homologues to human disease-related proteins in the yeast genome. *FEBS Lett.*, **426**, 7–16.

Banfi S, Borsani G, Rossi E *et al.* (1996) Identification and mapping of human cDNAs homologous to Drosophila mutant genes through EST database searching. *Nature Genet.*, **13**, 167–174.

Blackstock WP, Weir MP (1999) Proteomics: quantitative and physical mapping of cellular proteins. *Trends Biotechnol.*, **17**, 121–127.

Botstein D, White RL, Skolnick M, Davis RW (1980) Construction of a genetic linkage map in man using restriction fragment length polymorphism. *Am. J. Hum. Genet.*, **32**, 314–331.

Botstein D, Chervitz SA, Cherry JM (1997) Yeast as a model organism. *Science*, **277**, 1259–1260.

C. elegans **Sequencing Consortium** (1998) Genome sequence of the nematode *C. elegans*: a platform for investigating biology. *Science*, **282**, 2012–2017.

Cantor CR (1998) How will the human genome project improve our quality of life? *Nature Biotechnol.*, **16**, 212–213.

Cavalli-Sforza LL, Wilson AC, Cantor CR, Cook-Deegan RM, King MC (1991) Call for a worldwide survey of human genetic diversity: a vanishing opportunity for the Human Genome Project. *Genomics*, **11**, 490–491.

Chumakov I, Rigavit P, Guillou S *et al.* (1992) Continuum of overlapping clones spanning the entire human chromosome 21q. *Nature*, **359** (6394), 380–387.

Clark MS (1999) Comparative genomics: the key to understanding the Human Genome Project. *BioEssays*, **21**, 121–130.

Cohen D, Chumakov I, Weissenbach J (1993) A first generation physical map of the human genome. *Nature*, **366**, 698–701.

Cole ST, Brosch R, Parkhill J *et al.* (1998) Deciphering the biology of *Mycobacterium tuberculosis* from the complete genome sequence. *Nature*, **393**, 537–544.

Collins F, Guyer MS, Chakravarti A (1997) Variations on a theme: cataloging human DNA sequence variation. *Science*, **262**, 43–46.

Collins F, Patrinos A, Jordan E *et al.* (1998) New goals for the US Human Genome Project: 1998–2003. *Science*, **282**, 682–689.

Danesh J, Newton R, Beral V (1997) A human germ project? *Nature*, **389**, 21–24.

Deloukas P, Schuler GD, Gyapay G *et al.* (1998) A physical map of 30 000 human genes. *Science*, **282**, 744–746.

Donis-Keller H, Green P, Helms C *et al.* (1987) A genetic linkage map of the human genome. *Cell*, **51**, 319–337.

Dove A (1999) Proteomics: translating genomics into products? *Nature Biotechnol.*, **17**, 233–236.

Fire A, Xu S, Montgomery M, Kostas SA, Driver SE, Mello CC (1998) Potent and specific genetic interference by double-stranded RNA in *Caenorhabditis elegans*. *Nature*, **391**, 806–811.

Gardner MJ, Tettelin H, Carucci DJ *et al.* (1998) Chromosome 2 sequence of the human malaria parasite *Plasmodium falciparum*. *Science*, **282**, 1126–1128.

Garver KL, Garver B (1994) The Human Genome Project and eugenic concerns. *Am. J. Hum. Genet.*, **54**, 148–158.

Gershon D, Sobral BW, Horton B, Wickware P, Gavaghan H, Strobl M (1997) Bioinformatics in a post-genomics age. *Nature*, **389**, 417–422.

Goffeau A, Barrell BG, Bussey H *et al.* (1996) Life with 6000 genes. *Science*, **274**, 546–567.

Harding RM, Sajantila A (1998) Human genome diversity – a Project? *Nature Genet.*, **18**, 307–308.

Hudson TJ, Stein LD, Gerety SS *et al.* (1995) *Science*, **270**, 1945–1954.

Ichikawa H, Hosoda F, Arai Y, Shimizu K, Ohira M, Ohki M (1993) A *Not*I restriction map of the entire long arm of human chromosome 21. *Nature Genet.*, **4**, 361–365.

Knoppers BM (1999) Status, sale and patenting of human genetic material: an international survey. *Nature Genet.*, **22**, 23–25.

Lecrenier N, Foury F, Goffeau A (1998) Two-hybrid systematic screening of the yeast proteome. *BioEssays,* **20**, 1–6.

McKusick V (1989). HUGO news. The Human Genome Organization: history, purposes, and membership. *Genomics,* **5**, 385–387.

Meisler MH (1996) The role of the laboratory mouse in the human genome project. *Am. J. Hum. Genet.,* **59**, 764–771.

Murray JC, Buetow KH, Weber JL *et al.* (1994) A comprehensive human linkage map with centimorgan density. *Science,* **265**, 2049–2054.

Olsen GJ, Woese CR (1997) Archaeal genomics: an overview. *Cell,* **89**, 991–994.

Pennisi E (1997) Laboratory workhorse decoded. *Science,* **277**, 1432–1434.

Pennisi E (1998) A closer look at SNPs suggests difficulties. *Science,* **281**, 1787–1789.

Perrimon N (1998) New advances in Drosophila provide opportunities to study gene functions. *Proc. Natl Acad. Sci. USA,* **95**, 9716–9717.

Roush W (1997) A zebrafish genome project? *Science,* **275**, 923.

Rubin GM (1998) The *Drosophila* genome project: a progress report. *Trends Genet.,* **14**, 340–343.

Schafer AJ, Hawkins JR (1998) DNA variation and the future of human genetics. *Nature Biotechnol.,* **16**, 33–39.

Smith TF (1998) Functional genomics – bioinformatics is ready for the challenge. *Trends Genet.,* **14**, 291–293.

Spradling AC, Stern DM, Kiss I, Roote J, Laverty J, Rubin GM (1995) Gene disruptions using P transposable elements: an integral component of the *Drosophila* genome project. *Proc. Natl Acad. Sci. USA,* **92**, 10824–10830.

St John MA, Xu T (1997) Understanding human cancer in a fly? *Am. J. Hum. Genet.,* **61**, 1006–1010.

Tang CM, Hood DW, Moxon ER (1997) *Haemophilus* influence: the impact of whole genome sequencing on microbiology. *Trends Genet.,* **13**, 399–404.

Thomas SM, Davies ARW, Birtwistle NJ, Crowther SM, Burke JF (1996) Ownership of the human genome. *Nature,* **380**, 387–388.

Venter JC, Smith HO, Hood L. (1996) A new strategy for genome sequencing. *Nature,* **381**, 364–365.

Wang DG, Fan JB, Siao CJ *et al.* (1998) Large-scale identification, mapping and genotyping of single nucleotide polymorphisms in the human genome. *Science,* **280**, 1077–1082.

Weissenbach J, Gyapay G, Dib C, Vignal A, Morissette J, Millasseau P, Vaysseix G, Lathrop M (1992) A second generation linkage map of the human genome. *Nature,* **359**, 794–801.

Winzeler EA, Davis RW (1997) Functional analysis of the yeast genome. *Curr. Opin. Genet. Dev.,* **7**, 771–776.

Wodicka L, Dong H, Mittmann M, Ho M-H, Lockhart DJ (1997) Genome-wide expression monitoring in *Saccharomyces cerevisiae. Nature Biotechnol.,* **15**, 1359–1367.

Our place in the tree of life

By definition, the evolutionary origins of the human genome, and all genomes, are as old as life itself. The present chapter is not intended as an introduction to basic evolutionary theory or an overview of molecular evolutionary genetics *per se*. As a result, many fascinating areas are not covered, such as the idea that RNA used to be the primary information molecule before being superseded by DNA or the idea that the genetic code was initially a doublet code before evolving into the familiar triplet code, etc. The interested reader is advised to consult one of the more general molecular evolutionary genetics textbooks (see Further reading). Instead, this chapter is meant to focus on how comparative analyses of present day genomes have shed light on the evolutionary origin of human DNA. Much of the data are derived from comparisons of mammalian genomes, although comparison with more distant genomes is occasionally used to explain certain footprints of evolution, as in the origin of mitochondrial DNA and introns. A final section considers our recent evolutionary past and our uniqueness when compared with mammalian models, notably the mouse (an important model for understanding early human development and also human disease – see Chapter 21) and primates, our closest living relatives.

14.1 Evolution of the mitochondrial genome and the origin of eukaryotic cells

14.1.1 The mitochondrial genome most likely originated as a result of endocytosis of a prokaryotic cell by a eukaryotic cell precursor, but the nature of these cells is uncertain

In addition to the nuclear genome, mitochondria have a genome, as do the chloroplasts of plant cells. The organization and expression of mitochondrial and chloroplast genomes shows considerable similarities to that of prokaryotic cells (see below), suggesting that the evolution of eukaryotic cells involved a process in which a precursor to the eukaryotic cell (protoeukaryote) engulfed (endocytosed) some type of prokaryotic cell (the *symbiont*). Such a process is thought to have conferred a selective advantage for the resulting new cell and has been termed endosymbiosis.

The genomes of the engulfed prokaryotic cell and the host cell are envisaged to have given rise to the present day mitochondrial and nuclear genomes respectively (*Figure 14.1A*). However, the size of the mitochondrial genome of present day cells, especially in the case of animal cells, is very small. For example, the human mitochondrial genome is about 16 kb and has only 37 genes; the vast majority of mitochondrial proteins and functions are specified by genes in the nuclear genome (Section 7.1.1). Present day prokaryotic cells comprise two quite different types of cell: eubacteria and **archaea** (*Box 14.1*). They typically contain one or a few Mb of DNA and contain several hundreds or thousands of genes. As present day mitochondrial genomes are very much smaller than this, theories of mitochondrial genome origin envisage that many of the genes originally present in the engulfed cell were transferred to the genome of the host cell (*Figure 14.1B*; Doolittle, 1998). Currently there are two classes of hypothesis to explain the origin of mitochondria and the eukaryotic cell, exemplified by the serial endosymbiont hypothesis and the hydrogen hypothesis.

The serial endosymbiont hypothesis

The first version of the endosymbiont hypothesis was developed 30 years ago and was invoked to explain several features of the organization and expression of the mitochondrial genome which bore a resemblance to prokaryote genomes: small size, absence of introns, a very high percentage of coding DNA, a conspicuous lack of repeated DNA sequences and comparatively small prokaryotic-like rRNA genes. Phylogenetic analyses of rRNA sequences suggested that mitochondria were particularly closely related to the α subdivision of purple bacteria. Consequently, mitochondria were believed to

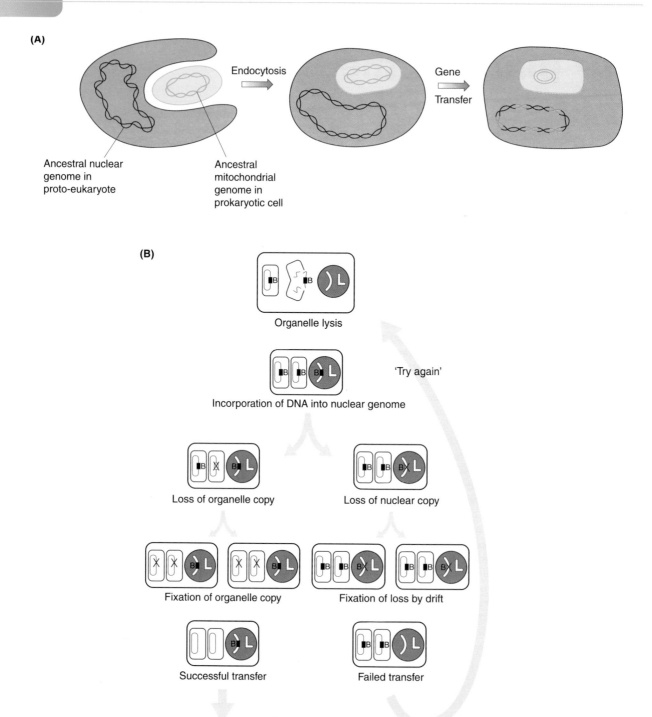

Figure 14.1: The human mitochondrial genome probably originated following endocytosis of a prokaryotic cell by a eukaryotic precursor cell.

(**A**) General endocytosis model whereby the mitochondrial genome is imagined to originate by endocytosis of a prokaryotic cell. In this particular example, the endocytosing cell does not have a nucleus but some models have imagined endocytosis by a protoeukaryotic cell with a nucleus. Following endocytosis, genes in the genome of the prokaryote are imagined to have been transferred to the precursor of the nuclear genome, leaving a much reduced mitochondrial genome. (**B**) One possible ratchet mechanism to explain gene loss from the genome of the endocytosed cell to the precursor of the nuclear genome. Redrawn from Doolittle (1998) *Trends Genet.*, **14**, pp. 307–311, with permission from Elsevier Science.

Box 14.1

The three kingdoms of life

The subdivision of cells into prokaryotes and eukaryotes as described in Section 2.1.1 is a simplification. During the last two decades or so, it has become clear that there are three kingdoms of life, the eukaryotes and two quite distinct kingdoms of prokaryotes:

■ **Eubacteria** (often now abbreviated to *bacteria*) are the commonly encountered prokaryotes which have traditionally been well-studied, e.g. Gram-negative and Gram-positive bacteria, cyanobacteria, etc.

■ **Archaea** (previously called *archaebacteria*, a term which is now considered inappropriate because of the increasing evidence for an evolutionary origin quite distinct from bacteria). They are single celled organisms typically found in extreme environments (e.g. hot springs, very high salt concentration, extremes of pH, etc.). Archaeal genomes resemble eubacterial genomes in form, but they are more related to eukaryotes than eubacteria in terms of the control of genetic information, e.g. promoters, enzymes involved in DNA replication, transcription, etc.

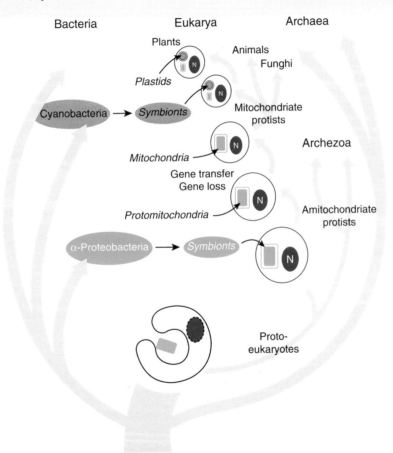

The three kingdoms of life and the origin of eukaryotes by endosymbiosis.

In this example, the mitochondria of eukaryotes are portrayed as having arisen by endocytosis of α-proteobacteria. The chloroplasts of plant cells are imagined to have originated by a subsequent endocytosis of cyanobacteria. Protists are single-celled eukaryotes. Reprinted from Doolittle (1998), *Trends Genet.*, **14**(8), pp. 307–311, with permission from Elsevier Science.

have originated as a result of endocytosis by anaerobic eukaryotic precursor cells of an aerobic eubacterium of this type (with an oxidative phosphorylation system). This was imagined to have occurred about 1.5 billion years ago when oxygen started to accumulate in significant quantities in the Earth's atmosphere; cells which

acquired the capacity for oxidative phosphorylation would have been at a strong selective advantage.

More recently, it has become clear that the nuclear genome of eukaryotes is an evolutionary chimera which is genetically related to both archaeal and eubacterial genomes. For example, eukaryotic genes involved in information transfer (replication, transcription, translation, etc.) are largely derived from archaea; however, operational genes (those involved in metabolism and biosynthesis of cofactors, amino acids, fatty acids, etc.) appear to be descended from eubacteria (Rivera *et al.*, 1998). The modern serial endosymbiont hypothesis envisages that eukaryotic cells evolved by a series of endocytoses. What is not clear in this model is when the prokaryotic cell containing the precursor mitochondrial genome was endocytosed, whether by a nucleated cell resembling a eukaryote, or by an archaeon (see Lopez-Garcia and Moreira, 1999 and Gray *et al.*, 1999 for references).

The hydrogen (syntrophy) hypothesis

The serial endosymbiont model considers that the precursor to the nuclear genome arose first and subsequently a precursor to the mitochondrial genome was captured by endocytosis, maybe even from a nucleated host cell. The hydrogen hypothesis, by contrast, considers that there could have been a *simultaneous* origin for the precursors of the nuclear and mitochondrial genomes, and that the respiration of ancestral mitochondria was anaerobic Martin and Muller, 1998; Lopez-Garcia and Moreira, 1999). In this hypothesis eukaryotes are suggested to have arisen by association of an anaerobic strictly autotrophic archaeon which was hydrogen-requiring (possibly a methanogen) with a hydrogen-producing eubacterium, such as an α-proteobacterium. The anticipated driving force for this association was a symbiotic metabolic association (syntrophy). Subsequently, to avoid pointless cycling of metabolites in its cytoplasm the host lost its autotrophic pathway and an irreversible heterotroph emerged containing ancestral mitochondria but no longer dependent on hydrogen. More efficient oxigenic respiration was then adopted by many such organisms and aerobic mitochondria evolved.

The hydrogen hypothesis was founded on several important observations. First, eukaryotes which do not possess mitochondria possess eubacterial-like metabolic enzymes (in addition to other known eubacterial-like genes). Secondly, hydrogenosomes (small membrane-bounded hydrogen-producing organelles found in some anaerobic protozoa and some types of fungi) appear to share a common ancestry with mitochondria (Gray *et al.*, 1999; Lopez-Garcia and Moreira, 1999). It is perhaps significant, too, that whereas most eukaryotes use histones to compact their nuclear DNA, the only prokaryotes that have histones and nucleosomes are the Euryarchaeota, the division of the Archaea that includes the hydrogen-consuming methanogens. This field is a fast-moving one and so interested readers are advised to consult recent review articles.

14.1.2 The mitochondrial genetic code most likely evolved as a result of reduced selection pressure in response to a greatly diminished coding capacity

The human mitochondrial genetic code is slightly different from the 'universal' genetic code that is used in the expression of polypeptides encoded by prokaryotic genomes, eukaryotic nuclear genomes and plant mitochondrial genomes (*Figure 1.22*). In addition, although it is identical to the genetic codes of other mammalian mitochondrial genomes, it shows some differences to the nonuniversal genetic code in the mitochondria of other eukaryotes, such as *Drosophila* and yeast cells. As described in the preceding section, theories of mitochondrial origins have envisaged that genes were transferred from the precursor mitochondrial genome to the nuclear genome. This may have occurred by successive processes of organelle lysis with incorporation of DNA into the nuclear genome, loss of organelle copies and fixation of gene loss by genetic drift (Doolittle, 1998). By whatever mechanism, the loss in genetic capacity could have relaxed the normal selection pressure that applies to large genomes.

From the above, one would expect that the original genome of the endocytosed prokaryotic cell would, like all large genomes, have been subject to strong conservative selection pressure. The universal genetic code would have been used because even slight alterations in the code could result in lack of function (or aberrant function) for large numbers of vitally important gene products, resulting in cell death. However, as the coding potential steadily diminished by gene transfer to the nuclear genome, there would have been progressively less selection pressure to conserve the original genetic code. Eventually a severely depleted genome resulted (only 13 genes in the human mitochondrial genome encode polypeptides). Slight altering of the otherwise universal genetic code could therefore be achieved without provoking disastrous consequences, because only a tiny number of polypeptides would be involved. It is also likely that the codons which have been altered (see *Figure 1.22*) have not been used extensively in locations where amino acid substitutions would have been deleterious.

14.2 Evolution of the eukaryotic nuclear genome: genome duplication and large-scale chromosomal alterations

As described in Section 14.1.1 the nuclear genome of eukaryotes is thought to have initially evolved as a mixture of archaeal genes (involved in information transfer) and eubacterial genes (involved in metabolism and other basic cellular functions). As eukaryotes developed into complex multicellular organisms, the number of genes and size of the nuclear genome increased and various other properties were altered, notably the amount of repetitive DNA and the fraction of coding DNA (see *Table 14.1*). The transition from the DNA of a typical simple eukaryotic cell precursor to the DNA of a human cell is therefore thought to have involved a huge increase in the size of the genome and a sizeable increase in gene number and in the percentage of noncoding and repetitive

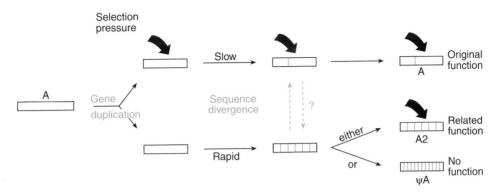

Figure 14.2: Gene duplication can lead to the acquisition of novel function or the formation of a pseudogene.

Duplication of gene A results in two equivalent gene copies. Selection pressure need be applied to only one gene copy (top) to maintain the presence of the original functional gene product. The other copy (bottom), will continue to be expressed but, in the absence of selection pressure to conserve its sequence, will accumulate mutations (vertical bars) relatively rapidly. It may acquire deleterious mutations and become a nonfunctional pseudogene which may continue to be expressed at the RNA level for some time, but which will eventually be transcriptionally silent (ψA). In some cases, however, the mutational differences may lead to a different expression pattern or other property that is selectively advantageous (A2). In the case of tandem gene duplication, subsequent sequence exchanges between the two copies (by mechanisms such as unequal crossover, see Section 9.3.2) will act as a brake on the rate of sequence divergence between the two gene copies.

DNA. Different mechanisms have been envisaged to contribute to the large increase in genome size:

■ *Rare duplications of the whole genome.*
■ *Frequent subgenomic duplication events resulting in gene and exon duplication.* Often such events occur at the subchromosomal level as a result of unequal crossover or unequal sister chromatid exchange, but interchromosomal exchanges are not uncommon, including retrotransposition, translocations and large-scale duplicative transpositions (see Section 14.2.3).
■ *Frequent subgenomic duplication events leading to increase in the amount of noncoding DNA.* An increase in the amount of noncoding DNA separating exons and genes is thought to have occurred principally by retrotransposition of repetitive elements such as *Alu* and LINE1 sequences.

In each case, the increase in genome size must have been accomplished without initially compromising the functions of the original DNA set. Instead, by providing additional genes, subsequent mutations could result in comparatively rapid sequence divergence: at each dupli-

cated gene locus, one gene is surplus to requirements and so can diverge rapidly because of the absence of selection pressure to conserve function. In some cases, such diverged genes may have acquired novel functions which could be selectively advantageous. In many cases, however, the additional gene sequences would be expected to acquire deleterious mutations and degenerate into nonfunctional pseudogenes (*Figure 14.2*).

14.2.1 Human genome evolution may have involved ancient genome duplication events, but the evidence has been obscured by subsequent chromosome and DNA rearrangements

Genome duplication (tetraploidization) is an effective way of increasing genome size and is responsible for the extensive polyploidy of many flowering plants. It can occur naturally when there is a failure of cell division after DNA duplication, so that a cell has double the usual number of chromosomes. Human somatic cells are normally diploid. However, if there is a failure of the first zygotic cell division, constitutional tetraploidy can result. Tetraploidy and other forms of polyploidy can be harmful and is often selected against. However, whole genome duplication via polyploidy has undoubtedly occurred relatively recently in maize, yeast, *Xenopus* and some types of fish. It is likely therefore that genome duplication occurred several times in the evolution of all eukaryotic lineages, including our own. Following genome duplication, an initially diploid cell could have undergone a transient tetraploid state; subsequent large-scale chromosome inversions and translocations, etc., could result in chromosome divergence and restore diploidy, but now with twice the number of chromosomes (*Figure 14.3*).

Table 14.1: Differences in DNA organization in the cells of simple and complex eukaryotes

Parameter	Yeast (*S. cerevisiae*)	*C. elegans*	Human
Number of cells	One	1000	10^{14}
Genome size	14 Mb	100 Mb	3000 Mb
Number of genes	6200	20 000	80 000?
Per cent coding DNA	~20%	~8%	~3%

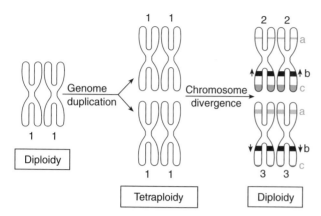

Figure 14.3: Genome duplication can lead to a transient tetraploid state before chromosome divergence restores diploidy.

Following duplication of a diploid genome, each pair of homologous chromosomes (e.g. chromosome 1) is now present as a pair of identical pairs. The resulting tetraploid state, however, can be restored to diploidy by chromosome divergence, e.g. by an interstitial deletion (upper panel, a), a terminal deletion (lower panel, c) or by an inversion (b).

If ancient tetraploidization events were rare in the evolution of the human genome, much intragenomic DNA shuffling would have occurred since the last such event. This means that the original evidence for tetraploidization events would be very largely obscured by subsequent chromosomal inversions, translocations, etc. Additionally, traces of gene duplication following genome duplication are likely to be frequently reduced by silencing of one member of each duplicated gene pair which then degenerates into a pseudogene. After hundreds of millions of years without any function, the nonprocessed pseudogenes generated following the last proposed genome duplication would have diverged so much in sequence as to be not recognizably related to the functional gene, even assuming they have not been lost during occasional rearrangements leading to gene deletion.

Genome duplication events during vertebrate evolution

In the case of vertebrates, two rounds of genome duplication have been envisaged at an early stage of vertebrate evolution, but the current evidence is fragmentary and its significance has been questioned (Skrabaneck and Wolfe, 1998). Gene numbers in different species have been taken to provide some evidence for two rounds of tetraploidization in vertebrates: invertebrates such as *C. elegans*, *Drosophila* and the sea squirt *Ciona intestinalis* are estimated to have about 15 000–20 000 genes, about one quarter that expected in mammalian genomes. In addition, many single-copy *Drosophila* genes have four vertebrate homologues and certain gene clusters appear to have been quadruplicated (see next section).

14.2.2 The existence of some paralogous chromosome segments has been alternatively viewed to reflect ancient genome duplications or subgenomic duplications

A major line of evidence for genome duplications in vertebrates is the existence of closely related gene clusters at different subchromosomal regions in a species, so-called **paralogous** chromosome segments (*Box 14.2* and *Figure 14.4*). Often such clusters contain genes that have been very highly conserved during evolution because they play crucial roles in early embryonic development. Some examples of quadruplicated gene clusters are known in the human genome and they have been taken as evidence of previous genome duplications. They include clusters containing fibroblast growth factor receptor genes and *HOX* genes (Skrabaneck and Wolfe, 1998).

HOX gene clusters

The most cited example of gene organization supporting two rounds of vertebrate genome duplication are the *HOX* genes, homeobox genes which are involved in specifying the anterior–posterior axis during early development. Humans and other mammals have four *HOX* gene clusters containing about nine such genes (*Figure 14.5*). The linear order of the genes in a cluster is thought to dictate the temporal order in which they are expressed during development and also their anterior limits of

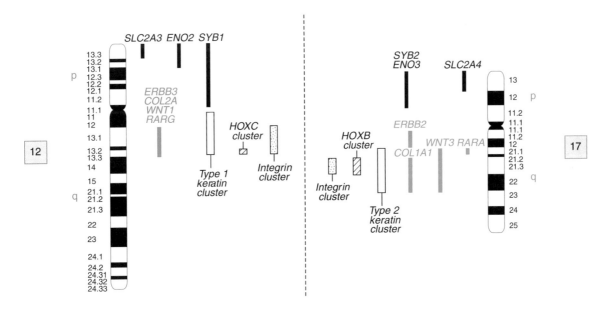

Figure 14.4: HSA12 and HSA17 appear to have paralogous chromosomal segments.

The approximate positions of some of the related genes mapping to human chromosomes 12 and 17 are indicated. The lengths of the bars mark the *maximum* uncertainty about the position of any member of a cluster; many of the genes are likely to be closely clustered.

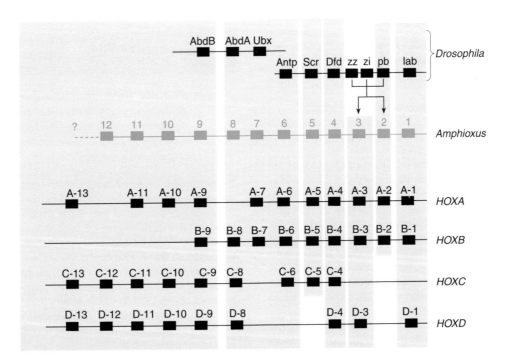

Figure 14.5: The organization of *HOX* gene clusters in mammals and *Amphioxus* suggests the possibility of one or two rounds of ancestral genome duplication.

Indicated paralogous groups consist of genes with very similar expression patterns and presumably similar functions. At the time of writing, 12 *HOX* genes had been isolated in *Amphioxus*, the invertebrate considered to be most closely related to vertebrates. The equivalent genes in *Drosophila* were presumably organized on a single cluster in an ancestral genome prior to a translocation which resulted in the *Ultrabithorax (Ubx)* and *Antennapedia (Antp)* clusters. The vertebrate ancestor presumably had 13 *HOX* genes but loss of individual genes following cluster duplications has led to each of the mammalian *HOX* clusters lacking one or more of the original genes. Boxes group genes with clearly related homeoboxes.

expression along the anterior–posterior axis (Ruddle *et al.*, 1994). The close similarity of the four clusters means that there is clear evidence for *paralogous HOX* genes, that is genes on different clusters which are more closely related to each other than they are to their neighbors (*Figure 14.5*). A single such *HOX* cluster exists in *Amphioxus*, with very close similarities to the mammalian *HOX* gene clusters (Garcia-Fernandez and Holland, 1994) As *Amphioxus* is thought to be the closest invertebrate relative of the vertebrates, the vertebrate ancestor may have had a single such cluster.

While the evidence above is consistent with two successive rounds of genome duplication during vertebrate evolution, contradictory evidence also exists. Analysis of the collagen genes which are closely linked to *HOX* clusters has suggested that the *HoxD* cluster branched off first from the ancestral lineage, followed by the *HoxA* cluster and finally *HoxB/HoxC*. This would require three separate duplication events and it may suggest that some of these steps were subgenomic duplications rather than whole genome duplications. Analysis of *Hox* clusters in other species also offers some support for subgenomic duplication. While the pufferfish has the expected four clusters, lamprey have only three such clusters. So either there has been some subgenomic duplication events, or whole clusters have been deleted. The latter possibility may also be suggested by the observation of seven clusters in zebrafish (Meyer and Malago-Trillo, 1999). Zebrafish may have undergone an additional recent genome duplication (as suggested by the presence of additional gene copies for many other types of gene) with subsequent loss of a single cluster.

14.2.3 There have been numerous major chromosome rearrangements during the evolution of mammalian genomes

In addition to whole genome duplication, a variety of different subgenomic DNA duplication events are possible, resulting from exchanges between nonhomologous chromosomes (chromosomal translocations), unequal exchanges between homologous chromosomes or the sister chromatids of a single chromosome, and DNA copy transposition events. Clearly, some of these mechanisms can also result in loss of genetic material. In addition, other mechanisms (chromosome inversions, simple DNA transpositions and balanced translocations) can result in no net gain or loss of material.

Mammalian genome evolution may have involved frequent subgenomic duplications and also rearrangements without net loss of DNA (Lundin, 1993). Small scale duplications can be expected to have occurred intrachromosomally by mechanisms such as unequal crossover and unequal sister chromatid exchange but interchromosomal **duplicative transposition** events may have been common too, where a segment of a chromosome duplicates and the copy is inserted elsewhere in the genome. For example, the pericentromeric regions of chromosomes are known to be unstable and duplications of pericentromeric regions fol-

lowed by insertion into other chromosomes may be frequent (Eichler, 1998). Other such examples could have occurred on a larger scale from a variety of different chromosomal regions. Even larger scale duplications could have resulted from ancestral whole chromosome duplications by **Robertsonian fusion** or subchromosomal duplications followed by **pericentric inversions**.

Comparisons of the present-day genome organization of humans and other mammals also suggest that large-scale rearrangements may have been frequent, and that karyotype and phenotype evolution can be uncoupled. For example, the Indian muntjac deer (*Muntiacus muntjak*) has only three types of (very large) chromosome, whereas its very close relative, the Chinese muntjac deer (*Muntiacus reevesi*), has 23 different chromosomes. The human and mouse karyograms are also quite different from each other and even the highly conserved X chromosome linkage group shows numerous differences in organization between the species (see *Figure 14.10*). The great apes are extremely closely related to humans but show clear cytogenetic differences as a result of several inversions, a translocation that has occurred exclusively in the human lineage and another that has occurred in the gorilla lineage (see *Figure 14.26*). Old World monkeys are also closely related to humans but, with the exception of the gibbons, numerous chromosome rearrangements have occurred since divergence from the human lineage.

14.3 Evolution of the human sex chromosomes

14.3.1 Despite their considerable structural differences, substantial blocks of sequence homology between the human X and Y chromosomes suggest a common origin

In mammals, pairs of homologous autosomal chromosomes are structurally virtually identical (homomorphic); chromosome pairing at meiosis is presumed to be facilitated by the high degree of sequence identity between homologs, albeit by a mechanism that is not understood. By contrast, the X and Y chromosomes of humans and other mammalian species are heteromorphic. The human X chromosome is a submetacentric chromosome which contains about 165 Mb of DNA, whereas the Y is acrocentric and is much smaller (containing about 60 Mb of DNA). The human X chromosome contains numerous important genes: on the basis of its size alone, it might be expected to contain about 4000 genes, but the comparative lack of CpG islands on the X chromosome (see *Figure 7.4*) indicates that the true figure may be substantially smaller. In marked contrast, the great bulk of the Y chromosome is genetically inert and is composed of constitutive heterochromatin consisting of different types of highly and moderately repetitive noncoding DNA. Only a very few functional genes are found on the Y chromosome, including some which have closely related homologues on the X chromosome, and several which are Y-specific and testis-expressed (*Figure 14.6*).

Despite being morphologically distinct, however, the X and Y chromosomes are able to pair during meiosis in male cells, and to exchange sequence information. Sequence exchanges occur within certain small regions of homology between the X and the Y chromosomes, known as **pseudoautosomal regions** because DNA sequences in these regions do not show strict sex-linked inheritance. In the human sex chromosomes, there are two pseudo-autosomal regions (Rappold, 1993; Ried *et al.,* 1998):

■ *The major pseudoautosomal region (PAR1)* extends over 2.6 Mb at the extreme tips of the short arms of the X and Y and is known to contain a dozen or so genes. It is the site of an **obligate crossover** during male meiosis which is thought to be required for correct meiotic segregation. This very small region is remarkable for its high recombination frequency (the sex-averaged recombination frequency is 28% which, for a region of only 2.6 Mb, is approximately 10 times the normal recombination frequency). The high

figure is, of course, mostly due to the obligatory crossover in male meiosis resulting in a crossover frequency approaching 50% (*Figure 14.7*). The boundary between the major pseudoautosomal region and the sex-specific region has been shown to map within the *XG* blood group gene, with the *SRY* male determinant gene occurring only about 5 kb from the boundary on the Y chromosome (*Figure 14.8*).

■ *The minor pseudoautosomal region (PAR2)* extends over 320 kb at the extreme tips of the long arms of the X and Y. Unlike the major pseudoautosomal region, crossover between the X and Y in this region is not so frequent, and is neither necessary nor sufficient for successful male meiosis. At the time of writing only two genes had been identified in this region: *IL9R* and *SYBL1*.

In addition to the two pseudoautosomal regions, the human X and Y chromosomes show substantial regions of homology elsewhere, including a variety of Xp–Yq and Xq–Yp homologies, as well as Xp–Yp and Xq–Yq homologies. The existence of such homologies suggests that the

(A)

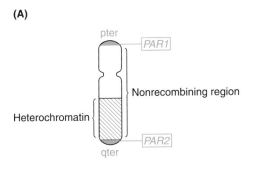

(B)

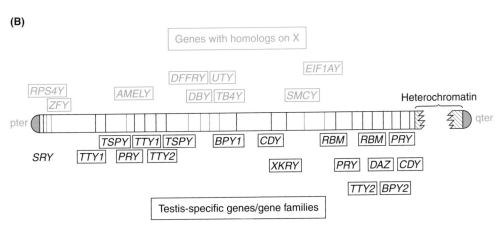

Figure 14.6: The great bulk of the Y chromosome is genetically inert but many of its few genes show testis-specific expression.

(**A**) Schematic illustration of the Y chromosome showing major and minor pseudoautosomal regions (the 2.6 Mb PAR1 and 320 kb PAR2 respectively) separated by the approximately 60 Mb nonrecombining region, much of which is heterochromatin. (**B**) Approximate positions of Y-linked genes. Note the very small number of genes (in 60 Mb of DNA one might expect about 1300–1500 genes normally; the heterochromatin region is thought to be devoid of transcriptionally active sequences). Of the few Y-linked genes the majority show testis-specific expression (with several occurring as gene families). See Lahn and Page (1997).

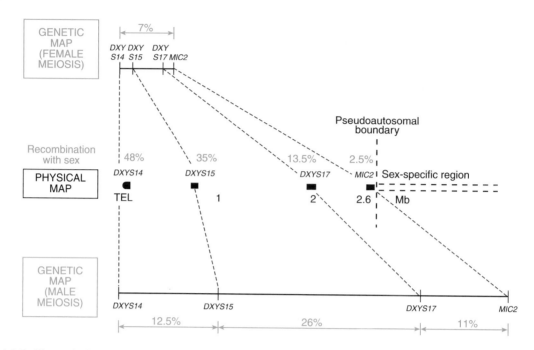

Figure 14.7: The major human pseudoautosomal region is characterized by a high overall recombination frequency and a large sex difference in recombination frequency.

The frequency of recombination with sex (i.e. of exchange between the X and Y chromosomes in male meiosis) shows a gradient along the length of the pseudoautosomal region – sequences such as *DXYS14* which are located at the telomere have an almost 50% chance of being exchanged between the X and the Y. Because of the obligate crossover in this very small region in male meiosis, the male genetic map is much larger than the corresponding female genetic map. For example, the recombination frequency between the markers *DXYS14* and *MIC2* is about 49.5% in male meiosis, but only 7% in female meiosis.

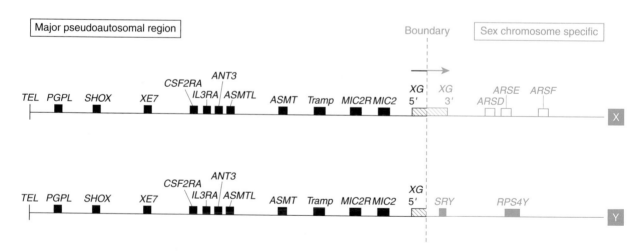

Figure 14.8: The boundary of the major human pseudoautosomal region occurs within the *XG* blood group gene.

The human major pseudoautosomal region (PAR1) is a 2.6 Mb region that is common to the tip of the short arms of the X chromosome (top) and Y chromosome (bottom). TEL – telomere. At the time of writing it was known to have about a dozen genes (Ried *et al.*, 1998). At the right hand side the first part of the long sex chromosome-specific regions are shown. *Note* that the major pseudoautosomal boundary occurs within the *XG* blood group gene on the X chromosome. On the Y chromosome, the *XG* gene is truncated: the promoter and first few exons are present in the pseudoautosomal region, but thereafter there are Y chromosome-specific sequences, including the *SRY* gene and the *RPS4Y* gene. See Weller *et al.* (1995).

two chromosomes have evolved from an ancestral homomorphic pair of chromosomes. Clearly, the two chromosomes have subsequently undergone substantial divergence, and sequences that are physically close on·one chromosome may have very widely spaced counterparts on the other (*Figure 14.9*). However, primate comparisons have shown that at least some of the existing X–Y homology results from very recent duplicative transposition events.

14.3.2 The evolution of the mammalian X chromosome has led to substantial species differences, both in chromosomal DNA organization and the pattern of X inactivation

Human–mouse divergence in gene order
Conservation of synteny between mouse and human

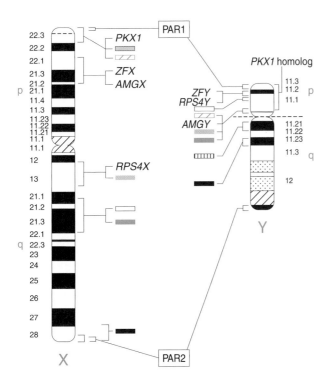

Figure 14.9: The human X and Y chromosomes show several regions of homology in addition to the common pseudoautosomal regions.

Note the contrasting spatial organization between different pairs of homology blocks. For example, a sequence block in the Yq11.21 region has a counterpart in the most distal Xp band, Xp22.3, whereas sequences in the nearby band Yq11.23 have homologs in the most distal Xq band, Xq28. The gene encoding ribosomal protein S4, *RPS4*, and the *ZFY* gene are clustered on distal Yp11.1 but have widely spaced homologs on the X, and the amelogenin gene, *AMGY,* is part of a cluster of homology regions at proximal Yp11.1 with widely scattered homologs on the X. Abbreviations: PAR1, major pseudoautosomal region; PAR2, minor pseudoautosomal region.

is most pronounced in the case of the X chromosome: almost the entire X linkage group appears to be conserved between the two species (known exceptions include three human pseudoautosomal genes with autosomal orthologs in mouse; see below). This remarkable conservation of synteny for X-linked genes also applies to other mammals and appears to be evolutionarily related to the development of a special form of dosage compensation: *X chromosome inactivation* (see Sections 2.2.3 and 8.5.6). Once established by evolutionary design, or accident (Ohno, 1973), X chromosome inactivation would be expected to ensure conservation of synteny because X–autosome translocations would be selected against (the normal 2:1 ratio of gene dosage for autosomal and X-linked genes would be destroyed). As expected, there is extensive conservation of synteny of the genes on the mouse and human X chromosomes. Nevertheless, there are major differences in gene order: fine mapping of X-linked DNA sequences in the two species indicates regions of homology which can only have been generated by a variety of different chromosomal inversions in the lineages leading to present day mice and humans (*Figure 14.10*).

Evolutionary instability of pseudoautosomal regions
The pseudoautosomal regions have not been well-conserved in evolution. There is no equivalent to the human PAR2 in mouse and various other mammals, including primate species. Of the two human genes known to be located in PAR2, *IL9R* has a mouse ortholog which is autosomal while the mouse ortholog of *SYBL1* lies close to the centromere of the X (*Figure 14.11*). There are also major species differences in the position of the boundary for the major pseudoautosomal region and in the gene content of this region. In humans the major pseudoautosomal boundary occurs within the *XG* gene (*Figure 14.8*). In mouse, however, the boundary lies within the *Fxy* gene, whose human ortholog *FXY* is located more proximally within the X chromosome-specific region (*Figure 14.11*; see Perry *et al.,* 1998).

Several of the genes in the human major pseudoautosomal region are autosomal in other species. This is consistent with the idea that there has been repeated addition of autosomal segments onto the pseudoautosomal regions of the X or the Y chromosome which are then recombined onto the other sex chromosome (Graves *et al.,* 1998; see also Section 14.3.3). The major pseudoautosomal region and neighboring regions are also thought to be comparatively unstable regions. Frequent DNA exchanges result in a high incidence of gene fusions, exon duplications and **exon shuffling** (see Ried *et al.,* 1998 and Sections 14.5.1 and 14.5.2). Genes within or close to the major pseudoautosomal region are also subject to rapid sequence evolution. The *SRY* gene located only 5 kb from the major pseudoautosomal boundary is very poorly conserved in evolution (Pamilo and O'Neil, 1997) and standard hybridization-based screening has failed to find mouse orthologs for many of the human genes that map within or close to the major pseudoautosomal region,

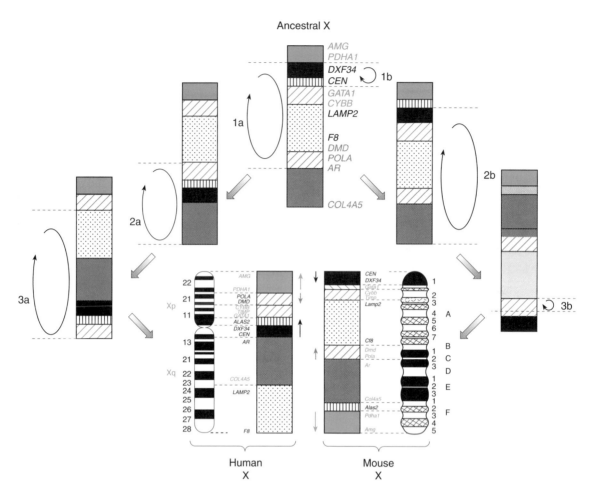

Figure 14.10: Several X chromosome inversions appear to have occurred since human–mouse divergence.

The present-day organization of human and mouse X chromosomes is shown at the bottom. A minimum of eight different homology blocks are defined by the presence in each of multiple orthologous genes and DNA sequences, of which only the most proximal and the most distal markers are shown. The remainder of the figure shows one possible explanation of how the existing chromosome organizations may have been derived by a series of inversions from a common ancestral X chromosome. Redrawn from Blair *et al.* (1994) with permission from Academic Press Inc. Subsequent studies have shown that the situation is even more complicated than portrayed here but the important point which is illustrated is that inversions have been common events in mammalian X chromosome evolution.

presumably because of very high levels of sequence divergence (see also Section 14.6.1).

Human–mouse divergence in X inactivation patterns

The rationale for X chromosome inactivation is to act as a dosage compensation mechanism for those X chromosome genes (the vast majority) which do *not* have homologs on the Y chromosome. However, a small minority of human X-linked genes do have functional homologs on the Y chromosome. Such genes, which are common to both the X and the Y, do not show a sex difference in dosage; as a result, they would be expected to escape X inactivation. Those genes in the major pseudoautosomal region which have been tested for X inactivation status, have all been shown to be expressed on both active and inactive X chromosomes. In the minor pseudoautosomal

region, the *IL9R* gene escapes X inactivation but surprisingly the *SYBL1* gene is subject to X inactivation and is not easily accommodated in proposed schemes for explaining how genes common to the X and Y came to be inactivated (Jegalian and Page, 1998). When present on the Y chromosome, *SYBL1* is methylated and not expressed. This type of **Y-inactivation** constitutes a novel way of maintaining equality in gene dosage between the two sexes.

In addition to the genes in the pseudoautosomal regions, several other human X-linked genes are known to escape X inactivation, including genes which map to proximal Xp and proximal Xq regions, while genes known to map to intermediate locations are often subject to X inactivation (*Figure 14.12*). In some cases where detailed gene mapping has been conducted, clear evidence exists for multigene domains outside the pseudoautosomal region which escape X inactivation, such as one at Xp11.2 (Miller and

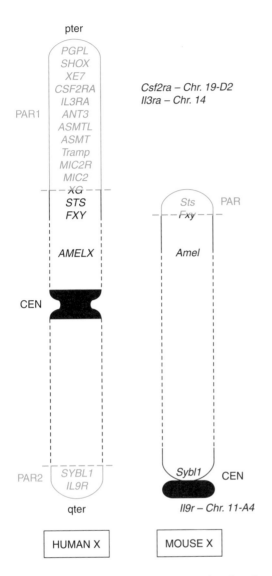

Figure 14.11: Mammalian pseudoautosomal regions have not been well-conserved during evolution.

The major pseudoautosomal region is a target for X-autosome translocations with periodic additions to it from autosomal sequences during evolution. It is also a region characterized by a high frequency of exon duplication and shuffling and rapid sequence divergence (Ried *et al.*, 1998). The example here illustrates the human–mouse differences for the pseudoautosomal regions on the X chromosome with pseudoautosomal regions and genes therein shown in blue. *Note* that many of the human pseudoautosomal region genes have autosomal counterparts in mice; the mouse orthologs of many of the other genes haven't been able to be found because of suspected large sequence divergence. There is only one mouse pseudoautosomal region which is located at the tip of the chromosome distal from the centromere. Although the mouse X is traditionally shown with the centromere (CEN) at the top, it is inverted here to emphasize that the mouse PAR resembles human PAR1 albeit with a much reduced set of genes and a boundary that interrupts the *Fxy* gene (whose human ortholog *FXY* is located more proximally.

Willard, 1998). As expected, many of the human genes which map outside the major pseudoautosomal region and escape X inactivation have functional homologs on the Y chromosome. Some, however, do not. For example, the *UBE1* and *SB1.8* genes escape inactivation but do not appear to have any homologs on the Y chromosome. Other genes such as the Kallman syndrome gene *KAL1*, and the steroid sulfatase gene *STS* do have homologs on the Y chromosome, but these are nonfunctional pseudogenes. It is likely, therefore, that for some genes sex difference in gene dosage is not a problem and is tolerated (Disteche, 1995). In the mouse, however, there are considerable differences in the pattern of X inactivation. For example, the human non-pseudoautosomal genes *ZFX*, *RPS4X* and *UBE1* all escape inactivation, but the murine homologs *Zfx*, *Rps4* and *Ube1X* (which unlike the human *UBE1* gene has a homolog on the Y) are all subject to X inactivation (*Figure 14.12*).

14.3.3 Comparison of genes in distantly related mammals suggests that much of the short arm of the human X chromosome has recently been acquired by X–autosomal translocation

Mammals are classified into two subclasses prototheria (the monotremes or egg-laying mammals) and *theria* which in turn are subdivided into two infraclasses: metatheria (marsupials) and eutheria, a group which includes placental mammals (*Figure 14.13*). Many eutherian X-linked genes are found to be X-linked in marsupials. However, genes mapping to a large part of human Xp (distal to Xp11.3) have orthologs on autosomes of both marsupials and monotremes. Because the prototherian divergence pre-dated the metatherian–eutherian divergence, the simplest explanation is that at least one large autosomal region was translocated to the X chromosome early in the eutherian lineage.

Translocation of autosomal genes on to the X chromosome will result in the formerly autosomal genes being subject to X inactivation. Not only is one X chromosome shut down in all female cells, but inactivation of the single X chromosome in male cells is required during spermatogenesis. However, certain genes on human Xp would be expected to be crucially important for cell function. For example, the *PDHA1* gene encodes the E1α subunit of the pyruvate dehydrogenase complex, an enzyme essential in aerobic energy metabolism. In marsupials the *PDHA1* gene is located on an autosome and so is expressed during spermatogenesis. In contrast, the *PDHA1* gene in humans (and other eutherian mammals) is X-linked, and is not expressed during spermatogenesis. However, a closely related gene, *PDHA2*, encodes a testis-specific human isoform which presumably has evolved in response to the silencing of X-linked genes during spermatogenesis. The *PDHA2* gene is intron-less and is thought to be an example of a functional processed gene, generated by reverse transcription from the mRNA of the *PDHA1* gene (see *Figure 7.13* for the general mechanism).

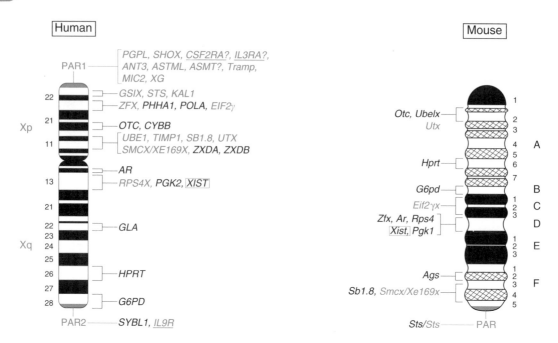

Figure 14.12: Genes that escape inactivation on the human X chromosome are widely distributed, but there are some notable differences in X-inactivation patterns in the mouse.

Genes that escape inactivation are shown in blue, those that are subject to inactivation are in black. Question marks denote genes presumed to escape X-inactivation but for which direct evidence is lacking. The *XIST/Xist* gene which is responsible for initiating X-inactivation is highlighted by boxes and is expressed only on the *inactivated X*, not the active X. The pattern of escape from X-inactivation is now thought to comprise small multigene domains at different positions on the X (Miller and Willard, 1998). The pseudoautosomal genes *CSF2RA*, *IL3RA* and *IL9R* are underlined to denote that the mouse orthologs are autosomal. *Note* that the mouse *Sts* (steroid sulfatase) locus may escape inactivation or be subject to inactivation in different mouse strains. *Note* differences in X-inactivation for the human-mouse orthologs: *UBE1 – Ube1x*; *RPS4X – Rps4*; *ZFX – Zfx*. PAR1, PAR2, major and minor pseudoautosomal regions in humans. PAR, the mouse pseudoautosomal region, which partly corresponds to the human PAR1.

14.3.4 Enforced lack of recombination has led to a severe loss of genetic capacity on the mammalian Y chromosome and thereafter to the development of the X-inactivation mechanism of dosage compensation

The heteromorphic mammalian sex chromosomes most likely evolved from homomorphic autosomes

Distinct sex chromosomes have been independently developed in many animals with disparate evolutionary lineages, including not only mammals, but birds (where the females are ZW, the *heterogametic* sex, and the males are ZZ, the *homogametic* sex), and certain species of fish, reptiles and insects. In each case, it is thought that the two sex chromosomes started off as virtually identical *autosomes*, except that one of them happened to evolve a major sex-determining locus (the *SRY* locus in humans; see *Figure 14.14*). Subsequent evolution resulted in the two chromosomes becoming increasingly dissimilar until, in many species, the Y was reduced to a tiny chromosome with only a very few functional genes (see *Figure 14.6*). There would appear to be evolutionary pressure to adopt the strategy of having two structurally and func-

tionally different sex chromosomes, and it seems that this pressure is gradually driving the Y chromosome to extinction. Eventually, one would expect a switch to a sex determination system where maleness is conferred simply by X : autosome gene dosage and XO individuals are male, as in the case of *Drosophila*.

Why should the X and Y chromosomes diverge, and why should the Y degenerate?

Clearly to maintain sex differences, recombination needs to be suppressed in the region of the major sex-determining locus (*SRY* in humans is located in a nonrecombining region just 5 kb proximal to the major pseudoautosomal region; see *Figure 14.8*). Additionally, environmental circumstances may have offered a selective advantage for breaking down recombination between the sex chromosomes. For example, one trigger could have been the development of *sexually antagonistic genes*, with alleles which may be of benefit to the heterogametic sex (XY), but harmful to the homogametic sex (XX). If such genes accumulate, then there will be a selective pressure to ensure that they are not transmitted to the homogametic sex, a restriction which can be met if they are present on a nonrecombining Y chromosome. Certainly, recombination between the present-day human X and Y chromo-

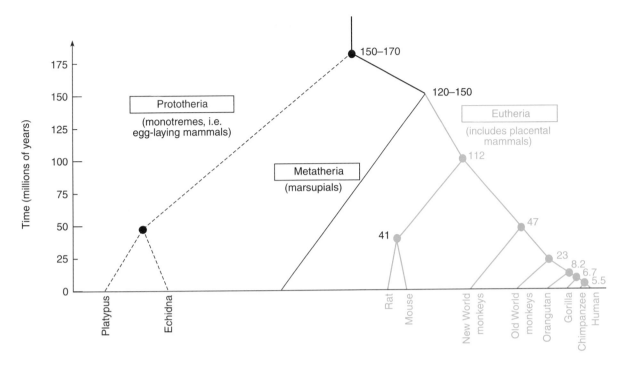

Figure 14.13: Mammalian phylogeny.

Numbers to the right refer to the approximate dates of divergence of the indicated lineages in millions of years. For example, the lineage giving rise to modern day humans is thought to have diverged from chimpanzees about 5.5 million years ago. Estimates for times for divergence of indicated lineages are based on the data reported by Kumar and Hedges (1998).

somes is very limited, being very largely confined to the tiny major pseudoautosomal region at the tips of the short arms (PAR1).

From the above the human Y chromosome can be viewed as an essentially asexual (nonrecombining) component within an otherwise sexual genome (the X chromosome can recombine along its length with a fully paired homolog in female meiosis). Population genetics predicts that a nonrecombining chromosome should degenerate by a process known as Muller's rachet. If the mutation rate is reasonably high the absence of mutations means that harmful mutations can gradually accumulate in genes on that chromosome over long evolutionary time scales: mutant alleles may drift to fixation as Y chromosomes with fewer mutants are lost by chance, or they may 'hitchhike' along with a favorable allele in a region protected from recombination. Once mutations accumulate in the nonrecombining Y, the loss of function of genes means that there is no selective pressure to retain that DNA segment and the chromosome will gradually contract by a series of deletions (*Figure 14.14*).

Y chromosome degeneration and the development of X chromosome inactivation

The evolution of the mammalian sex determination system shown in *Figure 14.14* is also inextricably interwoven with the evolution of the X-inactivation mechanism for

dosage compensation (see Ellis, 1998). In response to large-scale destruction of Y chromosome sequences by the process described above there would have been pressure to increase gene expression on the X chromosome. However, this would lead to excessive X chromosome gene expression in females which could cause reduced fitness. As a result a form of gene dosage compensation evolved whereby a single X chromosome was selected to be inactivated in female cells (X-inactivation).

14.4 Evolution of human DNA sequence families and DNA organization

14.4.1 Gene duplication is a mechanism for generating functional divergence that has frequently been used in the evolution of mammalian genomes

In addition to large-scale gene duplication events involving the whole genome or large chromosomal segments (Section 14.2), selective duplication of specific genes can occur by small scale duplications involving copy transposition events and also tandem gene duplication.

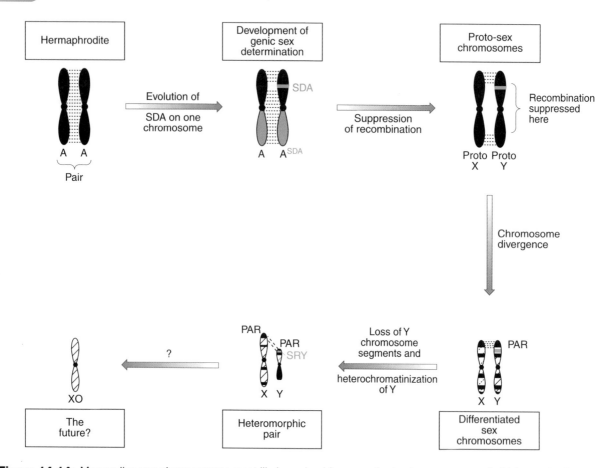

Figure 14.14: Mammalian sex chromosomes most likely evolved from a pair of autosomes, one of which acquired a sex-determining allele, leading to recombination suppression and chromosome differentiation.

One of a homologous pair of autosomes in an ancestral genome is envisaged to have evolved a sex determining allele (SDA). Thereafter, the need to avoid exchange of the SDA and possibly the evolution of *sexually antagonistic genes* led to suppression of recombination between the two chromosomes, except in small regions, known as pseudoautosomal regions (PARs). Lack of exchange between the homologs led to chromosome divergence. Because most of the Y is not involved in any recombination events, it degenerated by a series of chromosomal deletions. Present day human X and Y chromosomes retain small regions of homology outside the PARs, partly as a result of very recent X–Y transpositions. Possibly the inexorable pressure to reduce the Y means that eventually it will be completely eliminated and a mechanism of sex determination will evolve which is based simply on X : autosome gene dosage.

Mechanisms resulting in small scale gene duplication

Duplicative (or copy) transposition involves a duplication of a DNA sequence prior to transposition. Small-scale DNA transposition in mammalian genomes most often occurs through an RNA intermediate and frequently results in a moderate to large interspersed repeat family. Processed copies of genes transcribed by RNA polymerase II normally lack functional regulatory sequences present in the original gene and, with few exceptions, degenerate into pseudogenes (**processed pseudogenes**; see Section 7.3.5). The exceptions all appear to be sequences which are copied from X-linked genes and which show testis-specific gene expression (e.g. the pyruvate dehydrogenase genes, *PDHA2*; see Section 14.3.3). Genes transcribed by RNA polymerase III, however, often contain an internal promoter sequence. After transposition the internal promoter can be used to transcribe a new copy which may be able to transpose in turn, leading to eventually very high copy numbers. This is the way in which the *Alu* repeat family appears to have evolved, using the reverse transcriptase of LINE1 elements to make cDNA copies (see Section 7.4.5 and *Figure 14.25*).

Tandem gene duplication often occurs as a result of unequal crossover events or unequal sister chromatid exchanges. Numerous clustered human gene families show evidence of having acquired multiple members by this mechanism. In many cases, the duplicated genes degenerate into nonprocessed pseudogenes (Section 7.3.5). However, the transition between functioning duplicated gene and nonfunctional pseudogene may be a gradual one. This has given rise to the concept of the **expressed pseudogene**, a gene which is expressed at the mRNA level, or even at the polypeptide level, but which is nevertheless nonfunctional (Section 7.3.5). The absence of function means that selection pressure to conserve function will be relaxed and eventually mutations will accumulate, often leading to silencing of gene expression. Alternatively, the mutations may eventually result in the acquisition of different expression patterns and sometimes different functions (see *Figure 14.2*).

Acquisition of different expression patterns

Some diverged duplicated genes are known to be expressed predominantly in different environments. Sequence divergence in the different genes in the α-globin gene cluster and in the β-globin gene cluster may result in encoded products with slightly different biological properties. For example, the ε-, ζ- and γ-globin chains could possibly be especially suited to binding oxygen in the comparatively oxygen-poor environment of early development, whereas the α- and β-globin chains may be the preferred polypeptides in the environment of adult tissues.

Genes encoding different tissue-specific **isoforms** (alternative forms of the same protein) or **isozymes** (alternative forms of an enzyme) also appear to have evolved by gene duplication. For example, the enzyme alkaline phosphatase is encoded by at least four different genes which show tissue-specific differences in expression. Of these, three are clustered near the telomere of 2q: *ALPI* and *ALPP* encode alternative forms of the enzyme (87% protein sequence similarity) found in intestine and placenta, respectively, and *ALPPL* encodes a placental-like isozyme. A fourth member, *ALPL*, is located near the telomere of 1p and encodes an isozyme expressed in liver, bone, kidney and some other tissues, and is more distantly related to the intestinal and placental forms (57% and 52% sequence similarity, respectively). Note, however, that duplicated genes encoding subcellular-specific forms of the same protein are often located on different chromosomes, with gene duplication possibly arising from an ancestral genome duplication event. For example, in human liver there are two major isoforms of aldehyde dehydrogenase, a cytosolic and a mitochondrial form, which show 68% sequence identity over their 500 amino acid long sequences. The cytosolic and mitochondrial forms are, respectively, encoded by the *ALDH1* gene on chromosome 9q and the *ALDH2* gene on chromosome 12q. The two genes each have 13 exons and nine out of the 12 introns occur in homologous positions in the two coding sequences, strongly suggesting a common evolutionary origin by some kind of ancient gene duplication event (Strachan, 1992, pp. 32–33).

14.4.2 Concerted evolution occurs as a result of intragenomic (intraspecific) sequence exchanges within a DNA sequence family

In the case of certain gene and DNA sequence families, there may be a closer sequence relationship between individual family members in one species (paralogs) than that between orthologs in different species (**concerted evolution**). Thus, if we consider a specific gene family in two species, A and B, concerted evolution means that a family member in species A will be more closely related to other members of that family in species A than it will be to an ortholog or any other members of the same family in species B (*Figure 14.15*). Concerted evolution occurs because of various genetic mechanisms which cause sequence exchange between nonallelic DNA sequences within a genome. These mechanisms, which include *unequal crossover, unequal sister chromatid exchange* and *gene conversion*-like mechanisms (see Sections 9.3.2 and 9.3.3), are particularly prevalent in the case of tandemly repeated DNA sequences. For example, unequal crossover and unequal sister chromatid exchange can result in a specific repeat sequence spreading through an array of tandem repeats, and eventually replacing the other repeats, thereby resulting in sequence homogenization (see *Figure 9.8*). Because of meiotic recombination, the resulting effect can be transmitted to other genomes in a sexual population. As a result, concerted evolution may be observed between members of a DNA family *within a species*; sequence exchange between homologous sequences in the DNA from different animal species is essentially nonexistent.

The rDNA genes provide a useful example of sequence exchange between repeats within a cluster but, in addition and more unusually, sequence exchanges can occur between clusters. In the human genome the rDNA genes are organized as large clusters of tandem repeats containing 50–60 copies of an approximately 40-kb repeat unit (see *Figure 8.3*). The high degree of sequence homology between such large repeats facilitates frequent sequence exchanges between nonallelic repeats. Additionally, the clusters are located on the short arms of the acrocentric chromosomes 13, 14, 15, 21 and 22 which frequently exchange sequences by nonhomologous chromosome translocations. As a result, individual human rDNA genes are more similar to each other than they are to the rDNA genes of other primates.

14.4.3 Some gene families do not show strong evidence of concerted evolution: sequence homologies between orthologs may be greater than between different family members in one species

In general, members of a gene family or superfamily which are located in the same gene cluster show a higher degree of sequence homology than do members present on different clusters, and the degree of sequence homology is usually greatest between closely neighboring genes within a cluster. This is so for the following reasons:

- Gene duplication events leading to formation of any one cluster are often examples of recent tandem duplications, whereas the duplications that have given rise to the different clusters are often

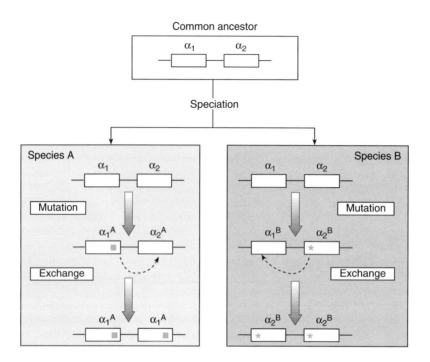

Figure 14.15: Concerted evolution occurs in DNA sequence families when a relatively high level of sequence exchange occurs between family members.

The family illustrated has two members α_1 and α_2 which arose by tandem duplication, and was inherited from a common ancestor by two species, A and B. Mutation leads to sequence divergence and the $\alpha_1^A - \alpha_2^A$ divergence would be expected to be similar to that for $\alpha_1^B - \alpha_2^B$. Indeed, if one copy were under selection pressure to maintain function, and the other diverged rapidly, one might expect the conserved copy in species A to be more similar to that in B than it would be to the other copy in A. However, if there are frequent sequence exchanges *within a species*, the copies in A will be more related to each other than to those in B and vice versa.

comparatively ancient and may have resulted from ancestral genome duplication events.

■ The evolution of gene clusters by a series of tandem duplications will tend to mean that closely neighboring genes are more likely to have originated by a recent tandem duplication than more distantly spaced genes in the same cluster.

■ Following gene duplication there may be two competing forces which affect the sequence identity between the duplicated genes: sequence divergence (the sequences of the duplicated genes may be identical initially but during evolution will gradually become different as a result of independent accumulation of mutations in the two genes); and sequence homogenization (periodic sequence exchanges between the two genes will tend to result in sharing of sequences between them and therefore maintain sequence identity). Such homogenization results from genetic mechanisms (unequal crossover, unequal sister chromatid exchange and gene conversion) which are much more prevalent in tandemly duplicated genes than in distantly located or nonsyntenic duplicated genes. As a result, the sequences of distantly spaced genes or nonsyntenic genes will have a tendency to diverge more rapidly than those of tandemly duplicated genes.

The globin gene superfamily provides some useful examples. Sequence homology between the genes and gene products from different clusters (e.g. α-globin and β-globin) is much less than between genes and gene products from a single cluster (*Figure 14.16* and *Table 14.2*). This is largely so because the different clusters are presumed to have originated early in evolution while gene duplications within clusters occurred comparatively recently. In the latter case, some duplication events are presumed to have occurred very recently, leading to duplicated genes which are almost identical. For example, the two human α-globin genes *HBA1* and *HBA2* encode identical products, and the products of the two γ-globin genes *HBG1* and *HBG2* differ by a single amino acid. In other cases, the duplicated genes within a cluster are clearly more diverged in sequence, presumably because the relevant duplication events occurred some time ago.

In contrast to tandemly repeated genes, intracluster sequence exchanges between globin genes are likely to be infrequent (except for the very recently duplicated genes, such as *HBA1* and *HBA2*). This is so because the different globin genes are small (1.6 kb) and the chromosomal DNA separating them is not well conserved. Additionally, the stringent developmental regulation of gene expression within a cluster presumably imposes a functional constraint, minimizing sequence exchanges

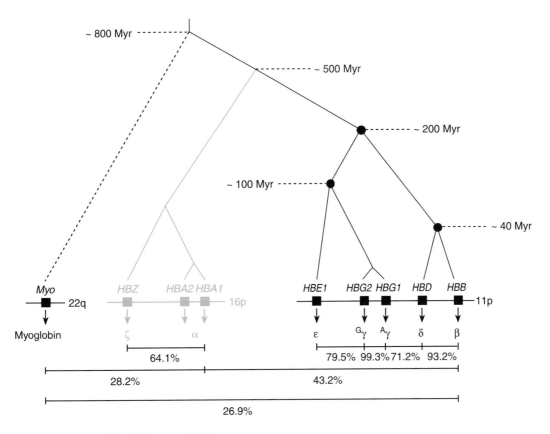

Figure 14.16: Evolution of the globin superfamily.

Globins encoded by genes within a cluster show a greater degree of sequence homology than those encoded by genes on different clusters (as do the genes themselves). The close relationship between genes within a cluster is thought to be due to comparatively recent gene duplication events, whereas the duplication events giving rise to the different clusters are much more ancient. *Note* that neighboring genes in a cluster may be very closely related as a result of very recent tandem gene duplication: the *HBA1* and *HBA2* genes encode identical α-globins, and the *HBG1* and *HBG2* encode γ-globins that differ by a single amino acid.

Table 14.2: Sequence variation in globin genes

	Sequence homology (%)			
	Coding DNA	5′ + 3′ UTS	Introns	Amino acid sequence
Human β-globin/chimp β-globin	100	<100[a]	<98.4	100
Human β-globin/rabbit β-globin	89.3	<79[b]	<67[b]	90.4
Human β-globin/mouse β-globin[c]	82.1	<66[b]	<61[b]	80.1
Human β-globin/human ε-globin	79.1	62	50	75.3

[a]5′ UTS only.
[b]Maximum homologies based on counting insertions or deletions of three or more nucleotides as single sites.
[c]Either β^{maj} or β^{min}.

within different types of gene in a cluster. As a result, the sequence homology between orthologs in distant species such as mouse and humans may be greater than that between genes on the same human cluster. For example, the sequence homology between human and rabbit β-globin genes or even between human and mouse β-globin genes is greater than that between the human β-globin and ε-globin genes (*Table 14.2*). The β–ε globin gene split most likely occurred some time before human–mouse divergence and the lack of frequent sequence exchanges within the human β-globin cluster has resulted in the considerable genetic distance between human β-globin and human ε-globin being maintained.

Table 14.3: Comparison of genome organization and gene expression in humans and mice

Parameter	Similarities/differences in humans and mice
Genome size and gene number	About 3000 Mb in both genomes; estimate of 65 000–80 000 human genes and the number in mice is expected to be close to this figure
Chromosome number and type	Humans: 22 pairs of autosomes, X and Y; most are metacentric or submetacentric (*Figure 2.18* and *Table 2.3*) Mice: 19 pairs of autosomes, X and Y, all acrocentric
CpG islands	Human genome: about 45 000 Mouse genome: about 37 000 (Antequara and Bird, 1993)
Conservation of synteny	Apart from the X chromosome (see below), is confined to subchromosomal regions which may be small or substantial (see *Figure 14.23*)
X chromosome genes	Almost complete conservation of synteny, but gene order can be quite different because of inversions (*Figure 14.10*)
Gene family organization	Very similar for individual families. Recent gene duplications have led to differences in gene number, e.g. there are two mouse β-globin genes and one α-globin gene, but only one human β-globin gene and two α-globin genes
Gene organization	Generally similar. Similar sizes of coding DNA (average $\sim$ 550 codons) and intron positions often conserved
Sequence homology	Coding DNA: usually about 70–90% homology, with polypeptide products usually within the 75–95% range. Noncoding DNA: usually extremely dissimilar
Gene expression	Very similar in general. For orthologous genes, can have differences in the choice of alternative promoters and patterns of splicing. Some differences in imprinting patterns for autosomal genes and many differences in X-inactivation status for X-linked genes (*Figure 14.12*)
Tandemly repeated noncoding DNA	Telomeric repeat sequences highly conserved. Other repeats not normally conserved in sequence, but similar overall patterns (location of satellite sequences at centromeres, widespread occurrence of microsatellite sequences, etc.)
Interspersed noncoding DNA	Some sequence conservation of interspersed repeats, notably the LINE-1 element. The human *Alu* and mouse *B1* repeats are both thought to have originated from 7SL RNA copies (see *Figure 14.25*) but are considerably diverged in sequence335

14.5 Evolution of gene structure

Eukaryotic genes are often larger and more complex than those from simple organisms. The process of gene elongation during evolution appears to have frequently involved repetition of existing amino acid sequences, often as a result of exon duplication. Additionally, the structure of many eukaryotic polypeptides suggests that frequent exchange of structural or functional protein domains has occurred at the gene level by **exon shuffling**, resulting in complex mosaic genes capable of specifying a variety of different protein modules.

14.5.1 Complex genes can evolve by intragenic duplication, often as a result of exon duplication

In addition to the other forms of DNA duplication dis-

cussed earlier, human genes, like other eukaryotic genes, often show evidence of intragenic DNA duplication which can be substantial. For example, many genes are known to encode polypeptides whose sequences are completely or mostly composed of large repeats, with sequence homology between the repeats being very high in some cases (see *Table 7.8*). In many cases, a repeat corresponds to a protein domain, and in some cases individual repeats are encoded as a result of exon duplication. Building of larger polypeptides by repetition of a previously designed protein module offers a variety of evolutionary advantages:

■   *Dosage repetition.* The ubiquitin-encoding genes, *UbB* and *UbC*, encode polypeptides containing, respectively, three and nine tandem repeats of the sequence for ubiquitin, a small protein with several functions, notably in proteolysis. Most of the proteins that are degraded in the cytosol are hydrolyzed in large protein complexes known as *proteasomes* but, before the proteins can be delivered to the proteasomes, they need to be tagged by covalent binding to a series of ubiquitin molecules, forming a multiubiquitin chain. Because many proteins are short lived, large amounts of **ubiquitin** molecules need to be synthesized. These genes may therefore have evolved to express many copies of the ubiquitin sequence by gene elongation through intragenic repetition, as opposed to the tandem gene duplication mechanism used in the case of genes encoding rRNA or histones. The large polypeptide precursor is then cleaved to generate multiple copies of the desired ubiquitin monomer.

■ *Structural extension.* Repeating domains may be particularly advantageous in the case of proteins that have a major structural role. An illustrative example is provided by the 41 exons of the *COL1A1* gene which encode the part of α1(I) collagen that forms a triple helix; each exon encodes essentially an integral number of copies (one to three) of an 18 amino acid motif which itself is composed of six tandem repeats of the structure Gly–X–Y where X and Y are variable amino acids.

■ *Domain divergence.* In most cases, intragenic duplication events have been followed by substantial nucleotide sequence divergence between the different repeat units. Such divergence presumably provides the opportunity of acquiring different, though related, functions. Sometimes the degree of sequence divergence between the repeats is such that the repeated structure may not be obvious at the sequence level. For example, in the case of the variable and constant domains of immunoglobulins, conservation of the secondary structure is much more apparent than that of the amino acid sequence (*Fig-*

ure 8.2). In some cases where the repeated structure is not obvious, statistical analysis can nevertheless reveal evidence for structural similarity.

14.5.2 Exon shuffling permits diverse combinations of structure and functional modules, and may be mediated by transposable elements

Many genes encode modules found in another type of gene. Fibronectin, a large extracellular matrix protein, contains multiple repeated domains encoded by individual exons or pairs of exons and is a good example of classical exon duplication (*Figure 14.17*). One of the repeated domains was subsequently found in tissue plasminogen activator. Like fibronectin, tissue plasminogen activator also contains other domains. They include a structural module characteristic of the epidermal growth factor precursor, and so-called kringle modules which have been found in other polypeptides such as prourokinase and plasminogen, etc. (*Figure 14.17*). Such observations have suggested the possibility of mechanisms permitting exon shuffling between genes (Patthy, 1994).

Intragenic exon duplication can be explained by a variety of mechanisms including unequal crossover, or unequal sister chromatid exchanges (see *Figure 9.7*). In order to avoid frameshifts, one would expect selective amplification of exons with a total number of nucleotides exactly divisible by three (i.e. exons which are flanked by introns of the same phase, such as 0,0, 1,1 and 2,2 exons; see *Box 14.3*). This is what is observed for exons that are duplicated within a gene and also for exons encoding modules shared by different genes (*Figure 14.17*).

Figure 14.17: Exon duplication and exon shuffling.

The fibronectin gene contains 12 copies of an exon encoding a finger module, which is also found in the products of other genes such as the tissue plasminogen activator (TPA) gene. In addition, it contains 15 copies of a pair of exons which together specify a module shared with cell surface receptors and other extracellular matrix proteins. Similarly, the epidermal growth factor precursor gene has 10 copies of an exon encoding a growth factor module which is also found in the TPA and prourokinase genes. The latter two genes also contain exons encoding *kringle* modules. Exon duplication within a gene and exon shuffling between genes selectively use exons where the total number of nucleotides is exactly divisible by three (exon groups 0,0; 1,1 and 2,2; see *Box 14.3*). Exon duplication could be mediated by unequal crossover and unequal sister chromatid exchanges, possibly assisted by interspersed repetitive sequences within introns. Exon shuffling may be mediated by transposable elements (see *Figure 14.18*).

Box 14.3

Intron groups and intron phases

Introns are heterogeneous entities with different functional capacities and notable structural differences. Depending on the extent to which they rely on extrinsic factors to engage in RNA splicing and on the nature of the splicing reaction, they can be classified into different intron groups. The structural differences include enormous length differences (unlike exons which appear to be much more homogeneous in length; see *Figure 7.7*) and some other characteristics, such as differences in positions within coding sequences. The latter characteristic, which is particularly relevant to gene expression, is recognized in the subdivision of introns into different intron phases or intron types.

Intron groups

- **Spliceosomal introns** are the conventional introns of eukaryotic cells. They are transcribed into RNA in the primary transcript and are excised at the RNA level during RNA processing by spliceosomes (see *Figure 1.16*). Only a few short sequences [those at the splice junctions and at the branch site (see *Figure 1.15*) and some regulatory sequences which are occasionally found] appear to be important for gene function. As a result, spliceosomal introns can tolerate large insertions and can be very long (e.g. intron 44 of the dystrophin gene is 140 kb long). Spliceosomal introns are likely to have arisen comparatively recently in evolution, and may have evolved from group II introns (Cavalier-Smith, 1991). By tolerating the insertion of mobile elements, they facilitated exon shuffling (Section 14.5.2).

- **Group I and II introns** have significant secondary structure and can catalyze their own excision without the requirement for a spliceosome (i.e. they are *self-splicing* or *autocatalytic introns*). They are found in both eubacteria and eukaryotes but are very restricted in their distribution, being found primarily in rRNA and tRNA genes, and in a few protein-coding genes found in some types of mitochondria, chloroplasts and bacteriophages. Both groups may also act as mobile elements, and mobile group II introns encode a reverse transcriptase-like activity which is strikingly similar to that of LINE-1 elements (Belfort, 1993). The two groups differ in the identity of conserved splicing signals and in the nature of the splicing reaction.

- **Archaeal introns** have been found only in tRNA and rRNA genes in *archaebacteria*. They have no conserved internal structure and, unlike group I and group II introns, are not self-splicing. Although they require proteins for the splicing mechanism, they do not, unlike spliceosomal introns, require *trans*-acting RNA molecules for the splicing reaction.

Intron phases

The *phase* of an intron refers to the position at which it interrupts a coding DNA sequence which specifies polypeptides (*note* that this term is therefore irrelevant for those introns which happen to interrupt an untranslated sequence). There are three types (see *Figure 14.19*):

- **Phase 0 introns** interrupt the coding sequence between adjacent codons. They are much more numerous than phase 1 or phase 2 introns (see below) and may represent the ancestral state.
- **Phase 1 introns** interrupt a codon between the first and second base positions.
- **Phase 2 introns** interrupt a codon between the second and third base positions.

Note that internal exons can be classified into various groups, depending on the phase of the two flanking introns. Exons where the total number of nucleotides is exactly divisible by three will fall into three groups (0,0; 1,1; and 2,2). *Exon duplication* within a gene and *exon shuffling* between genes involves such exons because when inserted they do not alter the translational reading frame, unlike the other six exon groups (0,1; 0,2; 1,0; 1,2; 2,0; and 2,1).

How do genes which are not necessarily closely related come to share sequences encoding very similar protein modules? One attractive possibility is *retrotransposon-mediated exon shuffling*. The most abundant retrotransposons in the human genome are the LINE1 (L1) elements which belong to the non-LTR class of retrotransposons (Section 7.4.6). Moran *et al.* (1999) developed an efficient L1 retrotransposition assay in cultured human cells and showed that L1 can insert into the intron of a gene and thence can make a copy of a downstream exon which can be inserted into another gene following another round of retrotransposition. This is possible because the L1 retrotransposition machinery has a weak specificity for its own 3′ end (and can act on other sequences including Alu sequences and processed pseudogenes which both lack reverse transcriptase). Because the L1 element's own poly (A) signal is weak, transcription of a L1 repeat within a gene often bypasses its own poly (A) sequence and uses instead a downstream poly (A) signal from the host gene. In so-doing it can make a copy of a host exon which can be stitched into another gene after another retrotransposition event (*Figure 14.18*).

14.5.3 The origin of spliceosomal introns is controversial but their phylogenetic distribution suggests that many introns have been inserted into genes comparatively recently in evolution

Following the discovery of split genes in 1977, the significance of spliceosomal introns has been intensely debated. The introns found in mammalian genes are large compared with those in other species and the intron sequences are not well conserved. Nevertheless, it is becoming apparent that many introns contain functionally important sequences involved in gene regulation and the sequences of some short introns have been considerably conserved in evolution (see *Table 14.2* for an example). Any proposed function for introns in human

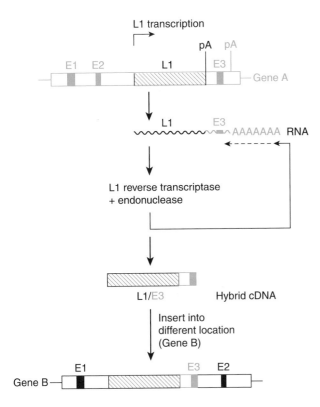

Figure 14.18: Exon shuffling between genes can be mediated by transposable elements.

The LINE1 (L1) sequence family contains members that actively transpose in the human genome. L1 elements have weak poly(A) signals and so transcription can continue past such a signal until another nearby poly(A) signal is reached as in the case of gene A at top. The resulting RNA copy can contain a transcript not just of L1 sequences but also of a downstream exon (in this case E3). The L1 reverse transcriptase complex can then act on the extended poly (A) sequence to produce a cDNA copy that contains both L1 and E3 sequences. Subsequent transposition into a new chromosomal location may lead to insertion of exon 3 into a different gene (gene B). See Moran *et al.* (1999).

genes cannot be a general one, however, because of the small minority which lack introns (see *Table 7.6*).

The evolution of spliceosomal introns and their relationship to exons have also been the subject of much controversy (Logsdon, 1998). Essentially there are two alternative positions:

- *The 'introns-early' view.* Different versions of the introns-early view exist but the *exon theory of genes* has been the most influential. It considers that exons are the descendants of ancient minigenes and spliceosomal introns are the descendants of self-splicing spacers which were located between the minigenes and *were present in primordial cells.* Exons have been considered as units which encode structural or functional domains, permitting evolution of

larger genes by exon shuffling, a strategy that was favored particularly in eukaryotes. By contrast, introns are imagined to have been effectively lost from archaea and eubacteria.

- *The 'introns-late' view.* This idea does not deny that exon shuffling between genes occurs but holds that split genes have arisen as a result of comparatively recent insertion of introns into genes. In this case, spliceosomal introns are thought to have descended from group II introns (see *Box 14.3*). The latter type of intron can function as a mobile element and are envisaged to have been introduced when a prokaryotic cell was endocytosed by a precursor to eukaryotic cells.

An important component of the exon theory of genes was the idea that exons in polypeptide-encoding genes represented functional or structural units. Exons consisting only of untranslated sequences (e.g. the first exon of the insulin gene; see *Figure 14.19*) cannot be accommodated in this view. Even in the case of coding exons, however, the evidence has been meager and in four major examples cited as evidence for the exon theory of genes, objective methods for detecting correspondence between exons and units of protein structure failed to identify any such correspondence (Stoltzfus *et al.*, 1994). Another line of evidence used to support the introns-early view has been the claim that only phase 0 introns are correlated with the structure of ancient proteins (de Souza *et al.*, 1998), but the phase correlations could instead reflect insertional bias.

The exon theory of genes was also supported by the apparent conservation of the positions of introns in genes known to have duplicated early in evolution, such as the globin genes (*Figure 14.20*). Against this view, the intron locations in numerous gene families (e.g. actin, myosin and tubulin families) are not conserved, suggesting instead that introns have been inserted recently, and in general phylogenetic studies have very strongly supported an introns-late model (Logsdon, 1998). One major problem with the introns-early model is that the requirement for introns to have been present since primordial times does not fit well with the very large number of positions where introns occur within a gene when different species are compared. It would mean not only that many original introns must have been lost subsequently from genes, but also that some genes must originally have had such a large number of introns that the corresponding exon sizes must have been tiny.

14.6 What makes us human? Comparative mammalian genome organization and the evolution of modern humans

The virtual universality of the genetic code, the high degree of conservation of key biochemical reactions, the huge evolutionary conservation of key developmental

HBB

| | 1 | 29 | 30 | 31 | 104 | 105 | 146 |

5'UTS   ATG GTG --- GGC–AG | gt-----ag | G-CTG---AGG | gt-----ag | CCC-----CAC | TAA 3'UTS

Val Gly Ar | 2   2 | g Leu Arg | 0   0 | Leu His

Phase 2   *Phase 0*

INS

| | | 1 | | 80 | 81 | | 110 |

5'UTS | gt--ag | ATG ------------ CAG G | gt---ag | TG --------- AAC TAG 3'UTS

V | 1   1 | al

Phase 1

Figure 14.19: Introns within polypeptide-encoding DNA can be grouped into three phases, according to the point of insertion.

Phase 0 introns do not interrupt codons unlike phase 1 and phase 2 introns. A phase 1 intron in the human insulin gene *INS* interrupts a codon specifying valine (position 81 in the preproinsulin precursor). A phase 2 intron in the human β-globin gene *HBB* interrupts a codon specifying arginine at position 30 in the mature polypeptide (the initial product has 147 amino acids but the initiator methionine is cleaved during processing; see *Figure 1.19*). Internal exon 2 of the *HBB* gene spans codon positions 30–104 and is an example of a 2,0 exon (see *Box 14.3*). Even if it encoded a specific structure/function module, it would not be eligible for exon duplication or exon shuffling because the number of nucleotides it contains is not exactly divisible by three. *Note* that some introns are found in the untranslated sequences, such as the first intron in the *INS* gene. This means that some exons (such as exon 1 of the *INS* gene) can be composed entirely of untranslated sequence.

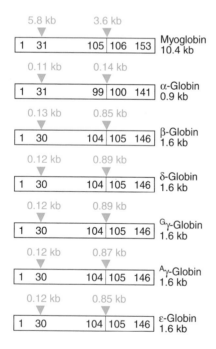

Figure 14.20: Members of the globin superfamily contain two introns which show reasonable conservation of positions, but not of size.

Boxes represent the mature polypeptides. Numbers contained within the boxes are the amino acid positions. Gene sizes are indicated to the right and intron sizes and locations are shown in blue.

processes – these are features which emphasize the close relationship of humans to species that are morphologically quite distinct and evolutionarily distantly related. So what is it that makes us different? While many of the fundamental features of human cells, genome organization and gene expression are common to all eukaryotes, mammalian-specific features can be identified, such as genomic imprinting and X inactivation. In addition, certain other components of the genome or aspects of its expression show still higher levels of specificity.

14.6.1 What makes us different from mice?

Increasingly we rely on extrapolation from mouse studies to infer the situation in humans. For example, our knowledge of gene expression patterns in early human development is minimal because of the lack of early stage embryos for study; instead we study readily available mouse embryos. Because of the power of transgenic technology and gene targeting in mouse embryonic stem cells, the mouse is the most commonly used animal model of human disease. The extrapolation from mouse studies to humans has been justified by the general assumption that genomic DNA organization and gene expression patterns of mice and humans have been highly conserved, despite the approximately 110 million years since the two lineages diverged from a common ancestor (Kumar and Hedges, 1998). Increasingly, however, there is greater appreciation of differences between the two species.

General aspects of genome organization

The genome sizes are comparable (3000 Mb of DNA), and both genomes can be divided into isochores (large chromosomal regions in which the base composition of the DNA is comparatively homogeneous but which is variable between isochores). The human isochore classes include two light (AT rich) classes L1 and L2, and three heavy (GC rich) classes H1, H2 and H3, but the mouse genome is comparatively lacking in the H3 isochore (Sabeur *et al.*, 1993). Cytogenetic analyses appear to reveal very different chromosome organizations: the mouse has 20 pairs of acrocentric chromosomes whereas there are 23 pairs of human chromosomes, most of which are metacentric or submetacentric. Nevertheless, comparison of high resolution mouse and human chromosome maps has indicated that *orthologous* chromosomal segments (e.g. those containing the major histocompatibility complex of mouse and humans) are located in regions where there is considerable similarity of cytogenetic banding patterns, albeit over relatively small chromosomal regions (Sawyer and Hozier, 1986).

Gene number

There appears to have been an erosion of CpG islands from the mouse genome: approximately 45 000 CpG islands are found in the human genome, but only 37 000 in the equivalent sized mouse genome (Antequara and Bird, 1993). This does not simply reflect a proportional reduction in gene number in the mouse genome because analysis of the sequence databases suggests that about 56% of human genes but only about 47% of mouse genes have CpG islands. On this basis, therefore, the total number of genes in humans and mice would appear to be much the same (about 80 000 when calculated from CpG island data). However, gene families often show different numbers of genes in different mammals (*Figure 14.21*). Such differences are expected to reflect complex processes of gene duplication and loss, with clear evidence of interlocus sequence exchanges (*Figure 14.22*).

In some cases, human genes do not appear to have any rodent orthologs, or if they exist there has been so much sequence divergence that they cannot be identified by standard hybridization methods even at low stringency. Examples include several of the genes mapping at or close to the major pseudoautosomal region such as *SHOX*, a locus for Leri–Weill syndrome, *ANT3*, *MIC2* and *KAL*, the Kallman syndrome gene. Similarly, there appear to be four human apolipoprotein (a) genes but none can be detected in rodent genomes (Lawn, 1996). The recent origin of the apolipoprotein (a) genes most likely occurred following a duplication of the related plasminogen locus. See also Ottolenghi and Vekemans (1998).

Gene distribution

Gene order has not been generally well conserved between human and mouse chromosomes. As in most mammals, human–mouse comparisons show a generally strong conservation of genes on the X chromosome. However, a few genes on the human X chromosome are known to have autosomal orthologs in mouse (see *Figure 14.11*). In addition, the general order of genes on the human and mouse X chromosomes is rather different, although conserved over subchromosomal regions (see *Figure 14.10*). For any one human or mouse autosome, orthologous regions are found on a variety of different chromosomes in the other species (*Figure 14.23*). However, again there is conservation of gene order over small to moderate sized subchromosomal regions. Such partial **conservation of synteny** (i.e. a group of linked genes in one species is paralleled by a linkage group between the orthologous genes in the other species) has proven to be very useful in identifying some human disease genes (see Section 15.4.3).

Gene organization and gene expression

The sizes of coding DNA in mouse and human genes are nearly identical, with an average size of perhaps about 550 codons (Makalowski and Boguski, 1998 and unpublished data). The respective polypeptide sequences show a high degree of sequence similarity, often within the 80–95% range. However, different classes of polypeptides may be extremely conserved (such as many gene products that are important in development or in crucially important cellular functions such as ribosomal function) while others, notably ligands and receptors that are important in host defense, can be much more divergent (*Figure 14.24*). The sequence similarity of coding DNA is generally a few per cent less than that for the polypeptide products (largely because of silent nucleotide substitutions, notably at the third base position of codons).

Species differences in gene expression include differences in RNA processing and the alternative usage of promoters (see also Ottolenghi and Vekemans, 1998). For example, the human aldolase A gene has an additional promoter which does not function in the rat ortholog and similar parallels are expected for some human and mouse orthologs. Other human–mouse differences include considerable differences in the pattern of X chromosome inactivation (see *Figure 14.12*) and also in the conservation of imprinting. For example, the mouse insulin-like growth factor receptor gene *Igfr* is imprinted and paternal alleles are not expressed, but a different pattern of polymorphic imprinting is found in humans.

Noncoding DNA

Introns and noncoding DNA flanking genes are generally so highly diverged that alignment of orthologous sequences from the two species can be extremely difficult unless the comparison is confined to sequences which are located close to exons. Thus, very short introns (less than 200 bp) can be aligned and compared (see *Table 14.2*) but larger introns (accounting for the great majority of introns) are progressively more difficult to compare because of the very high sequence divergence. However, a striking example of the conservation of noncoding DNA occurs in the TCR cluster, where sequencing of about 100 kb in mouse and humans

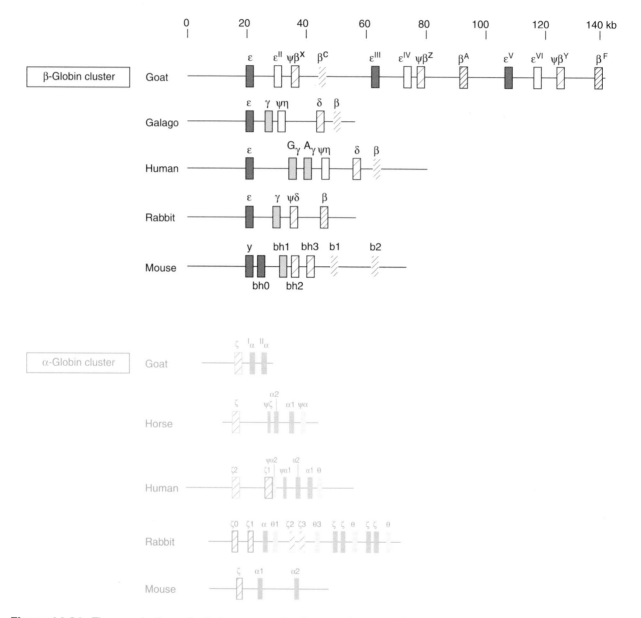

Figure 14.21: The organizations of orthologous gene families can show considerable differences in different mammals.

Shading of genes indicates proposed orthologous relationships, so that, for example, the horse ψα gene is orthologous to the θ gene of humans and rabbits. *Note* that the large number of genes in the goat β-globin cluster and the rabbit α-globin cluster reflects recent tandem triplication and quadruplication events respectively. The functional status of the θ genes is uncertain: the human θ gene may be an example of an expressed pseudogene (see *Box 7.3*) while in the rabbit the θ1 and θ2 genes are pseudogenes, and the other two θ genes are likely to be pseudogenes. Redrawn from Hardison and Miller (1993), © The Society for Molecular Biology and Evolution, PA.

reveals a sequence identity of approximately 70%, even though only about 6% of the DNA is coding DNA (Koop and Hood, 1994). This is likely to be related to the very unusual mechanisms for expressing TCR and immunoglobulin genes (Section 8.6).

Telomeric minisatellite DNA is conserved between mice and humans, and indeed the TTAGGG repeats are conserved throughout vertebrates, presumably because of selection pressure to ensure continued recognition by

the telomerase enzyme (*Figure 2.9*). However, highly repetitive DNA sequences in general are among the most rapidly diverging sequences because of the virtual absence of conservative selection pressure. For example satellite DNA sequences in the human and mouse genomes are quite different, and there is poor conservation of hypervariable minisatellites and microsatellites at orthologous locations in humans and mice. Nevertheless, there are several examples of apparent conservation of

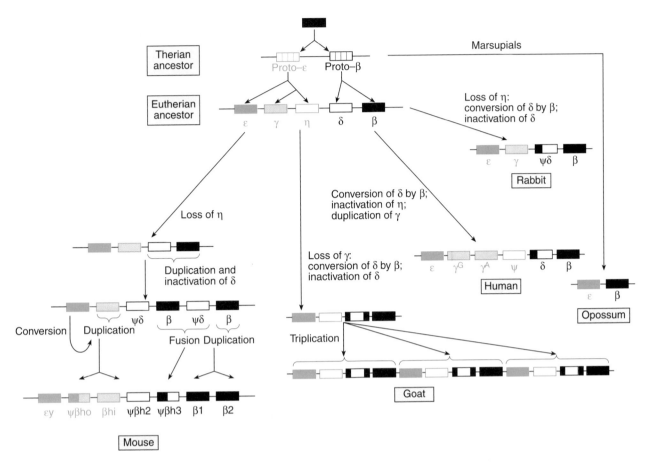

Figure 14.22: The evolution of the mammalian β-globin gene cluster has involved frequent gene duplications, conversions and gene loss or inactivation.

Note that, in addition to gene duplication and gene loss events, there are frequent examples where the sequence of one gene shows evidence of having been copied from the sequence of another gene. This is loosely described as conversion in this figure, but may have involved mechanisms other than gene conversion in some cases. For example, the conversion of δ by β in the lineage leading to rabbits could have involved an unequal crossover event of the type that most likely resulted in the production of the ψh3 gene in the mouse lineage. Redrawn from Tagle *et al.* (1992) with permission from Academic Press Inc.

intragenic microsatellites located within orthologs in human and mouse or rat (Stallings, 1995).

In general highly repeated interspersed elements are not well-conserved. The *Alu* repeat appears to have evolved as a processed pseudogene from transcripts of the 7SL RNA gene (Ullu and Tschudi, 1984; see *Figure 14.25*) and appears to be specific to primates. It does, however, have a type of counterpart in the mouse genome, the *B1* repeat, which also appears to have been generated from a 7SL RNA-like gene. The huge amplification in copy number apppears largely to have been generated some time ago in both species and the sequence divergence between the consensus sequences for the two types of repeat family is sufficiently high that probes can be made from them which permit distinction between the two genomes. The LINE-1 repeats are, however, conserved in humans and mouse (and throughout mammals), largely because of conservative selection pressure

to maintain the sequence of the large *ORF2* sequence which specifies the reverse transcriptase.

14.6.2 What makes us different from the great apes?

Traditional primate classifications places humans as the sole living members of the family Hominidae and the African great apes (gorillas, chimpanzees and bonobos i.e. pygmy chimpanzees) are placed together with Asian great apes (orangutans) within the subfamily Ponginae of the family Pongidae. However, this anthropocentric view has been strongly challenged by overwhelming evidence that the African great apes share their more recent common ancestry with humans than with orangutans (see Goodman, 1999). Nucleotide sequence data indicate that divergence of human–chimpanzee, human–gorilla and human–orangutan lineages occurred about 5.5, 6.7 and

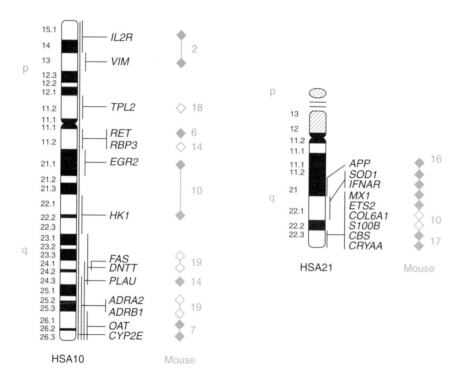

Figure 14.23: Conservation of orthologous human and mouse linkage groups is limited to subchromosomal regions.

Note that genes on human chromosome 10 (HSA10) have orthologs on at least seven different mouse chromosomes. Human chromosome 21 (HSA21) shows considerable conservation of synteny with mouse chromosome 16, but genes at the end of 21q have counterparts in mouse chromosomes 10 and 17. Note that the X chromosome is very highly conserved in mammals, but even then gene order can vary considerably in man and mouse (*Figure 14.10*). Adapted from O'Brien *et al.* (1993) *Nature Genet.*, **3**, pp. 103–112, with permission from Nature America Inc.

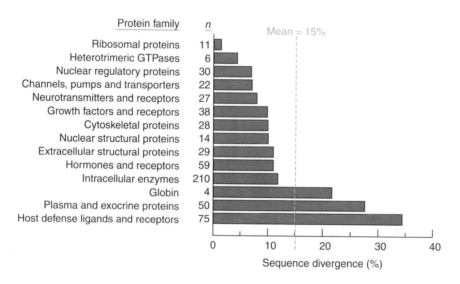

Figure 14.24: Amino acid sequence divergence between human and rodent orthologs.

The average human–rodent sequence divergence is shown for 14 protein families, representing a total of 603 proteins compared between human and mouse/rat. *Note* that certain types of protein, such as the ribosomal proteins, are very highly conserved between mice and humans. Others, notably ligands or receptors that function in host defense, show considerable sequence divergence. Redrawn from Murphy (1993) *Cell*, **73**, pp. 823–826, with permission from Cell Press.

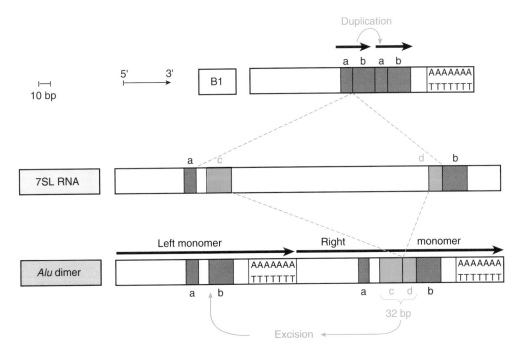

Figure 14.25: The human *Alu* repeat and the mouse *B1* repeat evolved from processed copies of the 7SL RNA gene.

Extensive homology of the *Alu* repeat sequences to the ends of the 7SL RNA sequence suggests that a polyadenylated copy of the 7SL RNA gene integrated elsewhere in the genome by a retrotransposition event (see *Figure 8.7*). In some cases, the integrated copies were able to produce RNA transcripts of their own, using the internal promoter of the 7SL RNA gene. At a very early stage, an internal segment (between c and d) was lost. Subsequently, a 32-bp central segment containing regions flanking the original deletion (c + d) was deleted to give a related repeat unit. Fusion of the two types of unit resulted in the classical *Alu* dimeric repeat, with the left (5′) monomer lacking a 32-bp sequence and the right (3′) monomer containing the 32-bp sequence. *Note* that in the human genome there are also multiple copies of a *free left Alu monomer* (*FLAM*) and a *free right Alu monomer* (*FRAM*). In the mouse, a similar process of copying from the 7SL RNA gene appears to have occurred with subsequent deletion of a large internal unit (between a and b), followed by tandem duplication of flanking regions (a + b).

8.2 million years ago respectively (Kumar and Hedges, 1998), and humans are now thought to be particularly closely related to chimpanzees and bonobos (Goodman *et al.*, 1998). The divergence into separate species may initially have been driven by small cytogenetic differences and/or mutations in key genes regulating gamete formation or regulation of early embryonic development. However, once speciation had been accomplished, the effective reproductive isolation meant that species-specific patterns of intragenomic sequence exchange could result in extending differences between species.

Genome organization and coding DNA
Cytogenetic comparisons of the great apes emphasize the very strong conservation of banding patterns (Yunis and Prakash, 1982). The only major structural differences are a number of pericentric and paracentric inversions, the recent fusion of two chromosomes to form human chromosome 2, and a reciprocal translocation between the gorilla chromosomes which correspond to human chromosomes 5 and 17. In addition, the extent of heterochromatinization is variable, with most gorilla chromosome arms and about half of the chimpanzee chromosome arms containing terminal heterochromatic G bands which are absent from human and orangutan

chromosomes (see *Figure 14.26*). Although the present information on comparative gene mapping in primates is sketchy, the available details show evidence of strong conservation of synteny (i.e. linked genes in humans are almost always linked in the great apes). However, large-scale organization at certain loci can differ. For example, a large part of the human Ig κ locus on 2p is duplicated (see *Figure 8.28*) but this is not the case in the corresponding chimpanzee and gorilla chromosomes (Ermert *et al.*, 1995).

When orthologous human and chimpanzee sequences are compared, the coding DNA typically shows 98–100% sequence identity (see *Table 14.2* for an example). Indeed, in some cases, specific alleles of certain human genes are more closely related to orthologs in chimpanzees than they are to other human alleles. For example, at the human HLA-DRβ locus, the alleles *HLA-DRB1*0302* and *HLA-DRB1*0701* are clearly closer in sequence to certain alleles of the orthologous chimpanzee (*Pan troglodytes*) gene *Patr-DRB* than they are to each other (*Figure 14.27*). Such observations are consistent with a comparatively ancient origin for such divergent alleles, predating man–chimpanzee divergence. Although there are as yet limited data, extrapolating from known human–mouse differences suggests that extremely few human genes

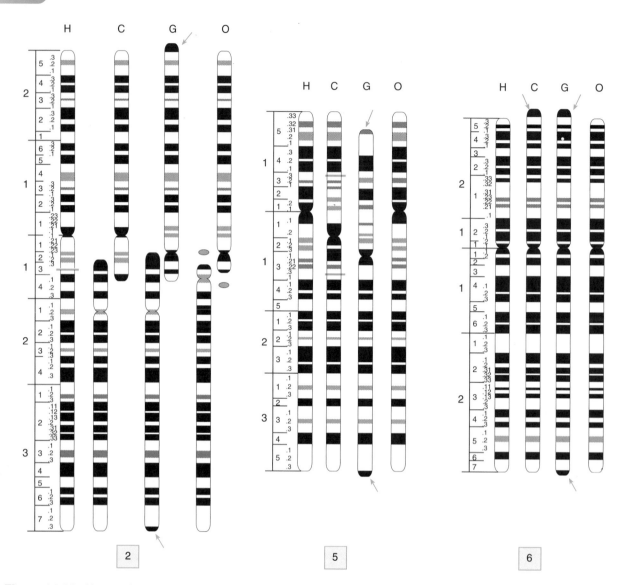

Figure 14.26: Human chromosome banding patterns are very similar to those of the great apes.

The ideograms represent selected primate chromosomes from 1000-band late prophase preparations. Human (H) chromosomes 2, 5 and 6 are illustrated together with the corresponding orthologs in chimpanzee (C), gorilla (G) and orangutan (O). Human chromosome 2 appears to have evolved by fusion of two primate chromosomes (with the point of fusion possibly at 2q13). *Note* the extremely similar structures for human chromosome 6 and its orthologs – the only readily visible differences are due to additional telomeric heterochromatin on the short arm of the chimpanzee ortholog, and on both arms of the gorilla ortholog (see arrows). In contrast, the orthologs of human chromosome 5 show appreciable differences. They include a pericentric inversion in the chimpanzee chromosome with breakpoints corresponding to human 5p13 and 5q13, and considerable differences in the gorilla ortholog (which has undergone a reciprocal translocation with a chromosome corresponding to human chromosome 17). Reprinted with permission from Yunis and Prakash (1982) *Science*, **215**, pp. 1525–1530. © 1982, American Association for the Advancement of Science.

could be expected to lack counterparts in the chimp and gorilla genomes and vice versa. In just about all cases, one would expect that these human-specific genes would have arisen by very recent gene duplication events so that both gene copies may be identical, and as such be rather unlikely to provide much input into reinforcing the observed anatomical and developmental differences between humans and the great apes. Nevertheless

human–primate differences do exist and the molecular basis for these differences are now beginning to be discovered (Gibbons, 1998).

Noncoding DNA
Noncoding DNA from humans and apes can show extremely high levels of sequence homology. For example, pairwise comparisons of orthologous noncod-

Table 14.4: Comparison of genome organization and gene expression in humans and the great apes

Parameter	Similarities/differences in humans and great apes
Genome size and gene number	About 3000 Mb in both genomes; estimate of 65 000–80 000 human genes and number in great apes expected to be extremely close to this figure
Chromosome banding patterns	Very similar. The only major differences are the fusion of two primate chromosomes to produce human chromosome 2, a reciprocal translocation in the gorilla lineage, and a few examples of paracentric and pericentric inversions (see *Figure 14.26*)
Conservation of synteny	Extensive
Gene family organization	Extremely similar for individual families. Very recent gene duplications may occasionally result in differences in gene number
Sequence homology	Coding DNA: average sequence homology about 98–100% Noncoding DNA: usually extremely similar ($\sim$ 98%), but some sequences are restricted to humans and others which are found in some apes appear to be missing from the human genome
Gene expression	Extremely similar in general
Tandemly repeated noncoding DNA	Telomeric repeat sequences highly conserved. Other repeats not normally conserved in sequence, but similar overall patterns (location of satellite sequences at centromeres, widespread occurrence of microsatellite sequences, etc.)
Interspersed noncoding DNA	The *Alu* repeat is found in the great apes and other primates, but human-specific subsets are known (see text)

ing sequences spanning more than 12.5 kb of the β-globin gene cluster showed sequence divergence of only 1.7, 1.8 and 3.3% in the case of human–chimp, human–gorilla and human–orangutan comparisons, respectively (Goodman *et al.*, 1998). However, highly repeated DNA families appear to be undergoing a more rapid evolution. Although a common alphoid sequence is conserved in all human and great ape chromosomes (Baldini *et al.*, 1993), the vast majority of human chromosome-specific alphoid sequences do not hybridize to the centromeres of the corresponding chimpanzee and gorilla chromosomes (Archidiacono *et al.*, 1995). In addition, a subterminal satellite DNA located adjacent to the telomeres of chimpanzee and gorilla chromosomes has no counterpart in

human and orangutan chromosomes (Royle *et al.*, 1994). This satellite most likely is the major component of the additional heterochromatic terminal G bands of chimpanzee and gorilla chromosomes (see *Figure 14.26*). Minisatellite and microsatellite sequences can also differ between humans and primates. Telomere sequences are conserved but hypervariable minisatellite sequences show transient evolution in the primate genomes – highly polymorphic human minisatellites often have monomorphic or minimal variability in the corresponding chromosomes of the great apes (Gray and Jeffreys, 1991). Microsatellites also show differences at orthologous positions in humans and other primates (Rubinsztein *et al.*, 1995).

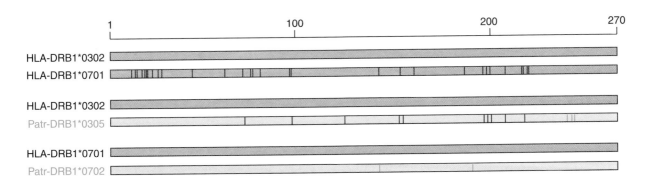

Figure 14.27: Some human alleles show greater sequence divergence than when individually compared with orthologous chimpanzee genes.

From a total of 270 amino acid positions, the *HLA-DRB*10302* and *HLA-DRB1*0701* alleles show a total of 31 differences (13%). Comparison of either allele with alleles at the orthologous chimpanzee locus (*Patr-DRB1*) identifies more closely related human–chimpanzee pairs, such as *HLA-DRB1*0701* and *Patr-DRB1*0702* (only two amino acid differences out of 270). This suggests that some present-day *HLA* alleles pre-date the human–chimpanzee split. See Gibbons (1995) for further details. Redrawn from Klein *et al.* (1993) *Scientific American*, **269**, pp. 675–680, with permission from Scientific American Inc.

Highly repetitive interspersed DNA can also show differences. Although the *Alu* repeat family is found in other primates, several different subfamilies have been recognized (Jurka and Miloslajevic, 1991, for a classification) and appear to have spread at different periods of primate evolution. The average age of the oldest subfamily, the *Alu* J repeats, was estimated at about 55 million years. This family, like other old subfamilies, is characterized by considerable divergence beween the members but comparatively close resemblance of the consensus to the 7SL RNA sequence. A small number of the *Alu* sequences belong to families which are extremely recent in evolutionary origin and contain members that are actively transposing. They include the Sb1 (previously alternatively known as the PV or HS subfamily) and Sb2 families which appear, on the basis of copy number, to be very largely human-specific (Zietkiewicz *et al.*, 1994).

14.6.3 DNA-based studies indicate that the genetic diversity in humans is very limited and that we are descended from individuals who lived in east Africa about 200 000 years ago

The limited genetic diversity in humans

Humans are unusual among primates in that we show much more limited genetic variability than our close relatives, the chimpanzees, and other apes. For example, the sequence of a 729 bp intron in the *ZFY* gene on the Y chromosome revealed no differences when the Y chromosomes of 38 different men were sampled, although there were many differences when referenced against the equivalent sequences from chimpanzees, gorillas and orangutans (Dorit *et al.*, 1995). Other studies on genes located in diverse genomic regions all confirm the low nucleotide diversity and therefore limited genetic variability of humans (Li and Sadler, 1991). By contrast, the genetic diversity in individual species of the great apes is very much higher. Chimpanzees show substantially more genetic variation in their nuclear genomes than humans do, and several chimpanzee and bonobo clades (and even single social groups) have retained substantially more mitochondrial variation than is seen in the entire human species (Gagneux *et al.*, 1999). These findings strongly suggest that at a recent time in our evolutionary past, the human population went through a genetic 'bottleneck' (i.e. a severe reduction in effective population size), so that a large part of the previously existing genetic variability was lost.

The great majority of the existing overall genetic variation in humans is represented by individual diversity within populations. By contrast, differences between racial groups accounts for only about 10% of the total variation (Barbujani *et al.*, 1997). Differences do occur, however, in the extent of genetic diversity within different populations, with African populations demonstrating the greatest diversity. These data are consistent with a recent genetic bottleneck followed by rapid population expansion from African populations (see also below).

DNA analyses on fossil remains

Research into our recent evolutionary past has traditionally been bedevilled by the incompleteness of the fossil record although exciting finds continue to be made about our recent ancestors (Culotta, 1999). DNA-based studies have provided a powerful alternative way of analysing fossils. Because the DNA from such ancient remains is present in small amounts and has been considerably degraded, PCR is used to amplify small overlapping portions of the DNA which can then be sequenced. Unfortunately, there is an age limit to this approach because of the chemical instability of DNA; previous hopes of large scale ancient DNA studies have been tempered by the realization that most samples more than 100 000 years old fail to yield amplifiable DNA.

Despite the above limitations, some important findings can be made. Krings *et al.* (1997) reported the successful amplification and sequencing of a portion of the hypervariable segment of the human mtDNA control region from a Neanderthal fossil expected to be from 30 000–100 000 years old. Neanderthals were a population of archaic humans who inhabited Europe and Western Asia between about 230 000 and 30 000 years ago and for part of this time they coexisted with modern humans but their relationship to modern humans was unclear. The careful study of Krings *et al.* (1997) showed that the neanderthal mtDNA sequence was clearly different from that of modern humans (about three times the average difference found between different humans but about half the average difference between humans and chimpanzees.). They were able to conclude that neanderthals went extinct without contributing any mtDNA to modern humans, and that the lineages leading to neanderthals and modern humans diverged about 550 000–690 000 years ago.

Reconstructing the history of human populations

A variety of DNA-based studies have been carried out on extant human and primate populations in order to infer our evolutionary past. DNA sequences or markers from selected loci are studied in usually large numbers of individuals and phylogenies are compiled based on the degree of relatedness of the individual samples (see von Haeseler *et al.*, 1995 and Jorde *et al.*, 1998 for the types of approach that are used). In many cases sequences or markers from mtDNA or from the nonrecombining portion of the Y chromosome have been used, or noncoding sequences in nuclear DNA. The absence of recombination in mtDNA/Y sequences makes interpretation of the data easier, and coalescence approaches are more easily applied to estimate the date of the common ancestor of the individuals sampled (*Figure 14.28*). Using this type of approach both mtDNA and Y chromosome DNA studies indicate that the common ancestor of modern humans can be traced back to somewhat less than about 200 000 years ago, This does not mean of course that only a single person (e.g. 'the mitochondrial Eve') was present at that time; instead, it simply means that the DNA of the other

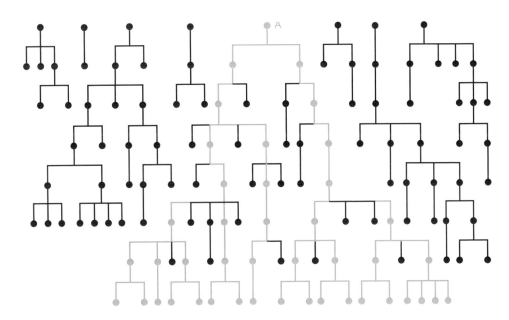

Figure 14.28: Coalescence analyses seek to trace back lineages until they *coalesce* in a single individual.

The example shows *uniparental inheritance* which can be tracked using mt DNA markers (inheritance through the maternal line) or by using markers from the nonrecombining portion of the Y chromosome (paternal inheritance). All extant populations (shown in blue) can ultimately be traced back to a single individual A. Many other people may have lived at the same time as this individual, but, unlike A, they and all individuals shown in black haven't transmitted their DNA to the present generation. Using this type of approach with mtDNA or Y markers a common ancestor of all living humans is estimated to have lived somewhat less than 200 000 years ago (see text).

people living at that time didn't get transmitted to the present human population (*Figure 14.28*).

As mtDNA and the nonrecombining Y sequences are inherited through the maternal line or paternal line respectively, the above analyses also give insights into the possibility of differential migration of the two sexes. Perhaps surprisingly, these type of studies have shown that female genes have geographically migrated more than

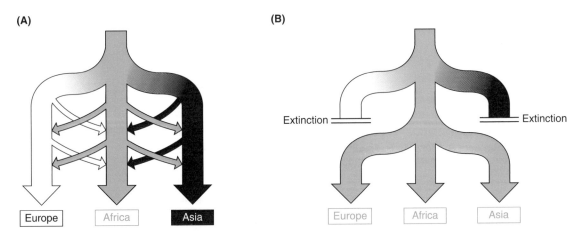

Figure 14.29: Competing hypotheses to explain the origins of modern humans.

(**A**) The *multiregional hypothesis.* Modern *Homo sapiens* is imagined to have evolved from more archaic forms over the course of a million years or so, at several different locations in the Old World. The high degree of genetic homogeneity was maintained by natural selection and by *gene flow* (small arrows) between different populations. (**B**) The *African replacement* ('*Out of Africa*') *hypothesis.* Modern humans arose in Africa approximately 100 000–200 000 years ago and dispersed throughout the Old World. In so doing, they replaced other populations who did not then contribute to the genes of modern humans. Reprinted from Jorde *et al* (1998) *BioEssays*, **20**, pp. 126–136 with permission from John Wiley & Sons, Inc.

male genes. One possible explanation is that men may typically have traveled greater distances in their lifetimes, but when it came time to settle down and have children they often went home to their birthplaces to be joined by spouses who may have had to migrate some distance from their birthplace.

Where have we come from: the African replacement ('Out of Africa') model versus the multiregional hypothesis

DNA-based studies have proved very valuable in studying recent human history (see Cavalli-Sforza *et al.*, Further Reading) but there has been a long-standing controversy regarding the geographical origins of modern humans. There is no doubt that African populations have the largest amount of genetic diversity but two competing hypotheses attempt to explain our origins:

■ The **multiregional hypothesis** (*Figure 14.29A*). This states that modern *Homo sapiens* evolved from more archaic forms over the course of a million years or so, at several different locations in the Old World. The

high degree of genetic homogeneity was maintained by natural selection and by **gene flow** between different populations.

■ The **African replacement hypothesis** (*Figure 14.29B*). This states that modern humans arose in Africa approximately 100 000–200 000 years ago and dispersed throughout the Old World to replace archaic human species completely.

The debate regarding these two competing hypotheses remains to be resolved. Evolutionary theory predicts that an older, 'source' population will typically have greater diversity than a population derived more recently from it and the very strong evidence for greater genetic diversity of African populations (Jorde *et al.*, 1998) is consistent with the African replacement hypothesis. However, there remains considerable uncertainty regarding various parameters such as gene flow patterns, and population sizes. Perhaps we will have a more definitive answer by the year 2030, say, when the entire genomes of almost everyone on the planet may have been sequenced using DNA chips!

Further reading

Cavalli-Sforza LL, Menozzi P, Piazza A (1994) *The History and Geography of Human Genes*. Princeton University Press, Princeton, NJ.

Doolittle WF (1999) Phylogenetic classification and the universal tree. *Science*, **284**, 2124–2128.

Jackson M, Strachan T, Dover GA (1996) *Human Genome Evolution*. BIOS Scientific Publishers, Oxford, UK.

Jones S, Martin R, Pilbeam D (1992) *The Cambridge Encyclopaedia of Human Evolution*. Cambridge University Press, Cambridge.

Li WH, Graur D (1991) *Fundamentals of Molecular Evolution*. Sinauer Associates, Sunderland, MA.

Nei M (1987) *Molecular Evolutionary Genetics*. Columbia University Press, New York.

The NCBI Taxonomy database at http://www.ncbi.nlm.nih.gov/Taxonomy/tax.html.

References

Antequara F, Bird A (1993) Number of CpG islands and genes in human and mouse. *Proc. Natl Acad. Sci. USA*, **90**, 11995–11999.

Archidiacono N, Antonacci R, Marzella R, Finelli P, Lonoce A, Rocchi M (1995) Comparative mapping of human alphoid sequences in great apes using fluorescence *in situ* hybridization. *Genomics*, **25**, 477–484.

Baldini A, Ried T, Shridhar V, Ogura K, D'Aiuto L, Rocchi M, Ward DC (1993) An alphoid DNA sequence conserved in all human and great ape chromosomes: evidence for ancient centromeric sequences at human chromosomal regions 2q21 and 9q13. *Hum. Genet*, **90**, 577–583.

Barbujani G, Magagni A, Minch E, Cavalli-Sforza LL (1997) An apportionment of human DNA diversity. *Proc. Natl Acad. Sci. USA*, **94**, 4516–4519.

Belfort M (1993) An expanding universe of introns. *Science*, **262**, 1009–1010.

Blair HJ, Reed V, Laval SH, Boyd Y (1994) New insights into the Man–mouse comparative map of the X chromosome. *Genomics*, **19**, 215–220.

Cavalier-Smith T (1991) Intron phylogeny: a new hypothesis. *Trends Genet.*, **7**, 145–148.

Culotta E (1999) A new human ancestor? *Science*, **284**, 572–573.

de Souza SJ, Long M, Klein RJ, Roy S, Lin S, Gilbert W (1998) Towards a resolution of the introns early/late debate: only phase zero introns are correlated with the structure of ancient proteins. *Proc. Natl Acad. Sci. USA*, **95**, 5094–5098.

Disteche CM (1995) Escape from X-inactivation in human and mouse. *Trends Genet.,* **11**, 17–22.

Doolittle WF (1998) You are what you eat: a gene transfer rachet could account for bacterial genes in eukaryotic nuclear genomes. *Trends Genet.,* **14**, 307–311.

Dorit RL, Akashi H, Gilbert W (1995) Absence of polymorphism at the ZFY locus on the human Y chromosome. *Science,* **268**, 1183–1185.

Eichler EE (1998) Masquerading repeats: paralogous pitfalls of the human genome. *Genome Res,* **8**, 758–762.

Ellis N (1998) The war of the sex chromosomes. *Nature Genet.,* **20**, 9–10.

Ermert K, Mitlohner H, Schempp W, Zachau HG (1995) The immunoglobulin kappa locus of primates. *Genomics,* **25**, 623–629.

Gagneux P, Wills C, Gerloff U, Tautz D, Morin PA, Boesch C, Fruth B, Hohmann G, Ryder OA, Woodruff DS (1999) Mitochondrial sequences show diverse evolutionary histories of African hominoids. *Proc. Natl Acad. Sci. USA,* **96**, 5077–5082.

Garcia-Fernandez J, Holland PW (1994) Archetypal organization of the Amphioxus *Hox* gene cluster. *Nature,* **360**, 563–566.

Gibbons A (1995) The mystery of humanity's missing mutations. *Science,* **268**, 35–36.

Gibbons A (1998) Which of our genes make us human? *Science,* **281**, 1432–1434.

Goodman M (1999) The genomic record of humankind's evolutionary roots. *Am. J. Hum. Genet.,* **64**, 31–39.

Goodman M, Porter CA, Czelusniak J *et al.* (1998) Towards a phylogenetic classification of primates based on DNA evidence complemented by fossil evidence. *Mol. Phylogenet. Evol.,* **9**, 585–598.

Graves JAM, Wakefield MJ, Toder R (1998) The origin and evolution of the pseudoautosomal regions of human sex chromosomes. *Hum. Mol. Genet.,* **7**, 1991–1996.

Gray IC, Jeffreys AJ (1991) Evolutionary transcience of hypervariable minisatellites in man and the primates. *Proc. Royal Soc. Lond. B,* **243**, 241–253.

Gray MW, Burger G, Lang BF (1999) Mitochondrial evolution. *Science,* **283**, 1476–1481.

Hardison R, Miller W (1993) Use of long sequence alignments to study the evolution and regulation of mammalian globin gene clusters. *Mol. Biol. Evol.,* **10**, 73–102.

Jegalian K, Page DC (1998) A proposed pathway by which genes common to mammalian X and Y chromosomes evolve to become X inactivated. *Nature,* **394**, 776–780.

Jorde LB, Bamshad M, Rogers AR (1998) Using mitochondrial and nuclear DNA markers to reconstruct human evolution. *BioEssays,* **20**, 126–136.

Jurka J, Miloslajevic A (1991) Reconstruction and analysis of human Alu genes. *J. Mol. Evol.,* **32**, 105–121.

Klein J, Takahata N, Ayala FJ (1993) MHC polymorphism and human origins. *Scientific American,* **269**, 78–83.

Koop BF, Hood L (1994) Striking sequence similarity over almost 100 kilobases of human and mouse T cell receptor DNA. *Nature Genet.,* **7**, 48–53.

Krings M, Stone A, Schmitz W, Krainitzki H, Stoneking M, Paabo S (1997) Neandertal DNA sequences and the origins of modern humans. *Cell,* **90**, 19–30.

Kumar S, Hedges SB (1998) A molecular timescale for vertebrate evolution. *Nature,* **392**, 917–920.

Lahn BT, Page DC (1997) Functional coherence of the human Y chromosome. *Science,* **278**, 675–680.

Lalioti MD, Scott HS, Buresi C *et al.* (1997) Dodecamer repeat expansion in cystatin B gene in progressive myoclonus epilepsy. *Nature,* **386**, 847–851.

Lawn RM (1996) How often has Lp(a) evolved? *Clin. Genet.,* **49**, 176–184.

Li WH, Sadler LA (1991) Low nucleotide diversity in man. *Genetics,* **129**, 513–523.

Logsdon Jr, JM (1998) The recent origins of spliceosomal introns revisited. *Curr. Opin. Genet. Dev.,* **8**, 637–648.

Lopez-Garcia P, Moreira D (1999) Metabolic symbiosis at the origin of eukaryotes. *Trends Biochem. Sci.,* **24**, 88–93.

Lundin LG (1993) Evolution of the vertebrate genome as reflected in paralogous chromosomal regions in Man and the house mouse. *Genomics,* **16**, 1–19.

Makalowski W, Boguski MS (1998) Evolutionary parameters of the transcribed mammalian genome: an analysis of 2,820 orthologous rodent and human sequences. *Proc. Natl Acad. Sci. USA,* **95**, 9407–9412.

Martin W, Muller M (1998) The hydrogen hypothesis for the first eukaryote. *Nature,* **392**, 37–41.

Meyer A, Malaga-Trillo E (1999) Vertebrate genomics: more fishy tales about Hox genes. *Curr. Biol.,* **9**, R210–R213.

Miller AP, Willard HF (1998) Chromosomal basis of X chromosome inactivation: identification of a multigene domain in Xp11.21–p11.22 that escapes X inactivation. *Proc. Natl Acad. Sci. USA,* **95**, 8709–8714.

Moran JV, DeBerardinis RJ, Kazazian HH Jr (1999) Exon shuffling by L1 retrotransposition. *Science,* **283**, 1530–1534.

Murphy PM (1993) Molecular mimicry and the generation of host defense protein diversity. *Cell,* **72**, 823–826.

O'Brien SJ, Womack JE, Lyons LA, Moore KJ, Jenkins NA, Copeland NG (1993) Anchored reference loci for comparative genome mapping in mammals. *Nature Genet.,* **3**, 103–112.

Ohno S (1973) Ancient linkage groups and frozen accidents. *Nature,* **244**, 259–262.

Ottolenghi C, Vekemans M (1998) Genetic divergence between mouse and humans: a useful direction for gene pathway analysis. *Teratology,* **58**, 82–87.

Pamilo P, O'Neill RJ (1997) Evolution of the Sry genes. *Mol. Biol. Evol.,* **14**, 49–55

Patthy L (1994) Exons and introns. *Curr. Opin. Struct. Biol.,* **4**, 383–392.

Perry J, Feather S, Smith A, Palmer S, Ashworth A (1998) The human *FXY* gene is located within Xp22.3: implications for evolution of the mammalian X chromosome. *Hum. Mol. Genet.,* **7**, 299–305.

Rappold GA (1993) The pseudoautosomal regions of the human sex chromosomes. *Hum. Genet.,* **92**, 315–324.

Ried K, Rao E, Schiebel K, Rappold GA (1998) Gene duplications as a recurrent theme in the evolution of the human pseudoautosomal region 1: isolation of the gene *ASMTL. Hum. Molec. Genet.,* **7**, 1771–1778.

Rivera MC, Jain R, Moore JE, Lake JA (1998) Genomic evidence for two functionally distinct gene classes. *Proc. Natl Acad. Sci. USA,* **95**, 6239–6244.

Royle NJ, Baird DM, Jeffreys AJ (1994) A subterminal satellite located adjacent to telomeres in chimpanzee is absent from the human genome. *Nature Genet.,* **6**, 52–56.

Rubinsztein DC, Amos W, Leggo J, Goodburn S, Jain S, Li SH, Margolis RL, Ross CA, Ferguson-Smith MA (1995) Microsatellite evolution – evidence for directionality and variation in rate between species. *Nature Genet.,* **10**, 337–343.

Ruddle FH, Barlels JL, Bentley KL, Kappen C, Murtha MT, Pendleton JW (1994) Evolution of Hox genes. *Annu. Rev. Genet.,* **28**, 423–442.

Sabeur G, Macaya G, Kadi F, Bernardi G (1993) The isochore patterns of mammalian genomes and their phylogenic implications. *J. Mol. Evol.,* **37**, 93–108.

Sawyer JR, Hozier JC (1986) High resolution of mouse chromosomes: banding conservation between Man and mouse. *Science,* **232**, 1632–1635.

Skrabanek L, Wolfe KH (1998) Eukaryote genome duplication – where's the evidence? *Curr. Opin. Genet. Dev.,* **8**, 694–700.

Stallings RL (1995) Conservation and evolution of (CT)n/(GA)n microsatellite sequences at orthologous positions in diverse mammalian genomes. *Genomics,* **25**, 107–113.

Stoltzfus A, Spenceer DF, Zuker M, Logsdon, Jr JM Doolittle WF (1994) Testing the exon theory of genes – the evidence from protein structure. *Science,* **265** (5169), 202–207.

Strachan T (1992) *The Human Genome,* pp. 32–33. Bios Scientific Publishers, Oxford.

Tagle DA, Stanhope MJ, Siemieniak DR, Benson P, Goodman M, Slightom JL (1992) The β globin gene cluster of the prosimian primate *Galago crassicaudatus*: nucleotide sequence determination of the 41 kb cluster and comparative sequence analyses. *Genomics,* **13**, 741–760.

Ullu E, Tschudi C (1984) Alu sequences are processed 7SL RNA genes. *Nature,* **312**, 171–172.

von Haeseler A, Sajantila A, Paabo S (1995) The genetical archaeology of the human genome. *Nature Genet.,* **14**, 135–140.

Weller PA, Critcher R, Goodfellow PN, German J, Ellis NA (1995) The human Y chromosome homolog of *XG*: transcription of a naturally truncated gene. *Hum. Mol. Genet.,* **4**, 859–868.

Yunis JJ, Prakash O (1982) The origin of man: a chromosomal pictorial legacy. *Science,* **215**, 1525–1530.

Zietkiewicz E, Richer C, Makalowski W, Jurka J, Labuda D (1994) A young Alu subfamily amplified independently in human and African great apes lineages. *Nucleic Acids Res.,* **22** (25), 5608–5612.

Identifying human disease genes

15

A more accurate, though less snappy title for this chapter might be 'Identifying genetic determinants of human phenotypes'. Some mendelian phenotypes, like red–green color blindness, may be regarded as normal variants rather than diseases. Nor would the many genetic variants that contribute in a minor way to our susceptibility to common nonmendelian diseases normally be called disease genes, since they are neither necessary nor sufficient for developing the disease. But for all genetically determined phenotypic variants, one can in principle use the methods described here to discover what DNA sequence variants are responsible. Such variants will be found in many but not all the 80 000 or so human genes. Some genes are indispensable to embryonic function, so that deleterious mutations result in embryonic lethality and go unrecorded in humans. In other cases, abolition of gene function may normally have no effect on the phenotype because other nonallelic genes also supply the same function (genetic redundancy).

15.1 Principles and strategies in identifying disease genes

Few areas have moved as fast as human disease gene identification. Before 1980, very few human genes had been identified as disease loci. The few early successes involved a handful of diseases with a known biochemical basis where it was possible to purify the gene product. In the 1980s, advances in recombinant DNA technology allowed a new approach, positional cloning, sometimes given the rather meaningless label 'reverse genetics'. The number of disease genes identified started to increase, but these early successes were hard won, heroic efforts. With the advent of PCR for linkage studies and mutation screening, it all became much easier. Now the human and other genome projects have made available a vast range of resources – maps, clones, sequences, expression data and phenotypic data. Identifying novel disease genes has become commonplace and is currently occurring on a weekly basis. Soon the land-

scape will change again, as the complete human genome sequence becomes available, so that all genes will in theory be accessible through databases.

Figure 15.1 summarizes some of the routes that have been followed to identify human disease genes. If the figure seems complicated, that is because there is no standard procedure for gene identification. All pathways converge on mutation testing in a candidate gene, but there is not one single entry point, and there is no unique pathway to the candidate gene. For discussion of the principles, we can divide the methods into those that do not require us to know the chromosomal location of the disease locus (Section 15.2) and those that depend on this knowledge (Section 15.3). In reality, groups trying to identify a disease gene will use several parallel approaches, with the emphasis shifting from one candidate and one line of attack to another in response to clues that emerge from the team's own results or new external data, and to new possibilities arising from technical developments. Most genes are identified by defining a candidate gene on the basis of both its chromosomal location and its properties (the positional candidate approach; Section 15.4).

15.2 Position-independent strategies for identifying disease genes

Historically, the first disease genes were identified by pure position-independent methods, simply because no relevant mapping information existed and the techniques were not available to generate it. However, methods based on sequence homology or functional complementation that are in principle position-independent, work much better when applied to predefined candidate subchromosomal regions, rather than to the whole human genome. Homology searches in particular become exceedingly powerful when combined with positional information. Their use in 'cybercloning' and positional candidate approaches to gene identification is considered in Section 15.4.

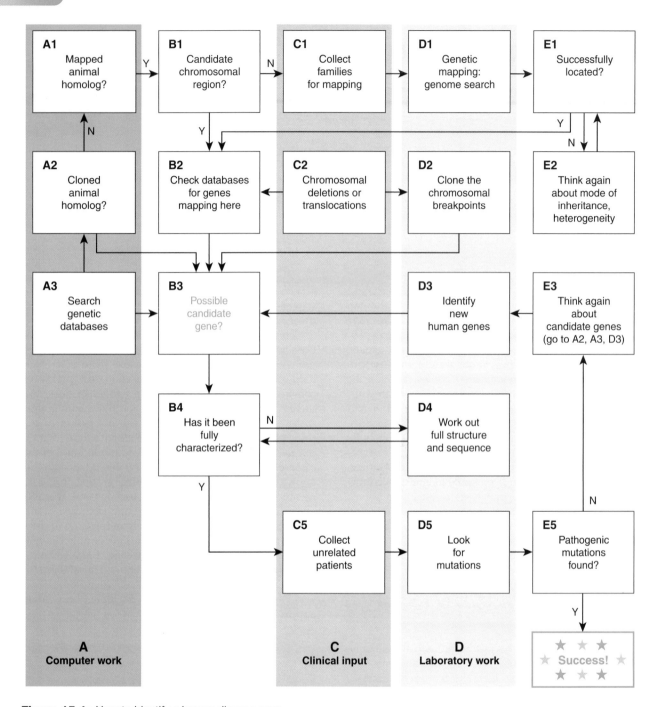

Figure 15.1: How to identify a human disease gene.

There is no single pathway to success, but the key step is to arrive at a plausible candidate gene, which can then be tested for mutations in affected people. Note the interplay between clinical work, laboratory benchwork and computer analysis. Database searching is becoming more and more crucial as information from genome projects accumulates.

15.2.1 Identification of a disease gene through knowledge of the protein product

If the biochemical basis of an inherited disease is known, it may be possible to purify and partially characterize some of the gene product. If this can be done, gene-specific oligonucleotides or specific antibodies can be generated that can be used to identify the gene.

Use of gene-specific oligonucleotides

This approach relies on the ability to isolate sufficient protein product to permit amino acid sequencing. Specific peptide bonds in the protein product can be cleaved using proteolytic enzymes such as trypsin (cuts at the carboxyl end of lysine or arginine residues) or reagents such as cyanogen bromide (cuts at the carboxyl end of methionine residues). The amino acid sequence of each resulting peptide can be determined by chemical sequencing. This involves a repeated series of chemical reactions in an automated amino acid sequencer. In each cycle, the peptide is exposed to a chemical that covalently bonds to the N-terminal amino acid and cleaves it off, allowing it to be identified by chromatography. Sequence overlaps identify overlapping peptides, enabling longer sequences to be assembled.

The resulting amino acid sequence is inspected to identify regions containing amino acids with minimal codon degeneracy (e.g. methionines and tryptophans are uniquely encoded by AUG and UGG codons, respectively). Once suitable regions have been identified, combinations of oligonucleotides are synthesized to correspond to all possible codon permutations. The resulting mix of partially degenerate oligonucleotides is labeled and used as a probe to screen cDNA libraries. As only one of the oligonucleotides in the mix will correspond to the authentic sequence, it is important to keep the number of different oligonucleotides low so as to increase the chance of identifying the correct target. Once a suitable cDNA clone is isolated, it can be used to screen a genomic DNA library in order to isolate genomic DNA clones for full characterization of the gene.

Identification of the hemophilia A gene (MIM 306700) followed this approach. Biochemical analysis of serum samples from patients had previously identified a genetic deficiency of blood clotting factor VIII. Purification of factor VIII from plasma is not straightforward, partly because it is present in very low quantities. One approach involved isolating small quantities of factor VIII from large volumes of pig blood by standard protein purification techniques. The purified product allowed the production of gene-specific oligonucleotide probes for library screening (*Figure 15.2*).

Library screening by hybridization can be tedious when a complex mixture of oligonucleotides is used, because the results are greatly influenced by the hybridization conditions. A more rapid alternative is to use partially degenerate oligonucleotides as PCR primers. One early strategy was to use two such sets corresponding to amino acid sequences from different regions of the protein as primers. By using total cDNA from a suitable source as a template, a specific cDNA could be amplified spanning the codons from the two different regions (*Figure 6.11*). However, this approach demands considerable prior information about the protein sequence. A more convenient alternative is to prepare cDNA and ligate it to vector DNA molecules. PCR can then be performed using one vector-specific primer and one primer composed of a panel of partially degenerate oligonucleotides.

Use of specific antibodies

If even small amounts of the normal protein product can be isolated, specific antibodies can be raised. The protein, or a peptide derived from it, is conjugated to a powerful immunogenic **hapten** such as keyhole limpet hemocyanin, and the compound molecule injected into a rabbit or mouse. The hapten activates B lymphocytes, and the protein or peptide of interest activates helper T

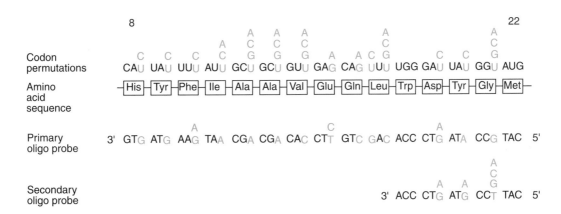

Figure 15.2: The factor VIII gene, the locus for hemophilia A, was cloned by product-directed oligonucleotide screening of DNA libraries.

The figure illustrates one way in which factor VIII DNA clones were obtained, following cleavage of purified porcine factor VIII protein into peptides and amino acid sequencing. The resulting sequences were inspected to identify regions with low codon redundancy. The top panel shows a sequence of 15 amino acids from His8 to Met22 in one of the peptides, with the possible codon permutations above (with variable nucleotides in color). This sequence was selected because of the generally low codon redundancy: two amino acids, Trp and Met, are specified by a single codon and another seven can be specified by just two alternative codons. A partially degenerate 45-bp antisense oligonucleotide probe was prepared and used as a primary hybridization probe to screen a porcine genomic DNA library, and thereafter secondary screening used a 15-bp antisense degenerate oligonucleotide probe corresponding to the sequence from Trp18 to Met22. The porcine factor VIII genomic clone was then used to screen human DNA libraries to identify the human gene (see Gitschier *et al.*, 1984).

lymphocytes, leading to production of antibodies. Mouse or rabbit antibodies that are specific for the desired protein or peptide can then be used in various ways to identify a corresponding cDNA.

An early approach was to enrich for mRNA encoding the protein product in a cell-free *in vitro* protein synthesis system. This was how the gene that is mutated in phenylketonuria (MIM 261600) was identified in 1982 (Robson *et al.*, 1982). Phenylketonuria was known to be caused by a lack of the enzyme phenylalanine hydroxylase (PAH). PAH enzyme was purified from rat liver, a known site of expression. Specific antibodies were raised and used to immunoprecipitate polysomes containing *PAH* mRNA. The purified mRNA was converted to cDNA, and a specific rat cDNA clone was isolated. This was then used as a probe to isolate the human cDNA from a human liver cDNA library.

This type of approach has been superseded by antibody screening of cDNA expression libraries. cDNA from a relevant tissue is cloned into an expression vector. Inserts within the recombinant DNA clones are expected to be expressed within the host cell to produce foreign polypeptides. Appropriate antibodies can then be used to screen colony filters from the library to identify clones encoding the product of interest.

15.2.2 Identification of a disease gene through knowledge of the DNA sequence

This most usually arises when the researcher is considering what diseases might be caused by mutations in a particular known gene. Alternatively, a novel human disease gene may be identified by homology, either to a paralogous human gene (Section 15.4.2) or to an orthologous gene in another species (Section 15.4.3). An interesting application of DNA sequence knowledge is the attempt to clone genes containing expanded trinucleotide repeats. As shown in *Box 16.7*, expanded trinucleotide repeats are known to cause several inherited neurological disorders. Often these disorders show anticipation – that is, the disease presents at an earlier age and with increased severity in successive generations. If a disease under investigation shows any of these features, it may be worth screening for triplet repeat expansions. The repeat expansion detection method of Schalling *et al.* (1993) permits detection of expanded repeats in unfractionated genomic DNA of affected patients, and methods have been developed for cloning any expanded repeats detected (Koob *et al.*, 1998). This approach was recently used in a completely position-independent way to identify a novel repeat expansion that causes a form of spinocerebellar ataxia (SCA8) (Koob *et al.*, 1999).

15.2.3 Identification of a disease gene through knowledge of its normal function

Functional cloning depends on expressing random fragments of human DNA in a cell or organism, and isolating any fragments that cause a desired change in function. The usual approach is a functional complementation assay, seeking fragments that correct a defect in the recipient. Examples include the following:

- *Functional complementation in mammalian cell lines*. For example, a variety of mammalian cell lines have been generated that are deficient in DNA repair. They (Robson *et al.*, 1982) show abnormal responses following exposure to UV irradiation or chemical mutagens. These mutant cells, or alternatively cells derived from patients with a DNA repair deficiency, can be transformed by fragments of normal human DNA or human chromosomes in order to produce a repair-competent phenotype. This was the way in which cDNA clones for the Fanconi's anemia group C (FACC; MIM 227645) gene were first obtained (Strathdee *et al.*, 1992). Similarly, the ability of transferred chromosomes or clones to correct the uncontrolled growth of tumor cell lines has been used to help locate and then identify tumor suppressor genes (Chapter 18).
- *Functional complementation in yeast*. Innumerable yeast mutants have been defined, and genetic analysis in yeast is particularly sophisticated because of the ease of performing homologous recombination. Some proteins have been so highly conserved during evolution that the human protein can complement a yeast mutant defective in the corresponding protein. This approach has been successful in identifying the human genes that specify various enzymes of purine and pyrimidine biosynthesis, and also some crucially important transcription factors.
- *Functional complementation in transgenic mice*. Occasionally a mouse gene has been identified by constructing transgenic mice, using nonmutant BAC clones from a candidate region, crossing them to mice carrying the mutation, and checking which transgene corrects the defect. This strategy was first used to identify a clock gene (Antoch *et al.*, 1997), and more recently as a crucial step in identifying the human *DFNB3* deafness gene (Probst *et al.*, 1998; *Figure 15.3*). *DFNB3* had been mapped to a location that corresponded in the mouse to the location of the deafness gene *shaker-2*. Transgenic mice were constructed using BACs from the *shaker-2* candidate region, and a BAC that corrected the *shaker-2* phenotype was identified. This led to identifying the *shaker-2* gene as an unconventional myosin, MYO15. The human *MYO15* gene was then isolated based on its close homology to the mouse gene, its position within the *DFNB3* candidate region confirmed, and mutations demonstrated in *DFNB3* affected people.
- *Isolation of activated oncogenes*. This is done by their effect on the growth of mouse 3T3 fibroblasts (see *Figure 18.4*).

If a patient has a disease because of a chromosomal deletion, identifying genes present in a normal person but absent in the patient would pinpoint the disease gene.

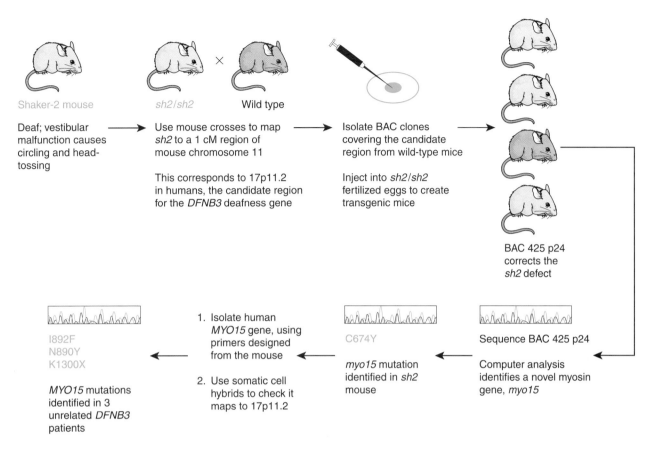

Figure 15.3: Functional complementation in transgenic mice as a tool for identifying a human disease gene.

The *shaker-2* mouse mutation was identified by finding a wild-type clone that corrected the defect. Human families with a similar phenotype that mapped to the corresponding chromosomal location proved to have mutations in the orthologous gene.

More generally, genes implicated in a disease may be expressed to a different degree in patients and controls (this depends on the type of mutation: missense mutations alter the function but not the expression of the mRNA, but many other types of mutation result in low or absent levels of mRNA – see Chapter 16). Methods that identify the differential presence or expression of a gene therefore provide a possible route to position-independent identification of a disease gene, although more usually they are one arm of a positional candidate strategy.

Subtraction cloning

Subtraction cloning can be used to select clones of the DNA that is deleted in an individual with a chromosomal deletion. Two DNA samples are compared, a normal 'test' DNA and a deleted 'driver' DNA. The test DNA is mixed with a large excess of driver DNA, denatured and re-annealed. By one means or another, double helices are selected in which both strands consist of test DNA. These preferentially represent sequences in the test DNA that are absent from the driver DNA. The most celebrated application of subtraction cloning was in identifying the dystrophin (*DMD*) gene (Section 15.3.4). The test DNA came from a normal individual, and the driver DNA from

a patient who had a deletion including the dystrophin gene. The test clones remaining after subtraction were enriched for DNA derived from the region missing in the affected patient.

In the historically important case of dystrophin, subtraction cloning directly yielded clones from the desired unknown gene, but this was an exceptional success. Subtraction cloning is a very difficult technique, which has seldom succeeded with genomic DNA. A more recent approach to the same problem is representational difference analysis (RDA; Lisitsyn, 1995). RDA uses several means, including selective PCR, to enrich sequences present in the test but not the driver DNA, and has been used to isolate regions amplified or deleted in cancer cells (Schutte *et al.*, 1995). Genes in such regions are positional candidate tumor suppressor or oncogenes (Chapter 18).

Subtractive hybridization works better with mRNA than genomic DNA and subtractive hybridization or mRNA differential display (Section 20.2.4) have been used to identify differentially expressed transcripts. In the future, expression arrays (*Figure 20.6*) could be efficient tools for such analyses, although they will not contain any novel uncharacterized genes. Generally the role of these techniques in gene identification is not to isolate the disease gene directly, but to produce collections of sequences that may be position-independent candidates for a disease because of their pattern of expression. These can then be screened further by positional criteria.

Subtractive hybridization has also been used in several projects to produce libraries of cDNAs specifically expressed in a certain tissue, by subtraction of a tissue-specific cDNA library against one or more nonspecific libraries (Swaroop *et al.*, 1991). Such subtraction libraries are a good source of candidate genes for diseases affecting just the tissue in question. Several disease genes have been identified by screening an appropriate subtraction library for sequences mapping to the same chromosomal location as the disease (see, for example, Yasunaga *et al.*, 1999). This is an example of the power of the positional candidate approach (Section 15.4).

15.3 In positional cloning, disease genes are identified using only knowledge of their approximate chromosomal location

At the opposite pole to the position-independent gene identification strategies, positional cloning identifies a disease gene based on no information except its approximate chromosomal location. The first successful gene identifications based only on positional information, published in 1986, marked a triumphant new era for human molecular genetics. One after another, genes for important disorders such as Duchenne muscular dystrophy, cystic fibrosis, Huntington disease, adult polycystic kidney disease, colorectal cancer, breast cancer, etc. were isolated. However, positional cloning can be desperately hard work, and by 1995 only about 50 inherited disease genes had been identified by this approach (Collins, 1995). The four examples discussed in Sections 15.3.4–15.3.7 (*Table 15.1*) illustrate typical approaches to positional cloning. Each of these disease genes was identified by first mapping the disease as closely as possible in affected families, and then identifying a novel candidate gene and showing that patients had mutations in that

gene. In all cases except Treacher Collins syndrome, there was some clue to help identify the correct candidate gene.

Positional cloning projects recapitulate the Human Genome Project in miniature. In both cases the researchers first produce high-resolution genetic and physical maps, then build clone contigs before identifying and sequencing transcripts. The only parts specific to positional cloning are selecting the candidate region and identifying pathogenic mutations in patients with the disease in question. Thus general progress in the Human Genome Project has had an enormous impact on positional cloning projects. Defining the right candidate region and testing candidate genes can still be long hard tasks (as *Figure 15.4* makes clear) but most of the intermediate stages can be achieved by intelligent use of existing Genome Project data.

15.3.1 The first step in positional cloning is to define the candidate region as tightly as possible

The methods of mapping mendelian and nonmendelian disease genes have been described in Chapters 11 and 12, respectively, while the use of loss of heterozygosity mapping to locate tumor suppressor genes is described in Section 18.5.3. Even with all the fruits of the Human Genome Project to hand, positional cloning can still be laborious and frustrating. More than any other factor, the size of the candidate region determines the work required, and so every effort is made to define as small a candidate region as possible. Regions larger than about 1 Mb of DNA present serious obstacles to positional cloning.

Often the initial localization from genetic mapping defines a candidate region of 10 Mb or more. The next step is to collect as many families as possible and establish a dense cover of polymorphic markers across the region. Suitable markers may be found by database searching, but if this fails then YACs, BACs and cosmids must be isolated from the candidate region and screened for polymorphisms. These might be microsatellites or single nucleotide polymorphisms, and depending on how well developed the physical map of the region is, the new markers might either be localized physically as

Table 15.1: Examples of disease genes identified by positional cloning

Disease	MIM no.	Map position	Gene	Approach
Duchenne muscular dystrophy (Section 15.3.4)	310200	Xp21.3	Dystrophin	(a) clone translocation breakpoints (b) clone sequences missing in a patient with a deletion
Cystic fibrosis (Section 15.3.5)	219700	7q31	CFTR	Linkage disequilibrium
Branchio-oto-renal syndrome (BOR) (Section 15.3.6)	113650	8q13	EYA1	Sequencing genomic clones; homology to *Drosophila* gene
Treacher Collins syndrome (Section 15.3.7)	154500	5q32–33.1	TCOF1	Transcript mapping

The disease gene was found within the minimal region defined by linkage analysis using the approaches shown. See text for details.

Candidate region

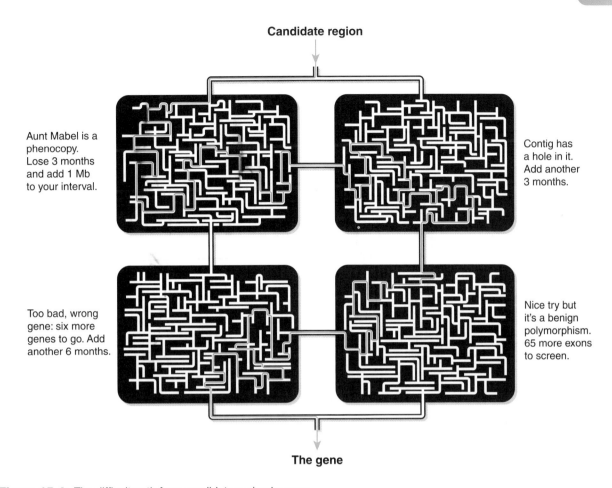

Aunt Mabel is a
phenocopy.
Lose 3 months
and add 1 Mb
to your interval.

Contig has
a hole in it.
Add another
3 months.

Too bad, wrong
gene: six more
genes to go. Add
another 6 months.

Nice try but
it's a benign
polymorphism.
65 more exons
to screen.

The gene

Figure 15.4: The difficult path from candidate region to gene.

One researcher's view of the frustrations of positional cloning. Image courtesy of Dr Richard Smith, University of Iowa.

sequence tagged sites (STS) on a contig, mapped physically using a radiation hybrid panel (Section 10.1.3), or mapped genetically in CEPH families (see Section 11.4.3). For mendelian conditions, where recombinants can be identified unambiguously, the limit of genetic resolution is reached when pairs of closely spaced markers define the positions of the closest recombinations on either side of the disease locus. This is decided by inspecting individual haplotypes rather than by statistical analysis (*Figure 15.5*). When mapping low-penetrance disease susceptibility loci, recombinants cannot be pinpointed in this way (Section 12.2.2), so all one can do is sharpen the lod score curve as far as possible by using the biggest possible dataset.

When single recombinants define the boundaries of a region that is to be searched, it is important to consider possible sources of error (see Section 11.5.1). Meticulous clinical diagnoses are imperative. Key recombinations are more reliable if they occur in unambiguously affected people – an unaffected individual may carry a nonpenetrant disease gene, which can lead to them being misinterpreted as recombinant when in fact they are nonrecombinant. Sometimes, despite good positive lod scores, there appear to be recombinants with every

marker tried. This is usually an indication that somebody has been diagnosed wrongly (labeled as affected when unaffected, or vice versa) or else that the disease gene in one or more of the families under investigation does not map to the candidate region. Alternatively, perhaps the markers are wrongly ordered on the genetic map.

Linkage disequilibrium may allow very high resolution mapping

Linkage disequilibrium (association at the population level of a particular marker allele with a disease) can allow genetic mapping to be taken to a very high resolution, as discussed in Section 12.4. Linkage disequilibrium has been enormously valuable for guiding positional cloning (the example of cystic fibrosis is discussed below) but not all diseases show it (European Consortium on MEN1, 1997). It is seen only when many of the apparently unrelated affected people in a population in fact derive their disease chromosome from a shared ancestor. Thus the easiest diseases to map very finely are those where most affected people carry the same ancestral mutation, and an ancestral haplotype can be defined, as illustrated in *Figure 12.5*. In such cases there is a price to be paid later, when candidate genes are being tested for mutations,

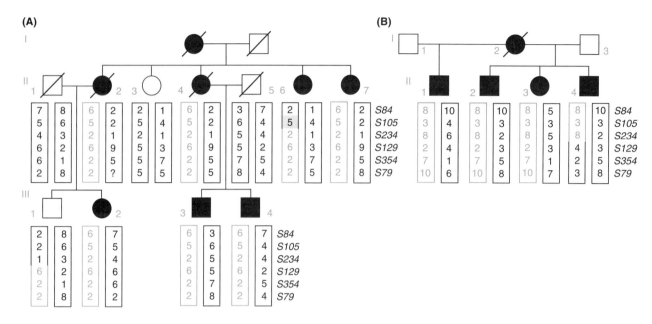

Figure 15.5: Crossover analysis seeks to map a gene by defining flanking proximal and distal recombinants.

The figure shows haplotype analysis in two pedigrees with a dominantly inherited skin disorder, Darier–White disease (MIM 124200), which had previously been mapped to 12q. Genotypes in II-1, II-2, II-4 and II-5 in pedigree (A) are inferred. (**A**) In this family the disease gene segregates with the marker haplotype 6-5-2-6-2-2 between *D12S84* (proximal) and *D12S79* (distal). A crossover in II-6 shows that the disease gene must map distal to *D12S84*. The positioning of *D12S105* is ambiguous because of presumed homozygosity for allele 5 in I-1 – compare the genotypes for II-3 and II-6. (**B**) In this family the disease gene must lie proximal to *D12S129*. The combined data indicate that the Darier's disease gene must map between the proximal marker *D12S84* and the distal marker *D12S129*. Reproduced from Carter *et al.* (1994) *Genomics*, **24**, 378–382, with permission from Academic Press.

because a diversity of mutations gives a much higher chance of spotting the correct gene. Plans to identify the genetic factors responsible for susceptibility to many common diseases rest almost entirely on the hope that association studies will pinpoint the location of the susceptibility genes. If it were to turn out that at most susceptibility loci many different variants predispose to the disease, then the whole endeavor would be in serious trouble.

15.3.2 Genes within the candidate region can be identified by a combination of database searching and transcript mapping

As shown in *Figure 15.1*, known genes from the candidate region can be found by database searching. If none of the known genes is a promising candidate, or if after testing no mutations can be found, then novel genes must be sought. The general methods for identifying unknown transcribed sequences from within a contig of genomic clones have been discussed in detail in Section 10.4, and are summarized briefly in *Box 15.1*. With the continuing progress of the Human Genome Project, the emphasis has moved strongly towards identifying genes from the databases. Once the human genome sequence is completed, it

---

Box 15.1

Transcript mapping: how to identify expressed sequences within genomic clones from a candidate region

Methods for transcript mapping are described in Section 10.4. In summary, these include:

- database searches to identify genes and expressed sequence tags (ESTs) known to map within or near to the candidate region;
- sequence analysis, to detect matches to unmapped ESTs in the databases;
- sequence analysis, to detect genomic sequences having characteristics of exons (*Figure 10.25*);
- cDNA library screening, using as probes genomic clones from the candidate region;
- cDNA selection, for ultra-sensitive detection of cDNAs derived from the candidate region (*Figure 10.24*);
- exon trapping, to find genomic sequences flanked by functional splice signals (*Figure 10.23*);
- zoo blotting, to seek evolutionarily conserved sequences (*Figure 10.21*);
- CpG island identification, to seek the regions of under-methylated DNA which often lie close to genes (*Figure 10.22*).

should in theory no longer be necessary to clone any human gene from scratch. However, present computational methods for identifying genes using genomic DNA sequence data (Section 10.4.5) are far from perfect, and for some time to come it may still be necessary to get one's hands dirty in the laboratory if one wishes to identify every expressed sequence from a candidate region.

Whenever cDNA libraries are to be screened, the question arises which libraries should be used. Often the pathology of the disease under study suggests a particular investigation. For example, when studying a neuromuscular disease it makes sense to start by screening muscle cDNA libraries. However, tissue-specific diseases are often caused by malfunction of widely expressed genes (Section 16.7.1), so if one library fails it is always worth screening others. Fetal brain is a popular choice because it has a particularly high number of expressed sequences.

15.3.3 Chromosomal aberrations can provide a useful short-cut to locating a disease gene

Researchers are constantly on the alert for special patients or observations that will short-cut the labor of pure positional cloning. Cancer studies in particular have relied on investigations of chromosomal abnormalities (Chapter 18), and identification of many other disease genes has

also been greatly helped by finding patients with a chromosomal abnormality (*Table 15.2*). Alert clinicians play a crucial role in identifying such patients (*Box 15.2*).

Translocations and inversions
If a person with an apparently balanced translocation or inversion is phenotypically abnormal, there are three possible explanations:

(i) the finding is coincidental;
(ii) the rearrangement is not in fact balanced – there is an unnoticed loss or gain of material;
(iii) one of the chromosome breakpoints causes the disease.

A chromosomal break can cause a loss-of-function phenotype if it disrupts the coding sequence of a gene, or separates it from a nearby regulatory region. Alternatively, it could cause a gain of function, for example by splicing regulatory sequences from one gene to distal coding sequences from another gene, causing inappropriate expression (this is rare in inherited disease but common in tumorigenesis, see Chapter 18). In either case, the breakpoint provides a valuable clue to the exact physical location of the disease gene. The clue is valuable but not infallible: sometimes breakpoints can alter expression of a gene located hundreds of kilobases away by affecting the structure of large-scale chromatin domains (*Box 15.3*).

The precise location of a chromosome breakpoint is

Table 15.2: The first ten years of positional cloning (selected highlights)

Year	Disease	MIM number	Location	Gene	Chromosome abnormality
1986	Duchenne muscular dystrophy	310200	Xp21.3	DMD	(a) del(X)(p21.3) (b) t(X;21)(p21.3:p13)
	Retinoblastoma	180200	13q14	RB	del(13)(q13.1q14.5)
1989	Cystic fibrosis	219700	7q31	CFTR	None
1990	Neurofibromatosis 1	162200	17q11.2	NF1	Balanced translocations t(1;17)(p34.3:q11.2) t(17;22)(q11.2:q11.2)
	Wilms' tumor	194070	11p13	WT1	del(11)(p14p13)
1991	Aniridia	106210	11p13	PAX6	t(4;11)(q22;p13) del(11)(p13)
	Familial polyposis coli	175100	5q21	APC	del(5)(q15q22)
	Fragile-X syndrome	309550	Xq27.3	FMR1	FRAXA fragile site
	Myotonic dystrophy	160900	19q13.3	DMPK	None
1993	Huntington's disease	143100	4p16	HD	None
	Tuberous sclerosis 2	191092	16p13	TSC2	Microdeletions in candidate region
	von Hippel–Lindau disease	193300	3p25	VHL	Microdeletions in candidate region
1994	Achondroplasia	100800	4p16	FGFR3	None
	Early-onset breast/ovarian cancer	113705	17q21	BRCA1	None
	Polycystic kidney disease	173900 601313	16p13.3	PKD1	t(16;22) (p13.3;q11.21)
1995	Spinal muscular atrophy	253300 600354	5q13	SMN1	None

The data illustrate how important cytogenetic abnormalities were for cloning disease genes. For TSC2 and VHL, microdeletions were identified only late in the refinement of the candidate region.

Box 15.2

Pointers to the presence of large-scale mutations

Clinicians can make a major contribution to positional cloning projects by finding patients who have large-scale mutations. Pointers include the following:

1. *A cytogenetic abnormality in a patient with the standard clinical presentation*
   If a disease gene has already been mapped to a certain subchromosomal location and then a patient with that disease is found who has a chromosome abnormality affecting that same location, the chromosome abnormality most probably caused the disease.

   ■ Patients with balanced translocations or inversions often have breakpoints located within the disease gene, or very close to it. Cloning their breakpoints often provides the quickest route to identifying the disease gene.
   ■ Patients with interstitial deletions may also be valuable; the breakpoints may be located some distance from the disease gene but, if the segment that is lost is small, mapping the breakpoints may enable the gene to be mapped to a small interval.

   Most such patients will have *de novo* mutations. Some

researchers feel that performing chromosome analysis on all patients with *de novo* mutations is a worthwhile expenditure of research effort. All researchers carefully monitor reports of patients with chromosomal abnormalities, hoping to discover cases showing features of the disease they are investigating.

2. *Additional mental retardation*
   Rare patients may have the expected phenotype, but in addition be severely mentally retarded. Although this may be coincidental, such cases can be caused by deletions that eliminate the disease gene plus additional neighboring genes. Large chromosomal deletions almost always cause severe mental retardation, reflecting the involvement of a high proportion of our genes in fetal brain development. When the patient has a *de novo* mutation, cytogenetic and molecular analysis is warranted.

3. *Contiguous gene syndromes*
   Very rarely a patient appears to suffer from several different genetic disorders simultaneously. This may be just very bad luck, but sometimes the cause is simultaneous deletion of a contiguous set of genes. Contiguous gene syndromes are described in Section 16.8.1; they are particularly well defined for X-linked diseases.

Box 15.3

Position effects – a pitfall in disease gene identification

In general genes appear to be arranged more or less at random on chromosomes, and the exact arrangement or order does not matter. In *Drosophila* however, it is well known that the local megabase-scale chromatin organization can affect gene expression – in particular, genes are silenced if placed within or close to heterochromatin. The same appears to be true in mice and men.

Studies of transgene expression (Section 21.2.3) show that correct tissue-specific gene expression can depend on sequences located hundreds of kilobases away from the coding sequence of a gene. Several human examples are known of translocation breakpoints affecting expression of a gene several hundred kilobases away. The examples of aniridia (MIM 106210) and the *PAX6* gene, and campomelic dysplasia (MIM 211970) and the *SOX9* gene were given in Section 8.5.1.

Thus balanced translocation breakpoints are not necessarily located within, or even very close to, the gene they inactivate, which reduces their value as tools for cloning disease genes.

rare-cutter restriction endonuclease and subjected to pulsed field gel electrophoresis (Section 10.2.2).

Deletions and duplications

Chromosomal deletions cause abnormalities due to loss of genes in males with X chromosome deletions, and reduced levels of dosage-sensitive gene products in people heterozygous for autosomal deletions (*Figure 16.9*). Cytogenetically visible deletions involve many megabases of DNA. Such large deletions often produce rather complex, nonspecific phenotypes, but if specific elements can be seen, the deletion may provide a pointer to a broad subchromosomal localization. In the past, subtraction cloning was attempted using such deletions, as described below for the dystrophin gene.

Small-scale deletions (microdeletions) are valuable for positional cloning. Deletions of tens or hundreds of kilobases of DNA are not uncommon in some disorders (Section 16.8.1). Often they are generated by unequal recombination between flanking repeat sequences (Section 9.5.4). Microdeletions can be identified by several methods.

■ *Noninheritance of marker alleles.* If a microdeletion eliminates a marker locus, individuals carrying the deletion will be hemizygous. On testing, they appear to be homozygotes and transmission of the disease chromosome is often accompanied by what appears to be nonmendelian segregation (*Figure 15.7; Figure 17.12* shows an X-linked example).

■ *PCR dosage analysis.* Any sequence-tagged site that

most easily defined by using FISH (*Figure 15.6*). Alternatively, different DNA clones from the relevant region can be used in turn to see if any can identify patient-specific restriction fragments, by hybridizing each clone to the patient's genomic DNA which has been digested with a

(A)

(B)

(C)

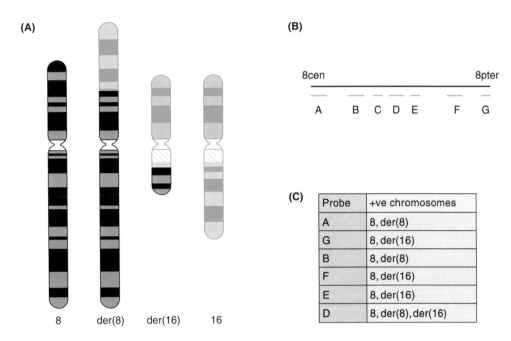

Probe	+ve chromosomes
A	8, der(8)
G	8, der(16)
B	8, der(8)
F	8, der(16)
E	8, der(16)
D	8, der(8), der(16)

8   der(8)   der(16)   16

Figure 15.6: Using fluorescence *in situ* hybridization to define a translocation breakpoint.

(A) Cytogenetically defined translocation t(8;16)(p22;q12). **(B)** physical map of part of the breakpoint region in a normal chromosome 8, showing approximate locations of seven clones. **(C)** Results of successive FISH experiments. The breakpoint is within the sequence represented in clone D. This result would normally be confirmed using clones from chromosome 16.

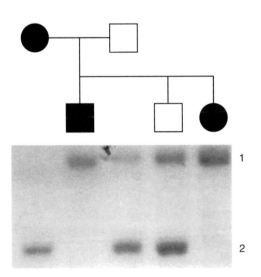

Figure 15.7: Deletions at a disease locus may result in noninheritance of closely linked markers.

The pedigree shows a family with type 2 neurofibromatosis (MIM 101000), with types for a TaqI RFLP at the closely linked neurofilament heavy chain (*NEFH*) locus. The two affected offspring inherit no NEFH allele from their mother. The likely cause of both the disease and the anomalous inheritance pattern is a deletion encompassing both the *NEFH* gene and *NF2* genes. This was confirmed by the analysis shown in *Figure 15.8*. Reproduced from Watson *et al.* (1993) by permission of Oxford University Press.

maps within a microdeletion will be present in half dosage in deletion carriers. Various quantitative or semiquantitative PCR methods can be used to detect the reduced dosage.

■ *FISH mapping.* Once suspected, a microdeletion can also be confirmed by this method. A DNA clone from the deleted interval is used as a hybridization probe against a metaphase chromosome preparation from the patient. For autosomal microdeletions, the probe should hybridize to only one of the two homologs (*Figure 15.8*).

■ *Hybridization-based restriction mapping.* Unlike the previous methods, which require markers that are deleted, this approach can be used with markers located several hundred kilobases away from the deletion. Genomic DNA from a panel of patients is digested with rare-cutter restriction endonucleases, and the fragments size-fractionated by pulsed field gel electrophoresis. DNA probes that map in the vicinity of the disease locus can then be tested in turn to see if they can detect abnormal patient-specific hybridization bands. If a probe detects abnormal size bands in a patient's DNA digested separately with two or more different enzymes, a large-scale mutation is indicated (*Figure 15.8*).

Duplications have not played any significant role in positional cloning. If unequal recombination between flanking repeats is a major cause of microdeletions, microduplications should be equally frequent, but in fact

(A)

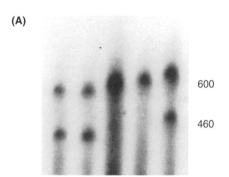

600

460

(B)

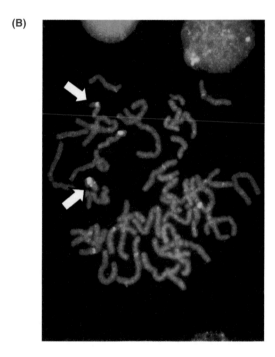

Figure 15.8: Microdeletions can be identified by restriction mapping using PFGE and by FISH.

The panels illustrate two sets of analyses to test whether affected members in the pedigree in *Figure 15.7* carried a small deletion of chromosome 22 in the vicinity of the NF2 locus. (**A**) PFGE analysis. A genomic DNA clone from the *NF2* gene region was hybridized to a Southern blot of genomic DNA from indicated family members, which had been digested with *Not*I and size-fractionated by PFGE. The 600-kb *Not*I fragment represents wild-type alleles. Note the additional, approximately 460-kb band found only in the affected individuals, resulting from a deletion which simultaneously eliminated the *NF2* gene. (**B**) FISH analysis. Metaphase chromosome preparations from an affected individual were hybridized with two DNA probes. A probe that hybridizes to repeat sequences found at the centromeres of chromosomes 14 and 22 produces a strong signal on the two homologs for each chromosome. A cosmid from the *NF2* gene region, however, hybridizes to only one of the two chromosome 22 homologs. This is a single copy probe so it gives a fainter signal (a dot from each chromatid) than the repeat sequence probe. Reproduced from Watson *et al.* (1993) *Hum. Mol. Genet.*, **2**, 701–704, by permission of Oxford University Press.

they are rarely observed. Probably most duplications are overlooked because they are nonpathogenic. If a duplication is associated with an abnormal phenotype, the cause is most likely to be a loss of function of a gene that is disrupted by the breakpoint. Occasionally there may be a dosage effect when a complete working gene is duplicated (*Figure 16.7*). Microduplications can be detected by careful dosage analysis, by long-range restriction mapping or by finding people who have three alleles of a marker.

15.3.4 Chromosomal deletions and translocations assisted positional cloning of the dystrophin gene

Duchenne muscular dystrophy (DMD, MIM 310200) was a major test-bed for positional cloning methods. Years of careful investigation of the pathological changes in affected muscle had failed to reveal the biochemical basis of DMD. In the early 1980s, several groups competed to clone the *DMD* gene, using different approaches. The pioneering work of these groups, overcoming formidable technical difficulties to clone an unprecedented gene, was probably the major inspiration for most subsequent positional cloning efforts. This work has been well reviewed by Worton and Thompson (1988).

The DMD gene was localized by linkage analysis and through X-autosome translocations

The *DMD* locus was mapped to Xp21 by linkage to a restriction fragment length polymorphism as long ago as 1982 (the first disease to be so mapped). Additional confirmation of this localization came from studies of rare affected females. These women, about 20 of whom have been described worldwide, occur sporadically in families with no history of DMD, and there is no evidence that they have inherited a conventional *DMD* mutation from either parent. Instead, they all carry balanced X-autosome translocations. Although each woman has a different autosomal breakpoint, and many different autosomes are involved, the X chromosome breakpoint is always at Xp21. The pathogenesis results from an unusual mechanism. X inactivation is random, but those cells in which the der(X) translocation chromosome was inactivated suffer genetic imbalances and die. In the cells that survive, the normal X is inactive, whereas the active der(X) does not produce any dystrophin because the translocation breakpoint has disrupted the dystrophin gene (*Figure 15.9*).

Isolation of the DMD gene by subtraction cloning

Kunkel's group in Boston used DNA from a boy 'BB' (Section 16.8.1) who had DMD and a cytogenetically visible Xp21 deletion. A technically very difficult subtraction cloning procedure (Section 15.2.3) was used to isolate clones from normal DNA that corresponded to sequences deleted in BB. Individual DNA clones in the subtraction library were then used as probes in Southern blot

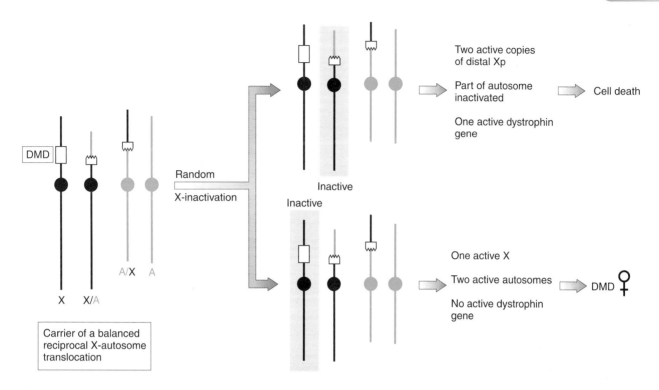

Two active copies of distal Xp

Part of autosome inactivated → Cell death

One active dystrophin gene

Inactive

Inactive

One active X

Two active autosomes → DMD ♀

No active dystrophin gene

Random X-inactivation

DMD

A/X   A

X   X/A

Carrier of a balanced reciprocal X-autosome translocation

Figure 15.9: Nonrandom X inactivation occurs in female DMD patients with Xp21-autosome translocations.

The translocation is balanced, but the X chromosome breakpoint disrupts the dystrophin gene. X inactivation is random, but cells which inactivate the translocated X die because of lethal genetic imbalance. The embryo develops entirely from cells where the normal X is inactivated, leading to a woman with no functional dystrophin gene. The resulting failure to produce any dystrophin causes DMD.

hybridization against DNA samples from normal people and DMD patients. One clone, pERT87-8, detected deletions in DNA from about 7% of cytogenetically normal DMD patients. It also detected polymorphisms that were shown by family studies to be tightly linked to DMD. These results showed that pERT87-8 was located much closer to the *DMD* gene than any previously isolated clones (in fact it was within the gene, in intron 13). Other nearby genomic probes were isolated by chromosome walking and used to screen muscle cDNA libraries. Given the low abundance of dystrophin mRNA and, as we now know, the small size and widely scattered location of the exons, finding cDNA clones was far from easy, but eventually clones were identified, and subsequently the whole remarkable dystrophin gene (see *Figure 8.13*) was characterized.

Isolation of the DMD gene by cloning a translocation breakpoint

While Kunkel's group was working on subtraction cloning, Worton's group in Toronto was successful with a different approach. One of the affected women described above had an X;21 translocation with a breakpoint in the short arm of chromosome 21. Knowing that 21p is occupied by arrays of repeated rRNA genes (Section 8.2.1), Worton's group prepared a genomic library and set out to

find clones containing both rDNA and X chromosome sequences. This led to isolation of *XJ* (X junction) clones which, in a similar way to Kunkel's pERT87-8 probe, detected deletions and polymorphisms. *XJ* turned out to be located in intron 17 of the dystrophin gene.

15.3.5 Linkage disequilibrium was an important aid to positional cloning of the cystic fibrosis gene

In 1985, studies of affected sib-pairs (see *Figure 12.2*) showed that the gene for CF (MIM 219700) was linked to a protein polymorphism of the enzyme paraoxonase. At that time, the chromosomal location of the paraoxonase gene was not known (this illustrates one of the big advantages of using DNA rather than protein polymorphisms for mapping). A rapid mapping effort located the paraoxonase gene to chromosome 7, and a variety of DNA markers were used to show that CF mapped to 7q31-q32. The *MET* oncogene was established as a proximal flanking marker, and an anonymous clone *D7S8* as a distal marker.

Despite an intensive world-wide search, no CF patients have been discovered with translocation, inversion or deletion breakpoints at 7q31-q32, nor did any microdeletions emerge during the progress of the research. Without large-scale mutations to help, identifying the CF gene

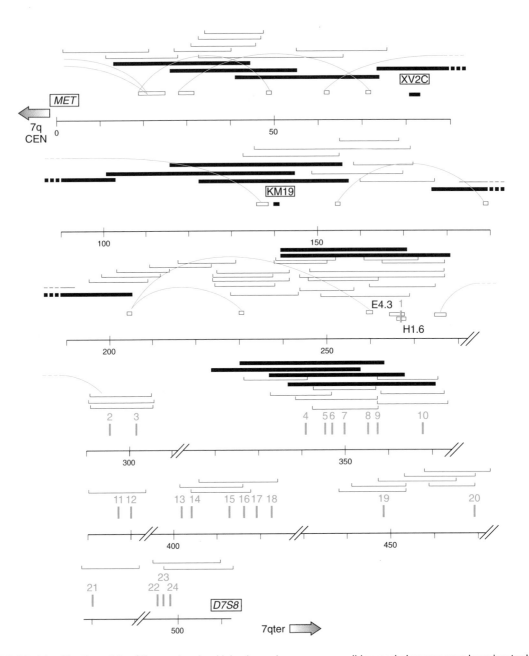

Figure 15.10: Identification of the CF gene involved laborious chromosome walking and chromosome jumping techniques.

Starting from the flanking markers *MET* (proximal) and *D7S8* (distal), an intervening region of about 500 kb was intensively mapped. Chromosome walking was used to identify overlapping λ and cosmid clones (short thin and long thick horizontal lines, respectively, above the restriction map). Chromosome jumping steps (color arcs) facilitated this process. After several false starts, the overlapping E4.3 and H1.6 clones, which contained evolutionarily conserved sequences (as detected by zoo blotting; see *Figure 10.21*), were used to isolate a cognate cDNA clone. The cDNA clone was then used to map back to λ genomic clones and the gene was shown to contain 24 exons. Gaps remained, however (e.g. between exons III and IV). The full structure of the gene was later shown to comprise 27 exons. Verification of the gene's involvement in CF was obtained by demonstrating patient-specific mutations (see text). Reproduced with permission from Rommens *et al.* (1989) *Science*, **245**, 1059–1065. Copyright 1989 American Association for the Advancement of Science.

proved an exceedingly arduous task, requiring extensive genetic mapping and exhaustive molecular characterization of the candidate region. In those days, before the Human Genome Project, generating clones covering the region between the flanking markers *D7S8* and *MET* was a major effort (*Figure 15.10*). The techniques used for this task included:

- *Chromosome walking* (Section 10.3.2). All this work was conducted before human YAC libraries were available, and so the chromosome walking used cosmid and phage λ libraries. Thus, individual steps were only about 10–20 kb, and frustrated researchers talked about chromosome crawling.

- *Chromosome microdissection.* Using micromanipulation techniques, a small chromosomal segment can be physically cut out from an individual chromosome in a spread on a microscope slide (Edstrom *et al.*, 1987). Extremely fine needles, or a laser beam, are used to perform the microdissection in a series of cells. The excised fragments, typically representing a single chromosomal band, are collected, pooled and the DNA extracted for use in constructing DNA libraries (Ludecke *et al.*, 1989). This technical tour de force has been useful for generating DNA clones from several disease-associated chromosomal regions. However, microdissection libraries are difficult to construct; contamination by extraneous DNA has been a problem and the clone complexity (the number of different DNA sequences) has often been poor. Now that large-scale YAC and BAC mapping has made clones from all chromosomal regions available, microdissection is no longer necessary for clone generation.

- *Chromosome jumping.* This now obsolete technique used circularization of large DNA fragments from the region of interest to hop from one genomic clone to another located several hundred kilobases away (Poustka *et al.*, 1987). Successful jumps in the candidate region provided new start points for chromosome walking.

Eventually most of the region was encompassed by DNA clones. Considering the formidable labor required to do this in 1985–88, makes one appreciate the impact of Human Genome Project resources on present-day positional cloning efforts. Linkage disequilibrium then provided valuable clues about the location of the CF gene. The mutation rate is very low and CF is maintained in the population largely by heterozygote advantage (*Box 3.6*). CF disease chromosomes in apparently unrelated people often derive from a distant common ancestor. The original flanking markers *D7S8* and *MET* show only extremely weak disequilibrium, but some of the newer markers such as *KM19* and *XV2.c* generated from clones within the candidate region showed strong disequilibrium with CF (*Table 12.1*). Linkage disequilibrium data can be hard to interpret (see *Figure 12.6*) but, in the case of CF, a gradient of steadily increasing disequilibrium pointed quite effectively to the location of the 5′ end of the gene.

Eventual isolation of the *CFTR* gene, unaided by large-scale mutations, required extensive screening of libraries for the elusive cDNA. A further small difficulty was encountered in producing convincing evidence that the gene eventually cloned was indeed the site of mutations causing CF. Because of the powerful linkage disequilibrium, it was expected that most CF mutant chromosomes in the population would share a great deal of ancestral sequence. Therefore, showing that a particular sequence change (the *F508del* mutation, *Table 17.3*) was present on 70% of CF chromosomes did not prove in a wholly convincing manner that *CFTR* was the CF gene, still less that *F508del* caused CF. *F508del* could have been simply a neutral variant inherited along with CF on the ancestral disease chromosome, especially since the sequence change left the reading frame intact.

The *F508del* mutation is present in the heterozygous state in 3–4% of phenotypically normal individuals (we would now identify them as CF carriers). The fact that *F508del* homozygotes were always severely affected was persuasive but not conclusive – *F508del* could have been in more or less complete linkage disequilibrium with the real CF mutation. Biochemical and pathological knowledge was important, in showing that the *CFTR* gene encoded an ion channel, and that the pathogenesis of CF was ultimately caused by defective regulation of chloride ion transport across apical membranes. The subsequent identification of minority disease alleles like *G542X*, where the expected effect on gene expression was more obviously deleterious, provided further confirmation that the true disease locus had been identified.

15.3.6 Positional cloning of the gene causing branchio-oto-renal syndrome was achieved by large-scale sequencing of clones from the candidate region

The autosomal dominant branchio-oto-renal (BOR) syndrome (MIM 113650: branchial fistulas, malformation of the external and inner ear with hearing loss; hypoplasia or absence of kidneys) was mapped to 8q13 following a clue from an affected patient who had a rearrangement of chromosome 8. The initial interval of 7 cM was refined to an interval of 470–650 kb by further mapping and delineation of a chromosomal deletion in the patient mentioned. P1 and PAC clones were isolated by screening genomic libraries with markers within or close to the candidate region, and gaps in the contig were filled by chromosome walking. The minimum tiling path (the smallest number of clones from which a contig can be built) across the candidate region involved 3 P1 and 3 PAC clones.

It was decided to isolate genes from the contig by large-scale sequencing of plasmid subclones. Checking the sequence against the EMBL and GenBank databases revealed homology between part of the sequence obtained and the *Drosophila* developmental gene *eyes absent* (*eya*). Further genomic sequence was then translated, and then searched for homologies to the deduced amino acid sequence of the *Drosophila eya* gene. This resulted in the identification of seven putative exons showing 69% identity and 88% similarity at the amino acid level to the putative eya protein (note that amino acid sequences are more likely to show detectable homology than nucleotide sequences). The human cDNA was then isolated from a 9-week total fetal mRNA library, and seven mutations in the gene, named *EYA1* were demonstrated in 42 unrelated BOR patients. Expression studies in the mouse demonstrated expression consistent with the developmental abnormalities of BOR syndrome.

This work (Abdelhak *et al.*, 1997) unambiguously identified *EYA1* as the human BOR gene, but the function of the gene product was not identified, nor was it clear why the *Drosophila* phenotype consists of reduced or absent compound eyes. As so often with positional cloning, identifying the gene was just the start of understanding the syndrome.

15.3.7 Identification of the gene causing Treacher Collins syndrome illustrates positional cloning in its purest form

Treacher Collins syndrome (MIM 154500) is an autosomal dominant disorder of craniofacial development with a variable phenotype including abnormalities of the external and middle ears, hypoplasia of the mandible and zygomatic complex and cleft palate. Linkage was initially established to markers at 5q31-q34. Because the markers in that region at the time (1991–92) were not very informative, new microsatellites were isolated and used to refine the candidate region to 5q32-33.1. A combined genetic and radiation hybrid map was constructed across this interval, and by 1994 the team had assembled a YAC contig. This was converted to a cosmid contig, and cDNA library screening and exon trapping were used to generate a transcript map. At least seven genes were identified in the critical region. Further rounds of marker isolation and crossover analysis produced a confusing picture of overlapping recombinations, but led eventually to isolation of a candidate incomplete cDNA from a placental library. Northern blotting and zoo blotting showed that the gene was widely expressed and conserved across species, but database searches revealed no significant homologies. The exon–intron structure was determined, and mutation analysis demonstrated five different mutations in unrelated patients.

Isolation of the *TCOF1* gene (Treacher Collins Syndrome Collaborative Group, 1996) illustrates positional cloning in its purest form. No relevant chromosomal abnormalities were found (there were four patients with TCS who had chromosomal translocations or deletions, but markers from each of the breakpoints showed no linkage to TCS in family studies, so presumably these cases were all coincidental). There is no linkage disequilibrium – not surprisingly, since 60% of cases are new mutations. The candidate region is gene-rich, so there were many possible candidates, and the gene eventually identified had no features that made it a particularly promising candidate. The gene product is now believed

to be a nucleolar phosphoprotein that is involved in some aspect of nucleolar trafficking. Why mutations should cause Treacher Collins syndrome is not yet known.

15.4 Positional candidate strategies identify candidate genes by a combination of their map position and expression, function or homology

The position-independent and positional cloning strategies described in the last two sections are in principle quite separate, but in reality most disease genes have been identified by a positional candidate strategy, using a combination of positional and nonpositional information.

- A purely position-independent approach will rarely succeed because molecular pathology is too complicated. Predictions of the biochemical function of an unknown disease gene are often proved wrong once the gene is isolated. Likewise, predicting the phenotypic effect of mutations in a known gene usually works only in very general terms. Mutations in rhodopsin (the pigment of the rods of the retina) will probably affect vision, but we cannot predict which of the many forms of hereditary retinal malfunction they might cause. Again, mutations in fibrillin (a connective tissue component) will probably cause a connective tissue disease – but which one of the hundreds known?

- A purely positional approach is inefficient because candidate regions identified by positional cloning usually contain dozens of genes. It can be very time-consuming to identify every transcript from the region, and excessively laborious to screen them all for mutations. Many are identified only as cDNAs or ESTs, so that before mutation screening can be started, their exon–intron structure must be determined. Thus it is essential to prioritize candidates. This requires position-independent information about their pattern of expression, likely function, or homology to genes implicated in relevant mutants in model organisms.

15.4.1 Criteria for selecting a candidate gene: expression pattern and function

From the list of genes that map to the candidate region, one would look for a gene that shows appropriate expression and/or appropriate function. Alternatively or additionally, as discussed in the following sections, one would look for homology to some other human or non-human gene that is known to have appropriate expression or function.

Appropriate expression pattern

A good candidate gene should have an expression pattern consistent with the disease phenotype. Expression need not be restricted to the affected tissue, because there are many examples of widely expressed genes causing a tissue-specific disease (Section 16.7.1), but the candidate should at least be expressed at the time and in the place where the pathology is seen. For example, neural tube defects are likely to involve genes that are expressed during the 3rd–4th weeks of human embryonic development, shortly before or during neurulation. The expression of candidate genes can be tested by RT-PCR or Northern blotting, but the best method for revealing the exact expression pattern is *in situ* hybridization against mRNA in tissue sections (*Figure 5.17*). For embryonic stages this is most conveniently performed using sections of mouse embryos at the equivalent developmental stages (7.5–9.5 embryonic days in the case of neural tube defects). The expression in human embryos is likely to be very similar, although this cannot be guaranteed, and centralized resources of staged human embryo sections have been established to allow the equivalent analyses to be performed where necessary on human embryos.

Appropriate function

Studying the pathology of a genetic disease only rarely gives information precise enough to allow position-independent identification of the disease gene, but it often allows good positional candidates to be selected. Rhodopsin and fibrillin, mentioned above, provide typical examples.

The gene for human rhodopsin was cloned in 1984, and it was mapped to 3q21-qter in 1986. Among disorders involving hereditary retinal degeneration are the various forms of retinitis pigmentosa (RP), which are marked by progressive visual loss resulting from clumping of the retinal pigment. Although rhodopsin was a possible candidate gene for some forms of RP, it was only one of many proteins that were known to be involved in phototransduction. However, in 1989, linkage analyses in a large Irish RP family mapped their disease gene to 3q in the neighborhood of rhodopsin. Rhodopsin was now a serious candidate gene, and patient-specific mutations in the rhodopsin (*RHO*) gene were identified within a year (see OMIM entry 180380).

The phenotype of Marfan syndrome (MFS, MIM 154700: excessive growth of long bones; lax joints; dislocation of lenses; liability to aortic aneurysms) suggested some abnormality in a connective tissue component. Linkage analysis mapped the MFS gene to 15q, and subsequently the gene for the connective tissue protein fibrillin was localized to 15q21.1 by *in situ* hybridization. Fibrillin was then an obvious positional candidate, and patient-specific mutations were soon demonstrated (see McKusick 1991 for discussion of the background).

Candidate genes may also be suggested on the basis of a close functional relationship to a gene known to be involved in a similar disease. The genes could be related by encoding a receptor and its ligand, or other interacting

components in the same metabolic or developmental pathway. For example, some of the genes implicated in Hirschsprung disease were identified using this logic, as described in Section 19.5.2.

15.4.2 Criteria for selecting a candidate gene: homology to a relevant human gene or EST

Preliminary identification of transcripts often comes from matching genomic sequence generated from the candidate region against unmapped ESTs in the databases. Finding a match suggests the presence of an exon in the genomic DNA, and may provide leads to identifying more of the gene or to guessing its function. Sometimes a gene in the candidate region turns out to be closely related to a known disease gene. If the diseases are similar, the new gene becomes a compelling candidate. Members of multigene families can be fairly readily assessed as positional candidates on this basis. For example, after fibrillin was identified as the gene mutated in Marfan syndrome, a second fibrillin gene was shown to map to 5q. This therefore became a candidate location for other Marfan-like phenotypes. A related condition, congenital contractural arachnodactyly (MIM 121050) was mapped to 5q and shown to be caused by mutations in the *FBN2* gene (Putnam *et al.*, 1995).

Selecting candidate disease genes by homology is often more successful using model organisms as described below than by considering human paralogs. Many diseases show extensive locus heterogeneity, and it is not usually the case that the different genes involved are related in any obvious way, either structurally or functionally (*Table 15.3*). One must always bear in mind the complexity of every human organ, tissue and developmental process. Each requires very many different genes and pathways, with the result that mutations in many unrelated genes can produce similar phenotypes.

15.4.3 Criteria for selecting a candidate gene: homology to a relevant gene in a model organism

Over the past decade it has become increasingly clear how far structural and functional homologies extend across even very distantly related species. Virtually every mouse gene has an exact human counterpart, and the same is probably true of other less well explored mammalian species. More surprisingly, extensive homologies can be detected between human genes and genes in zebrafish, *Drosophila*, the nematode worm *Caenorhabditis elegans* and even yeast. Even more than gene sequences, pathways are often highly conserved, so that knowledge of a developmental or control pathway in *Drosophila* or yeast can be used to predict the likely working of human pathways – although mammals often have several parallel paths corresponding to a single path in lower organisms. Being able to predict the possible protein–protein interactions governing a pathway assists experimental

Table 15.3: Functions of some genes implicated in human non-syndromic sensorineural hearing loss

Locus	Chromosomal location	Gene	Function
DFNA1	5q31	*DIAPH1*	Cytokinesis
DFNA2	1p34	*KCNQ4*	K^+ ion channel
DFNA9	14q12-q13	*COCH*	Uncertain
DFNA12, DFNB21	11q22-q24	*TECTA*	Structural component of tectorial membrane
DFNB1	13q12	*Cx26 (GJB2)*	Intercellular gap junction
DFNB2	11q13.5	*MYO7A*	Myosin (molecular motor)
DFNB3	17p11.2	*MYO15*	Myosin (molecular motor)
DFNB4	7q31	*PDS*	Chloride-iodide transporter
DFNB9	2p23	*OTOF*	Vesicle–membrane fusion
DFN3	Xq21.1	*POU3F4*	Transcription factor
12S RNA	Mitochondrion	12S ribosomal RNA	Energy generation by mitochondria

Many structurally and biochemically unrelated genes are implicated in hereditary hearing loss. This diversity means that we cannot predict the nature of the many as yet unidentified deafness genes in a position-independent way. However, knowing some of the genes and pathways involved helps prioritize positional candidates.

attempts to identify novel disease genes, for example by yeast two-hybrid screening (Section 20.4.1). Perhaps the most striking demonstration of the conservation of function between distant organisms is contained in a paper by Rincón-Limas *et al.* (1999) which shows that a wingless *Drosophila* mutant called *apterous* can be corrected by transfection with the human *apterous* homolog *Lhx2* – in other words, humans have a fully functional gene for making flies grow wings!

A very powerful means of selecting good candidates from among a set of human genes is therefore to search the databases for evidence of homologous genes in these well-studied model organisms, as described in Section 20.1.4. If a homolog is detected, one can see what is known about its function. Such data might include the pattern of expression and the phenotype of mutants. Additionally in the mouse, though not in nonmammalian species, the likely chromosomal location of the human ortholog can often be predicted from mouse mapping data, allowing prediction of as yet uncharacterized positional candidates.

Using clues from the mouse

Human–mouse phenotypic homologies provide particularly valuable clues towards identifying human disease genes for several reasons:

- Of the genetically well explored organisms, mice are much the closest to humans in evolutionary terms. Therefore orthologous gene mutations are more likely to produce similar phenotypes in humans and mice than in humans and lower organisms. Not infrequently, however, despite the close evolutionary relationship, the phenotypes are considerably different (see Section 21.4.6).
- Mouse phenotypic information often translates readily into positional candidate information. Backcross mapping (see *Box 15.4*) allows quick and accurate mapping in the mouse. Thus most mouse mutants have been mapped, or can easily be mapped, and there is considerable conservation of synteny

between humans and mice (see *Figure 15.11*). Once a chromosomal location for a gene of interest is known in mouse or humans, it is usually (though not always) possible to predict the likely location of that gene in the other species. Figure 15.12 shows the example of a mouse chromosomal region where prediction of the corresponding human location is easy, and an adjoining region where prediction could be difficult. A database of human–mouse map relationships is maintained at http://www.ncbi.nlm.nih.gov /Omim/Homology (DeBry & Seldin, 1996). *Table 15.4* shows an example of using mouse information to predict possible locations of human deafness genes.

- Exon sequences are usually well conserved between orthologous human and mouse genes. Once a human or mouse gene is isolated, probes or primers can be designed to screen DNA libraries from the other species in order to identify the orthologous gene.

An example: Waardenburg syndrome and the Splotch mouse

Waardenburg syndrome type 1 (WS1, MIM 193500) illustrates the value of human-mouse comparisons. A pedigree of this autosomal dominant but variable condition was shown in *Figure 3.5C*. The characteristic pigmentary abnormalities and hearing loss of WS1 are caused by absence of melanocytes from the affected parts (including the inner ear, where melanocytes are required in the stria vascularis of the cochlea in order for normal hearing to develop). Linkage analysis, aided by the description of a chromosomal abnormality in an affected patient, localized the gene for WS1 to the distal part of 2q. At this point, a likely mouse homolog emerged. The *Splotch* (Sp) mouse mutant has pigmentary abnormalities caused by patchy absence of melanocytes, and the *Sp* gene maps to a linkage group on mouse chromosome 1 that shows extensive conservation of synteny with distal human 2q.

Consideration of the pathogenesis provided further evidence that *WS1* and *Sp* are orthologous genes. The root cause of the phenotype lies in the embryonic neural crest,

Box 15.4

Mapping mouse genes

Several methods are available for easy and rapid mapping of phenotypes or DNA clones in mice. Together with the ability to construct transgenic mice (Chapter 21), they make the mouse especially useful for comparisons with humans. Methods include those described below:

Interspecific crosses (*Mus musculus/Mus spretus* or *Mus castaneus*)
The species have different alleles at many polymorphic loci, making it easy to recognize the origin of a marker allele. This is exploited in two ways:

■ *Constructing marker framework maps.* Several laboratories have generated large sets of F2 backcrossed mice. Any marker or cloned gene can be assigned rapidly to a small chromosomal segment defined by two recombination breakpoints in the collection of backcrossed mice. For example, the collaborative European backcross was produced from a *M. spretus/ musculus* (C57BL) cross. Five hundred F2 mice were produced by backcrossing with *spretus*, and 500 by backcrossing with C57BL. All microsatellites in the framework map are scored in every mouse.
■ *Mapping a new phenotype.* A cross must be set up specifically to do this but, unlike with humans, any number of F2 mice can be bred to map to the desired

resolution. *Musculus* × *castaneus* crosses are easier to breed than *musculus* × *spretus*.

Recombinant inbred strains
These are obtained by systematic inbreeding of the progeny of a cross, for example the widely used BXD strains are a set of 26 lines derived by over 60 generations of inbreeding from the progeny of a C57BL/6J × DBA/2J cross. They provide unlimited supplies of a panel of chromosomes with fixed recombination points. DNA is available as a public resource, and the strains function rather like the CEPH families do for humans (see Section 11.4.3). Recombinant inbred strains are particularly suited to mapping quantitative traits (see Chapter 19), which can be defined in each parent strain and averaged over a number of animals of each recombinant type. Compared with mice from interspecific crosses, it may be harder to find a marker in a given region which distinguishes the two original strains, and the resolution is lower because of the smaller numbers.

Congenic strains
These are identical except at a specific locus. They are produced by repeated backcrossing, and can be used to explore the effect of changing just one genetic factor on a constant background.

Silver (1995; see Further reading) gives an overview of mouse genetics, and Copeland and Jenkins (1991) describe the uses of interspecific crosses.

Table 15.4: Using mapped mouse mutations to predict possible locations of genes causing hearing loss in humans

Mouse mutant	Mouse map location	Predicted human map location(s)	Possible human ortholog
Wo	Chr 1, 25 cM	2q12; 2q31–q33; 6p11	
Sp (Pax3)	Chr 1, 54 cM	2p36	*WS1*
dr	Chr 1, 89 cM	1q23–q31	*DFNA7*
Lp	Chr 1, 92 cM	1q21–q23	*DFNA7*
fi	Chr 2, 34 cM	2q14–q37	
kr	Chr 2, 91 cM	20q11–q13	
Sig	Chr 6, 1 cM	7q21–q31	*DFNB4*
Hoxa1	Chr 6, 25 cM	7p14–p15	*DFNA5*
nv	Chr 7, 4 cM	19q13	*DFNA4*
hb	Chr 7, 65 cM	16p11–p13; 10q24–q26	
Fgf3	Chr 7, 70 cM	11q13.3	
ha1	Chr 10, 56 cM	12q22–q24.1	
Fu	Chr 17, 12 cM	6p21	
Tw	Chr 18, 9 cM	18q11–q12	
sy	Chr 18, 36 cM	5q31–q32; 18p11–q21	
Dc	Chr 19, 6 cM	11q12–q13	*DFNA2, DFNA11*

The left hand columns list a series of mouse mutants with structural abnormalities of the inner ear, and their location on the mouse map (chromosome, distance from centromere in cM). Using mouse–human synteny relationships, as illustrated in *Figure 15.11*, the likely locations of the human orthologs are predicted. In some cases, a human deafness gene has been mapped to the corresponding location (right hand column). Such information allows high-resolution mapping and gene cloning efforts in the two species to reinforce each other. Data from Hereditary Hearing Loss Homepage, http://dnalab-www.uia.ac.be/dnalab/hhh/. Reproduced by kind permission of the MRC Institute of Hearing Research, Nottingham, UK.

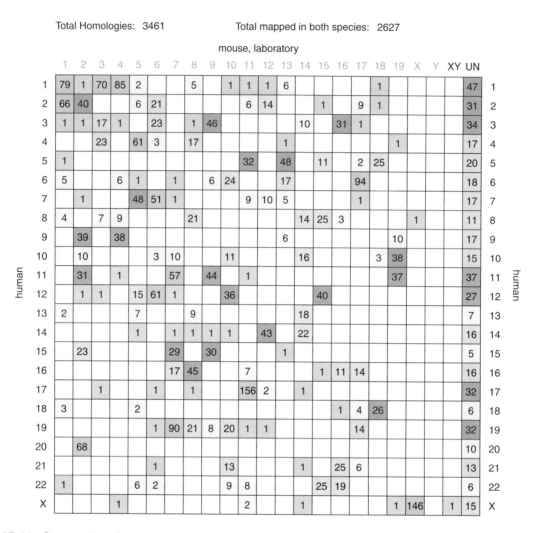

Total Homologies: 3461   Total mapped in both species: 2627

mouse, laboratory

human	1	2	3	4	5	6	7	8	9	10	11	12	13	14	15	16	17	18	19	X	Y	XY	UN	
1	79	1	70	85	2		5			1	1	1	6				1						47	1
2	66	40		6	21					6	14			1		9	1						31	2
3	1	1	17	1		23		1	46				10			31	1						34	3
4			23		61	3		17					1					1					17	4
5	1									32		48		11		2	25						20	5
6	5			6	1		1		6	24		17				94							18	6
7		1			48	51	1				9	10	5			1							17	7
8	4		7	9				21					14	25	3				1				11	8
9		39		38									6				10						17	9
10		10			3	10			11			16				3	38						15	10
11		31		1		57		44			1						37						37	11
12		1	1		15	61	1			36					40								27	12
13	2				7			9					18										7	13
14					1		1	1	1	1		43		22									16	14
15		23				29		30					1										5	15
16						17	45				7				1	11	14						16	16
17			1			1		1			156	2		1									32	17
18	3				2										1	4	26						6	18
19						1	90	21	8	20	1	1					14						32	19
20		68																					10	20
21						1				13				1		25	6						13	21
22	1			6	2					9	8					25	19						6	22
X				1							2			1				1	146			1	15	X

Figure 15.11: Conservation of synteny between human and mouse genetic maps.

The Oxford Grid for human and mouse shows an overall comparison of the two species. Each cell shows the number of orthologous genes mapping to particular chromosomes in mouse and man. Cells are color coded according to the number of orthologs mapped. The nonrandom distribution is obvious (Blake *et al.*, 1999). Reproduced with permission from the Mouse Genome Database, Mouse Genome Informatics, The Jackson Laboratory, Bar Harbor, Maine (http://www.informatics.jax.org/) (25 May 1999).

because melanocytes originate in the neural crest and migrate out to their final locations during embryonic development. Although heterozygous *Sp* mice resemble WS1 patients, homozygous *Sp* mice have neural tube defects, and have been studied for many years as a model for human neural tube defects.

A positional candidate gene emerged when the murine *Pax-3* gene was mapped to the vicinity of the *Sp* locus. *Pax-3* is one of a family of genes (*PAX* genes) that encode transcription factors containing the paired box DNA-binding motif, and it is expressed in mouse embryos in the developing nervous system, including the neural crest. The sequence of *Pax-3* was almost identical to the limited sequence which had previously been published for an unmapped human genomic clone, *HuP2*. Such observations prompted mutation screening of *Pax-3* and *HuP2* and led to identification of mutations in *Splotch*

mice and humans with WS1 (reviewed by Strachan & Read, 1994). As the underlying genes, *Pax-3* and *HuP2* were clearly orthologs, the *HuP2* gene was subsequently re-named *PAX3*.

Limitations of human–mouse homologies

Though enormously valuable as a guide to human–mouse homologies, conservation of synteny is not always sufficient to allow identification of positional candidates. This is illustrated by the *mi/MITF* locus, mutations in which cause another variant of Waardenburg syndrome, WS type 2 (MIM 193510) in man. The *mi* locus on mouse chromosome 6 has long been recognized as a likely candidate homolog of some form of WS, but attempts to predict the location of the human homolog failed. Genes mapping close to *mi* have human homologs mapping to 3p25, 3q21-q24 and 10q11.2. Each of these locations was

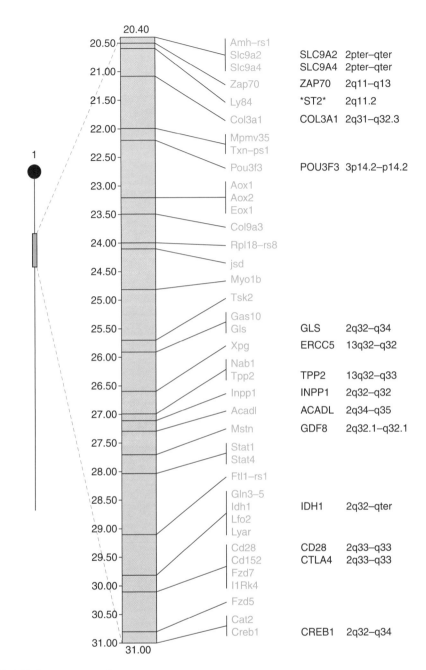

Figure 15.12: Detail of one mouse chromosome.

The map shows the part of mouse chromosome 1, from 48 to 62 cM from the centromere. Mapped mouse genes are shown in color. Where a human ortholog has been mapped, its human map location is shown. The distal part of the mouse map shows good conservation of synteny with the distal long arm of human chromosome 2, but in the proximal part of the mouse chromosome the relationship to human chromosomes is more complex. Mouse Chromosome 1 Linkage Map reproduced with permission from the Mouse Genome Database, Mouse Genome Informatics, The Jackson Laboratory, Bar Harbor, Maine (http://www.informatics.jax.org/) (25 May 1999). See Blake *et al.*, 1999.

tested for linkage to *WS2*, with negative results. Not until the human *mi* homolog, *MITF*, had been cloned and mapped by FISH to 3p14, and *WS2* in humans had been independently mapped by linkage to the same location (see *Figure 11.6*), was there sufficient evidence to begin a successful search for mutations in man (Tassabehji *et al.*, 1994).

Using clues from mutant phenotypes in lower organisms

Phenotypic homologies with lower organisms have mostly been used after a human gene and its orthologs have been identified, as part of the exploration of the gene's function. In a few cases, however, homologies have been used prospectively, to help identify a human

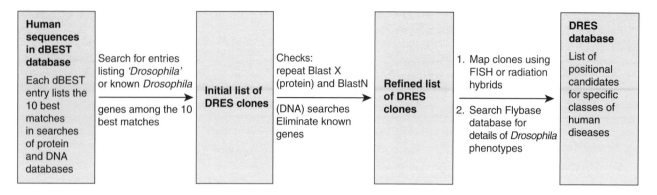

Figure 15.13: A systematic database search for *Drosophila* mutants that could be positional candidates for human disease genes (Banfi *et al.*, 1996).

disease gene. A particularly successful example was the identification of the *MSH2* and *MLH1* genes that are mutated in certain forms of hereditary colon cancer. These genes were identified after phenotypic resemblances led researchers to suspect that the cancers might be caused by mutations in the human homologs of yeast mismatch repair genes (Section 18.7.1).

An attempt to use *Drosophila* phenotypic information systematically to identify positional candidates for human diseases is the DRES database (Banfi *et al.*, 1996). The dbEST human EST database was searched for matches to *Drosophila* genes with known mutant phenotypes. Sixty six novel matches were detected. The map position of each EST was determined by both FISH and radiation hybrid mapping (*Figure 15.13*).

Sequence homologies with lower organisms

Just as in the general Human Genome Project, genomic or cDNA sequences generated in the exploration of a candidate disease region are routinely checked against data from lower organisms. A match of genomic sequence suggests the presence of a gene, and if the match is to a known gene in the lower organism, it suggests the nature of the human gene. The example of branchio-oto-renal syndrome, above, shows how useful this approach can be.

15.5 Confirming a candidate gene

Ultimately all approaches used to identify disease genes generate candidates, which then have to be tested individually to see if there is compelling evidence that mutations in them do cause the disease in question. Demonstrating that a candidate gene is likely to be the disease locus can be done by various means.

- *Mutation screening*. Screening for patient-specific mutations in the candidate gene is by far the most popular method, because it is generally applicable and comparatively rapid. The reasons why particular mutations may be found in certain diseases are dis-

cussed in Chapter 16, and the methods of testing for mutations are described in Chapter 17. Identifying mutations in several unrelated affected individuals strongly suggests that the correct candidate gene has been chosen, but formal proof requires additional evidence.

- *Restoration of normal phenotype* in vitro. For some disorders, usually ones where mutations cause loss of function, the phenotype is reversible. If a cell line that displays the mutant phenotype can be cultured from the cells of a patient, transfection of a cloned normal allele into the cultured disease cells may result in restoration of the normal phenotype by complementing the genetic deficiency.

- *Production of a mouse model of the disease* (Chapter 21). Once a putative disease gene is identified, a transgenic mouse model can be constructed. If the human phenotype is known to result from loss of function, gene targeting can be used to generate a germline knockout mutation in the mouse ortholog. If the human disease results from a gain of function, then attempts to make a transgenic mouse model normally involve introducing a disease allele into the mouse germline. The mutant mice are expected to show some resemblance to humans with the disease, although this expectation may not always be met even when the correct gene has been identified.

15.5.1 Principles of mutation screening to confirm a candidate gene

However promising a candidate gene appears to be for a disease, it must be shown to be mutated in affected people. Mutation screening entails testing DNA samples from a sizable panel of unrelated patients and control individuals. The first step is to design pairs of primers for PCR-amplifying portions of the coding DNA, either from a genomic DNA sample (if the exon/intron boundaries are known), or from cDNA generated by RT-PCR from mRNA of patients (*Figure 6.5*). The products of individual amplification reactions are then subjected to one or more of the mutation screening procedures described in Section 17.1.4 that are designed to detect unknown point mutations.

Mutation screening is often straightforward for diseases where a good proportion of patients carry independent mutations (typically, severe early onset dominant or X-linked recessive disorders) and where the disease phenotype results from loss of function of the gene. As explained in Section 16.4, if the correct gene is tested, a panel of DNA samples from unrelated patients will usually show a variety of different mutations, including some with an obviously deleterious effect on gene expression (nonsense mutations, frameshift mutations, etc.). *Figure 16.1* shows an example from the work on Waardenburg syndrome mentioned above. If the identified mutations are absent from control samples, then the conclusion that the gene being tested really is the locus for the disease becomes almost inescapable. However, other circumstances may make the identification of mutations and the interpretation of mutation screening more difficult.

- *Unsuspected locus heterogeneity.* Often mutations in several different genes can give almost identical phenotypes, so that a panel of unselected patient samples may have pathogenic mutations in different genes. If the candidate gene being tested is responsible for only a small proportion of cases, most samples will show no mutation in that gene. Ideally, one would use only samples from families with demonstrated linkage to the candidate region, but this may be impracticable. Family sizes for recessive and some dominant disorders are often too small for independent linkage analyses, and in some severe dominant disorders most patients present as sporadic cases without a family history.
- *Mutational homogeneity.* This problem was discussed above in connection with CF. Most apparently unrelated patients carry the same mutation, *F508del.* See Sections 16.3.3 and 17.1.2 for further discussion and examples of mutational homogeneity.
- Mutations are not unambiguously pathogenic. It may be difficult to identify missense mutations as being pathogenic as opposed to being neutral variants with no major effect on gene expression. Some guidelines to help decide whether a sequence change is pathogenic are given in *Box 16.4.*
- *Mutations may be hard to find.* Large genes are more difficult to screen for mutations, and sometimes mutations seem very hard to find. Current examples include the *NF1* and *PKD1* genes that are mutated in neurofibromatosis 1 (MIM 162200) and adult

polycystic kidney disease (MIM 173900) respectively. Mutations in the *F8C* gene causing severe hemophilia A seemed to be hard to find, until it was discovered that most of the missing mutations were large inversions which disrupted the gene (see *Figure 9.20*) but were not detected by the PCR methods normally used.

15.5.2 Once a candidate gene is confirmed, the next step is to understand its function

Identifying the gene involved in a genetic disease opens the way to several lines of investigation. The ability to identify mutations should immediately lead to improved diagnosis and counseling, as described in *Chapter 17*. Understanding the molecular pathology (why the mutated gene causes the disease; see *Chapter 16*) may also lead to insight into related diseases, and hopefully eventually to more effective treatment including perhaps gene therapy (*Chapter 22*).

A second line of enquiry concerns the normal function of the gene product. For example, until the Duchenne muscular dystrophy (DMD) gene was identified there was no knowledge of the way the contractile machinery of muscle cells is anchored to the sarcolemma. Analysis of functional domains and motifs (Mushegian *et al.,* 1997) and the search for experimentally manipulable homologs in the mouse, fruit fly, nematode and yeast are powerful tools for this work. This large topic is covered in *Chapter 20*; a foretaste of the sort of information that can be generated by database searching can be seen in the following information, taken from the Hereditary Hearing Loss Homepage database (http://dnalab-www.uia.ac.be/dnalab/hhh/) and describing the gene that was identified by positional cloning of an autosomal dominant hearing loss locus (*DFNA1*) in one large Costa Rican family (see OMIM entry 124900):

The human DFNA1 protein product DIAPH1, mouse *p140mDia*, and *Drosophila diaphanous* are homologs of *Saccharomyces cervisiae* protein *Bni1p*. The proteins are highly conserved overall. The genes encoding these proteins are members of the formin gene family, which also includes the mouse *limb deformity* gene, *Drosophila cappuccino*, *Aspergillus nidulans* gene *sepA*, and *S. pombe* genes *fus1* and *cdc12*. These genes are involved in cytokinesis and establishment of cell polarity. All formins share Rho-binding domains in their N-terminal regions, polyproline stretches in the central region of each sequence, and formin-homology domains in the C-terminal region.

Further reading

Silver LM (1995) *Mouse Genetics: Concepts and Applications.* Oxford University Press, Oxford.

References

Abdelhak S, Kalatzis V, Heilig R *et al.* (1997) A human homologue of the Drosophila eyes absent gene underlies Branchio-Oto-Renal (BOR) syndrome and identifies a novel gene family. *Nature Genet.*, **15**, 157–164.

Antoch MP, Song E-J, Chang A-M *et al.* (1997) Functional identification of the mouse circadian Clock gene by transgenic BAC rescue. *Cell*, **89**, 655–667.

Banfi S, Borsani G, Rossi E *et al.* (1996) Identification and mapping of human cDNAs homologous to Drosophila mutant genes through EST database searching. *Nature Genet.*, **13**, 167–174; website http//www.tigem.it/LOCAL/drosophila/dros.html

Blake JA, Richardson JE, Davisson MT, Eppig JT and the Mouse Genome Database Group (1999) The Mouse Genome Database (MGD): Genetic and genomic information about the laboratory mouse. *Nucleic Acids Res.*, **27**, 95–98.

Carter SA, Bryce SD, Munro CS *et al.* (1994) Linkage analyses in British pedigrees suggest a single locus for Darier disease and narrow the location to the interval between D12S105 and D12S129. *Genomics*, **24**, 378–382.

Collins FS (1995) Positional cloning moves from the perditional to the traditional. *Nature Genet.*, **9**, 347–350.

Copeland NG, Jenkins NA (1991) Development and applications of a molecular genetic linkage map of the mouse genome. *Trends Genet.*, **7**, 113–118.

DeBry RW, Seldin MF (1996) Human/mouse homology relationships. *Genomics*, **33**, 337–351. An updated electronic version is at http: //www.ncbi.nlm.nih.gov/Omim/Homology/

Edstrom J-E, Kaiser R, Rohme D (1987) Microcloning of mammalian metaphase chromosomes. *Meth. Enzymol.*, **151**, 503–516.

European Consortium on MEN1 (1997) Linkage disequilibrium studies in multiple endocrine neoplasia type 1 (MEN1). *Hum. Genet.*, **100**, 657–665.

Gitschier J, Wood WI, Goralka TM *et al.* (1984) Characterization of the human factor VIII gene. *Nature*, **312**, 326–330.

Koob MD, Benzow KA, Bird TD, Day JW, Moseley ML, Ranum LP (1998) Rapid cloning of expanded trinucleotide repeat sequences from genomic DNA. *Nature Genet.*, **18**, 72–75.

Koob MD, Moseley ML, Schut LJ *et al.* (1999) Untranslated CTG expansion causes a novel form of spinocerebellar ataxia (SCA8). *Nature Genet.*, **21**, 379–384.

Lisitsyn NA (1995) Representational difference analysis: finding the differences between genomes. *Trends Genet.*, **11**, 303–307.

Ludecke HJ, Senger G, Claussen U, Horsthemke B (1989) Cloning defined regions of the human genome by microdissection of banded chromosomes and enzymatic amplification. *Nature*, **338**, 348–350.

McKusick VA (1991) The defect in Marfan syndrome. *Nature*, **352**, 279–281.

Mushegian AR, Bassett DE, Boguski MS, Bork P, Koonin EV (1997) Positionally cloned human disease genes: patterns of evolutionary conservation and functional motifs. *Proc. Natl Acad. Sci. USA*, **94**, 5831–5836.

Poustka A, Pohl TM, Barlow DP, Frischauf AM, Lehrach H (1987) Construction and use of human chromosome jumping libraries from NotI-digested DNA. *Nature*, **325**, 353–355.

Probst FJ, Fridell RA, Raphael Y *et al.* (1998) Correction of deafness in *shaker-2* mice by an unconventional myosin in a BAC transgene. *Science*, **280**, 1444–1447. See also the accompanying paper, **Wang A, Liang Y, Fridell RA *et al.*** (1998) Association of unconventional myosin *MYO15* mutations with human nonsyndromic deafness *DFNB3*. *Science*, **280**, 1447–1451.

Putnam EA, Zhang H, Ramirez F, Milewicz DM (1995) Fibrillin-2 (FBN2) mutations result in the Marfan-like disorder, congenital contractural arachnodactyly. *Nature Genet.*, **11**, 456–458.

Rincón-Limas DE, Lu C-H, Canal I *et al.* (1999) Conservation of the expression and function of *apterous* orthologs in *Drosophila* and mammals. *Proc. Natl Acad. Sci. USA*, **96**, 2165–2170.

Robson KJH, Chandra T, MacGillivray RTA, Woo SLC (1982) Polysome immunoprecipitation of phenylalanine hydroxylase mRNA from rat liver and cloning of its cDNA. *Proc. Natl Acad. Sci. USA*, **79**, 4701–4705.

Rommens JM, Januzzi MC, Kerem B-S *et al.* (1989) Identification of the cystic fibrosis gene: chromosome walking and jumping. *Science*, **245**, 1059–1065.

Schalling M, Hudson TJ, Buetow KH, Housman DE (1993) Direct detection of novel expanded trinucleotide repeats in the human genome. *Nature Genet.*, **4**, 135–139.

Schutte M, da Costa LT, Hahn SA *et al.* (1995) Identification by representational difference analysis of a homozygous deletion in pancreatic carcinoma that lies within the BRCA2 region. *Proc. Natl Acad. Sci. USA*, **92**, 5950–5954.

Strachan T, Read AP (1994) PAX genes. *Curr. Opin. Genet. Dev.*, **4**, 427–438.

Strathdee CA, Gavish H, Shannan WR, Buchwald M (1992) Cloning of cDNAs for Fanconi's anaemia by functional complementation. *Nature*, **356**, 763–767.

Swaroop A, Xu J, Agarwal N, Weissman SM (1991). A simple and efficient cDNA library subtraction procedure: isolation of human retina-specific cDNA clones. *Nucleic Acids Res.*, **19**, 1954.

Tassabehji M, Newton VE, Read AP (1994) Waardenburg syndrome type 2 caused by mutations in the human microphthalmia (MITF) gene. *Nature Genet.*, **8**, 251–255.

Treacher Collins Syndrome Collaborative Group (1996). Positional cloning of a gene involved in the pathogenesis of Treacher Collins syndrome. *Nature Genet.*, **12**, 130–136.

Watson CJ, Gaunt L, Evans G, Patel K, Harris R, Strachan T (1993) A disease-associated germline deletion maps the type 2 neurofibromatosis (NF2) gene between the Ewing sarcoma region and the leukaemia inhibitory factor locus. *Hum. Mol. Genet.,* **2**, 701–704.

Worton RG, Thompson MW (1988) Genetics of Duchenne muscular dystrophy. *Annu. Rev. Genet.,* **22**, 601–629.

Yasunaga S, Grati M, Cohen-Salmon M *et al.* (1999) A mutation in *OTOF*, encoding otoferlin, a FER-1-like protein, causes *DFNB9*, a nonsyndromic form of deafness. *Nature Genet.,* **21**, 363–369.

Molecular pathology 16

16.1 Introduction

Molecular pathology seeks to explain why a given genetic change should result in a particular clinical phenotype. We have already reviewed the nature and mechanisms of mutations in Chapter 9 (briefly summarized in *Box 16.1*); this chapter is concerned with their effects on the phenotype. Molecular pathology requires us to work out the effect of a mutation on the quantity or function of the gene product, and to explain why the change is or is not pathogenic for any particular cell, tissue or stage of development.

Not surprisingly, given the complexity of genetic interactions, molecular pathology is at present a very imperfect science. The greatest successes to date have been in understanding cancer, where the phenotype to be explained, uncontrolled cell proliferation, is relatively simple. For most other genetic diseases we would like to explain complex clinical findings. Often these are the end result of a long chain of causation, and all too often they are not predictable or even readily comprehensible in our present state of knowledge. Nevertheless, as the emphasis of the Human Genome Project moves from cataloging genes to understanding their function, the study of molecular pathology has moved to center stage.

One of the major advantages of studying humans rather than laboratory organisms is that the healthcare systems worldwide act as a gigantic and continuous mutation screen. Any human phenotype that occurs with a frequency greater than 1 in 10^9 is probably already described somewhere in the literature, and for most inherited diseases where the gene responsible has been identified, many different mutations are known. We cannot do experiments on humans or breed them to order, but humans provide unique opportunities to observe the clinical effects of many different changes in a given gene. This generates hypotheses, which must then be tested in animals. Thus investigations of naturally occurring human mutations are complemented by studies of specific mutations in transgenic animals (see Chapter 21).

Box 16.1

The main classes of mutation

Deletions ranging from 1 bp to megabases.
Insertions including duplications.
Single base substitutions:
   Missense mutations replace one amino acid with another in the gene product;
   Nonsense mutations replace an amino acid codon with a stop codon;
   Splice site mutations create or destroy signals for exon-intron splicing.
Frameshifts can be produced by deletions, insertions or splicing errors.
Dynamic mutations (tandem repeats that often change size on transmission to children).
See *Table 16.1* for some examples, and Chapter 9 for more details and discussion of mechanisms.

16.2 There are rules for the nomenclature of mutations and databases of mutations

The preferred nomenclature of genes is laid down by the Genome Database Nomenclature Committee (http://www.gene.ucl.ac.uk/nomenclature/; printed version: White *et al.*, 1997). A valuable summary of genetic nomenclature for many different organisms including man was published as a supplement to *Trends in Genetics* in 1998 (see Further reading).

Mutations can be described in two ways: by their effects or by detailing the sequence change. *Box 16.2* shows one possible nomenclature for effects, currently more widely used for laboratory organisms than humans. *Box 16.3* summarizes the recommended conventions for describing sequence changes (Antonarakis *et al.*, 1998).

Box 16.2

A nomenclature for describing the effect of an allele

Null allele or **amorph**: an allele that produces no product.
Hypomorph: an allele that produces a reduced amount or activity of product.
Hypermorph: an allele that produces increased amount or activity of product.
Neomorph: an allele with a novel activity or product.
Antimorph: an allele whose activity or product antagonizes the activity of the normal product.

Box 16.3

Nomenclature for describing mutations
(see Antonarakis *et al*. (1998) for full details)

Amino acid substitutions
Use the one-letter codes: A, alanine; C, cysteine; D, aspartic acid; E, glutamic acid; F, phenylalanine; G, glycine; H, histidine; I, isoleucine; K, lysine; M, methionine; N, asparagine; P, proline; Q, glutamine; R, arginine; S, serine; T, threonine; V, valine; W, tryptophan; Y, tyrosine; X means a stop codon. 3-letter codes are also acceptable.
R117H or **Arg117His** – replace arginine 117 by histidine (the initiator methionine is codon 1).
G542X or **Gly542Stop** – glycine 542 replaced by a stop codon.

Nucleotide substitution
The A of the initiator ATG codon is +1; the immediately preceding base is −1. There is no zero. Give the nucleotide number followed by the change. If necessary use g. and c. to designate genomic and cDNA sequences. For changes within introns, when only the cDNA sequence is known in full, specify the intron number by IVSn or the number of the nearest exon position.
1162G > A – replace guanine at position 1162 by adenine.
621 + 1G > T or **IVS4 + 1G > T** – replace G by T at the first base of intron 4; exon 4 ends at nt 621.

Deletions and insertions
Use del for deletions and ins for insertions. As above, for DNA changes the nucleotide position or interval comes first, for amino acid changes the amino acid symbol comes first.
F508del – delete phenylalanine 508
6232–6236del or **6232–6236delATAAG** – delete 5 nucleotides (which can be specified) starting with nt 6232.
409–410insC – insert C between nt 409 and 410.

Systematic attempts are now being made to establish disease-specific databases of mutations (Krawczak and Cooper, 1997). These can be accessed through central points such as the Human Gene Mutation Database (http://www.uwcm.ac.uk/uwcm/mg/hgmd0.html). For some but not all genes, allelic variants are also listed in the OMIM database (http://www.hgmp.mrc.ac.uk/omim/). Cooper and Krawczak (1993) have performed a number of useful meta-analyses on diferent types of human mutation.

16.3 A first classification of mutations is into loss of function vs gain of function mutations

16.3.1 The convenient nomenclature of A and a alleles hides a vast diversity of DNA sequence changes

Over 750 different cystic fibrosis mutant alleles have been described, and a similar number of different mutations in the β-globin gene. There is no reason why these should all fit into a few tidy categories. In principle however, mutation of a gene might cause a phenotypic change in either of two ways:

■ the product may have reduced or no function (loss of function mutation – an amorph or hypomorph in the terminology of *Box 16.2*);

■ the product may do something positively abnormal (**gain of function mutation** – a hypermorph or neomorph).

Loss of function mutations most often produce recessive phenotypes. For most gene products the precise quantity is not crucial, and we can get by on half the normal amount. Thus most inborn errors of metabolism are recessive. For some gene products, however, 50% of the normal level is not sufficient for normal function, and **haploinsufficiency** produces an abnormal phenotype, which is therefore inherited in a dominant manner (see Section 16.4.3). Sometimes also a nonfunctional mutant polypeptide interferes with the function of the normal allele in a heterozygous person, giving a **dominant negative** effect (an antimorph in the terminology of *Box 16.2* – see Section 16.4.4).

Gain of function mutations usually cause dominant phenotypes, because the presence of a normal allele does not prevent the mutant allele from behaving abnormally. Often this involves a control or signaling system behaving inappropriately – signaling when it should not, or failing to switch a process off when it should. Sometimes the gain of function involves the product doing something novel – a protein containing an expanded polyglutamine repeat forming abnormal aggregates, for example.

Inevitably some mutations cannot easily be classified as either loss or gain of function. Has a permanently open ion channel lost the function of closing or gained the function of inappropriate opening? A dominant negative mutant allele has lost its function but also does something positively abnormal. Nevertheless, the distinction between loss of function and gain of function is a useful first tool for thinking about molecular pathology.

16.3.2 Loss of function is likely when point mutations in a gene produce the same pathological change as deletions

Purely genetic evidence, without biochemical studies, can often suggest whether a phenotype is caused by loss or gain of function. When a clinical phenotype results from loss of function of a gene, we would expect any change that inactivates the gene product to produce the same clinical result. We should be able to find point mutations which have the same effect as mutations that delete or disrupt the gene. Waardenburg syndrome Type 1 (MIM 193500) provides an example. As *Figure 16.1* shows, causative mutations in the *PAX3* gene include amino acid substitutions, frameshifts, splicing mutations, and in some patients complete deletion of the *PAX3* sequence. Since all these events produce the same clinical result, its cause must be loss of function of *PAX3*. Similarly, among diseases caused by unstable trinucleotide repeats (see *Box 16.7*), Fragile-X and Friedreich ataxia are occasionally caused by other types of mutation in their respective genes, pointing to loss of function, whereas Huntington disease is never seen with any other type of mutation, suggesting a gain of function.

16.3.3 Gain of function is likely when only a specific mutation in a gene produces a given pathology

Gain of function is likely to require a much more specific change than loss of function. The mutational spectrum in gain-of-function conditions should be correspondingly more restricted, and the same condition should not be produced by deletion or disruption of the gene. Likely examples include Huntington disease (see *Box 16.7*), and achondroplasia (MIM 100800: short-limbed dwarfism). Virtually all achondroplastics have one of two mutations in the fibroblast growth factor receptor gene *FGFR3*, both of which cause the same amino acid change, G380R (Bellus *et al.*, 1995). Other mutations in the same gene produce other syndromes (Section 16.7.3). For unknown reasons, the mutation rate for G380R is extraordinarily high, so that achondroplasia is one of the commoner genetic abnormalities, despite requiring a very specific DNA sequence change.

Mutational homogeneity is a good first indicator of a gain of function, but there are other reasons why a single mutation may account for all or most cases of a disease:

■   diseases where what one observes is directly related

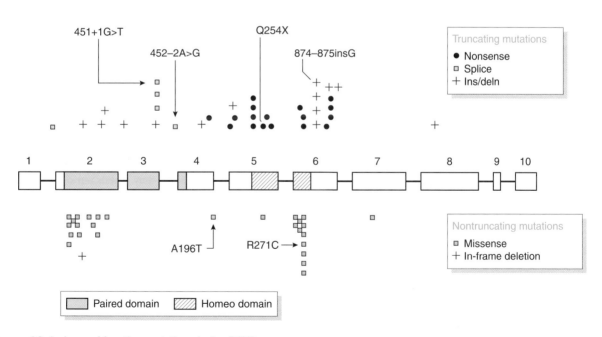

Figure 16.1: Loss of function mutations in the *PAX3* gene.

The 10 exons of the gene are shown as boxes, with the connecting introns not to scale. The shaded areas show the sequences encoding the two DNA-binding domains of the PAX3 protein. Note that mutations that completely destroy the structure of the PAX3 protein (drawn above the gene diagram) are scattered over at least the first six exons of the gene, but missense mutations (shown below the gene diagram) are concentrated in two regions, the 5′ part of the paired domain and the third helix of the homeodomain. A196T is believed to affect splicing. The other named mutations are mentioned in *Table 16.1*. The 874-875insG mutation introduces a seventh G into a run of six Gs; it has arisen independently several times and illustrates the relatively high frequency of slipped-strand mispairing (Section 9.5.1).

to the gene product itself, rather than a more remote consequence of the genetic change, may be defined in terms of a particular variant product, as in sickle cell disease (see *Box 16.5*);

- a molecular mechanism may make a certain sequence change in a gene much more likely than any other change – e.g. the CGG expansion in Fragile-X syndrome (see *Box 16.8*);
- there may be a **founder effect** – for example, certain disease mutations are common among Ashkenazi Jews, presumably reflecting mutations present in a fairly small number of founders of the present Ashkenazi population (Motulsky, 1995);
- selection favoring heterozygotes (Section 3.3.2) enhances founder effects and often results in one or a few specific mutations being common in a population.

16.3.4 Deciding whether a DNA sequence change in a gene is pathogenic can be difficult

Not every sequence variant seen in an affected person is necessarily pathogenic. If the genome-wide average heterozygosity of 0.0032 is applied to coding sequences, then screening a panel of 100 patients for mutations in a 3-kb coding sequence would reveal about 500 sequence changes. Even allowing for the much higher conservation of coding sequences, screening on such a scale will almost certainly reveal some rare nonpathogenic sequence variants, as well as pathogenic changes. If each variant occurs in one person in 10 000 in the population, it will not show up in any panel of normal controls of realistic size. How does one decide whether a sequence variant is pathogenic?

If the pathogenic mechanism is gain of function, then as explained above (Section 16.3.3), the mutation is likely to be very specific. Any sequence change different from the standard mutation is probably not pathogenic, at least for the disease in question. Loss of function mutations are usually much more heterogeneous. Only a functional test, either *in vitro* (Chapter 20) or *in vivo* (Chapter 21), can definitively show whether a DNA sequence change in a gene affects the function, but useful clues can be obtained by considering the nature of the sequence change (*Box 16.4*).

16.4 Loss of function mutations

16.4.1 Many different changes to a gene can cause loss of function

Not surprisingly, there are many ways of reducing or abolishing the function of a gene product (*Table 16.1* and *Figure 16.2*). Some of these have been discussed in Section 9.4. The hemoglobinopathies (*Box 16.5*) exemplify many of these mechanisms especially well. In fact, globin mutations can be found to illustrate virtually every process

---

Box 16.4

Guidelines for deciding whether a DNA sequence change is pathogenic

- Deletions of the whole gene, nonsense mutations and frameshifts are almost certain to destroy the gene function.
- Mutations that change the conserved GT...AG nucleotides flanking most introns affect splicing, and will usually abolish the function of the gene. The effects of other sequence changes on splicing are harder to predict, and should ideally be checked by RT-PCR or an *in vitro* splicing assay. See Section 9.4.5
- A missense mutation is more likely to be pathogenic if it affects a part of the protein known to be functionally important. For example, the missense mutations in *Figure 16.1*, all of which cause loss of function, are concentrated in the key DNA-binding domains of the PAX3 protein.
- Evolutionary conservation is a useful guide to function. Changing an amino acid is more likely to affect function if it is conserved across species (orthologs) or between members of a gene family (paralogs).
- Amino acid substitutions are more likely to affect function if they are nonconservative (replace a polar by a nonpolar amino acid, or an acidic by a basic one – see *Box 9.2*).
- A sequence change in a disease gene that is present in a *de novo* affected patient and not in the unaffected parents may well be pathogenic.

---

described in this book. Readers with a particular interest in these diseases are recommended to consult one of the excellent reviews of this topic (e.g. Weatherall *et al.*, 1995; see further reading).

When considering the likely result of a mutation on the gene product, some points to bear in mind are as follows:

- Small deletions and insertions have a much more drastic effect on the gene product if they introduce a frameshift (that is, if they add or remove a number of nucleotides that is not an exact multiple of 3). Deletions in the dystrophin (*DMD*) gene provide striking examples (*Figure 16.3*). Almost regardless of the size of the deletion, frameshifting deletions produce the lethal Duchenne muscular dystrophy, in which no dystrophin is produced, whereas nonframeshifting mutations cause the milder Becker form, in which dystrophin is present but abnormal.
- Nonsense mutations often trigger mRNA instability (see Section 9.4.6 and Hentze and Kulozik, 1999) rather than cause production of a truncated protein.
- Base substitutions in coding sequences may be pathogenic because of an effect on splicing or because they destroy an embedded signal (a nuclear localization signal, for example), rather than because of their effect on the amino acids encoded. Activation of a cryptic splice site is particularly hard to predict – see Section 9.4.5 and Berget (1995).

Table 16.1: Eleven ways to reduce or abolish the function of a gene product (see *Table 9.5* for a classification of mutations by their nature and location in the gene)

Change	Example
Delete: (i) the entire gene (ii) part of the gene	Most α-thalassemia mutations (*Figure 16.2*) 60% of Duchenne muscular dystrophy (*Figure 16.3*)
Insert a sequence into the gene	Insertion of LINE-1 repetitive sequence (see Section 9.5.6) into *F8C* gene in hemophilia A
Disrupt the gene structure: (i) by a translocation (ii) by an inversion	X-autosome translocations in women with Duchenne muscular dystrophy (*Figure 15.9*) Inversion in *F8C* gene (*Figure 9.20*)
Prevent the promoter working: (i) by mutation (ii) by methylation	β-Globin −29A → G mutation (*Table 17.2*) Fragile-X full mutation (*FMR1*) (*Box 16.8*)
Destabilize the mRNA: (i) by a polyadenylation site mutation · (ii) by nonsense-mediated RNA decay	α-globin AATAAA → AATAGA mutation Fibrillin mutations (*FBN1*)
Prevent correct splicing *Figure 9.11*: (i) by inactivating donor splice site (ii) by inactivating acceptor splice site (iii) by activating a cryptic splice site	*PAX3* 451 + 1G → T mutation (*Figure 16.1*) *PAX3* 452-2A → G mutation (*Figure 16.1*) β-Globin intron 1 −110G → A mutation (*Figure 9.13*)
Introduce a frameshift in translation	*PAX3* 874–875insG mutation (*Figure 16.1*)
Convert a codon into a stop codon	*PAX3* Q254X mutation (*Figure 16.1*)
Replace an essential aminoacid	*PAX3* R271C mutation (*Figure 16.1*)
Prevent post-transcriptional processing (Section 16.6.1).	Cleavage-resistant collagen N-terminal propeptide in Ehlers Danlos VII syndrome
Prevent correct cellular localization of product	F508del mutation in cystic fibrosis

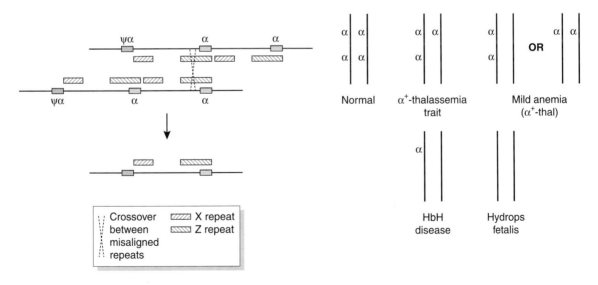

Figure 16.2: Deletions of α-globin genes in α-thalassemia.

Normal copies of chromosome 16 carry two active α-globin genes and an inactive pseudogene arranged in tandem. Repeat blocks (labeled X and Z) may misalign, allowing unequal crossover. The diagram shows unequal crossover between mis-aligned Z repeats producing a chromosome carrying only one active α gene. Unequal crossovers between X repeats have a similar effect. Unequal crossovers between other repeats (not shown) can produce chromosomes carrying no functional α gene. Individuals may thus have any number from 0 to 4 or more α-globin genes. The consequences become more severe as the number of α genes diminishes. See Weatherall *et al.*, 1995 for detail.

Box 16.5

Hemoglobinopathies

Hemoglobinopathies occupy a special place in clinical genetics for many reasons. They are by far the most common serious mendelian diseases on a worldwide scale. Globins illuminate important aspects of evolution of the genome (see *Figures 14.16, 14.20–22*) and of diseases in populations. Developmental controls are probably better understood for globins than for any other human genes (see *Figures 8.6, 8.22 and 8.23*). More mutations and more diseases are described for hemoglobins than for any other gene family. Clinical symptoms follow very directly from malfunction of the protein, which at 15 g per 100 ml of blood is easy to study, so that the relationship between molecular and clinical events is clearer for the hemoglobinopathies than for most other diseases.

Hemoglobinopathies are classified into two main groups:

■ **The thalassemias** are caused by inadequate quantities of the α or β chains. Alleles are classified into those

producing no product (α^0, β^0) and those producing reduced amounts of product (α^+, β^+). The underlying defects include examples of all the types detailed here in Section 16.4 (see also *Table 17.2*).

■ **Abnormal hemoglobins** with amino acid changes cause a variety of problems, of which sickle cell disease is the best known. The E6V mutation replaces a polar by a neutral amino acid on the outer surface of the β-globin molecule. This causes increased intermolecular adhesion, leading to aggregation of deoxyhemoglobin and distortion of the red cell. Sickled red cells have decreased survival time (leading to anemia) and tend to occlude capillaries, leading to ischemia and infarction of organs downstream of the blockage. Other amino acid changes can cause anemia, cyanosis, polycythemia (excessive numbers of red cells), methemoglobinemia (conversion of the iron from the ferrous to the ferric state), etc.

See Weatherall *et al*. (1995) for a comprehensive review.

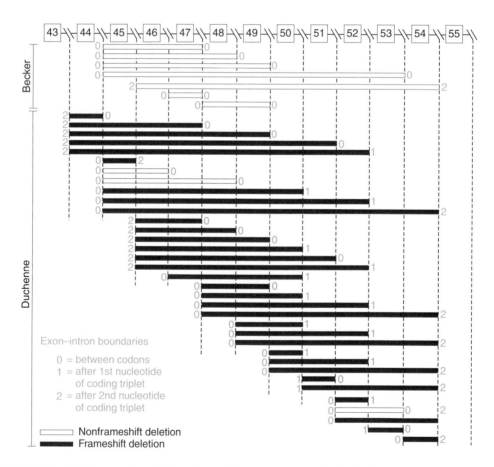

Figure 16.3: Deletions in the central part of the dystrophin gene associated with Becker and Duchenne muscular dystrophy.

Numbered boxes represent exons 43–55. Deletions that generate frameshifts cause the lethal DMD, while frame-neutral deletions cause the milder BMD. Western blots show no detectable dystrophin in muscle biopsies of DMD patients, but dystrophin of reduced molecular weight in BMD patients. See *Figure 14.9* and *Box 14.3* for more detail on exon/intron boundaries, and *Figure 17.11* to see how these deletions are characterized in the laboratory.

16.4.2 Epigenetic modification can abolish gene function even without a DNA sequence change

Heritable changes that do not depend on changes in a DNA sequence are called **epigenetic** (see Section 8.1). They may affect expression of a gene or the properties of its product. A set of mini-reviews in the 1 May 1998 issue of *Cell* (Vol 93, pp 301–337) discuss many of the diverse facets of epigenetics. Important epigenetic mechanisms include:

■ **DNA methylation**. Silencing of an intact gene by methylation of adjacent control sequences is a normal part of development, differentiation and X-inactivation – but methylation can occasionally cause pathogenic loss of function. In many tumors, for example, function of the *CDKN2A* tumor suppressor gene is abrogated by methylation of the promoter rather than by mutating its DNA sequence (Chapter 18). In Fragile-X syndrome, the *FMR1* gene is silenced by methylation, although in this case the methylation is triggered by a local DNA sequence change, expansion of a trinucleotide repeat (*Box 16.8*).

■ **Changes in chromatin configuration** as a result of chromosomal rearrangements can also up-regulate or silence expression of an intact gene – for example the MYC oncogene is over-expressed when a translocation places it in the transcriptionally active immunoglobulin region (*Figure 18.7*).

■ **Imprinting** (Sections 3.2.4, 8 and 8.5.4). Imprinted genes are a particularly intriguing example of epigenetic modification. Their expression is controlled by patterns of methylation that differ according to the parental origin of the gene. When either the imprinting mechanism malfunctions or the parental origin is not as expected, pathogenic loss of function or inappropriate expression can occur in intact genes. Several human diseases involve imprinted genes, the best known being Prader–Willi and Angelman syndromes (*Box 16.6*).

■ **Changes in protein conformation**. Sometimes it appears that a conformational change can propagate through a population of protein molecules, converting them from a stable native conformation into a new form with different properties. The process may be analogous to crystallization. The behavior of prion proteins (Prusiner *et al.*, 1998) is the most striking example, but natural processes of protein aggregation into subcellular structures might also be seen in this light.

Molecular pathology of Prader–Willi and Angelman syndromes

Prader–Willi syndrome (PWS) and Angelman syndrome (AS) are both caused by problems with differentially imprinted genes at 15q11-q13. They exemplify the complicated molecular pathology associated with clusters of imprinted genes.

■ **PWS** (MIM 176260: mental retardation, hypotonia, gross obesity, male hypogenitalism) is caused by loss of function of genes that are expressed only from the paternal chromosome.

■ **AS** (MIM 105830: mental retardation, lack of speech, growth retardation, hyperactivity, inappropriate laughter) is due to loss of function of a closely linked gene that is expressed only from the maternal chromosome. Some children with AS will have two functional copies of the PWS genes, and vice versa, but this over-expression does not appear to have any phenotypic effect.

As shown in the Table, a variety of events can lead to lack of a paternal (PWS) or a maternal (AS) copy of the relevant chromosome 15 sequences.

■ *De novo* deletions of 15q11-q13 are the commonest cause. Usually the deletions are large and remove the same sequences in both PWS and AS; rare cases with small deletions show that the PWS and AS critical regions are adjacent but not overlapping.

■ **Uniparental disomy** is detected when DNA marker studies show that a person with apparently normal chromosomes has inherited both homologs of a particular pair (No. 15 in this case) from one parent. The usual cause is trisomy rescue. A trisomy 15 conceptus develops to a multicell stage; normally it would die, but if a chance mitotic nondisjunction (Section 2.6.2)

produces a cell with only two copies of chromosome 15 at a sufficiently early stage of development, that cell can go on to make the whole of a surviving baby. Most trisomy 15 conceptuses are $15^M15^M15^P$. One time out of three, random loss of one chromosome 15 will produce a fetus with maternal uniparental disomy, 15^M15^M. Lacking any paternal 15, the fetus will have Prader–Willi syndrome.

■ AS may be caused entirely by lack of expression of the *UBE3A* gene, since some inherited cases have apparently normal chromosome structure and imprinting, but have point mutations in this gene. The cause of PWS may be more complex, because no simply inherited cases have been found.

■ Occasionally something has gone wrong with the mechanism of imprinting. Both chromosome 15 homologs carry the same parent-specific methylation pattern, although marker studies show they originate from different parents. Such cases are very interesting for the light they throw on the imprinting mechanism.

Origins of Prader–Willi and Angelman syndromes

Event	Proportion of PWS	Proportion of AS
Deletions	~75%	~75%
Uniparental disomy	~20%	~3%
Point mutations	not seen	~15% (in *UBE3A* gene)
Imprinting errors	~2%	~5%

16.4.3 Haploinsufficiency describes the case where a 50% reduction in the level of gene function causes an abnormal phenotype

Loss of function mutations tend to be recessive because heterozygotes often function perfectly normally. Sometimes this is because feedback loops compensate for the reduced dosage by increasing transcription or the activity of the gene product, but in many cases the cell and organism are able to function normally with only a 50% level of gene action. Only relatively few genes show **haploinsufficiency**; *Table 16.2* lists some examples.

One might reasonably ask why there should be dosage sensitivity for *any* gene product. Why has natural selection not managed things better? If a gene is expressed so that two copies make a barely sufficient amount of product, selection for variants with higher levels of expression should lead to the evolution of a more robust organism, with no obvious price to be paid. The answer is that in most cases this has indeed happened, which is why relatively few genes are dosage-sensitive. Sometimes perhaps, if the gene product is needed in large quantities, the total synthetic capacity of the cell, even at maximum transcription levels, may be insufficient if only one copy of the gene is present. An example may be elastin. In people heterozygous for a deletion or loss of function mutation of elastin, for the most part the elastic tissues (skin, lung, blood vessels) work normally, but often the aorta, a highly elastic tissue, shows some degree of narrowing just above the heart (supravalvular aortic stenosis), which may require surgery (see Section 16.8.1).

Certain gene functions, however, are inherently dosage-sensitive (Fisher and Scambler, 1994). These include:

- gene products that are part of a quantitative signaling system whose function depends on partial or variable occupancy of a receptor, DNA-binding site, etc.;
- gene products that compete with each other to determine a developmental or metabolic switch;
- gene products that co-operate with each other in interactions with fixed stoichiometry (such as the α and β globins or many structural proteins).

In each case the gene product is titrated against something else in the cell. What matters is not the correct absolute level of product, but the correct relative levels of interacting products. Genes whose products act essentially alone, such as many soluble enzymes of metabolism, seldom show dosage effects. Pathological effects caused by gene dosage depend on interactions, and so are subject to modification by changes elsewhere in the genome. Thus these dominant conditions often show highly variable expression (see Section 16.6.3).

16.4.4 Mutations in proteins that work as dimers or multimers sometimes produce dominant negative effects

A **dominant negative** effect occurs when a mutant polypeptide not only loses its own function, but also interferes with the product of the normal allele in a heterozygote. Dominant negative mutations cause more severe effects than deletion or nonsense mutations in the same gene. Some sort of physical association of the normal and mutant products is required for a dominant negative effect. Structural proteins that contribute to multimeric structures are vulnerable to dominant negative effects. Collagens provide a classic example.

Fibrillar collagens, the major structural proteins of connective tissue, are built of triple helices of polypeptide chains, sometimes homotrimers, sometimes heterotrimers, that are assembled into close-packed crosslinked arrays to form rigid fibrils. In newly synthesized polypeptide chains (preprocollagen), N- and C-terminal propeptides flank a regular repeating sequence (Gly–X–Y)$_n$, where either X or Y is usually proline, and the other is any amino acid. Three preprocollagen chains associate and wind into a triple helix under control of the C-terminal propeptide. After formation of the triple helix, the N- and C-terminal propeptides are cleaved off (*Figure 16.4*). A polypeptide that complexes with normal chains, but then wrecks the triple helix can reduce the yield of functional collagen to well below 50%. The molecular pathology of collagen mutations is very rich, and is discussed below (see Section 16.6.1, *Table 16.6B*).

Nonstructural proteins that dimerize or oligomerize also show dominant negative effects. For example, transcription factors of the b-HLH-Zip family (see *Figure 8.8*) bind DNA as dimers. Mutants that cannot dimerize often cause recessive phenotypes, but mutants that are able to sequester functioning molecules into inactive dimers give dominant phenotypes (Hemesath *et al.*, 1994). The ion channels in cell membranes provide another example of

Table 16.2: Phenotypes probably caused by haploinsufficiency (see text for details)

Condition	MIM no.	Gene	Notes
Alagille syndrome	118450	*JAG1*	See Section 16.8.1
Multiple exostoses	133700	*EXT1*	See Section 16.8.1
Tomaculous neuropathy	162500	*PMP22*	See Section 16.6.2
Supravalvular aortic stenosis	185500	*ELN*	See this Section
Tricho-rhino-phalangeal syndrome	190350	*TRPS1*	See Section 16.8.1
Waardenburg syndrome Type 1	193500	*PAX3*	See Section 16.3.2

multimeric structures that are subject to dominant negative effects (Section 16.6.1).

16.5 Gain of function mutations

16.5.1 Acquisition of a novel function is rare in inherited disease but common in cancer

Making random changes in a gene is quite likely to stop it working, but very unlikely to give it a novel function. The only mechanism that commonly generates novel functional genes is when a chromosomal rearrangement joins functional modules of two different genes (*Table 16.3*). Such exon-shuffling was no doubt important in evolution; for molecular pathology, it is most often noticed when it leads to cancer. Many acquired tumor-specific chromosomal rearrangements produce chimeric genes with novel activities that lead to uncontrolled cell proliferation (see *Table 18.3*). A rare case of an inherited point mutation conferring a novel function on a protein is the Pittsburg allele at the *PI* locus (MIM 107400; *Figure 16.5*).

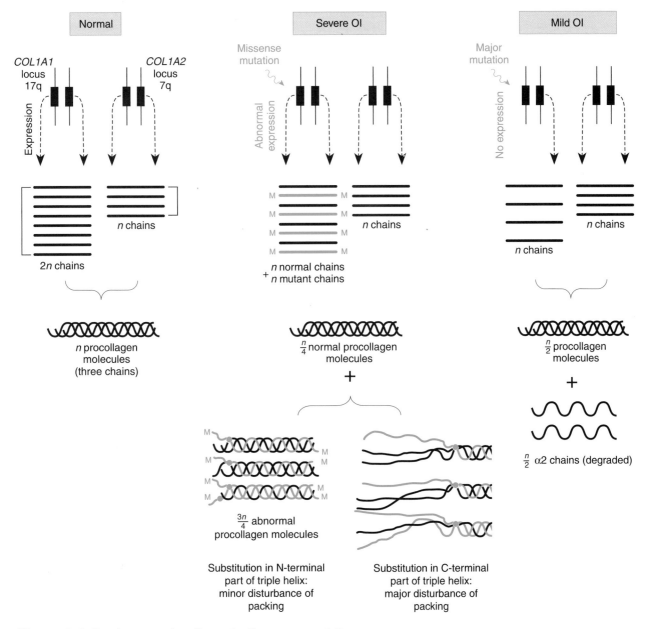

Figure 16.4: Dominant negative effects of collagen gene mutations.

Collagen fibrils are built of arrays of triple-helical procollagen units. The type I procollagen comprises two chains encoded by the *COL1A1* gene and one encoded by *COL1A2*. Null mutations in either gene have a less severe effect than mutations encoding polypeptides which cause the triple helix to be nonfunctional (see Dalgleish, 1997).

Box 16.7

Unstable expanding repeats – a novel cause of disease

Unstable expanding trinucleotide repeats were an entirely novel and unprecedented disease mechanism when first discovered in 1991, and they raise two major questions:

- What is the mechanism of the instability and expansion? This is discussed in *Section 9.5.2*
- Why do expanded repeats make you ill? Discussed here.

A hallmark of all these diseases is **anticipation** – that is, the age of onset is lower and/or the severity worse, in successive generations. Two different classes of expansion have been noted; the currently known examples are tabulated below. In some cases, intermediate-sized alleles are non-pathogenic but unstable, and readily expand to full mutation alleles (e.g. FRAXA repeats of 50–200 units); in other cases such alleles only very occasionally expand (e.g. HD alleles with 29–35 repeats). Data from OMIM and Andrews *et al.,* 1997. The list of diseases is likely to expand in the future. An expanded polyalanine tract in the *HOXD13* gene has been found in patients with synpolydactyly (MIM 186000); the normal gene has a run of 15 alanines and the pathogenic forms have 22–29 alanines. However, this does not seem to be another unstable expanding repeat. The expansion is probably the result of unequal crossing over, and at least in one family it has been stable for 7 generations (Akarsu *et al.,* 1996).

Highly expanded repeats cause loss of gene function
In Fragile-X syndrome and Friedreich ataxia, an enormously expanded repeat causes a loss of function by abolishing transcription. The same is true for the expanding 12-mer in juvenile myoclonus epilepsy. In each case, the disease is occasionally caused by different, more conventional, loss of function mutations in the gene. Such mutations produce the identical clinical phenotype to expansions, apart (presumably) from not showing anticipation. Other similar highly expanded repeats, such as *FRA16A* (expanded CCG repeat) or *FRA16B* (an expanded 33 bp minisatellite) are nonpathogenic, presumably because no important gene is located nearby.

Myotonic dystrophy may be different because no other mutation has ever been found in a myotonic dystrophy patient, so there must be something quite specific about the action of the CTG repeat. It may affect processing of the primary transcript in a specific way, or it may affect expression of a whole series of genes by altering chromatin structure in this gene-rich chromosomal region. The site of the expansion forms part of the CpG island of an adjacent gene, *DMAHP* (MIM 600963), and the expansion reduces expression of this gene. *SCA8* (Koob *et al.,* 1999) may have a similar mechanism.

The CAG repeats encode polyglutamine tracts within the gene product that cause it to aggregate within certain cells and kill them
Common features of the eight diseases caused by expansion of an unstable CAG repeat within a gene include:

- They are all late-onset neurodegenerative diseases, and except for Kennedy disease, are all dominantly inherited.
- No other mutation in the gene has been found that causes the disease.

Disease	MIM no.	Mode of inheritance	Location of gene	Location of repeat	Repeat sequence	Stable repeat no.	Unstable repeat no.
1. Very large expansions of repeats outside coding sequences							
Fragile-X site A (FRAXA)	309550	X	Xq27.3	5′UT	$(CGG)_n$	6–54	200–>1000
Fragile-X site E (FRAXE)	309548	X	Xq28	Promoter	$(CCG)_n$	6–25	>200
Friedreich ataxia (FA)	229300	AR	9q13–q21.1	Intron 1	$(GAA)_n$	7–22	200–1700
Myotonic dystrophy (DM)	160900	AD	19q13	3′UT	$(CTG)_n$	5–35	50–4000
Spinocerebellar ataxia 8	–	AD	13q21	Untranslated RNA	$(CTG)_n$	16–37	110–>500
Juvenile myoclonus epilepsy (JME)	254800	AR	21q22.3	Promoter	$(CCCCGC CCCGCG)_n$	2–3	40–80
2. Modest expansions of CAG repeats within coding sequences							
Huntington disease (HD)	143100	AD	4p16.3	Coding	$(CAG)_n$	6–35	36–>100
Kennedy disease (SBMA)	313200	XR	Xq21	Coding	$(CAG)_n$	9–35	38–62
Spinocerebellar ataxia 1 (SCA1)	164400	AD	6p23	Coding	$(CAG)_n$	6–38	39–83
Spinocerebellar ataxia 2 (SCA2)	183090	AD	12q24	Coding	$(CAG)_n$	14–31	32–77
Machado–Joseph disease (SCA3, MJD)	109150	AD	14q32.1	Coding	$(CAG)_n$	12–39	62–86
Spinocerebellar ataxia 6 (SCA6)	183086	AD	19p13	Coding	$(CAG)_n$	4–17	21–30
Spinocerebellar ataxia 7 (SCA7)	164500	AD	3p12–p21.1	Coding	$(CAG)_n$	7–35	37–200
Dentatorubral-pallidoluysian atrophy (DRPLA)	125370	AD	12p	Coding	$(CAG)_n$	3–35	49–88

Box 16.7 *(continued)*

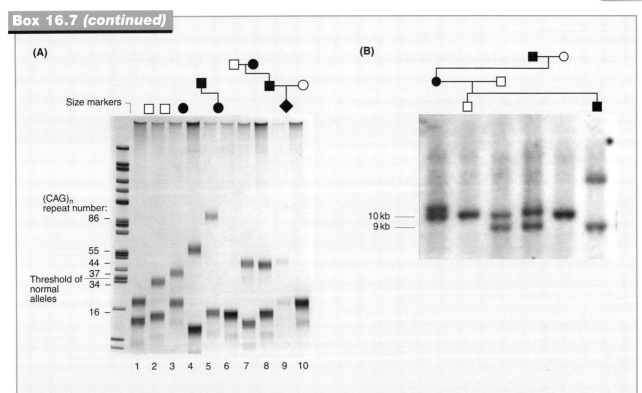

Laboratory diagnosis of trinucleotide repeat diseases.

(**A**) Huntington disease. A fragment of the gene containing the $(CAG)_n$ repeat has been amplified by PCR and run out on a polyacrylamide gel. Bands are revealed by silver staining. The scale shows numbers of repeats. Lanes 1, 2, 6 and 10 are from unaffected people, lanes 3, 4, 5, 7 and 8 are from affected people. Lane 5 is a juvenile onset case; her father (lane 4) had 45 repeats but she has 86. Lane 9 is an affected fetus, diagnosed prenatally. Courtesy of Dr Alan Dodge, St Mary's Hospital, Manchester. (**B**) Myotonic dystrophy. Southern blot of DNA digested with *Eco*RI. Bands of 9 or 10 kb (arrows) are normal variants. The grandfather has cataracts but no other sign of myotonic dystrophy. His 10 kb band appears to be very slightly expanded, but this is not unambiguous on the evidence of this gel alone. His daughter has one normal and one definitely expanded 10 kb band; she has classical adult onset myotonic dystrophy. Her son has a massive expansion and the severe congenital form of the disease. Courtesy of Dr Simon Ramsden, St Mary's Hospital, Manchester.

- ■ The expanded allele is transcribed and translated.
- ■ The trinucleotide repeat encodes a polyglutamine tract in the protein.
- ■ There is a critical threshold repeat size, below which the repeat is nonpathogenic and above which it causes disease.
- ■ The larger the repeat, above the threshold, the earlier is the age of onset (on average; predictions cannot be made for individual patients, but there is a clear statistical correlation).

The androgen receptor mutation in Kennedy disease provides clear evidence that CAG-repeat diseases involve a specific gain of function. Loss of function mutations in this gene are well known and cause androgen insensitivity or testicular feminization syndrome (MIM 300068), a failure of male sexual differentiation. The polyglutamine expansion, by contrast, causes a quite different neurodegenerative disease, although patients often also show minor feminization. The other CAG-repeat dis-

ease genes so far identified are widely expressed and encode proteins of unknown function. When the polyglutamine tract exceeds the threshold length the protein aggregates, forming an inclusion body that apparently kills the cell (Kim and Tanzi, 1998). The different clinical features of each disease reflect killing of different cells, presumably because of interactions with other cell-specific proteins. Neuronal cell death caused by protein aggregates is a common thread in the pathology of CAG repeat diseases, Alzheimer disease, Parkinson disease and the prion diseases; the mechanisms and their general significance remain to be discovered.

Laboratory diagnosis of expanded repeats
A single PCR reaction makes the diagnosis in the polyglutamine repeat diseases. Panel (A) in the figure shows an example from Huntington disease. The very large expansions in myotonic dystrophy (B) require Southern blotting.

Box 16.8

Laboratory diagnosis of fragile X

■ **Cytogenetic testing** – the fragile site is seen only when cells are grown under conditions of folate or thymidine depletion. For unknown reasons, it is never present in more than 50% of cells, and the frequency is often much lower, especially in carrier females. Diagnosing a carrier by cytogenetic testing requires analysis of a large number of cells.

■ **Molecular testing** – PCR will amplify the normal and premutation repeats, but Southern blotting is necessary to detect full expansions – the long G–C tract is refractory to PCR amplification. Careful inspection of the autoradiograph is needed to ensure that the diffuse smears from large mitotically unstable repeats are not missed.

The DNA of the inactivated X in a female, and the DNA of any X carrying a full mutation, is methylated. A combination of *Eco*RI and the methylation-sensitive restriction enzyme *Ecl*XI is used and the blot is hybridized to Ox1.9 or a similar probe (*Figure*). The X in a normal male (track 1) and the active normal X in a female (tracks 2, 3, 4, 6) give a small fragment (N). Unmethylated premutation alleles (P) give a slightly larger band – tracks 4 and 5 are female premutation carriers, track 7 a normal transmitting male. Methylated (inactive) X sequences do nut cut with *Ecl*XI, and give larger bands. Abbreviations: NM, methylated normal or premutation sequences; F, methylated full mutation sequences. (Courtesy of Dr Simon Ramsden, St Mary's Hospital, Manchester.)

16.5.2 Over-expression may be pathogenic

Gross over-expression of certain genes is common in cancer cells. The mechanisms by which somatic genetic changes produce over-expression include massive re-duplication of the gene or transposition of a gene normally expressed at low level into a highly active chromatin environment. These are discussed more fully in *Chapter 18*.

Inherited diseases are not often caused by constitutional over-expression of a single gene. Duplication of the *DSS* gene on Xp21.3 causes male to female sex reversal, probably as a direct result of the doubled dosage (Bardoni *et al.*, 1994). For the *PMP22* peripheral myelin protein gene, an increase in gene dosage from two to three copies is enough to produce Charcot–Marie–Tooth disease (see *Figure 16.7*). Such modest increases in gene expression are probably seldom pathogenic, although a similar degree of dosage sensitivity of unidentified genes must explain many features of chromosomal trisomies (Section 16.8.2). Overactivity of an abnormal gene product, with normal transcription and translation of the gene, can produce similar effects.

16.5.3 Qualitative changes in a gene product can cause gain of function

Although gains of truly novel functions are very rare in inherited disease, activating mutations that modify cellular signaling responses quite often produce dominant phenotypes. The G-protein coupled hormone receptors provide good examples. Many hormones exert their effects on target cells by binding to the extracellular domains of transmembrane receptors. Binding of ligand causes the cytoplasmic tail of the receptor to catalyse conversion of an inactive (GDP-bound) G-protein into an active (GTP-bound) form, and this relays the signal further by stimulating adenylyl cyclase. Some mutations cause receptors to activate adenylyl cyclase even in the absence of ligand.

■ Familial male precocious puberty (MIM 176410:

Table 16.3: Mechanisms of gain of function mutations

Malfunction	Gene	Disease	MIM no.
Overexpression	*PMP22*	Charcot–Marie–Tooth disease	118200
Receptor permanently 'on'	*GNAS1*	McCune–Albright disease	174800
Acquire new substrate	*PI* (Pittsburgh allele)	α_1-Antitrypsin deficiency	107400
Ion channel inappropriately open	*SCN4A*	Paramyotonia congenita	168300
Structurally abnormal multimers	*COL2A1*	Osteogenesis imperfecta	Various
Protein aggregation	*HD*	Huntington disease	143100
Chimeric gene	*BCR-ABL*	Chronic myeloid leukemia	151410

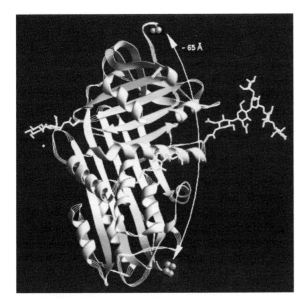

Figure 16.5: An inherited mutation causing a protein to gain a novel function.

The α_1-antitrypsin molecule inhibits elastase. Methionine 358 in the reactive center acts as a 'bait' for elastase; when the peptide link between Met358 and Ser359 is cleaved, the two residues spring 65 Å apart, as shown here. Elastase is trapped and inactivated. The Pittsburgh variant has a missense mutation M358R which replaces the methionine bait with arginine. This destroys affinity for elastase but creates a bait for thrombin. As a novel constitutively active antithrombin, the Pittsburgh variant produces a lethal bleeding disorder (Owen *et al.*, 1983). Image from University of Geneva ExPASy molecular biology worldwide web server.

onset of puberty by age 4 in affected boys) is found with a constitutively active luteinizing hormone receptor.

■ Autosomal dominant thyroid hyperplasia can be caused by an activating mutation in the thyroid stimulating hormone receptor (see MIM 275200).

■ Jansen's metaphyseal chondrodystrophy (MIM 156400: a disorder of bone growth) can be caused by a constitutively active parathyroid hormone receptor.

■ A constitutionally active $G_s\alpha$ protein (part of the G-protein) causes McCune–Albright syndrome or polyostotic fibrous dysplasia (PFD, MIM 174800). PFD is known only as a somatic condition in mosaics – probably constitutional mutations would be lethal. Depending on the tissues carrying the mutant cell line, the result is polyostotic fibrous dysplasia, café-au-lait spots, sexual precocity and other hyperfunctional endocrinopathies. Loss of function mutations of the same gene often underlie a different disease, Albright's hereditary osteodystrophy (see *Table 16.5*).

16.6 Molecular pathology: from gene to disease

The starting point in thinking about molecular pathology may be either a gene or a disease. These two approaches are considered separately in this section and the next, although of course a full understanding of molecular pathology would merge the two.

16.6.1 For loss of function mutations the phenotypic effect depends on the residual level of gene product

The DNA sequence changes described in *Table 16.1* can cause varying degrees of loss of function. Many amino acid substitutions have little or no effect, while some mutations will totally abolish the function. A mutation may be present in one or both copies of a gene. When both homologues are affected, they may be affected unequally – people with autosomal recessive conditions are often compound heterozygotes, with two different mutations. If both mutations cause loss of function, but to differing degrees, the least severe allele will dictate the level of residual function.

Figure 16.6 shows four possible relations between the level of residual gene function and the clinical phenotype.

(i) A simple recessive condition. People heterozygous for a mutation that totally abolishes gene function are phenotypically normal, provided their remaining allele is not significantly defective.

(ii) A dominant condition caused by haploinsufficiency. In reality, this simple situation is rare. If a 50%

reduction in gene product causes symptoms, a more severe reduction will probably have more severe effects.

(iii) A recessive condition with graded severity. Among many examples are:

■ mutants in the X-linked hypoxanthine guanine phosphoribosyl transferase (HPRT) gene. The extent of residual enzyme activity in mutants correlates well with the clinical phenotype of affected males (*Table 16.4*).

■ Reduced copy numbers of α-globin genes produce successively more severe effects. As shown in *Figure 16.2*, most people have four copies of the α-globin gene (αα/αα). People with three copies (αα/α-) are healthy; those with two (whether the phase is α-/α- or αα/--) suffer mild α-thalassemia; those with only one gene (α-/--) have severe disease, while lack of all four α genes (--/--) causes lethal hydrops fetalis.

(iv) Closely related to the situation in (c), decreasing residual function of a gene may extend the phenotype, perhaps causing a condition with a different clinical label. Depending on the position of the thresholds, several different situations can arise:

■ Several related recessive conditions may be caused by successive reductions in gene function at a single locus. For example, extracellular matrix is rich in sulfated proteoglycans like heparan sulfate and chondroitin sulfate, and defects in sulfate transport interfere with skeletal development. Loss of function mutations in the *DTDST* sulfate transporter cause three related autosomal recessive skeletal dysplasias,

diastrophic dysplasia (MIM 226600), atelosteogenesis II (MIM 256050) and achrondrogenesis Type 1B (MIM 600972), depending on the extent of loss (Hastbacka *et al.*, 1996).

■ Moderate reduction in function, caused by either haploinsufficiency or dominant negative effects, may produce a dominant condition, while very severe reduction in homozygotes may produce a recessive condition. The dominantly inherited Romano–Ward syndrome (MIM 192500: cardiac arrhythmia) is caused by dominant negative mutations in the *KVLQT1* K$^+$ channel; in transfected *Xenopus* oocytes the heterozygote ion channels have about 20% of normal activity. A simple loss of function mutation in the same gene has no clinical effect in heterozygotes (with 50% function) but causes the recessive Jervell and Lange–Nielsen syndrome (MIM 220400: heart

Table 16.4: Consequences of decreasing function of hypoxanthine guanine phosphoribosyl transferase

HPRT activity (% of normal)	Phenotype
>60	Normal
8–60	Neurologically normal; hyperuricemia (gout)
1.6–8	Neurological problem (choreoathetosis)
1.4–1.6	Lesch–Nyhan syndrome (choreoathetosis, self-mutilation) but intelligence normal
<1.4	Classical Lesch–Nyhan syndrome (MIM 308000; choreoathetosis, self-mutilation and mental retardation)

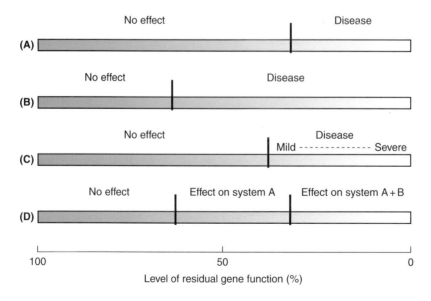

Figure 16.6: Four possible relationships between loss of function and clinical phenotype.

See text for discussion.

problems and hearing loss) in homozygotes. In the *Xenopus* assay, ion channels from JLN patients totally lack function (Wollnik *et al.*, 1997).

■ Mutations in the same gene can produce two or more dominant conditions, the milder one by simple haploinsufficiency, and more severe forms through dominant negative effects. This happens in the *COL1A1* or *COL1A2* genes that encode type I collagen (*Figure 16.4*). Mutations in these genes usually produce osteogenesis imperfecta (OI; brittle bone disease). Frameshifts and nonsense mutations produce type 1 OI, the mildest form, while amino acid substitutions in the Gly–X–Y repeated units are seen in the more severe types II, III and IV OI. The genotype–phenotype relationship is quite subtle. Substitution of glycine by a bulkier amino acid in the Gly–X–Y unit has a dominant negative effect by disrupting the close packing of the collagen triple helix. The helix is assembled starting at the C-terminal end, and substitution of glycines close to that end has a more severe effect than substitutions nearer the N-terminal end. Skipping of exon 6 (of *COL1A1* or *COL1A2*) has a quite different effect. The site for cleavage of the N-terminal propeptide is lost and abnormal collagen is produced that causes Ehlers Danlos syndrome Type VII (MIM 130060; laxity of skin and joints). A different function has been lost, and a different phenotype results.

16.6.2 Loss of function and gain of function mutations in the same gene will cause different diseases

We have seen that loss of function mutations in the *PAX3* gene cause the developmental abnormality Type 1 Waardenburg syndrome (*Figure 16.1*). A totally different phenotype is seen when an acquired chromosomal translocation creates a novel chimeric gene by fusing *PAX3* to another transcription factor gene, *FKHR* in a somatic cell. The gain of function of this hybrid transcription factor causes the development of the childhood tumor, alveolar rhabdomyosarcoma (see *Table 18.3*).

A striking example concerns the *RET* gene. *RET* encodes a receptor that straddles the cell membrane. When its ligand (GDNF) binds to the extracellular domain it induces dimerization of the receptors, which then transmit the signal into the cell via tyrosine kinase modules in their cytoplasmic domain. A variety of loss of function mutations – frameshifts, nonsense mutations and amino acid substitutions that interfere with the post-translational maturation of the RET protein – are one cause of Hirschsprung's disease (MIM 142623; intractable constipation caused by absence of enteric ganglia in the bowel – see Section 19.5.2). Certain very specific missense mutations in the *RET* gene are seen in a totally different set of diseases, familial medullary thyroid carcinoma and the related but more extensive multiple endocrine neo-

plasia type 2. These are gain of function mutations, producing receptor that reacts excessively to ligand or is constitutively active and dimerizes even in the absence of ligand. Curiously, some people with missense mutations affecting cysteines 618 or 620 suffer from both thyroid cancer and Hirschsprung disease – simultaneous loss and gain of function. This reminds us that loss of function and gain of function are not always simple scalar quantities; mutations may have different effects in the different cell types in which a gene is expressed.

Table 16.5 lists a number of cases where mutations in a single gene can result in more than one disease. Usually the gain of function mutant produces a qualitatively abnormal protein. Occasionally a simple dosage effect can be pathogenic. The peripheral myelin protein gene *PMP22* is an example. Unequal crossovers between repeat sequences on chromosome 17p11 create duplications or deletions of a 1.5 Mb region that contains the *PMP22* gene (*Figure 16.7*). Heterozygous carriers of the deletion or duplication have one copy or three copies, respectively, of this gene. People who have only a single copy suffer from hereditary neuropathy with pressure palsies or tomaculous neuropathy (MIM 162500), while as mentioned above, people with three copies have a clinically different neuropathy, Charcot–Marie–Tooth disease 1A (CMT1A; MIM 118220).

16.6.3 Variability within families is evidence of modifier genes or chance effects

Many mendelian conditions are clinically variable even between affected members of the same family who carry exactly the same mutation. Intrafamilial variability must be caused by some combination of the effects of other unlinked genes (modifier genes) and environmental effects (including chance events). Phenotypes depending on haploinsufficiency are especially sensitive to the effects of modifiers, as discussed above (Section 16.4.3). Waardenburg syndrome is a typical example: *Figure 16.1* shows the evidence that this dominant condition is caused by haploinsufficiency, and *Figure 3.5C* shows typical intrafamilial variation.

An example of how a modifier might work comes from an interesting family with apparent digenic inheritance of ocular albinism (Morell *et al.*, 1997). Tyrosinase is a key enzyme of melanocytes; deficiency leads to oculocutaneous albinism (MIM 203100). A common variant of the tyrosinase gene, R402Q, encodes an enzyme with reduced activity, but the residual activity is sufficiently high for even homozygotes to be phenotypically normal. However, in the family reported by Morell *et al.*, people carrying one or two copies of R402Q showed ocular albinism (a mild form of oculocutaneous albinism) when they also carried a mutation in *MITF*, a gene involved in differentiation of melanocytes. Mutations in *MITF* alone do not cause ocular albinism.

Intrafamilial variability is a big problem in genetic counseling because families contemplating childbearing want to know how severely affected a child would be.

Table 16.5: Examples of genes responsible for more than one disease

Gene	Location	Diseases	Symbol	MIM no.
PAX3	2q35	Waardenburg syndrome type 1	WS1	193500
		Alveolar rhabdomyosarcoma	RMS2	268220
CFTR	7p31.2	Cystic fibrosis	CF	219700
		Bilateral absence of vas deferens		
RET	10q11.2	Multiple endocrine neoplasia type 2A	MEN2A	171400
		Multiple endocrine neoplasia type 2B	MEN2B	162300
		Medullary thyroid carcinoma	FMTC	155420
		Hirschsprung disease	HSCR	142623
PMP22	17p11.2	Charcot-Marie-Tooth neuropathy type 1A	CMT1A	118220
		Tomaculous neuropathy	HNPP	162500
SCN4A	17q23.1-q25.3	Paramyotonia congenita	PMC	168300
		Hyperkalemic periodic paralysis	HYPP	170500
		Acetazolamide-responsive myotonia congenita		
PRNP	20p12-pter	Creutzfeldt-Jakob disease	CJD	123400
		Familial fatal insomnia	FFI	176640
GNAS1	20q13.2	Albright hereditary osteodystrophy	AHO	103580
		McCune-Albright syndrome	PFD	174800
AR	Xcen-q22	Testicular feminization syndrome	TFM	313700
		Kennedy disease	SBMA	313200

Thus there is a clinical as well as a scientific motivation to identify modifier genes. The work of Easton *et al.* (1993) on neurofibromatosis type 1 shows how statistical analysis of clinical phenotypes within large families can provide evidence for modifier genes. These might then be sought by a whole genome search, comparing clinically concordant and discordant relatives – but very large samples would probably be needed for success. The role of

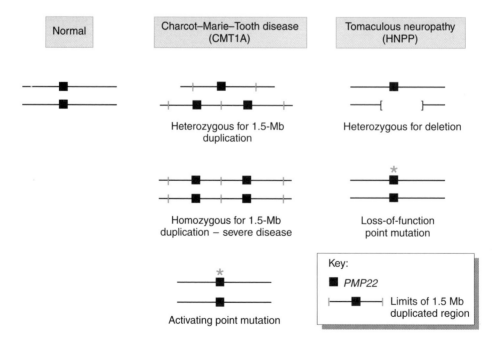

Figure 16.7: Gene dosage effects with the *PMP22* gene.

Most patients with Charcot–Marie–Tooth disease are heterozygous for a 1.5-Mb duplication at 17p11.2, including the gene for peripheral myelin protein, *PMP22* (black square). A patient homozygous for the duplication had very severe disease. Some patients have only two copies of the *PMP22* gene, but one copy carries an activating mutation. Deletion or loss-of-function mutation of the *PMP22* gene is seen in patients with tomaculous neuropathy (Patel and Lupski, 1994).

pure chance should also not be ignored, especially in conditions with a patchy phenotype. Examples include patchy depigmentation in Waardenburg syndrome, and the variable numbers of neurofibromata or polyps in neurofibromatosis type 1 (MIM 162200) and adenomatous polyposis coli (MIM 175100), respectively.

Candidate modifier genes may be suggested by knowledge of the biochemical interactions of the primary gene product, or from studies in mice, where the necessary genetic analysis is feasible. An example of the latter is the identification of phospholipase A as a major modifier of the number of tumors seen in a mouse model of polyposis coli. The human counterpart (*PLA2G2A*, MIM 172411) has been investigated as a modifier of the severity of adenomatous polyposis coli. Results to date are promising but not conclusive (lod score of 2). Whatever the final verdict, this work provides a model of how clinically important modifiers might be identified.

16.6.4 In mitochondrial diseases heteroplasmy and instability complicate the relationship between genotype and phenotype

Mitochondrial diseases (Section 3.1.5) can be caused by point mutations, deletions or duplications that abolish the function of genes in the densely packed mitochondrial genome (*Figure 7.2*). Cells typically contain thousands of mtDNA molecules. A major complication is that cells can be homoplasmic (every mtDNA molecule carries the causative mutation) or heteroplasmic (cells have a mixed population of normal and mutant mitochondrial DNA). Additionally, both the mutations and the heteroplasmy often seem to evolve with time within an individual. The same individual can carry both deletions and duplications, and the proportion can change with time (Poulton *et al.*, 1993).

Phenotype–genotype correlations are particularly hard to establish with mitochondrial diseases. The same sequence change is frequently seen in people with different syndromes, and attempts to explain severity of symptoms by differing degrees of heteroplasmy have not been convincing – for example, 60–70% of people with Leber's hereditary optic atrophy (MIM 535000: sudden irreversible loss of vision) have a substitution at nt 11778 of the mitochondrial genome, but some patients appear to be homoplasmic while others, no less severely affected, are heteroplasmic. Possible reasons for the difficulties include:

- heteroplasmy can be tissue-specific, and the tissue that is examined (typically blood or muscle) may not be the critical tissue in the pathogenesis;

- mtDNA is much more variable than nuclear DNA, and some syndromes may depend on the combination of the reported mutation with other unidentified variants;

- some mitochondrial diseases seem to be of a quantitative nature: small mutational changes accumulate

that reduce the energy-generating capacity of the mitochondrion, and at some threshold deficit clinical symptoms appear;

- many mitochondrial functions are encoded by nuclear genes (see *Box 7.1*), so that nuclear variation can be an important cause or modifier of mitochondrial phenotype.

The MITOMAP database of mitochondrial mutations (http://infinity.gen.emory.edu/mitomap.html) summarizes all the information on phenotypes and genotypes, and shows just how great is the challenge of relating them.

16.7 Molecular pathology: from disease to gene

Very often the starting point for thinking about molecular pathology is a disease rather than a gene. This approach gives an alternative viewpoint of genotype–phenotype correlations. The overall message is that one must not be naïve when speculating about the gene defect underlying a clinical syndrome.

16.7.1 The gene underlying a disease may not be the obvious one

Mutations leading to deficiency of a protein are not necessarily in the structural gene encoding the protein

Agammaglobulinemia (lack of immunoglobulins, leading to clinical immunodeficiency) is often mendelian. It is natural to assume the cause would be mutations in the immunoglobulin genes. But the immunoglobulin genes are located on chromosomes 2, 14 and 22, and agammaglobulinemias do not map to these locations. Many forms are X-linked. Remembering the many steps needed to turn a newly synthesized polypeptide into a correctly functioning protein (Section 1.5), this lack of one-to-one correspondence between the mutation and the protein structural gene should not come as any great surprise. Failures in immunoglobulin gene processing, in B-cell maturation, or in the overall development of the immune system will all produce immunodeficiency.

One gene defect can sometimes produce multiple enzyme defects

I-cell disease or mucolipidosis II (MIM 252500) is marked by deficiencies of multiple lysosomal enzymes. The primary defect is not in the structural gene for any of these enzymes, but in an N-acetylglucosamine-1-phosphotransferase that phosphorylates mannose residues on the glycosylated enzyme molecules. The phosphomannose is a signal that targets the enzymes to lysosomes; in its absence the lysosomes lack a whole series of enzymes.

Mutations often affect only a subset of the tissues in which the gene is expressed

The pattern of tissue-specific expression of a gene is a

poor predictor of the clinical effects of mutations. Tissues where a gene is not expressed are unlikely to suffer primary pathology, but the converse is not true. Usually only a subset of expressing tissues are affected. The *HD* gene is widely expressed, but Huntington disease affects only limited regions of the brain. The retinoblastoma (*RB1*) gene (Section 18.6.1) is ubiquitously expressed, but only the retina is commonly affected by inherited mutations. This is also strikingly seen in the lysosomal disorders. Gene expression is required in a single cell type, the macrophage, which is found in many tissues. But not all macrophage-containing tissues are abnormal in affected patients. Explanations are not hard to find:

■ Genes are not necessarily expressed only in the tissues where they are needed. Provided expression does no harm, there may be little selective pressure to switch off expression, even in tissues where expression confers no benefit.

■ Loss of a gene function will affect some tissues much more than others, because of the varying roles and metabolic requirements of different cell types and varying degrees of functional redundancy in the meshwork of interactions within a cell. The 'gatekeeper gene' concept from cancer genetics (Section 18.8.1) is likely to be applicable to many other cell functions and malfunctions, in addition to the cell turnover that goes wrong in cancer.

■ Any gain of function may be pathological for some cell types and harmless for others – see the example of the *RET* gene (Section 16.6.2).

16.7.2 Locus heterogeneity is the rule rather than the exception

Locus heterogeneity describes the situation where the same disease can be caused by mutations in several different genes. It is important to think about the biological role of a gene product, and the molecules with which it interacts, rather than expecting a one-to-one relationship between genes and syndromes. As we saw in Section 3.1.4, clinical syndromes often result from failure or malfunction of a developmental or physiological pathway; equally, many cellular structures and functions depend on multicomponent protein aggregates. If the correct functioning of several genes is required, then mutations in any of the genes may cause the same, or a very similar, phenotype.

Once again, the collagens (*Figure 16.4*; Section 16.6.1) provide good examples. We have seen that type I collagen, the major collagen of skin, bone, tendon and ligaments, is built of triple helices comprising two α(1) chains and one α(2). Mutations in either the *COL1A1* or *COL1A2* genes cause the same condition, dominant osteogenesis imperfecta. Type II collagen forms fibrils in cartilage and other tissues including the vitreous of the eye. It is made of homotrimeric helices of *COL2A1* chains. Different mutations in the *COL2A1* gene result in an overlapping spectrum of skeletal dysplasias including Stickler syn-

drome, spondyloepiphyseal dysplasia and Kniest dysplasia. A similar phenotype can result from mutations in the type XI collagen, which is a minor component of the type II fibril. In all these cases, which syndrome is produced depends on the overall effect on the final collagen fibrils, and not on which gene is mutated.

16.7.3 Mutations in different members of a gene family can underlie a series of related or overlapping syndromes

Mutations in the genes encoding fibroblast growth factor receptors exemplify the way phenotypes can depend on a network of interactions rather than a single linear pathway. The ten fibroblast growth factors govern important developmental processes through four cell surface receptors, FGFR1–4. Most tissues express multiple FGFRs, including splice variants of each. The FGF receptors are receptor tyrosine kinases that act in a similar manner to the RET protein described above: signal transduction requires receptor dimerization, and this can involve homodimers or heterodimers. FGFR mutants could produce an altered balance of splice forms, change the balance of homo- and heterodimers, reduce signaling by a dominant negative effect or produce constitutionally active dimers. Thus there is the potential for complex genetic effects.

Very specific mutations of the receptor genes are responsible for a series of dominant disorders of skeletal growth (*Figure 16.8*). Mutations in *FGFR2* on 10q25 are found in Crouzon, Jackson–Weiss, Pfeiffer and Apert syndromes, while different specific mutations in *FGFR3* at 4p16 produce achondroplasia, thanatophoric dysplasia types 1 and 2, hypochondroplasia, Crouzon syndrome with acanthosis nigricans and Muencke's coronal craniosynostosis. Some patients with Pfeiffer syndrome have a mutation in *FGFR1*. For clinical descriptions, references and an introduction to the molecular pathology of these syndromes, see OMIM and Wilkie (1997). The very specific nature of the mutations suggests a gain of function, and the achondroplasia, thanatophoric dysplasia and Crouzon mutants have been shown to produce receptors with varying degrees of constitutive (ligand independent) activation when transfected into certain types of cells (Naski *et al.*, 1996).

16.7.4 Clinical and molecular classifications are alternative tools for thinking about diseases, and each is valid in its own sphere

The connective tissue disorders caused by collagen gene mutations, which have been a recurring theme in this chapter, illustrate the difference between clinical and molecular classifications of diseases.

■ All mendelian diseases can be classified on a molecular basis (*Table 16.6B*), first by the locus involved and second by the particular mutant allele at that locus.

■ Genetic diseases can also be classified clinically

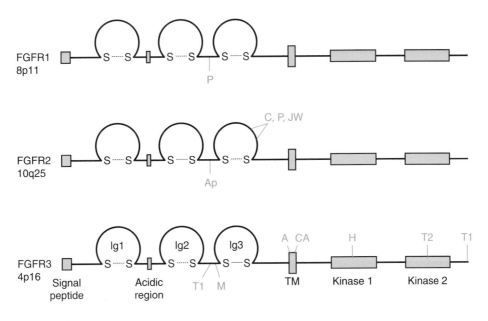

Figure 16.8: FGFR mutations.

Three of the four highly homologous fibroblast growth factor receptors are shown. Each receptor tyrosine kinase has three immunoglobulin-like extracellular domains (held by S–S bridges), a transmembrane domain, and paired intracellular tyrosine kinase domains. Very specific missense mutations are associated with a series of skeletal dysplasias (achondroplasia A, hypochondroplasia H, thanatophoric dysplasia Types 1 and 2, T1 and T2) and craniosynostosis syndromes (Apert Ap, Crouzon C, Jackson–Weiss JW, Muencke M, Pfeiffer P). Other mutations can cause the Beare–Stevenson cutis gyrata skin disease (CA). Some mutations in the Ig3 domain of FGFR2 are associated in different families with Crouzon, Jackson–Weiss and Pfeiffer syndromes.

according to the symptoms and the prognosis, as well as by the pathogenesis (*Table 16.6A*). Clinical categories defined in this way may not correspond exactly to a molecular classification, but they may be more useful for suggesting the prognosis and how the patient should be managed.

Clinical labels are not simply conventions. They evolve as knowledge of the underlying genetics advances – diseases are lumped together (Duchenne and Becker muscular dystrophy) or split (*BRCA1* breast cancer from sporadic breast cancer). A molecular classification is essential for molecular diagnosis, and it may allow more accurate counseling – for example, molecular analysis shows that unaffected parents who have more than one affected child with osteogenesis imperfecta are not carriers of a recessive form of OI, but germinal mosaics (see *Figure 3.8*). However, a full-blown molecular classification is not always clinically useful – for example, although OMIM lists nine loci causing Usher syndrome (recessive deafness and blindness), clinically it is only useful to distinguish three types, which vary in their severity. Thus a molecular classification illuminates rather than supersedes the clinical classification.

16.8 Molecular pathology of chromosomal disorders

16.8.1 Contiguous gene and microdeletion syndromes bridge the gap between single gene and chromosomal syndromes

If our 3000 Mb genome contains 50 000–100 000 genes, a deletion of a megabase or so, which is too small to be seen under the microscope, may still involve dozens of genes. An increasing number of well characterized clinical syndromes are proving to be caused by such microdeletions, or occasionally microduplications (*Table 16.7*). Once the cause is recognized, further cases can easily be diagnosed by fluorescence *in situ* hybridization (Section 10.1.4) using a probe from the deleted region.

X-chromosome contiguous gene syndromes
In males, X-chromosome microdeletions produce well-defined **contiguous gene syndromes** that show superimposed features of several different X-linked mendelian diseases. A classic case was the boy 'BB' who suffered from Duchenne muscular dystrophy (MIM 310200),

Table 16.6A: Clinical classification of the connective tissue diseases. OI, osteogenesis imperfecta (brittle bone disease); SED, spondylo-epiphyseal dysplasia; EDS, Ehlers–Danlos syndrome

Disease	MIM no.	Features
OI type I	166200	Mild–moderate bone fragility; blue sclerae; normal stature;
(divided on dental findings into IA, IB, IC)	(166240)	hearing loss (50%)
OI type II	166210	Very severe bone fragility; perinatal lethal
OI type III	(166230)	Moderate–severe bone fragility; progressive deformity; very short stature; often hearing loss
OI type IV (divided on dental findings into IVA, IVB)	166220	Mild–moderate bone fragility; normal sclerae; variable stature
SED	183900	Short stature, short neck, spondylo-epiphysal dysplasia
Stickler syndrome	108300	Mild SED, cleft palate, high myopia, hearing loss
Kniest dysplasia	156550	Disproportionate short stature; short neck; SED, etc.
EDS type VII	130060	Lax joints and skin

Table 16.6B: Molecular classification of the connective tissue diseases

Gene	Location	Mutations	Syndrome
COL1A1	17q22	Null alleles	OI type I
		Partial deletions; C-terminal substitutions	OI type II
		N-terminal substitutions	OI types I, III or IV
		Deletion of exon 6	EDS type VII
COL1A2	7q22.1	Splice mutations; exon deletions	OI type I
		C-terminal mutations	OI type II, IV
		N-terminal substitutions	OI type III
		Deletion of exon 6	EDS type VII
COL2A1	12q13	Point mutations	SED
		Nonsense mutation	Stickler syndrome
		Defect in conversion	Kniest dysplasia
		Missense	Achondrogenesis II, spondylo-meta-epiphyseal dysplasia
COL11A2	6p21.3	Splicing mutation	Stickler syndrome

chronic granulomatous disease (MIM 306400) and retinitis pigmentosa (MIM 312600), together with mental retardation (Francke *et al.*, 1985). He had a chromosomal deletion in Xp21 that removed a contiguous set of genes and incidentally provided investigators with the means to clone the genes whose absence caused two of his diseases, DMD and chronic granulomatous disease. Deletions of the tip of Xp are seen in another set of contiguous gene syndromes. Successively larger deletions remove more genes and add more diseases to the syndrome (Ballabio and Andria, 1992). Microdeletions are relatively frequent in some parts of the X chromosome (e.g. Xp21, proximal Xq) but rare or unknown in others (e.g. Xp22.1–22.2, Xq28). No doubt deletion of certain individual genes, and visible deletions in gene-rich regions, would be lethal.

Autosomal microdeletion syndromes

Not all syndromes that can be associated with microdeletions are true microdeletion syndromes (reviewed by Budarf and Emanuel, 1997). For example Alagille syndrome (MIM 118450) is seen in patients with a microdeletion at 20p11, but 93% of Alagille patients have no deletion. The cause of the syndrome in all cases is haplo-insufficiency for a single gene, *JAG1* located at 20p11, due to deletions or point mutations. True microdeletion syndromes (sometimes called segmental aneusomy syndromes) are caused by haploinsufficiency of several genes. Langer–Giedion syndrome (LGS, MIM 150230) is an example of an autosomal contiguous gene syndrome. LGS is caused by deletion of two adjacent dosage-sensitive genes in 8q24, and possibly a third gene causing mental retardation. Mutation of the *TRPS1* gene alone produces the typical face, bulbous nose and sparse hair of LGS; mutation of the adjacent *EXT1* gene produces multiple exostoses. LGS combines these features with mental retardation. However, autosomal microdeletions do not in general produce clear-cut contiguous gene syndromes representing the cumulative effect of all the deleted genes. Homozygous deletions are usually lethal, and in a heterozygote only the few dosage-sensitive genes will affect the phenotype (*Figure 16.9*). Thus it is a challenge to work out which of the deleted genes causes which aspect of the syndrome.

Williams syndrome (WLS; MIM 194050) provides an interesting example of these problems. People with WLS have a recognizable face, they are growth retarded, as infants they may have life-threatening hypercalcemia, and

Table 16.7: Syndromes often caused by autosomal chromosomal microdeletions

Syndrome	MIM no.	Chromosomal anomaly
Wolf–Hirschhorn	194190	Deletion of 4p16.3
Cri du chat	123450	Deletion of 5p15.2–p15.3
Williams	194050	Deletion of 7q11.23 including the elastin gene
WAGR (Wilms tumor, aniridia, genital anomalies, growth retardation)	194072	Deletion of 11p13 including *WT1* and *PAX6* genes
Prader–Willi	176270	Lack of paternal genes at 15q11–q13
Angelman	105830	Lack of maternal genes at 15q11–q13
Rubinstein–Taybi	180849	Deletion of 16p13.3
Miller–Diecker lissencephaly	247200	Deletion of 17p13.3
Smith–Magenis	182290	Deletion of 17p11.2
Alagille	118450	Deletion at 20p12.1–p11.23
Di George, velocardiofacial, Schprinzen ('Catch 22')	192430	Deletion at 22q11.21–q11.23

In some syndromes, such as Alagille syndrome, only a minority of patients have a deletion, suggesting that loss of function of a single dosage-sensitive gene causes the phenotype. In other cases, such as Williams syndrome, all patients with the full syndrome have a microdeletion, suggesting that more than one locus is involved in generating the phenotype.

they often have supravalvular aortic stenosis (SVAS). As mentioned above (Section 16.4.3), isolated SVAS sometimes occurs as an autosomal dominant condition (MIM 185500). Dominant SVAS was mapped to 7q11.23 and shown sometimes to result from deletion or disruption of the elastin gene. This provided the clue for identifying the microdeletion in WLS. Typically 1–2 Mb of DNA is deleted, including the elastin gene. This explains the SVAS

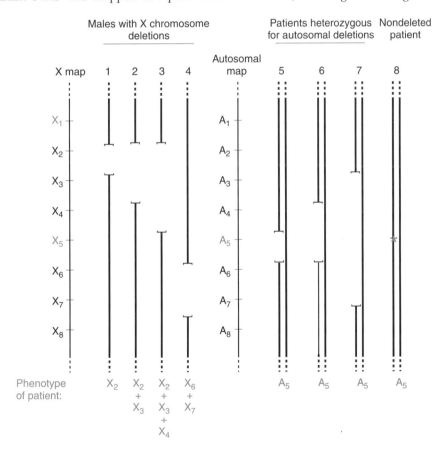

Figure 16.9: X-linked and autosomal microdeletion syndromes.

On the X chromosome, deletion of genes X_1 or X_5 is lethal in males. Patients 1–3 show a nested series of contiguous gene syndromes, patient 4 a nonoverlapping contiguous gene syndrome. On the autosome, only gene A_5 is dosage sensitive. Patients 5–7, with different sized deletions, all show the same phenotype as patient 8, who is heterozygous for a loss-of-function point mutation in the A_5 gene.

component of WLS. Potentially the facial features could also be caused by a deficiency of the connective tissue protein elastin, but this is evidently not the case because people with simple elastin mutations often have SVAS but do not have the characteristic Williams face. Several other genes have been identified that are deleted in WLS, but it is not easy to decide the role of each gene in the syndrome. This is especially true for the intriguing mental phenotype. People with WLS are usually moderately mentally retarded, to about the same extent as people with Down syndrome, but they have a very distinctive cognitive profile and personality (Tassabehji *et al.*, 1999). They are highly sociable, often musical, and talk remarkably well, but have a specific inability at manipulating shapes (visuospatial constructive ability). Identifying the relevant genes from among all those deleted in WLS might provide an entry to identifying genetic determinants of normal human cognition and behavior. Other microdeletion syndromes are also associated with specific behaviors, and so unraveling the constituent genes of these syndromes is an area of great current research interest.

16.8.2 The major effects of chromosomal aneuploidies may be caused by dosage imbalances in a few identifiable genes

Monosomies and trisomies probably owe their characteristic phenotypes to a few major gene effects superimposed on many minor disturbances of development.

Genes where a 50% increase of dosage has a major effect must be very uncommon, and so it should be possible to identify the few genes that cause the major features of, for example, Down syndrome (DS). Studies of patients with translocations show that the Down syndrome critical region is in 21q22.2; trisomy of other parts of chromosome 21 does not cause DS. At least two candidate genes for mental retardation have been identified from this region: *DYRK*, a gene whose *Drosophila* and mouse homologs (*minibrain*) produces dosage-sensitive learning defects (Smith and Rubin, 1997), and *DSCAM*, a brain-specific cell adhesion molecule (Yamakawa *et al.*, 1998).

The abnormalities of Turner syndrome might be due to haploinsufficiency at the earliest stages of development, before X inactivation takes place, but more probably stem from haploinsufficiency of genes that escape X inactivation and that have a functional Y counterpart. Only 18 such genes are known (Lahn and Page, 1997). Some people with partial deletions of Xp have a Turner or partial Turner phenotype, and their deletions can be used to try to pinpoint the genes responsible for particular features of the syndrome (see Zinn *et al.*, 1998). Other candidates are suggested by overlapping mendelian conditions – the pseudoautosomal region at the tip of Xp (Section 2.4.3) contains a homeobox gene *SHOX* that is very occasionally mutated in chromosomally normal people with short stature or with Leri–Weill dyschondreostosis (MIM 127300). Thus *SHOX* is a candidate for the short stature of Turner syndrome.

Further reading

Genetic Nomenclature Guide (1998) *Trends Genet.* Suppl.
Antonarakis SA (1998) Recommendations for a nomenclature system for human gene mutations. *Hum. Mutat.*, **11**,1–3.

References

Akarsu AN, Stoilov I, Yilmaz E, Sayli BS, Sarfarazi M (1996) Genomic structure of *HOXD13* gene: a nine polyalanine duplication causes synpolydactyly in two unrelated families. *Hum. Molec. Genet.*, **5**, 945–952.
Andrews SE, Goldberg YP, Hayden MR (1997) Rethinking genotype and phenotype correlations in polyglutamine expansion disorders. *Hum. Molec. Genet.*, **6**, 2005–2010.
Antonarakis SA (1998) Recommendations for a nomenclature system for human gene mutations. *Hum Mutat.*, **11**, 1–3.
Ballabio A, Andria G (1992) Deletions and translocations involving the distal short arm of the human X chromosome: review and hypotheses. *Hum. Molec. Genet.*, **1**, 221–227.
Bardoni B, Zanaria E, Guioli S, et al. (1994) A dosage sensitive locus at chromosome Xp21 is involved in male to female sex reversal. *Nat. Genet.*, **7**, 497–501.
Bellus GA, Hefferon TW, Ortiz de Luna RI et al. (1995) Achondroplasia is defined by recurrent G380R mutations of FGFR3. *Am. J. Hum. Genet.*, **56**, 368–373.
Berget S (1995) Exon recognition in vertebrate splicing. *J. Biol. Chem.*, **270**, 2411–2414.
Budarf ML, Emanuel BS (1997) Progress in the autosomal segmental aneusomy syndromes (SASs): single or multi-locus disorders? *Hum. Molec. Genet.*, **10**, 1657–1665.
Cooper DN, Krawczak M. (1993) *Human Gene Mutation*. Bios Scientific Publishers, Oxford.
Dalgleish R (1997) The human Type 1 collagen mutation database. *Nucleic Acids Res.*, **25**, 181–187. Update 1998: *Nucleic Acids Res.*, **26**, 253–255.

Easton DF, Ponder MA, Huson SM, Ponder BAJ (1993) An analysis of variation in expression of neurofibromatosis (NF) type 1 (NF1): evidence for modifying genes. *Am. J. Hum. Genet.,* **53**, 305–313.

Fisher E, Scambler P (1994) Two for joy, one for sorrow. *Nat. Genet.,* **7**, 5–7.

Francke U, Ochs HD, de Martinville B *et al.* (1985) Minor Xp21 chromosome deletion in a male associated with expression of Duchenne muscular dystrophy, chronic granulomatous disease, retinitis pigmentosa and McLeod syndrome. *Am. J. Hum. Genet.,* **37**, 250–267.

Hastbacka J, Superti-Furga A, Wilcox WR, Rimoin DL, Cohn DH, Lander ES (1996) Atelosteogenesis type II is caused by mutations in the diastrophic dysplasia sulfate-transporter gene (DTDST): evidence for a phenotypic series involving three chondrodysplasias. *Am. J. Hum. Genet.,* **58**, 255–262.

Hemesath TJ, Steingrimsson E, McGill G *et al.* (1994) Microphthalmia, a critical factor in melanocyte development, defines a discrete transcription factor family. *Genes Dev.,* **8**, 2770–2780.

Hentze MW, Kulozik AE (1999) A perfect message: RNA surveillance and nonsense-mediated decay. *Cell,* **96**, 307–310.

Kim TW, Tanzi RE (1998) Neuronal intranuclear inclusions in polyglutamine diseases: nuclear weapons or nuclear fallout? *Neuron,* **21**, 657–659.

Koob MD, Moseley ML, Schut LJ *et al.* (1999) Untranslated CTG expansion causes a novel form of spinocerebellar ataxia (SCA8). *Nature Genet.,* **21**, 379–384.

Krawczak M, Cooper DN (1997) The human gene mutation database. *Trends Genet.,* **13**, 121–122.

Lahn BT, Page DC (1997) Functional coherence of the human Y chromosome. *Science,* **278**, 675–680.

Morell R, Spritz RA, Ho L *et al.* (1997) Apparent digenic inheritance of Waardenburg syndrome type 2 (WS2) and autosomal recessive ocular albinism (AROA). *Hum. Molec. Genet.,* **6**, 659–664.

Motulsky AG (1995) Jewish diseases and origins. *Nat. Genet.,* **9**, 99–101.

Naski MC, Wang Q, Xu J, Ornitz DM (1996) Graded activation of fibroblast growth factor receptor 3 by mutations causing achondroplasia and thanatophoric dysplasia. *Nat. Genet.,* **13**, 233–237.

Owen MC, Brennan SO, Lewis JH, Carrell RW (1983) Mutation of α_1 antitrypsin to antithrombin: α_1 antitryspin Pittsburgh (358 Met→Arg), a fatal bleeding disorder. *New Engl. J. Med.,* **309**, 694–698.

Patel PI, Lupski JR (1994) Charcot–Marie–Tooth disease: a new paradigm for the mechanism of inherited disease. *Trends Genet.,* **10**, 128–33.

Poulton J, Deadman ME, Bindoff L, Morten K, Land J, Brown G (1993) Families of mtDNA re-arrangements can be detected in patients with mtDNA deletions: duplications may be a transient intermediate form. *Hum. Molec. Genet.,* **2**, 23–30.

Prusiner SB, Scott MR, DeArmond SJ, Cohen FE (1998) Prion protein biology. *Cell,* **93**, 337–348.

Smith DJ, Rubin EM (1997) Functional screening and complex traits: human 21q22.2 sequences affecting learning in mice. *Hum. Molec. Genet.,* **6**, 1729–1733.

Tassabehji M, Metcalfe K, Karmiloff-Smith A *et al.* (1999) Williams syndrome: using chromosomal microdeletions as a tool to dissect cognitive and physical phenotypes. *Am. J. Hum. Genet.,* **64**, 118–125.

Weatherall DJ, Clegg JB, Higgs DR, Wood WG. (1995) The Hemoglobinopathies. In: *The Metabolic and Molecular Basis of Inherited Disease* (eds Scriver CR, Beaudet AL, Sly WS, Valle D), 7th edn. McGraw Hill: New York.

White JA *et al.* (1997) Guidelines for Human Gene Nomenclature. *Genomics,* **45**, 468–471.

Wilkie AO (1997) Craniosynostosis: genes and mechanisms. *Hum. Molec. Genet.,* **6**, 1647–1656.

Wollnik B, Schroeder BC, Kubisch C, Esperer HD, Wieacker H, Jentsch T (1997) Pathophysiological mechanisms of dominant and recessive *KVLQT1* K$^+$ channel mutations found in inherited cardiac arrhythmias. *Hum. Molec. Genet.,* **6**, 1943–1949.

Yamakawa K, Huo Y-K, Haendel MA, Hubert R, Chen X-N, Lyons GE, Korenberg JR (1998) DSCAM: a novel member of the immunoglobulin superfamily maps in a Down syndrome region and is involved in the development of the nervous system. *Hum. Molec. Genet.,* **7**, 227–237.

Zinn AR, Tonk VS, Cen Z *et al.* (1998) Evidence for a Turner syndrome locus or loci at Xp11.2-p22.1. *Am. J. Hum. Genet.,* **63**, 1757–1766.

Genetic testing in individuals and populations

17

Geneticists have no monopoly on DNA-based diagnosis. For microbiologists and virologists, for example, PCR is a central tool for identifying pathogens. Hematologists, oncologists and other pathologists all use DNA testing as a basis for diagnosis. For reviews of a range of applications, see the book by Newton and Graham (Further reading). However, for the purposes of this chapter, we will define genetic testing as testing for mendelian factors. The factors indicate a person's risk of developing or transmitting a disease, or identify her, or indicate her relationship to somebody else.

Wherever possible, we shall use two of the most common mendelian diseases, cystic fibrosis and Duchenne muscular dystrophy, to illustrate the various testing methods. Between them these two diseases exemplify many of the situations that arise in genetic testing for mendelian disorders. As always in this book, we concentrate on the principles and not the practical details. The reader interested in specific procedures can find a series of 'best practice' guidelines for laboratory diagnosis of the commoner mendelian diseases at http//:www.cmgs.org. These have been drawn up by consensus workshops of the UK Clinical Molecular Genetics Society.

How much information a genetic test can give depends on the state of knowledge about the gene(s) involved, but in principle laboratory genetic diagnosis can be made in two essentially different ways.

- **Direct testing:** a sample (DNA, RNA, protein, etc.) from a *consultand* is tested to see whether or not he has a certain genotype – typically, a pathogenic mutation in a certain gene. The test is of an individual, and gives information about that individual.
- **Gene tracking:** linked markers are used in family studies to discover whether or not the consultand inherited the high-risk chromosome from a heterozygous parent. The test is of a family, and gives information about the segregation of a chromosomal segment in the family.

With each year that passes, the role of gene tracking shrinks and the applications of direct testing grow. However, direct testing is not always possible, and even when it is scientifically possible, it may not always be practical in the context of a routine diagnostic service.

17.1 Direct testing is like any other path lab investigation: a sample from the patient is tested to see if it is normal or abnormal

The optimal, though not always practical, method of laboratory genetic diagnosis is to test a person's gene or gene product directly to see whether the sequence is normal or mutant. We must of course know which gene to examine and we must know the relevant 'normal' (wild-type) sequence (*Figure 17.1*).

17.1.1 Direct testing can use a variety of methods, almost all based on PCR, applied to a wide range of sample types

Direct testing is almost always done by PCR, applying the methods described in Section 6.2. The few applications of Southern blotting include testing for major gene rearrangements or disruptions and for Fragile X and myotonic dystrophy full mutations (*Boxes 16.7 and 16.8*). The sensitivity of PCR allows us to use a wide range of tissue samples. These can include the following:

- **Blood samples** – the most widely used source of DNA from adults.
- **Mouthwashes or buccal scrapes** – being noninvasive, they are especially favored for population screening programs. Mouthwashes yield sufficient DNA for a few dozen tests, and by using whole genome amplification (Section 6.2.4) more extensive testing of a single sample may be possible.
- **Chorionic villus biopsy samples** – the best source of fetal DNA (better than amniocentesis specimens).

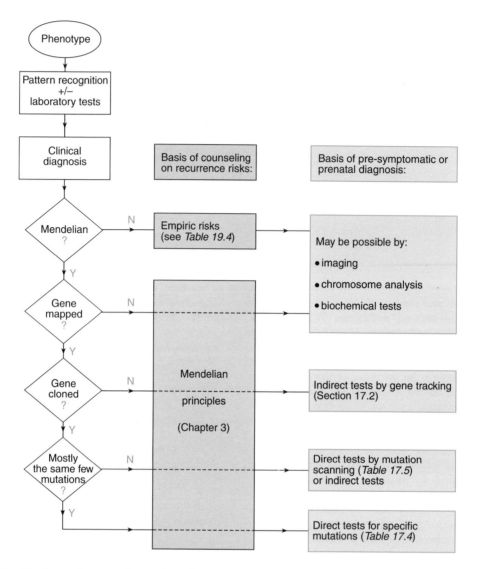

Figure 17.1: Genetic diagnosis, counseling and prediction.

The flow diagram shows how a clinical diagnosis is made starting from the patient's phenotype and then, depending on the mode of inheritance and the state of genetic knowledge, various means are deployed for counseling and predictive tests.

- One or two cells removed from eight-cell stage embryos, for pre-implantation diagnosis after *in vitro* fertilization.
- Hair, semen, etc. for criminal investigations.
- Archived pathological specimens, for typing dead people when no DNA has been stored, or testing tumors for genetic changes. Only short sequences, 250 bp or less, can be reliably amplified from fixed tissue specimens.
- Guthrie cards – these are the cards on which a spot of dried blood is sent to a laboratory for neonatal screening for phenylketonuria (PKU) in the UK and elsewhere; not all the blood spot is used for the screening test. They are a possible source of DNA from a dead child.

RNA has advantages over DNA, but is more difficult to obtain and handle

If a gene has to be scanned for unknown mutations, testing by RT-PCR (see *Figure 6.5*) offers several advantages. DNA testing usually involves amplifying and testing each exon separately, and this can be a major chore in a gene with many exons. Most of the mutation-scanning methods (see *Table 17.5*) can scan fragments larger than the average sized exon, so that an RT-PCR product can be examined using a smaller number of reactions. Also, only RT-PCR can reliably detect aberrant splicing, which is sometimes hard to predict from a DNA sequence change, or may be caused by activation of a cryptic splice site deep within an intron. However, RNA is much less convenient to obtain and work with. Samples must be

Table 17.1: Examples of diseases that show a limited range of mutations

Disease	Cause	Comments
Huntington disease, myotonic dystrophy	Gain of function mutation	Unstable expanded repeat (*see Box 16.7*)
Fragile X	Common molecular mechanism: expansion of an unstable repeat	See *Boxes 16.7 and 16.8*; other mutations occur, but are rare
Achondroplasia	Only G380R produces this particular phenotype; very high mutation rate	Two distinct changes, both causing G380R in *FGFR3* gene (*Figure 16.8*)
Sickle cell disease	Only this particular mutation produces the sickle-cell phenotype	E6V in *HBB* gene (see *Figure 5.11*)
α- and β-thalassemia	Selection for heterozygotes leads to different ancestral mutations being common in different populations	See *Figure 16.2* (α-thalassemia) and *Table 17.2* (β-thalassemia)
Cystic fibrosis	Common ancestral mutations in northern European populations; ancient heterozygote advantage	See *Table 17.3* and Section 3.3.2
Charcot–Marie–Tooth disease (HMSN1)	Common molecular mechanism: recombination between misaligned repeats	Duplication of 1.5 Mb at 17p11.2 (*Figure 16.7*); point mutations also occur
21-Hydroxylase deficiency	Common molecular mechanism: sequence exchange with adjacent closely related pseudogene	See *Figure 9.17*
Tay–Sachs disease	Founder effect in Ashkenazi Jews; ancient heterozygote advantage	Two common *HEXA* mutations in Ashkenazi: 4-bp insertion in exon11 (73%); exon 11 donor splice site G>C (15%)

See Section 16.3 for further discussion of the reasons why some diseases show a limited range of mutations, while others have extensive allelic heterogeneity.

handled with extreme care and processed rapidly to avoid degrading mRNA, and the gene of interest may not be expressed in readily accessible tissues. In addition, many mutations result in unstable mRNA (Section 9.4.6), so that the RT-PCR product from a heterozygous person may show only the normal allele.

Functional assays of proteins have a role in genetic testing

A protein-based functional assay might classify the products of a highly heterogeneous allelic series into two simple groups, functional and nonfunctional – which is, after all, the essential question in most diagnoses. The problem with functional assays is that they are specific to a particular protein. DNA technology by contrast is generic. This has obvious advantages for the diagnostic lab, but in addition it encourages technical development, since any new technique can be applied to many problems.

17.1.2 Some diseases show limited allelic heterogeneity, and genetic testing involves testing for specific mutations

For some diseases, most or all affected people have the same mutation, or one of a small number of different mutations. *Table 17.1* lists some examples. We saw why this should happen in Section 16.3.3; in brief it is when:

- the disease depends on a specific molecular mechanism;
- the nature of the gene is such that one particular mutation occurs repeatedly;
- affected people mostly carry the same ancestral disease mutation, or one of a limited number of ancestral mutations.

Testing for these diseases starts with checking for specific mutations, using one of the methods described in Section 17.1.3.

β-Thalassemia and cystic fibrosis illustrate different situations of limited mutational diversity

For both these conditions, a very large number of mutations in the relevant gene have been described, but in each case a handful of mutations account for the majority of cases in any particular population. With β-thalassemia different mutations are predominant in different populations (*Table 17.2*). DNA testing is not needed to diagnose carriers or affected people (orthodox hematology does this perfectly well) but it is the method of choice for prenatal diagnosis. Provided one has DNA samples from the

Table 17.2: The main β-thalassemia mutations in different countries

Population	Mutation	MIM no.*	Frequency (%)	Clinical effect
Sardinia	Codon 39 (C>T)	.0312	95.7	β^0
	Codon 6 (delA)	.0327	2.1	β^0
	Codon 76 (del C)	.0330	0.7	β^0
	Intron 1–110 (G>A)	.0364	0.5	β^+
	Intron 2–745 (C>G)	.0367	0.4	β^+
Greece	Intron 1–110 (G>A)	.0364	43.7	β^+
	Codon 39 (C>T)	.0312	17.4	β^0
	Intron 1–1 (G>A)	.0346	13.6	β^0
	Intron 1–6 (T>C)	.0360	7.4	β^+
	Intron 2–745 (C>G)	.0367	7.1	β^+
China	Codon 41/42 (delTCTT)	.0326	38.6	β^0
	Intron 2–654 (C>T)	.0368	15.7	β^0
	Codon 71/72 (insA)	.0328	12.4	β^0
	−28 (A>G)	.0381	11.6	β^+
	Codon 17 (A>T)	.0311	10.5	β^0
Pakistan	Codon 8/9 (insG)	.0325	28.9	β^0
	Intron 1–5 (G>C)	.0357	26.4	β^+
	619-bp deletion	–	23.3	β^+
	Intron 1–1 (G>T)	.0347	8.2	β^0
	Codon 41/42 (delTCTT)	.0326	7.9	β^0
US black African	−29 (A>G)	.0379	60.3	β^+
	−88 (C>T)	.0372	21.4	β^+
	Codon 24 (T>A)	.0369	7.9	β^+
	Codon 6 (delA)	.0327	0.8	β^0

In each country, certain mutations are frequent because of a combination of founder effects and selection favoring heterozygotes.
* HBB mutations are listed in OMIM entry 141900 (β-globin) under the number shown, e.g. codon 39 (C>T) is 141900.0312. Data courtesy of Dr J Old, Institute of Molecular Medicine, Oxford.

parents and knows their ethnic origin, the parental mutations can often be found using only a small cocktail of direct tests, after which the fetus can be readily checked.

In cystic fibrosis, by contrast, the F508del mutation is the commonest in all European populations, and is believed to be of ancient origin. However, the proportion of all mutations that are F508del varies, being generally high in the north and west of Europe and lower in the south. Direct testing for cystic fibrosis mutations divides into two phases. First, a limited number of specified mutations, always including F508del, are sought using the methods described in Section 17.1.3. As *Table 17.3* shows, there is no obvious natural cut-off in terms of diminishing returns on testing for specific mutations. If this phase fails to reveal the mutations, then if resources allow, a screen for unknown mutations may be instituted, using the methods described in Section 17.1.4, or alternatively gene tracking (see *Figure 17.14*) may be used. The impact of this diversity on proposals for population screening is discussed below (see *Figure 17.18*).

Some combination of founder effects and heterozygote advantage must be the cause of this relative mutational homogeneity in recessive disorders. However, surprisingly often, when a recessive disease is particularly common in a certain population, it turns out that more than one mutation is responsible. An example is Tay–Sachs disease among Ashkenazi Jews, where there are two common *HEXA* mutations (*Table 17.1*). It is difficult to explain this situation except by assuming there was substantial heterozygote advantage some time when the founding population was small.

17.1.3 Some genetic testing methods test for presence or absence of a specified DNA sequence change

Testing for the presence or absence of a known sequence change is a different and much simpler problem than scanning a gene for the presence of *any* mutation. Some of the main methods were described in Section 6.2.3, and are summarized in *Table 17.4*. Many variants of these and other methods have been developed as kits by biotechnology companies. Testing for a known change is useful for:

■ diseases with limited allelic heterogeneity, such as those discussed above;

■ diagnosis within a family. Mutation scanning methods may be needed to define the family mutation but, once it is characterized, other family members normally need be tested only for that particular mutation;

■ in research, for testing control samples. A common problem in positional cloning is that a candidate gene has been identified, and a patient has a sequence change in this gene. The question then arises, is this change pathogenic (confirming that this gene is indeed the disease gene), or might it be a nonpathogenic polymorphism (see Section 16.3.4)? One common approach is to screen a panel of 100 or so normal control samples for the presence of the change (which does not, of course, solve the problem of rare neutral variants).

Table 17.3: Distribution of CFTR mutations in 300 CF chromosomes from the Northwest of England

Mutation	Exon	Frequency (%)	Cumulative frequency (%)
F508del	10	79.9	79.9
G551D	11	2.6	82.5
G542X	11	1.5	84.0
G85E	3	1.5	85.5
N1303K	21	1.2	86.7
621+1G>T	4	0.9	87.6
1898+1G>A	12	0.9	88.5
W1282X	21	0.9	89.4
Q493X	10	0.6	90.0
1154insTC	7	0.6	90.6
3849+10kb (C>T)	Intron 19	0.6	91.2
R553X	10	0.3	91.5
V520F	10	0.3	91.8
R117H	4	0.3	92.1
R1283M	20	0.3	92.4
R347P	7	0.3	92.7
E60X	3	0.3	93.0
Unknown/private	–	7.0	100

F508del and a few of the other relatively common mutations are probably ancient and spread through selection favoring heterozygotes; the other mutations are probably recent, rare and highly heterogeneous. CF is more homogeneous in this population than in most others. See *Box 16.3* for nomenclature of mutations. Data courtesy of Dr Andrew Wallace, St Mary's Hospital, Manchester.

Testing for the presence or absence of a restriction site

When a base substitution mutation creates or abolishes the recognition site of a restriction enzyme, this allows a simple direct PCR test for the mutation (*Figure 6.6*). Although hundreds of restriction enzymes are known, they almost all recognize symmetrical palindromic sites, and many point mutations will not happen to affect such sequences. Also, sites for rare and obscure restriction enzymes are unsuitable for routine diagnostic use because the enzymes are expensive and often of poor quality. Sometimes, however, a diagnostic restriction site can be introduced by a form of PCR mutagenesis using carefully designed primers. *Figure 17.2* shows an example.

Use of allele-specific oligonucleotide (ASO) hybridization

Under suitably stringent hybridization conditions, these short synthetic probes hybridize only to a perfectly matched sequence (see *Figure 5.10*). *Figure 5.11* demonstrates the use of dot-blot hybridization with ASO probes to detect the single base substitution that causes sickle cell disease. For diagnostic purposes, a reverse dot-blot procedure has often been used. A screen for a series of defined cystic fibrosis mutations, for example, would use a series of ASOs specific for each mutant allele, spotted onto a single membrane which is then hybridized to

Table 17.4: Methods of testing for a specified mutation.

Method	Comments
Restriction digestion of PCR-amplified DNA; check size of products on a gel (*Figure 6.6*)	Only when the mutation creates or abolishes a natural restriction site, or one engineered by use of special PCR primers (*Figure 17.2*)
Hybridize PCR-amplified DNA to allele-specific oligonucleotides (ASO) on a dot–blot or gene chip	General method for specified point mutations; large arrays allow scanning for any mutation
PCR using allele-specific primers (ARMS test)	General method for point mutations; primer design critical (*Figure 17.3*) Can be adapted to chip technology. Can provide real-time quantitative readout, using TaqMan technology (*Figure 6.10*)
Oligonucleotide ligation assay (OLA)	General method for specified point mutations (*Figure 17.4*)
PCR with primers located either side of a translocation breakpoint	Successful amplification shows presence of the suspected deletion or specified rearrangement
Check size of expanded repeat	Dynamic repeat diseases (*Box 16.7*) only; large expansions require Southern blots, smaller ones can be done by PCR

See Section 6.2.3 for description of the principles.

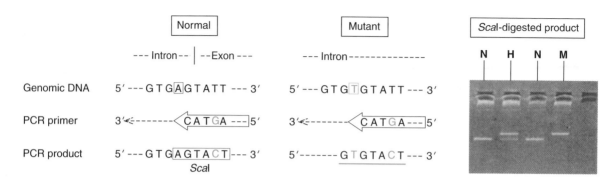

Figure 17.2: Introducing an artificial diagnostic restriction site.

An A>T mutation in the intron 4 splice site of the *FACC* gene does not create or abolish a restriction site. The PCR primer stops short of this altered base, but has a single base mismatch (blue G) in a noncritical position which does not prevent it hybridizing to and amplifying both the normal and mutant sequences. The mismatch in the primer introduces an AGTACT restriction site for *Sca*I into the PCR product from the normal sequence. The *Sca*I-digested product from homozygous normal (N), heterozygous (H) and homozygous mutant (M) patients is shown. Courtesy of Dr Rachel Gibson, Guy's Hospital, London.

labeled PCR-amplified test DNA. Recently reverse dot-blotting has developed from manually-spotted arrays of small numbers of ASOs to very large ASO arrays on 'gene chips' that can potentially detect all possible mutations in a gene (Section 17.1.4).

Allele-specific PCR amplification (ARMS test)

The principle of the ARMS (amplification refractory mutation system) method was shown in *Figure 6.9*. Paired PCR reactions are carried out. One primer (the common primer) is the same in both reactions, the other exists in two slightly different versions, one specific for the normal sequence and the other specific for the mutant sequence.

Additional control primers are usually included, to amplify some unrelated sequence from every sample as a check that the PCR reaction has worked. The location of the common primer can be chosen to give different sized products for different mutations, so that the PCR products of multiplexed reactions form a ladder on a gel. With careful primer design, the mutation-specific primers can also be made to give distinguishable products. For example, they can be labeled with different fluorescent or other labels, or given 5' extensions of different sizes. Multiplexed mutation-specific PCR is well suited to screening large numbers of samples for a given panel of mutations (*Figure 17.3*).

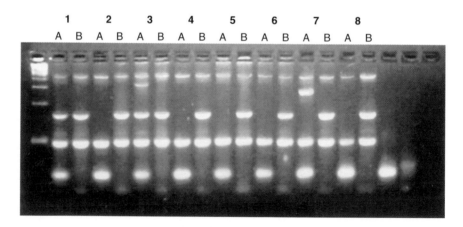

Figure 17.3: Multiplex ARMS test to detect 12 cystic fibrosis mutations.

Each sample is tested with two multiplex mutation-specific PCR reactions. Set A amplifies DNA containing the 1717- G>A, G542X, W1282X, N1303K, F508del or 3849+10kbC>T mutations; set B gives products with templates containing 621+1G>T, R553X, G551D, R117H, R1162X or R334W mutations and the normal allele of F508del. Each product is a different size, so the nature of any mutation is revealed by the position of the band in the gel. Each tube also amplifies two control sequences. Note that the normal alleles of mutations other than F508del are not tested for. Sample 1 is heterozygous for F508del; sample 3 is a compound heterozygote F508del / 1717-1 G>A; sample 7 includes G542X (homozygous or heterozygous). Courtesy of Dr Michelle Coleman, St Mary's Hospital, Manchester; data obtained using the Elucigene™ kit from Zeneca Diagnostics Ltd.

The ARMS principle can be adapted to allow the course of a quantitative PCR reaction to be followed in real time, using TaqMan or related fluorescer-quencher methods (Section 6.2.3).

Oligonucleotide ligation assay (OLA)

In the OLA test for base substitution mutations, two oligonucleotides are constructed that hybridize to adjacent sequences in the target, with the join sited at the position of the mutation. DNA ligase will covalently join the two oligonucleotides only if they are perfectly hybridized (Nickerson *et al.*, 1990). Various formats for the test are possible, for example as an ELISA reaction or, as in *Figure 17.4*, for analysis on a fluorescence sequencer.

17.1.4 Mutation scanning methods are used to check whether a gene carries any mutation

For the great majority of diseases there is extensive allelic heterogeneity, and genetic testing requires a search for any mutation anywhere within or near the relevant gene. The biggest current problem in laboratory genetic diagnosis is the lack of any quick, cheap and reliable method for doing this. *Table 17.5* lists a number of methods that can be used to seek mutations in a gene, and these are briefly described below; for laboratory details, see the book by Cotton *et al.* (Further reading). The table also summarizes their advantages and disadvantages, from the point of view of a diagnostic laboratory.

The requirements for routine diagnostic and for research use are rather different. Researchers usually want to screen a small number of samples for mutations in several candidate genes as quickly as possible. They are not worried if the methods require several days of benchwork or high levels of skill, but they do not want to invest much effort in fine-tuning a method for a particular gene. Sequencing and heteroduplex/SSCP analysis have been the methods most frequently used. Diagnostic laboratories grew up in a research culture and inherited the same attitudes, but increasingly now they need methods for testing a large number of samples for

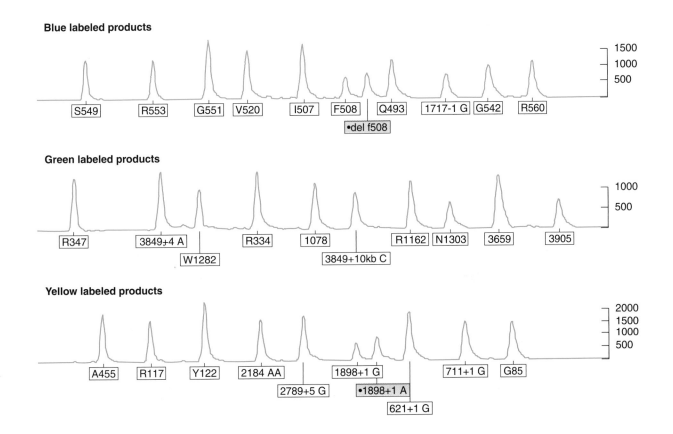

Figure 17.4: Using the oligonucleotide ligation assay to test for 31 known CF mutations.

After multiplex PCR, a multiple OLA is performed. Ligation oligonucleotides are designed so that products for each mutation and its normal counterpart can be distinguished by size and color of label. The trace is from a single gel track, with blue, green and yellow labeled products displayed separately. Ligation products from two mutations, F508del and the splice site mutation 1898 + 1G>A, are seen, showing that the sample comes from a compound heterozygote. Courtesy of Dr Andrew Wallace, St Mary's Hospital, Manchester.

Table 17.5: Methods for scanning a gene for mutations

Method	Advantages	Disadvantages
Southern blot, hybridize to cDNA probe	Only way to detect major deletions and rearrangements	Laborious, expensive Needs several μg of DNA
Sequencing	Detects all changes Mutations fully characterized	Expensive; can be hard to interpret
Heteroduplex gel mobility	Very simple, cheap	Sequences <200 bp only Limited sensitivity Does not reveal position of change
Denaturing HPLC	Quick, high throughput; quantitative	Expensive equipment Does not reveal position of change
Single-strand conformation polymorphism (SSCP) analysis	Simple, cheap equipment	Sequences <200 bp only Limited sensitivity Does not reveal position of change
Denaturing gradient gel electrophoresis (DGGE)	High sensitivity	Choice of primers is critical Expensive primers Does not reveal position of change
Dideoxy fingerprinting	High sensitivity	Complicated to interpret

Mismatch cleavage
  (i) chemical
 	
  (ii) enzymatic
	
	High sensitivity
	Shows position of change
	No nasty reagents
		
		Toxic chemicals
		Experimentally difficult
		Poor quality results

Protein truncation test (PTT)	High sensitivity for chain terminating mutations Shows position of change	Chain terminating mutations only Expensive Experimentally difficult Usually needs RNA

Oligonucleotide arrays (gene chips):
  (i) hybridization arrays
  (ii) minisequencing arrays
	Both types:
	  Quick
	  High throughput
	  Might detect and define all changes
		Both types:
		Novel technology still under development
		Expensive equipment
		Limited range of genes

The table summarizes the advantages and disadvantages of each method for use in a routine diagnostic service. Heteroduplex gel mobility and SSCP can be performed simultaneously on a single gel (see *Figure 17.6*).

mutations in just a few genes, with as near 100% sensitivity as possible. Time can be spent optimizing a test for a particular gene, provided the method once developed is quick and simple. Methods particularly suited to this approach include DGGE, two-dimensional gels, the protein truncation test, and many commercial kits.

For routine diagnostic use, all these methods suffer in varying degree from two limitations:

- They are quite laborious and expensive for use in a diagnostic service that needs to produce answers quickly and within a modest budget.
- They detect differences between the patient's sequence and the published 'normal' sequence – but they do not generally distinguish between pathogenic and chance nonpathogenic changes.

It seems likely that eventually oligonucleotide arrays of one sort or another (DNA chips) will replace most other methods for routine mutation scanning in the commoner diseases, and automated sequencing will be increasingly

used for the rarer diseases, where the investment to produce custom chips is not warranted. Widespread use of these methods will highlight the problem of deciding whether a sequence change is pathogenic or not (*Box 16.4*). Hopefully, advancing knowledge of gene function and the development of disease-specific mutation databases will make this task easier. Nevertheless, at least for the next few years, testing for unknown mutations in a service laboratory remains a considerable problem.

Sequencing, particularly of RT-PCR product, is increasingly the method of choice for mutation scanning

As automated fluorescence sequencers become standard items of laboratory equipment, sequencing becomes more and more attractive as a means of mutation scanning (*Figure 17.5*). Mutations picked up by other methods are often confirmed by sequencing, so it is tempting to do the sequencing straight away. Nevertheless, sequencing is expensive, especially for scanning many exons of a gene

(A) Patient

C A A A G A A A A A T C N T A A A C T C A

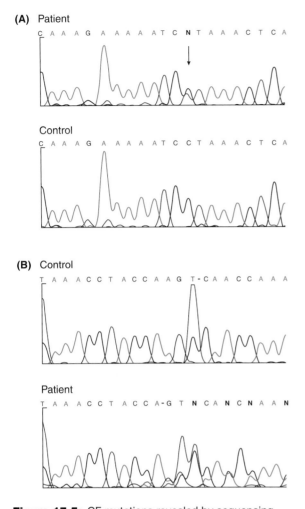

Control

C A A A G A A A A A T C C T A A A C T C A

(B) Control

T A A A C C T A C C A A G T - C A A C C A A A

Patient

T A A A C C T A C C A - G T N C A N C N A A N

Figure 17.5: CF mutations revealed by sequencing.

(A) A base substitution in exon 3. The double peak (arrow) in the upper trace shows the sample contains the mutation 332C>T (amino acid substitution P67L) in heterozygous form. Careful quality control is needed to allow heterozygous base substitutions to be reliably distinguished from noise. **(B)** A single base deletion 3659delC in exon 19 (lower trace). Sequence downstream of the deletion is confused, reflecting overlapping sequence of the two alleles in this heterozygote. The change would be confirmed by sequencing the reverse strand. Courtesy of Dr Andrew Wallace, St Mary's Hospital, Manchester.

in a genomic DNA sample, and much time can be wasted investigating artefacts, especially if the sequencing template is not of the highest quality.

Several methods rely on formation of heteroduplexes

Many tests use the properties of heteroduplexes to detect differences between two sequences. Most mutations occur in heterozygous form, and heteroduplexes can be formed simply by heating the test PCR product to denature it, and then cooling slowly. For homozygous mutations, or X-linked mutations in males, it is necessary to

add some reference wild-type DNA. Several properties of heteroduplexes can be exploited:

- Heteroduplexes often have abnormal mobility on nondenaturing polyacrylamide gels (*Figure 17.6*, lower panel). Special gels (Hydrolink™, MDE™) are supposed to improve the resolution. This is a particularly simple method to use. If fragments no more than 200 bp long are tested, insertions, deletions and most but not all single-base substitutions are detectable (Keen *et al.*, 1991).

- Heteroduplexes have abnormal denaturing profiles. This is exploited in denaturing gradient gel electrophoresis (DGGE, *Figure 17.7*) and denaturing high performance liquid chromatography (dHPLC). In both cases the mobility of a fragment changes markedly when it denatures. These methods require tailoring to the particular DNA sequence under test, and so are best suited to routine analysis of a given fragment in many samples. DGGE requires special primers with a 5′ poly(G;C) extension (a GC clamp). Once optimized, these methods have very high sensitivity. Two-dimensional DGGE gels have near 100% sensitivity for mutation detection (Dhanda

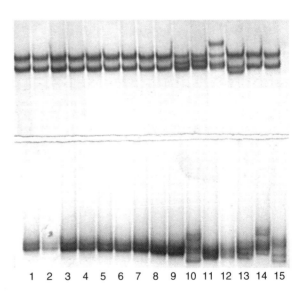

1 2 3 4 5 6 7 8 9 10 11 12 13 14 15

Figure 17.6: CF mutation screening by heteroduplex and SSCP analysis.

Exon 3 of the CFTR gene was PCR amplified from genomic DNA of nine unrelated patients and six control samples with known exon 3 mutations. After denaturation, the samples were loaded on to a nondenaturing polyacrylamide gel. Some of each product re-annealed to give double-stranded DNA. This runs faster in the gel and gives the bands seen in the lower panel. The single-stranded DNA runs more slowly in the same gel (upper panel). Variant SSCP and heteroduplex patterns can be seen with the control samples (mutations G85E, L88S, R75X, P67L, E60X and R75Q in lanes 10–15 respectively) but none of the patient samples shows an abnormal pattern for this exon. Courtesy of Dr Andrew Wallace, St Mary's Hospital, Manchester.

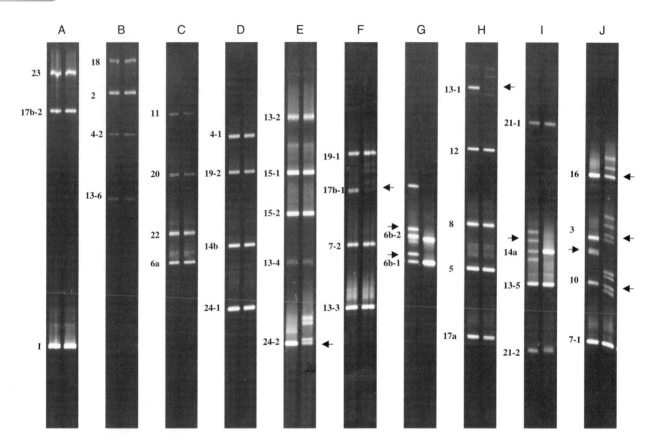

Figure 17.7: CF mutation detection by denaturing gradient gel electrophoresis.

Each exon of the *CFTR* gene (except exon 9, which in this protocol is sequenced directly) is PCR-amplified in one or more segments and run on 9% polyacrylamide gels containing a gradient of urea-formaldehyde denaturant. Bands contain exons as labeled. The band from any amplicon that contains a heterozygous variant splits into usually four sub-bands (arrows). Individual 1 (left lane in each panel) has variants in amplicons 6, 10 and 14; individual B (right lane) has variants in amplicons 3, 10, 13, 16, 17 and 24. Characterization of the variants showed that individual 1 was heterozygous for F508del (exon 10), and individual 2 was a compound heterozygote E60X (exon 3) / R1070Q (exon 17). Other variants were nonpathogenic. Courtesy of Dr Hans Scheffer, University of Groningen, Netherlands.

et al., 1998). As with any method that finds every sequence variant in a large gene, it is necessary to sort out pathogenic from nonpathogenic changes (*Figure 17.7*).

■ Mismatched bases in heteroduplexes are sensitive to cleavage by chemicals or enzymes. The chemical cleavage of mismatch (CCM) method (*Figure 17.8*) is a sensitive method for mutation detection, with the advantages that quite large fragments (over 1 kb) can be analyzed, and the location of the mismatch is pinpointed by the size of the fragments generated. Its disadvantages are that it uses very toxic chemicals, particularly osmium tetroxide (though this can be substituted by potassium permanganate), and some practice is needed before it works well. An alternative is enzymic cleavage of mismatches, which uses enzymes such as T4 phage resolvase or endonuclease VII to achieve the same result without the toxic chemicals. Unfortunately in most people's hands, the quality of the gels produced leaves much to be desired.

Single-strand conformational polymorphism (SSCP) analysis is one of the most popular methods for mutation scanning

Single-stranded DNA has a tendency to fold up and form complex structures stabilized by weak intramolecular bonds, notably base-pairing hydrogen bonds. The electrophoretic mobilities of such structures on nondenaturing gels will depend not only on their chain lengths but also on their conformations, which are dictated by the DNA sequence. For SSCP (*Figure 17.6*), amplified DNA samples (which may be RT-PCR products) are denatured and loaded on a nondenaturing polyacrylamide gel. Primers can be radiolabeled, or unlabeled products can be detected by silver staining. Control samples must be run, so that differences from the wild-type pattern can be noticed. SSCP is simple and adequately sensitive for fragments up to 200 bp long, but it does not reveal the nature or position of any mutation detected (Sheffield *et al.,* 1993). SSCP and heteroduplex analysis can be combined on a single gel, as in *Figure 17.6*. The precise pattern of

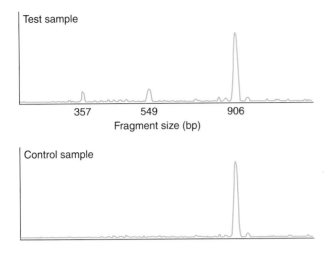

Figure 17.8: CF mutation detection by chemical cleavage of mismatch.

The test sample, PCR-amplified exon 13 of the *CFTR* gene, was denatured and allowed to renature. Because a mutation is present in heterozygous form, heteroduplexes are formed, and these are cleaved by treatment with OsO_4 (or $KMnO_4$) or hydroxylamine. Running on a fluorescence sequencer shows the full-length fragment (906 bp) plus cleavage products of 357 and 549 bp. Sequencing revealed the mutation 2184delA. Courtesy of Dr Julie Wu, St Mary's Hospital, Manchester.

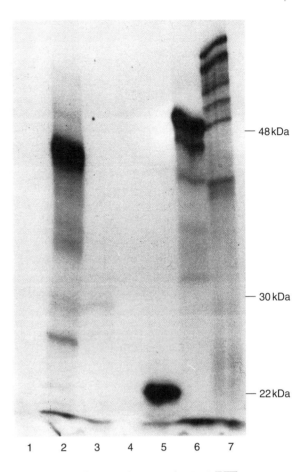

Figure 17.9: The protein truncation test (PTT).

Coding sequence without introns (cDNA or large exons in genomic DNA) is PCR amplified using a special forward primer that includes a T7 promoter, a eukaryotic translation initiator with an ATG start codon, and a gene-specific 3′ sequence designed so that the sequence amplified reads in-frame from the ATG. A coupled transcription–translation system is used to produce polypeptide from the PCR product, and the protein is checked for size by SDS-PAGE gel electrophoresis. A truncated polypeptide points to the presence of a premature stop codon. In this example, RT-PCR product from two males with DMD has been analyzed. Patient 1 exons 58–68 give normal (47 kDa) product only (lane 2) but exons 67–79 (lane 3) encode a 30 kDa truncated product (arrow) – the band is very faint in this example. In patient 2 exons 58–68 give a truncated product of 22 kDa (lane 5); exons 67–79 give normal (48 kDa) product only (lane 6). Sequencing revealed mutations 10431 + 1G>A and 9405C>A in the two patients. Photo courtesy of Drs Steve Abbs and Zandra Hatton, Medical and Molecular Genetics, Guy's Hospital, London.

bands seen is very dependent on details of the conditions. An elaboration of SSCP, **dideoxy fingerprinting**, analyzes each band in a sequencing ladder by SSCP, and is claimed to give 100% sensitivity (Sarkar *et al.*, 1992).

The protein truncation test (PTT) provides an efficient means to check for mutations that produce premature termination codons

The PTT (*Figure 17.9*) is a specific test for frameshifts, splice site or nonsense mutations that truncate a protein product (van der Luijt *et al.*, 1994). Clearly, the strength and weakness of the PTT is that it detects only certain classes of mutation. It would not be useful for cystic fibrosis, where only a minority of mutations introduce premature termination codons. But in Duchenne muscular dystrophy, adenomatous polyposis coli or *BRCA1*-related breast cancer, missense mutations are infrequent, and any such change found may well be coincidental and nonpathogenic. For such diseases, the PTT has several advantages. It conveniently ignores silent or missense base substitutions, and (like mismatch cleavage methods, but unlike SSCP) it reveals the approximate location of any mutation. Large exons, such as exon 15 of the *APC* gene (6.5 kb), can be tested using genomic DNA rather than by RT-PCR. Several variants have been developed to give cleaner results, usually by incorporating an immuno-precipitation step.

(A) Wild-type   GAGGTCGTATCCATGCCTTACAGTCCAGG

A>C mutant   GAGGTCGTATCC<u>C</u>TGCCTTACAGTCCAGG

Cell	Oligo	Wild-type Mismatch	Wild-type Hyb	Mutant Mismatch	Mutant Hyb
11A	AGGTCGTAT<u>a</u>CaTGCCTTAC	1	+	2	−
11G	AGGTCGTAT<u>g</u>CaTGCCTTAC	1	+	2	−
11C	AGGTCGTAT<u>c</u>CaTGCCTTAC	0	++	1	+
11T	AGGTCGTAT<u>t</u>CaTGCCTTAC	1	+	2	−
12A	GGTCGTATC<u>a</u>aTGCCTTACA	1	+	2	−
12G	GGTCGTATC<u>g</u>aTGCCTTACA	1	+	2	−
12C	GGTCGTATCCaTGCCTTACA	0	++	1	+
12T	GGTCGTATC<u>t</u>aTGCCTTACA	1	+	2	−
13A	GTCGTATCC<u>a</u>TGCCTTACAG	0	++	1	+
13G	GTCGTATCC<u>g</u>TGCCTTACAG	1	+	1	+
13C	GTCGTATCC<u>C</u>TGCCTTACAG	1	+	0	++
13T	GTCGTATCC<u>t</u>TGCCTTACAG	1	+	1	+
14A	TCGTATCCa<u>a</u>GCCTTACAGT	1	+	2	−
14G	TCGTATCCa<u>g</u>GCCTTACAGT	1	+	2	−
14C	TCGTATCCa<u>c</u>GCCTTACAGT	1	+	2	−
14T	TCGTATCCa<u>T</u>GCCTTACAGT	0	++	1	+

Cell	Wild-type 11	12	13	14		Mutant 11	12	13	14
A			██						
G									
C	██	██						██	
T				██					

(B) Wild-type   CTAGTTCGACGAGGTCGTATCCATGCCTTACAGTCCAGG

Target DNA   CTAGTTCGACGAGGTCGTATCC<u>C</u>TGCCTTACAGTCCAGG

Primer 1	5′ CTAGTTCGACGAGGTCGTA 3′	ddT-red
Primer 2	5′ TAGTTCGACGAGGTCGTAT 3′	ddC-blue
Primer 3	5′ AGTTCGACGAGGTCGTATC 3′	ddC-blue
Primer 4	5′ GTTCGACGAGGTCGTATCC 3′	ddC-blue
Primer 5	5′ TTCGACGAGGTCGTATCC<u>A</u> 3′	No label added
Primer 6	5′ TCGACGAGGTCGTATCC<u>A</u>T 3′	ddG-yellow (v. weak)
Primer 7	5′ CGACGAGGTCGTATCC<u>A</u>TG 3′	ddC-blue (weak)
Primer 8	5′ GACGAGGTCGTATCC<u>A</u>TGC 3′	ddC-blue
Primer 9	5′ ACGAGGTCGTATCC<u>A</u>TGCC 3′	ddC-red
Primer 10	5′ CGAGGTCGTATCC<u>A</u>TGCCT 3′	ddT-red
Primer 11	5′ GAGGTCGTATCC<u>A</u>TGCCTT 3′	ddA-green

Figure 17.10 (*Opposite*): Oligonucleotide arrays.

(**A**) Principles of mutation detection by hybridization to an oligonucleotide array. Oligonucleotides are arrayed in sets of four, with each set corresponding to part of the wild-type sequence and to each of the three possible base substitutions at a central position in the oligo. The mismatches to the wild-type sequence are underlined. The mutant sequence has an A>C substitution at position 13. Nucleotides with mismatches to the mutant sequence are in lower case. The number of mismatches and strength of hybridization to the wild-type and mutant sequences are shown on the right. Oligos with one mismatch hybridize weakly; those with two mismatches do not hybridize. When hybridized to the normal DNA, the sequence can be read off as a series of strongly hybridizing cells. With the mutant DNA, the chip will show an area of diminished hybridization with one strongly hybridizing cell marking the substituted base (lower panel). (**B**) Principles of minisequencing array. Each cell of the array contains an oligonucleotide anchored by its 5′end, corresponding to part of the target sequence. PCR-amplified target DNA is hybridized to the array. It acts as a template for extension of the primers. Each primer is extended by a single nucleotide, using color-labeled dideoxynucleotides. The result of using the A>C mutation as above is shown. The blue label added to primer 4 identifies the changed base. Primer 5 will fail to add any label because of the mismatched 3′ end.

High-density oligonucleotide (GeneChip®) arrays promise to make high-throughput mutation scanning feasible

Trials to date have involved custom-designed gene-specific arrays of 20–25-mer oligonucleotide probes, in either of two basic designs (*Figure 17.10*):

- **Hybridization chips** contain oligonucleotides matching all wild-type and single nucleotide substitution sequences in a gene. Test DNA is PCR-amplified, fluorescently labeled and hybridized to the array, either alone or (preferably) in competition with a reference wild-type sequence, labeled with a different color. Trials of these designs have achieved over 90% mutation detection in blind trials of *BRCA1* and *ATM* samples (Hacia, 1999). In general these designs detect homozygous base substitutions well, but miss a few heterozygous substitutions and have major problems with insertion mutations.

- **Minisequencing chips** use arrayed oligonucleotide primers with free 3′-OH groups. Unlabeled PCR-amplified test DNA is hybridized to the array, and DNA polymerase plus four differently labeled dideoxynucleotides are added. The test DNA acts as template for addition of a single labeled dideoxynucleotide to each array primer. As in an ARMS reaction, addition will occur only if the 3′ end of the primer exactly matches the template. The array could be made with primers specific not only for the wild-type sequence but also for all possible mutations. This technology has been less extensively tested than hybridization array technology because the chemistry used to produce most large arrays to date anchors the 3′ end of the oligonucleotide to the support, leaving a free 5′ end.

The emerging technology of gene chips promises to revolutionize mutation scanning, as well as other branches of human molecular genetics, but the revolution has not yet happened. Mutation detection arrays face a more difficult problem than expression arrays – the anchored probes are shorter, so specific hybridization is more difficult to achieve, and the test DNA must be PCR-amplified, unlike expression arrays, which can use total cellular poly(A) RNA. The current state of the art has been well reviewed by Hacia (1999).

17.1.5 Diagnosis of Duchenne muscular dystrophy: approaches to detecting structural rearrangements or point mutations in a giant gene that has high rates of mutation and recombination

As stated at the start of this chapter, we are using cystic fibrosis and Duchenne muscular dystrophy to illustrate many of the applications of genetic testing. Both involve large genes with extensive allelic heterogeneity, but beyond that, CF and DMD pose rather different sets of problems for DNA diagnosis (*Table 17.6*). Between them, they show many of the issues involved in testing for mendelian diseases.

A deletion screen reveals the majority of mutations in affected males

Although dystrophin mutations are extremely heterogeneous, 60–65% of all mutations are deletions of one or more exons (see *Figure 16.3*), and these preferentially affect certain exons. In affected males, two multiplex PCR reactions (that shown in *Figure 17.11* and one testing exons in the 5′ part of the gene) will reveal 98% of all deletions. Most deletions remove more than one exon; deletions that appear to affect noncontiguous exons and deletions of just a single exon may need confirming (because of the risk of spurious results due to PCR failure).

Detecting deletions in females, for carrier testing, is much more difficult because the normal X chromosome masks any deletion present on the other X. Sequences deleted on one X will show a 50% reduction in dosage compared to nondeleted sequences, and this can be detected by quantitative PCR or quantitative Southern blotting. Very careful work is needed to get a result sufficiently unambiguous that a woman could prudently base reproductive decisions on it. A principal requirement for quantitative PCR is to restrict the number of cycles, so that the reaction does not reach a plateau; methods that use competitive reactions are also more likely to be reliably quantitative. Fluorescence sequencing machines, with their inherently quantitative mode of operation and their ability to measure small quantities of product, are probably the best tool for detecting carriers of deletions and duplications (Yau *et al.*, 1996).

Table 17.6: The contrasting genetics of cystic fibrosis and Duchenne muscular dystrophy

Cystic fibrosis	Duchenne muscular dystrophy
Autosomal recessive	X-linked recessive
Loss of function mutations	Loss of function mutations
Fairly large gene:	Giant gene:

250 kb genomic DNA
   27 exons
   6.5 kb mRNA
	2400 kb genomic DNA
	   79 exons
	   14 kb mRNA

Almost all mutations are single nucleotide changes	65% of mutations are deletions encompassing one or more complete exons 5% duplications 30% nonsense, splice site, etc. mutations Missense mutations are very unusual
New mutations are extremely rare	New mutations are very frequent
Mosaicism is not a problem	Mosaicism is common
Little intragenic recombination	Recombination hotspot (12% between markers at either end of the gene)

Genetic testing in DMD and CF require different sets of approaches

If a deletion is segregating in a family, typing females for microsatellites mapping within the deletion may reveal apparent nonmaternity, where a mother has transmitted no marker allele to her daughter because of the deletion (see *Figure 17.12*). In such families nonmaternity proves a woman is a carrier, while heterozygosity (in the daughter or sister of a carrier) proves a woman is not a carrier. Several markers suitable for this purpose have been identified in the introns at deletion hotspots. A final option for detecting deletion carriers is fluorescence *in*

situ hybridization to metaphase chromosomes (Section 10.1.4), using a probe that does not hybridize to the deleted chromosome. Both the FISH and microsatellite methods work best in families where there is an affected male in whom the deletion can first be defined.

Mutation scanning in nondeletion families
If the multiplex screen fails to show a deletion, then defining the mutation is difficult. With 79 exons, screening the DNA exon by exon using SSCP or other standard

(A)

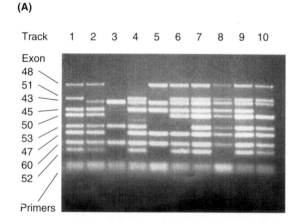

(B)

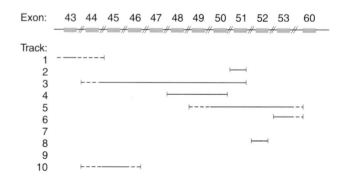

Figure 17.11: Multiplex deletion screen for dystrophin.

(A) Products of multiplex PCR amplification of nine exons, using samples from 10 unrelated patients with Duchenne/Becker muscular dystrophy. PCR primers have been designed so that each exon, with some flanking intron sequence, gives a different sized PCR product. Courtesy of Dr R. Mountford, Liverpool Women's Hospital. **(B)** Interpretation: solid lines show exons definitely deleted, dotted lines show possible extent of deletion running into untested exons. No deletion is seen in samples 7 and 9 – these patients may have point mutations, or deletions of exons not examined in this test. Exon sizes and spacing are not to scale.

(A)

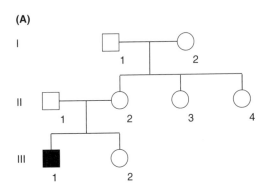

(B)

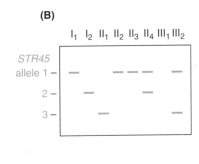

Figure 17.12: A family with Duchenne muscular dystrophy. **(A)** pedigree; **(B)** results of typing with the intragenic marker *STR45*.

The affected boy III-1 has a deletion which includes *STR45* (lane 7 of the gel is blank). His mother II-2 and his aunt II-3 inherited no allele of *STR45* from their mother I-2, showing that the deletion is being transmitted in the family. I-2 is apparently homozygous for this highly polymorphic marker (lane 2), but in fact is hemizygous. The other aunt II-4 and the sister III-2 are heterozygous for the marker, and therefore do not carry the deletion.

methods is excessively laborious. Missense mutations very seldom cause DMD, so the best method is probably to use the protein truncation test, as illustrated in *Figure 17.9*. Because the dystrophin gene consists of so many scattered small exons (averaging only 180 bp), PTT testing requires RNA. Dystrophin is primarily expressed in muscle, but in skilled hands low-frequency ectopic ('illegitimate') transcripts can be amplified from lymphocytes.

Gene tracking in nondeletion families

In the setting of a routine diagnostic service, if the multiplex deletion screen fails to show a deletion, or if there is no sample available from an affected male, the best answer may be gene tracking, as described below. DMD presents special problems for gene tracking because of the extremely high recombination frequency across the gene. Even intragenic markers show an average 5% recombination with the disease. Therefore it is prudent to use flanking markers for gene tracking, as in *Figure 17.13*.

New mutations and mosaicism in DMD

To complete the list of problems posed by DMD, there is a high frequency of new mutations. The mutation-selection equilibrium calculations in *Box 3.5* show that for any lethal X-linked recessive condition (f = 0), 1/3 of cases are fresh mutations. Therefore the mother of an isolated DMD boy has only a 2/3 chance of being a carrier. This greatly complicates the risk calculations that are necessary for interpreting gene tracking results. The interested reader should consult the book by Bridge (Further reading) for example calculations. Moreover, as shown in *Figure 3.7*, the first mutation carrier in a DMD pedigree is very often a mosaic (male or female). This raises yet more problems, both for risk estimation and for interpretation of the results of direct testing.

17.2 Gene tracking

Gene tracking was historically the first type of DNA diagnostic method to be widely used. Most of the mendelian diseases that form the bread-and-butter work of diagnostic laboratories went through a phase of gene tracking,

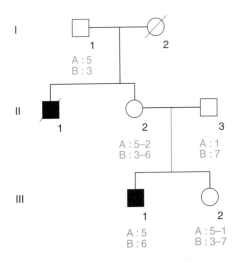

Figure 17.13: Gene tracking in Duchenne muscular dystrophy using flanking markers.

The family has been typed for two polymorphisms A and B that flank the dystrophin locus. III-2 can have inherited DMD only if she has one recombination between marker A and DMD and another between DMD and marker B. If the recombination fractions are θ_A and θ_B respectively, then the probability of a double recombinant is of the order $\theta_A\theta_B$, which typically will be well under 1%. III-1 has a recombination between marker locus A and DMD.

then moved on to direct tests once the genes were cloned. Huntington disease, cystic fibrosis and myotonic dystrophy are familiar examples. A similar progression is likely with any disease that is studied by the classic approach of linkage analysis followed by positional cloning. However, the progression is not inevitable. With some diseases, even though the gene has been cloned, mutations are hard to find. In some cases the known mutations are scattered widely over a large gene and, for others, mutation detection, for unknown reasons, has not so far been

very successful. Thus gene tracking using linked markers still has its place in modern molecular diagnosis.

17.2.1 Gene tracking involves three logical steps

Box 17.1 illustrates the essential logic of gene tracking. This logic can be applied to diseases with any mode of inheritance. Always there is at least one parent who could have passed on the disease allele to the proband, and who

---

Box 17.1

Gene tracking: four stages in the investigation of a late-onset autosomal dominant disease where direct mutation detection is not possible

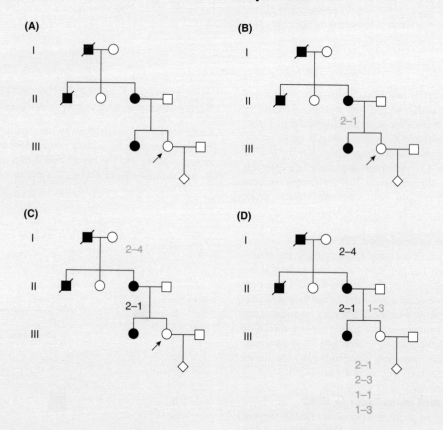

(A) III-2, who is pregnant, wishes a pre-symptomatic test to show whether she has inherited the disease allele.

(B) The first step is to tell the parent's two chromosomes apart. A marker, closely linked to the disease locus, is found for which II-3 is heterozygous.

(C) Next we must establish phase – that is, work out which marker allele in II-3 is segregating with the disease allele. I-2 is typed for the marker. II-3 must have inherited marker allele 2 from her mother, which therefore marks her unaffected chromosome. Her affected chromosome, inherited from her dead father, carries marker allele 1.

(D) By typing III-2 and her father we can work out which marker allele III-2 received from her mother. If she is 2–1 or 2–3, it is good news: she inherited marker allele 2 from her mother, which is the grandmaternal allele. If she types 1–1 or 1–3 it is bad news: she inherited the grandpaternal chromosome, which carries the disease allele.

Note that it is the segregation pattern in the family, and not the actual marker genotype, that is important: if III-2 has the same marker genotype, 2–1, as her affected mother, this is good and not bad news for her.

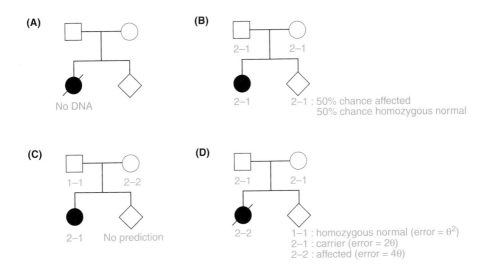

Figure 17.14: Gene tracking for prenatal diagnosis of an autosomal recessive disease.

Four families each have a child affected with a recessive disease. Direct mutation testing is not possible (either because the gene has not been cloned, or because the mutations could not be found). (**A**) No diagnosis is possible if there is no sample from the affected child. (**B**) If everybody has the same heterozygous genotype for the marker, the result is not clinically useful. (**C**) If the parents are homozygous for the marker, no prediction is possible with this marker. (**D**) Successful prediction. The error rates shown are the risk of predicting an unaffected pregnancy when the fetus is affected, or vice versa, if the marker used shows a recombination fraction θ with the disease locus. These examples emphasize the need for both an appropriate pedigree structure (DNA must be available from the affected child) and informative marker types.

may or may not have done so. The process always follows the three steps:

1. distinguish the two chromosomes in the relevant parent(s) – i.e. find a closely linked marker for which they are heterozygous;
2. determine phase – i.e. work out which chromosome carries the disease allele;
3. work out which chromosome the consultand received.

The prerequisites for gene tracking are:

1. the disease should be adequately mapped, so that markers can be used that are known to be tightly linked to the disease locus; and
2. the pedigree structure and sample availability must allow determination of phase.

Nowadays, informativeness of the marker is not usually a big problem. With over 10 000 highly polymorphic microsatellites mapped across the human genome, it is almost always possible to find an informative marker that maps close to the disease locus.

Figure 17.14 shows gene tracking for an autosomal recessive disease. The pedigrees emphasize the need for both an appropriate pedigree structure (DNA must be available from the affected child) and informative marker types. Even if the affected child is dead, if the Guthrie card (Section 17.1.1) can be retrieved, sufficient DNA for PCR typing can usually be extracted from the dried blood spot.

17.2.2 Recombination sets a fundamental limit on the accuracy of gene tracking

Because the DNA marker used for gene tracking is not the sequence that causes the disease, there is always the possibility of making a wrong prediction if recombination separates the disease and the marker. The recombination fraction, and hence the error rate, can be estimated from family studies by standard linkage analysis (Chapter 11). With almost any disease there should be a good choice of markers showing less than 1% recombination with the disease locus. This follows from the observations that one nucleotide in 500 is polymorphic, and that loci 1 Mb apart show approximately 1% recombination (Section 11.1.4). Ideally one uses an intragenic marker, such as a microsatellite within an intron. The special problem of the DMD recombination hotspot has been mentioned above.

Recombination between marker and disease can never be ruled out, even for very tightly linked markers, but the error rate can be greatly reduced by using two marker loci, situated on opposite sides of the disease locus. With such flanking or bridging markers, a recombination between either marker and the disease will also produce a marker–marker recombinant, which can be detected (e.g. III-1, *Figure 17.13*). If a marker–marker recombinant is seen in the consultand, then no prediction can be made about inheritance of the disease, but at least a false prediction has been avoided. Provided no marker–marker recombinant is seen, the only residual risk is that of double recombinants. The true probability

Box 17.2

Use of Bayes' theorem for combining probabilities

A formal statement of Bayes' theorem is:

$$P(H_i \mid E) = P(H_i).P(E \mid H_i) / \sum_i [P(H_i).P(E \mid H_i)]$$

$P(H_i)$ means the probability of the ith hypothesis, and the vertical line means 'given', so that $P(E \mid H_i)$ means the probability of the evidence (E) given hypothesis H_i.

The steps in performing a Bayesian calculation are as follows:

(i)   Set up a table with one column for each of the alternative hypotheses. Cover all the alternatives.

(ii)   Assign a **prior probability** to each alternative. The prior probabilities of all the hypotheses must sum to 1. It is not important at this stage to worry about exactly what information you should use to decide the prior probability, as long as it is consistent across the columns. You will not be using all the information (otherwise there would be no point in doing the calculation because you would already have the answer) and any information not used in the prior probability can be used later.

(iii)   Using one item of information not included in the prior probabilities, calculate a conditional probability for each hypothesis. The conditional probability is the probability of the information, given the hypothesis $P(E \mid H_i)$ [*not* the probability of the hypothesis given the information, $P(H_i \mid E)$]. The conditional probabilities for the different hypotheses do not necessarily sum to 1.

(iv)   If there are further items of information not yet included, repeat step (iii) as many times as necessary until all information has been used once and once only. The end result is a number of lines of conditional probabilities in each column.

(v)   Within each column, multiply together the prior and all the conditional probabilities. This gives a joint probability. The joint probabilities do not necessarily sum to 1 across the columns.

(vi)   If there are just two columns, the joint probabilities can be used directly as odds. Alternatively the joint probabilities can be scaled to give final probabilities which do sum to 1. This is done by dividing each joint probability by the sum of all the joint probabilities.

- the probability of disease–marker and marker–marker recombination;
- uncertainty, due to imperfect pedigree structure or limited informativeness of the markers, about who transmitted what marker allele to whom (see *Figure 11.4C* for an example);
- uncertainty as to whether somebody in the pedigree carries a newly mutant disease allele (see *Figure 3.7* for an example of this problem in DMD).

Two alternative methods are available for performing the calculation.

Bayesian calculations

Bayes' theorem provides a general method for combining probabilities into a final overall probability. The theory and procedure are shown in *Box 17.2*, and a sample calculation is set out in *Figure 17.15*. A very detailed set of calculations covering almost every conceivable situation in DNA diagnostics can be found in the book by Bridge (Further reading), which the interested reader should consult.

For simple pedigrees, Bayesian calculations give a quick answer, but for more complex pedigrees the calculations can get very elaborate. Few people feel fully confident of their ability to work through a complex pedigree correctly, although the attempt is a valuable mental exercise for teasing out the factors contributing to the final risk. An alternative is to use a linkage analysis program.

Using linkage programs for calculating genetic risks

At first sight it may seem surprising that a program designed to calculate lod scores can also calculate genetic risks – but in fact the two are closely related (*Figure 17.16*). Linkage analysis programs are general-purpose engines for calculating the likelihood of a pedigree, given certain data and assumptions. For calculating the likelihood of linkage we calculate the ratio:

$$\frac{\text{likelihood of data} \mid \text{linkage, recombination fraction } \theta}{\text{likelihood of data} \mid \text{no linkage } (\theta = 0.5)}$$

For estimating the risk that a proband carries a disease gene, we calculate the ratio:

$$\frac{\text{likelihood of data} \mid \text{proband is a carrier, recombination fraction } \theta}{\text{likelihood of data} \mid \text{proband is not a carrier, recombination fraction } \theta}$$

As in *Box 17.2*, the vertical line | means 'given'.

of a double recombinant is very low because of interference (Section 11.1.3). With a suitable choice of markers, the risk of an error due to unnoticed recombination is likely to be much smaller than the risk of a wrong prediction due to human error in obtaining and processing the DNA samples.

17.2.3 Calculating risks in gene tracking

Unlike direct testing, gene tracking always involves a calculation. Factors to be taken into account in assessing the final risk include:

17.3 Population screening

Population screening follows naturally from the ability to test directly for the presence of a mutation. Traditionally a distinction is drawn between screening and diagnosis. A screening test defines a high-risk group, who are then

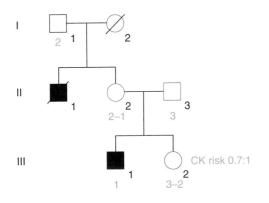

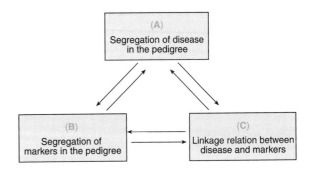

Figure 17.16: Use of linkage analysis programs for calculating genetic risks.

Given information on any two of these subjects, the program can calculate the third. For linkage analysis, the program is given (**A**) and (**B**), and calculates (**C**). For calculating genetic risks, the program is given (**B**) and (**C**), and calculates (**A**).

Hypothesis: III₂ is	A carrier	Not a carrier
Prior probability	1/2	1/2
Conditional (1): DNA result	0.05	0.95
Conditional (2): CK data	0.7	1
Joint probability	0.0175	0.475
Final probability	0.0175/0.4925	0.475/0.4925
	= 0.036	= 0.964

Figure 17.15: A Bayesian calculation of genetic risk.

III-2 wishes to know her risk of being a carrier of DMD, which affected her brother III-1 and uncle II-1. Serum creatine kinase testing (an indicator of subclinical muscle damage common in DMD carriers) gave carrier:noncarrier odds of 0.7:1. A DNA marker that shows on average 5% recombination with DMD gave the types shown. The risk calculation, following the guidelines in *Box 17.2*, gives her overall carrier risk as 3.6%.

given a definitive diagnostic test. DNA tests are rather different because there are no separate screening and diagnostic tests. However, proposals to introduce any population screening test still need to satisfy the same criteria (*Table 17.7*), regardless of the technology used.

17.3.1 Acceptable screening programs must fit certain criteria

What would screening achieve?

The most important single function of any screening program is to produce some useful outcome. It is quite unacceptable to tell people out of the blue that they are at risk of something unpleasant, unless the knowledge enables them to do something about the risk. Proposals to screen for genes conferring susceptibility to breast cancer or heart attacks must be assessed stringently against this

Table 17.7: Requirements for a population screening program

Requirement	Examples and comments
A positive result must lead to some useful action	▪ Preventive treatment, e.g. special diet for PKU ▪ Review and choice of reproductive options in CF carrier screening
The whole program must be socially and ethically acceptable	▪ Subjects must opt in with informed consent ▪ Screening without counseling is unacceptable ▪ There must be no pressure to terminate affected pregnancies ▪ Screening must not be seen as discriminatory
The test must have high sensitivity and specificity	▪ Tests with many false negatives undermine confidence in the program ▪ Tests with many false positives, even if these are subsequently filtered out by a definitive diagnostic test, can create unacceptably high levels of anxiety among normal people
The benefits of the program must outweigh its costs	▪ It is unethical to use limited health care budgets in an inefficient way

criterion. Predictive testing for HD might appear to break this rule – but it is offered only to people who are at high risk of HD and who are suffering such agonies of uncertainty that they request a predictive test, and persist despite counseling in which all the disadvantages are pointed out.

Ideally the useful outcome is treatment, as in neonatal screening for phenylketonuria. Increased medical surveillance is a useful outcome only if it greatly improves the prognosis. A special case is screening for carrier status, where the outcome is the possibility of avoiding the birth of an affected child. People unwilling to accept prenatal diagnosis and termination of affected pregnancies would not see this as a useful outcome, and in general should not be screened.

An ethical framework for screening

Ethical issues in genetic population screening have been discussed by a committee of distinguished American geneticists, clinicians, lawyers and theologians, and the reader is referred to their report for a very detailed survey (Andrews *et al.*, 1994). It is in the nature of ethical problems that they have no solutions, but certain principles emerge.

- Any program must be voluntary, with subjects taking the positive decision to opt in.
- Programs must respect the autonomy and privacy of the subject.
- People who score positive on the test must not be pressured into any particular course of action. For example, in countries with insurance-based health care systems, it would be unacceptable for insurance companies to put pressure on carrier couples to accept prenatal diagnosis, or financial pressure or inducements to terminate affected pregnancies.
- Information should be confidential. This may seem obvious, but it can be a difficult issue – we like to think that drivers of heavy trucks or jumbo jets have been tested for all possible risks. Societies with insurance-based health care systems have particular problems about the confidentiality of genetic data, since insurance companies will argue that they are penalizing low-risk people by not loading the premiums of high-risk people.

17.3.2 Specificity and sensitivity measure the technical performance of a screening test

Compared with the ethical problems, the technical questions in population screening are fairly simple. The performance of a test can be measured by its sensitivity and specificity (*Figure 17.17*).

Specificity of a test

Unexpectedly, perhaps, false positive test results can pose a more serious problem than false negatives. If the specificity is low, then a positive test result does not mean much. Even if the false positives can be filtered out sub-

	Affected	Not affected
+ve on test	a	b
–ve on test	c	d

Sensitivity of test = a/(a + c)

Specificity of test = d/(b + d)

Figure 17.17: Sensitivity and specificity of a screening test.

sequently by a diagnostic test, many people will have been worried unnecessarily. *Table 17.8* shows that if a test does have a significant false positive rate, then the specificity is hopelessly low except when testing for very common conditions. DNA tests are potentially valuable for population screening because, compared with biochemical tests, they should generate very few false positives. The most likely causes of false positives in DNA testing are laboratory or clerical errors.

Sensitivity of a test

A test must pick up a reasonable proportion of its intended target (i.e. the sensitivity must be high). While the specificity of DNA tests looks encouraging, the sensitivity usually depends on the degree of allelic heterogeneity. Unless a disease is unusually homogeneous (*Table 17.1*), it is not practicable to test for every conceivable mutation, especially in a large-throughput population screening program. Normally only a subset of mutations will be tested for. *Figure 17.18* shows how the choice of mutations can affect the outcome of a CF carrier screening program.

It is clear that simply testing for the commonest mutation, F508del, would not produce an acceptable program. Almost as many affected children would be born to couples negative on the screening program as would be detected by the screening. Whether or not such a program were financially cost-effective, it would surely be socially unacceptable. What constitutes an acceptable program is harder to define. One suggestion focuses on '+/−' couples (i.e. couples with one known carrier and the partner negative on all the tests). The partner still might be a carrier of a rare mutation. Professor LP Ten Kate suggested that an acceptable screening program is one in which the risk for such +/− couples is no higher than the general population risk before screening. That would require a sensitivity of about 95%.

17.3.3 Organization of a genetic screening program

Assuming the proposed program looks ethically acceptable and cost-effective, who should be screened? Three examples highlight some of the options.

Table 17.8: A test that performs well in the laboratory may be useless for population screening

Prevalence of condition	True positives in population screened	True positives detected by screening	True negatives in population screened	False positives detected by screening	Ratio of true:total positives
1/1000	1000	990	999 000	9990	0.09
1/10 000	100	99	999 900	9999	0.0098
1/100 000	10	10	999 990	10 000	0.001

In a laboratory trial on a panel of 100 affected and 100 control people, this hypothetical test was 99% accurate: it gave a positive result for 99% of true positives, and a negative result for 99% of true negatives. The table shows results of screening 1 million people. The great majority of all people positive on the test are false positives. Such a test is unlikely to be acceptable socially or viable financially for any mendelian disease (these typically affect less than 1 person in 1000).

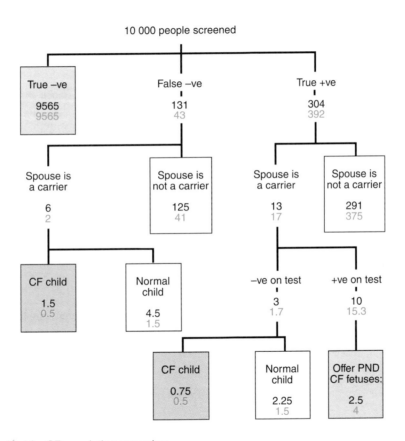

Figure 17.18: Flowchart for CF population screening.

Results of screening 10 000 people, 1:23 of whom is a carrier. If a person tests positive, his/her spouse is then tested. Black figures show results using a test which detects 70% of CF mutations (i.e. testing for F508del only); blue figures show results for a test with 90% sensitivity. Blue boxes represent cases which would be seen as successes for the screening program (regardless of what action they then take), gray boxes represent failures. PND, prenatal diagnosis.

Neonatal screening: screening for phenylketonuria

All babies in the UK are tested a few days after birth for PKU. A blood spot from a heel-prick is collected on a card (the Guthrie card) during a home visit and sent to a central laboratory. The phenylalanine level in the blood is measured by chromatography or a bacterial growth test. This is the screening test. Babies whose level is above a threshold are called in for a definitive diagnostic test.

Only a small proportion eventually turn out to have PKU. The lack of informed consent on the part of the infant is justified by the benefit it receives from dietary treatment (Smith, 1993).

Prenatal screening: screening for β-thalassemia

Carriers of β-thalassemia can be detected by conventional hematological testing, either before marriage or in the antenatal clinic. Carrier–carrier couples can be offered prenatal diagnosis by DNA analysis. Pre-implantation diagnosis or fetal stem-cell transplants may become alternatives to termination of affected pregnancies. Two ethnic groups in the UK have a high incidence of β-thalassemia: Cypriots and Pakistanis. Screening was quickly accepted by Cypriots in the UK but uptake has been slower among Pakistanis. The comparison illustrates the complex social questions surrounding genetic screening and the relevance of cultural background (Gill and Modell, 1998). Importantly, long-term studies of the Cypriot community show how the success of screening can be measured, not by counts of affected fetuses aborted, but by counts of couples having normal families. Before screening was available, many carrier couples opted to have no children; now they are using screening and having normal families (Modell *et al.*, 1984).

Population screening for carriers: proposals for cystic fibrosis

It is now technically feasible and financially worthwhile to screen northern European populations to detect CF carriers. Surveys in the UK suggest that most carrier–carrier couples would opt for prenatal diagnosis and would value the opportunity to ensure that they did not have affected children. This view might change if treatment becomes more effective, for example using gene therapy.

If a screening program is to be introduced, two sets of questions must be considered. How many mutations should the laboratory test for, and who should be offered the test? The problems raised by allelic heterogeneity have been discussed above (see *Figure 17.18*). On the question of who to screen, *Table 17.9* shows some possibilities considered in the UK. Naturally the way health care delivery is organized in each country will determine the range of possibilities. Preliminary results from controlled pilot studies suggest that none of the methods has had the negative effects (increased anxiety) sometimes predicted.

17.4 DNA profiling can be used for identifying individuals and determining relationships

We use the term DNA profiling to refer to the general use of DNA tests to establish identity or relationships. DNA fingerprinting is reserved for the technique invented by Jeffreys *et al.* (1985) using multilocus probes. For more detail on this whole area, the reader should consult the book by Evett and Weir (Further reading).

17.4.1 A variety of different DNA polymorphisms have been used for profiling

DNA fingerprinting using minisatellite probes

These probes contain the common core sequence of a hypervariable dispersed repetitive sequence GGGCA-GGAXG, first discovered by Jeffreys *et al.* (1985) in the myoglobin gene (see Section 7.4.2). When hybridized to Southern blots they give an individual-specific fingerprint of bands (*Figure 17.19*). Their chief disadvantage is that it is not possible to tell which pairs of bands in a fingerprint represent alleles. Thus, when matching DNA fingerprints, one matches each band individually by position and intensity. Other hypervariable repeated sequences have been used in the same way, for example those detected by the synthetic oligonucleotide $(CAC)_5$ (Krawczak and Schmidtke, 1998).

DNA profiling using single-locus minisatellite markers

Minisatellite probes recognize single-locus variable

Table 17.9: Possible ways of organizing population screening for carriers of CF

Group tested	Advantages	Disadvantages
Neonates	Easily organized	No consequences for 20 years Many families would forget the result Unethical to test children
School leavers	Easily organized Inform people before they start relationships	Difficult to conduct ethically Risk of stigmatization of carriers
Couples from physician's lists	Couple is unit of risk Stresses physician's role in preventitive medicine Allows time for decisions	Difficult to control quality of counseling
Women in antenatal clinic	Easily organized Rapid results	Bombshell effect for carriers Partner may be unavailable Time pressure on laboratory
Adult volunteers ('drop-in CF center')	Few ethical problems	Bad framework for counseling No targeting to suitable users May be inefficient use of resources

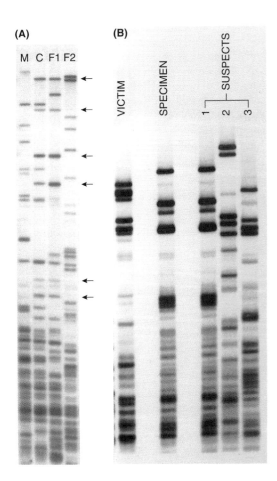

Figure 17.19: Legal and forensic use of DNA fingerprinting.

(**A**) A paternity test. 'Fingerprints' are shown from the mother (M), child (C) and two possible fathers (F1, F2). The DNA fingerprint of F1 contains all the paternal bands found in the child, whereas that of F2 contains only one of the paternal bands. (**B**) A rape case. The 'fingerprint' of suspect 1 exactly matches that from the semen sample S on a vaginal swab from the victim. As a result of this evidence, Suspect 1 was charged with rape and found guilty. Photograph courtesy of Cellmark Diagnostics, Abingdon, Oxfordshire.

tandem repeats on Southern blots. Each probe should reveal two bands in any person's DNA, representing the two alleles. Profiling is based on four to ten different polymorphisms. These probes allow exact calculations of probabilities (of paternity, of the suspect not being the rapist, etc.), if the gene frequency of each allele in the population is known. For matching alleles between different gel tracks, the continuously variable distance along the gel has to be divided into a number of 'bins'. Bands falling within the same bin are deemed to match. It is imperative that the criteria used for judging matches in each profiling test should be the same binning criteria that were used to calculate the population frequencies of each allele. The binning criteria can be arbitrary within

certain limits, but they must be consistent. Minor variations within repeated units of some minisatellites potentially allow an almost infinite variety of alleles to be discriminated, so that the genotype at a single locus might suffice to identify an individual (Jeffreys *et al.*, 1991).

DNA profiling using microsatellite markers

Microsatellite polymorphisms (Section 7.4.3) are based on short tandem repeats, usually di-, tri- or tetranucleotides. They have the advantages over minisatellites that they can be typed by PCR and that discrete alleles can be defined unambiguously by the precise repeat number. This avoids the binning problem and makes it easier to relate the results to population gene frequencies.

The use of Y-chromosome and mitochondrial polymorphisms

For tracing relationships to dead persons, Y-chromosome and mitochondrial DNA polymorphisms are especially useful because of their sex-specific pattern of transmission. An interesting example was the identification of the remains of the Russian Tsar and his family, killed by the Bolsheviks in 1917, by comparing DNA profiles of excavated remains with living distant relatives (Gill *et al.*, 1994).

17.4.2 DNA profiling can be used to determine the zygosity of twins

In studying nonmendelian characters (Chapter 19), and sometimes in genetic counseling, it is important to know whether a pair of twins are monozygotic (MZ, identical) or dizygotic (DZ, fraternal). Traditional methods depended on an assessment of phenotypic resemblance or on the condition of the membranes at birth (twins contained within a single chorion are always MZ, though the converse is not true). Errors in zygosity determination systematically inflate heritability estimates for non-mendelian characters, because very similar DZ twins are wrongly counted as MZ, while very different MZ twins are wrongly scored as DZ.

Genetic markers provide a much more reliable test of zygosity. The extensive literature on using blood groups for this purpose is summarized by Race and Sanger (1975). DNA profiling is nowadays the method of choice. The Jeffreys fingerprinting probe allows a very simple test – samples from MZ twins look like the same sample loaded twice, and samples from DZ twins show some differences. An error rate could be calculated from empirical data on band sharing by unrelated people, using some defined binning strategy (see above).

When single-locus markers are used, if twins give the same types, then for each locus, the probability that DZ twins would type alike is calculated. If the parents have been typed, this follows from mendelian principles; otherwise the probability of DZ twins typing the same must be calculated for each possible parental mating and weighted by the probability of that mating calculated from population gene frequencies. The resultant probabilities for each (unlinked) locus are multiplied, to give an overall likelihood P_I that DZ twins would give the same results with all the markers used. The probability that the twins are MZ is then:

$$P_m = m \, / \, [m + (1 - m)P_I]$$

where m is the proportion of twins in the population who are MZ (about 0.4 for like-sex pairs). Sample calculations are given in Appendix 4 of Vogel and Motulsky (Further reading).

17.4.3 DNA profiling can be used to disprove or establish paternity

Excluding paternity is fairly simple – if the child has a marker allele not present in either the mother or alleged father then, barring new mutations, the alleged father is not the biological father. Proving paternity is, in principle, impossible – one can never prove that there is not another man in the world who could have given the child that particular set of marker alleles. All one can do is establish a probability of nonpaternity that is low enough to satisfy the courts and, if possible, the putative father.

DNA fingerprinting probes have been widely used for this purpose (*Figure 17.19*). Bands must be binned according to an arbitrary but consistent scheme, as explained above, to decide whether or not each nonmaternal band in the child fits a band in the alleged father. Then if, say, 10/10 bands fit, the odds that the suspect, rather than a random man from the population, is the father are $1:p^{10}$, where p is the chance that a random man from that population would have a band matching a given band in the child. Even for $p = 0.2$, p^{10} is only 10^{-7}. Single-locus probes allow a more explicit calculation of the odds (*Figure 17.20*). A series of four to ten unlinked single-locus markers can give overwhelming odds favoring paternity if all the bands fit.

17.4.4 DNA profiling is a powerful tool for forensic investigations

DNA profiling for forensic purposes follows the same principles as paternity testing. Scene-of-crime material (bloodstains, hairs or a vaginal swab from a rape victim) are typed and matched to a DNA sample from the suspect. If the bands don't match, the suspect is excluded. One of the most powerful applications of DNA profiling is for preventing miscarriages of justice. If the bands do all match, the odds that the criminal is the suspect rather than a random member of the population can be calculated, based on the allele frequencies in the population. Of course, if the alternative were the suspect's brother, the odds would look very different. The fate of DNA evidence in courts provides a fascinating insight into the difference between scientific and legal cultures. There are at least three stumbling blocks for DNA data.

- The jury may simply not believe, or perhaps choose to ignore, the DNA data, as evidently happened in the OJ Simpson trial. A fascinating account of the DNA evidence is given by Weir (1995).
- The jury may be led into a false probability argument, the so-called Prosecutor's Fallacy. Suppose a suspect's DNA profile matches the scene-of-crime sample. The Prosecutor's Fallacy confuses two different probabilities:

  (i) the probability the suspect is innocent, given the match;
  (ii) the probability of a match, given that the suspect is innocent.

The jury should consider the first probability, not the second.

Using Bayesian notation (*Box 17.2*), with M = match, G = suspect is guilty, I = suspect is innocent, we want to calculate $P_I | M$, and not $P_M | I$. If the suspect were guilty, the samples would necessarily match: $P_M | G = 1$. Population genetic arguments might say there is a 1 in 10^6 chance that a randomly selected person would have the same profile as the crime sample: $P_M | I = 10^{-6}$. Suppose the guilty person could have been any one of 10^7 men in the local population. If there is no other evidence to implicate him, he is simply a random member of the population and the prior probability that he is guilty (before considering the DNA evidence) is $P_G = 10^{-7}$. The prior probability that he is innocent is $P_I = 1 - 10^{-7}$, ≈ 1. Baye's theorem tells us that

$$P_I | M = (P_I.P_M | I) \, / \, [(P_I.P_M | I) + (P_G.P_M | G)]$$
$$= 10^{-6} / \, (10^{-6} + 10^{-7})$$
$$= 1.0/1.1 = 0.9$$

The prosecutor would no doubt be happy to see the jury use 10^6 instead of 0.9 for the probability that the suspect is innocent! Given the Bayesian argument, it is clear that a forensic test needs $P_M | I$ to be 10^{-10} or less if it is to be able to convict a suspect on DNA evidence alone.

- Objections may be raised to some of the principles by which DNA-based probabilities are calculated.

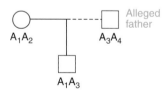

Figure 17.20: Using single-locus markers for a paternity test.

The odds that the alleged father, rather than a random member of the population, is the true father are $1/2:q_3$, where q_3 is the gene frequency of A_3. A series of n unlinked markers would be used and, if paternity were not excluded, the odds would be $(1/2)^n : q_A.q_B.q_C...q_N$.

(i) The multiplicative principle, that the overall probabilities can be obtained by multiplying the individual probability for each band or locus, depends on the assumption that bands are independent. If the population were actually stratified into reproductively isolated groups, each of whom tended to have a particular subset of bands or alleles, the calculation would be misleading. This is serious because it is the multiplicative principle that allows such exceedingly definite likelihoods to be given.

(ii) For single-locus markers, the probability depends on the gene frequencies. DNA profiling laboratories maintain databases of gene frequencies – but were these determined in an appropriate ethnic group for the case being considered? Taken to

extremes, this argument implies that the DNA evidence might identify the criminal as belonging to a particular ethnic group, but would not show which member of the group it was who committed the crime.

These issues have been debated at great length, especially in the American courts. Both objections are valid in principle, but the question is whether they make enough difference to matter. General opinion is that they do not. It would be ironic if courts, seeing opposing expert witnesses giving odds of correct identification differing a million-fold (10^5:1 versus 10^{11}:1), were to decide that DNA evidence is hopelessly unreliable, and turn instead to eyewitness identification (odds of correct identification < 50:50).

Further reading

Bridge PJ (1994) *The Calculation of Genetic Risks – Worked Examples in DNA Diagnostics.* Johns Hopkins University Press, Baltimore, MD.

Clinical Molecular Genetics Society Best Practice Guidelines for Molecular Genetics Services. http://www.cmgs.org

Cotton RGH, Edkins E, Forrest S (eds) (1998) *Mutation Detection: a Practical Approach.* IRL Press, Oxford.

Evett IW, Weir BS (1998) *Interpreting DNA Evidence: Statistical Genetics for Forensic Scientists.* Sinauer Associates, Inc., Sunderland, MA.

Newton CR, Graham A (1997) *PCR*, 2nd edn. BIOS Scientific Publishers, Oxford.

Vogel F, Motulsky AG (1996) *Human Genetics*, 3rd edn. Springer-Verlag, Berlin.

References

Andrews LB, Fullarton JE, Holtzman NA, Motulsky AG (1994) *Assessing Genetic Risks – Implications for Health and Social Policy.* National Academy Press, Washington, DC.

Dhanda RK, Smith RM, Scott CB, Eng C, Vijk J (1998) A simple system for automated two-dimensional electrophoresis: applications to genetic testing. *Genetic Testing*, **2**, 67–70.

Gill P, Ivanov PL, Kimpton C et al. (1994) Identification of the remains of the Romanov family by DNA analysis. *Nature Genet.*, **6**, 130–135.

Gill PS, Modell B (1998) Thalassaemia in Britain: a tale of two communities. Births are rising among British Asians but falling in Cypriots. *Br. Med. J.*, **317**, 761–762.

Hacia JG (1999) Resequencing and mutational analysis using oligonucleotide microarrays. *Nature Genet.*, **21** (suppl.) 42–47.

Jeffreys AJ, Wilson V, Thein LS (1985) Individual-specific fingerprints of human DNA. *Nature*, **314**, 67–73.

Jeffreys AJ, MacLeod A, Tamaki K, Neil DL, Monckton DG (1991) Minisatellite repeat coding as a digital approach to DNA typing. *Nature*, **354**, 204–209.

Keen J, Lester D, Inglehearn C, Curtis A, Bhattacharya S (1991) Rapid detection of single base mismatches as heteroduplexes on Hydrolink gels. *Trends Genet.*, **7**, 5.

Krawczak M, Schmidtke J (1998) *DNA Fingerprinting*, 2nd edn. BIOS Scientific Publishers, Oxford.

Modell B, Petrou M, Ward RH et al. (1984) Effect of fetal diagnostic testing on birth-rate of thalassaemia major in Britain. *Lancet*, **ii**, 1383–1386.

Nickerson DA, Kaiser R, Lappin S, Stewart J, Hood L (1990) Automated DNA diagnostics using an ELISA-based oligonucleotide ligation assay. *Proc. Natl Acad. Sci. USA*, **87**, 8923–8927.

Race RR, Sanger R (1975) *Blood Groups in Man*, 6th edn. Blackwell, Oxford.

Sarkar G, Yoon HS, Sommer SS (1992) Dideoxy fingerprinting (ddF): a rapid and efficient screen for the presence of mutations. *Genomics*, **13**, 441–443.

Sheffield VC, Beck JS, Kwitek AE, Sandstrom DW, Stone EM (1993) The sensitivity of single strand conformational polymorphism analysis for the detection of single base substitutions. *Genomics*, **16**, 325–332.

Smith I, MRC Working Party on Phenylketonuria (1993) Phenylketonuria due to phenylalanine hydroxylase deficiency: an unfolding story. *Br. Med. J.*, **306**, 115–119.

van der Luijt R, Khan PM, Vasen H, van Leeuwen C, Tops C, Roest PA *et al*. (1994) Rapid detection of translation-terminating mutations at the adenomatous polyposis coli (APC) gene by direct protein truncation test. *Genomics*, **20**, 1–4.

Weir BS (1995) DNA statistics in the Simpson matter. *Nature Genet.*, **11**, 365–368.

Yau SC, Bobrow M, Mathew CG, Abbs S (1996) Accurate diagnosis of carriers of deletions and duplications in Duchenne/Becker muscular dystrophy by fluorescent dosage analysis. *J. Med. Genet.*, **33**, 550–558.

Cancer genetics

18.1 Cancer is the natural end-state of multicellular organisms

Any population of organisms that shows hereditary variation in reproductive capacity will evolve by natural selection. Genotypes that reproduce faster or more extensively will come to dominate later generations, only to be supplanted in turn by yet more efficient reproducers. Exactly the same applies to the population of cells that constitutes a multicellular organism like man. Cellular proliferation is under genetic control, and if somatic mutation creates a variant that proliferates faster, the mutant clone will tend to take over the organism. Thus people have a natural tendency to turn into tumors.

Tumors, however, are not efficient at having babies and caring for them. At the level of the whole organism, there is powerful selection for mechanisms that prevent a person turning into a tumor, at least until she has produced and brought up her children. Thus we are ruled by two opposing sets of selective forces. One set of selective forces, however, is short term and the other long term. The evolution from a normal somatic cell to a malignant tumor takes place within the life of an individual, and has to start afresh with each new individual. But an organism with a good anti-tumor mechanism transmits it to its offspring, where it continues to evolve. A billion years of evolution have endowed us with sophisticated interlocking and overlapping mechanisms to protect us against tumors, at least during our reproductive life. Potential tumor cells are either repaired and brought back into line, or made to kill themselves (apoptosis). No single mutation can escape these mechanisms and convert a normal cell into a malignant one. Long ago, studies of the age-dependence of cancer suggested that on average 6–7 successive mutations are needed to convert a normal cell into an invasive carcinoma. In other words, only if half a dozen independent defenses are disabled by mutation can a normal cell convert into a malignant tumor.

The chance of a single cell undergoing six independent mutations is negligible, suggesting that cancer should be vanishingly rare. However, two general mechanisms exist that can allow the progression to happen (*Box 18.1*). Accumulating all these mutations nevertheless takes time, so that cancer is mainly a disease of post-reproductive life, when there is little selective pressure to improve the defenses still further.

18.2 Mutations in cancer cells typically affect a limited number of pathways

Not surprisingly, carcinogenic mutations usually affect the genes that control the birth (cell cycling) or death

---

Box 18.1

Two ways of making a series of successive mutations more likely

Turning a normal cell into a malignant cancer cell requires perhaps six specific mutations in the one cell. If a typical mutation rate is 10^{-7} per gene per cell, it is extremely unlikely that any one cell should suffer so many mutations (which is why most of us are alive). The probability of this happening to any one of the 10^{13} cells in a person is $10^{13} \times 10^{-42}$, or 1 in 10^{29}. Cancer nevertheless happens because of a combination of two mechanisms:

- Some mutations enhance cell proliferation, creating an expanded target population of cells for the next mutation (*Figure 18.1*).
- Some mutations affect the stability of the entire genome, at either the DNA or the chromosomal level, increasing the overall mutation rate.

Because cancers depend on these two mechanisms, they always develop in stages, starting with tissue hyperplasia or benign growths, while malignant tumor cells usually advertise their genomic instability by their bizarre karyotypes (*Figure18.6*).

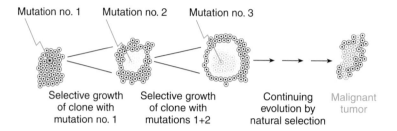

Figure 18.1: Multistage evolution of cancer.

Each successive mutation gives the cell a growth advantage, so that it forms an expanded clone, thus presenting a larger target for the next mutation.

(apoptosis) of cells. Two broad categories can be distinguished, although as always in biology, they are more tools for thinking about cancer than watertight exclusive classifications:

- **Oncogenes** (Sections 18.3 and 18.4). These are genes whose normal activity promotes cell proliferation. Gain of function mutations in tumor cells create forms that are excessively or inappropriately active. A single mutant allele may affect the phenotype of the cell. The non-mutant versions are properly called **proto-oncogenes**.
- **Tumor suppressor (TS) genes** (Section 18.5). TS gene products inhibit events leading towards cancer. Mutant versions in cancer cells have lost their function. Some TS gene products prevent cell cycle progression, some steer deviant cells into apoptosis, and others keep the genome stable and mutation rates low by ensuring accurate replication, repair and segregation of the cell's DNA. Both alleles of a TS gene must be inactivated to change the behavior of the cell.

By analogy with a bus, one can picture the oncogenes as the accelerator and the tumor suppressor genes as the brake. Jamming the accelerator on (a dominant gain of function of an oncogene) or having all the brakes fail (a recessive loss of function of a TS gene) will make the bus run out of control. Alternatively, a saboteur could simply loosen nuts and bolts at random (inactivate the TS genes that safeguard the integrity of the genome – see Section 18.7) and wait for a disaster to happen.

18.3 Oncogenes

18.3.1 Animal tumor viruses provided the first evidence of oncogenes

For many years it has been known that some animal leukemias, lymphomas and cancers are caused by viruses. A few human examples are also known (*Table 18.1*). Tumor viruses fall into three broad classes:

- **DNA viruses** normally infect cells lytically. They cause tumors by rare anomalous integrations into the DNA of non-permissive host cells (cells that do not support lytic infection). One way or another, integration of the viral genome implants the transcriptional activation or replication signals of the virus into the host genome and triggers cell proliferation. Some of the viral genes involved have been identified, such as those for the T-antigen of SV40 or E1A and E1B of

Table 18.1: Human and animal tumor viruses

Species	Disease	Virus	Type	Oncogene
Monkey	Sarcoma	SV40	DNA	T-antigen
Mouse	(Transformation *in vitro*)	Adenoviruses	DNA	*E1A, E1B*
Man	Cervical cancer	Papilloma virus HPV16	DNA	*E6, E7*
Man	Nasopharyngeal cancer	Epstein–Barr virus	DNA	*BNLF-1* (?)
Man	T-cell leukemia	HTLV-1, HTLV-2	RNA	–
Man	Kaposi sarcoma	HIV-1	RNA	–
Chicken	Sarcoma	Rous sarcoma virus	ATR	*src*
Rat	Sarcoma	Harvey rat sarcoma virus	ATR	*H-ras*
Mouse	Leukemia	Abelson leukemia virus	ATR	*abl*
Monkey	Sarcoma	Simian sarcoma virus	ATR	*sis*
Chicken	Erythroleukemia	Erythroleukemia virus	ATR	*erb-b*
Chicken	Sarcoma	Avian sarcoma virus 17	ATR	*jun*
Mouse	Osteosarcoma	FBJ osteosarcoma	ATR	*fos*
Cat	Sarcoma	McDonough feline sarcoma virus	ATR	*fms*
Chicken	Myelocytoma	Avian myelocytomatosis virus	ATR	*myc*

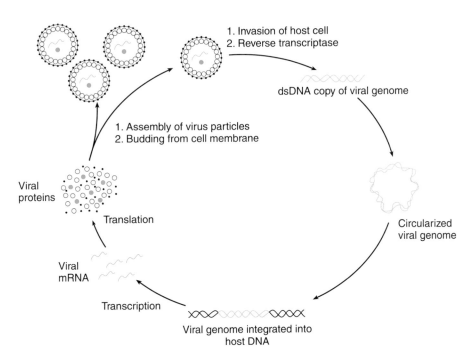

Figure 18.2: Retroviral life cycle.

The virus particle (top) contains the RNA genome and viral reverse transcriptase within an outer lipoprotein envelope and inner protein capsid. A double-stranded DNA copy of the viral genome integrates into the host DNA. Here it directs synthesis of viral RNA and proteins, which self-assemble and bud off from the cell membrane. The host cell is not killed.

adenoviruses. Unlike the classic retroviral oncogenes (see below) these genes are virus-specific and do not have exact cellular counterparts.

■ **Retroviruses** have a genome of RNA. They replicate via a DNA intermediate, which is made using a viral reverse transcriptase (*Figure 18.2*). These viruses do not normally kill the host cell (HIV is an exception), and only rarely transform it. The genome of a typical retrovirus consists of three genes, *gag*, *pol* and *env* (*Figure 18.3A*).

■ **Acute transforming retroviruses** are retrovirus particles which, unlike normal retroviruses, transform the host cell rapidly and with high efficiency. Their genomes include an additional gene, the oncogene (*Figure 18.3B*). Usually the oncogene replaces one or more essential viral genes, so that these viruses are replication-defective. To propagate them, they are grown in cells which are simultaneously infected with a replication-competent helper virus that supplies the missing functions. Studies of acute transforming retroviruses have revealed more than 50 different oncogenes.

18.3.2 An in vitro *transfection assay confirmed that cancer cells contain activated oncogenes*

An entirely independent way of discovering oncogenes came from a cell transformation assay. The NIH-3T3 mouse cell line readily undergoes transformation *in vitro*

– probably it has already acquired several of the successive genetic changes on the pathway to cancer, and one further change suffices to transform it. In the 3T3 test, the cells are transfected with random DNA fragments from human cancer cells. Potential oncogenes can then be identified by selecting transformants and recovering the human DNA present in them (*Figure 18.4*). Transformants are obtained when DNA from tumor cells is used, but not with DNA from nontumor cells. Thus tumor cells, even from nonviral tumors, contain activated oncogenes. This route led to the identification of essentially the same set of oncogenes as were found in acute transforming retroviruses.

18.3.3 *Oncogenes are mutated versions of genes involved in a variety of normal cellular functions*

It quickly became apparent that normal cells had counterparts of all the retroviral oncogenes (*Table 18.2*), and in fact that v-onc genes were transduced cellular genes. With a few exceptions, the v-onc gene products differ from their c-onc (proto-oncogene) counterparts by amino acid substitutions or truncations, which serve to activate the proto-oncogene.

Functional understanding of oncogenes began with the discovery in 1983 that the viral oncogene v-sis was derived from the normal cellular platelet-derived growth factor B (*PDGFB*) gene. Uncontrolled over-expression of a growth factor would be an obvious cause of cellular

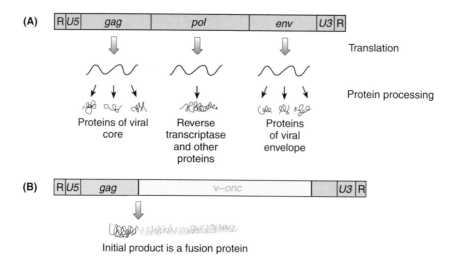

Figure 18.3: A normal and an acute transforming retrovirus.

The RNA genome has terminal repeats (R), subterminal unique sequences (U5, U3) and three genes, *gag*, *pol* and *env*. A complicated scheme of splicing and post-translational processing results in a variety of protein products. In an acute transforming retrovirus (bottom), one or more of the viral genes is replaced by a transduced cellular sequence, the oncogene. Initially it is translated into a fusion protein.

hyperproliferation. The roles of many cellular oncogenes (strictly speaking, proto-oncogenes) have now been elucidated (*Table 18.2*). Gratifyingly, they turn out to control exactly the sort of cellular functions that would be predicted to be disturbed in cancer. Five broad classes can be distinguished:

- secreted growth factors (e.g. *SIS*);
- cell surface receptors (e.g. *ERBB*, *FMS*);
- components of intracellular signal transduction systems (e.g. the *RAS* family, *ABL*);
- DNA-binding nuclear proteins, including transcription factors (e.g. *MYC*, *JUN*);
- components of the network of cyclins, cyclin-dependent kinases and kinase inhibitors that govern progress through the cell cycle (e.g. *MDM2*).

18.4 Activation of proto-oncogenes

Some of the best illustrations of molecular pathology in action are furnished by the various ways in which proto-oncogenes can become activated. Activation involves a gain of function. This can be quantitative (an increase in the production of an unaltered product) or qualitative (production of a subtly modified product as a result of a mutation, or production of a novel product from a chimeric gene created by a chromosomal rearrangement). These changes are dominant and normally affect only a single allele of the gene.

Activating mutations in oncogenes (unlike loss of function mutations in tumor suppressor genes, see below) are somatic events. Constitutional mutations would probably

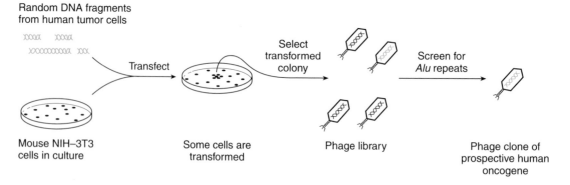

Figure 18.4: The NIH-3T3 assay.

Mouse 3T3 cells are transfected with random fragments of DNA from a human tumor. Any transformed cells (identified by their altered growth) are isolated, and a phage library constructed from their DNA. Phage are then screened for the human-specific *Alu* repeat to identify those containing human DNA, which potentially contains oncogenes.

Table 18.2: Viral and cellular oncogenes

Viral disease	v-onc	c-onc	Location	Function
Simian sarcoma	*v-sis*	*PDGFB*	22q13.1	Platelet-derived growth factor B subunit
Chicken erythroleukemia	*v-erb-b*	*EGFR*	7p13–q22	Epidermal growth factor receptor
McDonough feline sarcoma	*v-fms*	*CSF1R*	5q33	Macrophage colony-stimulating factor receptor
Harvey rat sarcoma	*v-ras*	*HRAS1*	11p15	Component of G-protein signal transduction
Abelson mouse leukemia	*v-abl*	*ABL*	9q34.1	Protein tyrosine kinase
Avian sarcoma 17	*v-jun*	*JUN*	1p32–p31	AP-1 transcription factor
Avian myelocytomatosis	*v-myc*	*MYC*	8q24.1	DNA-binding protein (transcription factor; see Rabbitts, 1994)
Mouse osteosarcoma	*v-fos*	*FOS*	14q24.3–q31	DNA-binding transcription factor

The viral genes are sometimes designated *v-src*, *v-myc* etc. and their cellular counterparts *c-src*, *c-myc* etc. The forms of the *c-onc* genes in normal cells are properly termed **proto-oncogenes**. Nowadays it is common to ignore these distinctions and simply use the term **oncogenes** for the normal genes. The abnormal versions can be described as activated oncogenes.

be lethal. We have met one exception to this: specific activating point mutations in the *RET* oncogene (Section 16.6.2) cause multiple endocrine neoplasia or familial thyroid cancer, and sometimes these mutations are inherited. It is very unusual to be able to build a functioning organism out of cells containing an activated oncogene. These *RET* mutations must affect the behavior of only very specific cells in very special circumstances. Note however that nonactivating mutations in proto-oncogenes may be inherited constitutionally, if their effect is unrelated to cancer. For example, inherited mutations that inactivate the *KIT* oncogene produce piebaldism (MIM 172800), while inherited loss-of-function mutations in *RET* predispose to Hirschsprung's disease (Section 19.5.2).

18.4.1 Activation of some oncogenes can occur by amplification

Many cancer cells contain multiple copies of structurally normal oncogenes. Breast cancers often amplify *ERBB2* and sometimes *MYC*; a related gene *NMYC* is usually amplified in late-stage neuroblastomas. Hundreds of extra copies may be present. They can exist as small separate chromosomes (*double minutes*) or as insertions within the normal chromosomes (*homogeneously staining regions, HSRs*). The genetic events producing HSRs may be quite complex because they usually contain sequences derived from several different chromosomes (reviewed by Pinkel, 1994). Similar gene amplifications are often seen in noncancer cells exposed to strong selective regimes – for example amplified dihydrofolate reductase genes in cells selected for resistance to methotrexate. In all cases the result is greatly to increase the level of gene expression.

Comparative genome hybridization (CGH) (Forozan *et al.*, 1997) can in principle reveal all regions of amplification in a single experiment, together with any regions of allele loss or aneuploidy, which may point to tumor suppressor genes (see below). The CGH test (*Figure 18.5*) uses a mixture of DNA from matched normal and tumor cells in competitive fluorescence *in situ* hybridization. With the aid of image-processing software, chromosomal

regions can be picked out where the ratio of FISH signal from normal and tumor DNA deviates from expectation. Depending on the direction of deviation, these mark regions of amplification or of allele loss in the tumor. The smallest alteration visible on standard CGH analysis is 5–10 Mb. By using high-density oligonucleotide arrays instead of metaphase chromosomes, CGH analysis can be carried to much higher resolution.

18.4.2 Some oncogenes are activated by point mutations

The *H-RAS1* gene (*Table 18.2*) is one of a family of *ras* genes that encode proteins involved in signal transduction from G-protein-coupled receptors. A signal from the receptor triggers binding of GTP to the RAS protein, and GTP-RAS transmits the signal onwards in the cell. RAS proteins have GTPase activity, and GTP-RAS is rapidly converted to the inactive GDP-RAS. Specific point mutations in *RAS* genes are frequently found in cells from a variety of tumors including colon, lung, breast and bladder cancers. These lead to amino acid substitutions that decrease the GTPase activity of the RAS protein. As a result, the GTP-RAS signal is inactivated more slowly, leading to excessive cellular response to the signal from the receptor.

18.4.3 Chromosomal translocations can create novel chimeric genes

Tumor cells typically have grossly abnormal karyotypes (*Figure 18.6*), with multiple extra and missing chromosomes, many translocations and so on. Most of these changes are random, and reflect a general genomic instability which is a normal part of carcinogenesis (see below). A huge research effort has been devoted to picking out tumor-specific changes superimposed on the background of random changes. Over 150 different tumor-specific breakpoints have now been recognized (Mitelman *et al.*, 1997), and they reveal an important common mechanism in tumorigenesis.

The   best-known   tumor-specific   rearrangement

(A)

(B)

(C)

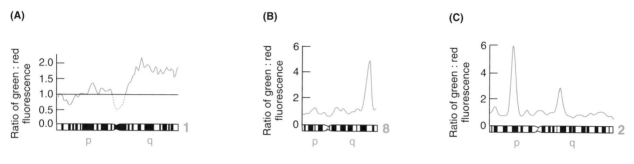

Figure 18.5: Comparative genome hybridization.

Tumor DNA and normal control DNA were labeled with green and red fluorescent labels respectively, then hybridized *in situ* together in equal quantities to chromosomes of a normal cell. The curves show computer-generated scans of the ratio of green:red fluorescence intensity along three chromosomes. **(A)** Chromosome 1: the breast cancer cell line 600PE carries an extra copy of 1q, and a possible interstitial deletion of proximal 1p34-p36. **(B)** Chromosome 8: the colon cancer cell line COLO320HSR shows amplification of the *MYC* region at 8q24. **(C)** Chromosome 2: the small cell lung cancer cell line NIH-H69 shows three regions of amplification. *NMYC* is amplified at 2p24; the content of the other two amplified regions is not known. Reprinted with permission from Kallioniemi *et al.* (1992) *Science*, **258**, pp. 818–821. Copyright 1992 American Association for the Advancement of Science.

produces the Philadelphia (Ph[1]) chromosome, a very small acrocentric chromosome seen in 90% of patients with chronic myeloid leukemia. This chromosome turns out to be produced by a balanced reciprocal 9;22 translocation. The breakpoint on chromosome 9 is within an intron of the *ABL* oncogene. The translocation joins most of the *ABL* genomic sequence onto a gene called

BCR (<u>b</u>reakpoint <u>c</u>luster <u>r</u>egion) on chromosome 22, creating a novel fusion gene (Chissoe *et al.*, 1995). This chimeric gene is expressed to produce a tyrosine kinase related to the *ABL* product but with abnormal transforming properties (*Figure 18.7A*).

Many other rearrangements are known that produce chimeric genes (*Table 18.3*). The products are normally

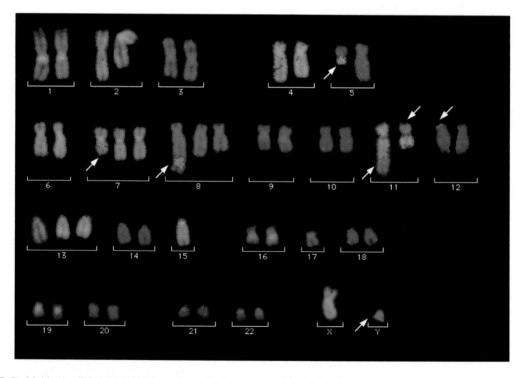

Figure 18.6: Multicolor FISH (M-FISH) karyotype of a human myeloid leukemia-derived cell line.

Note the numerous numerical and structural (arrowheads) abnormalities revealed by 24-color whole chromosome painting. These include t(5;15), der(7)t(7;15), der(8)ins(8;11), der(11)t(8;11), t(11;17) and der(Y)t(y;12) – this last is visible but hard to identify in this cell. Image Courtesy of Dr Lyndal Kearney, Institute of Molecular Medicine, Oxford. Tosi *et al.* (1999) *Genes, Chrom. Cancer*, **24**, pp. 213–221, copyright © John Wiley & Sons, Inc. Reprinted by permission of John Wiley & Sons, Inc.

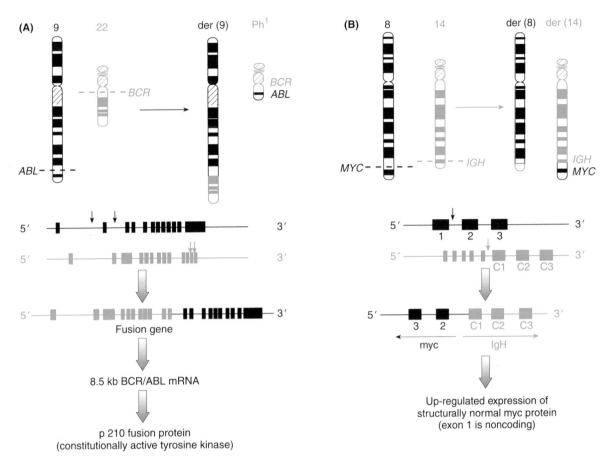

Figure 18.7: Chromosomal translocations which activate oncogenes.

(**A**) activation by qualitative change in the t(9;22) in chronic myeloid leukemia. The chimeric *BCR-ABL* fusion gene on the Philadelphia chromosome encodes a tyrosine kinase which does not respond to normal controls. See Chissoe *et al.* (1995) for more detail. (**B**) Activation by quantitative change in the t(8;14) in Burkitt's lymphoma. The *MYC* gene from chromosome 8 is translocated into the immunoglobulin heavy gene chain. In B-cells this region is actively transcribed, leading to over-expression of *MYC*.

transcription factors (or sometimes tyrosine kinases) which take their target specificity from one component gene, but couple it to an activation or ligand-binding domain from the other. This has been one of the most satisfying stories to emerge from cancer research, with several examples of clinical phenotypes being elegantly explained by a combination of cytogenetic and molecular genetic findings. The whole topic of chromosomal translocations in cancer and the underlying genetic events has been reviewed by Rabbitts (1994) and Sanchez-García (1997).

18.4.4 Oncogenes can be activated by transposition to an active chromatin domain

Burkitt's lymphoma is a childhood tumor common in malarial regions of Central Africa and Papua New Guinea. Mosquitoes and Epstein–Barr virus are believed to play some part in the etiology, but activation of the *MYC* oncogene is a central event. A characteristic chro-

mosomal translocation, t(8;14)(q24;q32) is seen in 75–85% of patients (*Figure 18.7B*). The remainder have t(2;8)-(p12;q24) or t(8;22)(q24;q11). Each of these translocations puts the *MYC* oncogene close to an immunoglobulin locus, *IGH* at 14q32, *IGK* at 2p12 or *IGL* at 22q11. Unlike the tumor-specific translocations shown in *Table 18.3*, the Burkitt's lymphoma translocations do not create novel chimeric genes. Instead, they put the oncogene in an environment of chromatin that is actively transcribed in antibody-producing B-cells. Usually exon 1 (which is noncoding) of the *MYC* gene is not included in the translocated material. Deprived of its normal upstream controls, and placed in an active chromatin domain, *MYC* is expressed at an inappropriately high level.

Many other chromosomal rearrangements put one or another oncogene into the neighborhood of either an immunoglobulin (*IGG*) or a T-cell receptor (*TCR*) gene (Rabbitts, 1994; Sanchez-García, 1997). Presumably the rearrangements arise by random malfunctioning of the recombinases that rearrange *IGG* or *TCR* genes during maturation of B and T cells (Section 8.6), and are then

Table 18.3: Chimeric genes produced by cancer-specific chromosomal rearrangements

Tumor	Rearrangement	Chimeric gene	Nature of chimeric product
CML	t(9;22)(q34;q11)	BCR-ABL	Tyrosine kinase
Ewing sarcoma	t(11;22)(q24;q12)	EWS-FLI1	Transcription factor
Ewing sarcoma (variant)	t(21;22)(q22;q12)	EWS-ERG	Transcription factor
Malignant melanoma of soft parts	t(12;22)(q13;q12)	EWS-ATF1	Transcription factor
Desmoplastic small round cell tumor	t(11;22)(p13;q12)	EWS-WT1	Transcription factor
Liposarcoma	t(12;16)(q13;p11)	FUS-CHOP	Transcription factor
AML	t(16;21)(p11;q22)	FUS-ERG	Transcription factor
Papillary thyroid carcinoma	inv(1)(q21;q31)	NTRK1–TPM3 (TRK oncogene)	Tyrosine kinase
Pre-B cell ALL	t(1;19)(q23;p13.3)	E2A-PBX1	Transcription factor
ALL	t(X;11)(q13;q23)	MLL-AFX1	Transcription factor
ALL	T(4;11)(q21;q23)	MLL-AF4	Transcription factor
ALL	t(9;11)(q21;q23)	MLL-AF9	Transcription factor
ALL	t(11;19)(q23;p13)	MLL-ENL	Transcription factor
Acute promyelocytic leukemia	t(15;17)(q22;q12)	PML-RARA	Transcription factor+retinoic acid receptor
Alveolar rhabdomyosarcoma	t(2;13)(q35;q14)	PAX3–FKHR	Transcription factor

Note how the same gene may be involved in several different rearrangements. For further details see Rabbitts (1994). CML, chronic myeloid leukemia; ALL, acute lymphoblastoid leukemia.

selected for their growth advantage. Predictably, these rearrangements are characteristic of leukemias and lymphomas, but not solid tumors.

18.5 Tumor suppressor genes

Cell fusion experiments show that the transformed phenotype can often be corrected *in vitro* by fusion of the transformed cell with a normal cell. This provides evidence that tumorigenesis involves not only dominant activated oncogenes, but also recessive, loss-of-function mutations in other genes. These other genes are the tumor suppressor (TS) genes. Sometimes TS genes are called antioncogenes, but that is an unhelpful name because it wrongly implies that they are all specific antagonists or inhibitors of oncogenes. Some may be, but like oncogenes, TS genes can have a variety of functions (see below).

TS genes have been discovered by three main routes:

■ positional cloning of the genes causing rare familial cancers;
■ defining chromosomal locations commonly deleted in tumor cells (by loss of heterozygosity analysis or comparative genomic hybridization);
■ testing tumors for mutations in genes known to be involved in cell cycle regulation.

The rare eye tumor, retinoblastoma, has been the main test-bed for defining the concepts and methods of TS gene research.

18.5.1 Retinoblastoma exemplifies Knudson's two-hit hypothesis

Retinoblastoma (MIM 180200) is a rare, aggressive childhood tumor of the retina. 60% of cases are sporadic and unilateral; the other 40% are inherited as an imperfectly penetrant autosomal dominant trait, which was mapped to 13q14. In familial retinoblastoma bilateral tumors are common. In 1971 AG Knudson proposed that two successive mutations ('hits') were required to turn a normal cell into a tumor cell (*Figure 18.8*), and that in familial forms one of the hits was inherited. A seminal study by Cavenee *et al.* (1983) both proved Knudson's hypothesis, and established the paradigm for laboratory investigations of TS genes.

Cavenee and colleagues sought evidence of somatic mutations at the *RB1* locus in sporadic retinoblastoma by typing surgically removed tumor material with a series of markers from chromosome 13. When they compared the results on blood and tumor samples from the same patients, they noted several cases where the constitutional (blood) DNA was heterozygous for one or more chromosome 13 markers, but the tumor cells were apparently homozygous. They reasoned that what they were seeing was one of Knudson's 'hits': loss of one functional copy of a tumor suppressor gene. Combining cytogenetic analysis with studies of markers from different regions of 13q, Cavenee *et al.* were able to suggest a number of mechanisms for the loss (*Figure 18.9*). Later studies confirmed this interpretation by showing that in inherited cases, it was always the wild-type allele that was lost in this way.

Loss of one marker allele but retention of the other is often seen because in both sporadic and inherited disease one hit is usually a point mutation while the other often involves loss of all or part of a chromosome. Inherited TS mutations are usually small-scale mutations – large chromosomal deletions would probably be lethal if carried in every cell of the body. Individual tumor cells may well be viable with a large deletion in heterozygous form, but

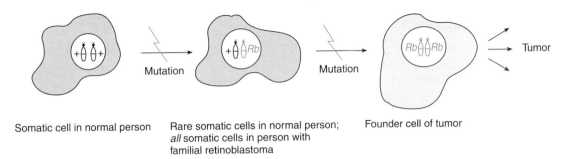

Figure 18.8: Knudson's two-hit hypothesis.

Suppose there are 1 million target cells and the probability of mutation is 10^{-5} per cell. Sporadic retinoblastoma requires two hits and will affect 1 person in 10 000 ($10^6 \times 10^{-5} \times 10^{-5} = 10^{-4}$), while the familial form requires only one hit and will be quite highly penetrant ($10^6 \times 10^{-5} => 1$).

large homozygous deletions are likely to be lethal even at the cell level. Thus both familial and sporadic tumors tend to retain markers surrounding a TS gene on one chromosomal homolog, but lose them from the other. However, if the wild-type allele is silenced by methylation rather than by deletion (Section 18.5.4), no loss of heterozygosity will be seen.

18.5.2 Rare familial cancers identify many TS genes

Following the example of retinoblastoma, many rare mendelian cancers are believed to involve TS genes via a two-hit mechanism (*Table 18.4*). As with retinoblastoma, mapping the genes in these rare families opens the way to positional cloning of TS genes. In many cases (*APC, NF2, PTC* for example) the TS gene identified in this way turns

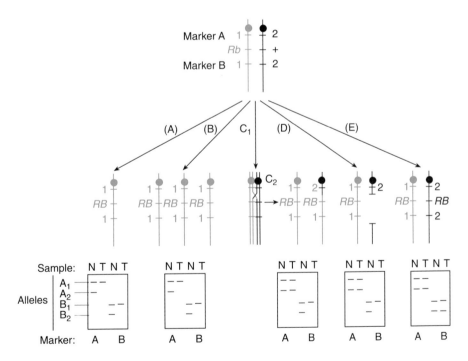

Figure 18.9: Mechanisms of loss of wild-type allele in retinoblastoma.

(**A**) Loss of a whole chromosome by mitotic non-disjunction. (**B**) Loss followed by reduplication to give (in the case studied by Cavenee *et al.*) three copies of the Rb chromosome. (**C**) Mitotic recombination proximal to the Rb locus (c_1), followed by segregation of both Rb-bearing chromosomes into one daughter cell (c_2); this was the first demonstration of mitotic recombination in humans, or indeed in mammals. (**D**) Deletion of the wild-type allele. (**E**) Pathogenic point mutation of the wild-type allele (adapted from Cavenee *et al.*, 1983). The figures underneath show results of typing normal (N) and tumor (T) DNA for the two markers A and B located as shown. Note the patterns of loss of heterozygosity.

Table 18.4: Rare familial cancers caused by TS gene mutations

Disease	MIM No.	Map location	Gene
Familial adenomatous polyposis coli	175100	5q21	APC
Hereditary non-polyposis colon cancer	120435, 120436	2p16, 3p21.3	MSH2, MLH1
Breast-ovarian cancer	113705	17q21	BRCA1
Breast cancer (early onset)	600185	13q12–q13	BRCA2
Li-Fraumeni syndrome	151623	17p13	TP53
Gorlin's basal cell nevus syndrome	109400	9q22–q31	PTC
Ataxia telangiectasia	208900	11q22–q23	ATM
Retinoblastoma	180200	13q14	RB1
Neurofibromatosis I (von Recklinghausen disease)	162200	17q12–q22	NF1
Neurofibromatosis 2 (vestibular schwannomas)	101000	22q12.2	NF2
Familial melanoma	600160	9p21	CDKN2A
von Hippel–Lindau disease	193300	3p25–p26	VHL

References to the genes and diseases may be found in OMIM under the numbers cited.

Box 18.2

Two-hit mechanisms may explain patchy mendelian phenotypes

It is possible that a two-hit mechanism similar to the Knudson model (*Figure 18.9*) may explain some non-cancer phenotypes that are mendelian in families but patchy in individuals. Why for example does piebaldism (MIM 172800) produce patches and spots of depigmented skin, rather than general albinism? Why does polycystic kidney disease (MIM 173900) produce a limited number of grossly dilated tubules in the kidney? In each case the inherited genetic defect is present in every cell, but only some show the phenotype. At least in the case of polycystic kidney disease, it has been demonstrated that the dilated tubules have lost the second copy of the *PKD1* gene (Qian *et al.*, 1996).

---

out to be important in the corresponding sporadic cancer, though this is not always the case – *BRCA1* mutations are not found in sporadic breast tumors.

18.5.3 Loss of heterozygosity (LoH) screening identifies locations of TS genes

By screening paired blood and tumor samples with markers spaced across the genome, we can discover candidate locations for TS genes (*Figure 18.10*). For meaningful results, a large panel of tumors must be screened with closely spaced markers. Highly polymorphic microsatellite markers are used in order to minimize the number of uninformative cases where the constitutional DNA is homozygous for the marker. Not all tumors will show the pattern of one small and one large abnormality that is needed to produce visible LoH. Advanced cancer cells often show LoH at as many as one quarter of all loci, so large samples are needed to tease out the specific changes from the general background chromosomal instability. Finally, most pathological tumor samples contain a mixture of intergrowing tumor and non-tumor (stromal) tissue, so that LoH shows as a decreased relative intensity (allelic imbalance) rather than total loss of the band from

one allele (*Figure 18.11A*). Thus screening for LoH is quite laborious. Comparative genome hybridization (*Figure 18.5*) allows much easier screening for large deletions, but lacks the resolution required to detect small deletions. Using the CGH principle of competitive hybridization, but to DNA microarrays rather than chromosome spreads, may solve this problem (Pinkel *et al.*, 1998).

Small homozygous deletions also occur, and are common at certain loci, such as *CDKN2A* (Section 18.6.3). Being small, homozygous deletions give an excellent pointer to the location of the TS gene. However, they also set a trap for the investigator. Markers that are homozygously deleted in the tumor tend to amplify from the contaminating stromal material, which is chromosomally normal. When there are overlapping deletions on the two homologs (*Figure 18.12*), the LoH results can appear to show evidence for two non-existent tumor suppressor genes, located either side of the one true gene. FISH is a good technique for confirming homozygous deletions. Alternatively, cell lines can be used where there is no problem of stromal contamination.

18.5.4 Tumor suppressor genes are often silenced epigenetically by methylation

Tumor supressor genes may be silenced by deletion (reflected in loss of heterozygosity) or by point mutations, but there is increasing evidence for a third mechanism – DNA methylation (Versteeg, 1997; Jones and Laird, 1999). As explained in Section 8.4, cytosines in CpG dinucleotides are liable to be methylated, and when the cytosine lies in a CpG island within the promoter region of a gene, methylation is often associated with lack of expression of the gene. Many tumors show genome-wide disturbances of the normal methylation pattern. More specifically, CpG island methylation has been demonstrated for several tumor suppressor genes in a variety of cancers (Jones and Laird, 1999). At least in the cases of the *CDKN2A, VHL, RB1* and *MLH1* genes, there is evidence that promoter methylation causes silencing of the gene. In one study, 84% of colorectal tumors with microsatellite

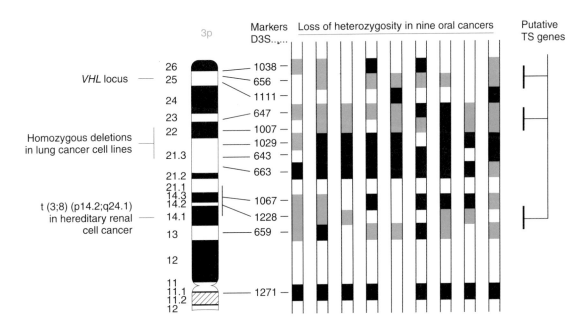

Figure 18.10: Possible tumor suppressor genes on chromosome 3p.

On the right are the results of typing constitutional and tumor DNA from a series of patients with oral tumors, using various markers (D3S1038, etc) from 3p (Wu *et al.*, 1994). Color signifies loss of heterozygosity (LoH), black signifies retention, and clear areas are where markers were uninformative because of constitutional homozygosity. Note the complex pattern of LoH in these tumors, which is typical of such studies. Three distinct regions of LoH are seen. These appear to correspond to regions (left side of diagram) previously identified by a translocation breakpoint in a family with dominant renal cell carcinoma, by homozygous deletions in some lung cancer cell lines, and by the location of a known tumor suppressor gene, *VHL*.

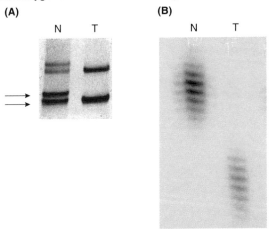

Figure 18.11: Genetic changes in tumors.

(**A**) Loss of heterozygosity. The normal tissue sample (N) is heterozygous for the marker D8S522 (arrows), while the tumor sample (T) has lost the upper allele. The bands higher up the gel are 'conformation bands', subsidiary bands produced by alternately folded sequences of each allele. Photograph courtesy of Dr Nalin Thakker, St Mary's Hospital, Manchester. (**B**) Microsatellite instability. The colorectal tumor sample (T) has an allele of the BAT26 marker that is not seen in the normal tissue (N) of the patient. BAT26 detects a poly(A) microsatellite that is especially sensitive to replication error repair defects. Courtesy of Loveena Verma, Medical Molecular Genetics Laboratory, University of Birmingham.

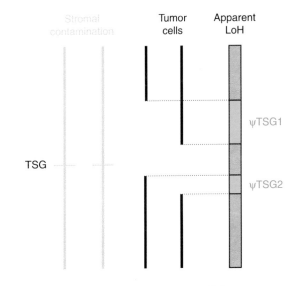

Figure 18.12: A pitfall in interpreting LoH data.

The true event in this tumor is homozygous deletion of TSG. Because of amplification of contaminating stromal material (pale lines), LoH data show retention of heterozygosity at the TSG locus, with LoH at two flanking regions (blue). The pattern suggests the existence of two spurious tumor suppressor loci, ψTSG1 and ψTSG2. The true situation would be revealed by fluorescence *in situ* hybridization, or by immunohistochemical testing for the TSG gene product.

instability showed methylation of the *MLH1* promoter (Herman *et al.*, 1998), and in cell lines from such tumors *MLH1* expression can be restored by treatment with the demethylating agent 5-aza-2'-deoxycytidine. Gene silencing by methylation is an example of an epigenetic mechanism (Section 16.4.2), a heritable change in gene expression that does not depend on a DNA sequence change. Standard techniques for mutation screening overlook changes in methylation, so their importance has probably been underestimated.

18.6 Control of the cell cycle

Any cell at any time has three choices of behavior: it can remain static, it can divide or it can die (apoptosis). Some cells also have the option of differentiating. Cells select one of these options in response to internal and external signals (*Figure 18.13A*). Oncogenes and tumor suppressor genes play key roles in generating and interpreting these signals.

Life would be very simple if the signal and response were connected by a single linear pathway (*Figure 18.13B*), but this seems never to be the case. Rather, multiple branching, overlapping and partially redundant pathways control the behavior of the cell (*Figure 18.13C*). Probably such complicated networks are necessary to confer stability and resilience on the extraordinarily complex machinery of a cell. Experimentally, unravelling the

precise genetic circuitry of the controls is exceedingly difficult, partly because of their complexity and partly because it is difficult to distinguish direct from indirect effects in transfection or knockout experiments. DNA array technology is the best hope for dealing with this problem. Hybridizing poly(A)$^+$ RNA to arrayed probes representing hundreds or thousands of genes allows significant changes to be picked out from the web of interactions. Normal and cancer cells can be compared, or transfected cells and controls, to get an overall snapshot of the state of the web in a certain cellular condition (see *Figure 20.6*).

Figure 18.14 shows part of the cell cycle control system that involves the products of three key genes, *RB1*, *TP53* and *CDKN2A*. One way or another, tumor cells must inactivate this control system – probably in fact they need to inactivate both the *RB1* and *TP53* arms of the system. Thus these three tumor suppressor genes are central players in carcinogenesis, and are among the most commonly altered genes in tumor cells. Each also has a role in inherited cancers.

18.6.1 Function of pRb, the RB1 gene product

The *RB1* gene is widely expressed, encoding a 110-kd nuclear protein, pRb. In normal cells pRb is inactivated by phosphorylation and activated by dephosphorylation. Active (dephosphorylated) pRb binds and inactivates the cellular transcription factor E2F1, function of which is required for cell cycle progression (*Figure 18.14*). For reviews of pRb and E2F1, see Weinberg (1995, 1996). The G$_1$–S checkpoint seems to be the most crucial in the cell cycle; 2–4 hours before a cell enters S-phase, pRb is phosphorylated. This releases the inhibition of E2F1 and allows the cell to proceed to S phase. Phosphorylation is governed by a cascade of cyclins, cyclin-dependent kinases and cyclin kinase inhibitors.

(A)

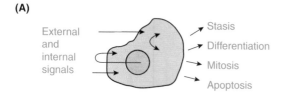

(B)

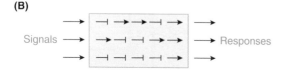

(C)

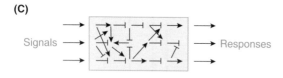

Figure 18.13: The options open to a cell, and how it chooses.

(A) In response to internal and external signals, a cell chooses between stasis, mitosis, apoptosis and sometimes differentiation. **(B)** An imaginary cell in which signals are linked to responses by linear unbranched pathways of stimulation ($\rightarrow$) or inhibition ($-|$). Human cells do not function like this. **(C)** In real cells signals feed into a complex network of partially redundant interactions, the outcome of which is not easy to predict analytically.

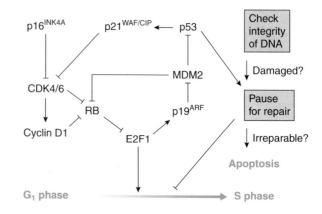

Figure 18.14: Controls on cell cycle progression and genomic integrity mediated by the *RB1*, *TP53* and *CDKN2A* gene products.

These controls form at least part of the G$_1$–S cell cycle checkpoint. p53 also has other activities.

RB1 gene mutations produce sporadic or inherited retinoblastoma, this being the classic example of Knudson's two-hit hypothesis (*Figure 18.8*). It is not clear why constitutional mutation of a gene so fundamental to cell cycle control should result specifically in retinoblastoma and a small number of other tumors, principally osteosarcomas. However, this is a common theme in molecular pathology: mutation of a gene produces a phenotypic effect in only a subset of the cells or tissues in which the gene is expressed and appears to have a function (*Section 16.7.1*). The product of the *MDM2* oncogene (which is amplified in many sarcomas) binds and inhibits pRb, thus favoring cell cycle progression. Several viral oncoproteins (adenovirus E1A, SV40-T antigen, human papillomavirus E7 protein) also bind and sequester or degrade pRb.

18.6.2 Function of p53, the TP53 gene product

p53 was first described in 1979 as a protein found in SV40-transformed cells, where it associated with the T-antigen. Later, the *TP53* gene which encodes p53 appeared as a dominant transforming gene in the 3T3 assay (*Figure 18.4*), and so was classed as an oncogene. Subsequently it transpired that while p53 from some tumor cells was oncogenic, p53 from normal cells positively suppressed tumorigenesis.

Loss or mutation of *TP53* is probably the commonest single genetic change in cancer. This reflects the central importance of p53, which has several functions in the cell. One is as a transcription factor. Tetramers of p53 bind DNA and can activate transcription of reporter genes placed downstream of a p53 binding site. However, p53 is believed to have a much broader role in the cell, which has been summarized as 'the guardian of the genome'. One of its guardian functions is to stop cells replicating damaged DNA (*Figure 18.14*). Normal cells with damaged DNA arrest at the G$_1$–S cell cycle checkpoint until the damage is repaired, but cells that lack p53 or contain a mutant form do not arrest at G$_1$. Replication of damaged DNA presumably leads to random genetic changes, some of which are oncogenic, similar to cells with a defective mismatch repair system (see below).

Probably related to this, p53 has a crucial role in cell death. In response to oncogenic stimuli, cells undergo apoptosis (programmed cell death). Apoptosis has come to occupy a central place in our understanding of the cancer process (reviewed by Fisher, 1994). It is one of the main higher-level controls that protect the organism against the consequences of the natural selection among its constituent cells described at the start of this chapter. Tumor cells lacking p53 do not undergo apoptosis, and so escape the control. p53 may be knocked out by deletion, by mutation or by the action of an inhibitor such as the *MDM2* gene product (which binds p53 and targets it for degradation; MDM2 also binds pRb, see above) or the E6 protein of papillomavirus.

Loss of heterozygosity assays confirmed the status of *TP53* as a tumor suppressor gene. *TP53* maps to 17p12,

and this is one of the commonest regions of loss of heterozygosity in a wide range of tumors. Tumors that have not lost *TP53* very often have mutated versions of it. To complete the picture of *TP53* as a TS gene, constitutional mutations in *TP53* are found in families with the dominantly inherited Li–Fraumeni syndrome (MIM 151623). Affected family members suffer multiple primary tumors, typically including soft tissue sarcomas, osteosarcomas, tumors of the breast, brain and adrenal cortex, and leukemia (*Figure 18.15*).

18.6.3 Function of CDKN2A and ARF, the CDKN2A gene products

The remarkable gene variously called *MTS1*, *INK4A* and *CDKN2A* at 9p13 encodes two structurally unrelated proteins (*Figure 18.16*). Exons 1α, 2 and 3 encode the CDKN2A (p16^{INK4A}) protein. A second promoter starts transcription further upstream at exon 1β. Exon 1β is spliced on to exons 2 and 3, but the reading frame is shifted, so that an entirely unrelated protein ARF (p19^{ARF}) (Alternative Reading Frame) is encoded.

Both gene products function in cell cycle control (*Figure 18.14*). CDKN2A functions upstream of the RB1 protein in control of the G$_1$–S cell cycle checkpoint. Cyclin-dependent kinases inactivate pRb by phosphorylation, but CDKN2A inhibits the kinases (Weinberg, 1995). Thus loss of CDKN2A function leads to loss of RB1 function and inappropriate cell cycling. The other product of the *CDKN2A* gene, ARF, mediates G$_1$ arrest by destabilizing MDM2 (Pomeranz *et al.*, 1998). MDM2 binds to p53 and induces its degradation. Thus ARF acts to maintain the level of p53. Loss of ARF function leads to excessive levels of MDM2, excessive destruction of p53, and hence loss of cell cycle control.

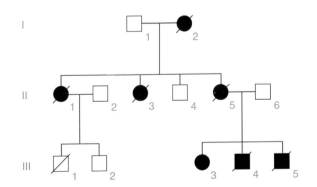

Figure 18.15: A typical pedigree of Li–Fraumeni syndrome.

Malignancies typical of Li–Fraumeni syndrome include bilateral breast cancer diagnosed at age 40 (I$_2$); a brain tumor at age 35 (II$_1$); soft tissue sarcoma at age 19 and breast cancer at age 33 (II$_3$); breast cancer at age 32 (II$_5$); osteosarcoma at age 8 (III$_3$); leukemia at age 2 (III$_4$); soft tissue sarcoma at age 3 (III$_5$). I$_1$ had cancer of the colon diagnosed at age 59 – this is assumed to be unrelated to the Li–Fraumeni syndrome. Pedigree from Malkin (1994).

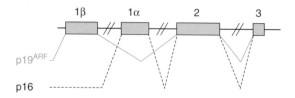

Figure 18.16: The two products of the *CDKN2A* gene.

This gene (also known as *MTS* and *INK4A*) encodes two completely unrelated proteins. CDKN2A, or p16^{INK4A}, is transcribed from exons 1α, 2 and 3, and ARF, or p19^{ARF}, from exons 1β, 2 and 3 – but with a different reading frame of exons 2 and 3. The two gene products are active in the RB1 and p53 arms of cell cycle control, respectively, as shown in *Figure 18.14*.

Inherited mutations, usually involving just the CDKN2A product, are seen in some families with multiple melanoma, but somatic mutations are very much more frequent. Homozygous deletion of the *CDKN2A* gene inactivates both the RB1 and the p53 arms of cell cycle control, and is a very common event in the development of many tumors. Other tumors have mutations that affect the CDKN2A but not the ARF product (e.g. inactivation of the 1α promoter by methylation). These tumors tend also to have p53 mutations, showing the importance of inactivating both arms of the control system shown in *Figure 18.14*.

18.7 Control of the integrity of the genome

We saw earlier (*Box 18.1*) that cancer can develop only if something happens to increase the vanishingly low probability of accumulating half a dozen specific mutations in a single cell. Some 'gatekeeper' mutations create expanded clones of cells as targets for subsequent mutations (*Figure 18.1*). Another class of genes that are commonly mutated in cancer cells are not directly involved in controlling the cell cycle. Instead, they have a general role as 'caretakers', ensuring the integrity of the genome. Loss of function mutations in these genes lead to a general genetic instability that has long been recognized as a feature of cancer cells – see the excellent review by Lengauer *et al.*, 1998. It can operate at either the nucleotide or the chromosomal level.

18.7.1 Nucleotide instability is manifest as defects in DNA replication or repair

Nucleotide excision repair defects
DNA that has been damaged by ionizing radiation, ultraviolet light or chemical mutagens contains single- or double-strand breaks and crosslinks that need to be repaired before the next round of replication (Section 9.6).

Table 18.5: Genes involved in DNA replication error repair

E. coli	Human	Location in man	% of HNPCC
MutS	*MSH2*	2p16	50–60%
	GTBP (MSH6)	2p16	v. low
MutL	*MLH1*	3p21.3	30–40%
	PMS1	2q31–q33	v. low
	PMS2	7p22	v. low

Loss of function mutations in some of the repair enzymes are seen in several cancer-prone syndromes, principally the various forms of xeroderma pigmentosum. These are inherited, constitutional mutations and the diseases are autosomal recessive, so that there is no somatic second hit on the repair gene. Instead, the somatic hit is DNA damage requiring repair, usually the result of UV light. XP patients are exceedingly sensitive to sunlight and develop many tumors on exposed skin.

Replication error repair defects
Loss of heterozygosity studies on colon cancers (Section 18.5.3) produced novel and unexpected results in some patients. Rather than lacking alleles present in the constitutional DNA, some tumor specimens appeared to contain extra, novel, alleles of the microsatellite markers used. LoH is a property of particular chromosomal regions, but the microsatellite instability (MSI or MIN) seemed to be general. Tumors could be classified into MSI$^+$ or MSI$^-$. MSI$^+$ tumors gained alleles for a good proportion of the markers tested, regardless of their chromosomal location. *Figure 18.11B* shows an example.

Microsatellite instability is a characteristic feature of the autosomal dominant hereditary non-polyposis colon cancer. HNPCC genes were mapped to two locations, 2p15–p22 and 3p21.3. In a wonderful example of lateral thinking, Fishel *et al.* (1993) related the MSI$^+$ phenomenon to so-called mutator genes in *E. coli* and yeast. These genes encode an error-correction system that checks the DNA for mismatched base pairs (*Figure 18.17*). Because the *E. coli* Dam system methylates adenine in GATC sequences, but not until some time after DNA replication, the system can recognize the newly-synthesized strand if it has not yet been methylated, and it can cut out and resynthesize the DNA surrounding a mismatch on this strand. Mutations in the genes that encode the MutHLS error-correction system lead to a 100 to 1000-fold general increase in mutation rates. Fishel and colleagues cloned the human homolog of one of these genes, MutS, and showed that it mapped to the location on 2p of one of the HNPCC genes and was constitutionally mutated in some HNPCC families. In all, five human homologs of the *E. coli* MutS or MutL genes have been implicated in cancer (Eshleman and Markowitz, 1996; Peltomaki and de la Chapelle, 1997; see *Table 18.5*.

Patients with HNPCC are constitutionally heterozygous for a loss-of-function mutation. Their normal cells still have a functioning mismatch repair system and do not show the MSI[+] phenotype. In a tumor, the second copy is lost by one of the mechanisms shown in *Figure 18.9*. Interestingly, these tumors are unusual in having relatively normal karyotypes – showing that while

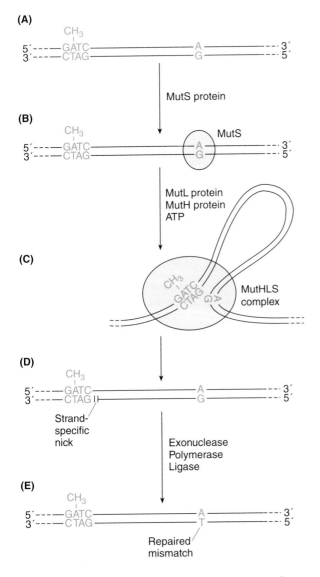

Figure 18.17: The MutHLS error correction system in *E. coli*.

A replication error introduces a mismatch (**A**). The MutS protein binds to mismatched base pairs (**B**). In an ATP-dependent reaction, a MutS–MutL–MutH complex is formed which probably brings any GATC sequence located within 1 kb either side of the mismatch into a loop (**C**). MutH makes a single-strand cut 5′ to the GATC sequence (**D**). The *E. coli* Dam methylation system methylates A in GATC, but in newly synthesized DNA only the template strand is methylated. MutH specifically cuts the unmethylated (newly synthesized) strand (**D**). Exonucleases, DNA polymerase and DNA ligase then strip back and repair the DNA (**E**). See Modrich (1995).

genetic instability is important in cancer, it may be achieved either through instability at the nucleotide level or by chromosomal instability.

Microsatellite instability is seen in about 13% of colorectal, gastric and endometrial carcinomas, but only occasionally in other tumors. Mismatch repair defects should affect all cells – why are they a feature of only a fairly restricted set of tumors? One explanation may be that in MSI[+] cells mutations in longer homopolymeric runs occur 1000 times as often as mutations of microsatellites. The TGFβ receptor II gene, which has an A_{10} run, has frameshifting mutations in 90% of MSI[+] cancers. Loss of TGFβ signalling seems to be an important step in development of colorectal cancer. In MSI[−] colorectal cancer, this system is often inactivated by loss of a downstream effector, *SMAD4* on chromosome 18 (White, 1998).

18.7.2 Chromosomal instability is a very common feature of cancer cells

Malignant tumor cells that do not show microsatellite instability usually have bizarrely abnormal karyotypes, with many losses, gains and rearrangements of chromosomes (*Figure 18.6*). DNA studies reinforce this picture of chromosomal instability: cells from a typical advanced colon, breast or prostate cancer show loss of heterozygosity at around one quarter of all loci. Only a few of the changes seem to be causally connected with the cancer; mostly they are a reflection of a general chromosomal instability.

Cell fusion experiments (Lengauer *et al.*, 1998) suggest that chromosomal instability is a dominant phenotype, and so presumably the result of a single mutant allele at the relevant locus. In yeast, mutations in many different genes can lead to chromosomal instability. The genes concerned are involved in chromosome condensation, functioning of the centromere and kinetochore, or checkpoints such as a spindle checkpoint that prevents chromatids separating until the chromosome is correctly aligned on the spindle. Identification of the corresponding human genes and investigation of their role in cancer has only just begun.

18.7.3 A DNA damage checkpoint that prevents cells containing damaged DNA from entering mitosis is often inactivated in cancer cells

Normal cells with unrepaired DNA damage do not enter mitosis. Details of this checkpoint mechanism are coming to light, mainly from experiments in yeast (Nurse, 1997; Weinert, 1998). In the presence of DNA damage the CDC2 cyclin-dependent kinase, which is the immediate controller of entry into mitosis, is inactive, and levels of p53 protein are raised. Among the mammalian genes involved in the damage checkpoint are *ATM* and maybe *ATR*, *BRCA1* and *BRCA2*.

ATM is the gene responsible for ataxia telangiectasia (AT, MIM 208900), a rare recessive combination of cerebellar ataxia, telangiectasia (dilation of blood vessels in the conjunctiva and eyeballs), immunodeficiency, growth retardation and sexual immaturity. AT patients have a strong predisposition to cancer. Homozygotes usually die of malignant disease before age 25, and there have been suggestions that heterozygotes have a raised risk of cancer – for example a 3.9-fold increased risk of breast cancer among women (Easton, 1994). AT affects about one person in 100 000 in the UK and USA, so the Hardy–Weinberg distribution (Section 3.3.1) suggests that one person in 158 of the population is heterozygous. If their raised risk of cancer is confirmed, this would represent a significant cancer risk at the population level.

In vitro, cells of AT patients show chromosomal instability and hypersensitivity to ionizing radiation or radiomimetic chemicals, even though DNA repair appears to be normal. Unlike normal cells, AT cells fail to accumulate p53 after irradiation. The *ATM* gene product has protein kinase activity, and activates p53 by phosphorylation of serine-15 (reviewed by Nakamura, 1998). Thus part of the DNA-damage checkpoint involves the ATM protein reacting to damage by activating p53. A related protein, ATR, appears to have an opposite effect: overexpression (rather than loss of function) of *ATR* inhibits the p53 response to DNA damage (Friend and Tapscott, 1998). Possibly the breast cancer genes, *BRCA1* and *BRCA2*, may also form part of the checkpoint mechanism. In all cases, if cells replicate damaged DNA, not only may mutations be propagated, but also DNA breaks or crosslinks may predispose to the chromosome deletions, translocations and inversions that are such a common feature of tumor cells.

18.7.4 Telomerase, the stability of chromosomal ends and the immortality of cancer cells

As we saw in Section 2.3.5, the ends of human chromosomes are protected by a repeat sequence $(TTAGGG)_n$, that is maintained by a special RNA-containing enzyme system, telomerase. Telomerase is present in the human germ line but is absent in most somatic tissues, and telomere length declines with time in normal somatic cells, a phenomenon which may contribute to the 'mitotic clock' that limits the number of divisions a cell can go through. There has been much excitement over the discovery that 90% of human primary tumors possess telomerase activity. Maybe this is the key to their immortality, and maybe an anti-telomerase agent would limit their mitotic potential. Indeed, ectopic expression of hTERT, the catalytic subunit of human telomerase, allows cells that were destined to senesce to multiply indefinitely – but in many cell types this also requires inactivation of the G_1–S controls shown in *Figure 18.14* (Weinberg, 1998).

18.8 The multistep evolution of cancer

Because cancers are the inevitable end-result of natural selection among the cells of an organism, rather than the result of a specific disease process, cancers of a given type do not all have mutations in a standard set of genes. Nevertheless, the requirements for genomic instability and successive clonal expansions at intermediate stages of the multistep evolution impose a certain regularity on the changes seen. In particular, the genetic changes involved in tumor induction may be different from, and more specific than, those involved in tumor progression.

18.8.1 Vogelstein's concept of 'gatekeeper' genes describes, even if it does not explain, the tissue specificity of many genetic cancers

The first mutation in the multistep evolution of a tumor is critical because it should confer some growth advantage on an otherwise normal cell. Sometimes the reason why a genetic change causes a particular type of cancer and not another is understandable – translocating the *MYC* oncogene into an immunoglobulin locus, for example, is likely to cause trouble for cells that express immunoglobulins at high levels. More often, however, no such connection is apparent. Why should mutations in the *RB1*, *TP53* and *MSH2* genes cause retinoblastoma, breast cancer and HNPCC, respectively, when the proteins encoded by these genes play important roles in virtually all cells? According to the gatekeeper hypothesis (Kinzler and Vogelstein, 1996), in a given renewing cell population one particular gene is responsible for maintaining a constant cell number. A mutation of a gatekeeper leads to a permanent imbalance of cell division over cell death, whereas mutations of other genes have no long-term effect if the gatekeeper is functioning correctly.

The tumor suppressor genes identified by studies of mendelian cancers are the gatekeepers for the tissue involved – *NF2* for Schwann cells, *VHL* for kidney cells, and so on. The reason why one particular widely expressed gene should be the gatekeeper for just one cell type must be buried in the complex networks of interactions that connect input to output in cell behavior (*Figure 18.13C*). An accumulation of minor differences leaves a certain gene with a predominant effect in one cell type but not another. For those cancers that do not have mendelian versions, maybe there is not a single gatekeeper.

18.8.2 The colorectal cancer model

Most colon cancer is sporadic. Familial cases fall into two categories:

■ **Familial adenomatous polyposis (FAP or APC**: MIM 175100) is an autosomal dominant condition in which the colon is carpeted with hundreds or

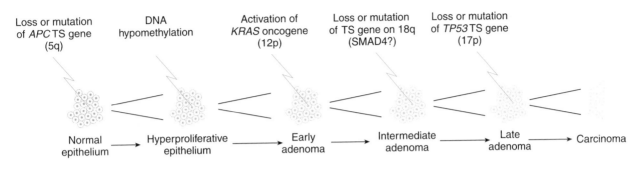

Figure 18.18: Fearon and Vogelstein's model for the development of colorectal cancer.

This is primarily a tool for thinking about how tumors develop, rather than a firm description. Every colorectal carcinoma is likely to have developed through the same progression of histological stages, but the underlying genetic changes are much less predictable. The sequence shown illustrates changes frequently involved in particular stages of progression. In HNPCC tumors, the early stages probably follow an alternative pathway. After loss of the wild-type MSH2 or MLH1 allele, which produces cells with deficient mismatch repair, specific genes including the TGFβ receptor II gene suffer frameshift mutations. The 100–1000-fold increase in mutation rate in these tumors must also speed progression through the later stages. See Fearon and Vogelstein (1990).

thousands of polyps. The polyps (adenomas) are not malignant, but if left in place, one or more of them is virtually certain to evolve into invasive carcinoma. The condition has been mapped to 5q21 and the gene responsible, *APC*, identified.

- **Hereditary non-polyposis colon cancer (HNPCC; MIM 120435, 120436)** is also autosomal dominant and highly penetrant, but unlike FAP there is no preceding phase of polyposis.

Malignant colorectal carcinomas develop from normal epithelium through microscopic dysplastic aberrant crypt foci to benign epithelial growths called adenomas. Adenomas can be classified into early (less than 1 cm in size), intermediate (more than 1 cm but without foci of carcinoma) or late (more than 1 cm and with foci of carcinoma). Adenomas develop into carcinomas, which eventually metastasize. Although there is not one invariant sequence of mutations in the development of every colorectal carcinoma, the most likely sequence is one where each successive step confers a growth advantage on the cell. Pointers to the commonest sequence include the following observations:

- In FAP, constitutional loss of one copy of the *APC* gene on 5q21 is sufficient to carpet the colon with adenomatous polyps. The very earliest detectable lesions in either sporadic colorectal cancer or FAP, dysplastic aberrant crypt foci, lack all APC expression. In FAP, about 1 epithelial cell in 10^6 develops into a polyp, a rate consistent with loss of the second *APC* allele being the determining event. The APC product binds β-catenin, and thus represses tran-

scription by the TCF7L2 transcription factor. Sporadic colorectal cancers sometimes have an intact APC gene but achieve the same effect by activating mutations of β-catenin, which therefore acts as an oncogene in this context.

- About 50% of intermediate and late adenomas, but only about 10% of early adenomas, have mutations in the *KRAS* oncogene (a relative of *HRAS, Table 18.2*). Thus *KRAS* mutations may often be involved in the progression from early to intermediate adenomas.
- About 50% of late adenomas and carcinomas show loss of heterozygosity on 18q. This is relatively uncommon in early and intermediate adenomas. It seems likely that the relevant gene is *SMAD4* rather than the initial candidate, *DCC* (see White, 1998).
- Colorectal cancers, but not adenomas, have a very high frequency of mutations in the *TP53* gene.

These are not the only changes seen in colorectal carcinomas, but they are the ones which can be most readily associated with specific stages, and they lead to the model shown in *Figure 18.18*. Analogous schemes could no doubt be constructed for other cancers if we had better knowledge of the early stages. The mutator genes *MSH2, MSH1*, etc., that are mutated in HNPCC are not directly involved in this pathway. Several genes have been identified that are frequently mutated in cancers with microsatellite instability (*TGFBR2, RAS, E2F-4, BAX*, etc.), and it seems likely that the mutator phenotype can either accelerate progress along the pathway shown in *Figure 18.18*, or facilitate alternative routes to malignant transformation.

Further reading

Lasko D, Cavenee W, Nordenskjöld M (1991) Loss of constitutional heterozygosity in human cancer. *Annu. Rev. Genet.,* **25**, 281–314.

References

Cavenee WK, Dryja TP, Phillips RA et al. (1983) Expression of recessive alleles by chromosomal mechanisms in retinoblastoma. *Nature,* **305**, 779–784.

Chissoe SL, Bodenteich A, Wang Y-F et al. (1995) Sequence and analysis of the human ABL gene, the BCR gene, and regions involved in the Philadelphia chromosomal translocation. *Genomics,* **27**, 67–82.

Easton DF (1994) Cancer risks in A-T heterozygotes. *Int. J. Radiat. Biol.,* **66**, S177–S182.

Eshleman JR, Markowitz SD (1996) Mismatch repair defects in human carcinogenesis. *Hum. Mol. Genet.,* **5**, 1489–1494.

Fearon ER, Vogelstein B (1990) A genetic model for colorectal tumorigenesis. *Cell,* **61**, 759–767.

Fishel R, Lescoe MK, Rao MRS et al. (1993) The human mutator gene homolog *MSH2* and its association with hereditary non-polyposis colon cancer. *Cell,* **75**, 1027–1038.

Fisher DE (1994) Apoptosis in cancer therapy: crossing the threshold. *Cell,* **78**, 539–542.

Forozan F, Karhu R, Kononen J, Kallioniemi A, Kallioniemi O-P (1997) Genome screening by comparative genomic hybridization. *Trends Genet.,* **13**, 405–409.

Friend SH, Tapscott SJ (1998) Sibling rivalry, arrested development and chromosomal mayhem. *Nat. Genet.,* **19**, 9–10.

Herman JG et al. (1998) Incidence and functional consequences of *hMLH1* promoter hypermethylation in colorectal carcinoma. *Proc. Natl Acad. Sci. USA,* **95**, 6870–6875.

Jones PA, Laird PW (1999) Cancer epigenetics comes of age. *Nature Genet.,* **21**, 163–167.

Kallioniemi A, Kallioniemi O-P, Sudar D, Rurovitz D, Gray JW, Waldman F, Pinkel D (1992) Comparative genomic hybridization for molecular cytogenetic analysis of solid tumors. *Science,* **258**, 818–821.

Kinzler KW, Vogelstein B (1996) Lessons from hereditary colorectal cancer. *Cell,* **87**, 159–170.

Lengauer C, Kinzler KW, Vogelstein B (1998) Genetic instabilities in human cancers. *Nature,* **396**, 643–649.

Malkin D. (1994) Germline p53 mutations and heritable cancer. *Annu. Rev. Genet.,* **28**, 443–465.

Mitelman F, Mertens F, Johansson B (1997) A breakpoint map of recurrent chromosomal rearrangements in human neoplasia. *Nature Genet.,* **15**, 417–474.

Modrich P (1995) Mismatch repair, genetic stability and tumour avoidance. *Phil. Trans. R. Soc. London Ser. B,* **347**, 89–95.

Nakamura Y (1998) ATM: the p53 booster. *Nature Med.,* **4**, 1231–1232.

Nurse P (1997) Checkpoint pathways come of age. *Cell,* **91**, 865–867.

Peltomaki P, de la Chapelle A (1997). Mutations predisposing to hereditary non-polyposis colorectal cancer. *Adv. Cancer Res.* **71**, 93–119.

Pinkel D (1994) Visualizing tumour amplification. *Nat. Genet.,* **8**, 107–108.

Pinkel D, Segraves R, Sudar D et al. (1998) High resolution analysis of DNA copy number variation using comparative genomic hybridization to microarrays. *Nature Genet.,* **20**, 207–211.

Pomeranz J, Schreiber-Agues N, Liégois NJ et al. (1998) The *Ink4a* tumor suppressor gene product, p19[Arf], interacts with MDM2 and neutralizes MDM2's inhibition of p53. *Cell,* **92**, 713–723.

Qian F, Watnick TJ, Onuchic LF, Germino GG (1996) The molecular basis of focal cyst formation in human autosomal dominant polycystic kidney disease type I. *Cell,* **87**, 979–987.

Rabbitts TH (1994) Chromosomal translocations in human cancer. *Nature,* **372**, 143–149.

Sanchez-García I (1997) Consequences of chromosomal abnormalities in tumor development. *Annu. Rev. Genet.,* **31**, 429–453.

Tosi S, Giudici G, Rambaldi A et al. (1999) Characterization of the human myeloid leukemia-derived cell line GF-D8 by multiplex fluorescence in situ hybridization, subtelomeric probes and comparative genomic hybridization. *Genes Chrom. Cancer,* **24**, 213–221.

Versteeg R (1997) Aberrant methylation in cancer. *Am. J. Hum. Genet.,* **60**, 751–754.

Weinberg RA (1995) The retinoblastoma protein and cell cycle control. *Cell,* **81**, 323–330.

Weinberg RA (1996) E2F and cell proliferation: a world turned upside down. *Cell,* **85**, 457–459.

Weinberg RA (1998) Bumps on the road to immortality. *Nature,* **396**, 23–24.

Weinert T (1998) DNA damage and checkpoint pathways: molecular anatomy and interactions with repair. *Cell,* **94**, 555–558.

White RL (1998) Tumor suppressing pathways. *Cell,* **92**, 591–592.

Wu CL, Sloan P, Read AP, Harris R, Thakker N (1994) Deletion mapping on the short arm of chromosome 3 in squamous cell carcinoma of the oral cavity. *Cancer Res.,* **54**, 6484–6488.

Complex diseases: theory and results

19

19.1 Deciding whether a nonmendelian character is genetic: the role of family, twin and adoption studies

19.1.1 The λ value is a measure of familial clustering

Nobody would dispute the involvement of genes in a character that consistently gives mendelian pedigree patterns or that is associated with a chromosomal abnormality. However, with nonmendelian characters, whether continuous (quantitative) or discontinuous (dichotomous), it is necessary to prove claims of genetic determination. The obvious way to approach this is to show that the character runs in families. The degree of family clustering of a disease can be expressed by the quantity λ_R, the risk to relative R of an affected proband compared with the population risk. Separate values can be calculated for each type of relative, for example λ_S for sibs. *Table 19.1* shows results from early studies of schizophrenia. Family clustering is evident from the raised λ values and, as

expected, λ values drop back towards 1 for more distant relationships – though in these data, not at the rate predicted for a purely genetic character.

19.1.2 The importance of shared family environment

Geneticists must never forget that humans give their children their environment as well as their genes. Many characters run in families because of the shared family environment – whether one's native language is English or Chinese, for example. One has therefore always to ask whether shared environment might be the explanation for a familial character. This is especially important for behavioral attributes like IQ or schizophrenia, which depend at least partly on upbringing, but it cannot be ignored even for physical characters or birth defects: a family might share an unusual diet or some traditional medicine that could cause developmental defects. Something more than a familial tendency is usually necessary to prove that a nonmendelian character is under genetic control. Unfortunately, these reservations are not always as clearly stated in the medical literature as they should

Table 19.1: Early studies of familial risk in schizophrenia

Dates	Studies	Relation	Incidence	λ^a
1928–62	14	Parents	336/7675 = 4.36%	5.45
			(corrected value[b] = 14.12%)	17.65[b]
1928–62	12	Sibs	724/8504 = 8.51%	10.6
1921–62	5	Children	151/1226 = 12.31%	15.4
1930–41	4	Uncles, aunts	68/3376 = 2.01%	2.5
1916–46	3	Half-sibs	10/311 = 3.22%	4.0
1926–38	5	Nephews, nieces	52/2315 = 2.25%	2.8
1928–38	4	Grandchildren	20/713 = 2.81%	3.5
1928–41	4	First cousins	71/2438 = 2.91%	3.6

[a] λ values are calculated assuming a population incidence of 0.8%.
[b] Correction allows for the fact that once schizophrenia has developed, people seldom have children.
Data assembled by Slater and Cowie (1970).

be. *Table 19.6* shows a demonstration of what can happen if shared family environment is ignored.

19.1.3 Twin studies suffer from many limitations

Francis Galton, who laid so much of the foundation of quantitative genetics, pointed out the value of twins for human genetics. Monozygotic (MZ) twins are genetically identical clones and will necessarily be concordant (both the same) for any genetically determined character. This is true regardless of the mode of inheritance or number of genes involved; the only exceptions (*Box 19.1*) are for characters dependent on postzygotic somatic genetic changes. Dizygotic (DZ) twins share half their genes on average, the same as any pair of sibs. Genetic characters should therefore show a higher concordance in MZ than DZ twins, and many characters do (*Table 19.2*).

Higher concordances in MZ compared with DZ twins might also be seen if the character is determined by environmental factors. For a start, half of DZ twins are of unlike sex, whereas all MZ twins are the same sex. Even if the comparison is restricted to same-sex DZ twins (as it is in the studies shown in *Table 19.2*), at least for behavioral traits the argument can be made that MZ twins are more likely to be very similar, to be dressed and treated the same, and thus to share more of their environment than DZ twins.

MZ twins separated at birth and brought up in entirely separate environments would provide the ideal experiment (Francis Crick once made the tongue-in-cheek suggestion that one of each pair of twins born should be donated to science for this purpose). Such separations happened in the past more often than one might expect because the birth of twins was sometimes the last straw for an overburdened mother. Fascinating television pro-

Table 19.2: Twin studies in schizophrenia

Study	Concordant MZ pairs	Concordant DZ pairs
Kringlen, 1968	14/55 (21/55)	4–10%
Fischer *et al.*, 1969	5/21 (10/21)	10–19%
Tienari, 1975	3/20 (5/16)	3/42
Farmer, 1987	6/16 (10/20)	1/21 (4/31)
Onstad *et al.*, 1991	8/24	1/28

The numbers show pairwise concordances, i.e. counts of the number of concordant (+/+) and discordant (+/−) pairs ascertained through an affected proband. Concordances can also be calculated probandwise, counting a pair twice if both were probands. This gives higher values for the MZ concordance. Probandwise concordances are thought to be more comparable with other measures of family clustering. Only the studies of Onstad and Farmer use the current standard diagnostic criteria, DSM-III. Figures in brackets are obtained using a wider definition of affected, including borderline phenotypes. For all references, see Fischer *et al.* (1969) and Onstad *et al.* (1991).

Box 19.1

Genetic differences between identical twins

All individuals, even monozygotic twins, differ in:

- their repertoire of antibodies and T-cell receptors (because of epigenetic rearrangements and somatic cell mutations);
- somatic mutations in general (*Chapter 18*);
- the numbers of mitochondrial DNA molecules (epigenetic partitioning);
- the pattern of X inactivation, if female.

grams can be made about twins reunited after 40 years of separation, who discover they have similar jobs, wear similar clothes and like the same music. As research material, however, separated twins have many drawbacks:

- There are very few of them, so any research is based on small numbers of arguably exceptional people.
- The separation was often not total – often they were separated some time after birth, and brought up by relatives.
- There is a bias of ascertainment – everybody wants to know about strikingly similar separated twins, but separated twins who are very different are not newsworthy.
- Even in principle, research on separated twins cannot distinguish intrauterine environmental causes from genetic causes. This may be important, for example in studies of sexual orientation ('the gay gene'), where some people have suggested that maternal hormones may affect the fetus *in utero* so as to influence its future sexual orientation.

Thus, for all their anecdotal fascination, separated twins have contributed relatively little to human genetic research.

19.1.4 Adoption studies are the most powerful way to disentangle genetic and environmental factors

If separating twins is an impractical way of disentangling heredity from family environment, adoption is much more promising. Two study designs are possible:

- find adopted people who suffer from a particular disease known to run in families, and ask whether it runs in their biological family or their adoptive family;
- start with affected subjects whose children have been adopted away from the family, and ask whether being adopted away from the family saves a child from the risk of the disease.

A celebrated (and controversial) study by Rosenthal and Kety (1968) used the first of these designs to test for genetic factors in schizophrenia. The diagnostic criteria used in this study have been criticized; there have also been claims (disputed) that not all diagnoses were made truly blind. However, an independent re-analysis using DSM-III diagnostic criteria (Kendler *et al.*, 1994) reached substantially the same conclusions. *Table 19.3* shows the results of a later extension of this study (Kety *et al.*, 1994).

The main obstacle in adoption studies is lack of information about the biological family, frequently made worse by the undesirability of approaching them with questions. A secondary problem is selective placement, where the adoption agency, in the interests of the child, chooses a family likely to resemble the biological family. Thus organizing a study may be difficult. Nevertheless, in countries where efficient adoption registers exist, this is unquestionably the most powerful method for checking how far a character is genetically determined. Quantitative characters can be similarly investigated by comparing the correlation between an adoptee and his/her biological or adoptive relatives. Sections 19.2 and 19.3 consider the main theory that has been used to explain how non-mendelian characters might be genetically determined.

19.2 Polygenic theory of quantitative traits

As we saw in Section 3.4.1, the early years of this century were marked by divergence between two traditions of genetics. The followers of Bateson saw genetics as the study of the transmission and segregation of mendelian genes, even if the phenotypic variants involved were rare or trivial, whereas the school of biometricians founded by Francis Galton saw genetics as the statistical study of evolutionarily important variation, normally in quantitative characters. Theoretical unification was achieved by Fisher in 1918, who demonstrated that the characters studied by the biometricians could be described in mendelian terms if they were **polygenic**, that is, governed by the simultaneous action of many gene loci. Any variable character that depends on the additive action of a large number of individually small independent causes will be distributed

in a normal (Gaussian) distribution in the population. This can be seen in the highly simplified example shown in *Figure 19.1*. We suppose the character depends on alleles at a single locus, then at two loci, then at three. As more loci are included we see two consequences:

(i) The simple one-to-one relationship between genotype and phenotype disappears: except for the extreme phenotypes, it is not possible to infer the genotype from the phenotype.
(ii) As the number of loci increases, the distribution looks increasingly like a Gaussian curve. Addition of a little environmental variation would smooth out the three-locus distribution into a good Gaussian curve.

A more sophisticated treatment allowing dominance and varying gene frequencies leads to the same conclusions. Since relatives share genes, their phenotypes are correlated, and Fisher's paper predicted the size of the correlation for different relationships.

19.2.1 Phenotypes where relatives share some but not all determinants will show regression to the mean, whether the determinants are genetic or environmental

A much misunderstood feature, both of biometric data and of polygenic theory, is regression to the mean. Suppose that variation in IQ were entirely genetically determined (we don't suppose this, but it is a convenient example to illustrate the arguments). On the simplest genetic assumptions, if one surveys mothers with an IQ of 120, their children would have an average IQ of 110, half way between the mothers' value and the population mean. Mothers with an IQ of 80 would have children with an average IQ of 90. Regression to the mean is often misinterpreted (see *Box 19.2*).

Figure 19.2 shows regression in our simplified two-locus model. For each class of mothers, the average IQ of

Table 19.3: An adoption study in schizophrenia

	Schizophrenia cases among biological relatives	Schizophrenia cases among adoptive relatives
Index cases (chronic schizophrenic adoptees)	44/279 (15.8%)	2/111 (1.8%)
Control adoptees (matched for age, sex, social status of adoptive family and number of years institutionalized)	5/234 (2.1%)	2/117 (1.7%)

The study involved 14 427 adopted persons aged 20–40 in Denmark. Forty seven of them were diagnosed as chronic schizophrenic. The 47 were matched with 47 nonschizophrenic control subjects from the same set of adoptees. Data of Kety *et al.* (1994).

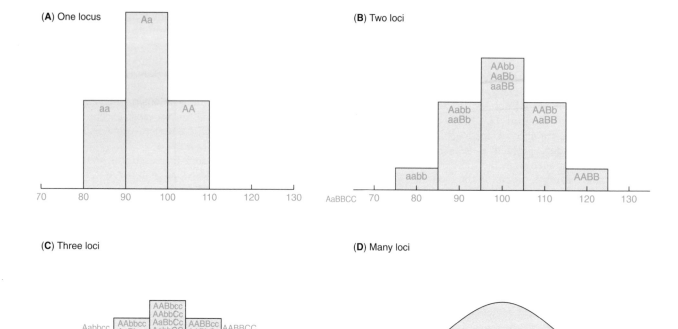

Figure 19.1: Successive approximations to a Gaussian distribution.

The charts show the distribution in the population of a hypothetical character that has a mean value of 100 units. The character is determined by the additive (co-dominant) effects of alleles. Each upper case allele adds 5 units to the value, and each lower case allele subtracts 5 units. All allele frequencies are 0.5. (**A**) The character is determined by a single locus. (**B**) Two loci. (**C**) Three loci: addition of a minor amount of 'random' (environmental or polygenic) variation produces the Gaussian curve (**D**).

Box 19.2

Two common misconceptions about regression to the mean

■   After a few generations everybody will be exactly the same.
■   Regression to the mean is a genetic phenomenon. If a character shows regression to the mean, it must be genetic.

Figure 19.2 shows that the first of these beliefs is wrong. In a simple genetic model:

■   the overall distribution is the same in each generation;
■   regression works both ways: for each class of children, the average for their mothers is half way between the children's value and the population mean. This may sound

paradoxical but it can be confirmed by inspecting, for example, the right-hand column of the bottom histogram in *Figure 19.2* (children of IQ 120). One-quarter of their mothers have IQ 120, half 110 and one-quarter 100, making an average of 110.

Regarding the second of these beliefs, regression to the mean is not a genetic mechanism but a purely statistical phenomenon. Whatever the determinants of IQ are (genetic, environmental or any mix of the two), if we take an exceptional group of mothers (for example, those with an IQ of 120), then these mothers must have had an exceptional set of determinants. If we take a second group who share half those determinants (their children, their sibs or either of their parents), the average phenotype in this second group will deviate from the population mean by half as much. Genetics provides the figure of one half, but not the principle of regression.

---

their children is half way between the mother's value and the population mean. However, this depends on a hidden assumption in this model, that there is random mating. For each class of mothers, the average IQ of their husbands is supposed to be 100. Thus the average IQ of the children is the mid-parental IQ, as common sense would

suggest. In the real world, highly intelligent women tend to marry men of above average intelligence (assortative mating) and we would not expect regression half way to the population mean, even if IQ were a purely genetic character.

The simplified model of *Figure 19.2* assumes there is no

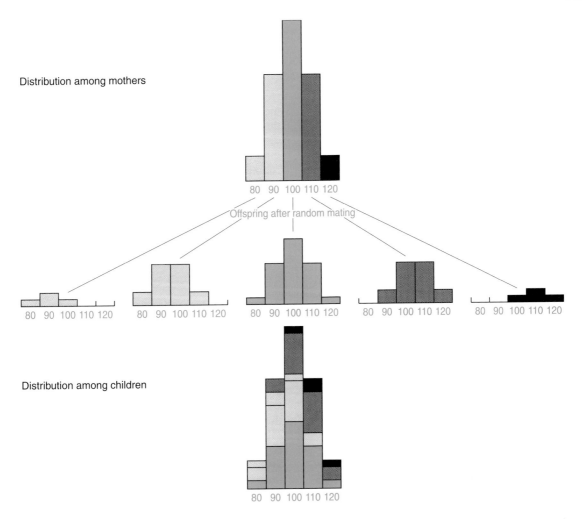

Figure 19.2: Regression to the mean.

The same character as in *Figure 19.1B*: mean 100, determined by co-dominant alleles A, a, B and b at two loci, all gene frequencies = 0.5. Top: distribution in a series of mothers. Middle: distributions in children of each class of mothers, assuming random mating. Bottom: summed distribution in the children. *Note* that: (i) the distribution in the children is the same as the distribution in the mothers; (ii) for each class of mothers, the mean for their children is half way between the mothers' value and the population mean (100); and (iii) for each class of children (bottom), the mean for their mothers is half way between the children's value and the population mean.

dominance. Each person's phenotype is the sum of the contribution of each allele at the relevant loci. If we allow dominance, the effect of some of a parent's genes will be masked by dominant alleles and invisible in their phenotype, but they can still be passed on and can affect the child's phenotype. Given dominance, the expectation for the child is no longer the mid-parental value. Our best guess about the likely phenotypic effect of the masked recessive alleles is obtained by looking at the rest of the population. Therefore, the child's expected phenotype will be displaced from the mid-parental value towards the population mean. How far it will be displaced depends on how important dominance is in determining the phenotype.

19.2.2 The heritability is the proportion of variance due to additive genetic effects

Gaussian curves are specified by only two parameters, the mean and the variance (or the standard deviation which is the square root of the variance). Variances have the useful property of being additive when they are due to independent causes (*Box 19.3*). Thus the overall variance of the phenotype V_P is the sum of the variances due to the individual causes of variation – the environmental variance V_E and the genetic variance V_G. V_G can in turn be broken down to a variance V_A due to simply additive genetic effects and an extra term V_D due to dominance effects. The **heritability** (h^2) of a trait is the proportion of the total variance which is genetic, that is V_G/V_P. For

Partitioning of variance

Variance of phenotype (V_P)
  = Genetic variance (V_G) + Environmental variance (V_E)
V_G = Variance due to additive genetic effects (V_A) +
  Variance due to dominant effects (V_D) → $V_P = V_A + V_D + V_E$
Heritability (broad) = V_G / V_P
Heritability (narrow) = V_A / V_P

animal breeders interested in breeding cows with higher milk yields, this is an important measure of how far a breeding program can create a herd in which the average animal resembles today's best. Strictly, V_G/V_P is the broad heritability. Dominance variance cannot be fixed by breeding, so the selection response is determined by the narrow heritability, V_A/V_P. Heritabilities of human traits are often estimated as part of segregation analysis (see Section 19.4 and *Table 19.5*).

The term 'heritability' is often misunderstood. Heritability is quite different from the mode of inheritance. The mode of inheritance (autosomal dominant, polygenic, etc.) is a fixed property of a trait, but heritability is not. 'Heritability of IQ' is shorthand for heritability of variations in IQ. Contrast the two questions:

■ *To what extent is IQ genetic?* This is a meaningless question.

■ *How much of the differences in IQ between people in a particular country at a particular time are caused by their genetic differences, and how much by their different environments and life histories?* This is a meaningful question, even if difficult to answer.

In different social circumstances, the heritability of IQ will differ. In an egalitarian society, we would expect IQ to have a higher heritability than in a society where access to education depended on accidents of birth. If everybody has equal opportunities, a number of the environmental differences between people have been removed. Therefore more of the remaining differences in IQ will be due to the genetic differences between people.

19.2.3 For many human behavioral traits, the simple partitioning of variance into environmental and genetic components is not applicable

Parents give their children both their genes and their environment, and genetic and environmental factors are often correlated. Genetic disadvantage and social disadvantage tend to go together. If genetic and environmental factors are not independent, V_P does not equal $V_G + V_E$; there are additional interaction variances. A proliferation of variances can rapidly reduce the explanatory power of the models, and in general this has been a difficult area in which to work.

19.3 Polygenic theory of discontinuous characters

Most of the classical 'polygenic' continuously variable characters like height or weight are of little interest to medical geneticists (although we discuss obesity in Section 19.5.7). Much more interesting are the innumerable diseases and malformations that tend to run in families, but that do not show mendelian pedigree patterns. A major conceptual tool in nonmendelian genetics was provided by Falconer's extension of polygenic theory to **dichotomous (discontinuous) characters** (those you either have or do not have).

19.3.1 Polygenic threshold theory can account for dichotomous non-mendelian characters

Falconer (1981) extended polygenic theory to discontinuous nonmendelian characters by postulating an underlying continuously variable susceptibility. You may or may not have a cleft palate, but every embryo has a certain susceptibility to cleft palate. The susceptibility may be low or high; it is polygenic and follows a Gaussian distribution in the population. Together with the polygenic susceptibility, Falconer postulated the existence of a threshold. Embryos whose susceptibility exceeds a critical threshold value develop cleft palate; those whose susceptibility is below the threshold, even if only just below, avoid cleft palate. Going back to the 'angels and devils' image of multifactorial disease in *Figure 3.11*, the threshold can be imagined as the neutral point of the balance. Changing the balance of factors tips it one way or the other.

For cleft palate, a polygenic threshold model seems intuitively reasonable (Fraser, 1980). All embryos start with a cleft palate. During early development the palatal shelves must become horizontal and fuse together. They must do this within a specific developmental window of time. Many different genetic and environmental factors influence embryonic development, so it seems reasonable that susceptibility should be polygenic. Whether the palatal shelves meet and fuse with ample time to spare, or whether they only just manage to fuse in time, is unimportant – if they fuse then a normal palate forms, and if they do not fuse then a cleft palate results. Thus there is a natural threshold superimposed on a variable developmental process.

19.3.2 Polygenic threshold theory helps explain how recurrence risks vary in families

Affected people have inherited an unfortunate combination of high-susceptibility genes. Their relatives who share genes with them will also, on average, have a raised susceptibility, the divergence from the population mean depending on the proportion of shared genes. Thus polygenic threshold characters tend to run in families (*Figure 19.3*). Parents who have had several affected children may just have been unlucky, but on average their susceptibility (and hence their children's susceptibility) must have been higher than parents with only one affected child. The threshold is fixed, but the average susceptibility, and hence the recurrence risk, rises with an increasing number of previous affected children.

Many supposed threshold conditions have different incidences in the two sexes. This implies sex-specific thresholds. Congenital pyloric stenosis, for example, is five times more common in boys than girls. The threshold must be higher for girls than boys, therefore relatives of an affected girl have a higher average susceptibility than relatives of an affected boy (*Figure 19.4*). The recurrence risk is correspondingly higher, although in each case the risk that a baby will be affected is five times higher if it is male (*Table 19.4*).

The mathematical properties of the Gaussian distribution predict the λ parameter (the ratio of incidence in relatives to incidence in the general population; Section 19.1.1). For first degree relatives, as a rough rule of thumb, the expected incidence is the square root of the population incidence. A polygenic threshold condition

Table 19.4: Recurrence risks for pyloric stenosis

Relatives of	Sons	Daughters	Brothers	Sisters
Male proband	19/296 (6.42%)	7/274 (2.55%)	5/230 (2.17%)	5/242 (2.07%)
Female proband	14/61 (22.95%)	7/62 (11.48%)	11/101 (10.89%)	9/101 (8.91%)

More boys than girls are affected, but the recurrence risk is higher for relatives of an affected girl. Data fit a polygenic threshold model with sex-specific thresholds (*Figure 19.4*). Data from Fuhrmann and Vogel (1976).

affecting 0.1% of newborns should affect about 1 in 30 of their sibs, parents or children. This prediction can then be compared with results of epidemiological surveys and any discrepancy used to calculate the heritability of the condition (*Figure 19.5*). Note, however, that the estimate of heritability depends on the assumption that the condition runs in families because of shared genes and not because of shared environment. As we saw in Section 19.1, proving that assumption is not always easy.

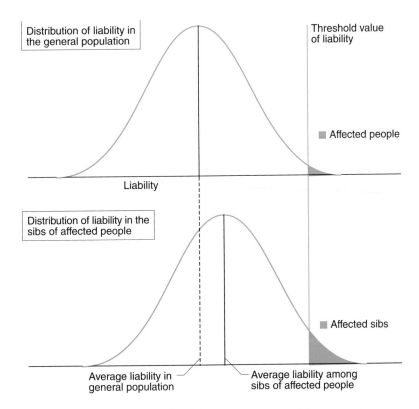

Figure 19.3: Falconer's polygenic threshold model for dichotomous nonmendelian characters.

Liability to the condition is polygenic and normally distributed (upper curve). People whose liability is above a certain threshold value are affected. Their sibs (lower curve) have a higher average liability than the population mean and a greater proportion of them have liability exceeding the threshold. Therefore the condition tends to run in families.

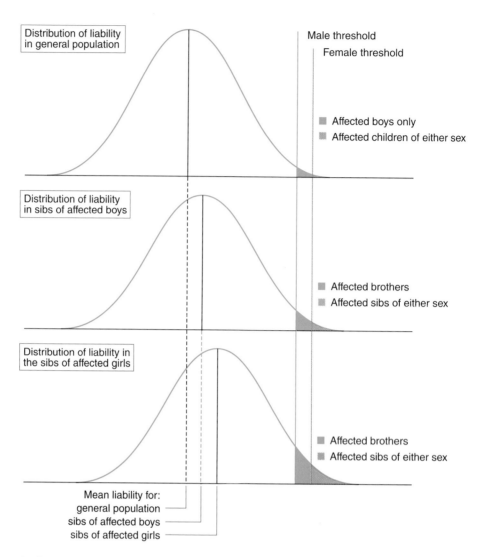

Figure 19.4: A polygenic character with sex-specific thresholds.

If a nonmendelian dichotomous character affects predominantly males, this is accommodated in multifactorial threshold theory by postulating a lower threshold for males than females. It follows that recurrence risks are higher for relatives of affected females, but the majority of those recurrent cases will be male. See *Table 19.4* for an example of data fitting this interpretation.

19.4 Segregation analysis allows analysis of characters that are anywhere on the spectrum between purely mendelian and purely polygenic

As we saw in *Figure 3.10*, pure mendelian and pure polygenic characters occupy opposite ends of a spectrum. In between are **oligogenic** traits governed by one or a few major susceptibility loci, possibly operating against a polygenic background, and possibly subject to major environmental influences. Segregation analysis is the main statistical tool for analyzing the inheritance of nonmendelian characters. It can provide evidence for a major

susceptibility locus and at least partly define its properties. By suggesting whether a complex disease is oligogenic or polygenic, segregation analysis can indicate whether it is worthwhile attempting to identify susceptibility genes by linkage or association studies, as outlined in Chapter 12.

19.4.1 Bias of ascertainment is often a problem with family data: the example of autosomal recessive conditions

Segregation analysis requires large datasets and is very sensitive to subtle biases in the way the data are collected. This can be illustrated by a mendelian example. Suppose

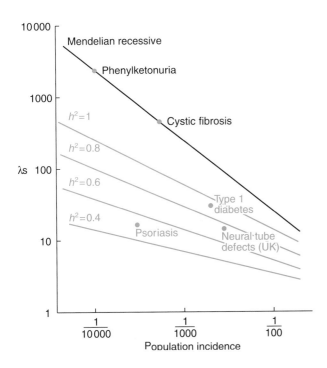

Figure 19.5: Familial clustering: relationship between population incidence and incidence in sibs of affected probands.

The value λ_S is the ratio of the incidence in sibs to the population incidence. The lines show the relationship for simple mendelian recessive or polygenic threshold inheritance. For a given population incidence, the higher the heritability, the greater is the value of λ_S. The graph can be used to estimate heritability from epidemiological data. From *The Genetics of Human Populations* by Cavalli-Sforza and Bodmer © 1978, 1971 by W. H. Freeman and Company. Used with permission.

one wishes to show that a condition is autosomal recessive. We could collect a set of families and check that the **segregation ratio** (the proportion of affected children) is 1 in 4. At first sight this would seem a trivial task, provided the condition is not too rare. But in fact the expected proportion of affected children in our family collection is not 1 in 4. The problem is **bias of ascertainment**.

Assuming there is no independent way of recognizing carriers, the families will be identified through an affected child. Thus the families shown on a white background in *Figure 19.6* will not be ascertained, and the observed segregation ratio in these two-child families is not 1/4 but 8/14. Families with three children, ascertained in the same way, would give a different segregation ratio, 48/111. The ratio for any given family size can be estimated from the truncated binomial distribution, a binomial expansion of $(\frac{1}{4} + \frac{3}{4})^n$ in which the last term (no affected children) is omitted. Experimental data can be corrected for this bias, most simply by the method of Li and Mantel (see *Box 19.4*).

The example above presupposes complete truncate ascertainment: we collect all families from some defined population who have at least one affected child. But this is not the only possible way of collecting families. We

might have ascertained affected children by taking the first 100 to be seen in a busy clinic (so that many more could have been ascertained from the same population by carrying on for longer). Under these conditions, a family with two affected children is twice as likely to be picked up as one with only a single affected child, and one with four affected is four times as likely. Single selection, where the probability of being ascertained is proportional to the number of affected children in the family, introduces a different bias of ascertainment, and requires a different statistical correction (see *Box 19.4*). We see that working out a segregation ratio requires data that have been collected in accordance with an explicit scheme of ascertainment, so that appropriate corrections can be applied.

19.4.2 Complex segregation analysis is a general method for estimating the most likely mix of genetic factors in pooled family data.

Analyzing data on the relatives of a large collection of people affected by a familial but non-mendelian disease is not a simple task. There could be both genetic and environmental factors at work; the genetic factors could be polygenic, oligogenic or mendelian with any mode of inheritance, or any mixture of these, while the environmental factors may include both familial and non-familial variables. In complex segregation analysis a whole range of possible genetic mechanisms, gene frequencies, penetrances, etc., are allowed, and the computer performs a maximum likelihood analysis to find the mix of parameter values which gives the greatest overall likelihood for the observed data. *Table 19.5* shows an example. As with lod score analysis (Chapter 11), the question asked is how much more likely the observations are on one hypothesis compared with another.

In the example of *Table 19.5*, the ability of specific models (sporadic, polygenic, dominant, recessive) to explain the data was compared with the likelihood calculated by a general model ('mixed model'), in which the computer

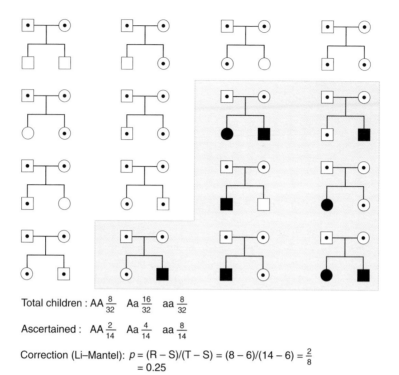

Total children : AA $\frac{8}{32}$ Aa $\frac{16}{32}$ aa $\frac{8}{32}$

Ascertained : AA $\frac{2}{14}$ Aa $\frac{4}{14}$ aa $\frac{8}{14}$

Correction (Li–Mantel): $p = (R - S)/(T - S) = (8 - 6)/(14 - 6) = \frac{2}{8}$
$$= 0.25$$

Figure 19.6: Bias of ascertainment.

Both parents are carriers of an autosomal recessive condition. Overall, one child in four is affected, but if families are ascertained through affected children, only the families shown in the blue area will be picked up, and the proportion of affected children is 8/14.

could freely optimize the mixture of single-gene, polygenic and random environmental causes. All models were constrained by overall incidences, sex ratios and probabilities of ascertainment estimated from the collected data. A single-locus dominant model is not significantly worse than the mixed model at explaining the data ($\chi^2 = 2.8$, $p = 0.42$), while models assuming no genetic factors, pure polygenic inheritance or pure recessive inheritance perform very badly. On the argument that simple explanations are preferable to complicated expla-

nations, the analysis suggests the existence of a major dominant susceptibility to Hirschsprung disease. Several such factors have now been identified (Section 19.5.2), though they probably account for only a minority of the susceptibility.

However clever the segregation analysis program, it can only maximize the likelihood across the parameters it was given. If a major factor is omitted, the result can be misleading. This was well illustrated by the data of McGuffin and Huckle (*Table 19.6*). They asked their

Table 19.5: Complex segregation analysis

Model	d	t	q	H	z	x	χ^2	p
Mixed	1.00	7.51	9.6×10^{-6}		0.01	0.15		
Sporadic							334	$<1 \times 10^{-5}$
Polygenic				1.00	1.00		78	$<1 \times 10^{-5}$
Major recessive locus	0.00	8.22	3.8×10^{-3}				35	$<1 \times 10^{-5}$
Major dominant locus	1.00	7.56	1.2×10^{-5}			0.19	2.8	0.42

Data are for families ascertained through a proband with long-segment Hirschsprung's disease. Parameters that can be varied are t (the difference in liability between people homozygous for the low-susceptibility and the high-susceptibility alleles of a major susceptibility gene, measured in units of standard deviation of liability), d (the degree of dominance of any major disease allele), q (the gene frequency of any major disease allele), H (the proportion of total variance in liability which is due to polygenic inheritance, in adults), z (the ratio of heritability in children to heritability in adults) and x (the proportion of cases due to new mutation). A single major locus encoding dominant susceptibility explains the data as well as a general model in which a mix of all mechanisms is allowed. Data from Badner *et al.* (1990).

Table 19.6: A recessive gene for attending medical school?

Model	d	t	q	H	χ^2	p
Mixed	0.087	4.04	0.089	0.008		
Sporadic					163	$<1 \times 10^{-5}$
Polygenic				0.845	14.4	<0.005
Major recessive locus	0.00	7.62	0.88		0.11	N.S.

Data of McGuffin and Huckle (1990) from a survey of medical students and their families. Meaning of symbols is as in *Table 19.5*. 'Affected' is defined as attending medical school. The analysis appears to support recessive inheritance, since this accounts for the data equally well as the unrestricted model. The point of this work was to illustrate how analysis of family data can produce spurious results if shared family environment is ignored (see text).

classes of medical students which of their relatives had attended medical school. When they fed the results through a segregation analysis program, it came up with results apparently favoring the existence of a recessive gene for attending medical school. Though amusing, this was not done as a joke, nor to discredit segregation analysis. The authors did not allow the computer to consider the likely true mechanism, shared family environment. The computer's next best alternative was mathematically valid but biologically unrealistic. The serious point McGuffin and Huckle were making was that there are many pitfalls in segregation analysis of human behavioral traits and incautious analyses can generate spurious genetic effects.

19.5 Seven examples illustrate the varying success of genetic dissection of complex diseases

There is no unified story in genetic analysis of complex diseases. We will now consider seven diseases that offer contrasting examples of genetic determination. The selection is more or less arbitrary – we could equally well have chosen other diseases – but our examples illustrate some typical situations. We have not concentrated on presenting success stories. In some cases there is little progress to report. Given the large efforts involved in every case, lack of success is almost as interesting as success.

19.5.1 Rhesus hemolytic disease: the interaction of two mendelian loci and obstetric history

Rhesus hemolytic disease of the newborn involves a pregnant woman mounting an immune attack against her baby's red cells. It runs in families but does not follow a mendelian pedigree pattern. The mechanism involves two mendelian loci interacting with the mother's history:

■ Hemolytic disease affects rhesus-positive babies of rhesus-negative mothers. The *D* allele at the *RH* locus encodes the D rhesus antigen, while the *d* allele is a gene deletion resulting in absence of D antigen. The matings at risk are dd mother × Dd or DD father.

■ Hemolytic disease is a risk only if the mother has previously been sensitized to D. Sensitization normally occurs at the time of birth of a previous rhesus-positive baby. Thus the firstborn child is rarely affected. Sometimes transplacental bleeds, miscarriages or unmatched blood transfusions may sensitize the mother and put her first baby at risk. Nowadays, sensitization is prevented by giving the mother an injection of anti-D antibody. The antibody clears rhesus-positive cells from her circulation before they can sensitize her.

■ If the mother and baby are ABO-incompatible, the mother's innate ABO antibodies will usually prevent sensitization.

Rhesus disease is not usually thought of as a complex disease. That is because we understand the genetics. If the mechanism had not been unraveled, we would probably be drawing Gaussian curves and thresholds.

19.5.2 Hirschsprung disease: a classical oligogenic disease, with several identified loci contributing to susceptibility, but none is either necessary or sufficient by itself

Hirschsprung disease (HSCR) is a congenital absence of ganglia in some or all of the large intestine or colon. The resulting lack of peristaltic action produces a grossly distended megacolon that is lethal neonatally unless the affected segment is removed. OMIM lists some rare syndromic forms caused by failure of the embryonic neural crest, but the common isolated HSCR is familial but non-mendelian. We have already seen (*Table 19.5*) how segregation analysis suggested a dominant susceptibility to isolated HSCR. One contributory locus, the *RET* oncogene on chromosome 10, was identified by linkage and mutation analysis following a clue from two HSCR patients with visible deletions of 10q11q21. The *RET* gene encodes a receptor tyrosine kinase, and in Section 16.6.2 we saw that loss of function mutations in *RET* predisposed to HSCR, while certain gain-of-function mutations had a totally different effect, producing multiple endocrine neoplasia. About 50% of patients with isolated

HSCR have a RET mutation, but so do some unaffected relatives.

Meanwhile a second HSCR susceptibility locus had been mapped to chromosome 13q in a huge multiply inbred Mennonite kindred. The report of this work (Puffenberger *et al*, 1994a) is well worth reading as an illustration of the complexities of mapping an imperfectly penetrant character in a pedigree that probably also contains a second unlinked disease susceptibility gene, and maybe phenocopies as well. Although the segregation analysis had suggested a dominant susceptibility, analysis of the Mennonite family was more compatible with recessive susceptibility. The gene responsible was identified as *EDNRB* (endothelin receptor B) by a positional candidate approach, based on phenotypic similarity to the *piebald-lethal* (s^l) mouse mutant. Interestingly, once the mutation in the family was defined (as W276C, Puffenberger *et al.*, 1994b) it turned out that affected people might have any genotype, but the penetrance was different, and sex-specific, for each genotype: 0.13 (M), 0.09 (F) for W/W, 0.33 (M), 0.08 (F) for W/C, and 0.85 (M), 0.60 (F) for C/C. This again illustrates the oligogenic character of Hirschsprung disease, with alleles at other loci modifying the effect of the *EDNRB* mutation.

Since *RET* and *EDNRB* both encode receptors, genes encoding their ligands (*GDNF* and *EDN3*) were natural candidate susceptibility genes. In all, five genes encoding components of these two biochemical pathways have been shown to contribute to isolated HSCR, but none of them shows high penetrance. Some patients carry mutations in more than one of these genes. Hofstra *et al.* (1997) discuss the difficulties of deciding which sequence changes contribute to the pathogenesis. Currently HSCR is probably the best example of an oligogenic disease, whose components are being revealed by a combination of mapping and analysis of functional candidates.

19.5.3 *Breast cancer: identifying a mendelian subset has led to important medical advances, but genetic susceptibility may be unimportant in the common sporadic form*

Although the common cancers are usually sporadic, the existence of 'cancer families' has been known for many years. When several relatives suffer the same rare cancer, like vestibular schwannomas (see NF2, MIM 101000), a mendelian syndrome is readily suspected, and investigation of such families has led to the identification of the tumor suppressor genes described in Section 18.5. Breast cancer is common, and so when several relatives have breast cancer, it is less clear whether this is just bad luck or a true cancer family. However, some breast cancer families have additional unusual features. Some families suffer breast cancer plus a variety of other rare cancers; this is the Li-Fraumeni syndrome (MIM 151623), which is a distinct condition caused by mutations in the *TP53* gene (Section 18.6.2), and not considered further here. Other families have just breast, or breast plus ovarian cancer,

but with an unusually early age of onset, frequent bilateral tumors and occasional affected males. Investigation of these families has led to identification of the *BRCA1* and *BRCA2* genes.

A large-scale segregation analysis of 1500 families (Newman *et al.*, 1988) supported the view that 4–5% of breast cancer, particularly early-onset cases, might be attributable to inherited factors. Families with near-mendelian pedigree patterns were collected for linkage analysis (*Figure 19.7*; see MIM 113705 for details of the linkage work). In 1990 a susceptibility locus, named *BRCA1*, was mapped to 17q21. The mean age at diagnosis in 17q-linked families was below 45. Later-onset families gave negative lod scores. The localization was soon confirmed by other groups, and over the next 2 years analysis of many families defined a candidate region 8 cM long in males and 17 cM long in females, located between markers D17S588 and D17S250. Meanwhile in 1994 a linkage search in 15 large families with breast cancer not linked to 17q mapped the *BRCA2* locus to 13q12 (see MIM 600185). It appeared that *BRCA1* might account for 80–90% of families with both breast and ovarian cancer, but only 50% of families with breast cancer alone. Male breast cancer was seen mainly in *BRCA2* families. A survey of 257 families with four or more cases of breast cancer suggested that 52% had mutations in *BRCA1*, 32% in *BRCA2* and 16% in other unidentified genes (Ford *et al.*, 1998).

A hectic race ensued to identify the two genes. *BRCA1* was cloned in 1994 and *BRCA2* in 1995 – see the MIM entries (113705 and 600185) for details. Early data suggested that a woman carrying a *BRCA1* mutation had an 85–90% chance of developing breast cancer, and a 40% risk of ovarian cancer. The risk of breast (but not ovarian) cancer for a woman with a *BRCA2* mutation is similar. However, these analyses were based on women from the families that had been used for mapping and mutation screening. Later, when mutations were found in women not selected for family history, far more nonpenetrant cases were found among their relatives. In Ashkenazi Jewish women, three particular mutations are rather frequent (185delAG and 5382insC in *BRCA1*, 6174delT in *BRCA2*). A study of relatives of Ashkenazi women with breast cancer, not selected on the basis of family history, suggested a lifetime risk for carriers of these mutations of 36% rather than the 85–90% usually quoted (Fodor *et al.*, 1998). Whatever the cause of this particular discrepancy, it is likely that there are mild, low penetrance mutations that would be missed in the initial studies but would be found if population screening were implemented. Until the natural history is better understood, population screening runs the risk of generating data that cannot be interpreted. For example, what advice would one give a woman with no family history who was found by population screening to carry a novel *BRCA1* mutation?

Both genes encoded large novel proteins that eventually turned out to be transcriptional co-activators, probably with additional roles in DNA repair. They behave as tumor suppressor genes, in that the inherited mutations cause loss of function (Section 16.3), and tumors from familial

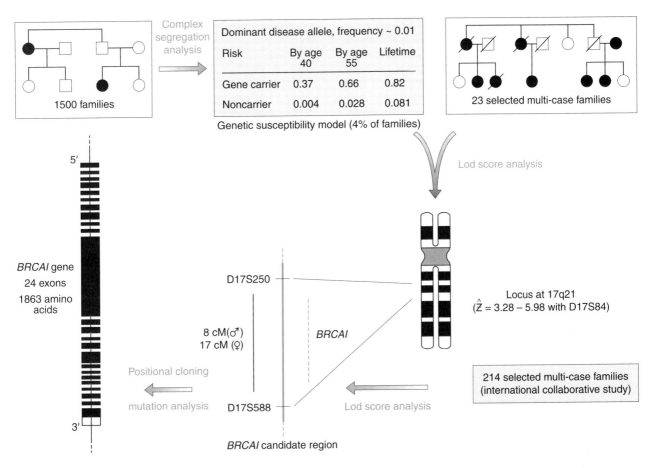

Figure 19.7: How the *BRCA1* gene was found.

This is an example of the successful use of mapping by lod score analysis followed by positional cloning to identify a gene conferring susceptibility to a common disease.

cases lose the wild-type allele. However, the story of breast cancer offers a striking contrast to colon cancer. In colon cancer, the *APC* gene was identified through the study of a rare mendelian form of the disease, but then turned out to be mutated in the common forms as well (Section 18.8.2). *BRCA1* and *BRCA2*, by contrast, are almost never mutated in sporadic breast cancer. Evidently other mechanisms and pathways are responsible for the sporadic late-onset breast cancer that affects about 8% of women in the UK. To a first approximation, breast cancer can be seen as two different diseases, a fairly rare mendelian early onset form, and a common late onset form in which no strong genetic factors have been demonstrated.

19.5.4 Alzheimer's disease: genetic factors are important both in the common late-onset form and in the rare mendelian early-onset forms, but they are different genes, acting in different ways

Alzheimer's disease (AD; MIM 104310) affects about 5% of persons over the age of 65, and about 20% over the age

of 80. There is progressive loss of memory, followed by disturbances of emotional behavior and general cognitive deterioration. Post mortem examination of the brain reveals loss of neurons with many amyloid-containing plaques. Degenerating neurons contain characteristic neurofibrillary tangles. Rarely, the onset is at a much younger age. The clinical and pathological features are identical in early-onset and late-onset AD, but early-onset disease is sometimes mendelian and autosomal dominant, whereas late-onset AD is non-mendelian and shows only modest familial clustering. Standard lod score analysis in dominant early-onset families allowed mapping and subsequently cloning of three genes, APP at 21q21, presenilin-1 at 14q24 and presenilin-2 at 1q42 (see MIM 104300, 104311 and 600579, respectively). Although mutations at these three loci account for only about 10% of AD cases with onset before age 65, they are particularly seen in the rare families afflicted by highly penetrant dominant AD with strikingly early onset. Anybody wanting an insight into what such disease means for a family should read *Hannah's Heirs* by Daniel Pollen (see Further reading).

Multicase late-onset families showed evidence of linkage

to chromosome 19 when analyzed by the affected pedigree member method (Section 12.2.3). There is no linkage in late-onset families to the loci implicated in early-onset disease. Searches for other susceptibility loci are being pursued through association studies; with such a late-onset disease it is not possible to use the standard forms of TDT, which require parental DNA (see, for example, Kehoe *et al.*, 1999). Eventually the susceptibility locus on chromosome 19 was identified as *APOE* (Apolipoprotein E, MIM 107741, located at 19q13.2). ApoE has three common forms with frequencies (in Caucasians) of about 0.08 (E2), 0.77 (E3) and 0.15 (E4). It turns out that both familial and sporadic late-onset AD are strongly associated with the E4 allele. E2 is associated with resistance to AD, and E4 with susceptibility. In cross-sectional studies of people over 65, E3/E4 people have about three times the risk, and E4/E4 people about 14 times, compared to E3 homozygotes. Thus, in contrast to breast cancer, not only the rare mendelian forms but also the common sporadic form has a substantial degree of genetic determination. ApoE appears to account for about 50% of the susceptibility to late-onset AD. However, E4 is neither necessary nor sufficient for AD. About half of all people with E4 will never develop AD, no matter how long they live, while many AD patients lack E4. A longitudinal study (Meyer *et al.*, 1998) has suggested that E4 may govern the age of onset rather than susceptibility. Once AD has started, its progress is no different in people with or without E4.

While the association of E4 with late-onset AD is beyond dispute, the mechanism is unclear. The E4 allele may be transcribed at a rather lower level than E3, and E2 at a rather higher level. The E2/E3/E4 determinants are amino acid substitutions R112C and R158C, which seem unlikely to affect the level of transcription directly. If the level of transcription is the true susceptibility factor, then its determinants may lie in the *APOE* promoter, and must be in linkage disequilibrium with the E2/E3/E4 polymorphism. Once identified, these determinants may prove more powerful predictors of risk. ApoE genotyping is technically simple, but it is a classic example of a test that is better not carried out. The statistical uncertainty allows neither confirmation of a clinical diagnosis nor useful predictive testing in healthy persons, and there is no preventive action that can be taken in response to a positive predictive test. Of course, insurance companies selling long-term care policies may see this in a different light.

19.5.5 Type 1 diabetes mellitus: large-scale affected sib-pair analysis defined susceptibility loci, and linkage disequilibrium is being used to try to define the genetic factors involved

Family and twin studies showed that diabetes tended to run in families, but for many years diabetes remained, as it was famously described, the geneticists' nightmare. The first step towards understanding the genetics was to distinguish the different types of diabetes (*Table 19.7*). Type 1 and Type 2 diabetes are different diseases, with different causes, different natural histories and different genetics. Type 1 diabetes (insulin-dependent diabetes, IDDM) is caused by autoimmune destruction of pancreatic β-cells, typically affecting young people and requiring lifelong insulin treatment. Type 2 diabetes (noninsulin-dependent diabetes, NIDDM) is caused by some combination of reduced insulin secretion and end-organ unresponsiveness. Age, obesity, physical inactivity and genetic factors all play contributory roles in the complex and doubtless heterogeneous pathogenesis of NIDDM. Maturity-onset diabetes of the young (MODY) is a rare condition which, being mendelian, is susceptible to standard linkage analysis. Several MODY genes have been identified (see MIM 600496), and mild mutations in the MODY genes might be susceptibility factors for NIDDM.

Once Type 1 diabetes was recognized as a separate disease, its genetics could be investigated (see MIM 222100). There is fairly strong familial clustering ($\lambda_s = 15$, MZ twin concordance ~ 30%). An association with certain HLA alleles was already established in the 1970s. In the UK, about 95% of patients with type 1 diabetes have the HLA-DR3 and/or DR4 antigens, compared with 45–54% of the general population. This was the first demonstration of a major genetic factor in diabetes, and it accounts for about half of the genetic predisposition. The strong linkage disequilibrium within the major histocompatibility complex makes it difficult to identify the primary determinant of HLA-linked susceptibility. Several different haplotypes are associated with susceptibil-

Table 19.7: Clinical classification of diabetes

Type 1 diabetes	Type 2 diabetes	MODY
Juvenile onset	Maturity onset (> 40 years)	Juvenile onset
0.4% of UK population	6% of US population	Rare
Requires insulin	Usually controllable by oral hypoglycemics	As type 2 diabetes
No obesity	Strong association with obesity	No obesity
Familial:	Familial:	Familial:
MZ concordance 30%	MZ twin concordance 40–100%	autosomal
sib risk 6–10%	sib risk 30% (maybe subclinical)	dominant?
Associated with	No HLA association	No HLA association
HLA-DR3 and DR4		

ity or resistance, but it turns out that haplotypes associated with a low risk of diabetes all carry an allele at the *DQB1* locus in which amino acid 57 is aspartic acid, while high-risk haplotypes have *DQB1* alleles with some other amino acid at this position. In one study, 96% of diabetics but only 20% of controls were homozygous non-ASP at DQB1 position 57. Exactly how the absence of aspartic acid at position 57 of the *DQB1* product predisposes to diabetes is not clear.

Early work on IDDM also identified a second susceptibility locus, *IDDM2*, close to the structural gene for insulin. Linkage disequilibrium mapping has revealed the actual determinant as a 14 bp minisatellite repeat upstream of the gene. Susceptibility is associated with short repeats (26–63 repeat units). Long repeats (140–210 units) cause the insulin gene to be transcribed at rather higher level in the developing thymus; it is argued that this increases the efficiency of deletion of insulin-reactive T-cell clones during development of the immune system, and so reduces the risk of an autoimmune attack. Both the *HLA-DQB* work and the *INS* findings underscore the likely difference between the type of subtle genetic changes that may underlie susceptibility to a common disease and the gross changes seen in mendelian diseases (see Section 3.4.3).

IDDM has been the major test-bed for the strategy of mapping susceptibility loci by large-scale analysis of affected-sib-pairs, followed by linkage disequilibrium mapping within confirmed regions. For a good flavor of the initial affected sib pair part of the work, and of the subsequent linkage disequilibrium mapping, see the papers by Davies *et al.* (1994) and Merriman *et al.* (1997). *Table 19.8* lists the putative loci reported by various groups. Of these, IDDM 1, 2, 4, 5, 8 and 12 are confirmed according to the Lander and Kruglyak criteria (Section 12.5.3). In some cases faith in a candidate location is reinforced by the proximity of a plausible candidate gene, or by reference to *idd* loci mapped in the NOD (nonobese

diabetic) mouse, which is a good model of IDDM (Ghosh *et al*, 1993).

19.5.6 Multiple sclerosis: much work but very modest progress

Dyment *et al.* (1997) give a good brief review of the epidemiology and genetics of multiple sclerosis (MS). The prevalence is about 1 in 1000 in Northern Europeans, and there is strong familial aggregation (λ_s = 20–40). The pathology involves autoimmune destruction of the myelin sheath that insulates nerve fibers, and as with many autoimmune diseases, there is a well-established population association with an HLA haplotype, in this case DRB1*1501, DQA1*0102, DQB1*0602. Three independent large affected sib pair studies were reported in the August 1996 issue of *Nature Genetics* (*Table 19.9*; summarized by Sawcer *et al.*, 1997). They make interesting comparison.

■ **Cambridge (UK) series**. 143 affected sib pairs were typed with 311 markers. Nineteen autosomal regions gave sharing significant at a pointwise 5% significance level, and six regions showed sharing at a level that would be expected by chance only once in a genome screen (the criterion for 'suggestive linkage', see Section 12.5.3). Thirty-two markers from these six regions, plus another three regions defined by candidate genes, were tested in a second sample of 108 sib pairs. 17q22 and 6p21 (the HLA region) gave maximum lod scores of 2.7 and 2.8, and the data implied λ_s of 1.7 and 1.5 respectively. Neither of the other two studies suggested any susceptibility locus on 17q. Genes with λ_s of 5.0 and 2.0 were excluded from 93% and 55% respectively of the genome.

■ **US/French series**. 52 families including 81 affected sib pairs and a number of other affected relatives were typed with 443 markers. As in the Cambridge

Table 19.8: Type 1 diabetes susceptibility loci suggested by affected sib pair (ASP) or transmission disequilibrium (TDT) analysis.

Locus	MIM no.	Location	Status
IDDM1	222100	6p21	λ_s = 3.1; determinant is HLA-DQB
IDDM2	125852	11p15	λ_s = 1.3; determinant is a VNTR upstream of INS gene
IDDM3	600318	15q26	λ_s = 1.4; detected in two studies
IDDM4	600319	11q13	λ_s = 1.6; confirmed in three screens
IDDM5	600320	6q24–q27	λ_s = 1.2; confirmed in two studies; homolog of mouse *idd5*?
IDDM6	601941	18q21	ASP and TDT evidence in one very large study.
IDDM7	600321	2q31–q33	λ_s = 1.3; seen in three ASP studies and TDT shows association
IDDM8	600883	6q25–q27	λ_s = 1.8; confirmed in two studies
IDDM10	601942	10p11–q11	ASP and TDT data in two studies from one group
IDDM11	601208	14q24–q31	Seen in one study
IDDM12	600388	2q33	Confirmed. Linkage disequilibrium with CTLA4
IDDM13	601318	2q34	Seen in one study. Same as *IDDM7* and/or *12*?
IDDM15	601666	6q21	Yet another 6q locus, seen in one study
IDDM17	603266	10q25	In one large Bedouin family

Data from the OMIM entries and papers cited therein; λ_s values are from Luo *et al.* (1995).

Table 19.9: Results of three whole-genome searches for susceptibility loci for multiple sclerosis

	Cambridge series	US / French series	Canadian series
Stage 1 screen	143 ASPs 311 markers	52 families including 81 ASPs; 443 markers	100ASPs 257 markers
MLS > 1*	1p36, 2p13, 3p14–p21 4q35, 14q32, 19q13	5q13–q23, 7q32–q34, 11p15, 12q24–qter, 19q13	2p16, 3q21–q24, 11q22.3, Xp21–p11.4
Suggestive linkage	1cen, 5cen, 7p21–p15, 12p13–p12, 17q22, 22q13	7q21–q22	5p (D5S406)
Stage 2 screen	108 ASP 6 regions of suggestive linkage tested	23 families including 45 ASP; data reported only for 6p21	(a) 44 ASP (b) 78 ASP Tested for 5p and 6p21
Overall result	MLS 2.8 for 6p21 MLS 2.7 for 17q22	MLS 3.6 for 6p21	MLS 0.65 for 6p21 MLS 1.6 for 5p

* The US–French study used three different statistical tests; loci in the 'MLS >1' row passed two tests and those shown as 'suggestive' passed all three tests. See text for details.

study, 19 chromosomal regions gave results with a pointwise significance < 0.05, based on a combination of ASP and affected pedigree member linkage (using IBS, not IBD criteria). Among the 19 candidate regions was 6p21, and excess allele sharing in this region was confirmed in a second set of 23 multiple-case families.

■ **Canadian series.** 100 affected sib pairs were studied with 257 markers. Five loci gave MLS values > 1 (expectation on the null hypothesis = 3.6 loci). Eighty eight percent of the genome was excluded for $\lambda_S = 3.0$. One region, on chromosome 5, gave a 2-point MLS of 4.24, but this fell to 1.8 on multipoint analysis. This region, together with the HLA region (which showed no evidence of excess allele sharing in the initial dataset) was studied in two additional datasets comprising 44 and 78 affected sib pairs. The chromosome 5 region showed modest evidence of linkage (MLS 1.8) in one of the two additional datasets, whilst HLA-linked markers showed modestly but not significantly increased allele sharing in both additional datasets.

The most striking feature of these studies is the overall low level of significance of the results. A follow-up study by the US-French group (Fontaine *et al.*, 1998) tested six candidate regions in a total of 261 affected sib pairs. Maximum lod scores in each region were 1.66 at 1p35, 0.22 at 2p14, 0.24 at 3p14, 0.00 at 5q12, 0.68 at 19q13 and 0.00 at Xq23.

These studies make it virtually certain that no single locus confers as much as 10% of the overall susceptibility in the populations tested. Thus although the λ_S value for MS is quite high at 20–40, it appears that MS lies towards the polygenic end of the spectrum of diseases, and it will not be easy to confirm susceptibility loci by ASP analysis. Association studies such as the TDT may hold more promise. Interestingly, linkage in multicase Finnish families has produced evidence of susceptibility at three candidate loci, HLA, MBP (myelin basic protein on 18q23)

and markers on 5p12–p14 at a position corresponding to the mouse *Eae2* locus (this governs susceptibility to experimental allergic encephalitis, a possible model of MS). The 5p location corresponds roughly (but not precisely) to the location highlighted by the Canadian study, but neither the genome scans described above nor directed linkage analysis by other groups have supported MBP. Overall, it seems likely that MS is marked by considerable genetic heterogeneity and that no strongly predisposing alleles are widely prevalent.

19.5.7 Obesity: genetic analysis of a quantitative trait

As a contentious emotive issue, obesity rivals schizophrenia, with one school of thought, prevalent among thin people, blaming poor self-discipline and moral laxity, and the stout party laying the blame on constitutional predisposition. Obesity is a major public health concern in advanced countries, and is of great interest to biotechnology companies, who see the fortune to be made from an effective slimming pill. For our purposes, the principal interest of obesity is as an example of a quantitative trait.

Simply measuring weight does not give a good quantitative variable for analysis because it confuses tall thin people with short fat people. Body mass index (weight in kg)/(height in m)^2, is a much better measure. The mean BMI in the US in 1983 was 22. Other investigators have tried to use measures closer to the underlying physiology, for example percent adipose tissue or serum leptin concentration (see below). In each case, some cut-off could be used to define obesity, for example BMI 25–29.9 = Grade 1, BMI 30–40 = Grade II and BMI > 40 = Grade III, with adjustments for age and sex. However, such an approach is arbitrary and loses data. It is much better to analyze the quantitative variable directly. Such quantitative measures are the bread and butter of animal and plant breeding, and the standard human linkage programs for both parametric analysis (e.g. MLINK, Section 11.5.2) and non-parametric analysis (e.g. GENEHUNTER, Section 12.2.4) are set up to handle quantitative variables. Comuzzie *et al.* (1997) describe how the covariance between relatives can be analyzed as a function of the identity by descent around a marker locus. Paterson *et al.* (1988) showed how a genome-wide linkage trawl could map **quantitative trait loci (QTL)** in the tomato, and the present generation of investigations of obesity attempt to replicate this success in humans.

As always with nonmendelian phenotypes, the first task is to check whether there is any genetic component in the causation. Obesity tends to run in families, but this could be the effect of shared family environment. However, twin and adoption studies suggest that around 70% of the variance in BMI can be attributed to genetic factors. Several groups have conducted whole-genome searches for loci predisposing to obesity (see Hager *et al.*, 1998 and references therein). Possible candidate loci are suggested by several rare mendelian diseases that are associated with obesity, for example Bardet–Biedl syndrome (MIM 209900). Other candidates come from the recent discovery of leptin and its receptor, which constitute a hormone system by which adipose tissues signal their state and regulate appetite. Mice with mutations affecting the leptin system are morbidly obese, and rare human examples have been described (see MIM 164160 for references). Predictably for this stage of the research, the published studies show little agreement about candidate regions, although some promising lod scores are reported. We chose obesity here as a nice example to illustrate the methods and approaches in human QTL analysis – unfortunately the best results are probably hidden in the files of biotechnology companies, not to be revealed until something patentable has been achieved.

19.6 Applications of genetic insights into complex diseases

We have seen in the previous section that the search for disease susceptibility genes has so far produced rather modest results. The general consensus of researchers is that the difficulties are caused by lack of statistical power, and are remediable by testing more samples with more markers. For a less optimistic view, see Weiss, *Genetic Variation and Human Disease* (1993). Assuming markers of susceptibility to a complex disease can be defined, identifying them must advance our understanding of the pathogenesis. Such understanding does not automatically lead to cures or improved treatment – a condition may still be incurable even if we understand it well, and purely symptomatic treatment is sometimes very effective – but at the very least, it must be better to understand the conditions one is trying to deal with. There would be clear benefits to pharmaceutical and epidemiological research.

Pharmaceutical companies are looking for two main benefits.

- Identifying the pathogenic processes underlying a clinical condition can suggest new targets for drug development.
- Patients respond differently to many drugs. With any given drug, a percentage of patients may fail to respond at all, and others may show unwanted side effects. These differences are probably largely genetically determined. If different genetic susceptibility factors precipitate the same disease in different patients, they may respond optimally to different treatments. This is likely to be particularly true for conditions like hypertension, where the pathology is ongoing. Diseases like Type 1 diabetes may be less promising – whatever the causes of the autoimmune attack on pancreatic beta cells, once it is underway we have to treat the consequences. Such individual-specific effects will be especially noticeable in complex diseases, because an important insight from theory is that a given polygenic phenotype can be produced by several different genotypes (see *Figures 3.11* and *19.1*). This influence of genotype is on top of the effects that polymorphisms in metabolic enzymes may have on the pharmacodynamics or side effects of a drug. The pharmaceutical industry expects genotype-specific drugs and genotype-specific dosage regimens to be an important part of 21st century medicine.

Epidemiological research should benefit from an ability to define genetic susceptibility. A typical investigation of environmental risk factors would follow a cohort of people, and retrospectively try to detect the difference between the environments of those who did and did not develop the disease. But the people who did not develop the disease may be a mixture of those who avoided the environmental trigger and those who encountered the trigger but escaped disease because they were not genetically susceptible. The analysis would be far more sensitive if a cohort of susceptible people could be defined.

A much more open question is the degree to which the new insights will lead to successful preventive medicine. Two extreme positions can be characterized:

- **The head-in-clouds position** – medicine in the 21st century will be focused on these susceptibility factors. Identifying them will allow us to move from a 'diagnose and treat' to a 'predict and prevent' healthcare system. Population screening will define people's susceptibility, and drugs or lifestyle changes will prevent disease.
- **The head-in-sand position** – most common diseases don't have mendelian subgroups. Most disease susceptibility factors are not strong enough to be usefully predictive for individuals. In the light of experience with smoking and diet, it is unrealistic to think many people will modify their lifestyle to avoid modest genetic risks.

Technical developments, especially DNA chips would make population screening easier, but one always needs to ask whether such screening would be cost-effective and ethically acceptable. The necessary conditions were discussed in detail in Section 17.3. The most essential single point, more important than any technical issue, is that identifying somebody as at risk should lead to some useful action. Usually this means action to avoid the risk. In general, one is talking about lifestyle changes – prophylactic drug treatment would be justifiable only if it could be targeted at a small number of high-risk people, and such groups are characteristic of mendelian rather than

the common complex diseases. Recommendations to change one's lifestyle provide a health-conscious minority of individuals with a valued option to avoid risk, but their overall impact on public health, on present evidence, is disappointingly low. Without compulsion, only simple and effortless changes that substantially reduce a large risk have much chance of improving public health significantly. In the end, the problem with both the head-in-clouds and the head-in-sand positions is that they over-generalize. Many common diseases will probably not yield major alterable risk factors, but if a few common diseases do, then the research will have been worthwhile.

Further reading

Kety SS, Rowland LP, Sidman RL, Matthysse SW (eds) (1983) *Genetics of Neurological and Psychiatric Disorders.* Raven Press, New York.
Pollen DA (1996) *Hannah's Heirs,* expanded edition. Oxford University Press, Oxford.
Rosenthal D, Kety SS (1968) *The Transmission of Schizophrenia.* Pergamon Press, Oxford.
Weiss KM (1993) *Genetic Variation and Human Disease: Principles and Evolutionary Approaches. Studies in Biological Anthropology Vol 11.* Cambridge University Press, New York.

References

Badner JA, Sieber WK, Garver KL, Chakravarti A (1990) A genetic study of Hirschsprung disease. *Am. J. Hum. Genet.,* **46**, 568–580.
Cavalli-Sforza L, Bodmer W (1971) *The Genetics of Human Populations.* Freeman, San Francisco.
Comuzzie AG, Hixson JE, Almasy L et al. (1997). A major quantitative trait locus determining serum leptin levels and fat mass is located on human chromosome 2. *Nature Genet.,* **15**, 273–276.
Davies JL, Kawaguchi Y, Bennett ST et al. (1994) A genome-wide search for human type 1 diabetes susceptibility genes. *Nature, 371,* 130–136.
Dyment DA, Sadnovich AD, Ebers GC (1997). Genetics of multiple sclerosis. *Hum. Molec. Genet.,* **6**, 1693–1698.
Falconer DS. (1981) *Introduction to Quantitative Genetics,* 2nd edn. Longman, London.
Fischer M, Harvald B, Hauge M (1969) A Danish twin study of schizophrenia. *Br. J. Psychiatr.,* **115**, 981–990.
Fisher RA (1918) The correlation between relatives under the supposition of mendelian inheritance. *Trans. R. Soc. Edin.,* **52**, 399–433.
Fodor FH, Weston A, Bleiweiss IJ et al. (1998). Frequency and carrier risk associated with common *BRCA1* and *BRCA2* mutations in Ashkenazi Jewish breast cancer patients. *Am. J. Hum. Genet.,* **63**, 45–51.
Fontaine B, Clanet M, Babron MC et al. (1998). Analysis of six previously identified regions of linkage in multiple sclerosis (MS). *Am. J. Hum. Genet.,* **63**, A289.
Ford D, Easton DF, Stratton M et al. (1998). Genetic heterogeneity and penetrance analysis of the *BRCA1* and *BRCA2* genes in breast cancer families. *Am. J. Hum. Genet.,* **62**, 676–689.
Fraser FC (1980) Evolution of a palatable multifactorial threshold model. *Am. J. Hum. Genet.,* **32**, 796–813.
Fuhrmann W, Vogel F (1976) *Genetic Counselling.* Springer, New York.
Ghosh S, Palmer SM, Rodrigues NR et al. (1993) Polygenic control of autoimmune diabetes in nonobese diabetic mice. *Nature Genet.,* **4**, 404–409.
Hager J, Dina C, Francke S et al. (1998) A genome-wide scan for human obesity genes reveals a major susceptibility locus on chromosome 10. *Nature Genet.,* **20**, 304–308.
Hofstra RMW, Osinga J, Buys CHCM (1997) Mutations in Hirschsprung disease: when does a mutation contribute to the phenotype?. *Eur. J. Hum. Genet.,* **5**, 180–185.
Kehoe P, Wavrant-De Vrieze F, Crook R et al. (1999). A full genome scan for late-onset Alzheimer disease. *Hum. Molec. Genet.,* **8**, 237–246.
Kendler KS, Gruenberg AM, Kinney DK (1994) Independent diagnoses of adoptees and relatives as defined by DSM-III in the provincial and national samples of the Danish Adoption Study of Schizophrenia. *Arch. Gen. Psychiatr.,* **51**, 456–468.
Kety SS, Wender PH, Jacobsen B, Ingraham LJ, Jansson L, Faber B, Kinney DK (1994) Mental illness in the biological and adoptive relatives of schizophrenic adoptees. Replication of the Copenhagen Study in the rest of Denmark. *Arch. Gen. Psychiatr.,* **51**, 442–455.

Luo DF, Bui MM, Muir A, McLaren NK, Thomson G, She J-X (1995) Affected sib-paired mapping of a novel susceptibility gene to insulin-dependent diabetes mellitus (IDDM8) on chromosome 6q25–q27. *Am. J. Hum. Genet.*, **57**, 911–919.

McGuffin P, Huckle P (1990) Simulation of Mendelism revisited: the recessive gene for attending medical school. *Am. J. Hum. Genet.*, **46**, 994–999.

Merriman T, Twells R, Merriman M *et al.* (1997) Evidence by allelic association-dependent methods for a type 1 diabetes polygene (*IDDM6*) on chromosome 18q21. *Hum. Mol. Genet.*, **6**, 1003–1010.

Meyer MR, Tschanz JT, Norton MC *et al.* (1998). ApoE genotype predicts when – not whether – one is predisposed to develop Alzheimer disease. *Nature Genet.*, **19**, 331–332.

Newman B, Austin MA, Lee M, King MC (1988) Inheritance of human breast cancer: evidence for autosomal dominant transmission in high-risk families. *Proc. Natl Acad. Sci. USA*, **85**, 3044–3048.

Onstad S, Skre I, Torgersen S, Kringlen E (1991) Twin concordance for DSM-III-R schizophrenia. *Acta Psychiatr. Scand.*, **83**, 395–401.

Paterson AH, Lander ES, Hewitt JD, Peterson S, Lincoln SE, Tanksley SD (1988) Resolution of quantitative traits into mendelian factors by using a complete linkage map of restriction fragment length polymorphisms. *Nature,* **335**, 721–726.

Puffenberger E, Kauffman E, Bolk S *et al.* (1994a) Identity-by-descent and association mapping of a recessive gene for Hirschsprung disease on human chromosome 13q22. *Hum. Mol. Genet.*, **3**, 1217–1225.

Puffenberger EG, Hosoda K, Washington SS, Nakao K, deWit D, Yanagisawa M, Chakravarti A (1994b) A missense mutation of the endothelin-B receptor gene in multigenic Hirschsprung's disease. *Cell*, **79**, 1257–1266

Sawcer S, Goodfellow PN, Compston A (1997) The genetic analysis of multiple sclerosis. *Trends Genet.*, **13**, 234–239.

Slater E, Cowie V (1970) *Genetics of Mental Disorders*. Oxford University Press, Oxford.

Studying human gene structure, expression and function using cultured cells and cell extracts

20

The structure, expression and function of human genes can be studied by a variety of methods. As described in this chapter, some well-established and recently developed methods use cell extracts or cultured cells to explore gene expression and gene function. In addition, genetically manipulated animals have been very important for studying gene function and for devising models of human disease. Experimental approaches using genetically modified animals are described in Chapter 21.

20.1 Gene structure and transcript mapping studies

Genomic or cDNA clones containing an uncharacterized gene have been identified by different routes (*Box 20.1*). An initial priority in defining gene structure is to obtain a full-length cDNA sequence and identify translational initiation and termination sites, and polyadenylation site(s). Exon/intron structure can then be determined by referencing the cDNA sequence against sequences of cognate genomic DNA clones. Subsequently, attempts may be made to complete gene characterization at the genomic level by sequencing of promoter regions, 5' and 3' flanking sequences and intron sequences.

20.1.1 Completion of full-length cDNA sequences can be achieved by assembling overlapping cDNA clones and by RT-PCR based procedures

Many of the clones found in cDNA libraries contain partial length sequences. A frequent method of constructing cDNA libraries uses an oligo(dT) primer to bind to the 3' poly(A) tail of the mRNA and prime first strand synthesis by reverse transcriptase (see *Figure 4.8*). However, very large mRNAs can be a problem because of the difficulty in completing cDNA synthesis, and such libraries typically show a general 3' end bias: many of the cDNA clones contain comparatively short sequences containing only the 3' end of the cDNA. One way of addressing this problem has been to construct randomly primed cDNA libraries: a degenerate hexanucleotide primer of the type commonly used for labeling DNA *in vitro* is used to prime first strand synthesis instead of the traditional oligo(dT) primer. The hexanucleotide primers hybridize at essentially random sites along the RNA, and can generate cDNA sequences of a sufficient size to be useful when cloned.

cDNA contig assembly

One way of achieving a full-length cDNA sequence is to screen different cDNA libraries and then map the extent of overlap between the inserts of positive clones (by sequencing or by PCR/hybridization-based mapping). As a result, a series of overlapping cDNA clones can be established, a cDNA clone contig (for an example, see the contig for the cystic fibrosis cDNA in Riordan *et al.*, 1989). Full or selected clone sequencing will define a consensus sequence which may provide a full length cDNA sequence.

Because of the recent huge proliferation of cDNA sequences in electronic databases it may be possible to obtain nearly full length cDNA sequences for a previously uncharacterized gene without actually doing any sequencing! Instead, cDNA contig assembly is achieved by a type of 'electronic PCR', and all that is required is a small starting sequence representing an expressed part of the gene of interest, and a computer. Sophisticated computer programs such as the EST blast program developed at Glaxo-Wellcome (electronic reference 1) permit an integrated approach to sequence database searching and sequence analysis in order to compile and extend a consensus cDNA sequence starting from a small reference sequence (see also Section 20.1.4).

Extending cDNA sequences by standard RT-PCR and RACE

Full-length copies of low abundance mRNAs may be difficult to obtain by conventional cloning, and so rapid PCR-based methods have proved to be useful alternatives. Standard reverse transcriptase (RT)-PCR methods are often used where sequence information has been obtained for dispersed exons at the level of genomic

Box 20.1

Obtaining gene clones for studying human gene structure, expression and function

Human gene clones can be obtained by cell-based or PCR-based DNA cloning of either genomic DNA or cDNA. PCR may provide a rapid way of getting an initial gene clone, but cell-based DNA cloning is normally used to provide the large quantities of starting DNA needed for comprehensive studies of gene structure and function.

cDNA enrichment cloning

The genomic DNA of all nucleated human cells consists of basically the same collection of sequences, but the mRNA populations may be very different. Total cDNA from different cell types is enriched for certain gene sequences at the expense of others. For example, α-globin and β-globin mRNA are abundant in erythrocyte mRNA and so globin cDNA sequences are abundant in erythrocyte cDNA libraries. PCR-based methods have also been devised to selectively amplify cDNAs corresponding to subsets of transcriptionally active genes (e.g. see mRNA differential display, Section 20.2.4).

Protein-directed cloning

It may be possible to isolate and purify a specific protein from a cell or from extracellular fluids in the case of secreted proteins. A partially purified protein preparation can be injected into rabbits or mice in order to raise specific antibodies, which can then be used to screen an expression cDNA library (see *Figure 4.18*). Alternatively, the amino acid sequence of a polypeptide is determined (see Stryer, 1995). The amino acid sequence is inspected to design gene-specific oligonucleotides which can then be used as hybridization probes for isolating a corresponding cDNA or genomic DNA clone (see *Figure 15.2*).

Location-directed cloning

Physical mapping and positional cloning projects can involve attempts to identify gene clones at specific subchromosomal locations (Section 15.3).

Homology-based cloning

Individual human genes may contain sequences or sequence motifs that are related to other human genes or to genes in other species, notably mammalian species. A previously isolated gene (from another species, or a human gene) can be labeled and used as a probe to isolate novel human genes by screening human cDNA or genomic libraries. Conserved sequence motifs can also be used to design PCR amplification methods for identifying novel members of gene families. In this case, partially degenerate oligonucleotide primers are used, often corresponding to short sequences of amino acids that are well conserved between the products of the different members of a gene family. Amplification with such primers will lead to a mixture of different products which can then be cloned in cells to identify and isolate novel gene sequences (e.g. Gavin and McMahon, 1993).

Random clone characterization

The above methods have been and continue to be very useful for isolating gene clones. They are now, however, being overtaken by large-scale characterization of random cDNA clones and construction of a human gene map (Section 13.2.3).

DNA. For example, random shotgun sequencing of large genomic clones (BACs etc) may result in sequences that are expected to be exons of the same gene but may be some distance apart from each other at the cDNA level. Or different exons can be isolated by *exon trapping* (Section 10.4.2). In such cases, primers can be designed for specific exons and used in RT-PCR to amplify portions of the cDNA sequence separating two such exons.

One popular method of obtaining full-length cDNA sequences is the RACE (rapid amplification of cDNA ends) technique (Frohman *et al.*, 1988). RACE-PCR is an anchor PCR modification of RT-PCR. The rationale is to amplify sequences between a single previously characterized region in the mRNA (cDNA) and an anchor sequence that is coupled to the 5′ or the 3′ end. A primer is designed from the known internal sequence and the second primer is selected from the relevant anchor sequence (see *Figure 20.1*).

20.1.2 Transcription start sites can be mapped by nuclease S1 protection and by primer extension

Important regulatory sequences are often located close to transcription start sites. Although 5′ RACE-PCR can per-

mit rescue of sequences corresponding to the 5′ end of a mRNA (and therefore, the transcriptional start site), two major methods are preferentially used to define the transcriptional start site.

Nuclease S1 protection

The endonuclease S1 is an enzyme from the mold *Aspergillus oryzae* which cleaves single-stranded RNA and DNA but not double-stranded molecules. In order to map the transcription start site for a gene, a genomic DNA clone suspected of containing the start site is required. The DNA clone is then digested with a suitable restriction endonuclease to generate a fragment that is expected to contain the transcription start site. As shown in *Figure 20.2A*, hybridization to the cognate mRNA and S1 nuclease digestion defines the distance of the transcription start site from the unlabeled end of the restriction fragment. If more precise localization is required, the labeled DNA fragment in the heteroduplex can be sequenced by a chemical method of DNA sequencing (see Maxam and Gilbert, 1980). Note that, in much the same way, nuclease S1 mapping can also be used to map other boundaries between coding and noncoding DNA such as exon/intron boundaries (see below) and the 3′ end of a transcript.

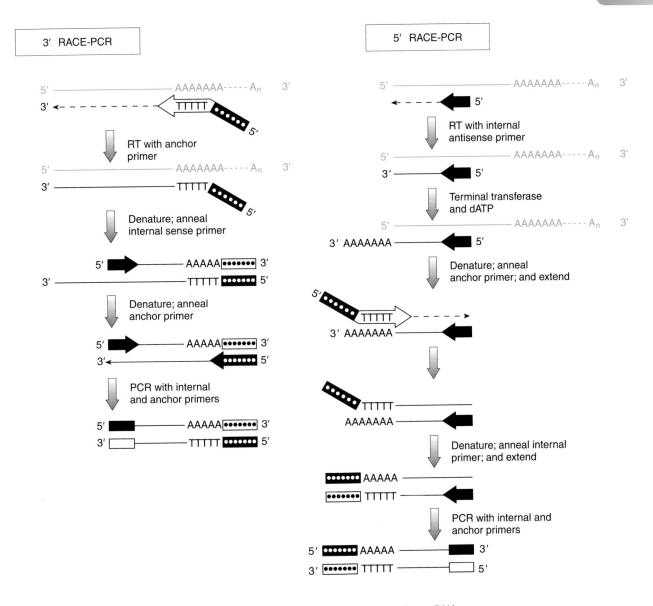

Figure 20.1: RACE-PCR facilitates the isolation of 5′ and 3′ end sequences from cDNA.

A preliminary step in RACE (*rapid amplification of cDNA ends*)-PCR involves the introduction of a specific sequence at either the 3′ or the 5′end by what is effectively a form of *5′-add-on mutagenesis* (Section 6.4.2). (**A**) 3′ RACE-PCR uses a starting antisense primer with a specific 5′ extension sequence (*anchor sequence*, often >15 nucleotides long) which becomes incorporated into the cDNA transcript at the reverse transcriptase step. An internal sense primer is then used to generate a short second strand ending in a sequence complementary to the original anchor sequence. Thereafter, PCR is initiated using the internal sense primer and an anchor sequence primer. (**B**) 5′ RACE-PCR. Here an internal antisense primer is used to prime synthesis from a mRNA template (blue) of a partial first cDNA strand (black). A poly(dA) is added to the 3′ end of the cDNA using terminal transferase. Second strand synthesis is primed using a sense primer with a specific extension (anchor) sequence. This strand is used as a template for a further synthesis step using the internal primer in order to produce a complementary copy of the anchor sequence. PCR can then be accomplished using internal and anchor sequence primers.

Primer extension

The method is very similar to the nuclease S1 protection method. In this case, the chosen restriction fragment must be shorter than the mRNA and the overhang is filled in using reverse transcriptase (*Figure 20.2B*). Again a more accurate location is possible by using the chemical sequencing method of Maxam and Gilbert (1980) to sequence the labeled DNA strand.

20.1.3 Exon/intron boundaries can be mapped by a variety of different methods

An early requirement in investigating gene structure is to isolate suitable genomic DNA clones covering the length of the gene. The availability of both genomic and cDNA clones then provides the opportunity of mapping

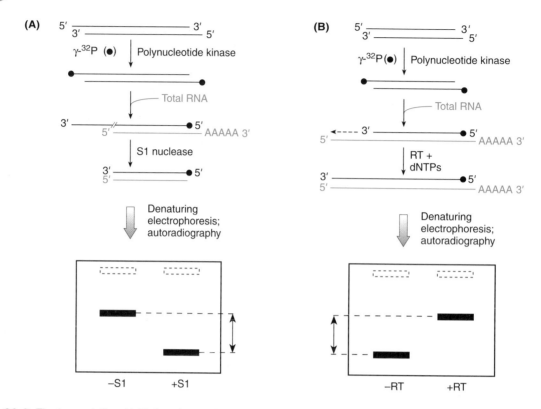

Figure 20.2: The transcriptional initiation site can be mapped by nuclease S1 protection or primer extension assays.

(**A**) Nuclease S1 protection assay. A restriction fragment from the 5' end of a cloned gene is suspected of containing the transcription initiation site. It is end-labeled at the 5' ends, then denatured and mixed with total RNA from cells in which the relevant gene is thought to be expressed. The cognate mRNA can hybridize to the antisense DNA strand to form an RNA–DNA heteroduplex. Subsequent treatment with nuclease S1 results in progressive cleavage of the overhanging 3' DNA sequence until the point at which the DNA is hybridized to the 5' end of the mRNA. Size-fractionation on a denaturing electrophoresis gel can identify the size difference between the original DNA and the DNA after nuclease S1 treatment. (**B**) Primer extension assay. In this case, the restriction fragment suspected of containing the transcriptional initiation site is deliberately chosen to be small. Hybridization with a cognate mRNA will then leave the mRNA with an overhanging 5' end. The DNA can serve as a primer for reverse transcriptase (RT) to extend its 3' end until the 5' end of the mRNA is reached. The size increase after reverse transcriptase treatment (+RT) compared with before treatment (−RT) maps the transcription initiation site. More precise mapping is possible in both methods by sequencing the DNA following treatment with S1 or RT.

exon/intron boundaries. Not all human genes have introns (see *Table 7.6*). Where introns are present, however, they are often comparatively large. The presence of introns can usually be inferred from comparative mapping of cognate genomic and cDNA clones. Once the full cDNA sequence has been established, sequencing primers can be designed from various segments of the cDNA and used in *cycle DNA sequencing* with denatured cosmid clones as the DNA sequencing templates (see Section 6.3.3). The sequences obtained should cross an exon/intron boundary, unless the exon is very large, in which case additional sequencing primers may be required.

Other methods of mapping exon/intron boundaries include nuclease S1 protection (see previous section) and PCR-based *genomic walking* methods. The latter includes techniques such as *bubble linker PCR* or *inverse PCR* (see *Figures 10.16* and *10.15*), which can be used with an exon-specific primer to amplify a short stretch of neighboring intronic sequence from a YAC clone template (or even

from total genomic DNA) and then sequence across the exon/intron boundaries. This was the way, for example, in which comprehensive exon/intron boundaries were established for the dystrophin gene (Roberts *et al.*, 1993).

20.1.4 Computer-based sequence analysis programs and database searches can illuminate aspects of gene structure, function and evolution

Once an accurate cDNA sequence has been obtained, a simple computer program can be used to translate all three translational reading frames in order to identify a long open reading frame (ORF), which would be interpreted as the frame used to generate a polypeptide product. In some cases, such as the *H19* and *XIST* genes, no significant ORF may be found, indicating that the gene does not encode a polypeptide. An alternative possible explanation, in some cases, may be that the gene specifies a very short polypeptide product, the translational reading frame for which may not be easily distinguished from the other two frames. Computer-based sequence analyses may clarify the likeliest possibility (see below).

Once the correct reading frame has been identified, the translational termination site can be identified easily by the presence of one of the recognized termination codons. Thereafter, possible polyadenylation sites can usually be distinguished by the presence of the well-conserved AATAAA sequence (often located ~600–800 bp downstream of the termination codon in mammalian cDNAs), which defines the end of the 3′ UTR. However, the identification of the translational start point and the extent of the 5′ UTR may be less straightforward. In the former case, ATG (which specifies methionine) is found in the vast majority of cases *but not all* – rare alternatives which have been found include ACG (Thr), CTG (Leu) and GTG (Val). Even if the initiation codon is ATG, it should also be noted that the N-terminal amino acid of the mature polypeptide may not be methionine (see the example of β-globin in *Figure 1.19*).

Inspection of the nucleotide sequence at the 5′end of a cDNA or in the corresponding sequence of a genomic DNA clone, may reveal a stop codon in the same reading frame as the presumed coding sequence. Clearly, the initiating codon must lie downstream from this and, unless there are several possible codons for methionine in this area, the identification of a methionine codon downstream of this and forming part of a large ORF is usually straightforward (Kozak, 1996). If, however, there are several possible ATG codons, the initiator codon may have to be identified by characterization of the poly-

peptide product in a suitable expression cloning system. Other important sequence motifs are most easily identified by computer-based sequence analysis.

Sequence analysis programs

Simple computer programs exist for constructing detailed restriction maps of given nucleotide sequences, and for searching for repetitive sequences (see electronic reference 2). Tandemly repeated small segments, such as triplet repeats, are usually evident by cursory inspection of the sequence. However, large repeats, distantly spaced repeats and repeats that have undergone substantial divergence may not be readily apparent until analyzed by computer. The output of such searches is often in the form of a **dot-matrix analysis**. A dot is marked on the chart wherever a string of nucleotides or amino acids at one position in the gene or protein matches a string at another position. Matching can be defined as identity, or as matching above a pre-set value. Internal repeats generate diagonal lines of contiguous dots superimposed on the random background. Dot-matrix analysis can be used to compare one sequence with another (*Figure 20.3*).

Database homology searches

Powerful computer programs have been devised to permit searching of nucleic acid and protein sequence databases for significant sequence matching (**sequence homology**) with a test sequence under investigation. Popularly used programs are the different **BLAST** and FASTA programs (see Ginsburg, 1994 and *Table 20.1*). Programs such as these use algorithms to identify optimal sequence alignments and typically display the output as

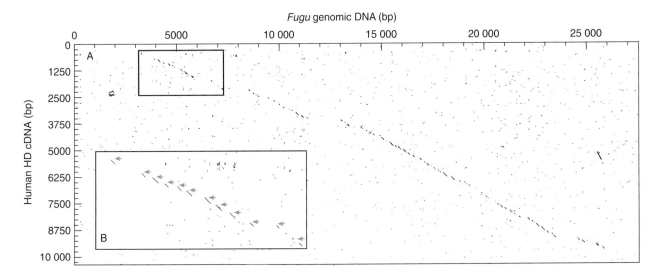

Figure 20.3: Dot-matrix analysis is a useful graphical method for comparing sequences.

The example shows comparison of the human Huntington's disease cDNA sequence with the genomic DNA sequence for the homologous gene in the pufferfish. The lines along the diagonal correspond to the regions of homology permitted within the parameters of the program, and define exons in this case. The insert (**B**) shows a tenfold increase in resolution of the first 12 exons (boxed in **A**) of the gene (indicated by arrows). *Note* that although the *Fugu* HD gene spans only 23 kb of genomic DNA compared to 170 kb for the human gene, all 67 exons are conserved. Reproduced from Baxendale *et al.* (1995) *Nature Genet.*, **10**, pp. 67–75, with permission from Nature America Inc.

Table 20.1: BLAST and FASTA programs for sequence comparisons

Program	Compares
FASTA	A nucleotide sequence against a nucleotide sequence database, or an amino acid sequence against a protein sequence database
TFASTA	An amino acid sequence against a nucleotide sequence database translated in all six reading frames
BLASTN	A nucleotide sequence against a nucleotide sequence database
BLASTX	A nucleotide sequence translated in all six reading frames against a protein sequence database
EST BLAST	A cDNA/EST sequence against cDNA/EST sequence databases
BLASTP	An amino acid sequence against a protein sequence database
TBLASTN	An amino acid sequence against a nucleotide sequence database translated in all six reading frames

Note that because the design of comparable programs such as FASTA and BLASTN is different, they may give different results (Ginsburg, 1994). All of the above programs are accessible through the Internet from various centers, such as the European Bioinformatics Institute (http://www.ebi.ac.uk/) and the US National Center for Biotechnology Information (http://www.ncbi.nih.gov/).

a series of pairwise comparisons between the test sequence (query sequence) and each related sequence which the program identifies in the database (subject sequences).

Different approaches can be taken to calculate the optimal sequence alignments. For example, in nucleotide sequence alignments the algorithm devised by Needleman and Wunsch (1970) seeks to maximize the number of matched nucleotides. In contrast, in other programs such as that of Waterman *et al.* (1976) the object is to minimize the number of mismatches. Pairwise comparisons of sequence alignments are comparatively simple when the test sequences are very closely matched and have similar, preferably identical, lengths. When the two sequences that are being matched are significantly different from each other and especially when there are clear differences in length due to deletions/insertions, considerable effort may be necessary to calculate the optimal alignment (*Figure 20.4A*).

If the nucleotide sequence under investigation is a coding sequence, then nucleotide sequence alignments can be aided by parallel amino acid sequence alignments using the assumed translational reading frame for the coding sequence. This is so because there are 20 different amino acids but only 4 different nucleotides. Pairwise alignments of amino acid sequences can also be aided by taking into account the chemical *subclasses* of amino acids. *Conservative substitutions* are nucleotide changes which result in an amino acid change but where the new amino acid is chemically similar to the replaced amino acid and typically belongs to the same subclass (*Box 9.2*). As a result, algorithms used to compare amino acid sequences typically use a scoring matrix in which pairs of scores are arranged in a 20 × 20 matrix where higher scores are accorded to identical amino acids and to ones which are of similar character (e.g. isoleucine and leucine) and lower scores are given to amino acids that are of different character (e.g.

(A) GATATTATCACTGGAGCCT GGC AGGAGCT
```
  * * *   * * * *   * * * * * * * * * *   *   * * * * * * *           or
```
GATTTTATGACTGGAGCCT – GA AGGAGCT

GATATTATCACTGGAGCCT GGC AGGAGCT
```
  * * *   * * * *   * * * * * * * * * *   *           * * * * * * *
```
GATTTTATGACTGGAGCCT GA – AGGAGCT

(B) Score = 56.2 bits (133), Expect = 2e – 08
Identities = 39/120 (32%), Positives = 58/120 (48%), Gaps = 9/120 (7%)

```
Query: 165  AKLLIKHDSNIGIPDVEGKIPLHWAANHKDPSAVHTVRCILDAAPTESLLNWQDYEGRTP  224
            A+LL++HD++        G  PLH A +H +   + V+ +L  +       W Y    TP
Sbjct: 548  AELLLEHDAHPNAAGKNGLTPLHVAVHHNN---LDIVKLLLPRGGSPHSPAWNGY---TP  601

Query: 225  LHFAVADGNLTVVDVLTSYE-SCNITSYDNLFRTPLHWAALLGHAQIVHLLLERNKSGTI  283
            LH A     +V   L Y  S  N  S   +  TPLH AA GH ++V LLL +   +G +
Sbjct: 602  LHIAAKQNQIEVARSLLQYGGSANAESVQGV--TPLHLAAQEGHTEMVALLLSKQANGNL  659
```

Figure 20.4: Nucleotide and amino acid sequence alignments.

(A) The difficulty in sequence alignments. This example illustrates a difficulty in finding a correct alignment between two nucleotide sequences which are clearly related. At the sequence GGC shown in blue at the top, there is uncertainty as to the best alignment with the corresponding sequence GA in the bottom sequence. **(B)** Sequence identity and sequence similarity. This worked example illustrates a BLASTP output. The Swissprot protein database was screened with a query sequence which represented amino acids 165–283 of a protein predicted from the newly discovered inversin gene, a gene involved in left–right axis specification. The highest score was detected when aligned with a mouse ankyrin sequence. The program considers not just *sequence identity* (39 of the 120 positions, or 32%, have identical residues in the two sequences, shown as blue letters), but also *sequence similarity* (here indicated as 'positives') whereby an additional 19 positions have chemically similar amino acids (shown as blue +).

isoleucine and aspartate; see Henikoff and Henikoff, 1992). The typical output gives two overall results for % sequence relatedness, often termed % sequence identity (matching of identical residues only) and % sequence similarity (matching of both identical residues and ones that are chemically related; *Figure 20.4B*).

Depending on the length of sequences under investigation a variety of different levels of sequence homology searching can be conducted.

Homolog and domain searching

A full-length or partial DNA sequence (or the interpreted amino acid sequence) for a human gene can be entered into a program that compares it against all other sequences stored in electronic sequence databases. Such comparisons will detect sequences that are significantly related to a substantial component (*domain*) or even all of the sequence of interest (homolog), either in the same species or in different species. Searching across large databases often provides the first indication of the function of a sequence, as in the case of the normal function of the neurofibromatosis type 2 gene (*NF2*). Homology searching of databases using the newly established NF2 cDNA sequence identified related sequences, notably those of moesin, ezrin and radixin (see, for example, Rouleau *et al.*, 1993). These proteins were known to act as structural links between cell membrane proteins and intermediate filament proteins, thereby providing some initial clues to the function of the *NF2* gene product.

Motif searching

This means analysing a sequence for the presence of a short sequence (*motif*) that is indicative of a specific functional characteristic of the gene being studied. This can be conducted at both nucleotide and amino acid sequence level:

- **Nucleotide motifs.** The test sequence can be queried for the presence of specific regulatory *cis*-acting sequences including general and tissue-specific regulatory elements (*Table 8.2*), response elements (*Table 8.4* and *Figure 8.9*), etc.
- **Amino acid motifs.** Information on subcellular or extracellular destination of a predicted protein sequence can be obtained by using specialized computer programs (e.g. electronic reference 3). Such programs screen for a variety of different motifs such as signal peptides, nuclear localization sequences, glycosylation signals or for the presence of likely transmembrane domains etc. (see *Table 1.7*). In addition, specific amino acid motifs may be uncovered by computer analysis which are associated with a particular function or family of genes with related functions. This is possible because there are 20 different amino acids and so short sequences may be of considerable significance. For example, the DEAD sequence (one letter code for Asp–Glu–Ala–Asp) would occur by chance about once in 170 human polypeptides. This

motif is found, along with some other motifs, in a family of RNA helicases (see *Figure 7.9*).

20.2 Studying gene expression using cultured cells or cell extracts

20.2.1 Principles of expression mapping

Obtaining information on patterns of gene expression can be done at different levels and using a variety of different technologies (*Figure 20.5*). Two important parameters are as follows:

- **Expression resolution.** Some methods are designed simply to track the gross expression of a gene in RNA extracts or protein extracts. Such low resolution expression patterns are usually attempted as a first pass approach. In addition to being able to sample expression in different tissues etc., they may provide useful information on product size and on possible isoforms. In contrast, high resolution expression can be obtained using methods which track expression patterns within a cell, or within groups of cells and tissues which are spatially organized in a manner representative of the normal *in vivo* organization.
- **Throughput.** Some methods are designed to obtain expression data for one or a very small number of genes at a time. Other methods can simultaneously track the expression of many genes, and in some cases, all of the genes in an organism (whole genome expression screening). In the latter case, complete sequencing of a genome permits identification of all genes thereby permitting simultaneous analysis of the expression of every single gene using DNA chip technology.

20.2.2 Nucleic acid hybridization is a popular tool for surveying gene expression at the RNA level and can be used to provide whole genome expression screens

Once a genomic or cDNA clone becomes available, it can be labeled and used as a probe to track the expression of that gene at the RNA level. Different methods are available to study RNA expression using RNA or cDNA extracts, or fixed tissues and cells.

Northern blot hybridization

This approach affords low resolution expression patterns by hybridizing a gene or cDNA probe to total RNA or poly (A)$^+$ RNA extracts prepared from different tissues or cell lines. Because the RNA is size-fractionated on a gel, it is possible to estimate the size of transcripts (see *Figure 5.13*). The presence of multiple hybridization bands in one lane may indicate the presence of differently sized isoforms.

	RESOLUTION	THROUGHPUT	EXAMPLES
RNA	High	Low	Tissue *in situ* hybridization Cellular *in situ* hybridization
	Low	Low	Northern blot hybridization RNA dot blot hybridization RT-PCR Ribonuclease protection assay
	Low	High	DNA microarray hybridization Differential display Serial analysis of gene expression (SAGE)
PROTEIN	High	Low	Immunocytochemistry Immunofluorescence microscopy
	Low	Low	Immunoblotting (western blotting)
	Low	High	2-D gel electrophoresis Mass spectrometry

Figure 20.5: Expression mapping can be conducted at different levels.

Tissue in situ hybridization

High resolution spatial expression patterns of RNA in tissues and groups of cells are normally obtained by **tissue *in situ* hybridization**. Usually, tissues are frozen or embedded in wax and then cut using a microtome to give very thin sections (e.g. 5 μm thick) which are mounted on a microscope slide. Hybridization of a suitable gene-specific probe to the tissue on the slide can then give detailed expression images representative of the distribution of the RNA in the tissue of origin (*Figure 5.17*). Often, the tissues used include embryonic tissues which have the advantage that their miniature size permits expression screening of many tissues in a single section.

In order to improve sensitivity, it is customary to use an antisense riboprobe, that is, a labeled antisense RNA probe which can be generated by *in vitro* transcription from a cDNA cloned in a suitable vector (see *Figure 5.4*) and which will hybridize specifically to the sense mRNA of the relevant gene. Sometimes, however, the gene in question is a member of a gene family and other closely related gene sequences are expressed. If so, some effort may be required to prepare a suitable locus-specific probe. This involves comparing the sequences, where available, of other members of the gene family and designing a probe to represent regions that are not well conserved between the family members. In some difficult cases it may be necessary to use **oligonucleotide probes**. If a gene is known to be expressed to give different isoforms e.g. by alternative splicing, oligonucleotide probes may be the only easy way of studying expression of the individual isoforms.

Studying RNA expression in microdissected cell populations and single cells

Using suitably labeled probes, specific RNA sequences can be tracked within *single cells* to identify sites of RNA processing, transport and cytoplasmic localization. By using quantitative fluorescence *in situ* hybridization and digital imaging microscopy, it has even been possible to visualize single RNA transcripts *in situ* (Femino *et al.*, 1998). Recently, powerful new methods have been developed which use lasers to microdissect tissue to produce pure cell populations from sources such as tissue biopsies and stained tissue, and even single cells (see Simone *et al.*, 1998; Schutze and Lahr, 1998). Such developments will allow a variety of gene expression analyses (see below) to be focused on single cells, or on homogeneous cell populations which will be more representative of the *in vivo* state than cell lines.

Multiplex and whole genome expression screens using DNA chip technology

Gene expression screens have been transformed by the capacity to prepare high density miniaturized arrays of oligonucleotides or DNA clones on glass surfaces ('DNA chips'; see Section 5.4.3). To conduct a multiplex gene expression screen, a microarray needs to be designed to contain cDNA clones or gene-specific oligonucleotides (designed from transcribed gene regions) to represent every gene of interest. To investigate expression of the chosen genes represented in such arrays, RNA is prepared from the selected target cells, converted into cDNA using standard reverse transcriptase procedures, labeled with a fluorescent tag and hybridized to the microarray.

The labeled cDNA is of course a very heterogeneous population of cDNA sequences corresponding to the RNA transcribed from all genes expressed in the target cells. A gene that is strongly expressed in the target cells will therefore be well-represented in the total cDNA probe; one that is poorly expressed will be poorly represented in the labeled cDNA. When the labeled cDNA is hybridized to the microarray, the intensity of signal remaining bound to individual cDNA clones or oligonucleotides represents how well the respective sequence is represented in the labeled cDNA, and hence is a measure of gene expression (*Figure 20.6*).

The potential of DNA chip technology for surveying gene expression is enormous. In the case of some genomes which have been completely sequenced, it affords the possibility of whole genome expression screening whereby the expression of every single gene in an organism can be monitored simultaneously. Already this is possible in the case of the yeast genome which comprises over 6000 genes (Wodicka *et al.*, 1997; see

Section 13.4.1). In the case of the human genome, essentially all genes should be known by 2002 and it is to be expected that improvements in DNA chip technology will permit whole human genome expression screening shortly thereafter. Numerous applications can be envisaged and will be particularly important in understanding functional networks linking our genes, the most important challenge in the post-sequencing era. The list of possible applications below is not meant to be exhaustive, but illustrates only selected examples:

- *Inducible gene expression* – investigation of how cells respond to environmental changes, such as exposure to different concentrations of a specific hormone, or other signaling molecule, or to heat shock, etc.
- *Tissue- and cell type-specific gene expression.*
- *Developmental stage-specific gene expression.*
- *Gene expression during differentiation* – investigation of how gene expression patterns are altered during differentiation.
- *Gene expression during tumorigenesis* – cells can be sampled at different recognized stages during the progression to cancer (see *Figure 18.18*).

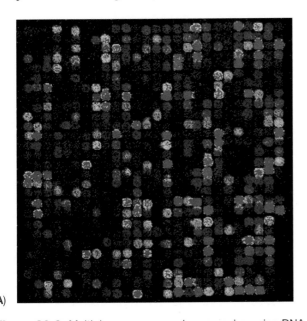

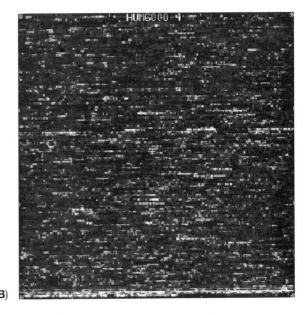

(A)　　　　　　　　　　　　　　　　　　　　(B)

Figure 20.6: Multiplex gene expression screening using DNA chips.

(A) An 82K microarray. Shown is a composite image of a DNA microarray containing 82 944 total features printed on a silylated microscope slide (CEL). The test spotting solution containing 1.0 pmole/μl Cy3-labeled 21-mer oligonucleotide suspended in micro-spotting solution (TeleChem), was deposited as 0.4 nl droplets at 120 μm center-to-center spacing with a PixSys 5500 gridding robot (Cartesian) equipped with a Stealth surface contact printhead and pins (TeleChem). Fluorescent scanning was accomplished with a ScanArray 3000 (GSI Lumonics) set at 100% PMT and 70% laser power in channel 1. The 16-bit tiff data are presented in a rainbow color palette for ease of viewing. Image kindly provided by Dr Mark Schena, Stanford University.
(B) Oligonucleotide microarray. A fluorescence image of a 1.28 cm × 1.28 cm array containing over 65 000 different oligonucleotide probes for 1641 unique genes (each gene is represented by 20 oligonucleotide pairs, with each pair including a faithful gene sequence oligonucleotide and a related mismatched oligonucleotide). The array also includes RNA standards that are used as quantitative references (β-actin, glyceraldehyde-3-phosphate dehydrogenase, transferrin receptor and one phage and three bacterial RNAs). The image was obtained following overnight hybridization of a sample made directly from human tissue, and was from one of four such arrays, collectively sampling 6500 different human genes. The image was kindly provided by Affymetrix, Inc. (Santa Clara, CA). Both images reprinted with permission from Strachan *et al.* (1997) *Nature Genet.*, **16**, 127, Nature America, Inc.

20.2.3 Serial analysis of gene expression (SAGE) is a method for multiplex gene expression screening which depends on very short sequence tags located at specific sites within individual transcripts

Although microarray technology has huge potential, it suffers from the drawback that it is very expensive and the hardware to produce DNA chips is not readily available. An alternative approach which permits simultaneous tracking of the expression of large numbers of transcripts is the SAGE (serial analysis of gene expression) method devised by Velculescu *et al.* (1995). SAGE is a much more accessible method because it does not require any sophisticated equipment to track gene expression. It is based on two principles:

■ *A short nucleotide sequence tag can uniquely identify the transcript from an individual gene provided it is from a defined position within the transcript.* For example, although the total number of human genes is expected to be of the order of 100 000, a sequence tag of only 9 bp can, in principle, distinguish $4^9 = 262\ 144$ different transcripts.

■ *Concatenation of short sequence tags allows the efficient analysis of transcripts in a serial manner.* The tags from different transcripts can be covalently linked together within a single clone and the clone can then be sequenced to identify the different tags in that clone.

To perform a SAGE analysis, poly $(A)^+$ RNA is first extracted from the source to be investigated and converted to cDNA using a biotinylated oligo (dT) primer. The resulting cDNA is cleaved with a frequent 4 base-pair cutter restriction nuclease (termed the anchoring enzyme) and liberated 3′ end fragments, which will contain a biotin group, are bound to a streptavidin bead. The streptavidin-bound cDNA is split into two fractions, and the two fractions are separately ligated *en masse* to two double-stranded oligonucleotide adaptors (**linkers**) A and B, each designed to contain the following: an over-

hang which is complementary in sequence to overhanging ends generated by the anchoring enzyme; a five base recognition sequence for an enzyme which is termed the tagging enzyme and which is a type IIS restriction nuclease (one which unlike standard type II restriction nucleases cleaves at a defined distance up to 20 bp from its recognition site); sufficient additional sequence to design a specific PCR primer.

The linkered cDNA is cleaved with the chosen tagging enzyme and so the only portions of the original cDNA sequences which will remain bound to the linkers are short sequence tags (*Figure 20.7*). The two pools of linkers are then ligated to each other and PCR amplification is carried out using primers specific for linkers A and B. The amplified product will be a complex population of sequences each containing two adjacent sequence tags (ditags) derived from the sequences individually coupled to linkers A and B. The individual ditags can then be released by cleavage with the anchoring enzyme and the different ditags can be concatenated by simple ligation and then cloned. Sequencing of the resulting clones can then produce long read-outs of sequential ditags of fixed length (*Figure 20.7*).

Although a considerable fraction of human genes remain to be characterized, SAGE provides a measure of transcript quantitation since genes that are expressed strongly in a specific tissue will frequently be recovered as sequence tags. In the original experiment, Velculescu *et al.* (1995) reported recovery of 840 sequence tags from pancreatic cDNA. Of these, 498 represented 77 different transcripts and the most abundant transcripts were all produced from genes known to have pancreatic function (e.g. procarboxypeptidase A1 was represented 64 times, pancreatic trypsinogen 2 was represented 46 times, etc.).

SAGE can be expected to be useful in comparative expression studies to identify differences in gene expression between two or more cellular sources of RNA. A big advantage of the method is its speed: all genes expressed at 0.5% of total mRNA in one tissue but not abundant in another could be found in a single day (approximately 1000 sequence tags from each tissue).

Figure 20.7 (Opposite): Multiplex gene expression screening using the SAGE method.

The basis of the method is to reduce each cDNA molecule to a representative short sequence tag (about nine nucleotides long). Individual tags are then joined together (*concatenation*) into a single long DNA clone as shown at the very bottom of the diagram where the numbers above the sequence tags represent a specific cDNA from which the tag was derived. Sequencing of the clone provides information on the different sequence tags which can identify the presence of corresponding mRNA sequences. The mRNA is converted to cDNA using an oligo (dT) primer with an attached biotin group and the biotinylated cDNA is cleaved with a frequently cutting restriction nuclease (the anchoring enzyme, AE; in this example it is *Nla*III which cuts immediately after the G in the 4 bp sequence CATG). The resulting 3′ end fragments which contain a biotin group are then selectively recovered by binding to streptavidin-coated beads, separated into two pools and then individually ligated to one of two double-stranded oligonucleotide linkers, A and B. The two linkers differ in sequence except that they have a 3′ CTAG overhang and immediately adjacent to it, a common recognition site for a type IIS restriction nuclease which will serve as the tagging enzyme (TE). In this example it is *Fok*I which recognizes the sequence GGATG (outlined in a box), but cleaves at 9/13 nucleotides downstream. Cleavage with *Fok*I generates a 9 bp sequence tag from each mRNA and fragments from the separate pools can be brought together to form 'ditags' then concatenated as shown.

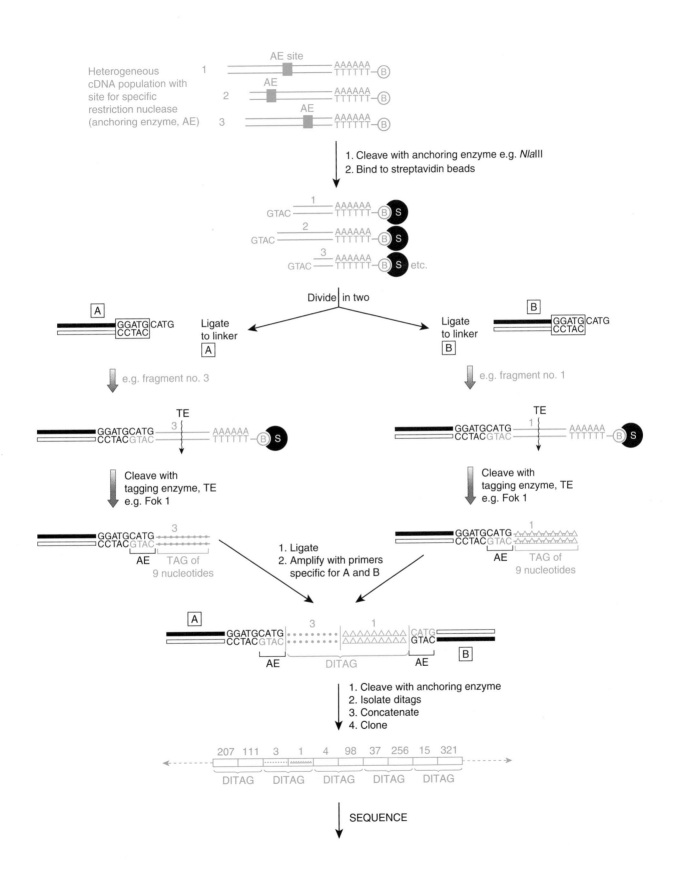

20.2.4 RT-PCR and mRNA differential display are popular PCR-based methods for studying gene expression

As described in Chapter 6, the great advantages of PCR are its speed, sensitivity and simplicity. Although it is not so suited to providing spatial patterns of expression (in the way that tissue *in situ* hybridization, for example, does), it can provide rapid gross patterns of expression which may be valuable.

Studying gene expression by conventional RT-PCR

RT-PCR can provide rough quantitation of the expression of a particular gene, which can be very useful in the case of cell types or tissues that are not easy to access in great quantity, such as early stage preimplantation human embryos, e.g. see Daniels *et al.* (1995). The extreme sensitivity of PCR means that RT-PCR can also be used to study expression in single cells. In addition, RT-PCR can be useful for identifying and studying different **isoforms** of an RNA transcript. For example, different mRNA isoforms may be produced by alternative splicing and can be identified when exon-specific primers identify extra amplification products in addition to the expected products (for an example, see Pykett *et al.*, 1994).

Multiplex gene expression using mRNA differential display

By using partially degenerate PCR primers, it is possible to devise modified RT-PCR methods for studying the expression of many genes simultaneously. **mRNA differential display** is one such technique. It uses a modified oligo(dT) primer which has a different single nucleotide or dinucleotide at the 3′ end causing it to bind to the poly(A) tail of a *subset* of mRNAs (Liang *et al.*, 1993). For example, if the oligonucleotide TTTTTTTTTTTCA is used as a primer, it will preferentially prime cDNA synthesis from those mRNAs where the dinucleotide TG precedes the poly(A) tail. The second primer which is used is usually an arbitrary short sequence (often 10 nucleotides long but, because of mismatching, especially at the 5′ end, it can bind to many more sites than expected for a decamer). The resulting amplification patterns are deliberately designed to produce a complex ladder of bands when size-fractionated in a long polyacrylamide gel (*Figure 20.8*).

Unlike DNA chip technology, mRNA differential display is a form of multiplex gene expression screening where the identities of the genes that are being tracked are not evident. Nevertheless it is useful in *comparative* gene expression studies to identify how gene expression alters in cells at different physiological or developmental stages. This allows identification of a small subset of genes whose expression patterns are different between the cell types. Although the method is prone to false positive differences, it can be useful in cloning genes which are differentially expressed in two or more cell populations. To do this, specific PCR bands displayed in one source but absent from another are isolated from the gel and the DNA is submitted to further PCR cycles (for an example, see Aiello *et al.*, 1994). Other methods for cloning genes which are differentially expressed in different tissue or cell sources include *subtractive hybridization* and *representation difference analysis* (see Section 15.2.4).

20.2.5 Antibodies permit highly specific screens for protein expression

Because of their exquisite diversity and sensitivity in detecting proteins, antibodies have numerous applications in research, and their therapeutic potential is considerable (see Section 22.1.3). Traditionally, antibodies have been isolated by immunizing animals, but increasingly genetically engineered antibodies are being used (*Box 20.2*). Antibodies have been used to detect proteins by different methods, and different labeling systems can be used.

Antibody labeling and detection systems

Antibodies can be labeled in different ways and, as with nucleic acid labeling and detection, antibodies can be used in either direct or indirect detection systems. In direct detection methods, the purified antibody is labeled appropriately with a reporter molecule (e.g. fluorescein, rhodamine, biotin etc. – see also Section 5.1.2) and then used directly to bind the target protein. In indirect detection systems, the primary antibody is used as an intermediate molecule and is not linked directly to a labeled group. Once bound to its target, the primary antibody is in turn bound by a secondary reagent (e.g. protein A or a secondary antibody) which is conjugated to a reporter which may be a fluorochrome, an enzyme (e.g. horseradish peroxidase, alkaline phosphatase, β-galactosidase, etc.) or colloidal gold. Direct detection systems have the disadvantage that a large variety of primary antibodies may need to be conjugated to reporter molecules, whereas the indirect systems offer the use of readily available commercial affinity-purified secondary antibodies. Labeling-detection systems for use with specific methods of tracking protein expression are outlined in *Table 20.2*.

Immunoblotting (Western blotting)

This method is designed to survey gross protein expression using cell extracts which are fractionated according to size. Usually this is achieved by one-dimensional SDS–PAGE, a form of polyacrylamide gel electrophoresis in which the mixture of extracted proteins is first dissolved in a solution of sodium dodecyl sulfate (SDS; an anionic detergent that disrupts nearly all noncovalent interactions in native proteins). Mercaptoethanol or dithiothreitol is also added to reduce disulfide bonds. Following electrophoresis, the fractionated proteins can be visualized by staining with a suitable dye (e.g. Coomassie blue) or a silver stain. Two-dimensional gels may also be used: the first dimension involves isoelectric focusing, that is separation according to charge in a pH

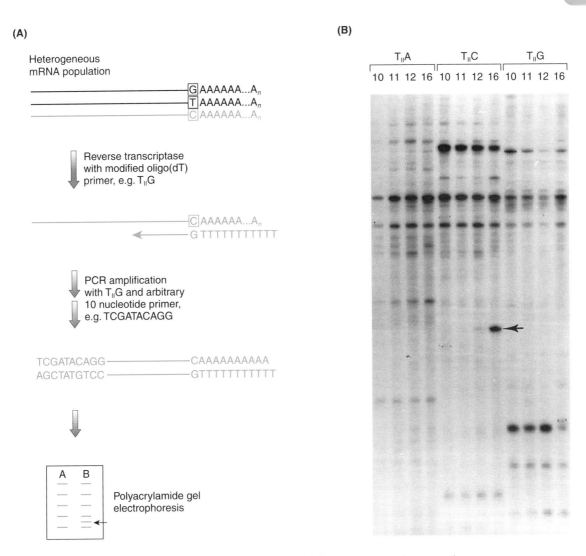

Figure 20.8: mRNA differential display is a rapid method of multiplex gene expression screening.

(**A**) Schematic representation of differential display. In this case, total RNA from two or more cell types is reverse transcribed using a modified oligo(dT) primer which in this particular example has 11 T residues and a single G at its 3′ end ($T_{11}G$). It should preferentially transcribe from mRNA sequences where a C precedes the poly(A) tail. Amplification is carried out using an arbitrary primer and products are size-fractionated on a polyacrylamide gel. Differences in amplification bands between the RNA sources compared (here, two sources A and B) indicate differential expression. (**B**) Differential display to identify genes differentially expressed at different stages of heart development. This example compares expression in the mouse heart at embryonic days 10, 11, 12 and 16. Three sets of reaction conditions were used, with a single arbitrary primer and either a $T_{11}A$, $T_{11}C$ or $T_{11}G$ primer. The figure shows a section of the gel where several bands can be seen to change in intensity at the different developmental stages, including a particularly prominent increase in expression for one band at embryonic day 16 (shown by the arrow). The photograph was kindly provided by Andy Curtis and David Wilson, University of Newcastle upon Tyne.

Table 20.2: Antibody labeling-detection for tracking protein expression

Label	Detection method	Application
Iodine-125	X-ray film	Immunoblotting
Enzyme	Chromogenic substrate detected by eye	Immunoblotting; immunocytochemistry
Biotin	Avidin or streptavidin coupled to various labels	Immunoblotting; immunocytochemistry
Fluorochrome	Fluorescence microscopy (*Figure 5.5*)	Immunocytochemistry; immunofluorescence microscopy

Box 20.2

Obtaining antibodies

Traditional methods of obtaining antibodies

Antibodies to human gene products have traditionally been obtained by repeatedly injecting suitable animals (e.g. rodents, rabbits, goats, etc.) with a suitable **immunogen**. Two types of immunogen are commonly used:

- **Synthetic peptides.** The amino acid sequence (as inferred from the known cDNA sequence) is inspected and a synthetic peptide (often 20–50 amino acids long) is designed. The idea is that, when conjugated to a suitable molecule (e.g. keyhole limpet hemocyanin), the peptide will adopt a conformation that resembles that of the corresponding segment of the native polypeptide. This approach is relatively simple, but success in generating suitably specific antibodies is far from assured and difficult to predict.

- **Fusion proteins.** An alternative approach is to insert a suitable cDNA sequence into a modified bacterial gene contained within an appropriate expression cloning vector. The rationale is that a hybrid mRNA will be produced which will be translated to give a fusion protein with an N-terminal region derived from the bacterial gene and the remainder derived from the inserted gene (see figure). The N-terminal bacterial sequence is often designed to be quite short, but may nevertheless confer some advantages. For example it can provide a signal sequence to ensure secretion of the fusion protein into the extracellular medium, thereby simplifying its purification, and it may protect the foreign protein from being degraded within the bacterium. Because the fusion protein contains most or all of the desired polypeptide sequence, the probability of raising specific antibodies may be reasonably high.

If the animal's immune system has responded, specific antibodies should be secreted into the serum. The antibody-rich serum (antiserum) which is collected contains a heterogeneous mixture of antibodies, each produced by a different B lymphocyte (*because immunoglobulin gene rearrangements are cell-specific as well as cell type (B lymphocyte)-specific*, see Section 8.6). The different antibodies recognize different parts (**epitopes**) of the immunogen (polyclonal antisera). A homogeneous preparation of antibodies can be prepared, however, by propagating a clone of cells (originally derived from a single B lymphocyte). Because B cells have a limited life-span in culture, it is preferable to establish an immortal cell line: antibody-producing cells are fused with cells derived from an immortal B-cell tumor. From the resulting heterogeneous mixture of hybrid cells, those hybrids that have both the ability to make a particular antibody and the ability to multiply indefinitely in culture are selected. Such hybridomas are propagated as individual clones, each of which can provide a permanent and stable source of a single type of **monoclonal antibody**.

The above methods of raising antibodies are not always guaranteed to produce suitably specific antibodies. An alternative approach for tracking the subcellular expression of a protein of interest is to use a previously obtained antibody to track it as a result of binding to an artificially coupled epitope (epitope tagging). In this procedure a recombinant DNA construct is generated by coupling a sequence that encodes an epitope for which a previously obtained antibody is available, to the coding

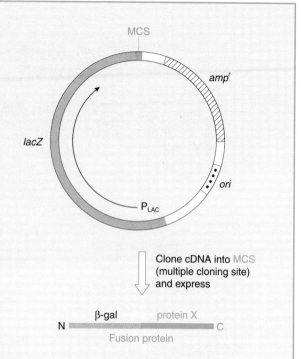

Figure: Fusion proteins are often designed as immunogens for raising antibodies.

In this example, the plasmid vector has an origin of replication (*ori*) and an ampicillin resistance gene (*amp*^{r}) for growth in *E. coli*. The multiple cloning site (MCS) is located immediately adjacent to a *lacZ* gene which can encode β-galactosidase with transcription occurring from the *lacZ* promoter (P$_{LAC}$) in the direction shown by the arrow. A cDNA sequence from a gene of interest (gene X) is cloned in a suitable orientation into the MCS. Expression from the *lacZ* promoter will result in a β-galactosidase–X fusion protein. This can be used as an immunogen to raise antibodies to protein X. A popular alternative is to use GST fusion proteins, where glutathione S transferase is coupled to the protein of interest. The fusion protein can be purified easily by affinity chromatography using glutathione agarose columns.

sequence of the protein of interest in much the same way as in the figure above except that in this case the vector system is designed to be expressed in mammalian cells or other cells in which it is intended to investigate expression. Expression of this construct in the desired cells can be monitored by using the antibody specific for the epitope tag to track the protein. Commonly used epitope tags are shown in the table below.

Sequence of tag	Origin	Location	mAb
DYKDDDDK	synthetic FLAG	N, C terminal	anti-FLAG M1
EQKLISEEDL	human c-Myc	N, C terminal	9E10
MASMTGGQQMG	T7 gene 10	N terminal	T7.Tag Ab
QPELAPEDPED	HSV protein D	C terminal	HSV.Tag Ab
RPKPQQFFGLM	substance P	C terminal	NC1/34
YPYDVPDYA	influenza HA1	N, C terminal	12CA5

gradient, and the second dimension, at right angles to the first, involves size-fractionation by SDS-PAGE (see Stryer, 1995). In this case the fractionated proteins are transferred ('blotted') to a sheet of nitrocellulose and then exposed to a specific antibody (see *Figure 20.9*).

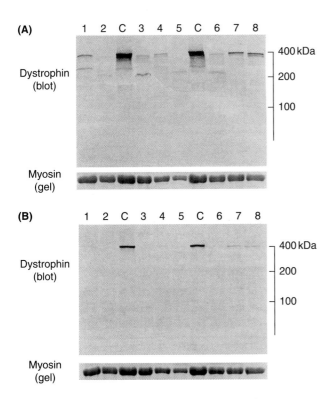

Figure 20.9: Immunoblotting (western blotting) detects proteins that have been size-fractionated on an electrophoresis gel.

Immunoblotting involves detection of polypeptides after size-fractionation in a polyacrylamide gel and transfer ('blotting') to a membrane. This example illustrates its application in detecting dystrophin using two antibodies. The Dy4/6D3 antibody is specific for the rod domain and was generated by using a fusion protein immunogen (see *Box 20.2*). The Dy6/C5 antibody is specific for the C-terminal region and was generated by using a synthetic peptide immunogen. Reproduced from Nicholson *et al.* (1993) with permission from the BMJ Publishing Group. The photograph was kindly provided by Dr Louise Anderson, University of Newcastle upon Tyne.

Immunocytochemistry (immunohistochemistry)

This technique is concerned with studying the overall expression pattern at the protein level, within a tissue or other multicellular structure. It can therefore be regarded as the protein equivalent of the tissue *in situ* hybridization methods used to screen RNA expression. As in the latter case, tissues are typically either frozen or embedded in wax and then cut into very thin sections with a microtome before being mounted on a slide. A suitably specific antibody is allowed to bind to the protein in the tissue section and can produce expression data that can be related to histological staining of neighboring tissue sections (*Figure 20.10*).

Immunofluorescence microscopy

This method is used when investigating the *subcellular location* for a protein of interest. A suitable fluorescent dye, such as fluorescein or rhodamine is coupled to the desired antibody, enabling the relevant protein to be localized within the cell by fluorescence microscopy (*Figure 5.5*).

Ultrastructural studies

Higher resolution still of the intracellular localization of a gene product or other molecule is possible using electron microscopy. The antibody is typically labeled with an electron-dense particle, such as colloidal gold spheres.

20.2.6 The green fluorescent protein has provided a powerful new approach to tracking gene expression and subcellular localization of proteins in animal cells

Green fluorescent protein (GFP) is a 238 amino acid protein originally identified in the jellyfish *Aequoria victoria*. Similar proteins are expressed in many jellyfish and appear to be responsible for the green light that they emit, being stimulated by energy obtained following oxidation of luciferin or another photoprotein (Tsien, 1998). When the green fluorescent protein (GFP) gene was cloned and **transfected** into target cells in culture, expression of GFP in heterologous cells was also marked by emission of the green fluorescent light. This means that GFP by itself can act as a functional fluorophore. It can therefore serve as a unique reporter since it does not require other agents such as antibodies, cofactors, enzyme substrates etc. As a result, GFP can readily be followed by conventional

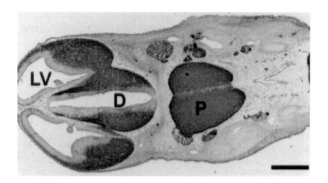

Figure 20.10: Immunocytochemistry.

In this example, β-tubulin expression was screened in a transverse section of the brain of a 12.5 embryonic day mouse. The antibody detection system used identifies expression ultimately as a brown color reaction based on horseradish peroxidase/3,3' diaminobenzidine. The underlying histology was revealed by counterstaining with toluidine blue. Abbreviations: LV, lateral ventricle; D, diencephalon; P, pons. Figure kindly provided by Steve Lisgo, University of Newcastle upon Tyne, UK.

fluorescence microscopy (*Figure 5.5*) and also confocal fluorescence microscopy (see *Box 20.3*) and has become a popular tool for tracking gene expression in animal cells.

Use as a reporter gene for tracking gene expression in vivo

An early application was as a **reporter gene** for monitoring gene expression *in vivo* in response to regulation by a coupled promoter sequence (see Section 20.3.1). This has

been successfully done in some organisms such as in *C. elegans* (Chalfie *et al.*, 1994) where the cuticle has generally impeded access to substrates required for detecting other reporter genes. A powerful characteristic of GFP as a reporter is that it can be imaged in live cells, as well as in fixed ones. This has permitted imaging of dynamic processes in cells. In mammalian cells, however, GFP expression driven by endogenous promoters is often rather weak, although mutant versions of GFP have been designed to permit optimal expression.

Use as a fusion tag for monitoring protein localization

The most successful and popular applications have used GFP as a tag in a fusion protein where it is coupled to the protein whose expression is to be tracked. In such cases, the principal aim is to investigate the subcellular localization of the protein under investigation. GFP itself is not localized within cells and in most cell types the fluorescence of GFP appears to be homogeneously spread all over the nucleus, cytoplasm and distal cell processes. Genetic engineering can be used to produce vectors containing a GFP coding sequence into which a coding sequence for an uncharacterized protein, X, can be cloned. The resulting GFP–X fusion construct can be transfected into suitable target cells such as cultured mammalian cells and expression of the GFP-X fusion protein can be monitored to track the subcellular location of the protein (*Figure 20.11*).

20.3 Identifying regulatory sequences through the use of reporter genes and DNA–protein interactions

20.3.1 The use of reporter genes and deletion analysis provides a first approach to identifying regulatory gene sequences

In addition to studying gene expression *per se*, it is important to be able to study the control of gene expression. This can be done in different ways. One way of (tentatively) identifying regulatory elements is to use comparative sequencing in different species coupled with computer-based sequence analysis (Section 13.4.1). A more general approach, however, is to use an artificial gene expression system and to study how deleting different segments of DNA upstream of the gene, or occasionally in the first intron, affects gene expression.

Clearly, human cells are the most appropriate system for studying the expression of human genes, but this then leaves the problem of how to follow expression of the transfected human gene in the presence of an endogenous homolog which may be expressed in the same cells. As a result, therefore, it is usual to clone the presumptive regulatory sequences into a vector which

---

Box 20.3

Confocal fluorescence microscopy

To examine a tissue using conventional light microscopy, it needs to be sliced into thin sections; the thinner the section, the sharper the image. But in the process of preparing sections information on the third dimension (the physical relationship between neighboring sections) gets lost. Confocal microscopy can be used to study comparatively thick specimens by making *optical sections* in the third dimension, instead of physical sections. This is achieved by the ability to obtain focused images at defined planes in the third dimension, which can be selected by the operator. A confocal microscope is normally used with fluorescence optics but instead of illuminating the whole specimen, the light is focused on a specific point at a specific *depth* in the specimen, usually by passing laser light through a pinhole. The illuminated specimen fluoresces and the emitted fluorescence is collected and passed through a second pinhole aperture placed next to a detector. This pinhole aperture used in detection is *confocal* with the pinhole aperture used in illumination and so the only fluorescent image which is in focus will be that which is emitted from the point at which the illuminating light was initially focused.

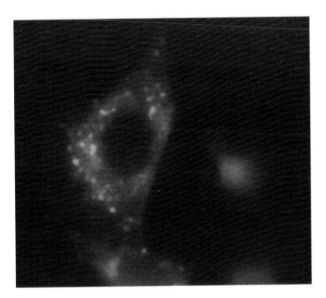

Figure 20.11: Tagging with the green fluorescent protein provides a powerful way of tracking protein expression.

This example shows a live transiently transfected HeLa cell expressing a GFP-tagged Batten disease protein. The coding sequence of the Batten disease gene, *CLN3*, was cloned into a GFP expression vector, pEGFP-N1, so that a fusion protein was produced. The latter consisted of the Batten disease protein with a GFP sequence coupled to its C terminus (a *GFP tag*) coupled to the Batten disease protein. This cell is an example of a small proportion of HeLa cells expressing CLN3p/GFP in a vesicular punctate pattern distributed throughout the cytoplasm. These and other analyses indicate that the Batten disease protein is a Golgi integral membrane protein. Reproduced from Kremmidiotis *et al.* (1999) *Hum. Mol. Genet.*, **8**, pp. 523–531, with permission from Oxford University Press.

Box 20.4

Methods for transferring genes into cultured animal cells

A variety of methods can be used to transfer genes into human and animal cells, but they can be grouped into two classes:

- **Transduction.** This describes virus-mediated gene transfer. Certain animal viruses naturally infect human and mammalian cells. They include both DNA viruses (SV40, adenovirus, etc.) and RNA viruses (e.g. HIV and other **retroviruses**). Modifications of these viruses can be used as vectors to transfer exogenous genes into suitable target cells at high efficiency.

- **Transfection.** This describes nonviral mediated gene transfer. Note that the term **transfection** is analogous to the process of **transformation** in bacteria. The latter term was not applied to animal cells because of its association with an altered phenotype and unrestrained growth. Gene transfer can be carried out by various methods, including the assistance of :

(a) Artificial lipid vesicles (**liposomes** – see Section 22.2.3) to which the DNA binds. The liposomes can fuse with the plasma membrane and so permit access to the cell interior;

(b) Calcium phosphate. The calcium phosphate and DNA form co-precipitates on the surface of the target cells. The high concentration of the DNA on the plasma membrane may increase the efficiency of transfection.

(c) **Electroporation.** A popular method whereby an electric shock is used to cause temporary membrane depolarization in the target cells, thereby assisting passage of large DNA molecules.

In some cases, the transferred genes can integrate into the chromosomes of the host cell and be stably inherited. In other cases they remain as extrachromosomal replicons (**episomes**) which may be capable of transient expression only.

contains a **reporter gene** downstream of the cloning site. The reporter gene is deliberately designed to be one which is not found in human cells and which can be assayed simply. Commonly used reporter genes include the bacterial *CAT* (chloramphenicol acetyl transferase) gene, derived from the Tn9 transposon of *E. coli*, the β-galactosidase gene and the firefly luciferase gene. Luciferase has the advantage of providing a very sensitive assay: it catalyzes the oxidation of luciferin with the emission of yellow–green light which can be detected easily and at low levels.

In such artificial expression systems, the vector is designed so that expression of the reporter gene is controlled almost entirely by the introduced upstream human sequences. The vector–reporter gene construct is then *transfected* into cultured target cells by standard methods (*Box 20.4*). If all the necessary elements are present for expression of the human gene, high level expression of the reporter gene will result when the con-

struct is transfected into an appropriate type of cultured human cell (*Figure 20.12*). Often the test construct is simply transfected into HeLa cells, a well-established cell line derived from a human cervical carcinoma. However, if the gene is known to be expressed predominantly in a certain type of cell, say hepatocytes, it is usual to use a more appropriate recipient cell line, a hepatoma cell line in this case. A series of progressive deletion constructs can then be made using the enzyme *exonuclease III* from *E. coli* which is inactive on single-stranded DNA but progressively cleaves from the 3′ end of double-stranded DNA. This allows mapping of sequences which control gene expression. Alternatively, a series of PCR amplification products can be designed covering the regions of interest and then cloned into a suitable expression vector (*Figure 20.13*).

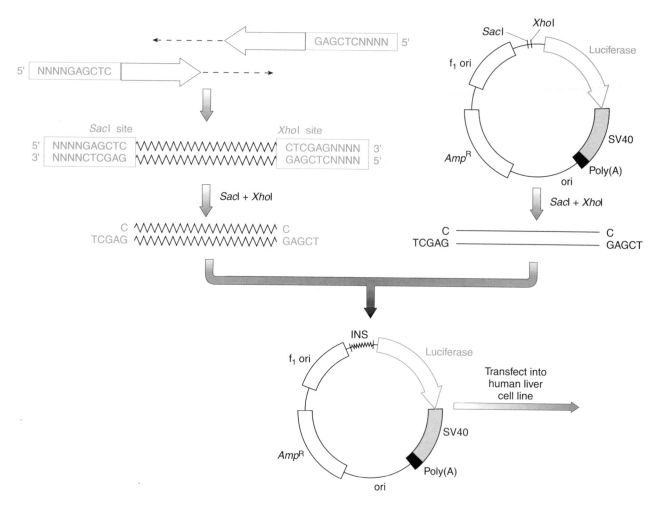

Figure 20.12: Regulatory sequences can be mapped by cloning into an expression vector containing a promotorless reporter gene.

The expression vector pGL2-Basic is a eukaryotic expression vector designed to assay for promoter sequences. It contains a promoterless luciferase reporter gene, a downstream SV40 polyadenylation signal 'poly(A)', a conventional origin of replication (ori), an f1 replication origin for producing single-stranded DNA if required, and an ampicillin resistance gene. The figure shows the way in which PCR amplification products corresponding to sequences upstream of the human factor VIII gene (*F8*) were cloned into this vector by Figueiredo and Brownlee (1995). The PCR primers were modified at their 5′ ends by a 10-nucleotide extension (in blue) which is not related to the target sequences, but is simply designed to include a recognition sequence for *Sac*I (GAGCTC) or for *Xho*I (CTCGAG). Amplification with these primers constitutes an example of *add-on mutagenesis*: the 'foreign' decanucleotide sequence becomes incorporated into the amplification product. The PCR products were cloned by double digestion with *Sac*I and *Xho*I and ligation to similarly cut vector DNA. The recombinants were transfected into human liver-derived cell lines and the inserts were assayed for *cis*-acting regulatory sequences which, in the presence of complementary *trans*-acting factors provided by the cell, could drive expression of the luciferase gene (see *Figure 20.13*).

20.3.2 DNase footprinting, gel retardation, and methylation interference assays can identify protein-binding sites on a DNA molecule

The deletion analysis described above usually provides only a rough indication of the location of a sequence that regulates gene expression. Because such control sequences can specifically bind regulatory proteins, another way of identifying them involves screening for

sequences that can specifically bind proteins. Often this involves using synthetic oligonucleotides corresponding to sequences from a region thought to contain a regulatory sequence. Different methods can be used as follows.

DNase I footprinting

When a protein binds specifically to a DNA sequence, only a few nucleotides of the DNA are involved in DNA–protein contacts. The bound protein, however, renders

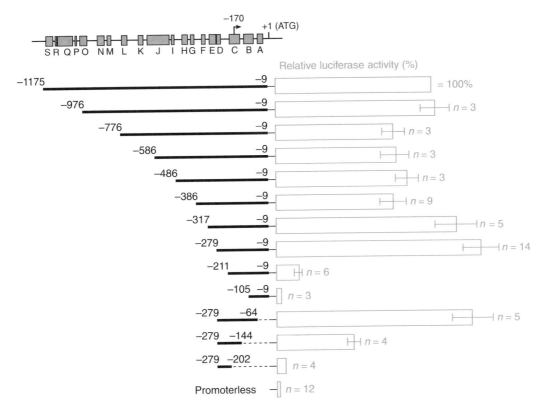

Figure 20.13: Deletion analysis of the human factor VIII gene promoter region.

Solid bars to the left indicate variously sized sequences upstream of the human factor VIII gene (*F8*) which had been cloned into the expression vector pGL2-Basic (see *Figure 20.12*). Coordinates −1175 to −9 are referenced against the initiator methionine, arbitrarily labeled as +1. Boxes on top indicate sequences upstream which were suggested to be protein-binding sites on the basis of *DNase I footprinting assays* (see *Figure 20.14*). Boxes on the right indicate the level of luciferase activity relative to the intact sequence, based on *n* replicate experiments. On the basis of observed luciferase activity for the different expression clones, the deletion mapping shows that all the necessary elements for maximal promoter activity are located in the region from −279 to −64, including protein-binding sites B, C and D. Reproduced from Figueiredo and Brownlee (1995) *J. Biol. Chem.*, **270**, pp. 11828–11838, with permission from The American Society for Biochemistry and Molecular Biology.

the underlying DNA segment relatively resistant to cleavage by pancreatic *deoxyribonuclease I* (DNase I), when compared with naked DNA. Short cloned DNA fragments (usually a few hundred bases long) are used as targets and are end-labeled at one end only. The cloned fragments are individually incubated in the presence or absence of a protein extract (often a nuclear extract from human HeLa cells) and subsequently exposed briefly to low concentrations of DNase I. Such *partial digestion* conditions ensure that, for any one DNA fragment, each DNA molecule is cut only rarely and at a random position if no protein is bound. The digestion products are then size-fractionated on long denaturing polyacrylamide gels, prior to autoradiography. Control samples will show a series of bands corresponding to DNA fragments of every possible length. The corresponding test lanes will, however, reveal gaps where no fragments are seen (foot-

prints), as the DNase I was not able to cut at these positions because of steric inhibition by the bound protein (*Figure 20.14*).

Gel retardation assay

This technique is alternatively known as the electrophoretic mobility shift assay (EMSA), and its rationale is that binding of a protein to a DNA fragment reduces its mobility during gel electrophoresis. A DNA fragment from a genomic clone or a corresponding synthetic oligonucleotide, which is suspected of containing regulatory sequences, is end-labeled and then mixed with a protein extract. The resulting preparation is size-fractionated by PAGE in parallel with a control sample in which the DNA was not mixed with the protein. DNA fragments which have bound protein will be identifiable as low mobility bands. Since DNA is a highly charged

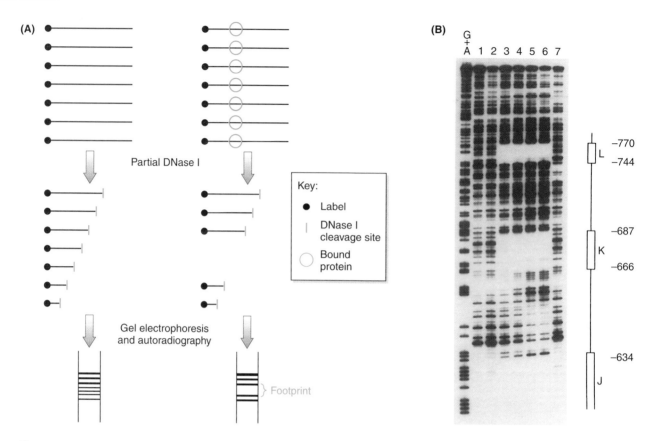

Figure 20.14: DNase I footprinting identifies protein-binding regions in a DNA molecule by their ability to confer resistance to cleavage by DNase I.

(A) Basis of method. DNA molecules are labeled at a single end. If the DNA is not complexed with protein (left), **partial digestion** with pancreatic DNase I can result in a ladder of DNA fragments of different sizes because the location of a cleavage site (blue vertical bars) on an *individual* DNA molecule is essentially random. If, however, they contain a protein-binding site, then incubation with a suitable protein extract may result in binding of protein at a specific region (right). Subsequent partial digestion with pancreatic DNase I does not produce random cleavage: the bound protein protects the underlying DNA sequence from cleavage. As a result, the spectrum of fragment sizes is skewed against certain fragment sizes, leaving a gap when size-fractionated on a gel, a DNase I 'footprint'. **(B)** Practical example. The example shows DNase I footprinting analysis of the human factor VIII promoter using liver cell nuclear extracts. Three protein-binding sites were identified in the region spanning −800 to −600: J, K and L (see also *Figure 20.13*). Lanes were: G + A, Maxam and Gilbert chemical sequencing reaction; *1* and *2*, control reactions, lacking nuclear extract; *3–6*, test reactions with increasing amounts of liver nuclear extract; *7*, control reaction in which the nuclear extract had been heated for 10 min at 75°C in order to denature the protein. Reproduced from Figueiredo and Brownlee (1995) *J. Biol. Chem.*, **270**, pp. 11828–11838, with permission from The American Society for Biochemistry and Molecular Biology.

molecule, there is the opportunity for nonspecific binding, and controls are required whereby a nonspecific competitor DNA is added in increasing concentrations (*Figure 20.15*).

Methylation interference assay

This technique complements the two outlined above by defining which guanines are implicated in the protein-binding site (see Piccolo *et al.*, 1995, for a typical application).

Once a DNA region has been identified as capable of binding protein, comparison of its sequence to other known protein-binding sequences may indicate the iden-

tity of the binding protein. A more general screening method for identifying a protein which binds to a given stretch of DNA has been described as southwestern screening. The method involves cloning cDNA into a suitable expression vector such as λgt11. The resulting expression library is plated out on to nutrient agar in large culture dishes, resulting in well-separated phage plaques in a lawn of bacteria. A nitrocellulose membrane is placed on top of the agar surface for a brief period, then peeled off to give an imprint of phage material originally present in the plaques (this so-called plaque-lift procedure is the phage equivalent of the colony blot procedure used to transfer separated bacterial colonies to a nitro-

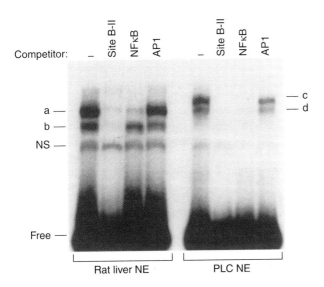

Figure 20.15: Protein-binding regions in a DNA molecule can be identified by gel retardation assays and the identity of the protein can be verified by competitive binding studies.

The figure shows investigation of a protein-binding site upstream of the factor VIII gene (site B; see *Figure 20.13*). This site had been identified by DNase I footprinting analysis and the gel retardation assay (electrophoretic mobility shift assay; EMSA) confirmed its existence. A sequence in the B site resembles the binding site for the ubiquitous transcription factor NF-κB. The EMSA assay used a nuclear protein extract (NE) from rat liver or from a human liver-derived cell line (PLC) and a labeled oligonucleotide from the factor VIII upstream site B (B-II). Assays were conducted in the absence or presence of an excess of competitor oligonucleotide. Most of the label runs near the bottom of the gel at the position of the free B-II oligonucleotide. Upper bands indicate the presence of complexes of protein bound to the labeled B-II oligonucleotide: a and b, specific complexes with rat nuclear extract; c and d, specific complexes with the human liver cell extract; NS, nonspecific protein binding. *Note* that when using the human liver cell nuclear extract, complex formation is strongly inhibited by both the competitor B-II oligonucleotide (control) and also by an oligonucleotide representing the consensus NF-κB binding sequence (PLC NE). Further experiments using purified NF-κB instead of the nuclear extracts verified the result. Reproduced from Figueiredo and Brownlee (1995) *J. Biol. Chem.*, **270**, pp. 11828–11838, with permission from The American Society for Biochemistry and Molecular Biology.

cellulose or nylon membrane; see *Figure 5.18*). Because the foreign cDNA sequences in the phage are expressed to give protein, expressed fusion proteins are absorbed on to the nitrocellulose membrane. The membrane is incubated with a radiolabled *duplex* DNA oligonucleotide representing the known protein-binding sequence. Under suitably stringent conditions, the DNA probe will bind selectively to protein expressed by an individual plaque, leading to identification of a specific binding protein (Old and

Primrose, 1994). See also the yeast one hybrid technique in Section 20.4.1.

20.4 Investigating gene function by identifying interactions between a protein and other macromolecules

After a polypeptide-encoding gene has been characterized at the nucleotide level and a predicted polypeptide sequence has been uncovered, further investigations focus on expression, regulation of expression and the biological function of the gene. Clues to function may sometimes be apparent from the nucleotide and/or amino acid sequences (through homology searches and motif sequence analysis; see Section 20.1.4), or from the expression patterns.

More direct approaches to investigate function seek to eliminate or inhibit expression of the gene and then study any alteration in the phenotype that results. Methods such as antisense technology have been devised for eliminating expression of a specific predetermined gene within cultured cells, but they are often technically difficult and the information that is obtained on gene function is often limited (see Section 22.3.2). More powerful methods for investigating the function of a human gene rely on extrapolation from mouse. The equivalent (**orthologous**) gene is identified in mouse and **gene targeting** is used to knock out its expression to see if its silencing produces some type of mutant phenotype (Section 21.3). An additional powerful approach is to express a gene sequence of interest to produce a protein and then use that protein in some way to identify other proteins or nucleic acids to which it can bind.

20.4.1 The yeast two-hybrid system is a powerful screen for identifying protein–protein associations

The function of a gene may not be obvious even after comprehensive study of its structure and expression patterns. In such cases, it may be advantageous to attempt to identify those proteins with which it may interact specifically, especially if these turn out to be ones that have previously been well studied and whose functions are known. Different approaches can be taken, including physical methods and library-based methods (see Phizicky and Fields, 1995). One such method is co-immunoprecipitation. This classical physical method, which requires a suitably specific antibody for the protein in question, is simple: cell lysates are generated, antibody is added, the antigen is precipitated and washed, and bound proteins are eluted and analyzed. If a polyclonal antibody is used, it is important to show that the co-precipitated proteins have been precipitated by the expected antibody, rather than a contaminating one. It also needs to be established that the antibody itself does not recognize the coprecipitated protein.

Classical methods such as coimmunoprecipitation are often not so effective. However, recently powerful methods have been developed to identify proteins that interact with a protein of interest. One widely used library-based method is the **yeast two-hybrid system**, also called the *interaction trap system* (Fields and Sternglanz, 1994). As well as identifying proteins that bind to a protein under study, this method can also be used to delineate domains or residues crucial for interaction. Proteins that physically bind to one another are detected by their ability, when bound, to activate transcription of a reporter gene.

The key to the two hybrid method is the observation that transcription factors which activate the expression of target genes require two separate domains, a DNA binding domain and an activation domain. In natural transciption factors, the DNA binding and activation domains are typically part of the same molecule (see Section 8.2.3). However, an active transcription factor can equally be made out of two proteins that associate together, one of which carries the DNA binding domain while the other carries the activation domain. The object of the two hybrid and derivative techniques is to use a target protein or a target nucleic acid as bait for specific recognition by an interacting protein which is fused to a necessary transcription factor component. As a result of the specific interaction, a transcription factor activity is formed which allows expression of a reporter gene.

To use the two-hybrid system, standard recombinant DNA methods are used to produce a fusion gene which encodes the protein under study coupled to one of the two domains necessary for the transcription factor function. Cells are cotransfected with this gene and also with a library of fusion genes where random cDNA sequences are coupled to coding sequence of the other transcription factor domain. The target cells are engineered to carry a reporter gene which can be activated when the appropriate transcription factor is formed by protein–protein association between the binding domain and the activation domain. These two domains do not interact directly, however. Instead, they are brought together when the proteins that are bound to each domain interact (*Figure 20.16A*).

The two-hybrid approach has been very successful for investigating mammalian gene function and is currently being applied on a whole genome scale to investigate protein–protein networks in yeast. In addition it has spawned some derivative methods designed to investigate protein interactions with other molecules. They include:

- **the one-hybrid system**. Here, the object is to identify proteins that bind to a target *cis*-acting regulatory element, usually a DNA sequence but sometimes an RNA sequence. In the former case, a sequence consisting of at least three tandem copies of a known target element (E) is inserted upstream of a reporter gene and integrated into the yeast genome to make a new reporter strain. The yeast strain is then trans-

formed with a cDNA library in which random cDNA sequences are coupled to a cDNA sequence encoding an activation domain needed for the reporter gene to be transcribed. The rationale is that one of the activation domain fusion proteins may be able to recognize and bind specifically to the *cis*-acting elements upstream of the reporter gene, thereby bringing the activation domain into close proximity. The resulting transcription of the reporter gene is the cue to identify this interaction (*Figure 20.16B*).

- **three-hybrid systems**. In addition to two proteins separately containing an activation domain and a binding domain, a third molecule is required which can be an RNA or protein molecule (Sengupta *et al.*, 1996).

20.4.2 Phage display is a form of expression cloning which has many applications, including the capacity to identify proteins that bind to a protein of interest

Phage display is a form of expression cloning of foreign genes using phage (Clackson and Wells, 1994). Genetic engineering techniques are used to insert foreign DNA fragments into a suitable phage coat protein gene. The modified gene can then be expressed as a fusion protein which is incorporated into the virion and displayed on the surface of the phage which, however, retains infectivity (fusion phage). If an antibody is available for a specific protein, phage displaying that protein can be selected by preferential binding to the antibody: affinity purification of virions bearing a target determinant can be achieved from a 10^8-fold excess of phage not bearing the determinant, using even minute quantities of the relevant antibody. Initially, phage display involved the use of filamentous phages such as fd, f_1, M13, etc., where the foreign gene was incorporated into a gene specifying a minor coat protein such as the gene III protein (*Figure 20.17*).

Several useful applications have been devised:

- *Antibody engineering*. Phage display is proving a powerful alternative source of constructing antibodies, including humanized antibodies, bypassing immunization and even hybridoma technology (Winter *et al.*, 1994; see Section 22.1.3).
- *General protein engineering*. Phage display is a powerful adjunct to random mutagenesis programs as a way of selecting for desired variants from a library of mutants.
- *Studying protein–protein interactions*. This is a library-based method which can be used to identify proteins that interact with a given protein. In the same way that antibodies can be used in affinity screening, a desired protein or any other molecule to which a protein can bind can be used as the selective agent. The protein can select fusion phage which display any other proteins that significantly bind to it.

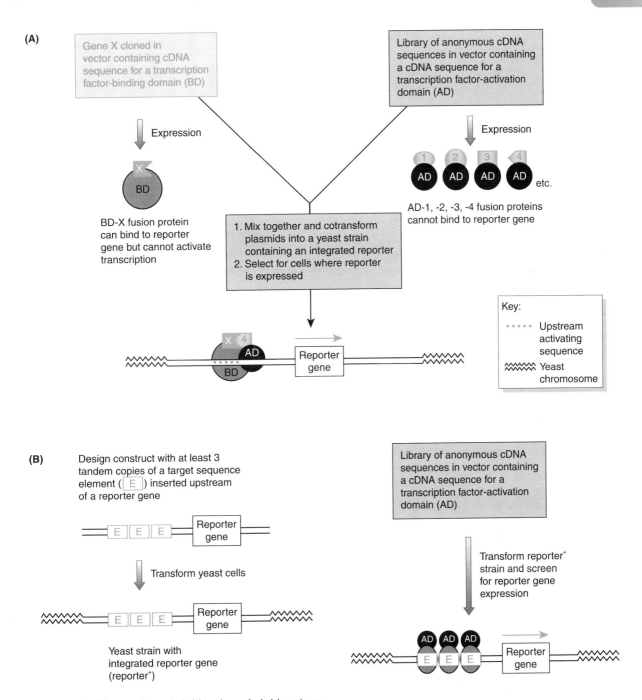

Figure 20.16: The yeast two-hybrid and one-hybrid systems.

(A) The yeast two-hybrid system. The reporter gene which is integrated into the yeast genome will only be expressed when a suitable transcription factor activity is available. This can happen when two fusion proteins are brought together: one that contains a binding domain (BD) and one that contains an activation domain (AD). The target protein X is fused to the binding domain which recognizes upstream activating sequences (promoter) adjacent to the reporter gene. The activation domain library shown on top right can express fusion proteins containing an activation domain coupled to one of many uncharacterized proteins (shown here as 1, 2, 3, 4, etc.). One of these may happen to interact with protein X (protein no. 4 in this example). Such an interaction will now bring the activation domain adjacent to the reporter gene and cause it to be expressed, thereby identifying the appropriate transformants. (B) The yeast one-hybrid system. Here the reporter yeast strain has been manipulated so that it contains several tandem copies of a sequence element of interest and the rationale is to identify a protein that normally interacts with this sequence element. Transformation of this strain by an activation domain fusion protein library will result in the expression of one of many different possible proteins coupled to an activation domain (just as in the two-hybrid scheme above). If one of these binds specifically to the target element, the reporter gene will be expressed. Variants of the one-hybrid system can also be designed to look for proteins that bind specifically to a target RNA sequence.

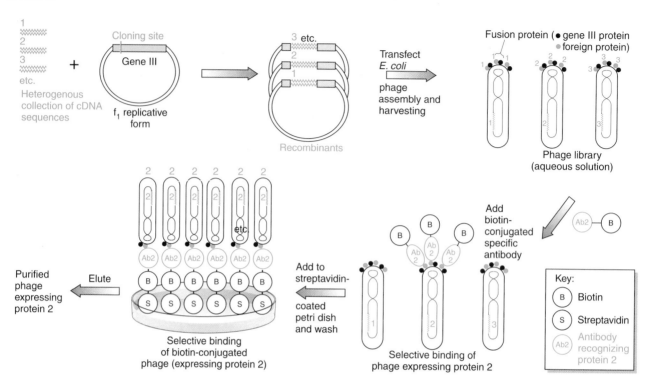

Figure 20.17: Phage display.

Phage display is a form of expression cloning which involves cloning cDNA into phage vectors and expressing foreign proteins on the phage surface. The figure illustrates one popular approach whereby the DNA is cloned into gene III of the filamentous phage f1 (or M13, etc.), a gene which encodes a minor phage coat protein. The cloning site is usually designed by site-specific mutagenesis to occur at a position corresponding to the extreme N-terminal sequence of the gene III protein. Following transfection of *E. coli,* phage assembly, extrusion from the cells and phage harvesting (see *Figure 4.17* for the general scheme of cloning using filamentous phage vectors), a phage library is produced. Recombinants with inserts which do not produce a frameshift in the reading frame may often be expressed to give a **fusion protein** in which the N-terminal component consists of a foreign protein sequence. An antibody which specifically recognizes one of the foreign protein sequences can then be used to bind specifically to the phage which displays the sequence, leading to its purification. Such affinity purification permits identification of cDNA sequences encoding an uncharacterized protein of interest (see Parmley and Smith, 1988).

Further reading

Farzaneh F, Cooper DN (eds) (1995) *Functional Analysis of the Human Genome.* BIOS Scientific Publishers, Oxford.
Spector DL, Goldman RD, Leinwand LA (1998) *Cells: a laboratory manual.* Cold Spring Harbor Laboratory Press, Cold Spring Harbor.

Electronic references

1. http://www.hgmp.mrc.ac.uk/ESTBlast
2. These and many other useful molecular biology programs have been compiled at various centers such as Pedro's BioMolecular Research Tools at http://www-biol.univ-mrs.fr/pedro/research_tools.html
3. Programs such as PSORT (protein sorting signals), TMPred (transmembrane prediction etc.). Available, for example, at Pedro's BioMolecular Research Tools at http://www-biol.univ-mrs.fr/pedro/research_tools.html

References

Aiello LP, Robinson GS, Lin Y-W, Nishio Y, King GL (1994) Identification of multiple genes in bovine retinal pericytes altered by exposure to elevated levels of glucose by using mRNA differential display. *Proc. Natl Acad. Sci. USA*, **91**, 6231–6235.

Baxendale S, Abdulla S, Elgar G et al. (1995) Comparative sequence analysis of the human and pufferfish Huntington's disease genes. *Nature Genet.*, **10**, 67-75.

Chalfie M, Tu Y, Euskirchen G, Ward WW, Prasher DC (1994) Green fluorescent protein as a marker for gene expression. *Science*, **263**, 802–805.

Clackson T, Wells JA (1994) *In vitro* selection from protein and people libraries. *Trends Biotechnol.*, **12**, 173–184.

Daniels R, Kinis T, Serhal P, Monk M (1995) Expression of myotonin protein kinase gene in preimplantation human embryos. *Hum. Mol. Genet.*, **4**, 389–393.

Femino AM, Fay FS, Fogarty K, Singer RH (1998) Visualization of single RNA transcripts *in situ*. *Science,* **280**, 585–590.

Fields S, Sternglanz R (1994) The two-hybrid system: an essay for protein–protein interactions. *Trends Genet.*, **10**, 286–292.

Figueirdo MS, Brownlee GG (1995) CIS-acting elements and transcription factors involved in the promoter activity of the human factor VIII gene. *J. Biol. Chem.*, **270**, 11828–11838.

Frohman MA, Dush MK, Martin GR (1988) Rapid production of full length cDNAs from rare transcripts: amplification using a single gene-specific oligonucleotide primer. *Proc. Natl Acad. Sci. USA*, **85**, 8998–9002.

Gavin BJ, McMahon AP (1993) Cloning developmentally regulated gene families. *Meth. Enzymol.*, **225**, 653–663.

Ginsburg M (1994) In: *Guide to Human Genome Computing* (MJ Bishop ed.), pp. 215–248. Academic Press, New York.

Henikoff S, Henikoff JG (1992) Amino acid substitution matrices from protein blocks. *Proc. Natl. Acad. Sci. USA,* **89**, 10915–10919.

Kozak M (1996) Interpreting cDNA sequences: some insights from studies on translation. *Mamm. Genome,* **7**, 563–574.

Kremmidiotis G, Lensink IL, Bilton RL, Woollat E, Chataway TK, Sutherland GR, Callen DF (1999) The Batten disease gene product (CLN3p) is a Golgi integral membrane protein. *Hum. Mol. Genet.*, **8**, 523–531.

Liang P, Averboukh L, Pardee AB (1993) Distribution and cloning of eukaryotic mRNAs by means of differential display: refinements and optimization. *Nucleic Acids Res.*, **21**, 3269–3275.

Maxam AM, Gilbert W (1980) Sequencing end labeled DNA with base-specific chemical cleavages. *Methods Enzymol.,* **65**, 499–560.

Needleman SB, Wunsch CD (1970) A general method applicable to the search of similarities in the amino acid sequences of two proteins. *J. Mol. Biol.,* **48**, 443–453.

Nicholson LVB, Johnson MA, Bushby KMD et al. (1993) Integrated study of 100 patients with Xp21 linked muscular dystrophy using clinical, genetic, immunochemical, and histopathological data. Part 2. Correlations within individual patients. *J. Med. Genet.*, **30**, 737–744.

Old RW, Primrose SB (1994) *Principles of Gene Manipulation. An Introduction to Genetic Engineering*, 5th edn. Blackwell Scientific Publications, Oxford.

Parmley SF, Smith GP (1988) Antibody-selectable filamentous fd phage vectors: affinity purification of target genes. *Gene*, **73**, 305–318.

Phizicky E, Fields S (1995) Protein–protein interactions: methods for detection and analysis. *Microbiol. Rev.*, **59**, 94–123.

Piccolo S, Bonaldo P, Vitale P, Volpin D, Bressan GM (1995) Transcriptional activation of the alpha 1 (VI) collagen gene during myoblast differentiation is mediated by multiple GA boxes. *J. Biol. Chem.*, **270**, 19583–19590.

Pykett MJ, Murphy M, Harnish PR, George DL (1994) The neurofibromatosis 2(NF2) turn or suppressor gene encodes multiple alternatively spliced transcripts. *Hum. Mol. Genet.*, **3**, 559–564.

Riordan JR, Rommens JM, Kerem B et al. (1989) Identification of the cystic fibrosis gene: cloning and characterization of complementary DNA. *Science,* **245**, 1066–1073.

Roberts RG, Coffey AJ, Bobrow M, Bentley DR (1993) Exon structure of the human dystrophin gene. *Genomics*, **16**, 536–538.

Rouleau G, Merel P, Lutchman M et al. (1993) Alteration in a new gene encoding a putative membrane-organising protein causes neurofibromatosis type 2. *Nature*, **363**, 515–521.

Schutze K, Lahr G (1998) Identification of expressed genes by laser-manipulated manipulation of single cells. *Nature Biotechnol.*, **16**, 737–742.

Sengupta DJ, Zhang B, Kraemer B, Pochart P, Fields S, Wickens M (1996) A three-hybrid system to detect RNA–protein interactions in vivo. *Proc. Natl Acad. Sci. USA*, **93**, 8496–8501.

Simone NL, Bonner RF, Gillespie JW, Emmert-Buck MR, Liotta LA (1998) Laser-capture microdissection: opening the microscopic frontier to molecular analysis. *Trends Genet.*, **16**, 272–276.

Stryer L (1995) *Biochemistry*, 4th edn. WH Freeman, New York.

Tsien RY (1998) Green fluorescent protein. *Ann. Rev. Biochem.*, **67**, 509–554.

Velculescu VE, Zhang L, Vogelstein B, Kinzler KW (1995) Serial analysis of gene expression. *Science,* **270**, 484–487.

Waterman MS, Smith TF, Beyer WA (1976) Some biological sequence metrics. *Adv. Math.,* **20**, 367–387.

Winter G, Griffiths AD, Hawkins RE, Hoogenboom HR (1994) Making antibodies by phage display technology. *Annu. Rev. Immunol.,* **12**, 433–455.

Wodicka L, Dong H, Mittmann M, Ho M-H, Lockhart DJ (1997) Genome-wide expression monitoring in *Saccharomyces cerevisiae. Nature Biotechnol.,* **15**, 1359–1367.

Genetic manipulation of animals

21.1 An overview of genetic manipulation of animals

Experimental animals have been used in biomedical research for decades. In many cases, aspects of physiology and biochemistry have been investigated, and artificial manipulations have often been confined to examining the effect of altering the animal's environment or some aspect of its phenotype. Some animals, notably *Drosophila* and mice, have been particularly amenable to genetic analyses and traditional genetic manipulation of animals has involved carefully selected breeding experiments or exposure of animals to powerful chemical or radioisotopic mutagens. A new era in animal research was ushered in during the early 1980s when successful experiments designed to genetically modify animals by inserting foreign DNA were first reported. These new methodologies were expected to have many advantages for research but two major areas have benefited:

■ *Gene function.* While the use of cultured cells and cell extracts can be extremely valuable in studying gene expression and function, the ability to insert genes into whole animals or to selectively delete or alter single predetermined genes in an animal provides enormous power in studying gene function.

■ *Animal models of disease.* Nature has provided some animal models of disease and some have been generated by random mutagenesis programmes, *but not in a predetermined way.* The new technologies held the promise of altering at will even single genes within a living animal in such a way as to mimic mutations faithfully in an analogous gene in humans, thereby providing a higher chance of resembling human disease phenotypes.

In order to create genetically modified animals, it is necessary to modify the DNA of germline cells so that the modified DNA is heritable. As a result, certain cells that have the capacity to differentiate into the different cells of an adult animal (or at least to give rise to germ line cells) were considered to be the optimal targets for introducing foreign DNA. The fertilized oocyte is one such cell, being totipotent. Other target cells are cells of very early stage embryos, including **embryonic stem (ES) cells.** Although such cells are postzygotic they represent a stage in development where there has been incomplete separation of the *soma* and the *germline.* Such cells are therefore capable of giving rise to both somatic and germline cells.

When a foreign DNA molecule is artifically introduced into the cells of an animal, a **transgenic animal** is produced. The foreign DNA molecule is called a transgene and may contain one or many genes. By inserting a transgene into a fertilized oocyte or cells from the early embryo, the resulting transgenic animal may be able to transmit the foreign DNA stably in its germline. Many different types of transgenic animals have been created including transgenic *Drosophila*, transgenic frogs, transgenic fish and a variety of transgenic mammals including mice, rats and various livestock animals. The technology of transgenesis and its applications are considered in Section 21.2.

Although transgenes often integrate into the host chromosomes without affecting the expression of any endogenous genes, occasionally the integration event alters endogenous gene expression (insertional mutation), producing a recognizable phenotype. This constitutes a form of *in vivo* mutagenesis, albeit at an *unselected* target gene. **Gene targeting** was developed as a method of *in vivo* mutagenesis in which the mutation is introduced into a *preselected* endogenous gene. This can be achieved in somatic cells, but gene targeting in cultured ES cells is particularly powerful because it can lead to the construction of an animal in which all nucleated cells contain a mutation at the desired locus (see Section 21.3 below). In mammals, gene targeting has been possible only in mice but research on ES cells from other species may extend the capacity for gene targeting in the near future. Note that in some cases gene targeting is used to produce a subtle mutation and as a result the ES cells used for blastocyst implantation do not contain any foreign sequences.

The resulting mice may be described as genetically modified but not as transgenic.

The ability to produce transgenic mice and particularly the ability to perform specific changes in a predetermined gene by gene targeting has permitted the design of many new animal models of human disease (Section 21.4). Another experimental approach involving genetic manipulation of animals has had a major impact recently. In 1997 a new era in mammalian genetics was heralded when the procedure of somatic cell nuclear transfer permitted 'cloning' of an adult mammal for the first time. This involved transfer of the nucleus from an adult cell into an enucleated oocyte and the technology has subsequently been applied as an alternative route to generating transgenic animals (Section 21.5).

21.2 The creation and applications of transgenic animals

Of the different transgenic animals that have been made thus far, transgenic *Drosophila*, transgenic frogs, and transgenic fish have been very important for understanding aspects of gene function and development in these species. Transgenic sheep and other transgenic livestock animals have been produced largely to serve as bioreactors, whole-animal expression cloning systems in which introduced genes are expressed to give large amounts of therapeutic or commercially valuable gene products (see Section 22.1.2). But it has been transgenic mice that have been the most useful to biomedical research, both in providing animal models of disease and in permitting the most useful analyses of mammalian gene function.

21.2.1 Transgenic animals can be produced following transfer of cloned DNA into fertilized oocytes and cells from very early stage embryos

Transgenesis involves transfer of foreign DNA into totipotent or pluripotent embryo cells (either fertilized oocytes, cells of the very early embryo or cultured embryonic stem cells) followed by insertion of the transferred DNA into host chromosomes. If the foreign DNA integrates into the chromosomes of a fertilized oocyte, the developing animal will be *fully transgenic* since all nucleated cells in the animal should contain the transgene. If chromosomal integration occurs later, at a postzygotic stage, the animal will be a mosaic, with some cells containing the transgene and some others lacking it. If the transgene is present in germline cells it can be passed through sperm or egg cells into some of the animal's progeny, and PCR-based tests can be used to quickly screen for the presence of the transgene. Progeny that are transgene positive can be expected to be fully transgenic; all their cells should have developed from a fertilized oocyte containing the transgene.

Pronuclear microinjection

To obtain transgenic mice by this route, females are superovulated, mated to fertile males and sacrificed the next day. Fertilized oocytes are recovered from excised oviducts. The DNA of interest is then microinjected using a micromanipulator into the male pronucleus of individual oocytes (*Figure 21.1*). Surviving oocytes are reimplanted into the oviducts of foster females and allowed to develop into mature animals (see Gordon, 1992).

During this procedure, the microinjected DNA (transgene) randomly integrates into chromosomal DNA, usually at a single site, although rarely two sites of integration are found in a single animal. Individual insertion sites typically contain multiple copies of the transgenes integrated into chromosomal DNA as head-to-tail concatemers (it is not unusual to find 50 or more copies at a single insertion site). As a result of chromosomal integration, the transgenes can be passed on to subsequent generations in mendelian fashion: if the foreign DNA has integrated at the one-cell stage, it should be transmitted to 50% of the offspring.

Transfer into pre- or postimplantation embryos

Cells from very early stage embryos may be totipotent or at least pluripotent and can provide a route for foreign DNA to enter the germline. DNA can be transferred into unselected cells of very early embryos, as described in this section or into cell lines derived from embryonic stem cells, as described in Section 21.2.2.

One method that allows foreign DNA to integrate into the chromosomes of the target cells uses **retroviruses**, RNA viruses which naturally undergo an intermediate DNA form prior to integrating into cellular genomes (see *Figure 18.2*). Infection of preimplantation mouse embryos with a retrovirus such as Moloney murine leukemia virus or injection of the retrovirus into early postimplantation mouse embryos results in mosaic offspring. Retroviruses should integrate rarely and at random into accessible cells, and the use of replication- defective retroviruses provides heritable markers for clonal descendants of the target cell (unlike wild-type viruses which spread from cell to cell). This approach has been used, therefore, for studying cell lineage using reporter genes.

In *Drosophila*, efficient chromosomal integration is possible by using sequences from a *Drosophila* transposable element known as the P element to permit insertion of single copies of a gene at random in the genome. The gene or DNA fragment to be inserted is first manipulated so as to be flanked by the two terminal sequences of the P element. The modified DNA is then microinjected into a very young *Drosophila* embryo along with a separate plasmid containing a gene encoding a transposase. In the presence of the transposase the terminal P element sequences allow the intervening DNA fragment to transpose and as a result of the ensuing transposition events, the injected DNA often enters the germline in a single copy.

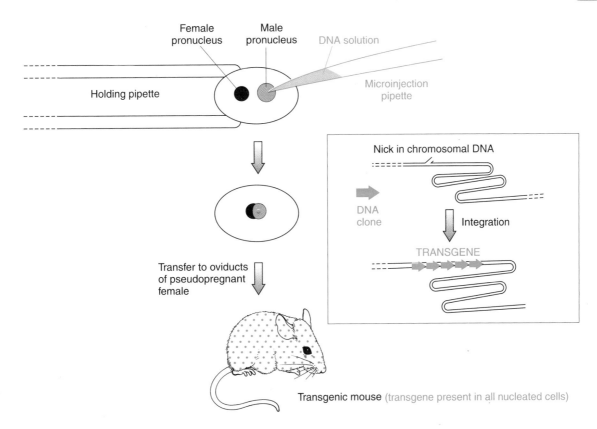

Figure 21.1: Construction of transgenic mice by pronuclear microinjection.

Very fine glass pipettes are constructed using specialized equipment: one, a holding pipette, has a bore which can accommodate part of a fertilized oocyte, and thereby hold it in place, while the microinjection pipette has a very fine point which is used to pierce the oocyte and thence the *male* pronucleus (because it is bigger). An aqueous solution of the desired DNA is then pipetted directly into the pronucleus. The introduced DNA clones can integrate into chromosomal DNA at nicks, forming transgenes, usually containing multiple head-to-tail copies. Following withdrawal of the micropipette, surviving oocytes are reimplanted into the oviducts of *pseudopregnant* foster females (which have been mated with a vasectomized male; the mating act can initiate physiological changes in the female which stimulate the development of the implanted embryos). Newborn mice resulting from development of the implanted embryos are checked by PCR for the presence of the desired DNA sequence (Gordon, 1992).

21.2.2 Cultured embryonic stem (ES) cells provide a powerful route to genetic modification of the germline

The microinjection of foreign DNA into fertilized oocytes is technically difficult and not suited to large-scale production of transgenic animals or to sophisticated genetic manipulation. A popular alternative, but one which has so far been restricted to the construction of genetically modified mice, involves transferring the foreign DNA initially into cultured **embryonic stem (ES) cells**. Mouse ES cells are derived from 3.5–4.5 day postcoitum embryos and arise from the *inner cell mass* of the blastocyst (see *Figure 21.2*). The ES cells can be cultured *in vitro* and retain the potential to contribute extensively to all of the tissues of a mouse, *including the germline,* when injected back into a host blastocyst and reimplanted in a pseudopregnant mouse.

The developing embryo is a **chimera**: it contains two populations of cells derived from different zygotes, those of the blastocyst and the implanted ES cells. If the two strains of cells are derived from mice with different coat colors, chimeric offspring can easily be identified (see *Figure 21.2*). Use of genetically modified ES cells results in a partially transgenic mouse. Because the injected ES cells can form all or part of the functional germ cells of the chimera, it is possible to derive fully transgenic mice. This is usually accomplished by screening the offspring of matings between chimeras (usually males) and mice with a coat color recessive to that of the strain from which the ES cells were derived (see *Figure 21.2*).

The big advantage of ES cells is that they can be grown readily in culture. This means that a variety of genetic manipulations can be conducted in cultured ES cells. Importantly, *the desired genetic modification can be verified in tissue culture,* before injecting the genetically modified cells into a blastocyst prior to implantation. For example, the desired gene can be ligated to a marker gene, such as

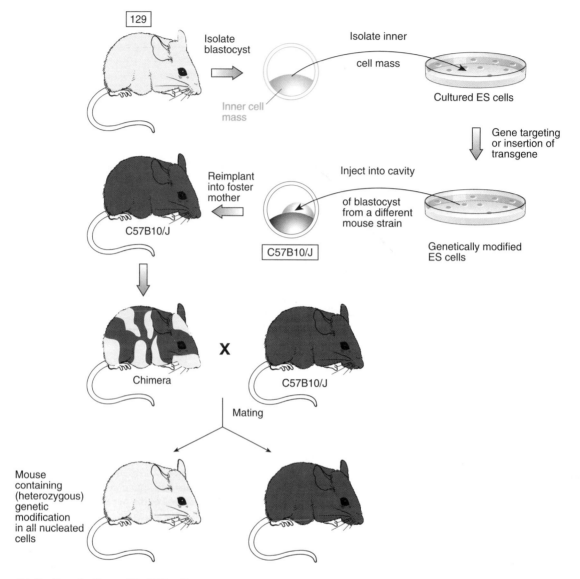

Figure 21.2: Genetically modified ES cells as a route for transferring foreign DNA or specific mutations into the mouse germline.

Cells from the *inner cell mass* were cultured following excision of oviducts and isolation of blastocysts from a suitable mouse strain (129). Such embryonic stem (ES) cells retain the capacity to differentiate into, ultimately, the different types of tissue in the adult mouse. ES cells can be genetically modified while in culture by insertion of foreign DNA or by introducing a subtle mutation. The modified ES cells can then be injected into isolated blastocysts of another mouse strain (e.g. C57B10/J which has a black coat color that is recessive to the agouti color of the 129 strain) and then implanted into a pseudopregnant foster mother of the same strain as the blastocyst. Subsequent development of the introduced blastocyst results in a **chimera** containing two populations of cells (including germline cells) which ultimately derive from different zygotes (normally evident by the presence of differently colored coat patches). Backcrossing of chimeras can produce mice that are heterozygous for the genetic modification. Subsequent interbreeding of heterozygous mutants generates homozygotes.

the *neo* gene, enabling a positive selection for cells that have been successfully transfected (see *Box 10.1*). The presence of the desired gene can also be verified quickly by a PCR-based assay. ES cells also offer the huge advantage of *gene targeting* by homologous recombination, a method which can permit a programed selective alteration of a single *predetermined* gene and also highly spe-

cific ways of chromosome engineering. Such approaches are extremely powerful for understanding gene function (Section 21.3) and also for creating animal models of disease (Section 21.4).

The ES cell approach to constructing transgenic mice was made possible by the successful establishment in the early 1980s of stable cell lines from isolated mouse ES cells. ES cells were not so readily identifed in other mammals, although there have been some important recent successes (see *Box 21.1*).

21.2.3 Transgenic animals have been used for a variety of studies investigating mammalian gene expression and function

Transgenic animals have been extremely important for analyzing human genes (Hanahan, 1989; Camper *et al.*, 1995; Theuring, 1995), and have helped greatly in our understanding of a variety of fundamental biological processes, notably in immunology, neurobiology, cancer and developmental studies. The following list is far from exhaustive but is intended to illustrate some major types of application:

- *Investigating gene expression and its regulation.* Although evidence for *cis*-acting regulatory elements is often inferred initially from studies using cultured cells, they need to be validated in whole animal studies. Transgenes consisting of the presumptive

regulatory sequence(s) coupled to a reporter gene, such as *lacZ*, provide a sensitive method of detecting gene expression and a powerful way of investigating regulation of gene expression. Long-range control of gene expression is often investigated using YAC transgenes, see Section 21.2.4.

- *Investigating gene function by targeted gene inactivation.* Specific genes can be inactivated by a gene targeting procedure to introduce a transgene into the target gene (**insertional inactivation**). The effect on the phenotype of creating a null mutation in the gene of interest can provide powerful clues to gene function.

- *Investigating dosage effects and ectopic expression.* In some cases, valuable information can be gained by over-expressing a transgene (e.g. Schedl *et al.*, 1996) or by expressing it *ectopically* (the transgene is coupled to a tissue-specific promoter which causes expression in cells; the phenotypic consequence may provide valuable clues to function).

- *Cell lineage ablation.* Transgenes can be designed consisting of a tissue-specific promoter coupled to a sequence encoding a toxin, for example diphtheria

---

Box 21.1

Isolation and manipulation of mammalian embryonic stem cells

Mouse ES cells

The embryo proper derives from the inner cell mass (ICM) of a blastocyst (*Box 2.2*). Mouse embryonic stem (ES) cells were first isolated from blastocysts by Evans and Kaufman (1981) and Martin (1981). The procedure involves placing a 4.5 day preimplantation embryo (blastocyst) on a monolayer of feeder cells. The latter provide a matrix for attachment and also secreted protein factors which inhibit newly formed ES cells from differentiating. Following attachment of the blastocyst, ICM cells proliferate. At an appropriate time, the ICM is physically removed with a micropipette, dispersed into small clumps of cells and seeded onto new feeder cells. Colonies are examined under the microscope for characteristic morphology, picked, dispersed into single cells and reseeded onto feeder layers. Eventually cells with a uniform morphology can be isolated, and cell lines can be established.

The term 'ES cell' was introduced to distinguish these embryo-derived stem cells from embryonal carcinoma (EC) cells, which had been derived from teratocarcinomas. In general ES cells have a less restricted developmental potential than EC cells. Both classes of cells are pluripotent and ES cells can give rise to all adult cell types, including germline cells. They are, however, not totipotent in the sense that the fertilized oocyte is: if implanted in a uterus they are unable to give rise to an embryo. In some cases EC cells have been reported to contribute to the germline in chimeras, but when mouse ES cells are injected into an isolated blastocyst from a different strain and implanted in a pseudopregnant foster mother (see *Figure 21.2*) they more consistently contribute to chimeras and particularly to the germline. It is this property which has made mouse ES cells so valuable for research. It should be noted, however, that the majority of ES cells used to form successful germline chimeras have been from a single

mouse strain (129) in combination with a single host embryo strain (C57BL/6).

ES cells in other mammals

The search for ES cells in other mammals has occupied researchers for many years, and although ES cells have been isolated from several other species in addition to the mouse, the great success in forming germline chimeras in mouse has not been paralleled thus far in other species. Recently, human embryonic stem cells have been isolated from cells of the blastocyst (Thomson *et al.*, 1998) and also from **primordial germ cells** (the cells of the early embryo that will eventually differentiate into sperm and oocytes; Shamblott *et al.*, 1998). For ethical reasons, these cells can never be used to form germline chimaeras, but their discovery opens up exciting new prospects for both basic and applied research (see Solter and Gearhart, 1999). Much work needs to be done, not least in finding ways for these undifferentiated cells to be turned into cells of a desired type. Once this can be achieved, however, several exciting possibilities can be envisaged including therapeutic applications:

- *Basic developmental biology.* There are numerous basic research issues which can be investigated in identifying the nature and control of cell commitment and cell differentiation.

- *Drug development.* Derived human cell lines can be fully characterized and used for screening drugs.

- *Cell and tissue transplantation.* Large quantities of disease-free cells might be produced for transplants, e.g. neuronal cells for treating Parkinson's disease, pancreatic beta cells for treatment of diabetes, and cells for various heart, brain and nerve grafts.

toxin subunit A or ricin. When the promoter becomes active at the appropriate stage of tissue differentiation, the toxin is produced and kills the cells. Thus, certain cell lineages in the animal can be eliminated (cell ablation) and the phenotypic consequences monitored.

■ *Investigating gain of function.* In principle, any mammalian gene that produces a dominant negative effect or gain of function can be investigated by introduction of an appropriate transgene. In some cases, this can provide proof of a suspected biological function. A classical example concerns the *Sry* gene. A variety of different genetic analyses had implicated this gene as a major male-determining gene but convincing proof was obtained using a transgenic mouse approach. The experiment consisted of transferring a cloned *Sry* gene into a fertilized 46,XX mouse oocyte. As a result of this artificial intervention, the resulting mouse, which nature had intended to be female, turned out to be male (Koopman *et al.*, 1991).

■ *Modeling human disease.* Insertional inactivation is often used to model loss of function mutations whilst gain of function mutations can often be modeled by inserting a mutant transgene (see Section 21.4).

21.2.4 The use of YAC transgenes and inducible promoters has greatly extended the applications of transgenic animals

Inducible promoters

For many applications it is desirable to have a transgene expressed under the control of a tissue-specific promoter or an inducible promoter (see Section 8.2.4). In the former case, genetic engineering can be used to splice known tissue-specific promoters from cloned genes to the gene of interest. In some cases, coupled regulatory elements can confer both position-independent and tissue-specific expression as with sequence elements from the β-globin locus control region (Grosveld *et al.*, 1987).

Various attempts have also been made to create inducible transgenic mice (e.g. by using heavy metal ions to induce expression of an integrated gene which has a coupled metallothionein promoter, etc.). Generally, the use of inducible promoters has been hampered by 'leakiness' in gene expression and by relatively low levels of induction, and they have often been applicable to a limited range of tissues. More recently, however, more promising systems have been developed. For example, methods employing tetracycline-regulated inducible expression have permitted construction of both highly inducible transfected cells (with much greater efficiency than the constitutive system) and transgenic mice (Shockett *et al.*, 1995). In the latter case, the expression of a reporter gene, such as the luciferase gene, can be controlled by altering the concentration of tetracycline in the drinking water of the animals.

YAC transgenics

Early studies of gene expression and regulation in transgenic animals involved transfer of small genes. However, expression of small transgenes often fails to follow the normal temporal and spatial patterns of expression or match the expression level of the endogenous homolog. An increasing number of human genes are known to be very large (see *Figure 7.7*). Even in the case of small genes, important regulatory elements that are required for correct expression may be located many kilobases upstream of the coding sequence (see *Figure 8.23*). In order to be able to study the expression and regulation of a human gene under the control of its own *cis*-acting regulatory elements, it was therefore necessary to establish transfection conditions which would allow the transfer of large DNA clones.

A major breakthrough in transgenic studies was the development of so-called **YAC transgenics** (Lamb and Gearhart, 1995). The first report to be published described transfer of a 670 kb YAC containing the human *HPRT* (hypoxanthine phosphoribosyltransferase) gene into mouse ES cells (Jakobovits *et al.*, 1993). This was accomplished by spheroplast fusion (i.e. fusion of ES cells with YAC-containing yeast cells that have been stripped of the hard cell wall; see Section 4.3.4). Fragments from the yeast genome can integrate at the same time, however, and so alternative methods have sought to purify an individual YAC by size-fractionation on a preparative gel using pulsed-field gel electrophoresis (assuming that the YAC migrates at a position in the gel that is different from any yeast chromosome). The purified YAC can be inserted into a fertilized oocyte by pronuclear microinjection (see above). This method is, however, limited to small YACs: the DNA of large YACs is more likely to fragment following microinjection with very fine micropipettes. Alternatively, purified YACs have been transferred into ES cells by using **liposomes**, artificial lipid vesicles that are used to transport molecules into a cell following fusion of the lipid coat with the plasma membrane of the recipient cell (see *Figure 22.6*).

YAC transgenics have permitted study of large genes such as the 400 kb human *APP* gene, a gene known to contribute to Alzheimer's disease, encoding the amyloid protein precursor. A YAC transgene containing *APP* showed tissue- and cell type-specific expression patterns closely mirroring that of the endogenous mouse gene (for this and other examples, see Lamb and Gearhart, 1995). Long-range gene regulation mechanisms (locus control regions, imprinting and other chromatin domain effects; see Chapter 8) can be modeled. An interesting application has been in the production of fully human antibodies in the mouse by transfer of human YACs containing large segments of the human heavy and kappa light chain immunoglobulin loci into mouse ES cells, and thence the creation of transgenic mice able to produce human antibodies (see Mendez *et al.*, 1997 and *Figure 21.3*). Finally, YAC transgenics may also find a role in modeling disease caused by large-scale gene dosage imbalance (see Section 21.4.4).

1. Isolate YACs containing sequences from human IgH (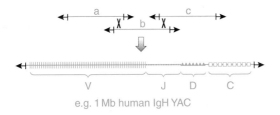) and human IgK chain (⊏ᴷ⊐) loci

2. Allow YACs from each locus to undergo homologous recombination in yeast cells to generate YACs wth large inserts

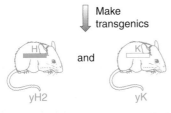

e.g. 1 Mb human IgH YAC

3. Fuse yeast cell spheroplasts containing large IgYAC to mouse ES cells

Make transgenics

and

yH2   yK

Mice with single human Ig transloci

4. Cross these mice with the DI strain which has endogenous IgH (◼◼ᴴ) and IgK (⊏ᴷ⊐) loci knocked out by gene targeting

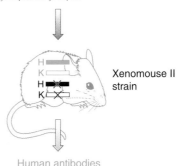

yH2 × DI   yK2 × DI

yH2; DI   yK2; DI

5. Cross yH2; DI to yK2; DI

Xenomouse II strain

Human antibodies

Figure 21.3: Use of YAC transgenesis to construct a mouse with a human antibody repertoire.

YACs containing Ig sequences are obtained by screening YAC libraries with suitable Ig probes. The recovery of YACs with comparatively small inserts meant that there was a need to artificially construct larger YACs by homologous recombination in yeast cells. Spheroplasts are created by treating the yeast cells so that the outer cell wall is stripped off, making the cells more amenable to cell fusion. See Mendez *et al.* (1997) for further details of the procedures used to construct these mice.

Transchromosomic animals

Ultimately, even YACs have upper limits for the size of foreign inserts that can be transferred. Mammalian artificial chromosomes have also been generated, including first generation human artificial chromosomes (Harrington *et al.*, 1997; Ikeno *et al.*, 1998). Such systems will have the capacity of transferring hundreds and possibly thousands of genes into transgenic animals, although a preferred route may be by using nuclear transfer technology (Section 21.5) rather than ES cells. Recently, however, transfer of whole chromosomes or chromosome fragments into ES cells has been possible by microcell-mediated chromosome transfer (see Sections 10.1.1 and 10.1.3). Using this approach Tomizuka *et al.* (1997) were able to transfer human chromosomes or chromosome fragments derived from normal fibroblasts into mouse ES cells. The resulting chimeric transchromosomic mice were viable, and the chromosome fragments appeared to show functional expression and could be transmitted through the germline.

21.3 Use of mouse embryonic stem cells in gene targeting and gene trapping

21.3.1 Gene targeting by homologous recombination in ES cells can be used to produce mice with a mutation in a predetermined gene

Gene targeting involves engineering a mutation in a *pre-selected* gene within an intact cell. It can therefore be viewed as a form of artificial **site-directed *in vivo* mutagenesis** (as opposed to the various methods of site-directed *in vitro* mutagenesis described in Section 6.4). The mutation may result in inactivation of gene expression (a **'knock-out' mutation**), or altered gene expression, and so can be useful for studying gene function (see below). In addition, the same method can be used to 'correct' a pathogenic mutation by restoring the normal phenotype, and so has therapeutic potential (see Section 22.3).

Gene targeting typically involves introducing a mutation by homologous recombination. A cloned gene (or gene segment) closely related in sequence to endogenous target gene is transfected into the appropriate cells. In some of the cells, homologous recombination occurs between the introduced gene and its chromosomal homolog. Gene targeting by homologous recombination has been achieved in some somatic mammalian cells, such as myoblasts. However, the most important application involves mouse ES cells: once a mutation has been engineered into a specific mouse gene within the ES cells, the modified ES cells can then be injected into the blastocyst of a foster mother and eventually a mouse can be produced with the mutation in the desired gene in all nucleated cells (Capecchi, 1989; Melton, 1994).

Homologous recombination in mammalian cells is a very rare occurrence (unlike in yeast cells, for example, where it occurs naturally at high frequencies, enabling sophisticated genetic manipulation). The frequency of homologous recombination is increased, however, when the degree of sequence homology between the introduced DNA and the target gene is very high. As a result, the introduced DNA clone is a *mouse* sequence which should preferably be isogenic (derived from the same mouse strain as the strain of mouse from which the ES cells were derived). Even then, the frequency of genuine homologous recombination events is very low and may be difficult to identify against a sizeable background of random integration events.

To assist identification of the desired homologous recombination events, the targeting vector (often a plasmid vector) contains a marker gene, such as the *neo* gene (see *Box 10.1*), which permits selection for cells that have taken up the introduced DNA. PCR assays are used to screen for evidence of a homologous recombination event (by using a marker-specific primer plus a primer derived from a sequence present in the target gene but absent from the introduced homologous gene segment). The targeting construct is transferred into cultured mouse ES cells by electroporation, a method in which pulses of high voltage are delivered to cells, causing temporary relaxation of the selective permeability properties of the plasma membranes. Two basic approaches have been used:

- **Insertion vectors** target the locus of interest by a single reciprocal recombination, causing insertion of the entire introduced DNA including the vector sequence (*Figure 21.4A*). This is the most reliable way of causing a knock-out mutation.
- **Replacement vectors** are designed to replace some of the sequence in the chromosomal gene by a homologous sequence from the introduced DNA (*Figure 21.4B*). This can occur as a result of a double reciprocal recombination or by gene conversion. The replacement method can inactivate a gene when the introduced sequence contains one or more premature termination codons or lacks critical coding sequences. It can also be used to correct a pathogenic mutation.

The replacement vector approach, as well as the insertion vector method, often leaves foreign sequences at the tar-

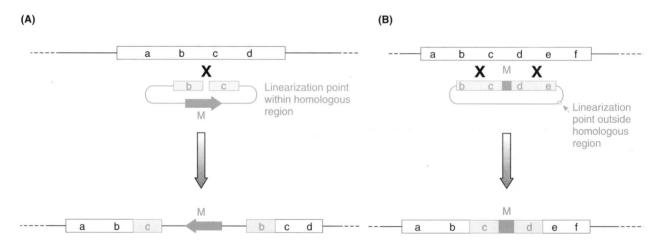

Figure 21.4: Gene targeting by homologous recombination can inactivate a predetermined chromosomal gene within an intact cell.

(**A**) Insertion vector method. The introduced vector DNA (blue) is cut at a unique site within a sequence which is identical or closely related to part of a chromosomal gene (black). Homologous recombination (X) can occur, leading to integration of the entire vector sequence including the marker gene (M). *Note* that the letters do not represent exons but are simply meant to indicate linear order within the gene. (**B**) Replacement vector method. In this case, the marker gene is contained within the sequence homologous to the endogenous gene, and the vector is cut at a unique location outside the homologous sequence. A double recombination or gene conversion event (X X) can result in replacement of internal sequences within the chromosomal gene by homologous sequences from the vector, including the marker gene.

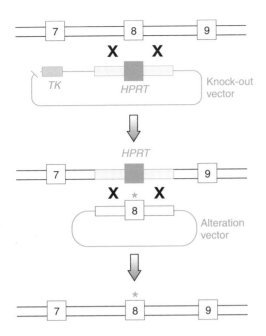

Figure 21.5: Double replacement gene targeting can be used to introduce subtle mutations.

Both the methods in *Figure 21.4* result in introduction of a substantial amount of exogenous sequence within the endogenous gene. To introduce a subtle mutation without leaving residual exogenous sequence, a double replacement method with positive and negative selection can be used (Melton, 1994). Exons in the endogenous gene are represented as numbered large boxes, and introns as long thin boxes. In order to introduce a subtle mutation, such as a single nucleotide substitution in exon 8, a replacement knock-out vector is used with a marker gene (e.g. the *HPRT* gene) flanked by homologous sequences from introns 7 and 8, and a second marker, such as the herpes simplex thymidine kinase (*TK*) gene outside the homologous region. Gene conversion, or double crossover within the flanking intron sequences, can lead to replacement of exon 8 by the *HPRT* gene, and can be selected for if a mutant *HPRT⁻* ES cell is used. A positive–negative selection system can be used: selection in the first step is for *HPRT⁺ TK⁻* cells. Cells containing random vector integrations will contain the *TK* gene and can be killed with the thymidine analog *gancyclovir* (see *Figure 22.13*). The second replacement involves introducing an altered exon 8 with a point mutation (*) to replace the *HPRT* gene and can be screened by identifying *HPRT⁻* cells. *Note* that mice engineered in this way cannot be described as transgenic because of the lack of foreign sequences in the germline.

get locus. In some cases, however, a more subtle mutation is required. For example, it may be desirable to investigate the effect of changing a single codon. Various two-step recombination techniques can be used to accomplish this method, and the resulting mouse, although genetically modified lacks any foreign sequences and so can no longer be described as transgenic (see *Figure 21.5* and Melton, 1994).

Gene targeting in mice is popularly used for producing artificial mouse models of human disease (Section 21.4). In addition, it provides a powerful general method of studying gene function. The gene in question is selectively inactivated, producing a **'knock-out' mouse**, and the effect of the mutation on the development of the mouse is monitored carefully. Sometimes there is little or no phenotypic consequence after inactivating a gene that would be expected to be crucially important, such as some genes which encode a transcription factor known to be expressed in early embryonic development. The lack of a phenotype in such cases is often thought to be due to **genetic redundancy** (another gene is able to carry out the function of the gene that has been knocked out). As a result, in some cases double or even triple gene knockouts have been carried out to analyze gene function, as in the case of some of the *Hox* genes (see, for example, Manley and Capecchi, 1997).

A useful example of investigating functional redundancy concerns studies of the mouse *Engrailed* genes, *En-1* and *En-2*. Both of these genes are homeobox genes which had been considered to play crucial roles in brain formation. *En-1* knock-outs have serious abnormalities but surprisingly *En-2* knock-outs have only minor problems. Expression of the *En-1* gene is switched on 8–10 hours before that of the *En-2* product, suggesting that perhaps the *En-1* product can compensate for the lack of *En-2* product in *En-2* knock-outs. To test for the possibility of functional redundancy, Hanks *et al.* (1995) used a variant of the knock-out procedure known as the **'knock-in'** technique. In this case the transgene used to knock-out the target endogenous gene is itself designed to be expressed under the control of the *cis*-acting elements of the knocked-out gene. A transgene containing an *En-2* coding sequence was used to knock-out the endogenous *En-1* gene. In so doing, the introduced *En-2* sequence came under control of the *En-1* regulatory sequences and was expressed before the endogenous *En-2* gene was switched on (see *Figure 21.6*). The resulting *En-1* knock-out mouse had a normal phenotype, demonstrating that the knocked-in *En-2* gene was functionally equivalent to *En-1* (Hanks *et al.*, 1995).

21.3.2 Site-specific recombination systems, notably the Cre–loxP system, extend the power of gene targeting

Several site-specific recombination systems from bacteriophages and yeasts have been characterized and are promising tools for genome engineering (Kilby *et al.*, 1993). Thus far, the Cre–*loxP* recombination system from bacteriophage P1 has been the most widely used. The natural function of the **Cre** (causes recombination) recombinase is to mediate recombination between two *loxP* sequences that are in the same orientation (the *loxP* sequence consists of 34 bp and comprises two inverted 13 bp repeats separated by a central asymmetric 8 bp spacer; *Figure 21.7*). As a result of recombination, the intervening sequence between the two *loxP* sites is

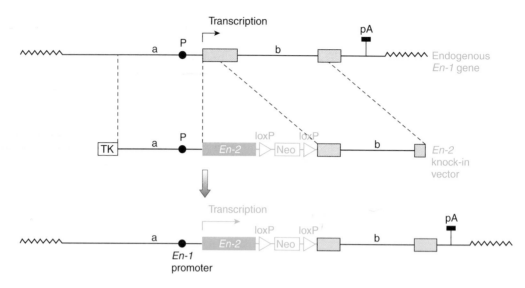

Figure 21.6: The knock-in method replaces the activity of one chromosomal gene by that of an introduced gene.

The *En-1* gene shown at top has two exons and coding sequences are shown by filled boxes. Its promoter (P) and polyadenylation site (pA) are also shown. The gene targeting vector ('knock-in vector') contains cloned *En-1* gene sequences comprising an upstream sequence *a* which contains the *En-1* promoter, and an internal segment *b* which spans the 3' end of exon 1, the single intron and the 5' coding sequence of exon 2. Separating these two sequences is the coding sequence of the *En-2* gene. Two marker cassettes include a thymidine kinase (TK) gene and a neomycin-resistance gene (Neo) both driven by a phosphoglycerate kinase promoter (chosen because phosphoglycerate kinase is expressed in ES cells). The Neo gene is flanked by loxP sequences. The targeting procedure results in replacement of endogenous sequences *a* and *b* but the 5' coding sequence of the *En-1* gene is deleted. The knocked-in *En-2* gene comes under the control of the *En-1* promoter (Hanks *et al.*, 1995). *Note* that the term 'knock-in' has also been applied to any procedure where an endogenous gene is inactivated by insertion of a new gene which is then expressed, even if the latter is only meant to serve as a reporter gene such as *lacZ*.

excised (see *Figure 4.15*). Using gene targeting, *lox*P sequences can be stitched into a desired gene or chromosomal location, and the subsequent provision of a gene encoding the Cre product can result in an artificially generated site-directed recombination event (see Chambers, 1994). As described below, several applications can be envisaged (Lobe and Nagy, 1998).

Tissue- and cell type-specific knock-outs

Some genes are vital to early development and simple knock-out experiments are generally not helpful because death ensues at the early embryonic stage. To overcome this problem, methods have been developed to inactivate expression of the target gene in only selected, predetermined cells of the animal (**conditional knock-out**). The animal can therefore survive and the effect of the knock-out can be studied in a tissue or cell type of interest.

An early example of a conditional knock-out sought

Figure 21.7: Structure of the *lox*P recognition sequence.

Note that the central 8 bp sequence which is flanked by the 13 bp inverted repeats is asymmetric and confers orientation.

to study the role of DNA polymerase β (an enzyme that is essential for embryonic development) in T lymphocytes (Gu *et al.*, 1994). The gene targeting procedure replaced part of the endogenous gene by an introduced homologous gene segment flanked by *lox*P sequences. Mice carrying this targeted mutation were then mated with a strain of mice which carried a *Cre* transgene gene under the control of a T cell-specific promoter. Offspring with the *lox*P-flanked pol β sequences plus the *Cre* transgene were identified and survived to adulthood. The Cre product was expressed only in T cells, leading to inactivation of the target gene in these cells by excision of DNA polymerase β gene segment between the two *lox*P sequences (see *Figure 21.8* for the general method).

Tissue- and cell type-specific gene activation

This approach is the opposite to that described above: it involves selective *activation* of a gene in certain cells of the animal to switch on a foreign gene only in predetermined cells of the animal (Barinaga, 1994).

Chromosome engineering

Another important recent development is a strategy for chromosome engineering in ES cells which relies on sequential gene targeting and Cre–*lox*P recombination. Gene targeting is used to integrate *lox*P sites at the desired chromosomal locations and, subsequently, transient

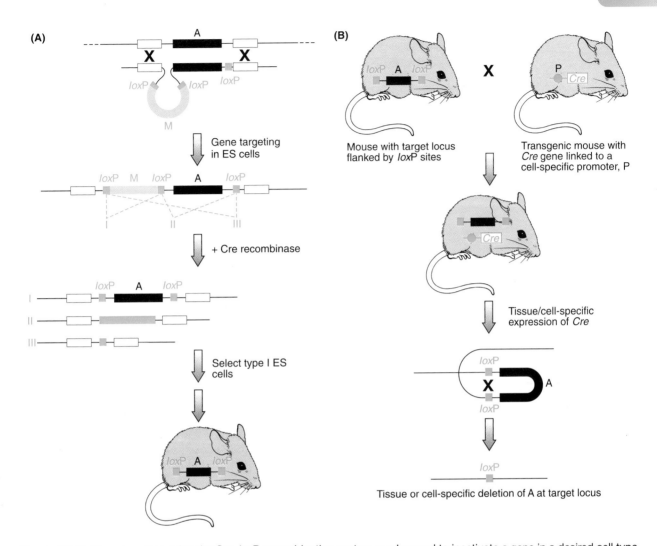

Figure 21.8: Gene targeting using the Cre–loxP recombination system can be used to inactivate a gene in a desired cell type.

(**A**) Illustration of a standard homologous recombination method using mouse ES cells, in which three *lox*P sites are introduced along with a marker M at a target locus A (typically a small gene or an internal exon which if deleted would cause a frameshift mutation). Subsequent transfection of a *Cre* recombinase gene and transient expression of this gene results in recombination between the introduced *lox*P sites to give different products. Type I recombinants are used to generate mice in which the target locus is flanked by *lox*P sites. Such mice can be mated with previously constructed transgenic mice (**B**) which carry an integrated construct consisting of the *Cre* recombinase gene linked to a tissue-specific promoter. Offspring which contain both the *lox*P-flanked target locus plus the *Cre* gene will express the *Cre* gene in the desired tissue type, and the resulting recombination between the *lox*P sites in these cells results in tissue-specific inactivation of the target locus A.

expression of Cre recombinase is used to mediate a selected chromosomal rearrangement (Ramirez-Solis *et al.*, 1995; Smith *et al.*, 1995; *Figure 21.9*). Chromosome engineering strategies of this type offer the exciting possibility of creating novel mouse lines with specific chromosomal abnormalities for genetic studies.

The multiple targeting and selection steps in ES cells used in the above chromosome engineering methods can be avoided using the novel approach of Herault *et al.* (1998). This targeted meiotic recombination method takes advantage of the homologous chromosome pairing that occurs naturally during meiosis at the first cell division. A transgene is designed to express Cre recombinase under

the control of a *Sycp1* promoter (the *Sycp1* gene encodes the SCYP1 protein which is part of the synaptonemal complex that facilitates crossing over). As a result Cre recombinase is produced in male spermatocytes during the zygotene to pachytene stages when chromosome pairing occurs.

21.3.3 Gene trapping in mouse ES cells allows an efficient approach to functional analysis

Large-scale approaches to investigating animal gene function have principally relied on exposing animals to

(A) 1. Use sequential gene targeting to introduce
loxP site (▶) plus a marker gene (Ⓜ) into two
desired locations on different chromosomes

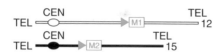

2. Expose loxP-containing chromosomes
to Cre recombinase

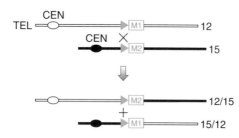

(B) 1. Use sequential gene targeting to introduce
loxP site plus a marker gene into two
desired locations flanking chromosomal
region to be deleted (⟋⟍⟋)

2. Allow to undergo intrachromosomal
'recombination' in presence of Cre
recombinase

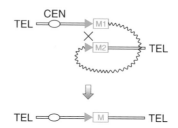

Figure 21.9: Chromosome engineering can be accomplished using cre–loxP systems.

(A) Use of targeted insertion of loxP sites to facilitate a chromosomal translocation. See Smith et al. (1995) for a practical example. **(B)** Use of targeted insertion of loxP sites to permit modeling of a microdeletion by intrachromosomal recombination. See Ramirez-Solis et al. (1995) for a practical example, and for other examples of intrachromosomal recombinations.

high doses of radiation, or to potent chemical mutagens such as *ethylnitrosurea* (*ENU*) or *ethyl methylsulfonate* (*EMS*). In mouse mutagenesis programs, for example, males are typically exposed to high levels of a suitable mutagen to induce a high frequency of mutation in sperm DNA. The progeny of irradiated mice are then screened for obvious phenotypic abnormalities. Such mutagenesis screens have been very useful in producing novel mutants, but a major problem is that the mutations occur essentially at random. Identification of the structural change(s) in the DNA of a single mutant animal may therefore require a laborious positional cloning approach.

A mutation that is induced by inserting a known foreign DNA sequence (transgene) into the mouse genome has a major advantage over one induced by chemical mutagens or X rays: *it leaves a sequence tag at the locus which is mutated.* As a result, rapid molecular characterization of the mutated locus is possible. Mouse ES cells provide a way of introducing such mutations into the germ line, and so random insertional mutagenesis in ES cells using transgenes was considered a useful way of producing a large number of mouse mutants which could quickly be characterized at the molecular level. Unlike gene targeting, the transgene should not be related to endogenous sequences, but should integrate essentially at random by **nonhomologous recombination**. However, because large sections of the mouse genome are noncoding, many random insertion events may not result in gene disruption.

In order to improve the efficiency of recovering mutations that are likely to have a phenotypic effect (by altering a gene or its expression) the **gene trap** approach was

devised (see Evans *et al.*, 1997). The underlying principle is that the transgene which is inserted into ES cells contains a defective reporter gene or marker gene *which lacks some component needed for gene expression.* The reporter or marker gene is designed to be expressed only after it inserts into a gene (within an intron or exon) or at a promoter. When it integrates at such positions it can acquire the expression element that it lacks.

Different gene trap strategies are possible. In some cases, the reporter lacks a functional promoter and so relies on chance integration next to an appropriate *cis*-acting sequence element that can activate its transcription. Other approaches have used a marker gene coupled to a suitable promoter but lacking a downstream polyadenylation signal. Here, the marker gene is designed to be expressed after integrating into a host cell gene such that a fusion RNA product is made that utilizes 3' host sequences in order to acquire a poly (A) tail (see *Figure 21.10*). Already, this type of approach has been applied on a large scale, with a recent report describing disruption and sequence identification of 2000 genes in mouse embryonic stem cells (Zambrowicz *et al.*, 1998).

21.4 Creating animal models of disease using transgenic technology and gene targeting

Animal models of human disease are crucially important to medical research. They allow detailed examination of

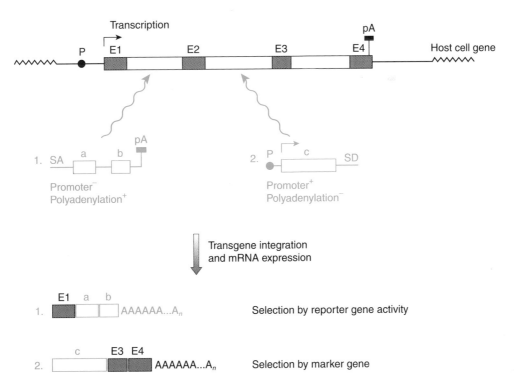

Figure 21.10: Gene trapping uses an expression-defective transgene to select for chromosomal integration events that occur within or close to a gene.

A specimen host cell gene with 4 exons (E1–E4) is shown at top together with its promoter (P) and polyadenylation signal (pA). Two possibilities for a gene trap vector are shown. Transgene 1 includes a reporter gene which lacks a promoter. It has two exonic sequences *a* and *b*, a polyadenylation signal (pA) and an upstream sequence which contains a **splice acceptor sequence (SA)**. In this example, transgene 1 integrates into intron 1 of the host cell gene. During transcription from the endogenous promoter the splice acceptor sequence will help the transgene exons to be spliced to the sequence from the first exon of the endogenous gene, producing a fusion transcript. Selection is based on this transcript having a functional reporter activity (see Evans *et al.*, 1997). Transgene 2 has a marker gene (drug-resistance, e.g. puromycin *N*-acetyl transferase) coupled to a promoter which works in ES cells (usually a phosphoglycerate kinase promoter) and a downstream **splice donor sequence**, but lacks a polyadenylation signal. Again, integration is intended to permit expression but this time from the transgene's promoter and the splice donor sequence helps the RNA transcript to be spliced to downstream host exons. The location of the integrated transgene can be identified by RACE-PCR (see *Figure 20.1*). See Zambrowicz *et al.* (1998) for a detailed example.

the physiological basis of disease. They also offer a front-line testing system for studying the efficacy of novel treatments before conducting clinical trials on human subjects. Although many individual human disorders do not have a good animal model, animal models exist for some representatives of all the major human disease classes: genetically determined diseases, disease due to infectious agents, sporadic cancers and autoimmune disorders (see Leiter *et al.*, 1987; Darling and Abbott, 1992; Clarke, 1994; Bedell *et al.*, 1997). Some animal models of human disease originated spontaneously; others have been generated artificially by a variety of different routes (*Table 21.1*).

Until recently, the great majority of available animal disease models were ones which arose spontaneously or had been artificially induced by random mutagenesis using exposure to high doses of mutagenic chemicals or X-rays (next section). More recently gene targeting and transgenic technologies have provided direct ways of

obtaining animal models of disease, and targeted mutations in the mouse have been particularly valuable. Interestingly, it has become increasingly clear that disease phenotypes due to comparable mutations in human and mouse gene homologs often show considerable differences (Section 21.4.5).

21.4.1 Animal models of disease occurring spontaneously or as a result of exposure to chemical mutagens or radiation may be difficult to identify

Spontaneous animal disease models

Mutant human phenotypes, especially those associated with obvious disease symptoms, are subject to intense scrutiny: most individuals who suffer from a disorder seek medical advice. If they present with a previously

Table 21.1: Classes of animal models of disease

Origin	Type	Comments
(A) Spontaneous	Germline mutation → inherited disorder Somatic mutation → cancer	See *Table 20.4*.
(B) Artificial intervention or artificially generated	Selective breeding to obtain strains that are genetically susceptible to disease	
	Infect animal strain with relevant microbial pathogen	
	Manipulate environment to induce disease without causing mutation	e.g. injection of pristane, a synthetic adjuvant oil, into the dermis of rats has produced a model of arthritis.
	In vivo mutagenesis using a strong mutagen such as X-rays or powerful chemical mutagens such as ethyl nitrosurea (ENU)	Large chemical mutagenesis programs have been established to produce mutant mice and zebrafish (Section 21.4.1)
	Genetic modification of fertilized egg cells or cells from the early embryo and subsequent animal breeding (*transgenic* and *gene targeting* technologies)	Sections 21.2 and 21.3

undescribed phenotype their case may well be referred to experts who often will document the phenotype in the medical literature. Given the motivation of both affected individuals and their families, physicians and interested medical researchers, and the large population size for screening (current total global population is ~6 billion individuals), there is a remarkably effective screening process for mutant human phenotypes. In contrast, many animal disease phenotypes will go unrecorded. Only a small percentage of the animal population is in captivity, and recording of spontaneous mutant phenotypes is largely dependent on examination of animal colonies bred for research purposes and, to a lesser extent, live-stock and pet populations. Only mutants with obvious external anomalies are likely to be noticed.

Despite the difficulty in identifying spontaneous animal mutants, a number of animal phenotypes have been described as likely models of human diseases (see *Table 21.2* for some examples). In some cases, the animal mutant phenotype closely parallels the corresponding clinical phenotype, but in others there is considerable divergence because of species differences in biochemical and developmental pathways (see Erickson, 1989; Erickson, 1996; Wynshaw-Boris, 1996; see also Section 21.4.5). Additionally, phenotypic differences may result because of different classes of mutation at orthologous loci.

Table 21.2: Examples of spontaneous animal mutants

Animal mutant	Phenotypic features and molecular pathogenesis
NOD mouse	Diabetic, but without being obese. Mimics human insulin-dependent diabetes mellitus
mdx mouse	X-linked muscular dystrophy due to mutations in mouse dystrophin gene. The original *mdx* mutant has a nonsense mutation but phenotype is much milder than Duchenne muscular dystrophy (DMD)
Hemophiliac dog	Missense mutation in canine factor IX gene causes complete loss of function. Human homolog is hemophilia B
Watanabe heritable hyperlipidemic (WHHL) rabbit	Hyperlipidemic as a result of a deletion of four codons of the low density lipoprotein receptor gene (*LDLR*); human homolog is familial hypercholesterolemia
Atherosclerotic pigs	Marked hypercholesterolemia. Normal LDL receptor activity, but variant apolipoproteins, including apolipoprotein B
Splotch mouse	Abnormal pigmentation; phenotypic overlap with Waardenburg syndrome suggested that it was an animal model for this disease, and confirmed by identification of mutations in homologous human and murine *PAX* genes
NF damselfish	Extensive neurofibromas suggest that it could be a homolog of human neurofibromatosis type I

The potential of animals for modeling human disease

Primates *should* provide the best animal models of human disease because they are so closely related to us (see Section 14.6.2). Humans and great apes show extensive developmental, anatomical, biochemical and physiological similarities. Primates are expensive to breed, however, and the population sizes in captivity are very small. Public sensitivity also plays a part: while some are opposed, in principle, to all animal experimentation, those that feel it is justified in the interests of medical research are generally more comfortable with experimentation on small laboratory animals such as mice and rats. Most importantly, primates are not well suited to experimentation: they are comparatively long lived and less fecund than rodents, and so breeding experiments are more difficult to organize. Given that novel therapeutic approaches are rapidly being developed for a range of human disorders (see Chapter 22), the long delay required to perform test experiments on primates has prompted the study of alternative models.

Mice have been the most widely used animal models of human disease. They are small and can be maintained in breeding colonies comparatively cheaply. They have a short lifespan ($\sim$2–3 years), a short generation time ($\sim$3 months) and are prolific (an average female will produce four to eight litters of six to eight pups). Because they can be bred easily, complex breeding programs can be arranged to produce recombinant inbred strains and **congenic strains** (Taylor, 1989 and *Box 15.4*) and their short generation time and lifespan means that the effects of transmitting a pathogenic mutation through several generations can be monitored relatively easily. As a result, the genetics of the laboratory mouse have been studied extensively for decades, and the phenotypes of many mutants have been recorded (Lyon and Searle, 1989). Most such mutants have originated spontaneously within breeding colonies. A few have also been produced artificially, initially by X-ray/chemical mutagenesis, but increasingly by

gene targeting. Mapping of the mouse mutants is facilitated by interspecific back-cross mapping (Avner *et al.,* 1988, and *Box 15.4*) and by the availability of numerous polymorphic markers ($>$5000 dinucleotide repeat markers have been mapped). Because regions which show conservation of synteny between mouse and humans have been well documented (see, e.g., *Figure 14.23*), this information is useful in identifying genuinely homologous single gene disorders in mouse and man.

Rats are comparatively large and have been more amenable to physiological, pharmacological and behavioral studies, especially in cardiovascular and neuropsychiatric studies. They have a longer generation time (11 weeks), and breeding colonies are more expensive. Some classes of human disorders (e.g. hypertension, behavioral disorders, etc.) have no good mouse models and instead have relied on rat models. A dense map of genetic markers is rapidly being constructed.

Zebrafish genetics has developed only extremely recently: a genetic linkage map and efficient mutagenesis and saturation screening procedures were first published in 1994. However, mutant screening is facilitated in zebrafish because the embryos are transparent. Large-scale chemical mutagenesis screens have recently been conducted, identifying many mutants with abnormalities of developing internal organs such as the heart (Mullins, 1994). Inevitably, because of the considerable evolutionary divergence between humans and zebrafish, the relevance of zebrafish mutants to human disease will be expected to be confined to disorders affecting pathways that are highly conserved during evolution. In addition to certain developmental disorders, diseases affecting very highly conserved proteins or pathways may be modeled and zebrafish models have recently been reported for disorders of heme biosynthesis (Brownlie *et al.,* 1998; Wang *et al.,* 1998).

Random mutagenesis using chemicals and irradiation

Classical methods of producing animal mutants have involved controlled exposure to mutagenic chemicals, notably ethyl nitrosurea (ENU) and ethyl methylsulfonate (EMS) or to high doses of X-rays. Large numbers of *Drosophila* and mouse mutants have been obtained by this method and, very recently, efficient large-scale mutagenesis screens have been conducted on both mice and also zebrafish (which offers some advantages as an animal model; see *Box 21.2*). A major problem with chemically-induced and irradiation-induced mutations, however, is that they are generated essentially at random. In order to identify a mutant phenotype of interest, a laborious screen for mutants needs to be conducted by close examination of the phenotypes following mutagenesis. The mutant phenotypes which have been described in these studies, as well as for spontaneous mutants, show a clear bias towards phenotypes with obvious external abnormalities, simply because of the ease of identifying them. Nevertheless,

several important models of human disease have been created using such methods.

21.4.2 Mice have been widely used as animal models of human disease largely because specific mutations can be created at a predetermined locus

Spontaneous and artificially produced disease phenotypes have been described in a wide range of animal species with differing potentials for modeling human disease. In some cases, the species may be too evolutionarily remote from humans to provide useful disease models. For example, numerous *Drosophila* mutants have been generated and studies of some developmental mutants have enabled the identification of human genes that are important in development, but they cannot serve as useful models of human disease. Other species, such as the zebrafish, which are also evolutionarily remote from us, offer some advantages as model organisms and models have been reported for some human diseases (*Box 21.2*).

Mammals would be expected to provide better disease models but, for a variety of reasons, our closest relatives, the great apes, have not been very useful in providing disease models. Instead, other mammals, notably mice, have been used widely to model human disease (see *Box 21.2*).

In the case of animal disease models which are artificially induced by exposure to mutagenic chemicals or radiation, or which originate spontaneously, there is little or no artificial control over the resulting phenotype and frequently, the identification of an animal disease model is serendipitous. The great advantage of transgenic/gene-targeted mouse models of disease is that *specific disease models can be constructed to order*. Provided that the relevant gene clones are available, including mutant genes in some cases, mice can be generated with a desired alteration in a chosen target gene. All the major classes of disease, inherited disorders, cancers, infectious diseases and autoimmune disorders can be modeled in this way (*Table 21.3*; Smithies, 1993; Clarke, 1994; Bedell *et al.*, 1997). In most cases, the transgenic/gene targeting approaches have been used to model single gene disorders but, increasingly, attempts are being made to produce mouse models of complex genetic diseases, such as Alzheimer's disease, atherosclerosis and essential hypertension effects.

21.4.3 Single gene disorders resulting from loss of function and gain of function mutations can be conveniently modeled by gene targeting and by integration of mutant genes respectively

Modeling loss of function mutations by gene targeting in mice

Many disease phenotypes, including those of essentially all recessively inherited disorders and many dominantly inherited disorders, are thought to result from loss of gene function. The simplest way of modeling the disease for single gene disorders of this type is to make a knock-out mouse. The first step is to isolate the orthologous mouse gene and to use a segment of it to knock out the endogenous gene in mouse ES cells using gene targeting. Following injection of the genetically modified ES cells into the blastocyst of a foster mother, and continued development, founder mice are obtained with the targeted mutation in a sizeable proportion of their germ cells. These mice can be interbred and the offspring can be screened for the presence of the desired mutation, and for the presence of the wild-type allele using PCR assays of cells collected from tail bleeds.

The gene targeting event is intended to create a null allele (where there is complete absence of gene expression), but sometimes the result may be a 'leaky' mutation and the mutant allele retains some gene expression. This may explain why gene targeting has produced mouse models of a disease with divergent phenotypes. For example, the mouse model of cystic fibrosis described by Snouwaert *et al.* (1992) had a severe phenotype while that reported by Dorin *et al.* (1992) had a mild phenotype because of a 'leaky' mutation. Differences in phenotype may also occur because of *modifier genes* using different mouse strains (see Section 21.4.5).

Modeling gain of function mutations by insertion of a mutant gene

This general experimental design has been used frequently in conjunction with the pronuclear microinjection technique of gene transfer. The disease to be modeled must be one where the presence of an introduced DNA is itself sufficient to induce pathogenesis, and can include inherited gain of function mutations, oncogenes, etc. To model such disorders, it is necessary to clone a mutant

Table 21.3: Examples of transgenic or gene-targeted mouse models of human disease

Human disease or abnormal phenotype	Gene	Method of constructing model
Cystic fibrosis	*CFTR*	Insertional inactivation by gene targeting
β-Thalassemia	*HBB* (β-globin)	Insertional inactivation by gene targeting
Hypercholesterolemia and atherosclerosis	Apolipoprotein genes, e.g. *APOE*	Insertional inactivation by gene targeting
Gaucher's disease		Insertional inactivation by gene targeting
Fragile-X syndrome	*FMR1*	Insertional inactivation by gene targeting
Gerstmann–Sträussler–Scheinker (GSS) protein gene with missense mutation syndrome	Prion protein gene (*PRNP*)	Integration of mutant mouse prion gene
Spinocerebellar ataxia type 1 (SCA1)	*SCA1* (ataxin)	Integration of mutant human ataxin gene with expanded triplet repeat
Alzheimer's disease	*APP* (β-amyloid precursor protein)	Integration of mutant full-length *APP* cDNA under control of a platelet-derived growth factor promoter

gene or, if necessary, design one by *in vitro* mutagenesis. The mutant gene is then simply inserted as a transgene, e.g. by microinjection into fertilized oocytes. Because there is no requirement for the introduced mutant gene to integrate at a specific location, human mutant genes will suffice although, in some cases, mouse mutant genes have been used. The two examples below illustrate this approach.

- An early example was intended to assess whether a leucine substitution found at codon 102 in a prion protein gene in a patient with Gerstmann–Sträussler–Scheinker (GSS) syndrome was pathogenic. An analogous mutation was artificially designed in a cloned mouse prion protein gene and the mutant gene was then injected into fertilized oocytes to produce transgenic mice. The mice went on to develop spontaneous neurodegeneration, reminiscent of that found in the human syndrome (Hsiao *et al.*, 1990). A variety of other experiments using prion protein transgenes have been very helpful in understanding prions (Gabizon and Taraboulos, 1997).

- Expanded triplet repeats causing neurodegenerative disorders (see *Box 16.7*) constitute another type of gain of function mutation. Spinocerebellar ataxia type 1 (SCA1) is a dominantly inherited disorder which results from unstable expansion of a CAG triplet repeat in the ataxin gene. It is characterized by degeneration of cerebellar Purkinje cells, spinocerebellar tracts and some brainstem neurons. Transgenic mice were produced by introduction of one of two transgenes driven by a Purkinje cell-specific promoter: the normal human ataxin gene (*SCA1*), and a mutant ataxin gene containing an expanded CAG repeat. Both types of transgene were stable in parent to offspring transmissions, but only those with the expanded allele developed ataxia and Purkinje cell degeneration, confirming the gain of function hypothesis (Burright *et al.*, 1995).

21.4.4 Considerable effort is currently being devoted to constructing mouse models of cancers and other complex genetic disorders

Modeling human cancers
(See Ghebranious and Donehower, 1998; Macleod and Jacks, 1999.)

- *Gain of function.* Disease due to inappropriate inactivation of a proto-oncogene can be modeled by constructing a transgenic mouse: the appropriate oncogene is introduced into the mouse genome by simple transgene integration.

- *Loss of function.* Disease due to inactivation of tumor suppressor genes can be modeled by constructing knock-out mice through gene targeting. For example, several models have been generated by inactivating the mouse homologs of the *TP53* and *RB1* genes but the phenotypes show only broad similarity to the

homologous human phenotypes, respectively Li–Fraumeni syndrome and retinoblastoma (see also Section 21.4.5).

Modeling chromosomal disorders
Existing mouse models for human chromosomal disorders are sparse. In some cases this is due to insufficient conservation of synteny between the two species. Taking the example of Down syndrome (trisomy 21), human chromosome 21 shares a large region of genetic homology with mouse chromosome 16 (see *Figure 14.23*), but trisomy 16 (Ts16) mice are not good models because they die *in utero*. The Ts16 mouse could never be expected to be a good model of human trisomy 21: it is not trisomic for all human chromosome 21 genes (the genes in the distal 2–3 Mb of human chromosome 21 have orthologs on mouse chromosomes 17 and 10) and is trisomic for some genes which have human orthologs on chromosomes other than chromosome 21.

In order to produce a better Down syndrome model in the mouse, attention has focused on the Down syndrome *critical region* at 21q21.3–q22.2 (deduced from observing the phenotypes of rare Down syndrome patients with partial trisomy 21). Within this region, the human *mini-brain* gene at 21q22.2 may be an important contributory locus to the associated learning defects: transgenic mice in which a 180 kb YAC containing the 100 kb human *minibrain* gene but apparently no other gene, develop learning defects (Smith *et al.*, 1997). A segmental trisomy 16 mouse, Ts65Dn, obtained by standard methods of irradiating mice was shown by Reeves *et al.* (1995) to have learning and behavior deficits. Other more recently produced models are discussed by Kola and Hertzog (1998).

Future efforts in modeling chromosomal disorders can be expected to take advantage of gene targeting using Cre–*lox*P. As described in Section 21.3.2 this system offers tremendous potential for genome engineering and can be used to engineer chromosome translocations at defined positions on preselected chromosomes. YAC transgenics can also be expected to be important in investigating over-expression of genes in other chromosomal disorders and in disorders resulting from aberrant gene dosage for large regions, such as Charcot–Marie–Tooth type 1A, which is due to overexpression as a result of a 1.5-Mb duplication in the *PNP22* gene region (see *Figure 16.7*).

Modeling complex diseases
Increasingly, the focus in human genetics is moving towards understanding the pathogenesis of complex genetic diseases such as atherosclerosis, essential hypertension, diabetes, etc. Such disorders have a complex etiology with multiple genetic and environmental components. Some valuable animal models have been produced for some of these disorders but gene targeting approaches are expected in the future to provide additional badly needed models (Smithies and Maeda, 1995). As long as suitably promising genes can be identified as being involved in the pathogenesis, then breeding

experiments can be used to bring different combinations of disease genes together, and the effect of different genetic backgrounds in different strains of mice and of different environmental factors can be assessed. This approach may not be so daunting as it sounds because, increasingly, many complex disease phenotypes are considered to be due mostly to the combination of only a very few major susceptibility genes. For example, a digenic model of spina bifida occulta was generated serendipitously in offspring obtained by crossing a mouse heterozygous for the *Patch* mutation (*pdgfrb*; platelet-derived growth factor receptor) with a mouse homozygous for the *undulated* mutation (*pax-1*) (see Helwig *et al.*, 1995).

21.4.5 Mouse models of human disease may be difficult to construct because of a variety of human/mouse differences

It is not uncommon for spontaneous or artificially generated mouse models of disease to show phenotypes that are considerably different from the homologous human disorders. For example, gene targeting to inactivate several mouse tumor suppressor genes has often produced disappointing mouse models, as in the case of *TP53* and *RB1* (retinoblastoma) knock-outs. There may be problems in achieving the desired type of mutation. For example, in gene targeting intended null mutations may be offset in some cases by exon skipping or some other form of 'leaky' transcription, or the expression of a transgene may be affected by various parameters causing an unexpected phenotype. Setting aside these possibilities, there are several areas where differences between mice and humans could be expected to result in divergent disease phenotypes for mutations in orthologous genes (Erickson, 1989; Erickson, 1996; Wynshaw-Boris, 1996):

■ *Biochemical pathways.* Although biochemical pathways in mammals are generally well conserved, some differences are known between the pathways of humans and mice. The human retina appears to depend heavily on the accurate function of the *Rb* gene product, but other vertebrate retinas do not. As a result, spontaneous retinoblastoma mouse mutants have not been described, and retinoblastoma is not a feature of *Rb1* knock-out mice. Another example may be provided by ganglioside degradation pathways and while deficiency of the hexosaminidase gene *HEXA* results in the severe Tay–Sachs disorder, inactivation of the mouse homolog *Hexa* results in abnormal accumulation of ganglioside in neurons but without motor deficits or learning deficits (Wynshaw-Boris, 1996).

■ *Developmental pathways.* The differences in human and mouse developmental pathways are not well understood but are expected to be significant for some organ systems, such as in brain development.

■ *Absolute time.* Because of the huge difference in the average lifespans of mice and humans, certain human disorders in which the disease is of late onset may possibly be difficult to model in mice.

■ *Genetic background: the importance of modifier genes.* Most human populations are outbred. Laboratory strains of mice, however, are very inbred. Often a particular phenotype can vary considerably in different strains of mice because of differences in their alleles at other loci (**modifier genes**), which can interact with the locus of interest.

A useful example of the importance of genetic background is the *Min* (*multiple intestinal neoplasia*) mouse which was generated by ENU mutagenesis and results from mutations in the mouse *Apc* gene. Mutations in the orthologous human gene, *APC*, cause adenomatous polyposis coli and related colon cancers and the *Min* mouse has been regarded as a good model for such disorders. The phenotype of the *Min* mouse is, however, dramatically modified by the genetic background. For example the number of colonic polyps in mice carrying APC^{Min} is strikingly dependent on the strain of mouse. Similar phenotypic variability is found in human families where different members of the same family may have strikingly different tumor phenotypes although they possess identical mutations in the *APC* gene. Some of the variability could be due to environmental factors, but the involvement of modifier genes had been strongly suspected. The *Min* mouse provides a well-defined genetic system for mapping and identifying modifier genes (Dietrich *et al.*, 1993; MacPhee *et al.*, 1995).

21.5 Manipulating animals by somatic cell nuclear transfer

21.5.1 Principles and practice of animal cloning

The term clones indicates genetic identity and so can describe genetically identical molecules (DNA clones), genetically identical cells or genetically identical organisms. *Animal clones* occur naturally as a result of *sexual reproduction*. For example, genetically identical twins are clones who happened to have received exactly the same set of genetic instructions from two donor individuals, a mother and a father. A form of animal cloning can also occur as a result of artificial manipulation to bring about a type of *asexual reproduction*. The genetic manipulation in this case uses nuclear transfer technology: a nucleus is removed from a donor cell then transplanted into an oocyte whose own nucleus has previously been removed. The resulting 'renucleated' oocyte can give rise to an individual who will carry the nuclear genome of only one donor individual, unlike genetically identical twins. The individual providing the donor nucleus and the individual that develops from the 'renucleated' oocyte are usually described as 'clones', but it should be noted that they share only the same nuclear DNA; they do not share the same mitochondrial DNA, unlike genetically identical twins.

Nuclear transfer technology was first employed in embryo cloning, in which the donor cell is derived from an early embryo, and has been long established in the case of amphibia. However, it was only comparatively recently when McGrath and Solter reported nuclear transplantation in the mouse embryo and paved the way for modern mammalian cloning (see Fulka *et al.*, 1998). Subsequently, nuclear transplantation was conducted successfully in the eggs of domestic animals, including sheep and cows.

Unlike embryo cloning the prospect of cloning *adults* had seemed remote. During their development from (ultimately) the fertilized oocyte, adult somatic cells undergo an extensive series of cell division and differentiation steps. Until recently it was thought that these processes were accompanied by irreversible modifications to the genome. Even in frogs transplantation of an adult cell nucleus had never been reported to give rise to an adult animal; instead, the renucleated embryos underwent early development but thereafter failed to develop to term. A variety of early experiments in mice were also unsuccessful before the landmark study of Wilmut *et al.* (1997) reported successful cloning of an adult sheep. For the first time, an adult nucleus had been reprogrammed to become totipotent once more, just like the genetic material in the fertilized oocyte from which the donor cell had ultimately developed.

In the Wilmut *et al.* study, the donor cells were derived from a cell line established from adult mammary gland cells and were fused to an enucleated metaphase II-arrested oocyte (*Figure 21.11A*). The donor cells were deprived of serum before use, forcing them to exit the cell cycle into a quiescent stage, G_0 (Stewart, 1997). A certain degree of gene silencing is a characteristic feature of the nuclei of G_0 cells. As egg cells are normally fertilized by transcriptionally inactive sperm cells, G_0 cells may be more amenable to full genetic reprogramming. Another consideration is the degree of chromosome condensation and of access to chromatin 'remodeling factors' such as transcription factors in the oocyte. In any event, the cloning was extremely inefficient: out of a total of 434 oocytes that were submitted to the procedure, only 29 developed to the transferable stage and of these only one developed to term, being born as the now famous Dolly (*Figure 21.11B*). Subsequent doubts about the exact origin of the donor cell and whether Dolly really was an *adult* clone (as opposed to a contaminant fetal cell) have been allayed by genetic testing of Dolly and the adult mammary gland donor cells. Importantly, successful animal cloning has also been achieved by other groups with comparatively high success in cloning of adult mice (Wakayama *et al.*, 1998) and cows (Kato *et al.*, 1998).

21.5.2 The successful cloning of an adult animal has major implications for research, medicine and society

The report by Wilmut *et al.* (1997) has generated enormous attention, in the scientific and general press, both because of its novelty and the significance for future work. In particular, the possible extrapolation to cloning of humans has generated a great deal of controversy.

Basic research

Successful cloning of adult animals has forced us to accept that genome modifications once considered irreversible can be reversed and that the genomes of adult cells can be reprogammed by factors in the oocyte to make them totipotent once again. Research investigations into the control of gene expression during development and basic processes of somatic differentiation, somatic mutation, aging and repair processes will undoubtedly benefit from animal cloning, especially cloning of mice. Other more recent studies are now forcing us to reconsider the *potency* of other cells. For example, adult mouse neural stem cells transplanted into an irradiated host animal have very recently been shown to develop into a variety of blood cell types (myeloid, lymphoid and early hematopoietic cells) and so the developmental potential of stem cells is not restricted to the differentiated elements of the tissue in which they reside (Bjornson *et al.*, 1999).

Cloning of livestock and transgenic animals

The successful cloning of adult sheep and cows is clearly attractive to people who wish to perpetuate prized livestock, racehorses, pets and endangered species. In addition, transgenic animals can be cloned. The traditional route for making a transgenic animal is by pronuclear microinjection (Section 21.2.1). But this may be rather inefficient. Transgenic sheep and other livestock have been produced to serve as **bioreactors**, sources of medically valuable products such as human insulin (see Section 22.1.2). However, in the case of transgenic sheep, for example, only 2–4% of the founder animals born by implanting eggs which have been microinjected with a transgene turn out to be transgenic. Producing founder transgenic animals by nuclear transfer should be more efficient and will allow more sophisticated genetic modifications. An early success was achieved by Dr Wilmut's group who used fetal sheep cells containing a factor IX transgene as donor cells to generate transgenic sheep and this has been followed by cloning of transgenic cattle (Pennisi, 1998).

Human cloning

The most contentious issue in cloning animals is, of course, the potential extrapolation to cloning humans (Shapiro, 1997; Johnson, 1998). Clearly, the technology is still poorly developed and the comparatively high incidence of spontaneous abortions, perinatal losses and anomalous births observed in animal cloning would make the prospect of human cloning unappealing at present. In many countries, existing legislation would also preclude attempts at human cloning. For example, in the UK it is a criminal offence to experiment with human embryos without a licence, which will not be granted under any circumstances for experiments with embryos more than 14 days old.

(A)

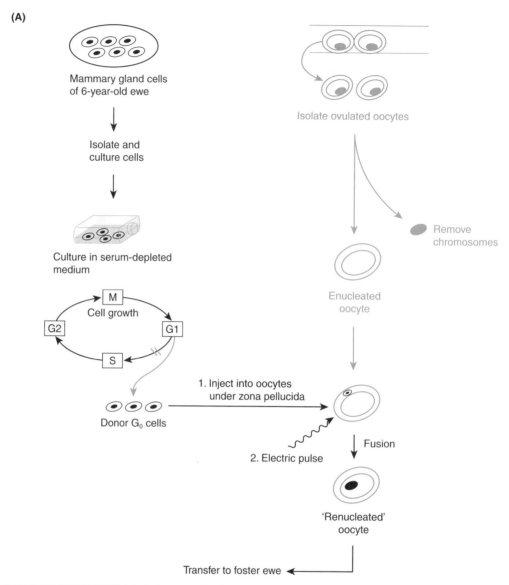

Mammary gland cells
of 6-year-old ewe

Isolate and
culture cells

Culture in serum-depleted
medium

M

Cell growth

G2

G1

S

Donor G_0 cells

1. Inject into oocytes
under zona pellucida

2. Electric pulse

Isolate ovulated oocytes

Remove
chromosomes

Enucleated
oocyte

Fusion

'Renucleated'
oocyte

Transfer to foster ewe

Figure 21.11: A sheep called Dolly was the outcome of
the first successful attempt at animal cloning.

(**A**) Experimental strategy used by Wilmut *et al.* (1997).
During culture in serum-depleted medium the normal cell
cycle is interrupted and cells move from the G_1 phase of cell
growth to a quiescent stage, G_0 where there is no cell
division. This quiescent state was thought to facilitate
successful nuclear transfer (see text). *Note* that in the
Wilmut *et al.* (1997) study the nuclear transfer occurred by
cell fusion, but in other cases different strategies have been
used. For example, in the successful cloning of adult mice
reported by Wakayama *et al.* (1998), a very fine needle was
used to take up the donor cell nucleus with minimal
contamination by donor cell cytoplasm. The donor cell was
quickly, but very gently, microinjected into the enucleated
oocyte. (**B**) Dolly with her first born, Bonnie. Original photo
kindly provided by the Roslin Institute.

Technological improvements in animal cloning will undoubtedly occur, however, and if the procedure were eventually to become both efficient and comparatively risk-free, there could be considerable pressure to apply nuclear replacement technology to human cells. Some applications need not involve human reproductive cloning. For example, nuclear replacement could be used to avoid transmission of inherited diseases derived from the mitochondria. Here, an unfertilized egg taken from an individual with mitochondrial disease could act as the donor with the nucleus being transferred into an enucleated egg from a donor containing normal mitochondria. The reconstructed egg could then be fertilized *in vitro*.

The use of nuclear transfer technology for human reproductive cloning is, inevitably, more contentious. For some infertile couples or women, for example, it could provide a welcome method of having children. However, the expectation that could be placed on such a child could be damaging to that individual because the parent(s) and later the child may be especially conscious of genetic identity between individuals *whose ages are quite different*. Unlike identical twins whose development proceeds in parallel, for example, a cloned child could be only too aware of how he/she might develop in later life by observing a parent who was essentially genetically identical. Against this, many would argue that a person's character and capability is not determined exclusively be his/her genetic endowment; the environment also has a powerful role to play.

Further reading

Galli-Taliadoris LA, Sedgwick JD, Wood SA, Korner H (1995) Gene knock-out technology: a methodological overview for the interested novice. *J. Immunol. Methods*, **181**, 1–15.

Kolata G (1997) *Clone: the road to Dolly and the path ahead*. William Morrow and Company, New York.

Kuhn R, Schwenk F (1997) Advances in gene targeting methods. *Curr. Opin. Immunol.*, **9**, 183–188.

Popko B (ed.) (1998) *Mouse Models of Human Genetic Neurological Disease*. Plenum Press, New York.

Shastry BS (1998) Gene disruption in mice: models of development and disease. *Mol. Cell. Biochem.*, **181**, 163–179.

Sikorski R, Peters R (1997) Transgenics on the internet. *Nature Biotechnol.*, **15**, 289.

TBASE: a transgenic/targeted mutation database at http: //www.gdb.org/Dan/tbase. html

References

Avner P, Amar L, Dandolo L, Guenet JL (1988) Genetic-analysis of the mouse using interspecific crosses. *Trends Genet.*, **4**, 18–23.

Barinaga M (1994) Knockout Mice: Round Two. *Science*, 265, 26–28.

Bedell MA, Jenkins NA, Copeland NG (1997) Mouse models of human disease. Part II. Recent progress and future directions. *Genes Dev.*, **11**, 11–43.

Bjornson CRR, Rietze RL, Reynolds BA, Magli MC, Vescovi AL (1999) Turning brain into blood: a hematopoietic fate adopted by adult neural stem cells in vivo. *Science*, 283, 534–537.

Brownlie A et al. (1998) Positional cloning of the zebrafish *sauternes* gene: a model for congenital sideroblastic anaemia. *Nature Genet.*, **20**, 244–250.

Burright EN, Clark HB, Servadio A et al. (1995) SCA1 transgenic mice – a model for neurodegeneration caused by CAG trinucleotide expansion. *Cell*, **82**, 937–948.

Camper SA, Saunders TL, Kendall SK et al. (1995) Implementing transgenic and embryonic stem-cell technology to study gene-expression, cell–cell interactions and gene-function. *Biol. Reprod.*, **52**, 246-257.

Capecchi M (1989) The new mouse genetics: altering the genome by gene targeting. *Trends Genet.*, **5**, 70–76.

Chambers CA (1994) TKO'ed: Lox, stock and barrel. *BioEssays*, **16**, 865–868.

Clarke AR (1994) Murine genetic models of human disease. *Curr. Opin. Genet. Dev.*, **4**, 453–460.

Darling SM, Abbott CM (1992) Mouse models of human single gene disorders. 1. Nontransgenic mice. *BioEssays*, **14**, 359–366.

Dietrich WF, Lander ES, Smith JS et al. (1993) Genetic identification of mom-1, a major modifier locus affecting min-induced intestinal neoplasia in the mouse. *Cell*, **75**, 631–639.

Dorin JR, Dickinson P, Alton EW et al. (1992) Cystic-fibrosis in the mouse by targeted insertional mutagenesis. *Nature*, **359**, 211–215.

Erickson RP (1989) Why isn't a mouse more like a man? *Trends Genet.*, **5**, 1–3.

Erickson RP (1996) Mouse models of human genetic disease: which mouse is more like a man? *Bioessays*, **18**, 993–998.

Evans MJ, Kaufman MH (1981) Establishment in culture of pluripotential cells from mouse embryos. *Nature*, **292**, 154–156.

Evans MJ, Carlton MBL, Russ AP (1997) Gene trapping and functional genomics. *Trends Genet.*, **13**, 370–374.

Fulka Jr J, First NL, Loi P, Moor RM (1998) Cloning by somatic cell nuclear transfer. *BioEssays,* **20,** 847–851.

Gabizon R, Taraboulos A (1997) Of mice and (mad) cows – transgenic mice help to understand prions. *Trends Genet.,* **13,** 264–269.

Ghebranious N, Donehower LA (1998) Mouse models in tumor suppression. *Oncogene,* **17,** 3385–3400.

Gordon JW (1992) Production of transgenic mice. *Methods. Enzymol.,* **225,** 747–771.

Grimes B, Cooke H (1998) Engineering mammalian chromosomes. *Hum. Mol. Genet.,* **7,** 1635–1640.

Grosveld F, van Assendelft GB, Greaves DR, Kolias G (1987) Position-independent, high-level expression of the human beta-globin gene in transgenic mice. *Cell,* **51,** 975–985.

Gu H, Marth JD, Orban PC, Mossmann H, Rajewsky K (1994) Deletion of a DNA-polymerase-beta gene segment in T-cells using cell-type-specific gene targeting. *Science,* **265,** 103–107.

Hanahan D (1989) Transgenic mice as probes into complex systems. *Science,* **246,** 1265–1275.

Hanks M, Wurst W, Anson-Cartwright L, Auerbach AB, Joyner AL (1995) Rescue of the *En-1* mutant phenotype by replacement of *En-1* with *En-2. Science,* **269,** 679–682.

Harrington JJ, Van Bokkelen G, Mays RW, Gustashaw K, Willard HF (1997) Formation of de novo centromeres and construction of first-generation human artificial microchromosomes. *Nature Genet.,***15,** 345–355.

Helwig U, Imai K, Schmahl W, Thomas BE, Varnum DS, Nadeau JH, Balling R (1995) Interaction between undulated and patch leads to an extreme form of spina-bifida in double-mutant mice. *Nature Genet.,* **11,** 60–63.

Herault Y, Rassoulzadegan M, Cuzin F, Duboule D (1998) Engineering chromosomes in mice through targeted meiotic recombination (TAMERE). *Nature Genet.,* **20,** 381–384.

Hsiao KK, Scott M, Foster D, Groth DF, Dearmond SJ, Prusiner SB (1990) Spontaneous neurodegeneration in transgenic mice with mutant prion protein. *Science,* **250,** 1587–1590.

Ikeno M, Okazaki T, Grimes B, Saitoh K, Cooke H, Masumoto H (1998) Generation of human artificial chromosomes from a YAC containing alphoid DNA with CENP-B boxed and human telomeres. *Nature Biotechnol.,* **16,** 431–439.

Jakobovits A, Moore AL, Green LL *et al.* (1993) Germ-line transmission and expression of a human-derived yeast artificial chromosome. *Nature,* **362,** 255–258.

Johnson M (1998) Cloning humans? *Bioessays,* **19,** 737–739.

Kato Y, Tani T, Sotomaru Y, Kurokawa K, Kato J, Doguchi H, Yasue H, Tsunoda Y (1998) Eight calves cloned from somatic cells of a single adult. *Science,* **282,** 2095–2098.

Kilby NJ, Snaith MR, Murray JAH (1993) Site-specific recombinases – tools for genome engineering. *Trends Genet.,* **9,** 413–421.

Kola I, Hertzog PJ (1998) Down syndrome and mouse models. *Curr. Opin. Genet. Dev.,* **8,** 316–321.

Koopman P, Gubbay J, Vivian N, Goodfellow P, Lovell-Badge R (1991) Male development of chromosomally female mice transgenic for Sry. *Nature,* **351,** 117–121.

Lamb BT, Gearhart JD (1995) Yac transgenics and the study of genetics and human disease. *Curr. Opin. Genet. Dev.,* **5,** 342–348.

Leiter EH, Beamer WG, Shultz LD, Barker JE, Lane PW (1987) Mouse models of genetic diseases. *Birth Defects,* **23,** 221–257.

Lobe CG, Nagy A (1998) Conditional genome alteration in mice. *Bioessays,* **20,** 200–208.

Lyon MF, Searle AG (1989) *Genetic Variants and Strains of the Laboratory Mouse,* 2nd edn. Oxford University Press, Oxford.

Macleod KF, Jacks T (1999) Insights into cancer from transgenic mouse models. *J. Pathol.,* **187,** 43–60.

MacPhee M, Chepenik KP, Liddell RA, Nelson KK, Siracusa LD, Buchberg AM (1995) The secretory phospholipase-a2 gene is a candidate for the mom1 locus, a major modifier of apc(min)-induced intestinal neoplasia. *Cell,* **81,** 957–966.

Manley NR, Capecchi MR (1997) Hox group 3 paralogous genes act synergistically in the formation of somitic and neural crest-derived structures. *Dev. Biol.,* **192,** 274–288.

Martin GR (1981) Isolation of a pluripotent cell line from early mouse embryos cultured in medium conditioned by teratocarcinoma stem cells. *Proc. Natl Acad. Sci. USA,* **78,** 7634–7638.

Melton DW (1994) Gene targeting in the mouse. *BioEssays,* **16,** 633–638.

Mendez MJ, Green LL, Corvalan JRF *et al.* (1997) Functional transplant of megabase human immunoglobulin loci recapitulates human antibody response in mice. *Nature Genet.,* **15,** 146–156.

Mullins MC, Hammerschmidt M, Haffter P, Nusslein-Volhard C (1994) Large-scale mutagenesis in the zebrafish: in search of genes controlling development in a vertebrate. *Curr. Biol.,* **4,** 189–202.

Pennisi E (1998) After Dolly, a pharming frenzy. *Science,* **279,** 646–648.

Ramirez-Solis R, Liu P, Bradley A (1995) Chromosome engineering in mice. *Nature,* **378,** 720–724.

Reeves RH, Irving NG, Moran TH *et al.* (1995) A mouse model for Down syndrome exhibits learning and behaviour deficits. *Nature Genet.,* **11,** 177–183.

Schedl A, Ross A, Lee M, Engelkamp D, Rashbass P, van Heyningen V, Hastie ND (1996) Influence of PAX6 gene dosage on development: overexpression causes severe eye abnormalities. *Cell,* **86,** 71–82.

Shamblott MJ, Axelman J, Wang S *et al.* (1998) Derivation of pluripotent stem cells from cultured human primordial germ cells. *Proc. Natl Acad. Sci. USA,* **95,** 13726–13731.

Shapiro HT (1997) Ethical and policy issues of human cloning. *Science*, **277**, 195–196.

Shockett P, Difillppantonio M, Hellman N, Schatz DG (1995) A modified tetracycline-regulated system provides autoregulatory, inducible gene-expression in cultured-cells and transgenic mice. *Proc. Natl Acad. Sci. USA*, **92**, 6522–6526.

Smith AJH, De Sousa MA, Kwabi-Addo B, Heppell-Parton A, Impey H, Rabbitts P (1995) A site-directed chromosomal translocation induced in embryonic stem-cells by cre-*lox*p recombination. *Nature Genet.*, **9**, 376–385.

Smith DJ, Stevens ME, Suclanagunta SP *et al.* (1997) Functional screening of 2 Mb of human chromosome 21q22.2 in transgenic mice implicates minibrain in learning defects associated with Down syndrome. *Nature Genet.*, **16**, 28–36.

Smithies O (1993) Animal models of human genetic diseases. *Trends Genet.*, **9**, 112–116.

Smithies O, Maeda N (1995) Gene targeting approaches to complex genetic diseases: Atherosclerosis and essential hypertension. *Proc. Natl Acad. Sci. USA*, **92**, 5266–5272.

Snouwaert JN, Brigman KK, Latour AM *et al.* (1992) An animal-model for cystic-fibrosis made by gene targeting. *Science*, **257**, 1083–1088.

Solter D, Gearhart J (1999) Putting stem cells to work. *Science*, **283**, 1468–1470.

Stewart C (1997) Nuclear transplantation. An udder way of making lambs. *Nature*, **385**, 769–771.

Taylor BA (1989) In: *Genetic Variants and Strains of the Laboratory Mouse,* 2nd edn (eds MF Lyon, AG Searle), pp. 773–796. Oxford University Press, Oxford.

Theuring F (1995) In: *Functional Analysis of the Human Genome* (eds F Farzaneh, DN Cooper), pp. 185–205. BIOS Scientific Publishers, Oxford.

Thomson JA, Itskovitz-Elder J, Shapiro SS *et al.* (1998) Embryonic stem cell lines derived from human blastocysts. *Science*, **282**, 1145–1147.

Tomizuka K, Yoshida H, Uejima H *et al.* (1997) Functional expression and germline transmission of a human chromosome fragment in chimaeric mice. *Nature Genet.*, **16**, 133–143.

Wakayama T, Perry AC, Zuccotti M, Johnson KR, Yanagimachi R (1998) Full-term development of mice from enucleated oocytes injected with cumulus cell nuclei. *Nature*, **394**, 369–374.

Wang H, Long Q, Marty SD, Sassa S, Lin S (1998) A zebrafish model for hepatoerythropoietic porphyria. *Nature Genet.*, **20**, 239–243.

Wilmut I, Schnieke AE, McWhir J, Kind AJ, Campbell KHS (1997) Viable offspring derived from fetal and adult mammalian cells. *Nature*, **385**, 810–813.

Wynshaw-Boris A (1996) Model mice and human disease. *Nature Genet.*, **13**, 259–260.

Zambrowicz BP, Friedrich GA, Buxton EC, Lilleberg SL, Person C, Sands AT (1998) Disruption and sequence identification of 2,000 genes in mouse embryonic stem cells. *Nature*, **392**, 608–611.

Gene therapy and other molecular genetic-based therapeutic approaches

22

22.1 Principles of molecular genetic-based therapies and treatment with recombinant proteins or genetically engineered vaccines

22.1.1 Principles of molecular genetic-based approaches to treating disease

Once a human disease gene has been characterized, molecular genetic tools can be used to dissect gene function and explore the biological processes involved in the normal and pathogenic states. The resulting information can be used to design novel therapies using conventional drug-based approaches. In addition, molecular genetic technologies have recently provided a variety of novel therapeutic approaches that can be categorized into two broad groups, depending on whether the therapeutic agent is a gene product/vaccine or genetic material.

Recombinant proteins and genetically engineered vaccines

Here the therapy is to deliver proteins or vaccines which have been produced by genetic engineeering instead of traditional methods. Methods involve:

- *expression cloning of normal gene products* – cloned genes are expressed in microorganisms or transgenic livestock in order to make large amounts of a medically valuable gene product;
- *production of genetically engineered antibodies* – antibody genes are manipulated so as to make novel antibodies, including partially or fully *humanized antibodies,* for use as therapeutic agents;
- *production of genetically engineered vaccines* – includes novel cancer vaccines and vaccines against infectious agents.

Gene therapy

The term *gene therapy* describes any procedure intended to treat or alleviate disease by genetically modifying the cells of a patient. It encompasses many different strategies and the material transferred into patient cells may be genes, gene segments or oligonucleotides. The genetic material may be transferred directly into cells within a patient (*in vivo* gene therapy), or cells may be removed from the patient and the genetic material inserted into them *in vitro*, prior to transplanting the modified cells back into the patient (*ex vivo* gene therapy). Because the molecular basis of diseases can vary widely, some gene therapy strategies are particularly suited to certain types of disorder, and some to others. Major disease classes include:

- *infectious diseases* (as a result of infection by a virus or bacterial pathogen);
- *cancers* (inappropriate continuation of cell division and cell proliferation as a result of activation of an oncogene or inactivation of a tumor suppressor gene or an apoptosis gene – see Chapter 18);
- *inherited disorders* (genetic deficiency of an individual gene product or genetically determined inappropriate expression of a gene);
- *immune system disorders* (includes allergies, inflammations and also autoimmune diseases, in which body cells are inappropriately destroyed by immune system cells).

A major motivation for gene therapy has been the need to develop novel treaments for diseases for which there is no effective conventional treatment. Gene therapy has the potential to treat all of the above classes of disorder. Depending on the basis of pathogenesis, different gene therapy strategies can be considered (*Box 22.1* and *Figure 22.1*). One, rather arbitrary, subdivision of gene therapy approaches is as follows:

- *Classical gene therapy.* The rationale of this type of approach is to deliver genes to appropriate target cells with the aim of obtaining optimal expression of the introduced genes. Once inside the desired cells in the patient, the expressed genes are intended to do one of the following:

General gene therapy strategies
(see also *Figure 22.1*)

Gene augmentation therapy (GAT)

For diseases caused by loss of function of a gene, introducing extra copies of the normal gene may increase the amount of normal gene product to a level where the normal phenotype is restored (see *Figure 22.1*). As a result GAT is targeted at clinical disorders where the *pathogenesis is reversible*. It also helps to have no precise requirement for expression levels of the introduced gene and a clinical response at low expression levels. GAT has been particularly applied to autosomal recessive disorders where even modest expression levels of an introduced gene may make a substantial difference. Dominantly inherited disorders are much less amenable to treatment: gain-of-function mutations are not treatable by this approach and, even if there is a loss-of-function mutation, high expression efficiency of the introduced gene is required: individuals with 50% of normal gene product are normally affected, and so the challenge is to increase the amount of gene product towards normal levels.

Targeted killing of specific cells

This general approach is popular in cancer gene therapies. Genes are directed to the target cells and then expressed so as to cause cell killing.

■ *Direct cell killing* is possible if the inserted genes are expressed to produce a lethal toxin (suicide genes), or a gene encoding a prodrug is inserted, conferring susceptibility to killing by a subsequently administered drug. Alternatively, selectively lytic viruses can be used.

■ *Indirect cell killing* uses immunostimulatory genes to provoke or enhance an immune response against the target cell.

Targeted mutation correction

If an inherited mutation produces a dominant-negative effect, gene augmentation is unlikely to help. Instead the resident mutation must be corrected. Because of practical difficulties, this approach has yet to be applied but, in principle, it can be done at different levels: at the gene level (e.g. by **gene targeting** methods based on homologous recombination); or at the RNA transcript level (e.g. by using particular types of therapeutic ribozymes or therapeutic RNA editing).

Targeted inhibition of gene expression

If disease cells display a novel gene product or inappropriate expression of a gene (as in the case of many cancers, infectious diseases, etc.), a variety of different systems can be used specifically to block the expression of a single gene at the DNA, RNA or protein levels. Allele-specific inhibition of expression may be possible in some cases, permitting therapies for some disorders resulting from dominant negative effects.

(a) produce a product that the patient lacks;

(b) kill diseased cells directly, e.g. by producing a toxin which kills the cells;

(c) activate cells of the immune system so as to aid killing of diseased cells.

■ *Nonclassical gene therapy.* The idea here is to inhibit the expression of genes associated with the pathogenesis, or to correct a genetic defect and so restore normal gene expression.

Current gene therapy is exclusively *somatic* gene therapy, the introduction of genes into somatic cells of an affected individual. The prospect of human germline gene therapy raises a number of ethical concerns, and is currently not sanctioned (see Section 22.6.1).

22.1.2 Recombinant pharamaceuticals can be produced by expression cloning in microorganisms or transgenic livestock

Advantages of obtaining medically valuable reagents by expression cloning

Once a human gene has been cloned, large amounts of the purified product can be obtained by *expression cloning* (Section 4.4.2). Often the desired gene is expressed in bacterial cells, which have the advantage that they can be cultured easily in large volumes. Using this approach, large amounts of recombinant pharmaceuticals can be generated. Expression cloning may provide the only product-based therapeutic route in those cases where

---

Figure 22.1 *(Opposite):* Five approaches to gene therapy.

Of the five illustrated approaches, four have been used in clinical trials. Gene augmentation therapy by simple addition of functional alleles has been used to treat several inherited disorders caused by genetic deficiency of a gene product. Artificial cell killing and immune system-assisted cell killing have been popular in the treatment of cancers. The former has involved transfer to cells of genes encoding toxic compounds (**suicide genes**), or **prodrugs** (reagents which confer sensitivity to subsequent treatment with a drug). Targeted inhibition of gene expression is particularly suitable for treating infectious diseases and some cancers. Targeted gene mutation correction, the repair of a genetic defect to restore a functional allele, is the exception: technical difficulties have meant that it is not sufficiently reliable to warrant clinical trials. The example shows correction of a mutation in a mutant gene by **homologous recombination**, but mutation correction may also be possible at the RNA level (see *Figure 22.10*). ODN, oligodeoxynucleotide; TFO, triplex-forming oligonucleotide.

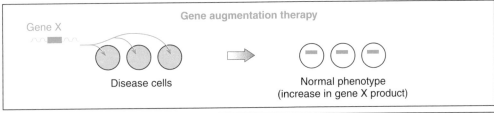

Gene augmentation therapy

Gene X

Disease cells

Normal phenotype
(increase in gene X product)

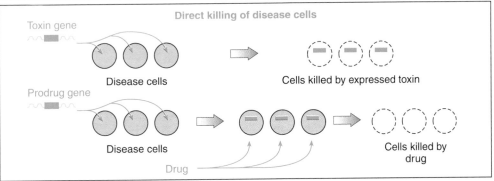

Direct killing of disease cells

Toxin gene

Disease cells

Cells killed by expressed toxin

Prodrug gene

Disease cells

Drug

Cells killed by drug

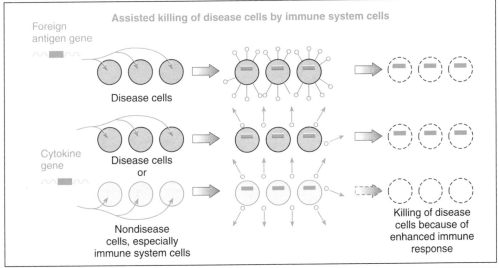

Assisted killing of disease cells by immune system cells

Foreign antigen gene

Disease cells

Cytokine gene

Disease cells

or

Nondisease cells, especially immune system cells

Killing of disease cells because of enhanced immune response

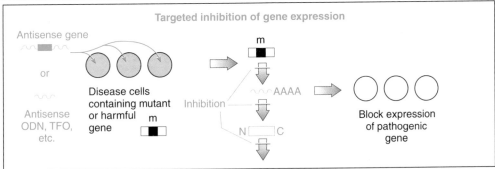

Targeted inhibition of gene expression

Antisense gene

or

Antisense ODN, TFO, etc.

Disease cells containing mutant or harmful gene

m

Inhibition

AAAA

N ☐ C

Block expression of pathogenic gene

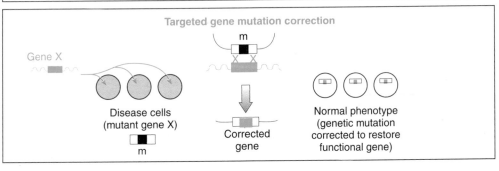

Targeted gene mutation correction

Gene X

Disease cells
(mutant gene X)

m

X   X

Corrected gene

Normal phenotype
(genetic mutation corrected to restore functional gene)

Box 22.2

Treatment using conventional animal or human products can be hazardous.

Animal products have been prepared by standard biochemical purification techniques and used to treat human patients. In many cases the treatment is based on the animal product having a function identical to that of a product which the patient lacks, e.g. insulin prepared from cows or pigs has traditionally been used to treat diabetes. Such procedures have been justified by easy access to large amounts of the appropriate animal tissues but are potentially hazardous. For example, if the animal product differs from the human one, say in terms of amino acid differences, glycosylation differences, etc. it can produce unwanted side-effects in highly immunoreactive individuals. In addition, there is the potential for introducing unwanted pathogens that might have co-purified with the product.

  Human products obtained by conventional routes can be hazardous too. Hemophilia A is a genetic deficiency of the blood clotting factor VIII which has been treated by supplementation with human factor VIII. In the past, the factor VIII was purified from the serum of unscreened human donors and many hemophiliacs consequently developed AIDS. Another example concerns treating growth hormone deficiency patients with human growth hormone. Some patients treated in this way developed Creutzfeldt–Jakob disease, a rare neurological disorder which is a human counterpart of 'mad cow disease') because the hormone had been extracted from large numbers of pooled cadaver pituitaries which had not been suitably screened.

Table 22.1: Examples of pharmaceutical products obtained by expression cloning

Product	For treatment of
Blood clotting factor VIII	Hemophilia A
Blood clotting factor IX	Hemophilia B
Erythropoietin	Anemia
Insulin	Diabetes
Growth hormone	Growth hormone deficiency
Tissue plasminogen activator	Thrombotic disorders
Hepatitis B vaccine	Hepatitis B
α-Interferon	Hairy cell leukemia; chronic hepatitis
β-Interferon	Multiple sclerosis
γ-Interferon	Infections in patients with chronic granulomatous disease
Interleukin-2	Renal cell carcinoma
Granulocyte colony-stimulating factor (G-CSF)	Neutropenia following chemotherapy
DNase (deoxyribonuclease)	Cystic fibrosis

biochemical purification of the product from a human or animal source is difficult or impossible. Safety risks are minimal, unlike those associated with products obtained from conventional human or animal sources (*Box 22.2*).

  Recombinant human insulin was first marketed in 1982 and, subsequently, a number of other cloned human gene products of medical interest have been produced commercially (see *Table 22.1*). Treatment with the products of cloned genes is not free from risks, however. For example, patients who completely lack a normal product may mount a vigorous immune response to the administered pharmaceutical product as in the case of some patients with severe hemophilia A who have been treated with recombinant factor VIII.

Transgenic livestock as a source of medically valuable products

Expression cloning often involves the use of microorganisms, but this approach may not always be suitable. For example, expression of a human gene in a bacterial cell can give a product that shows differences from the normal human product: the polypeptide may have the same sequence of amino acids but patterns of glycosylation may be different. This may mean that the gene product is not particularly stable in a human environment, or it may provoke an immune response, or its biological function

may be less effective than desired. In addition, the cost of producing purified recombinant pharmaceuticals may be rather high.

  The disadvantages of microbial expression cloning has prompted consideration of alternatives. In particular, increasing attention has been paid to constructing **transgenic** livestock (see Lubon, 1998), where the post-translational processing systems are more similar to analogous human systems. For example, a cloned human gene can be fused to a sheep gene specifying a milk protein then inserted into the genome of the sheep germline. The resultant transgenic sheep can secrete large quantities of the fusion protein in its milk. Transgenic pigs have also been designed to express human proteins in their milk, as in the case of human factor VIII (Paleyanda *et al.*, 1997).

22.1.3 Genetically engineered antibodies and vaccines have great therapeutic promise

Antibody engineering

Antibodies are natural therapeutic agents which are produced by B lymphocytes. In each B-cell precursor, a cell-specific rearrangement of antibody gene segments occurs so that *individual* B cells produce different antibodies (Section 8.6.2). Additional diversity is provided by other mechanisms, including frequent somatic mutation events. As a result, each one of us has a population of B cells which *collectively* ensures a huge repertoire of different antibodies as a defense system against a diverse array of foreign antigens. The antibody may be thought of as an *adaptor molecule*: it contains binding sites for foreign antigen at the variable (V) end, and at the constant (C) end it has binding sites for effector molecules. Binding of an antibody may by itself be sufficient to neutralize some toxins and viruses, but it is more common for the antibody to trigger the complement system and cell-mediated killing.

Artificially produced therapeutic antibodies are designed to be *monospecific* (they recognize a single type of antigenic site) and can recognize specific disease-associated antigens, leading to killing of the disease cells (see Berkower, 1996). Notable targets for such therapy are cancers (especially lymphomas and leukemias); infectious disease (using antibodies raised against antigens of the relevant pathogen); and autoimmune disorders (where antibodies recognize inappropriately expressed host cell antigens). A favorite way of producing immortal monospecific antibodies is to fuse individual antibody-producing B lymphocytes from an immunized mouse or rat with cells derived from an immortal mouse B-lymphocyte tumor. From the fusion products, hybridomas, a heterogenous mixture of hybrid cells which have the ability to make a particular antibody and to multiply indefinitely in culture, are selected. The hybridomas are propagated as individual clones, each of which can provide a permanent and stable source of a single type of **monoclonal antibody (mAb)**.

Until recently, the therapeutic antibody approach was not straightforward. Although rodent monoclonal antibodies (mAbs) can be created against human pathogens and cells, they normally have limited use in the clinic. This is because rodent mAbs have a short half-life in human serum and they can elicit an unwanted immune response in patients (producing human antirodent antibodies). In addition, only some of the different classes can trigger human effector functions. The generation of human mAbs would avoid these problems but has been difficult to achieve using standard hybridoma technology. Once immunoglobulin genes had been cloned, however, the possibility of designing artificial combinations of immunoglobulin gene segments arose (antibody engineering). Because different exons encode different domains of an antibody molecule, domain swapping could be done easily at the DNA level by artificially shuffling exons between different antibody genes (see Winter and Harris, 1993).

Chimeric and humanized antibodies

One immediate goal of antibody engineering was the production of chimeric and humanized antibodies, which are rodent–human recombinant antibodies (Winter and Harris, 1993; *Figure 22.2*). Humanizing of rodent antibodies could allow access to a large pool of well-characterized rodent mAbs for therapy, including those with specificities against human antigens that are difficult to elicit from a human immune response. Early versions contained the variable domains of a rodent antibody attached to the constant domains of a human antibody, a so-called chimeric (V/C) antibody. The immunogenicity of the rodent mAb is reduced, while allowing the effector functions to be selected for the therapeutic application. A further stage of humanizing antibodies is possible. The essential antigen-binding site is a subset of the variable region characterized by hypervariable sequences, the *complementarity-*

determining regions (CDRs). Accordingly, second generation humanized antibodies were CDR-grafted antibodies: the hypervariable antigen-binding loops of the rodent antibody were built into a human antibody, creating a humanized antibody. Chimeric V/C antibodies and CDR-engrafted antibodies have been constructed against a wide range of microbial pathogens and against human cell surface markers, including tumor cell antigens. Their clinical potential is considerable (see *Table 22.2*).

Fully human antibodies

Two approaches have been taken towards the construction of fully human antibodies (Vaughan *et al.*, 1998):

- *Phage display technology*. This technology bypasses hybridoma technology, and even immunization. Instead, antibodies are made *in vitro* by mimicking the selection strategies of the immune system (see Section 20.4.3).
- *Transgenic mice*. One powerful strategy involves transferring yeast artificial chromosomes containing large segments of the human heavy and light chain immunoglobulin loci into mouse embryonic stem cells, and subsequent production of transgenic mice. For example, Mendez *et al.* (1997) report the construction of transgenic mice containing 1.02 Mb human Ig heavy chain and 0.8 Mb human Ig light chain loci (*transloci*; see *Figure 21.3*). Such mice contain a very considerable portion of the human V gene segment repertoire and the human immunoglobulin transloci are able to undergo the normal program of rearrangement and hypermutation to generate human antibodies.

Table 22.2: Examples of the clinical potential of humanized antibodies

Target	Clinical potential
CDw52	Lymphomas, systemic vasculitis, rheumatoid arthritis
CD3	Organ transplantation
CD4	Organ transplants, rheumatoid arthritis, Crohn's disease
IL-2 receptor	Leukemias and lymphomas, organ transplants, graft-versus-host disease
TNF-α	Septic shock
HIV	AIDS
RSV	Respiratory syncytial virus infection
HSV	Neonatal, ocular and genital herpes infection
Lewis-Y	Cancer
p185^{HER2}	Cancer
PLAP	Cancer
CEA	Cancer

TNF, tumor necrosis factor; HIV, human immunodeficiency virus; RSV, Rous sarcoma virus; HSV, herpes simplex virus; p185^{HER2}, human epidermal growth factor receptor 2; PLAP, placental alkaline phosphatase; CEA, carcinoembryonic antigen. Derived from Winter and Harris (1993).

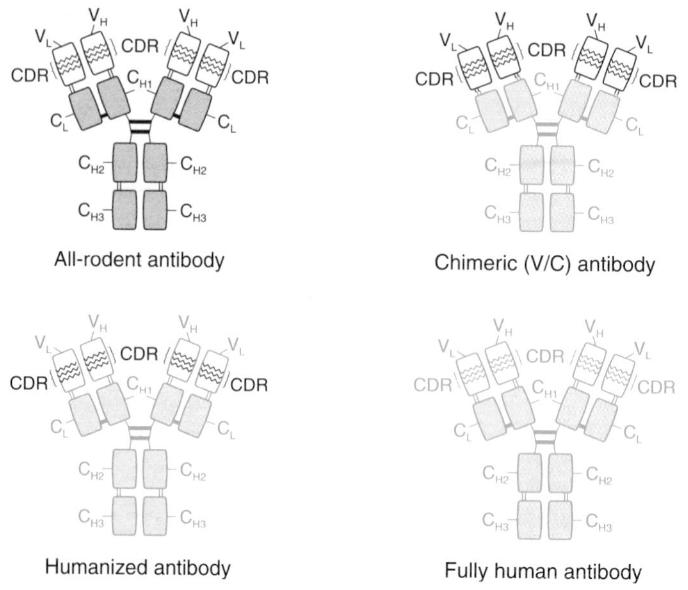

Figure 22.2: Genetically engineered antibodies.

Antibodies consist of two light (L) chains, each containing a variable domain (V_L) and a constant domain (C_L), plus two heavy chains, containing a variable domain (V_H) and three constant domains (C_{H1}, C_{H2} and C_{H3}). The variable domains contain hypervariable regions involved in antigen recognition, known as *complementarity determining regions (CDRs)*. Genetic engineering allows swapping of coding DNA segments to construct recombinant antibodies which are part human and part mouse, e.g. by replacing rodent constant domains with human constant domains (*chimeric antibodies*). If the variable domains are also made human except for the CDRs, *humanized antibodies* are produced. *Fully human antibodies* can be produced by phage display or transgenic technology. See Vaughan *et al.* (1998).

Genetically engineered vaccines

Recombinant DNA technology is also being applied to the construction of novel vaccines. Several different strategies are being used (Liu, 1998; Pardoll, 1998):

- *Nucleic acid vaccines.* These are typically bacterial plasmids containing genes encoding pathogen or tumor antigens which are delivered in saline solution by direct intramuscular injection. They normally carry a strong viral promoter which drives the expression of the gene of interest directly in the injected host. An alternative method of gene transfer uses a 'gene gun' to deliver gold beads onto which the DNA has been precipitated (see also Section 22.2.3). In the last few years, abundant evidence has been obtained that DNA vaccines can be effective (see Donnelly *et al.*, 1997).
- *Genetic modification of antigen.* This can be achieved, for example, by fusion with a cytokine gene to increase antigenicity.
- *Genetic modification of viruses.* Viral vector systems have been used to deliver the genes of heterologous pathogens. DNA vectors based on alphaviruses (which have single-stranded RNA genomes) have been the focus of much recent attention.
- *Genetic modification of microorganisms.* One way is to disable an organism, e.g. by removing genes required for pathogenesis or survival. This is a genetic method of *attenuation* so that a live vaccine

can be used without undue risk. Another approach involves inserting an exogenous gene that will be expressed in bacteria or parasites.

Note that some gene therapy approaches, such as adoptive immunotherapy, are effectively forms of genetically engineered vaccination (Section 22.5.2).

22.2 The technology of classical gene therapy

22.2.1 Genes can be inserted into the cells of patients by direct and indirect routes, and the inserted genes can integrate into the chromosomes or remain extrachromosomal

An essential component of classical gene therapy is that cloned genes have to be introduced and expressed in the cells of a patient in order to overcome the disease. Practically, this usually involves targeting the cells of diseased tissues. However, deliberate targeting of unaffected cells may be preferred in some approaches:

- *Immune system-mediated cell killing.* In many gene therapies the target cells are healthy immune system cells, and the idea is to enhance immune responses to cancer cells or infectious agents (Section 22.5.1).

■ *Delivery of gene products from cells at a remote location.* Genes may be targeted initially to one type of tissue while the gene products may be delivered to a remote location. For example, the myonuclei in muscle fibers have the advantage of being very long lived. Genetically engineered myoblasts therefore have the potential to ameliorate some nonmuscle diseases through long-term expression of exogenous genes which encode a product secreted into the blood stream (see for example, Jiao *et al.*, 1993).

Two major general approaches are used in the transfer of genes for gene therapy: transfer of genes into patient cells outside of the body (*ex vivo*) or inside the body (*in vivo*).

Ex vivo *gene transfer*

This initially involves transfer of cloned genes into cells grown in culture. Those cells which have been transformed successfully are selected, expanded by cell culture *in vitro*, then introduced into the patient. To avoid immune system rejection of the introduced cells, autologous cells are normally used: the cells are collected initially from the patient to be treated and grown in culture before being reintroduced into the same individual (see *Figure 22.3*). Clearly, this approach is only applicable to tissues that can be removed from the body, altered genetically and returned to the patient where they will engraft and survive for a long period of time (e.g. cells of the hematopoietic system and skin cells). Note that this type of gene therapy involves transplantation of autologous genetically modified cells and so can be considered a modified form of cell therapy (see *Box 22.3*).

In vivo *gene transfer*

Here the cloned genes are transferred directly into the tissues of the patient. This may be the only possible option in tissues where individual cells cannot be cultured *in vitro* in sufficient numbers (e.g. brain cells) and/or where cultured cells cannot be re-implanted efficiently in patients. Liposomes and certain viral vectors are increasingly being employed for this purpose. In the latter case, it is often convenient to implant vector-producing cells (VPCs), cultured cells which have been infected by the recombinant retrovirus *in vitro*: in this case the VPCs

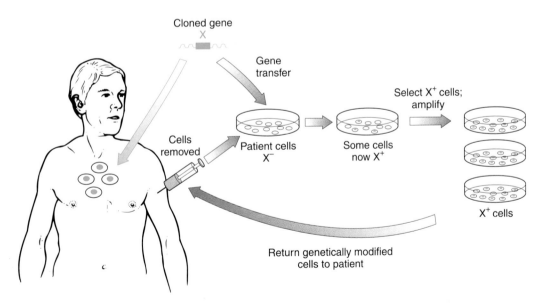

Figure 22.3: *In vivo* and *ex vivo* gene therapy.

In vivo gene therapy (blue arrow) entails the genetic modification of the cells of a patient *in situ*. *Ex vivo* gene therapy (black arrows) means that cells are modified outside the body before being implanted into the patient. The figure shows the usual situation where autologous cells are used, i.e. cells are removed from the patient, cultured *in vitro*, before being returned to the patient. Occasionally, however, the cells that are implanted do not belong to the patient but are allogeneic (from another human source) in which case HLA matching is routinely required to avoid immune rejection.

transfer the gene to surrounding disease cells. As there is no way of selecting and amplifying cells that have taken up and expressed the foreign gene, the success of this approach is crucially dependent on the general efficiency of gene transfer and expression.

Principles of gene transfer

Classical gene therapies normally require efficient transfer of cloned genes into disease cells so that the introduced genes are expressed at suitably high levels. In principle, there are numerous different physicochemical and biological methods that can be used to transfer exogenous genes into human cells. The size of DNA fragments that can be transferred is in most cases comparatively very limited, and so often the transferred gene is not a conventional gene. Instead, an artificial minigene may be used: a cDNA sequence containing the complete coding DNA sequence is engineered to be flanked by appropriate regulatory sequences for ensuring high level expression, such as a powerful viral promoter. Following gene transfer, the inserted genes may integrate into the chromosomes of the cell, or remain as extrachromosomal genetic elements (**episomes**).

Genes integrated into chromosomes

The advantage of integrating into a chromosome is that the gene can be perpetuated by chromosomal replication following cell division (*Figure 22.4*). As progeny cells also contain the introduced genes, long-term stable expression may be obtained. As a result, gene therapy using this approach may provide the possibility of a cure for some disorders. For example, in tissues composed of actively dividing cells, the key cells to target are **stem cells** (a minority population of undifferentiated precursor cells which gives rise to the mature differentiated cells of the tissue; see Section 2.2.2). This is so because stem cells not only give rise to the mature tissue cells, but during this procedure they also renew themselves. As a result, they are an immortal population of cells from which all other cells of the tissue are derived. High efficiency gene transfer into stem cells, and subsequent stable high level expression of a suitable introduced gene, can therefore provide the possibility of curing a genetic disorder.

Chromosomal integration has its disadvantages, however, because normally the insertion occurs almost randomly. As a result, the location of the inserted genes can

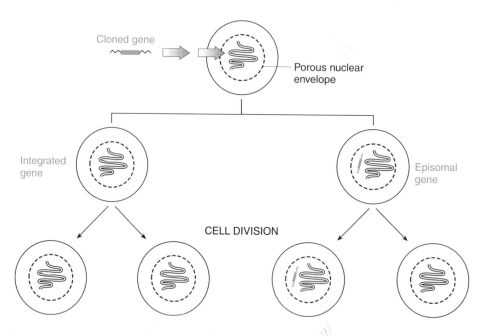

Figure 22.4: Exogenous genes that integrate into chromosomes can be stably transmitted to all daughter cells, unlike episomal (extrachromosomal) genes.

The figure illustrates two possible fates of genes that have been transferred into nucleated cells. If the cells are actively dividing, any genes which integrate stably into chromosomal DNA can be replicated under the control of the parent chromosome (during the S phase of the cell cycle). Following each cell division, an integrated gene will be stably inherited by both daughter cells. As a result, all cells that descend from a single cell in which stable integration took place, will contain the integrated gene. Gene therapy involving chromosomal integration of exogenous genes offers the possibility of continued stable expression of the inserted gene and a permanent cure, but carries certain risks, notably the possibility that one of the integration events may result in cancer (see text). By contrast, episomal genes which do not integrate but replicate extrachromosomally (under the control of a vector origin of replication) may not segregate to all daughter cells during subsequent mitoses. As a result, this type of approach has been particularly applied in gene therapies where the target tissue consists of nondividing cells (see text).

vary enormously from cell to cell. In many cases, inserted genes may not be expressed, e.g. they may have integrated into a highly condensed heterochromatic region. In some cases the integration event can result in death of the host cell (the insertion may occur within a crucially important gene, thereby inactivating it). Such an event has consequences only for the single cell in which the integration occurred. A greater concern is the possibility of cancer: an integration event in one of the many cells that are targeted could disturb the normal expression patterns of genes that control cell division or cell proliferation. For example, the integration can cause activation of an oncogene or it could inactivate a tumor suppressor gene or a gene involved in apoptosis (programmed cell death). *Ex vivo* gene therapy at least offers the opportunity for selecting cells where integration has been successful, amplifying them in cell culture and then checking the phenotypes for any obvious evidence of neoplastic transformation, prior to transferring the cells back into the patient.

Nonintegrated genes

Some gene transfer systems are designed to insert genes into cells where they remain as extrachromosomal elements and may be expressed at high levels (see *Table 22.3*). If the cells are actively dividing, the introduced gene may not segregate equally to daughter cells and so long-term expression may be a problem. As a result, the possibility of a cure for a genetic disorder may be remote: instead, repeated treatments involving gene transfer will be necessary. In some cases, however, there may be no need for stable long-term expression. For example, cancer gene therapies often involve transfer and expression of genes into cancer cells with a view to killing the cells. Once the malignancy has been eliminated, the therapeutic gene may no longer be needed.

22.2.2 Most gene therapy protocols have used mammalian viral vectors because of their high efficiency of gene transfer

The method chosen for gene transfer depends on the nature of the target tissue and whether transfer is to cultured cells *ex vivo* or to the cells of the patient *in vivo*. No one gene transfer system is ideal; each has its limitations and advantages. However, mammalian virus vectors have been the preferred vehicle for gene transfer because of their high efficiency of *transduction* into human cells (Anderson, 1998).

Oncoretroviral vectors

Retroviruses are RNA viruses which possess a reverse transcriptase function, enabling them to synthesize a complementary DNA form. Following infection (transduction), retroviruses deliver a nucleoprotein complex (preintegration complex) into the cytoplasm of infected cells. This complex reverse transcribes the viral RNA genome and then integrates the resulting DNA copy into a single site in the host cell chromosomes (*Figure 18.2*). Retroviruses are very efficient at transferring DNA into cells, and the integrated DNA can be stably propagated, offering the possibility of a permanent cure for a disease. Because of these properties, retroviruses were considered the most promising vehicles for gene delivery and currently about 60% of all approved clinical protocols utilize retroviral vectors.

The retrovirus vectors that have traditionally been used in gene therapy are derived from simple retroviruses (oncoretroviruses), notably murine leukemia virus. Unlike adenoviruses they can only be produced at relatively low titers and so it is not possible to get a large number of vector particles to the desired cell type *in vivo*. Since all the viral genes are removed from the vector, the

Table 22.3: Properties of major methods of gene transfer used in gene therapy and their applications

Features	Oncoretroviral	Adenoviral	Adeno-associated	Lentiviral	Liposomes
Maximum insert size	7–7.5 kb	>30 kb	4.0 kb	7–7.5 kb	Unlimited
Chromosomal integration	Yes	No; episomal	Yes/No	Yes	Very low frequency
Duration of expression *in vivo*	Short	Short	Long	Long	Short
Stability	Good	Good	Good	Untested	Very good
Route of gene delivery	*Ex vivo*	*Ex vivo* and *in vivo*	*Ex vivo* and *in vivo*	*Ex vivo* and *in vivo*	*Ex vivo* and *in vivo*
Concentration (particles per ml)	$>10^8$	$>10^{11}$	$>10^{12}$	$>10^8$	Unlimited
Ease of preparation and scale up	Pilot scale up, up to 20–50 litres	Easy to scale up	Difficult to purify; difficult to scale up	Not known	Easy to scale up
Host immunological response	Few problems	Extensive	Not known	Few problems	None
Pre-existing host immunity	Unlikely	Yes	Yes	Unlikely, except maybe AIDS patients	No
Safety concerns	Possibility of insertional mutagenesis	Inflammatory response, toxicity	Inflammatory response, toxicity	Possibility of insertional mutagenesis	None

Reproduced from Verma and Somia (1997) *Nature*, **389**, p. 242, with permission from Macmilan Magazines.

viruses cannot replicate by themselves. They can accept inserts of up to 8 kb of exogenous DNA and require a variety of packaging systems to enclose the viral genome within viral particles (simple injection of retroviral vectors is usually inappropriate for *in vivo* gene therapy because they can generally be killed by human complement).

Oncoretroviruses can only transduce cells that divide shortly after infection: the preintegration complex is excluded from the nucleus and can only reach the host cell chromosomes when the nuclear membrane is fragmented during cell division. This therefore limits potential target cells. Only certain blood cells (but *not* stem cells) and the cells lining the gastrointestinal tract are continually in division; other cell types undergo division but not continually and some important cell targets never divide (e.g. mature neurons). The property of transducing only dividing cells can, however, be beneficial to gene therapy for cancers of tissues that normally have nondividing cells: the actively dividing cancer cells can be selectively infected and killed without major risk to the nondividing cells of the normal tissue (see Section 22.5.3).

Adenovirus vectors

Adenoviruses are DNA viruses that produce infections of the upper respiratory tract and have a natural tropism for respiratory epithelium, the cornea and the gastrointestinal tract. Adenovirus vectors have been the second most popular delivery system in gene therapy (with extensive applications in gene therapy for cystic fibrosis and certain types of cancer) and have several advantages as gene delivery vectors. They are human viruses which can be produced at very high titers in culture, and they are able to infect a large number of different human cell types including nondividing cells. Entry into cells occurs by

receptor-mediated endocytosis (*Figure 22.5;* see also below) and transduction efficiency is very high (often approaching 100% *in vitro*). They are large viruses and so have the potential for accepting large inserts.

They also have some major disadvantages. The inserted DNA does not integrate, and so expression of inserted genes can be sustained over short periods only. The first generation recombinant adenoviruses used in cystic fibrosis gene therapy trials showed that transgene expression declined after about 2 weeks and was negligible after only 4 weeks (see below). Because they can infect virtually all human cells, adenovirus vectors may conceivably pose a risk in some therapies that are designed to kill cancer cells without causing toxicity to normal surrounding cells. Most importantly, first generation adenovirus vectors can generate unwanted immune responses, causing chronic inflammation.

Many of these difficulties have been addressed in the construction of second generation adenovirus vectors. For example, all of the adenovirus genes have been deleted from some newer adenovirus vectors ('gutless vectors') which then require the assistance of a helper virus. Such a virus provides certain viral functions *in trans* (e.g. enzymes involved in viral DNA replication etc.) which are essential for productive infection (including viral DNA replication, viral assembly and infection of new cells) by certain natural viruses, such as AAV, or artificially disabled viruses.

The risk of immune response to these vectors is negligible. This is an important consideration given the need to administer treatment frequently (because of the inability of adenovirus to integrate into chromosomal DNA). They also have the advantage that they can accept much larger inserts (up to 35 kb). Unfortunately, however,

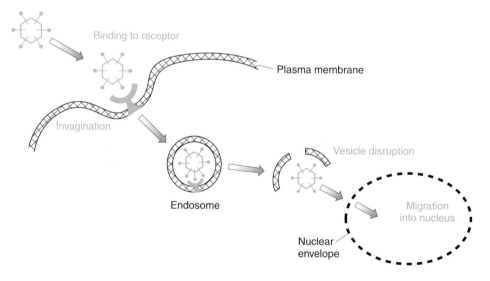

Figure 22.5: Adenoviruses enter cells by receptor-mediated endocytosis.

Binding of viral coat protein to a specific receptor on the plasma membrane of cells is followed by **endocytosis**, a process in which the plasma membrane invaginates and then pinches off to form an intracellular vesicle (**endosome**). Subsequent vesicle disruption by adenovirus proteins allows virions to escape and migrate towards the nucleus where viral DNA enters through pores in the nuclear envelope. Adapted from Curiel (1994) with permission from the New York Academy of Sciences.

deletion of adenoviral genes can also be counterproductive. Deletion of the E3 region removes the capacity to encode a protein that protects the virus from immune surveillance mechanisms in the host. In addition, fully disabled adenoviral vectors have much lower transduction efficiencies.

Adeno-associated virus vectors

Adeno-associated viruses (AAVs) are a group of small, single-stranded DNA viruses which cannot usually undergo productive infection without co-infection by a helper virus, such as an adenovirus or herpes simplex virus. In the absence of co-infection by a helper virus, unmodified human AAV integrates into chromosomal DNA, usually at a specific site on 19q13.3-qter. Subsequent superinfection with an adenovirus can activate the integrated virus DNA, resulting in progeny virions. AAV vectors can only accommodate inserts up to 4.5 kb, but they have the advantage of providing the possibility of long-term gene expression because they integrate into chromosomal DNA. They also provide a high degree of safety: because 96% of the parental AAV genome has been deleted, the AAV vectors lack any viral genes and recombinant AAV vectors only contain the gene of interest.

Herpes simplex virus vectors

HSV vectors are tropic for the central nervous system (CNS) and can establish lifelong latent infections in neurons. They have a comparatively large insert size capacity (>20 kb) but are nonintegrating and so long-term expression of transferred genes is not possible. Their major applications are expected to be in delivering genes into neurons for the treatment of neurological diseases, such as Parkinson's disease, and for treating CNS tumors.

Lentiviruses

The lentivirus family, which includes HIV (human immunodeficiency virus), are complex retroviruses that infect macrophages and lymphocytes. Unlike oncoretroviruses, lentiviruses are able to transduce nondividing cells. In the case of HIV, for example, the preintegration complex contains nuclear localization signals that permit its active transport through nuclear pores into the nucleus during interphase. Because of their ability to infect nondividing cells and to integrate into host cell chromosomes, considerable efforts are now being devoted to making lentivirus vectors for gene therapy (Naldini, 1998).

22.2.3 Concerns over the safety of recombinant viruses have prompted increasing interest in nonviral vector systems for gene therapy

Increasingly, concern has been expressed regarding the safety of viral vector systems. The recombinant viruses which are used for *ex vivo* gene therapy are designed to be disabled: typically some viral genes required for viral replication are deleted, and the therapeutic genes that are to be transferred are inserted in their place. The resulting replication-incompetent viruses are then intended to infect individual cells. In the case of retrovirus vectors, chromosomal integration is still possible but, like other replication-incompetent virus vectors, they should not be able to undergo a productive infection in which they replicate, assemble new virions and infect new cells. However, there is the remote possibility that the introduced viruses can recombine with endogenous retroviruses, resulting in recombinant progeny that can undergo productive infection. Additionally, adenoviruses are generally nonintegrating and the repeated injections that may be required may provoke severe inflammatory responses to the recombinant adenoviruses as has happened recently in a gene therapy trial for cystic fibrosis. Increasingly, therefore, attention has been focused toward studying alternative methods of gene transfer (Kay *et al.*, 1997).

Liposomes

Liposomes are spherical vesicles composed of synthetic lipid bilayers which mimic the structure of biological membranes. The DNA to be transferred is packaged *in vitro* with the liposomes and used directly for transferring the DNA to a suitable target tissue *in vivo* (*Figure 22.6*). The lipid coating allows the DNA to survive *in vivo*, bind to cells and be endocytosed into the cells. Cationic liposomes (where the positive charge on liposomes stabilize binding of negatively charged DNA), have become popular vehicles for gene transfer in *in vivo* gene therapy (see Huang and Li, 1997 for references). Unlike viral vectors, the DNA/lipid complexes are easy to prepare and there is no limit to the size of DNA that is transferred. However, the efficiency of gene transfer is low, and the introduced DNA is not designed to integrate into chromosomal DNA. As a result, expression of the inserted genes is transient.

Direct injection/particle bombardment

In some cases, DNA can be injected directly with a syringe and needle into a specific tissue, such as muscle. This approach has been considered, for example, in the case of DMD, where early studies investigated intramuscular injection of a dystrophin minigene into a mouse model, *mdx* (Acsadi *et al.*, 1991). An alternative direct injection approach uses particle bombardment ('gene gun') techniques: DNA is coated on to metal pellets and fired from a special gun into cells. Successful gene transfer into a number of different tissues has been obtained using this approach. Such direct injection techniques are simple and comparatively safe. However, there is poor efficiency of gene transfer, and a low level of stable integration of the injected DNA. The latter property is particularly disadvantageous in the case of proliferating cells, and would necessitate repeated injections. It may be less of a problem in tissues such as muscle which do not regularly proliferate, and in which the injected DNA may continue to be expressed for several months.

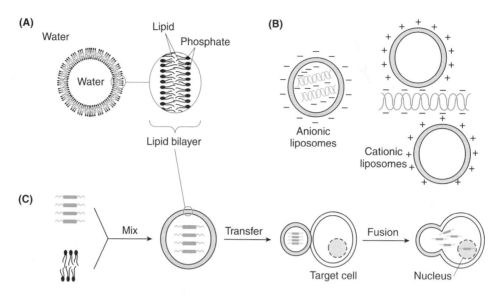

Figure 22.6: *In vivo* liposome gene delivery.

(**A**) and (**B**) Structure of liposomes. Liposomes are synthetic vesicles which can form spontaneously in aqueous solution following artificial mixing of lipid molecules. In some cases, a phospholipid bilayer is formed, with hydrophilic phosphate groups located on the external surfaces and hydrophobic lipids located internally (left). In other cases there is a multilamellar lipid envelope. Anionic liposomes have a negative surface charge and when the lipid constitutents are mixed with negatively charged DNA molecules (see panel C below), the DNA is internalized. Cationic liposomes have a surface positive charge and DNA molecules bind to the surface of liposomes. (**C**) Use of liposomes to transfer genes into cells. This figure illustrates the use of anionic liposomes to transfer internally located DNA into cells. The plasma membranes of cells are fluid structures whose principal components are phospholipids, and so mixing of cells and liposomes can result in occasional fusion between the lipid bilayer of the liposome and the plasma membrane. When this happens the cloned genes can be transferred into the cytoplasm of a cell, and can thence migrate to the nucleus by passive diffusion through the pores of the nuclear envelope. Note that, in practice, cationic liposomes have been more widely used for transferring DNA into cells.

Receptor-mediated endocytosis

The DNA is coupled to a targeting molecule that can bind to a specific cell surface receptor, inducing endocytosis and transfer of the DNA into cells. Coupling is normally achieved by covalently linking polylysine to the receptor molecule and then arranging for (reversible) binding of the negatively charged DNA to the positively charged polylysine component. For example, hepatocytes are distinguished by the presence on the cell surface of asialoglycoprotein receptors which clear asialoglycoproteins from the serum. Coupling of DNA to an asialoglycoprotein via a polycation such as polylysine can target the transfer of exogenous DNA into liver cells. The complexes can be infused into the liver either via the biliary tract or vascular bed, whereupon they are taken up by hepatocytes.

A more general approach utilizes the transferrin receptor which is expressed in many cell types, but is relatively enriched in proliferating cells and hemopoietic cells (*Figure 22.7*). Gene transfer efficiency may be high but the method is not designed to allow integration of the transferred genes. A further problem has been that the protein–DNA complexes are not particularly stable in serum. Additionally, the DNA conjugates may be entrapped in endosomes and degraded in lysosomes, unless

previously co-transferred with, or physically linked to, an adenovirus molecule (see *Figure 22.5*).

22.3 Therapeutics based on targeted inhibition of gene expression and mutation correction *in vivo*

22.3.1 Principles and applications of therapy based on targeted inhibition of gene expression *in vivo*

One way of treating certain human disorders is to selectively inhibit the expression of a predetermined gene *in vivo*. In principle, this general approach is particularly suited to treating cancers and infectious diseases, and some immunological disorders. In these cases, the basis of the therapy is to knock out the expression of a specific gene that allows disease cells to flourish, without interfering with normal cell function. For example, attention could be focused on selectively inhibiting the expression of a particular viral gene that is necessary for viral replication, or an inappropriately activated oncogene.

In addition to the above, targeted inhibition of gene

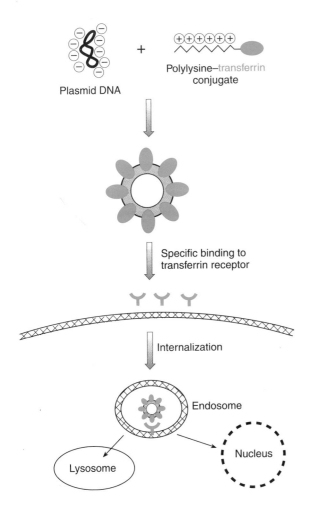

Figure 22.7: Gene transfer via the receptor-mediated endocytosis pathway.

The negatively charged plasmid DNA can bind reversibly to the positively charged polylysine attached to the transferrin molecule. During this process, the DNA is condensed into a compact circular toroid with the transferrin molecules located externally and free to bind to cell surface transferrin receptors. Following initial endosome formation, a portion of the endocytosed conjugates can migrate to the nucleus, although a very significant fraction is alternatively transferred to lysosomes where the DNA is degraded. The efficiency of transfer can be increased by the further refinement of coupling an inactivated adenovirus to the DNA–transferrin complex: following endocytosis and transport to lysosomes, the added adenovirus causes vesicle disruption (see *Figure 22.5*), allowing the DNA to avoid degradation and to survive in the cytoplasm. Adapted from Curiel (1994) with permission from The New York Academy of Sciences.

expression may offer the possibility of treating certain dominantly inherited disorders. If a dominantly inherited disorder is the result of a loss-of-function mutation, treatment may be difficult using conventional gene augmentation therapy: given that heterozygotes with 50% of normal gene product can be severely affected, very

efficient expression of the introduced genes would be required for the gene therapy to be successful. However, dominantly inherited disorders which arise because of a **gain-of-function mutation** may not be amenable to simple addition of normal genes. Instead, it may be possible, in some cases, to inhibit specifically the expression of the mutant gene, while maintaining expression of the normal allele. Such allele-specific inhibition of gene expression would be facilitated if the pathogenic mutation results in a significant sequence difference between the alleles.

The expression of a selected gene might be inhibited by a variety of different strategies. One possible type of approach involves specific *in vivo* mutagenesis of that gene, altering it to a form that is no longer functional. Gene targeting by homologous recombination offers the possibility of site-specific mutagenesis to inactivate a gene (Section 21.2.3). However, this technique has only very recently become feasible with normal diploid somatic cells and is still very inefficient. Instead, methods of blocking the expression of a gene without mutating it have been preferred. In principle, this can be accomplished at different levels: at the DNA level (by blocking transcription); at the RNA level (by blocking post-transcriptional processing, mRNA transport or engagement of the mRNA with the ribosomes); or at the protein level (by blocking post-translational processing, protein export or other steps that are crucial to the function of the protein).

Various techniques for selectively inhibiting expression of a specific gene have been devised, and include examples where expression is inhibited at all three major levels (see *Figure 22.8*):

- *Targeted inhibition of expression at the DNA level.* Under certain conditions, DNA can form triple-stranded structures, as occurs naturally in the case of a portion of the mitochondrial genome (Section 7.1.1). The rationale of triple helix therapeutics is to design a gene-specific oligonucleotide that will have a high chance of base-pairing with a defined double-stranded DNA sequence of a specific target gene in order to inhibit transcription of that gene (Vasquez and Wilson, 1998). Binding of the single-stranded oligonucleotide to a pre-existing double helix occurs by Hoogsteen hydrogen bonds and certain bases are preferred. The most stable of such bonds are formed by a G binding to the G of a GC base pair and a T binding to the A of an AT base pair.
- *Targeted inhibition of expression at the RNA level.* Antisense therapeutics involves binding of gene-specific oligonucleotides or polynucleotides to the RNA; in some cases, the binding agent may be a specifically engineered ribozyme, a catalytic RNA molecule that can cleave the RNA transcript (Section 22.3.2).
- *Targeted inhibition of expression at the protein level.* Oligonucleotide aptamers and intracellular antibodies can be designed to specifically bind to and inactivate a selected polypeptide/protein (Section 22.3.3).

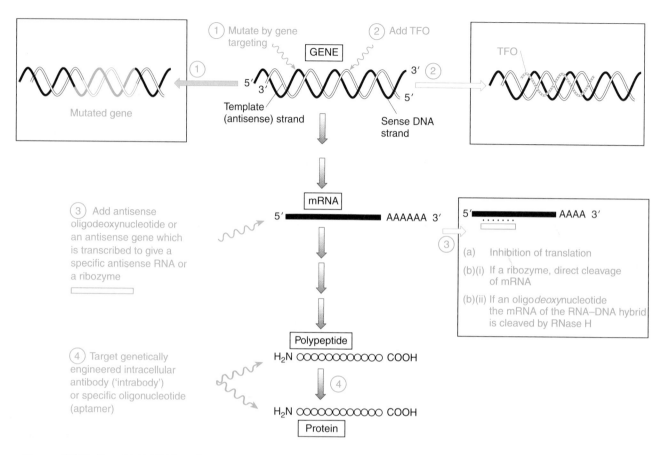

Figure 22.8: Targeted inhibition of gene expression *in vivo*.

Gene therapy based on selective inhibition of a predetermined gene *in vivo* can be achieved at several levels. In principle, it is possible to mutate the gene via homologous recombination-mediated gene targeting to a nonfunctional form (1). In practice, however, it is more convenient to block expression: at the level of transcription by binding a gene-specific triplex-forming oligonucleotide (TFO) to the promoter region (2), or at the mRNA level by binding a gene-specific antisense oligonucleotide or RNA (3). In the latter case, an antisense gene is normally provided which can encode a simple antisense RNA or a ribozyme (*Figure 22.9*). In each case, the binding interferes with the ability of the mRNA to direct polypeptide synthesis, and may ensure its destruction: a bound oligodeoxynucleotide makes the mRNA susceptible to cleavage by RNase H, while a bound ribozyme cleaves the RNA directly. The technology for specific inhibition at the polypeptide/protein level (4) is less well developed but is possible using genes which encode intracellular antibodies or oligonucleotide aptamers which specifically bind to the polypeptide and inhibit its function.

22.3.2 *Antisense oligonucleotides or polynucleotides can bind to a specific mRNA, inhibiting its translation and, in some cases, ensuring its destruction*

During transcription, only one of the two DNA strands in a DNA duplex, the **template strand** (or **antisense strand**), serves as a template for making a complementary RNA molecule. As a result, the base sequence of the single-stranded RNA transcript is identical to the other DNA strand (the **sense strand**), except that U replaces T. Any oligonucleotide or polynucleotide which is complementary in sequence to an mRNA sequence can therefore be considered to be an antisense sequence.

Binding of an antisense sequence to the corresponding mRNA sequence would be expected to interfere with translation, and thereby inhibit polypeptide synthesis. Indeed, naturally occurring antisense RNA is known to provide a way of regulating the expression of genes in some plant and animal cells, as well as in some microbes. Synthetic oligonucleotides can be designed to be complementary in sequence to a specific mRNA and, when transferred into cells, show evidence of inhibition of expression of the corresponding gene, which can occur at different levels not just translation. Antisense therapeutics is the application of antisense technology to block expression of a disease-causing gene such as a viral gene or a cancer gene with the aim of combating disease.

Antisense oligodeoxynucleotides

The use of artificial antisense oligodeoxynucleotides is

often favored, simply because they can be synthesized so simply. They can be transferred efficiently into the cytoplasm of cells using liposomes, and can migrate rapidly to the nucleus by passive diffusion through the pores of the nuclear envelope. *Note* that even although antisense oligonucleotides migrate to the nucleus, they do not bind the double-stranded DNA because they are specifically designed to have a low chance of binding to double-stranded DNA, unlike the oligonucleotides used in triple helix technologies which are deliberately designed for this purpose.

Antisense oligodeoxyribonucleotides (ODNs) are preferred as they are generally less vulnerable to nuclease attack than oligoribonucleotides. Nevertheless, to protect against degradation by cellular exonucleases it is still usual to modify the oligonucleotides at their 3' or 5' ends e.g. by introducing more resistant phosphorothioate bonds where sulfur atoms are linked to phosphate groups instead of the normal oxygen atoms. Antisense ODNs are also preferred because they have the additional advantage of inducing the destruction of an mRNA to which they bind. This is so because an ODN–mRNA hybrid, like all DNA–RNA hybrids, is vulnerable to attack by a specific class of intracellular ribonuclease, RNase H which selectively cleaves the RNA strand.

General optimism about the power of antisense technology has been sufficient for it to be used in several clinical trials (Wagner, 1994 and *Table 22.7*). However, formidable technical challenges remain to be overcome, including unexpected 'nonantisense effects'. Folding of target RNAs and/or their association with specific proteins in the cell often means that the antisense molecule is unable to bind to its target (see Branch, 1998).

Peptide nucleic acids

Peptide nucleic acids (PNAs) are artificially constructed by attaching the bases found in nucleic acids to a pseudopeptide backbone. The normal phosphodiester backbone is entirely replaced with a polyamide (peptide) backbone composed of 2-aminoethyl glycine units. As a result, PNAs have improved flexibility compared to DNA or RNA, which permits more stable hybridization to DNA or RNA (by Watson–Crick hydrogen bonding). They are also more resistant to nuclease attack and may therefore be useful alternatives to conventional antisense oligonucleotides (see Corey, 1997).

Ribozymes

Some RNA molecules are able to lower the activation energy for specific biochemical reactions, and so effectively function as enzymes (**ribozymes**). They contain two essential components: target recognition sequences (which base-pair with complementary sequences on target RNA molecules), and a catalytic component which cleaves the target RNA molecule while the base-pairing holds it in place. The cleavage leads to inactivation of the RNA, presumably because of subsequent recognition by intracellular nucleases of the two unnatural ends. Examples include human ribonuclease P and various ribozymes obtained from plant viroids (virus-like particles).

Genetic engineering can be employed to custom design the recognition sequence so that the ribozyme now contains sequences that are complementary to a specific mRNA molecule, and appropriately modified hammerhead ribozymes may be useful in gene therapy (*Figure 22.9*). A particularly promising application has been envisaged in treating dominant genetic disorders by specifically targeting the mutant allele (Phylactou *et al.*, 1998). Ribozymes such as the group I intron ribozyme can also be used to repair mutant RNA molecules (see Section 22.3.4)

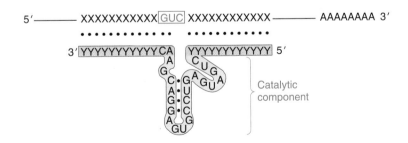

Figure 22.9: Genetically engineered hammerhead ribozymes.

The hammerhead ribozyme is a constituent of some plant viroids (virus-like particles) and is so called because of the shape of its catalytic component. It is *trans*-acting and cleaves specific target RNA molecules, whose recognition sequence contains a centrally located triplet: GUC (boxed) or a close variant. Recognition is achieved by two sequences which flank the catalytic component and which permit base-pairing to the target sequence. In order to custom design an artificial ribozyme that will cleave a predetermined mRNA sequence, it is simply necessary to choose a suitable GUC (or variant triplet) in the target and then design the ribozyme to contain the usual catalytic component flanked by sequences (YYYYY . . .) which are complementary to the sequences flanking the chosen triplet in the target sequence (XXXXX . . .). Because of the comparative lability of RNA, gene therapy experiments using ribozymes to inhibit gene expression may involve synthesis and transfer of genes encoding the desired ribozyme.

22.3.3 Artificially designed intracellular antibodies (intrabodies), oligonucleotide aptamers and mutant proteins can inhibit the function of a specific polypeptide

Intracellular antibodies (intrabodies)

Antibody function is normally conducted extracellularly. Once synthesized, they are normally secreted into the extracellular fluid, or are transported to the surface of the B cell to act as an antigen receptor. Recently, however, it has been possible to design genes encoding intracellular antibodies, or intrabodies. Intrabodies are engineered to have a single chain by coupling the variable domain of the heavy chain to the variable domain of the light chain through a peptide linker, thereby preserving the affinity of the parent antibody. Intrabodies can be directed to a particular cell compartment where they can bind to and inactivate a specific cell molecule such as a disease-causing protein, and so they have been envisaged to have potential for treating certain diseases, such as infectious diseases (Rondon and Marasco, 1997).

Oligonucleotide aptamers

Oligonucleotide **aptamers** are oligonucleotides which can bind to a specific protein sequence of interest. A general method of identifying aptamers is to start with partially degenerate oligonucleotides, and then simultaneously screen the many thousands of oligonucleotides for the ability to bind to a desired protein. The bound oligonucleotide can be eluted from the protein and sequenced to identify the specific recognition sequence. Transfer of large amounts of a chemically stabilized aptamer into cells can result in specific binding to a predermined polypeptide, thereby blocking its function. Currently, the therapeutic potential of this technology has yet to be realized (Osborne *et al.*, 1997).

Mutant proteins

Naturally occurring gain-of-function mutations can involve the production of a mutant polypeptide that binds to the wild-type protein, inhibiting its function. In many such cases, the wild-type polypeptides naturally associate to form multimers, and incorporation of a mutant protein inhibits this process (see Section 16.5). In some cases, gene therapy may be possible by designing genes to encode a mutant protein that can specifically bind to and inhibit a predetermined protein, such as a protein essential for the life-cycle of a pathogen. For example, one form of gene therapy for AIDS involves artificial production of a mutant HIV-1 protein in an attempt to inhibit multimerization of the viral core proteins (see Section 22.5.4).

22.3.4 Artificial correction of a pathogenic mutation in vivo is possible, in principle, but is very inefficient and not readily amenable to clinical applications

Certain disorders are not easy targets for conventional gene therapy. For example, dominantly inherited disorders where a simple mutation results in a pathogenic gain of function cannot be treated by gene augmentation therapy, and targeted inhibition of gene expression may be difficult to achieve. An alternative to conventional gene therapy involves repair of a mutant sequence *in vivo*. In principle, this can be done by a variety of different experimental strategies at both the level of the mutant gene or its transcript (Woolf, 1998).

Therapeutic repair at the DNA level

One possible approach is to achieve correction of the genetic defect by therapeutic gene targeting. However, while there have been substantial advances in our understanding of homologous recombination-based gene targeting in the cells of complex eukaryotes, the efficiency of homologous recombination remains extraordinarily low in such systems and there are formidable challenges in applying this technology to *in vivo* gene therapy. An alternative, recently developed gene targeting method, which uses chimeric RNA/DNA oligonucleotides has, however, been claimed to be a much more efficient way of inducing site-directed mutagenesis *in vivo* (Kren *et al.*, 1998). Other possibilities for therapeutic DNA repair utilize triple helix formation and *peptide nucleic acids* (see Section 22.3.2 and Woolf, 1998).

Therapeutic repair at the RNA level

An alternative approach to gene targeting is to repair the genetic defect at the RNA level. One possibility is to use a therapeutic ribozyme (Rossi, 1998). One method envisages using a class of ribozyme known as group I introns, which are distinguished by their ability to fold into a very specific shape, capable of both cutting and splicing RNA. If a transcript has, for example, a nonsense or a missense mutation, it may be possible to design specific ribozymes that can cut the RNA upstream of the mutation and then splice in a corrected transcript, a form of trans-splicing (see *Figure 22.10*). Thus far, this technology is in its infancy, and catalytic efficiency needs to be improved.

Another possibility is therapeutic RNA editing. This involves using a complementary RNA oligonucleotide to bind specifically to a mutant transcript at the sequence containing the pathogenic point mutation, and an RNA editing enzyme, such as double-stranded RNA adenosine deaminase, to direct the desired base modification. Again this technology is in its infancy and formidable technical difficulties need to be overcome before clinical applications can be envisaged.

22.4 Gene therapy for inherited disorders

Over the last two decades molecular genetic technologies have been spectacularly successful in identifying and characterizing novel disease genes, and in devising novel diagnostic tests for inherited disorders. In contrast, the dream of successfully applying molecular genetic tech-

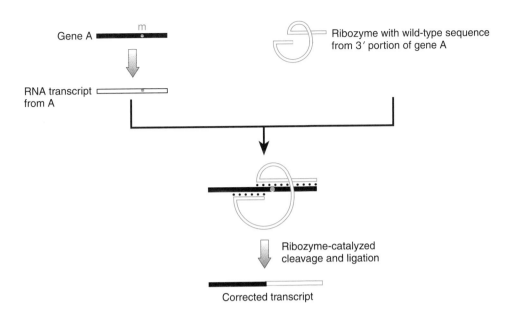

Figure 22.10: Some ribozymes also have the potential of repairing mutations in mRNA.

Group I introns are a class of self-splicing intron (see *Box 14.3*). The RNA transcript acts as a ribozyme by catalyzing the cleavage of the RNA and subsequent splicing. They possibly could be used as therapeutic agents capable of repairing certain mutations at the mRNA level (Cech, 1995). In this example, gene A has a missense or a nonsense mutation (m) and is transcribed to give a mutant RNA. The therapeutic ribozyme is designed so that its flanking recognition sequences are complementary to the 3′ end of the wild-type mRNA sequence for gene A, encompassing the location of the mutation. The ribozyme is designed to cleave the mutant mRNA at a position upstream of the mutation site. Subsequent ligation of the 5′ end of the mutant mRNA to the 3′ end sequence carried by the ribozyme can result in repair of the mutation at the mRNA level.

nologies on a large scale to curing, or even treating disease has remained unfulfilled. A new era was heralded when the first gene therapy trial for an inherited disease began in 1990, but exciting though this prospect was, reviews of clinical trials have shown that the initial enthusiasms were misplaced (Ross *et al.*, 1996; Knoell and Yiu, 1998). Even now, gene therapy has not *cured* any patient and there is precious little evidence for any significant clinical benefit in the trials that have been conducted thus far: any amelioration of the diseases that have been treated have been modest and very short-lived. Instead, there is now widespread recognition of the limitations of the current technologies and the need for safer and more efficient gene delivery systems (Verma and Somia, 1997; Anderson, 1998).

While recognizing that current gene delivery methods are not very effective, there remains considerable optimism that this is a temporary difficulty that can be overcome by future technological improvements. However, some genetic disorders may not be so easy to treat as others. Common nonmendelian genetic diseases may involve a complex interplay between different genetic loci and/or environmental factors, and so possible gene therapy approaches may not be straightforward. Single gene disorders where individuals are severely affected and where there is no effective treatment, are more obvious

candidates for gene therapy. Within the single gene disorder category, however, differing pathogeneses means that certain single gene disorders will be more amenable to gene therapy approaches than others (*Table 22.4*).

22.4.1 Recessively inherited disorders are conceptually the easiest inherited disorders to treat by gene therapy

Those disorders where the disease results from a simple deficiency of a specific gene product are generally the most amenable to treatment: high level expression of an introduced normal allele should be sufficient to overcome the genetic deficiency. Recessively inherited disorders have been of particular interest as candidates for gene therapy because the mutations are almost always simple loss-of-function mutations. Affected individuals have deficient expression from both alleles and so the disease phenotype is due to complete or almost complete absence of normal gene expression. Heterozygotes, however, have about 50% of the normal gene product and are normally asymptomatic. Additionally, there is, in at least some cases, wide variation in the normal levels of gene expression, so that a comparatively small percentage of the average normal amount of gene product may be sufficient to restore the normal phenotype. It is also often observed

Table 22.4: Factors governing the amenability of single gene disorders to gene therapy approaches

Factor	Most amenable	Least amenable
Mode of inheritance function	Recessive: affecteds usually have no or extremely little gene product, so that even low level expression of introduced genes can have an effect	Dominant: even where the mutation is a loss-of-mutation most affected people are heterozygotes, with at least 50% of the normal gene product already present
Nature of mutation product, etc.	Loss of function: can be treated simply by gene augmentation therapy (see *Box 20.2*)	Gain of function – novel mutant protein or toxic mutant may not be treated by simply adding normal genes. Instead, may need specifically to block expression of gene or repair genetic defect
Accessibility of target cells and amenability to cell culture	Readily accessible tissues, e.g. blood, skin, etc. Cells that can be cultured readily and reinserted in the patient permit *ex vivo* gene transfer	Tissues that are difficult to access (e.g. brain), or to derive cell cultures which can be reimplanted (thereby excluding *ex vivo* gene therapy)
Size of coding DNA	Small coding DNA size means easy to insert into vector e.g. β-globin = ~0.5 kb	Large coding DNA; may be difficult to insert into suitable vector
Control of gene expression	Loose control of gene expression with wide variation in normal expression levels, e.g. ADA expression (Section 22.4.2)	Tight control of gene expression, e.g. in the case of β-globin (Section 22.4.1)

that the severity of the phenotype of recessive disorders is inversely related to the amount of product that is expressed (see *Table 16.4*). As a result, even if the efficiency of gene transfer is low, modest expression levels for an introduced gene may make a substantial difference. This is quite unlike dominantly inherited disorders where heterozygotes with loss-of-function mutations have 50% of the normal gene product and may yet be severely affected.

Although recessively inherited disorders are, in principle, amenable to gene augmentation therapy, certain disorders are less amenable than others. In addition to the question of accessibility of the disease tissue, some disorders may be difficult to treat for other reasons. A good example is provided by β-thalassemia which results from mutations in the β-globin gene, *HBB*. This is a severe disorder affecting hundreds of thousands of people worldwide, and superficially would appear to be an excellent candidate for gene therapy: the gene is very small and has been characterized extensively, the disorder is recessively inherited and affects blood cells. An initial attempt at gene therapy for this disorder in 1980 failed, largely because of inefficient gene transfer and poor expression of the introduced β-globin genes. Even though we now know much about how this gene is expressed, there have been no subsequent gene therapy attempts. This is due to the problem of the very tight control of gene expression required following insertion of a normal β-globin gene into the desired cells: the amount of β-globin product made must be equal to the amount of α-globin. If too much β-globin were to be made, the imbalance between β-globin and α-globin chains would result in an α-thalassemia phenotype.

22.4.2 The first gene therapy trial for an inherited disease was initiated in 1990

The first gene therapy trial for an inherited disorder was initiated on 14 September 1990. The patient, Ashanthi DeSilva, was just 4 years old and was suffering from a very rare recessively inherited disorder, adenosine deaminase (ADA) deficiency. ADA is involved in the purine salvage pathway of nucleic acid degradation, and is a housekeeping enzyme which is synthesized in many different types of cell. An inherited deficiency of this enzyme has, however, particularly severe consequences in the case of T lymphocytes, one of the major classes of immune system cells. As a result, ADA⁻ patients suffer from severe combined immunodeficiency. This severe disorder was particularly amenable to gene therapy for a variety of reasons: the *ADA* gene is small, and had previously been cloned and extensively studied; the target cells are T cells which are easily accessible and easy to culture, enabling *ex vivo* gene therapy; the disorder is recessively inherited and, importantly, gene expression is not tightly controlled (enzyme levels in the normal population show huge differences between healthy individuals). The observation that allogeneic bone marrow transplantation can cure the disorder suggested that engraftment of T cells alone may be sufficient, and transfer of normal *ADA* genes into ADA⁻ T cells was noted to result in restoration of the normal phenotype.

Alternative treatments for ADA deficiency do exist. Indeed, the treatment of choice is bone marrow transplantation from a perfectly HLA-matched sibling donor, which provides a cure in about 80% of cases. For children where this is not an option, an alternative is enzyme replacement therapy, consisting of weekly intramuscular injections of ADA conjugated to polyethylene glycol (PEG). PEG stabilizes the ADA enzyme, allowing it to survive and function in the body for days. Inevitably, however, enzyme replacement therapy does not provide full immune reconstitution and so life expectancy is still likely to be shortened (T cells are required for mounting effective immune responses against invading micro-organisms, and in preventing cancer).

The novel ADA gene therapy approach involved essentially four steps:

(i)   cloning a normal ADA gene into a retroviral vector;
(ii) transfecting the ADA recombinant into cultured ADA$^-$ T lymphocytes from the patient;
(iii) identifying the resulting ADA$^+$ T cells and expanding them in culture;
(iv) re-implanting these cells in the patient (see *Figure 22.11*).

This approach was never going to be a cure; instead it was designed to be a form of treatment which would need to be repeated on many occasions. Successful treatment would require high efficiency gene transfer into bone marrow stem cells and high levels of expression. The trouble here is that human bone marrow stem cells are very difficult to isolate and insertion of retroviral vectors into such cells is very inefficient. Enrichment for such cells is possible using the monoclonal antibody CD34 which selectively binds a population of cells that includes totipotent stem cells and subsequent gene therapy trials

conducted on neonates used retroviral transduction of selected CD34$^+$ umbilical cord blood cells.

All patients in the ADA gene therapy trials were treated in parallel by conventional enzyme replacement therapy using PEG-ADA. The combined gene therapy plus enzyme replacement therapy appeared to give initially promising results (as assayed by various measures of antibody and T-cell function, and a dramatic decrease in infections compared with the incidence before treatment). However, cessation of the parallel PEG-ADA treatment led to a decline in immune function despite the persistence of ADA$^+$ T lymphocytes (Kohn *et al.*, 1998). The inescapable conclusion is that improved gene transfer and expression will be needed before ADA gene therapy can be successful.

22.4.3 Since the pioneering work on ADA deficiency, gene therapy trials have been initiated for a few inherited disorders

Gene therapy has been initiated for only a comparatively few inherited disorders in addition to ADA deficiency (see *Table 22.5* for examples). Different recessively inherited disorders have been targets for *in vivo* or *ex vivo* gene augmentation therapy and, in the one case where a dominantly inherited disorder has been treated, familial hypercholesterolaemia, the patients had the homozygous form of the disease. The following examples are simply illustrative of current progress and difficulties.

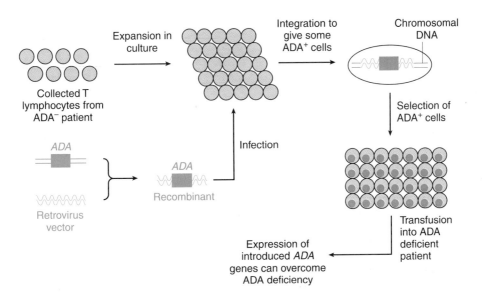

Figure 22.11: *Ex vivo* gene augmentation therapy for adenosine deaminase (ADA) deficiency.

Note that identification of suitably transformed cells is helped by having an appropriate selectable marker in the retrovirus vector, such as a *neo*^{R} gene which confers resistance to the neomycin analog G418 (see *Box 10.1*). Following infection, the target cells can be cultured in a medium containing G418 to select for the presence of retroviral sequences, and then assayed by PCR for the presence of the inserted ADA gene. Suitable ADA$^+$ cells can then be expanded in culture before being reintroduced into the patient.

Table 22.5: Examples of gene therapy trials for inherited disorders

Disorder	Cells altered	Gene therapy strategy
ADA deficiency	T cells and hemopoietic stem cells	*Ex vivo* GAT using recombinant retroviruses containing an *ADA* gene
Cystic fibrosis	Respiratory epithelium	*In vivo* GAT using recombinant adenoviruses or liposomes to deliver the *CFTR* gene
Familial hypercholesterolemia	Liver cells	*Ex vivo* GAT using retrovirus to deliver the LDL receptor gene (*LDLR*)
Gaucher's disease glucocerebrosidase	Hemopoietic stem cells	*Ex vivo* GAT using retroviruses to deliver the gene (*GBA*)

GAT, gene augmentation therapy.

Familial hypercholesterolemia (FH)

This disorder is caused by a dominantly inherited deficiency of low density lipoprotein (LDL) receptors, which are normally synthesized in the liver, and is characterized by premature coronary artery disease. About 50% of heterozygous affected males die by 60 years of age, unless treated. Because FH is such a common single gene disorder, homozygotes are occasionally seen. They suffer precocious onset of disease and increased severity, with death from myocardial infarction commonly occurring in late childhood. The first, and only, gene therapy for FH was initiated on the first of five patients with the homozygous form of the disease in 1992. The liver, being a solid internal organ, may not seem to be an ideal choice for targeting gene therapy, and its major cell population, the differentiated hepatocyte, is refractory to infection with retroviruses, the most widely used vector system. However, hepatocytes can be cultured *in vitro* and, under such conditions, are susceptible to retroviral infection.

Ex vivo gene therapy became a possibility when animal experiments showed that cultured hepatocytes could be injected via the portal venous system – the veins which drain from the intestine directly into the liver – after which they appear to seed in the liver. The gene therapy involved surgical removal of a sizeable portion of the left lobe of the patient's liver, disaggregation of the liver cells and plating in cell culture prior to infection with retroviruses containing a normal human *LDLR* gene (Grossman *et al.*, 1994). The genetically modified cells were infused back into the patient through a catheter implanted into a branch of the portal venous system. The patient's LDL/high density lipoprotein (HDL) ratio subsequently declined from 10–13 before gene therapy to 5–8, and such improvement was maintained over a long period. However, because of the invasiveness of the procedure and limited effectiveness of therapy, subsequent gene therapies are not intended until there is a significant advance in gene transfer efficiency.

Cystic fibrosis

Cystic fibrosis is an autosomal recessive disorder that results in defective transport of chloride ions through epithelial cells, and results from mutations in a gene, *CFTR*, which encodes a cAMP-regulated chloride chan-nel. The primary expression of the defect is in the lungs: a sticky mucus secretion accumulates which is prone to chronic infections. Because there are no methods to culture lung cells routinely in the laboratory, *in vivo* gene therapy approaches have been adopted. As respiratory epithelial cells are differentiated, retroviral vectors cannot be used. Instead, gene therapy trials have used adenovirus vectors or liposomes to transfer a suitably sized *CFTR* minigene, either through a bronchoscope or through the nasal cavity.

The first adenovirus-based protocol began in 1993 and, although preliminary data have confirmed gene transfer into respiratory epithelium *in vivo*, there have been major concerns regarding the safety of the procedure. The first patient to be treated with a high dose of recombinant adenovirus experienced transient pulmonary infiltrates and alterations in vital signs, before recovering uneventfully. This experience prompted recognition of the need to confirm the maximum tolerated adenovirus dose. The liposome-based gene therapy trials are regarded as safer procedures, but the efficiency of gene transfer is much lower. Despite an impressive amount of research, CF gene therapy remains ineffective (Boucher, 1999).

Duchenne muscular dystrophy

DMD is a severe X-linked recessive disorder: affected males suffer progressive muscle deterioration, are confined to a wheelchair in their teens and die usually by the third decade. The target tissue is skeletal muscle, and initial interest in treatment for this disorder focused on cell therapy because of the unique cell biology of muscle (Miller and Boyce, 1995). As well as muscle fibers (or myofibers – very long, post-mitotic, multinucleate cells), skeletal muscle contains mononucleate myoblasts which are normally quiescent but can divide and subsequently fuse with myofibers to repair muscle damage. Although implanting normal or genetically modified myoblasts into diseased muscles appeared attractive, difficulties have been evident with this approach in humans, despite promising pilot studies with myoblast transfer in mice.

Suitable gene therapy approaches have also been difficult to conceive, largely because of the lack of a suitable gene transfer system. Oncoretroviral vectors cannot be used because adult skeletal muscle fibers are postmitotic and hence not susceptible to oncoretroviral infection. Adenovirus vectors have been used to deliver genes to muscle fibers *in vivo* and, although the postmitotic state of muscle nuclei allows the expression to persist, the need for expression to continue over the course of a lifetime (which would be required for successful therapy) remains doubtful. A final problem is the sheer size of the dystrophin coding sequence (~14 kb), although a very large central segment appears not to be crucially important.

In addition to simple gene replacement strategies, alternative methods have been considered including up-regulation of genes encoding proteins that may have a compensatory function. The principle of therapeutic reactivation of fetal genes has been considered for β-thalassemia and sickle cell anemia. Here the idea is to offset genetic deficiencies in β-globin production by re-activation of other globin genes which are largely expressed during the fetal period, such as the γ-globin genes (β-thalassemia and sickle cell anemia patients may have a mild form of the disease if they produce unusually high levels of the HbF fetal haemoglobin; Olivieri and Weatherall, 1998). A similar strategy has been considered for Duchenne muscular dystrophy. Dystrophin has a close relative, utrophin, which is highly expressed during the fetal period and so there is the possibility that up-regulation of utrophin may confer a protective effect. Encouraging results have been obtained in mice where expression of utrophin transgenes in mice with dsytrophin deficiency leads to major improvements in muscle function (Deconinck *et al.*, 1997). Now, considerable effort is being devoted to standard drug-finding approaches to identify small molecules that naturally upregulate the utrophin gene with a view to administering them as drugs in future treatments.

22.5 Gene therapy for neoplastic disorders and infectious disease

22.5.1 General principles of gene therapy for neoplastic disorders and infectious disease

Cancer gene therapies

Many different approaches can be used for cancer gene therapy (see *Table 22.6*) and, in marked contrast to the few gene therapy trials for inherited disorders, numerous cancer gene therapy trials are currently being conducted (*Table 22.7*). This reflects partly the severity of the disorders that are being treated and the considerable funding for cancer research, and partly reflects the comparative ease in applying treatments based on targeted killing of disease cells, by introducing genes that encode toxins, etc. or by provoking enhanced immune responses. In a few cases, the gene therapy approach has focused on targeting single genes, such as *TP53* gene augmentation therapy and delivery of antisense *KRAS* genes in the case of some forms of non-small-cell lung cancer. In most cases, however, targeted killing of cancer cells has been conducted without knowing the molecular etiology of the cancer. Thus far, some significant advances have been made against local and metastatic tumor growth, but effective therapy awaits development of more effective methods to transfer and express transgenes or to induce antitumor responses (Hall *et al.*, 1997).

Gene therapy for infectious disorders

The gene therapy approaches for treating infectious disorders are slightly different. In common with cancer

---

Table 22.6: Potential applications of gene therapy for the treatment of cancer

General approaches

Artificial killing of cancer cells
Insert a gene encoding a toxin (e.g. diphtheria A chain) or a gene conferring sensitivity to a drug (e.g. herpes simplex thymidine kinase) into tumor cells

Stimulate natural killing of cancer cells
Enhance the immunogenicity of the tumor by, for example, inserting genes encoding foreign antigens or cytokines
Increase antitumor activity of immune system cells by, for example, inserting genes that encode cytokines
Induce normal tissues to produce antitumor substances (e.g. interleukin-2, interferon)
Production of recombinant vaccines for the prevention and treatment of malignancy (e.g. BCG-expressing tumor antigens)

Protect surrounding normal tissues from effects of chemotherapy/radiotherapy
Protect tissues from the systemic toxicities of chemotherapy (e.g. multiple drug resistance type 1 gene)

Tumors resulting from oncogene activation
Selectively inhibit the expression of the oncogene
Deliver gene-specific antisense oligonucleotide or ribozyme to bind/cleave oncogene mRNA
Inhibit transcription by triple helix formation following delivery of a gene-specific oligonucleotide
Use of intracellular antibodies or oligonucleotide aptamers to specifically bind to and inactivate the oncoprotein

Tumors arising from inactivation of tumor suppressor
Gene augmentation therapy
Insert wild-type tumor suppressor gene

Table 22.7: Examples of cancer gene therapy trials

Disorder	Cells altered	Gene therapy strategy

Brain tumors	Tumor cells *in vivo*
	Tumor cells *ex vivo*
	Hematopoietic stem
	  cells *ex vivo*
		Implanting of murine fibroblasts containing recombinant retroviruses
		  to infect brain cells and ultimately deliver HSV-tk gene
		DNA transfection to deliver antisense *IGF1*

Breast cancer	Fibroblasts *ex vivo* Hematopoietic stem cells *ex vivo*	Retroviruses to deliver *MDR1* gene Retroviruses to deliver *IL4* gene
Colorectal cancer	Tumor cells *in vivo* Tumor cells *ex vivo*	Retroviruses to deliver *MDR1* gene Liposomes to deliver genes encoding HLA-B7 and β_2-microglobulin
Malignant melanoma	Fibroblasts *ex vivo* Tumor cells *in vivo* Tumor cells *ex vivo* Fibroblasts *ex vivo*	Retroviruses to deliver *IL2* or *TNF* gene Retroviruses to deliver *IL2* or *IL4* genes Liposomes to deliver genes encoding HLA-B7 and β_2-microglobulin Retroviruses to deliver *IL2* gene
Myelogenous leukemia	T cells/tumor cells *ex vivo*	Retroviruses to deliver *IL4* gene
Neuroblastoma	Tumor cells	Retroviruses to deliver *TNFA* gene
Non-small cell lung cancer	Tumor cells Tumor cells *in vivo*	Retroviruses to deliver HSV-tk gene Retroviruses to deliver antisense *KRAS*
Ovarian cancer	Tumor cells *in vivo* Tumor cells *ex vivo*	Retroviruses to deliver wild-type TP53 gene Retroviruses to deliver HSV-tk gene
Renal cell carcinoma	Hematopoietic stem cells *ex vivo* Tumor cells *ex vivo*	Retroviruses to deliver *MDR1* gene Retroviruses to deliver *IL2* or TNF genes
Small cell lung cancer	Fibroblasts *ex vivo*	Retroviruses to deliver *IL4* gene
Solid tumors	Tumor cells *ex vivo* Tumor cells *in vivo*	DNA transfection to deliver *IL2* gene Liposomes to deliver genes encoding HLA-B7 and β_2-microglobulin

gene therapy, strategies can involve provoking a specific immune response or specific killing of infected cells. An increasingly popular additional approach targets the life-cycle of the infectious agent, reducing its ability to undergo productive infection. Some infectious agents are genetically comparatively stable. Others, however, may be undergoing rapid evolution and (much as in the case of cancer cells) present problems for any general therapy. The classic example is AIDS, where the infectious agent, HIV-1, appears to mutate rapidly.

22.5.2 Ex vivo *cancer gene therapies frequently involve attempts to recruit immune system cells to destroy the tumor cells*

Gene transfer into tumor-infiltrating lymphocytes

One of the earliest gene therapy protocols used a population of immune system cells for specifically targeting a foreign protein to a tumor. The therapy could be considered to be a form of adoptive immunotherapy (see below) because a gene encoding a cytokine, tumor necrosis factor-α (TNF-α), was transferred into tumor-infiltrating lym-

phocytes (TILs) in an effort to increase their antitumor efficacy. The TIL population is a natural population of T lymphocytes which can seek out and infiltrate tumor deposits, such as metastatic melanomas. TNF-α is a protein naturally produced by T lymphocytes which, if infused in sufficient amounts in mice, can destroy tumors. However, it is a toxic substance and intravenous infusion of TNF has significant adverse side-effects in humans. An attractive alternative was to use TILs as cellular vectors for transferring the toxic protein directly to tumors. The gene therapy approach that was used, therefore, involved retroviral-mediated transfer of a TNF gene to a TIL population which had initially been obtained from an excised tumor and then grown in culture. Subsequent transfusion of the genetically modified TILs into a patient with metastatic melanoma was expected to result in the TILs 'homing in' on the melanomas, expression of the introduced TNF gene and tumor regresssion (*Figure 22.12*). However, the trial has been marked by comparatively poor efficiency of gene transfer into human TILs and a down-regulation of cytokine expression by the TILs.

Adoptive immunotherapy by genetic modification of tumor cells

Animal studies in which murine tumor cells were geneti-

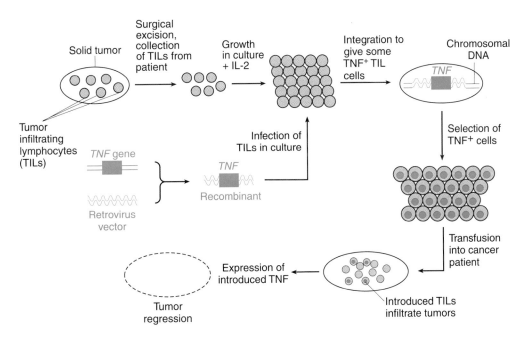

Figure 22.12: Genetic modification of cultured tumor-infiltrating lymphocytes can be used to target therapeutic genes to a solid tumor.

This approach has been used in an attempt at *ex vivo* gene therapy for metastatic melanoma. The tumor-infiltrating lymphocytes (TIL) appear to be able to 'home in' to tumor deposits. In this example, they act as cellular vectors for transporting to the melanomas a retrovirus recombinant which contains a gene specifying the anti-tumor cytokine TNF-α (tumor necrosis factor-α). Problems with the efficiency of gene transfer into the TILs and down-regulation of cytokines limited the success of this approach.

cally modified by the insertion of genes encoding various cytokines [several different interleukins (ILs), TNF-α, interferon (IFN)-γ, granulocyte–macrophage colony-stimulating factor (GM-CSF)] and then re-implanted in mice gave cause for encouragement. In each case, the genetically altered tumor cells either never grew, or grew and then regressed. In addition, most of the treated mice were then systemically immune to reimplantation of non-modified tumors. However, the results were much less satisfactory when animals with established, sizeable tumors were treated. Nevertheless, the idea of modifying a patient's own tumor cells for use as a vaccine (adoptive immunotherapy) caught on, and human gene therapy trials have been approved for the insertion of cytokine genes using retrovirus vectors for treating a wide variety of cancers (see *Table 22.7* and Ockert *et al.*, 1999).

In each case, the idea is to immunize the patients specifically against their own tumors by genetically modifying the tumor with one of a variety of genes that are expected to increase the host immune reactivity to the tumor. In addition to cytokine genes, other genes such as foreign HLA antigen genes have been transferred to tumors for the same general reason. Insertion of genes encoding HLA-B7 into tumors of patients lacking HLA-B7 is intended to provoke an immune response to the tumors as a consequence of the presence on the tumor cell surface of the effectively foreign HLA-B7 antigen (see *Table 22.7*

for some examples). Such a response is hoped to provide subsequent immunity against the same type of tumor even in the absence of the HLA-B7 antigen.

Adoptive immunotherapy by genetic modification of fibroblasts

One problem with *ex vivo* therapy for tumors is the difficulty in growing tumor cells *in vitro*: less than 50% of tumor cell lines grow in long-term culture. As an alternative, fibroblasts, which are much easier to adapt to long-term tissue culture, have been targeted in some cases. For example, transfer of genes encoding the cytokines IL-2 and IL-4 into skin fibroblasts grown in culture provides the basis of some clinical trials for treatment of breast cancer, colorectal cancer, melanoma and renal cell carcinoma. The IL-2- and IL-4-secreting fibroblasts are then mixed with irradiated autologous tumor cells and injected subcutaneously. In such cases, the hope is that the local production and secretion of cytokines by the transferred fibroblasts will induce a vigorous immune response to the nearby irradiated tumor cells and thereby result in a systemic anticancer immune response.

Other immunological approaches

Two other *ex vivo* gene therapy strategies use immunological approaches to tumor destruction. One involves transferring an antisense insulin-like growth factor-1

(*IGF1*) gene into tumor cells in order to block production of IGF-1 (Anthony *et al.*, 1998). Animal studies have shown that when tumor cells modified in this way are reimplanted *in vivo*, they provoke an immune response which can lead to destruction of nonmodified tumors, but the basis of immunological destruction is not known. A second approach involves the insertion of a co-stimulatory molecule such as B7-1 or B7-2, molecules which are normally present on lymphocytes, being required for full T-lymphocyte activation (see Putzer *et al.*, 1997).

22.5.3 In vivo *gene therapy may be the only feasible approach for some cancers*

Currently, a variety of different gene therapy approaches are being used involving genetic modification of tumor cells *in vivo*. In some cases, adoptive immunotherapy approaches are being employed, as in the case of increasing the immunogenicity of melanoma, colorectal tumors and a variety of solid tumors by the direct injection of liposomes containing a gene which encodes HLA-B7. The tumor cells take up the liposomes by phagocytosis and express the foreign HLA-B7 antigen transiently on their cells. More recent modifications include the additional insertion of a gene encoding the conserved light chain of HLA antigens, β_2-microglobulin.

A second approach has been the use of retrovirus-mediated transfer of a gene encoding a prodrug, a reagent that confers sensitivity to cell killing following subsequent administration of a suitable drug. In one recent example, the target cells were brain tumor cells, notably recurrent glioblastoma multiforme, and the retroviruses were provided in the form of murine fibroblasts that are producing retroviral vectors (retroviral vector-producing cells or VPCs). The cells were directly implanted into multiple areas within growing tumors using stereotactic injections guided by magnetic resonance imaging (*Figure 22.13*). Once injected, the VPCs continuously produce retroviral particles within the tumor mass, transferring genes into surrounding tumor cells. Although retroviruses are not normally used for *in vivo* gene therapy because of their sensitivity to serum complement, they are comparatively stable in this special environment and have the advantage that, since they only infect actively dividing cells, the tumor cells are a target, but not nearby brain cells (which are usually terminally differentiated).

The prodrug gene that was transferred is a HSV gene which encodes thymidine kinase (HSV-tk). HSV-tk confers sensitivity to the drug gancyclovir by phosphorylating it within the cell to form gancyclovir monophosphate which is subsequently converted by cellular kinases to gancyclovir triphosphate. This compound inhibits DNA polymerase and causes cell death (see *Figure 22.13*). Such therapy appears to benefit from a phenomenon known as the bystander effect: adjacent tumor cells that have not taken up the HSV-tk gene may still be destroyed. This is thought to be due to diffusion of the gancyclovir triphos-

phate from cells which have taken up the HSV-tk gene, perhaps via gap junctions.

22.5.4 Gene therapy for infectious disorders is often aimed at selectively interfering with the life-cycle of the infectious agent

Current gene therapy trials for infectious disorders are conspicuously targeted at treating AIDS patients. The infectious agent for this usually fatal disorder is a class of retrovirus known as *HIV-1* which can infect helper T lymphocytes, a crucially important subset of immune system cells (see *Figure 22.14*). Two features of HIV-1 make it especially deadly: it eventually kills the helper T cells (thereby rendering patients susceptible to other infections), and the provirus tends to persist in a latent state before being suddenly activated (the lack of virus production during the latent state complicates antiviral drug treatment). A major problem is that the HIV genome is mutating at a very high rate.

In principle, a variety of gene therapy strategies can be envisaged for treating AIDS. As in the case of cancer gene therapy, infected cells can be killed directly (by insertion of a gene encoding a toxin or a prodrug; see above) or indirectly, by enhancing an immune response against them. For example, this can involve transferring a gene that encodes an HIV-1 antigen, such as the envelope protein gp120, and expressing it in the patient in order to provoke an immune response against the HIV-1 virus, or the patient's immune system can be boosted by transfer and expression of a gene encoding a cytokine, such as an interferon. Another general approach, which is applicable to all disorders caused by infectious agents, is to find a means of interfering with the life-cycle of the infectious agent.

Gene therapy strategies designed to interfere with the HIV-1 life-cycle

A wide variety of such strategies are available (Gilboa and Smith, 1994). Inhibition has been envisaged at three major levels:

- *Blocking HIV-1 infection*. HIV-1 normally infects T lymphocytes by binding of the viral gp120 envelope protein to the CD4 receptor on the cell membrane. Transfer of a gene encoding a soluble form of the CD4 antigen (sCD4) into T lymphocytes or hemopoietic cells and subsequent expression will result in circulating sCD4. If the levels of circulating sCD4 are sufficiently high, binding of sCD4 to the gp120 protein of HIV-1 viruses could be imagined to inhibit infection of T-lymphocytes without compromising T lymphocyte function.

- *Inhibition at the RNA level*. The production of HIV-1 RNA can be selectively inhibited by standard antisense/ribozyme approaches (see Section 22.3.2), and also by the use of RNA decoys. The latter strategy exploits unique regulatory circuits which operate

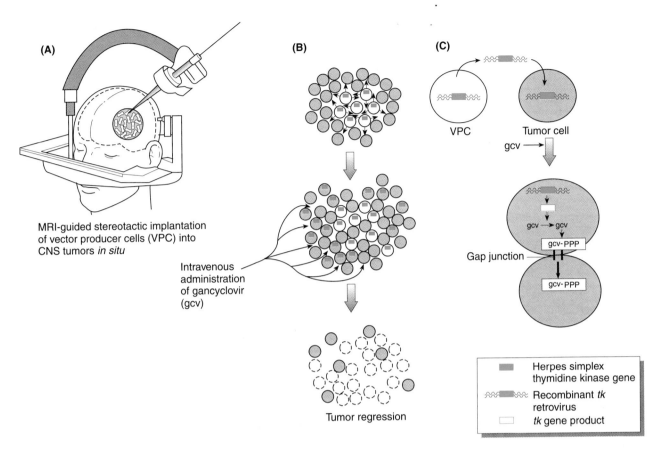

(A) MRI-guided stereotactic implantation of vector producer cells (VPC) into CNS tumors *in situ*

Intravenous administration of gancyclovir (gcv)

(B) Tumor regression

(C) VPC   Tumor cell   gcv →

gcv → gcv
gcv-PPP
Gap junction   gcv-PPP

Herpes simplex thymidine kinase gene

Recombinant *tk* retrovirus

tk gene product

Figure 22.13: *In vivo* gene therapy for brain tumors.

This example shows a strategy for treating glioblastoma multiforme *in situ* using a delivery method based on magnetic resonance imaging-guided stereotactic implantation of retrovirus vector-producing cells (VPCs). The retroviral vectors produced by the cells were used to transfer a gene encoding a prodrug, herpes simplex thymidine kinase (HSV-tk), into tumor cells. This reagent confers sensitivity to the drug gancyclovir: HSV-tk phosphorylates gancyclovir (gcv) to a monophosphorylated form gcv-P and, thereafter, cellular kinases convert this to gancyclovir triphosphate, gcv-PPP, a potent inhibitor of DNA polymerase which causes cell death. Because retroviruses infect only dividing cells, they infect the tumor cells, but not normal differentiated brain cells. The implanted VPCs transferred the HSV-tk gene to neighboring tumor cells, rendering them susceptible to killing following subsequent intravenous administration of gancyclovir. In addition, it was found that uninfected cells were also killed by a bystander effect: the gancyclovir triphosphate appeared to diffuse from infected cells to neighboring uninfected cells, possibly via gap junctions. Reproduced in part from Culver and Blaese (1994) with permission from Mary Ann Liebert Inc.

during HIV replication. Two key HIV regulatory gene products are tat and rev which bind to specific regions of the nascent viral RNA, known as TAR and RRE respectively (*Figure 22.14*). Artificial expression of short RNA sequences corresponding to TAR or RRE will generate a source of decoy sequences which can compete for binding of tat and rev, and possibly thereby inhibit binding of these proteins to their physiological target sequences.

■ *Inhibition at the protein level.* There are numerous different strategies. One strategy involves designing intracellular antibodies (see Section 22.3.3), against HIV-1 proteins, such as the envelope proteins. Another involves introducing genes that encode dominant-negative mutant HIV proteins which can bind to and inactivate HIV proteins (transdominant

proteins). For example, transdominant mutant forms of the gag proteins have been shown to be effective in limiting HIV-1 replication, possibly by interfering with multimerization and assembly of the viral core (Gilboa and Smith, 1994).

22.6 The ethics of human gene therapy

All current gene therapy trials involve treatment for somatic tissues (somatic gene therapy). Somatic gene therapy, in principle, has not raised many ethical concerns. Clearly, every effort must be made to ensure the

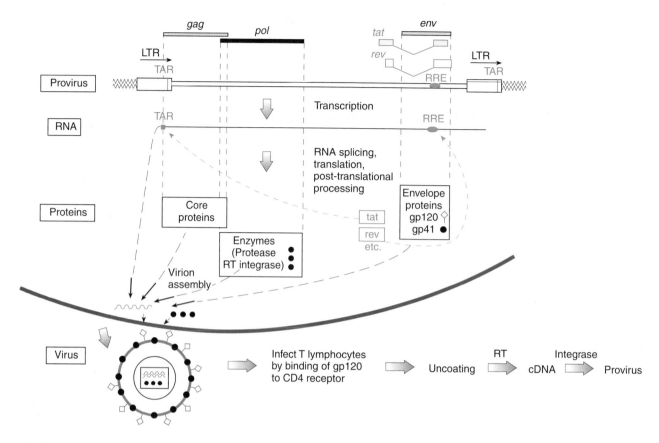

Figure 22.14: The HIV-1 virus life-cycle.

The HIV-1 virus is a retrovirus which contains two identical single-stranded viral RNA molecules and various viral proteins within a viral protein core, which itself is contained within an outer envelope. The latter contains lipids derived from host cell plasma membrane during budding from the cell, plus viral coat proteins gp120 and gp41. Penetration of HIV-1 into a T lymphocyte is effected by specific binding of the gp120 envelope protein to the CD4 receptor molecules present in the plasma membrane. After entering the cell, the viral protein coat is shed, and the viral RNA genome is converted into cDNA by viral reverse transcriptase (RT). Thereafter a viral integrase ensures integration of the viral cDNA into a host chromosome. The resulting **provirus** (see top) contains two long terminal repeats (LTRs), with transcription being initiated from within the upstream LTR. For the sake of clarity, the figure only shows some of the proteins encoded by the HIV-1 genome. In common with other retroviruses, are the *gag* (core proteins), *pol* (enzymes) and *env* (envelope proteins) genes. Tat and rev are regulatory proteins which are encoded in each case by two exons, necessitating RNA splicing. The tat protein functions by binding to a short RNA sequence at the extreme 5′ end of the RNA transcript, known as TAR (*trans*-acting response element); the rev protein binds to an RNA sequence, RRE (rev response element), which is encoded by sequence transcribed from the *env* gene.

safety of the patients, especially since the technologies being used for somatic gene therapy are still at an undeveloped stage. However, confining the treatment to somatic cells means that the consequences of the treatment are restricted to the individual patient who has consented to this procedure. Many, therefore, view the ethics of somatic gene therapy to be at least as acceptable as, say, organ transplantation, and feel that ethical approval is appropriate for carefully assessed proposals. Patients who are selected for such treatments have severely debilitating, and often life-threatening, disease for which no effective conventional therapy is available. As a result, despite the obvious imperfections of the technology, it may even be considered to be unethical to refuse such treatment. The same technology has the potential, of

course, to alter phenotypic characters that are not associated with disease, such as height for instance. Such **genetic enhancement**, although not currently considered, can be expected to pose greater ethical problems; attempts to produce genetically enhanced animals have not been a success and in some cases have been spectacular failures (Gordon, 1999).

Germline gene therapy, involving the genetic modification of germline cells (e.g. in the early zygote), is considered to be entirely different. It has been successfully practised on animals (e.g. to correct β-thalassemia in mice). However, thus far, it has not been sanctioned for the treatment of human disorders, and approval is unlikely to be given in the near future, if ever (see next section).

22.6.1 Human germline gene therapy has not been practised because of ethical concerns and limitations of the technology for germline manipulation

The lack of enthusiasm for the practice of germline gene therapy can be ascribed to three major reasons.

The imperfect technology for genetic modification of the germline

Germline gene therapy requires modification of the genetic material of chromosomes, but vector systems for accomplishing this do not allow accurate control over the integration site or event. In somatic gene therapy, the only major concern about lack of control over the fate of the transferred genes is the prospect that one or more cells undergoes neoplastic transformation. However, in germline gene therapy, genetic modification has implications not just for a single cell: accidental insertion of an introduced gene or DNA fragment could result in a novel inherited pathogenic mutation.

The questionable ethics of germline modification

Genetic modification of human germline cells may have consequences not just for the individual whose cells were originally altered, but also for all individuals who inherit the genetic modification in subsequent generations. Germline gene therapy would inevitably mean denial of the rights of these individuals to any choice about whether their genetic constitution should have been modified in the first place (Wivel and Walters, 1993). Some ethicists, however, have considered that the technology of germline modification will inevitably improve in the future to an acceptably high level and, provided there are adequate regulations and safeguards, there should then be no ethical objections (see, for example, Zimmerman, 1991). At a recent scientific research meeting in the USA some scientists have also come out in support of such a development (Wadman, 1998).

From the ethical point of view, an important consideration is to what extent technologies developed in an attempt to engineer the human germline could subsequently be used not to treat disease but in genetic enhancement. There are powerful arguments as to why germline gene therapy is pointless (see next section). There are serious concerns, therefore, that a hidden motive for germline gene therapy is to enable research to be done on germline manipulation with the ultimate aim of germline-based genetic enhancement. The latter could result in **positive eugenics** programs, whereby planned genetic modification of the germline could involve artificial selection for genes that are thought to confer advantageous traits.

The implications of human genetic enhancement are enormous. Future technological developments may make it possible to make very large alterations to the human germline by, for example, adding many novel genes using human artificial chromosomes (Grimes and Cooke, 1998).

Some people consider that this could advance human evolution, possibly paving the way for a new species, *homo sapientissimus*. To have any impact on evolution, however, genetic enhancement would need to be operated on an unfeasibly large scale (Gordon, 1999).

Even if positive eugenics programs were judged to be acceptable in principle and genetic enhancement were to be practised on a small scale, there are extremely serious ethical concerns. Who decides what traits are advantageous? Who decides how such programs will be carried out? Will the people selected to have their germlines altered be chosen on their ability to pay? How can we ensure that it will not lead to discrimination against individuals? Previous **negative eugenics** programs serve as a cautionary reminder. In the recent past, for example, there have been horrifying eugenics programs in Nazi Germany, and also in many states of the USA where compulsory sterilization of individuals adjudged to be feebleminded was practised well into the present century.

The questionable need for germline gene therapy

Germline genetic modification may be considered as a possible way of avoiding what would otherwise be the certain inheritance of a known harmful mutation. However, how often does this situation arise and how easy would it be to intervene? A 100% chance of inheriting a harmful mutation could most likely occur in two ways. One is when an affected woman is **homoplasmic** for a harmful mutation in the mitochondrial genome (see Section 16.6.4) and wishes to have a child. The trouble here is that, because of the multiple mitochondrial DNA molecules involved, gene therapy for such disorders is difficult to devise.

A second situation concerns inheritance of mutations in the nuclear genome. To have a 100% risk of inheriting a harmful mutation would require mating between a man and a woman both of whom have the same recessively inherited disease, an extremely rare occurrence. Instead, the vast majority of mutations in the nuclear genome are inherited with at most a 50% risk (for dominantly inherited disorders) or a 25% risk (for recessively inherited disorders). *In vitro* fertilization provides the most accessible way of modifying the germline. However, if the chance that any one zygote is normal is as high as 50 or 75%, gene transfer into an unscreened fertilized egg which may well be normal would be unacceptable: the procedure would inevitably carry some risk, even if the safety of the techniques for germline gene transfer improves markedly in the future. Thus, screening using sensitive PCR-based techniques would be required to identify a fertilized egg with the harmful mutation. Inevitably, the same procedure can be used to identify fertilized eggs that lack the harmful mutation. Since *in vitro* fertilization generally involves the production of several fertilized eggs, it would be much simpler to screen for normal eggs and select these for implantation, rather than to attempt genetic modification of fertilized eggs identified as carrying the harmful mutation.

Further reading

Lemoine N, Cooper D (1998) *Gene Therapy*. BIOS Scientific Publishers, Oxford.
Thomas A (ed.) (1998) Therapeutic horizons. *Nature*, **392**, (suppl.), 1–35.

References

Acsadi G, Dickson G, Love DR *et al.* (1991) Human dystrophin expression in MDX mice after intramuscular injection of DNA constructs. *Nature*, **352**, 815–818.

Anderson WF (1998) Human gene therapy. *Nature*, **392** (Suppl.), 25–30.

Anthony, DD, Pan YX, Wu SG, Shen F, Guo YJ (1998) Ex vivo and in vivo IGF-I antisense RNA strategies for treatment of cancer in humans. *Adv. Exp. Med. Biol.*, **451**, 27–34.

Berkower I (1996) The promise and pitfalls of monoclonal antibody therapeutics. *Curr. Opin. Biotechnol.*, **7**, 622–628.

Boucher RC (1999) Status of gene therapy for cystic fibrosis lung disease. *J. Clin. Invest.*, **103**, 441–445.

Branch AD (1998) A good antisense molecule is hard to find. *Trends Biochem. Sci.*, **23**, 45–50.

Cech T (1995) Group I introns: new molecular mechanisms for mRNA repair. *Biotechnology*, **13**, 323–326.

Corey DR (1997) Peptide nucleic acids: expanding the scope of nucleic acid recognition. *Trends Biotechnol.*, **15**, 224–229.

Culver KW, Blaese RM (1994) Gene therapy for cancer. *Trends Genet.*, **10**, 174–178.

Curiel DT (1994) High-efficiency gene transfer mediated by adenovirus–polylysine–DNA complexes. In: *Annals of the New York Academy of Sciences* (Huber and Lazo, eds), vol. 716, pp. 36–41. The New York Academy of Sciences, New York.

Deconinck N, Tinsley JM, De-Backer F, Fisher R, Kahn D, Phelps S, Davies KE, Gillis JM (1997) Expression of truncated utrophin leads to major functional improvements in dystrophin-deficient muscles of mice. *Nature Med.*, **3**, 1216–1221.

Donnelly JJ, Ulmer JB, Shiver JW, Liu MA (1997) DNA vaccines. *Annu. Rev. Immunol.*, **15**, 617–648.

Gage FH (1998) Cell therapy. *Nature*, **392** (Suppl.), 18–24.

Gilboa E, Smith C (1994) Gene therapy for infectious diseases: the AIDS model. *Trends Genet.*, **10**, 139–144.

Gordon JW (1999) Genetic enhancement in humans. *Science*, **283**, 2023–2024.

Grimes B, Cooke H (1998) Engineering mammalian chromosomes. *Hum. Mol. Genet.*, **7**, 1635–1640.

Grossman M, Raper SE, Kozarsky K, Stein EA, Engelhardt JF, Muller D, Lupien PJ, Wilson JM (1994) Successful ex vivo gene therapy directed to liver in a patient with familial hypercholesterol-aemia. *Nature Genet.*, **6**, 335–341.

Hall SJ, Chen S-H, Woo SLC (1997) The promise and reality of cancer gene therapy. *Am. J. Hum. Genet.*, **61**, 785–789.

Huang L, Li S (1997) Liposomal gene delivery: a complex package. *Nature Biotechnol.*, **15**, 620–621.

Jiao S, Gurevich V, Wolff JA (1993) Long-term correction of rat model of Parkinsons-disease by gene-therapy. *Nature*, **362**, 450–453.

Kay MA, Liu D, Hoogerbrugge PM (1997) Gene therapy. *Proc. Natl. Acad. Sci. USA.*, **94**, 12744–12746.

Knoell DL, Yiu IM (1998) Human gene therapy for hereditary diseases: a review of trials. *Am. J. Health Syst. Pharm.*, **55**, 899–904.

Kohn DB, Hershfield MS, Carbonaro D *et al.* (1998) T lymphocytes with a normal ADA gene accumulate after transplantation of transduced autologous umbilical cord blood CD34$^+$ cells on ADA-deficient SCID neonates. *Nature Med.*, **4**, 775–780.

Kren BT, Bandyopahyay P, Steer CL (1998) *In vivo* site-directed mutagenesis of the factor IX gene in rat liver and in isolated hepatocytes by chimeric RNA/DNA oligonucleotides. *Nature Med.*, **4**, 285–290.

Liu MA (1998) Vaccine developments. *Nature Med.* (Suppl.), **4**, 515–519.

Lubon H (1998) Transgenic animal bioreactors in biotechnology and production of blood proteins. *Biotechnol. Annu. Rev.*, **4**, 1–54.

Mendez MJ, Green LL, Corvalan JRF, Jia XC, Maynard-Currie CE, Yang X-D (1997) Functional transplant of megabase human immunoglobulin loci recapitulates human antibody response in mice. *Nature Genet.*, **15**, 146–156.

Miller JB, Boyce FM (1995) Gene therapy by and for muscle cells. *Trends Genet.*, **11**, 163–165.

Naldini L (1998) Lentiviruses as gene transfer agents for delivery to non-dividing cells. *Curr. Opin. Biotechnol.*, **9**, 457–463.

Ockert D, Schmitz M, Hampl M, Rieber EP (1999) Advances in cancer immunotherapy. *Immunol. Today*, **20**, 63–65.

Olivieri NF, Weatherall DJ (1998) The therapeutic reactivation of fetal hemoglobin. *Hum. Mol. Genet.*, **7**, 1655–1658.

Osborne SE, Matsumura I, Elington AD (1997) Aptamers as therapeutic and diagnostic reagents: problems and prospects. *Curr. Opin. Chem. Biol.,* 1, 5–9.

Paleyanda RK, Velander WH, Lee TK et al. (1997) Transgenic pigs produce functional human factor VIII milk. *Nature Biotechnol.,* **15**, 971–975.

Pardoll DM (1998) Cancer vaccines. *Nature Med.,* **4** (suppl.), 525–531.

Phylactou LA, Kilpatrick MW, Wood MJA (1998) Ribozymes as therapeutic tools for genetic disease. *Hum. Mol. Genet.,* **7**, 1649–1653.

Putzer BM, Hitt M, Muller WJ, Emtage P, Gauldie J, Graham FL (1997) Interleukin 12 and B7-1 costimulatory molecule expressed by an adenovirus vector act synergistically to facilitate tumor regression. *Proc. Natl Acad. Sci. USA.,* **94**, 10889–10894.

Rondon IJ, Marasco WA (1997) Intracellular antibodies (intrabodies) for gene therapy of infectious disease. *Annu. Rev. Microbiol.,* **51**, 257–283.

Ross G, Erickson R, Knorr D et al. (1996) Gene therapy in the United States: a five year report. *Hum. Gene Ther.,* **7**, 1781–1790.

Rossi JJ (1998) Ribozymes to the rescue: repairing genetically defective mRNAs. *Trends Genet.,* **14**, 295–298.

Smith A (1998) Cell therapy: in search of pluripotency. *Curr. Biol.,* **8**, R802–R804.

Vasquez KM, Wilson JH (1998) Triplex-directed modification of genes and gene activity. *Trends Biochem. Sci.,* **23**, 4–9.

Vaughan TJ, Osbourn JK, Tempest PR (1998) Human antibodies by design. *Nature Biotechnol.,* **16**, 535–539.

Verma IM, Somia N (1997) Gene therapy – promises, problems and prospects. *Nature,* **389**, 239–242.

Wadmann (1998) Germline gene therapy 'must be spared excessive regulation'. *Nature,* **392**, 317.

Wagner RW (1994) Gene inhibition using antisense oligodeoxynucleotides. *Nature,* **372**, 333–335.

Winter G, Harris W (1993) Humanized antibodies. *Immunol. Today,* **14**, 243–246.

Woolf TD (1998) Therapeutic repair of mutated nucleic acid sequences. *Nature Biotechnol.,* **16**, 341–344.

Wivel NA, Walters L (1993) Germ-line gene modification and disease prevention: some medical and ethical perspectives. *Science,* **262**, 533–538.

Zimmerman B (1991) Human germ-line therapy: the case for its development. *J. Med. Philos.,* **16**, 593–612.

Glossary

Wherever possible, reference is made to a figure or box that illustrates or expands the topic. Words in italics refer to glossary entries.

5′ add-on mutagenesis: a form of PCR mutagenesis in which the 5′ end of a primer is designed to introduce into the amplification product a DNA sequence not present in the target DNA. See *Figure 6.20*.

Allele: one of several alternative forms of a gene or DNA sequence at a specific chromosomal location (*locus*). At each autosomal locus an individual possesses two alleles, one inherited from the father and one from the mother.

Allele-specific oligonucleotide (ASO): a synthetic oligonucleotide, often about 20 bases long, which hybridizes to a specific target sequence and whose hybridization can be disrupted by a single base pair mismatch under carefully controlled conditions. ASOs are often labeled and used as allele-specific hybridization probes (see *Figure 6.9*), or as allele-specific primers in PCR. See *ARMS*.

Allelic association: particular alleles at two or more neighboring loci show allelic association if they occur together with frequencies significantly different from those predicted from the individual allele frequencies. Often called *linkage disequilibrium*.

Allelic exclusion: the mechanism whereby only one of the two immunoglobulin alleles in B lymphocytes, or T-cell receptor alleles in T lymphocytes, is expressed. More generally, any naturally occurring mechanism that causes only one allele to be expressed – see *Box 8.6*.

Alphoid DNA (or **α-satellite DNA**): a class of satellite DNA with an average repeat length of about 170 bp; found at centromeres. See Section 7.4.1.

Alternative splicing: the natural use of different sets of splice junction sequences to produce more than one product from a single gene. See Section 8.3.2, *Box 8.4*.

Alu-PCR: PCR using a primer that anneals to *Alu* repeats to amplify DNA located between two oppositely oriented *Alu* sequences. Used as a method of obtaining a fingerprint of bands from an uncharacterized human DNA. See *Figure 10.18*.

Amplimers: PCR primers.

Anchored PCR: PCR where one primer anneals to an oligonucleotide linker which has been ligated onto the end of an uncharacterized DNA fragment. See *Figures 6.12, 6.13*.

Aneuploidy: a chromosome constitution with one or more chromosomes extra or missing from a full euploid set.

Annealing: the association of complementary DNA (or RNA) strands to form a double helix.

Anonymous DNA: DNA not known to have a function.

Anticipation: a phenomenon in which the age of onset of a disorder is reduced and/or the severity of the phenotype is increased in successive generations. Characteristic of *dynamic mutations*.

Anticodon: the 3-base sequence in a tRNA molecule that base-pairs with the codon in mRNA. See *Figures 1.7B, 1.20, Table 1.5*.

Antisense oligonucleotide: a synthetic oligonucleotide designed to be complementary to a naturally occurring mRNA molecule. Formation of a double helix can prevent expression of the mRNA. See Section 22.3.2.

Antisense RNA: a transcript from the antisense strand of a gene. Naturally occurring antisense RNAs may negatively regulate gene expression.

Antisense strand (template strand): the DNA strand of a gene which, during transcription, is used as a template by RNA polymerase for synthesis of mRNA. See *Figure 1.12*.

Apoptosis: programmed cell death.

Archaeon (pl. **archaea**): a single-celled prokaryote superficially resembling a bacterium, but with molecular features indicative of a third kingdom of life. See *Box 14.1*.

ARMS (Amplification refractory mutation system): allele-specific PCR. See *Figure 6.9*.

Assortative mating: marriage between people of similar phenotype or genotype (e.g. tall people tend to marry tall people, deaf people tend to marry deaf people;

some people prefer to marry relatives). Assortative mating can produce a non-Hardy–Weinberg distribution of genotypes in a population.

Autonomously replicating sequence (ARS): in yeast, a DNA sequence having an origin of replication. See *Figure 2.8*, Section 4.3.4.

Autosome: any chromosome other than the sex chromosomes, X and Y.

Autozygosity: in an inbred person, homozygosity for alleles identical by descent.

Autozygosity mapping: a form of genetic mapping for autosomal recessive disorders in which affected individuals are expected to have two identical disease alleles by descent. See *Figure 11.8*.

Bacterial artificial chromosome (BAC): a recombinant plasmid in which inserts up to 300 kb long can be propagated in bacterial cells. See Section 4.3.3.

Barr body (sex chromatin): a blob of condensed chromatin visible at the edge of the nucleus in some interphase cells of females that represents the inactive X chromosome.

Bias of ascertainment: distortions in a set of data caused by the way cases are collected – for example, severely affected people are more likely to be ascertained than mildly affected people. See Section 19.4.1.

Biological fitness: see *Fitness*.

Biotin–streptavidin: the very high affinity of streptavidin for biotin allows biotin-labeled molecules to be isolated efficiently from a mixture. See, for example, *Figure 10.24*.

Bivalent: the four-stranded structure seen in prophase I of meiosis, comprising two synapsed homologous chromosomes. See *Figure 2.14*.

BLAST: a family of programs that search sequence databases for matches to a query sequence. See Section 20.1.4.

Blunt-ended: a DNA fragment having no single-stranded extensions.

Boundary elements: sequences that define the boundaries of coordinately regulated chromatin domains in chromosomes. See *Box 8.2*.

Branch site: in mRNA processing, a rather poorly defined sequence (consensus CTRAY; R = purine, Y = pyrimidine) located 10–50 bases upstream of the splice acceptor, containing the adenosine at which the lariat-splicing intermediate is formed. See *Figure 1.15*.

Bubble-linker PCR (vectorette-PCR): a form of *anchored PCR* that uses a linker containing an unpaired region (vectorette) to amplify adjacent uncharacterized sequences. See *Figure 10.16*.

Cap: a specialized chemical group that cells add to block the 5' end of mRNA. See *Figure 1.17*.

cDNA (complementary DNA): DNA synthesized by the enzyme reverse transcriptase using mRNA as a template, both experimentally (see *Figures 4.8* and *6.5*) or *in vivo* (see *Figures 7.13* and *14.18*).

cDNA selection: a hybridization-based method for retrieving genomic clones that have counterparts in a cDNA library. See *Figure 10.24*.

CentiMorgan (cM): a unit of genetic distance equivalent to a 1% probability of recombination during meiosis. See *Figure 11.3* for the correspondence between genetic and physical distances.

CentiRay (cR): a mapping unit when using radiation hybrids, which is dependent on the intensity of the irradiation. A distance of 1 cR_{8000} represents 1% frequency of breakage between two markers after exposure to a dose of 8000 rad. See *Figure 10.3*.

Central dogma: a summary of the usual information flow in organisms, from DNA to RNA to protein.

Centric fusion: strictly, formation of an abnormal chromosome by fusion of two chromosome arms at the centromeres. Loosely applied to the common human Robertsonian translocations, which are in fact produced by exchange between the proximal short arms of acrocentric chromosomes (*Figure 2.21*).

Centromere: the primary constriction of a chromosome, separating the short arm from the long arm, and the point at which spindle fibers attach to pull chromosomes apart during cell division. See Section 2.3.2.

Chiasma (pl. chiasmata): a visible crossover between paired homologous chromosomes in prophase I of meiosis. See *Figure 2.14*.

Chimera: an organism derived from more than one zygote. See *Figure 3.9*.

Chimeric insert: a clone insert consisting of two or more originally noncontiguous DNA sequences.

Chromatid: from the end of S phase of the cell cycle (*Figure 2.2*) until anaphase of cell division, chromosomes consist of two sister chromatids. Each contains a complete double helix and the two are exact copies of each other.

Chromatin fiber: the 30 nm coiled coil of DNA and histones that is believed to be the basic conformation of chromatin. See *Figure 2.7*.

Chromosome painting: fluorescence labeling of a whole chromosome by a *FISH* procedure in which the probe is a cocktail of many different DNA sequences from a single chromosome. See *Box 10.2*.

Chromosome walking: a method of isolating sequences adjacent on the chromosome to a characterized clone. See *Figure 10.14*.

Cis-acting: *cis*-acting regulatory sequences control the activity of a gene only when it is part of the same DNA molecule or chromosome.

Clone fingerprinting: involves generating a pattern of fragments from a clone by restriction digestion or *Alu-PCR* (*Figure 10.20*).

Coding DNA: DNA that encodes the amino acid sequence of a polypeptide, or a functional mature RNA which does not specify a polypeptide (see *Table 7.4*).

Codon: a nucleotide triplet (strictly in mRNA, but by extension, in genomic coding DNA) that specifies an amino acid or a translation stop signal. See *Figure 1.22*.

Coefficient of selection (s): $s = 1 - f$, where f is the biological *fitness* of a genotype.

Comparative genome hybridization (CGH): use of competitive fluorescence *in situ* hybridization to detect chromosomal regions that are amplified or deleted, especially in tumors. See *Figure 18.5*.

Competent: bacterial cells in a competent state are able to take up high molecular weight DNA.

Complementary strands: two nucleic acid strands are said to be complementary in sequence if they can form sufficient base pairs so as to generate a stable double-stranded structure.

Complementation: two alleles complement if in combination they restore the wild-type phenotype (see *Box 3.2*). Normally, alleles complement only if they are at different loci, although some cases of interallelic complementation occur.

Concatemer: multiple copies of the same sequence joined tandemly end to end.

Concerted evolution: the process whereby individual members of a DNA family within one species are more closely related to each other than to members of the same type of DNA family in other species. See *Figure 14.15*.

Conditional knock-out: an engineered mutant that causes loss of gene function under some circumstances (for example, raised temperature), or in some cells, but not others. See Section 21.3.2.

Congenic strains: strains of a laboratory animal that are identical except at one specific locus (*Box 15.4*).

Conservation of synteny: when genetic maps of two organisms are compared, synteny is conserved if loci that are located on the same chromosome in one organism are also located together on a single chromosome in the other. See *Figure 15.11*.

Conservative substitution: a mutation causing a codon to be replaced by another codon that specifies a different amino acid, but one which is related in chemical properties to the original amino acid.

Constitutional: an abnormality or a mutation of a genotype that was present in the fertilized egg, and is therefore present in all cells of a person.

Constitutive expression: a state where a gene is permanently active. Mutations that result in inappropriate constitutive expression are often pathogenic.

Constitutive heterochromatin: chromatin that is always heterochromatic (principally at centromeres of chromosomes).

Contig: a list or diagram showing an ordered arrangement of cloned overlapping fragments that collectively contain the sequence of an originally continuous DNA strand.

Contiguous gene (segmental aneusomy) syndrome: a syndrome caused by deletion of a contiguous set of genes, several or all of which contribute to the phenotype (see Section 16.8.1).

Continuous character: a character like height, which everybody has, but to a differing degree – as compared to a *dichotomous character* like polydactyly, which some people have and others do not have.

Copy number: the number of different copies of a particular DNA sequence in a genome.

Cosmid: a vector constructed by inserting the *cos* sequences of lambda phage into a plasmid. Permits cloning of inserts 30–46 kb long. See *Figure 4.14*.

Cot-1 DNA: a fraction of DNA consisting largely of highly repetitive sequences. Obtained from total genomic DNA by selecting for rapidly reassociating DNA sequences during renaturation of DNA. See *Box 5.3*.

CpG dinucleotide: the sequence 5′ CG 3′ within a longer DNA molecule. CpG dinucleotides are targets of a specific DNA methylation system in mammals that is important in control of gene expression.

CpG island: short stretch of DNA, often <1 kb, containing frequent unmethylated *CpG dinucleotides*. CpG islands tend to mark the 5′ ends of genes. See *Box 8.5*.

Cre-lox system: A technique for generating predefined chromosomal deletions. *Cre* is a bacteriophage P1 gene whose product facilitates recombination between *lox*P sequences. So called because it <u>c</u>reates <u>re</u>combination. See *Figures 4.15* and *21.8*.

Cryptic splice site: a sequence in pre-mRNA with some homology to a splice site. Cryptic splice sites may be used as splice sites when splicing is disturbed or after a base substitution mutation that increases the resemblance to a normal splice site. See *Figure 9.11*.

Degenerate oligonucleotides (or primers): a panel of synthetic oligonucleotides designed so that collectively they correspond to various codon permutations for a given sequence of amino acids.

Denaturation: dissociation of complementary strands to give single-stranded DNA and/or RNA.

Dichotomous character: a character like polydactyly, which some people have and others do not have – as opposed to a *continuous character* like height, which everybody has, but to differing degree.

Dideoxynucleotides (ddNTPs): synthetic nucleotides lacking 2′ and 3′ hydroxyl groups. They act as chain terminators during DNA replication. See *Figure 6.15*.

Differential display: see *mRNA differential display*.

Diploid: having two copies of each type of chromosome; the normal constitution of most human somatic cells.

Displacement (D) loop region: in mitochondrial DNA, a short triple-stranded region. See *Figure 7.2*.

DNA chip: a microarray of oligonucleotides or cDNA clones fixed on a glass surface. They are commonly used in a form of reverse hybridization assay to test for sequence variation in a known gene, or to profile gene expression in an mRNA preparation. See *Figures 5.20* and *20.6*.

DNA fingerprinting: a method which produces a pattern of hybridizing bands in a Southern blot that can be used to identify a person for legal or forensic purposes (*Figure 17.19*).

DNA footprinting: see *Footprinting*.

DNA ligase: an enzyme able to form a phosphodiester bond between adjacent but unlinked nucleotides in a double helix. See also *Ligation*.

DNA profiling: using genotypes at a series of polymorphic loci to recognize a person, usually for legal or forensic purposes. See Section 17.4.

DNAse I-hypersensitive sites: regions of chromatin that are rapidly digested by DNAse I. They are believed to be important long-range control sequences. See Section 8.5.2.

Dominant: (in human genetics) describes any trait that is expressed in a heterozygote. See also *Semi-dominant.*

Dominant negative mutation (antimorph): a mutation which results in a mutant gene product that can inhibit the function of the wild-type gene product in heterozygotes. See, for example, *Figure 16.4.*

DOP-PCR (degenerate oligonucleotide primer PCR): PCR using a mixture of closely related oligonucleotides rather than a single species as primer. See *Figure 6.11.*

Dosage compensation: any system that equalizes the amount of product produced by genes present in different numbers. In mammals, describes the X-inactivation mechanism that ensures equal amounts of X-encoded gene products in XX and XY cells. See *Figure 2.6.*

Dot-matrix analysis: a technique for comparing two DNA sequences. See *Figure 20.3.*

Downstream: in the 3' direction when a gene sequence is written in the conventional way (showing the sense strand, in the 5' to 3' direction).

Duplicative (copy or replicative) transposition: transposition of a copy of a DNA sequence while the original remains in place.

Dynamic mutation: an unstable expanded repeat that changes size between parent and child. See *Box 16.7.*

Ectopic transcription: see *Illegitimate transcription.*

Embryonic stem cells (ES cells): undifferentiated, pluripotent cells derived from an embryo. Mouse ES cell lines are commonly used as a vehicle for transferring foreign DNA into the germline in order to generate transgenic mice. See Section 21.3.

Empiric risks: risks calculated from survey data rather than from genetic theory. Genetic counseling in most nonmendelian conditions is based on empiric risks. See Section 3.4.4.

Enhancer: a set of short sequence elements which stimulate transcription of a gene and whose function is not critically dependent on their precise position or orientation. See *Box 8.2.*

Epigenetic: heritable (from mother cell to daughter cell, or sometimes from parent to child), but not produced by a change in DNA sequence. DNA methylation is the best understood mechanism.

Episome: any DNA sequence that can exist in an autonomous extra-chromosomal form or can be integrated into the chromosomal DNA of the cell. Often used to describe self-replicating and extra-chromosomal forms of DNA.

Epitope: a part of an antigen with which a particular antibody reacts. See *Box 20.2.*

ES cells: see *Embryonic stem cells.*

EST: see *Expressed sequence tag.*

Euchromatin: the fraction of the nuclear genome which contains transcriptionally active DNA and which, unlike *heterochromatin,* adopts a relatively extended conformation.

Eugenics: 'improving' a population by selective breeding from the 'best' types (positive eugenics) or preventing 'undesirable' types from breeding (negative eugenics).

Eukaryote: an animal or plant, having cells with a membrane-bound nucleus and organelles. May be multicellular (e.g. *metazoa*) or unicellular (*protist*). See *Figure 2.1.*

Euploidy: the state of having one or more complete sets of chromosomes with none extra or missing; the opposite of aneuploidy.

Exclusion mapping: genetic mapping with negative results, showing that the locus in question does not map to a particular location. Particularly useful for excluding a possible *candidate gene* without the labor of mutation screening.

Exon: a segment of a gene that is represented in the mature RNA product. Individual exons may contain coding DNA and/or noncoding DNA (untranslated sequences). See *Figures 1.14* and *1.19.*

Exon shuffling: evolution of a gene by combining exons from other pre-existing genes. See *Figures 14.17* and *14.18.*

Exon skipping: alternative splicing in which splice junction sites that are normally used in RNA splicing are by-passed. See *Figure 9.11.*

Exon trapping: a technique for detecting sequences within a cloned genomic DNA that are capable of splicing to exons within a specialized vector. See *Figure 10.23.*

Expressed pseudogene: a copy of a gene that is transcribed but not functional. See *Box 7.3.*

Expressed sequence tag (EST): a short (typically 100–300) bp partial cDNA sequence. See Section 13.2.3.

Expression cloning: cloning of cDNAs in specialized vectors permitting expression of a gene product from the insert. See Section 4.4.2.

FACS (fluorescence activated cell sorter): a machine that can separate cells, or individual chromosomes in suspensions, according to their ability to bind fluorescent dyes. Used, among other purposes, for *flow cytometry.* See *Figure 10.7.*

Facultative heterochromatin: chromatin that may exist as euchromatin or heterochromatin, depending on the state of the cell.

FISH: see *Fluorescence* in situ *hybridization.*

Fitness (f): in population genetics, a measure of the success in transmitting genotypes to the next generation. Also called biological fitness or reproductive fitness. f always lies between 0 and 1.

Flow cytometry (flow sorting): fractionation of individual chromosomes according to size and base composition in a fluorescence-activated chromosome (or cell) sorter. See *Figure 10.7.*

Fluorescence *in situ* hybridization (FISH): *in situ* *hybridization* using a fluorescently labeled DNA or

RNA probe. A key technique in modern molecular genetics – see *Figures 5.17, 10.5* and *10.6*.

Fluorophore: a fluorescent chemical group, used for labeling probes etc. See *Box 5.2*.

Footprinting: a method of identifying sequences within a cloned DNA molecule that can specifically bind protein molecules, such as transcription factors. See *Figure 20.14*.

Founder effect: high frequency of a particular allele in a population because the population is derived from a small number of founders, one or more of whom carried that allele.

Frameshift mutation: a mutation that alters the normal translational *reading frame* of a mRNA by adding or deleting a number of bases that is not a multiple of three.

Functional genomics: analysis of gene function on a large scale, by conducting parallel analyses of gene expression/function for large numbers of genes, even all genes in a genome.

Fusion (hybrid) gene: a gene containing coding sequence from two different genes, usually created by unequal crossover (*Figure 9.17*) or chromosomal translocations (*Figure 18.7A*).

Fusion protein: the product of a natural or engineered *fusion gene*: a single polypeptide chain containing amino acid sequences that are normally part of two or more separate polypeptides. See *Box 20.2*.

Gain-of-function mutation: a mutation that causes inappropriate expression or function of the gene product, rather than simply loss of function. See Sections 16.3, 16.5.

Gene conversion: a naturally occurring nonreciprocal genetic exchange in which a sequence of one DNA strand (acceptor sequence) is altered to become identical to the sequence of another DNA strand (donor sequence). See *Figures 9.10* and *9.17*.

Gene dosage: the number of copies of a gene. Abnormal dosage of some genes (dosage-sensitive genes) can cause developmental abnormalities. See Sections 16.4.3 and 16.5.2.

Gene frequency: the proportion of all alleles at a locus that are the allele in question. See Section 3.3. Strictly the term should be allele frequency, but the use of gene frequency is too well established now to change.

Gene pool: the totality of genes, either alleles at a given locus or over all loci, in a population.

Gene targeting: a form of *in vivo* mutagenesis whereby the sequence of a predetermined gene is selectively modified within an intact cell. See Section 21.3.

Gene trap: a method of selecting *transgene* insertions that have occurred into a gene. See *Figure 21.10*.

Genetic distance: distance on a genetic map, defined by recombination fractions and the *mapping function*, and measured in centiMorgans. See Section 11.1.

Genetic enhancement: the possible application of molecular genetic technologies to alter the human *germ line* and as a result alter the <u>normal</u> phenotype in some way that is considered to be beneficial.

Genetic load: the relative decrease in the average fitness, compared to what it would be if all individuals had the fittest genotype, i.e. $(w_{max} - w_{av})/w_{max}$. Genetic load is made up of *mutational load* and segregational load.

Genetic redundancy: performance of the same function in parallel by genes at more than one locus, so that loss of function mutations at one locus do not cause overall loss of function.

Genome: the total genetic complement of an organism or virus. For cellular organisms, the set of different DNA molecules, e.g. the human genome comprises 25 different DNA molecules (see *Table 7.1*).

Genome-wide *p* value: in testing for linkage or association, the probability on the null hypothesis of observing the statistic in question anywhere in a screen of the whole genome – see also *pointwise p value*. See Section 11.3.4.

Genotype: the genetic constitution of an individual, either overall or at a specific locus.

Germ cells (or **gametes**): sperm cells and egg cells.

Germ line: the germ cells and those cells which give rise to them; other cells of the body constitute the soma.

Germinal (gonadal or **gonosomal) mosaic:** an individual who has a subset of germ line cells carrying a mutation that is not found in other germ line cells. See *Figure 3.8*.

Haploid: describing a cell (typically a gamete) which has only a single copy of each chromosome (i.e. 23 in man).

Haploinsufficiency: a locus shows haploinsufficiency if producing a normal phenotype requires more gene product than the amount produced by a single copy. See Section 16.4.3 and *Table 16.2*.

Haplotype: a series of alleles found at linked loci on a single chromosome.

Hardy–Weinberg distribution: the simple relationship between gene frequencies and genotype frequencies that is found in a population under certain conditions. See *Box 3.3*.

Hemimethylation: DNA methylation on only one strand of the double helix. A transient state occurring after DNA replication. Hemimethylation allows proofreading enzymes to work out which strand is the original when they find a replication error. See *Figures 8.17* and *18.17*.

Hemizygous: having only one copy of a gene or DNA sequence in diploid cells. Males are hemizygous for most genes on the sex chromosomes. Deletions occurring on one autosome produce hemizygosity in males and in females.

Heritability: the proportion of the causation of a character that is due to genetic causes. See *Box 19.3* and *Figure 19.5* for examples of how heritability is calculated.

Heterochromatin: a chromosomal region that remains highly condensed throughout the cell cycle and shows little or no evidence of active gene expression. See *Figure 2.18*.

Heteroduplex: double-stranded DNA in which there is some mismatch between the two strands. See Section 17.1.4 for methods of detecting heteroduplexes.

Heteroplasmy: mosaicism, usually within a single cell, for mitochondrial DNA variants. See Section 3.1.5 and *Box 9.3*.

Heterozygous: an individual is heterozygous at a locus if (s)he has two different alleles at that locus.

Heterozygote advantage: the situation when somebody heterozygous for a mutation has a reproductive advantage over the normal homozygote. Sometimes called overdominance. Heterozygote advantage is the reason why several severe recessive diseases remain common (*Box 3.6*).

Homologous chromosomes (homologs): the two copies of a chromosome in a diploid cell. Unlike sister chromatids, homologous chromosomes are not copies of each other; one was inherited from the father and the other from the mother.

Homologous genes (homologs): two or more genes whose sequences are significantly related because of a close evolutionary relationship, either between species (*orthologs*) or within a species (*paralogs*).

Homoplasmy: of a cell or organism, having all copies of the mitochondrial DNA identical. See *Box 9.3*.

Homozygous: an individual is homozygous at a locus if (s)he has two identical alleles at that locus. The exact meaning depends on the stringency with which identity is established. For clinical purposes a person is often described as homozygous AA or aa if they have two normally functioning or two pathogenic alleles, regardless of whether they are in fact completely identical. Homozygosity for alleles *identical by descent* is called *autozygosity*.

Hotspot: a sequence associated with an abnormally high frequency of recombination or mutation.

Housekeeping gene: a gene whose expression is essential for the function of most or all types of cell.

Hybrid cell panel: a collection of *somatic cell hybrids* or *radiation hybrids* used for physical mapping.

Hybridization assay: testing for the presence of a given sequence in a DNA or RNA sample by mixing single DNA (or RNA) strands from a known probe with those of the poorly characterized target sample, then allowing complementary strands to anneal. See *Box 5.4*.

Hybridization stringency: the degree to which mismatches are tolerated in a *hybridization assay*. High stringency is achieved by using a high temperature and low salt concentration.

Identity by descent (IBD): alleles in an individual or in two people that are identical because they have both been inherited from the same common ancestor, as opposed to *identity by state*.

Identity by state (IBS): coincidental possession of alleles that appear identical. The alleles may or may not be truly identical. See *Figure 12.1*.

Illegitimate (ectopic) transcription: low-level transcription of a gene in a cell in which the gene is not normally expressed. Probably a near-universal phenomenon.

Imprinting: determination of the expression of a gene by its parental origin. See Section 8.5 and *Box 16.6*.

In situ **hybridization:** hybridization of a labeled nucleic acid to a target DNA or RNA which is typically immobilized on a microscopic slide.

In vitro **mutagenesis:** introduction of a predefined mutation into a cloned sequence. See *Figures 6.19* and *6.20*.

Inbreeding: marrying a blood relative. The term is comparative, since ultimately everybody is related. The coefficient of inbreeding is the proportion of a person's genes that are *identical by descent*.

Informative meiosis: in linkage analysis, a meiosis is informative if the genotypes in the pedigree allow us to decide whether it is recombinant or not (for a given pair of loci). See *Box 11.2*.

Insertional inactivation: inactivation of a gene by insertion of a foreign DNA sequence within it – a transposon (*Figure 7.16*) or a transgene (*Figure 21.1*), for example. See Section 9.5.6.

Insulator: see *Boundary element*.

Interference: in meiosis, the tendency of one crossover to inhibit further crossing-over within the same region of the chromosomes. See Section 11.1.3.

Interphase: all the time in the cell cycle when a cell is not dividing.

Intron: noncoding DNA which separates neighboring exons in a gene. During gene expression introns are transcribed into RNA but then the intron sequences are removed from the pre-mRNA by splicing (see *Figure 1.14*). Can be classified according to the mechanism of splicing or, if separating coding DNA sequences, their precise location within codons (see *Box 14.3*).

Isochromosome: an abnormal symmetrical chromosome, consisting of two identical arms, which are normally either the short arm or the long arm of a normal chromosome.

Isoforms/isozymes: alternative forms of a protein/enzyme.

Karyogram: an image showing the chromosomes of a cell sorted in order and arranged in pairs, such as *Figure 2.17*.

Karyotype: a summary of the chromosome constitution of a cell or person, such as 46,XY. See *Boxes 2.5* and *2.6* for nomenclature.

Knock-in mutation: a targeted mutation that replaces activity of one gene by that of an introduced gene (usually an allele). See *Figure 21.6*.

Knock-out mutation: the targeted inactivation of a gene within an intact cell.

Lagging strand: in DNA replication, the strand that is synthesized as Okazaki fragments (see *Figure 1.9*).

Leading strand: in DNA replication, the strand that is synthesized continuously (see *Figure 1.9*).

Library: a collection of clones. See *Figures 4.7* and *4.8*.

Ligase: see DNA ligase.

Ligation: formation of a 3′–5′ phosphodiester bond between nucleotides at the ends of two molecules (intermolecular ligation) or the two ends of the same molecule (intramolecular ligation, cyclization).

LINE (long interspersed nuclear element): a class of repetitive DNA sequences that make up about 10% of the human genome (*Table 7.12*). Some are active transposable elements. See *Figures 7.16* and *7.18*.

Linkage: the tendency of genes or other DNA sequences at specific loci to be inherited together as a consequence of their physical proximity on a single chromosome.

Linkage disequilibrium: see *allelic association*.

Linker (or adaptor oligonucleotide): a double-stranded oligonucleotide which is ligated to a DNA fragment of interest to assist cloning or PCR amplification. See, for example, *Figures 6.12* and *20.7*.

Liposome: a synthetic lipid membrane designed to transport a molecule of interest into a cell. See *Figure 22.6*.

Locus: a unique chromosomal location defining the position of an individual gene or DNA sequence.

Locus control region (LCR): a stretch of DNA containing regulatory elements which control the expression of genes in a gene cluster that may be located tens of kilobases away. See *Figure 8.23*.

Locus heterogeneity: determination of the same disease or phenotype by mutations at different loci. A major problem for linkage analysis. See Section 3.1.4.

Lod score (z): a measure of the likelihood of genetic linkage between loci. The log (base 10) of the odds that the loci are linked (with recombination fraction θ) rather than unlinked. For a mendelian character, a lod score greater than +3 is evidence of linkage; one that is less than −2 is evidence against linkage. See *Box 11.3* and *Figure 11.5*.

Manifesting heterozygote: a female carrier of an X-linked recessive condition who shows some clinical symptoms, presumably because of skewed X-inactivation. See Section 3.1.2.

Mapping function: a mathematical equation describing the relation between recombination fraction and *genetic distance*. The mapping function depends on the extent to which *interference* prevents close double recombinants. See Section 11.1.3.

Marker: a **genetic marker** is a polymorphic DNA or protein sequence deriving from a single chromosomal location, which is used in genetic mapping. See *Box 12.1*. A **marker chromosome** is an extra chromosome of unidentified origin.

Matrilineal inheritance: transmission from just the mother, but to children of either sex; the pattern of mitochondrial inheritance. See *Figure 3.4*.

Mean heterozygosity: of a marker, the likelihood that a randomly selected person will be heterozygous. A measure of the usefulness of the marker for linkage analysis (see Section 11.2.2).

Mendelian: of a pedigree pattern, conforming to one of the archetypic patterns shown in *Figure 3.2*; a character will give a mendelian pedigree pattern if it is determined at a single chromosomal location, regardless of whether or not the determinant is a gene in the molecular geneticist's sense.

Metaphase: the stage of cell division (mitosis or meiosis) when chromosomes are maximally contracted and lined up on the equatorial plane (metaphase plate) of a cell. See *Figures 2.10, 2.11* and *2.15*.

Metazoon (pl. metazoa): a multicellular animal as opposed to a unicellular protozoon.

Microarray: see *DNA chip*.

Microsatellite: small run (usually less than 0.1 kb) of tandem repeats of a very simple DNA sequence, usually 1–4 bp, for example $(CA)_n$. The primary tool for genetic mapping during the 1990s. See *Figures 6.7* and *6.8*.

Microsatellite instability: a phenomenon characteristic of certain tumor cells, where during DNA replication the repeat copy number of microsatellites is subject to random changes. Abbreviated to MIN, MSI or RER (replication error). See *Figure 18.11*.

MIM number: the catalog number for a gene or mendelian character, as listed in Victor McKusick's <u>M</u>endelian <u>I</u>nheritance in <u>M</u>an, available as a book and electronically (*OMIM*).

Minisatellite DNA: an intermediate size array (often 0.1–20 kb long) of short tandemly repeated DNA sequences. See *Table 7.11*. Hypervariable minisatellite DNA is the basis of DNA fingerprinting and many *VNTR* markers.

Mismatch repair: a natural enzymic process that replaces a mis-paired nucleotide in a DNA duplex (most likely present because of an error in DNA replication) to obtain perfect Watson–Crick base-pairing. See *Figure 18.17*.

Missense mutation: a nucleotide substitution that results in an amino acid change. See *Box 9.2*.

Modifier gene: a gene whose expression can influence a phenotype resulting from mutation at another locus. See Section 16.6.3.

Monoallelic expression: expression of only one of the two copies of a gene in a cell, because of X-inactivation, imprinting or other epigenetic change, or because of the gene rearrangements that take place with immunoglobulin and T-cell receptor genes. See *Box 8.6* for examples.

Monoclonal antibody (mAb): pure antibodies with a single specificity, produced by hybridoma technology, as distinct from polyclonal antibodies that are raised by immunization. See *Box 20.2*.

Mosaic: an individual who has two or more genetically different cell lines derived from a single zygote. The differences may be point mutations, chromosomal changes, etc. See *Figure 3.9*.

mRNA differential display: a PCR-based technique for comparing the mRNA species that are expressed in two related sources of cells to pick out differentially expressed genes. See *Figure 20.8*.

Multigene family: a set of evolutionarily related loci within a genome, at least one of which can encode a functional product. See Section 7.3.

Mutational load: the loss of fitness in a population due to deleterious recessive mutations. The mutational load and the segregational load (due to loss of homozygotes where there is heterozygote advantage) are the two components of the *genetic load*.

Mutator gene: a gene with an error-checking function. When such a gene is disabled by mutation, there is a general increase in the mutation rate in a cell. See Section 18.7.1.

Nondisjunction: failure of chromosomes (sister chromatids in mitosis or meiosis II; paired homologs in meiosis I) to separate (disjoin) at anaphase. The major cause of numerical chromosome abnormalities. See Section 2.6.2.

Nonhomologous recombination: recombination between sequences that either have no homology or have limited local homology. A major cause of insertions and deletions, at the genetic or chromosomal level. See for example *Figures 9.7* and *16.2*.

Nonpenetrance: the situation when somebody carrying an allele that normally causes a dominant phenotype does not show that phenotype. Due to the effect of other genetic loci or of the environment. A pitfall in genetic counseling. *Figure 3.5B* shows an example.

Nonsense mutation: a mutation that occurs within a codon and changes it to a stop codon. See *Box 9.2*.

Northern blot: a membrane bearing RNA molecules that have been size-fractionated by gel electrophoresis, used as a target for a hybridization assay. See *Figure 5.13*.

Nucleolar organizer region (**NOR**): the satellite stalks of human chromosomes 13, 14, 15, 21 and 22. NORs contain arrays of ribosomal DNA genes and can be selectively stained with silver. Each NOR forms a nucleolus in telophase of cell division; the nucleoli fuse in interphase.

Nucleosome: a structural unit of chromatin. See *Figure 2.7*.

Oligogenic: determined by a small number of genes acting together.

OMIM: O̱n-line M̱endelian I̱nheritance in M̱an, the central database of human genes and mendelian characters (http://www.ncbi.nlm.nih.gov/omim/ or http://www.hgmp.mrc.ac.uk/omim/). *MIM numbers* are the index numbers for entries in OMIM.

Oncogene: a gene involved in control of cell proliferation which, when overactive can help to transform a normal cell into a tumor cell. See *Table 18.2*. Originally the word was used only for the activated forms of the gene, and the normal cellular gene was called a *proto-oncogene*, but this distinction is now widely ignored.

Open reading frame (**ORF**): a significantly long sequence of DNA in which there are no termination codons in at least one of the possible reading frames. Six reading frames are possible for a DNA duplex because each strand can have three reading frames. See Section 20.1.4.

Ortholog: one of a set of homologous genes in different species (e.g. *PAX3* in humans and *Pax3* in mice). See *Box 14.2*.

Overdominant: phenotypes showing *heterozygote advantage*. A term used in population genetics.

Palindrome: a DNA sequence such as ATCGAT that reads the same when read in the 5'⇒3' direction on each strand. DNA–protein recognition, for example by restriction enzymes, often relies on palindromic sequences.

Paracentric inversion: inversion of a chromosomal segment that does not include the centromere. See *Figure 2.20*.

Paralog: one of a set of homologous genes within a single species. See *Box 14.2*.

Partial digestion: digestion, usually of DNA by a restriction enzyme, that is stopped before all target sequences have been cut. The object is to produce overlapping fragments. See *Figure 4.7*.

Penetrance: the frequency with which a genotype manifests itself in a given phenotype.

Pericentric inversion: inversion of a chromosomal segment that includes the centromere. See *Figure 2.20*.

Phage display: an expression cloning method in which foreign genes are inserted into a phage vector and are expressed to give polypeptides that are displayed on the surface (protein coat) of the phage. See *Figure 20.17*.

Phase: of linked markers – the relation (coupling or repulsion) between alleles at two linked loci. If allele A1 is on the same physical chromosome as allele B1, they are in coupling; if they are on different parental homologs they are in repulsion. See *Figure 11.4*. **Of the cell cycle** – G_1, S, G_2, M and G_0 phases (see *Figure 2.2*). **Of an intron** – a term used to classify introns in coding sequences according to the position at which they interrupt the message (see *Box 14.3*).

Phenotype: the observable characteristics of a cell or organism, including the result of any test that is not a direct test of the *genotype*.

Physical distance: distance between genes or sequences measured in kilobases, megabases or (in radiation hybrid mapping) centiRays.

Physical mapping: a method of mapping genes or DNA sequences on a chromosome which does not rely on meiotic segregation (*genetic mapping*).

Point mutation: a mutation causing a small alteration in the DNA sequence at a locus. The meaning is a little imprecise: when being compared to chromosomal mutations, the term 'point mutation' might be used to cover quite large (but submicroscopic) changes within a single gene, whereas when mutations at a single locus are being discussed, 'point mutations' would mean the substitution, insertion or deletion of just a single nucleotide.

Pointwise *p* value: in linkage analysis, the probability on the null hypothesis of exceeding the observed value of the statistic at one given position in the genome. Compare with the *genome-wide* p *value*. See Sections 11.3.4 and 12.5.2.

Polyadenylation: addition of typically 200 A residues to the 3′ end of a mRNA. The poly(A) tail is important for stabilizing mRNA. See *Figure 1.18*.

Polygenic character: a character determined by the combined action of a number of genetic loci. Mathematical polygenic theory (see Sections 19.2 and 19.3) assumes there are very many loci, each with a small effect.

Polymorphism: strictly, the existence of two or more variants (alleles, phenotypes, sequence variants, chromosomal structure variants) at significant frequencies in the population. Looser usages among molecular geneticists include (1) any sequence variant present at a frequency >1% in a population, (2) any nonpathogenic sequence variant, regardless of frequency.

Polyploid: having multiple chromosome sets as a result of a genetic event that is abnormal (e.g., constitutional or mosaic triploidy, tetraploidy, etc.) or programmed (e.g. some plants and certain human body cells are naturally polyploid).

Primary transcript: the initial RNA product of transcription, before intronic sequences are spliced out. See *Figure 1.14*.

Primer: a short oligonucleotide, often 15–25 bases long, which base-pairs specifically to a target sequence to allow a polymerase to initiate synthesis of a complementary strand. Suitable primers are crucial for PCR, RT-PCR and DNA sequencing. See *Figures 6.1* and *6.14*.

Primer extension: a method of identifying the transcription initiation site. See *Figure 20.2*.

Primordial germ cells: cells in the fetus which give rise eventually to the sperm or egg cells.

Prior probability: in Bayesian statistics, the initially estimated probability of an outcome before all relevant information has been taken into account. See *Box 17.2* and *Figure 17.15*.

Probe: a *known* DNA or RNA fragment (or a collection of different known fragments) which is used in a *hybridization assay* to identify closely related DNA or RNA sequences within a complex, *poorly understood* mixture of nucleic acids (the *target*). In standard hybridization assays, the probe is labeled (*Figure 5.8*), but in reverse hybridization assays the target is labeled (see *Box 5.4*).

Processed pseudogene (**retropseudogene**): a pseudogene which lacks intronic sequences and flanking sequences of the related functional gene. Originates by reverse transcription. See *Figure 7.13* and *Box 7.3*.

Prokaryote: a unicellular organism, either a bacterium or an *archaeon*, where the cell has a simple internal structure. See *Figure 2.1*.

Promoter: a combination of short sequence elements to which RNA polymerase binds in order to initiate transcription of a gene. See *Figure 1.13*.

Proofreading: an enzymic mechanism by which DNA replication errors are identified and corrected. See *Figure 18.17*.

Protein truncation test: a method of screening for chain-terminating mutations by artificially expressing a mutant allele in a coupled transcription–translation system. See *Figure 17.9*.

Proteome: the complete protein repertoire of an organism (named by analogy to the *genome*, the complete gene repertoire).

Protist: any unicellular eukaryote. Can resemble an animal (protozoon), or a plant.

Proto-oncogene: a cellular gene which can be converted by activating mutations into an *oncogene*. The term *oncogene* is now widely used for both the normal and activated forms of such genes. See *Table 18.2*.

Provirus: a phage or viral genome that has integrated into the chromosomal DNA of a cell. See *Figures 4.10* and *22.14*.

Pseudoautosomal region: a region at the tips of the sex chromosomes marked by homology between the X and Y chromosomes which may be involved in recombination during male meiosis. See *Figure 14.7*.

Pseudogene: a DNA sequence which shows a high degree of sequence homology to a nonallelic functional gene but which is itself nonfunctional. Different classes of pseudogene are described in *Box 7.3*.

Quantitative trait locus (QTL): a locus important in determining the phenotype of a continuous character. Section 19.5.7 discusses the search for QTL underlying human obesity.

Radiation hybrid: in human physical mapping, a rodent cell that contains numerous small fragments of human chromosomes. Produced by fusion with a lethally irradiated human cell. Radiation hybrid panels allow very rapid mapping of *STSs*. See *Figure 10.4*.

Rare cutter: a restriction nuclease which cuts DNA infrequently because the sequence it recognizes is large and/or contains one or more CpGs. Examples are *Not*I, *Sac*II and *Bss*HII. See *Table 4.1*.

Reading frame: during translation, the way the continuous sequence of the mRNA is read as a series of triplet codons. There are three possible reading frames for any mRNA, and the correct reading frame is set by correct recognition of the AUG initiation codon.

Reassociation kinetics: The rates at which complementary DNA strands reassociate. Highly repetitive sequences reassociate rapidly; single copy sequences reassociate slowly. See Section 5.2.2.

Recessive: a character is recessive if it is manifest only in the homozygote.

Recombinant: In linkage analysis, a person who inherits from a parent a combination of alleles that is the result of a crossover during meiosis. See *Figure 11.1*. **Recombinant DNA** is DNA containing covalently linked sequences with different origins, for example a vector with an insert.

Recombination fraction: for a given pair of loci, the proportion of meioses in which they are separated by recombination. Usually signified as θ. θ values vary between 0 and 0.5. See Section 11.1.

Renaturation (reannealing, reassociation): reformation of double helices from complementary single strands. The basis of all *hybridization assays*.

Repetitive DNA: a DNA sequence that is present in many identical or similar copies in the genome. The copies can be tandemly repeated or dispersed. See *Tables 7.11* and *7.12*.

Replication fork: the point of bifurcation when a DNA double helix is being replicated. Two replication forks proceeding outwards from a single starting point create a replication bubble (*Figure 1.10*).

Replication slippage: a mistake in replication of a tandemly repeated DNA sequence, that results in the newly synthesized strand having extra or missing repeat units compared to the template. See *Figure 9.5*.

Reporter gene: a gene used to test the ability of an upstream sequence joined on to it to cause its expression. Putative *cis*-acting regulatory sequences can be coupled to a reporter gene and transfected into suitable cells to study their function. Alternatively, transgenic mice or other organisms can be made with a promotorless reporter gene integrated at random into its chromosomes, so that expression of the reporter indicates the presence of an efficient promoter (*Figure 21.10*).

Reporter molecule: a molecule whose presence is readily detected (for example, a fluorescent molecule) that is attached to a DNA sequence we wish to monitor. See, for example, *Figure 5.7*.

Response elements: sequence usually located a short distance upstream of promoters that makes gene expression responsive to some chemical in the cellular environment. *Table 8.4* lists some examples.

Restriction fragment length polymorphism (RFLP): a polymorphic difference in the size of allelic restriction fragments as a result of the polymorphic presence or absence of a particular restriction site. RFLPs can be assayed by *Southern blotting* (*Figures 5.14* and *5.15*) or PCR (*Figure 6.6*).

Restriction map: a diagram showing the positions of restriction sites within a DNA sequence. In short sequences these can be used as landmarks. See *Figure 4.9*.

Restriction site: a short DNA sequence, often 4–8 bp long and usually *palindromic*, which is recognized by a restriction endonuclease. See *Table 4.1*.

Retrotransposon (retroposon): a transposable DNA element that transposes by means of an RNA intermediate. Retroposons encode a *reverse transcriptase* that acts on the RNA transcript to make a cDNA copy, which then integrates into chromosomal DNA at a different location. See *Box 7.4* and *Figures 7.17* and *7.18*.

Retrovirus: an RNA virus with a reverse transcriptase function, enabling the RNA genome to be copied into cDNA prior to integration into the chromosomes of a host cell. See *Figures 18.2* and *18.3*.

Reverse transcriptase: an enzyme that can make a DNA strand using an RNA template. Used to make cDNA libraries (*Figure 4.8*) and for RT-PCR (Section 20.2.4). Reverse transcription is an essential part of the retroviral life cycle (*Figure 18.2*) and is also encoded by some LINE-1 elements (*Figure 7.18*).

Riboprobe: a labeled RNA probe prepared by *in vitro* transcription from a cloned DNA sequence. See *Figure 5.4*.

Ribozyme: a natural or synthetic catalytic RNA molecule. See *Figures 22.9* and *22.10*.

RNA editing: a natural process in which specific changes occur post-transcriptionally in the base sequence of an RNA molecule. Occurs rarely in human genes. See *Figure 8.16*.

Robertsonian fusion: a chromosomal rearrangement that converts two acrocentric chromosomes into one metacentric. See *Figure 2.21*. Sometimes called *centric fusion*, although the point of exchange is actually in the proximal short arm, and not at the centromere.

RT-PCR (reverse transcriptase-PCR): a PCR reaction that amplifies a cDNA made by reverse transcription of a mRNA. See Section 20.2.4.

Satellite: this word has two different meanings in genetics: (i) **satellites on chromosomes** are stalked projections variably present on the short arms of human acrocentric chromosomes (13, 14, 15, 21, 22). (ii) **satellite DNA** originally described a DNA fraction that forms separate minor bands on density gradient centrifugation because of its unusual base composition. The DNA is composed of arrays of tandemly repeated DNA sequences. See Section 7.4.1 and *Table 7.11*.

Secondary structure: regions of a single-stranded nucleic acid or protein/polypeptide molecule where chemical bonding occurs between distantly spaced nucleotides or amino acids resulting in complex structures. Secondary structure is often due to intra-strand hydrogen bonding (*Figures 1.7* and *1.24*).

Segregation ratio: the proportion of offspring who inherit a given gene or character from a parent.

Semiconservative: DNA replication is semiconservative because each daughter duplex contains one old and one newly synthesized strand (*Figure 1.8*).

Semidominant: describes a mutation which, in the heterozygote, produces a phenotype intermediate (but not necessarily halfway) between the wild type and the homozygote. A term widely used in mouse genetics, but better avoided, at least in human genetics, since dominance is a property of a character and not of an allele.

Sense strand: the DNA strand of a gene that is complementary in sequence to the template (antisense) strand, and identical to the transcribed RNA sequence (except that DNA contains T where RNA has U). Quoted gene sequences are always of the sense strand, in the 5′–3′ direction. See *Figure 1.12*.

Sequence homology: a measure of the similarity in the sequences of two nucleic acids or two polypeptides (see *Figure 20.4B*).

Sequence tagged site (STS): any unique piece of DNA for which a specific PCR assay has been designed, so that any DNA sample can be easily tested for its presence or absence. See *Box 10.3*.

Sib-pair analysis: a form of linkage analysis in which markers are tested for linkage to a disease or phenotypic trait by measuring the extent to which affected sib pairs share marker haplotypes. See *Figure 12.2.*

Signal sequence (leader sequence): a sequence of about 20 aminoacids at the N-terminus of a polypeptide that controls its destination within or outside the cell. See Section 1.5.4.

Silencer: combination of short DNA sequence elements which suppress transcription of a gene. See *Box 8.2.*

Silent (synonymous) mutation: a mutation that changes a codon but does not alter the amino acid encoded. See *Box 9.2.* Such mutations may still have effects on mRNA splicing or stability.

SINE (short interspersed repetitive element): a class of moderate to highly repetitive DNA sequence families, of which the best known in humans is the *Alu* repeat family (*Figure 7.18*). See *Table 7.12.*

Sister chromatid: two chromatids present within a single chromosome and joined by a centromere. Nonsister chromatids are present on different but homologous chromosomes.

Sister chromatid exchange (SCE): a recombination event involving sister chromatids. Since sister chromatids are duplicates of each other, such exchanges should have no effect unless they are *unequal*. However, an increased frequency of SCEs is evidence of DNA damage.

Site-directed mutagenesis: production of a specific pre-determined change in a DNA sequence. Can be done *in vitro* on cloned DNA (section 6.4) or *in vivo* by homologous recombination (Section 21.3).

Slipped strand mispairing (replication slippage): a process in which the complementary strands of a double helix pair out of register at a tandemly repeated sequence. The resulting daughter DNA strands will have extra or missing repeat units. See *Figure 9.5.*

SNP (single nucleotide polymorphism): any polymorphic variation at a single nucleotide. SNPs include RFLPs, but also other polymorphisms that do not alter any restriction site. Although less informative than microsatellites, SNPs are more amenable to large-scale automated scoring.

snRNA: small nuclear RNAs, often part of the RNA splicing mechanism. See Section 1.4.1 and *Table 7.4.*

Somatic cell: any cell in the body except the gametes.

Somatic cell hybrid: an artificially constructed cell in which chromosomes have been stably introduced from cells of different species. See *Figure 10.1.*

Southern blot: transfer of DNA fragments from an electrophoretic gel to a nylon or nitrocellulose membrane (filter), in preparation for a hybridization assay. See *Figure 5.12.*

Splice acceptor site: the junction between the 3′ end of an intron and the start of the next exon. Consensus sequence y_{11}nyagR (Y = pyrimidine, R = purine, upper case = exon). See *Figure 1.15.*

Splice donor site: the junction between the end of an exon and the start (5′ end) of the downstream intron. Consensus sequence AGgtragt (R = purine, upper case = exon). See *Figure 1.15.*

Spliceosome: a ribonucleoprotein complex used in RNA splicing. See *Figure 1.16.*

Splicing: normally RNA splicing, in which RNA sequences transcribed from introns are excised from a primary transcript and those transcribed from exons are spliced together in the same linear order as the exons (see *Figure 1.14*). Natural DNA splicing is a much rarer event but occurs in B and T lymphocytes (see *Figure 8.27* and *8.28*).

Splicing enhancer sequences: sequences that increase the probability that a nearby potential splice site will actually be used.

SSCP or SSCA: single stranded conformation polymorphism or analysis, a commonly used method for point mutation screening. See *Figure 17.6.*

Stem cell: a cell which can act as a precursor to differentiated cells but which retains the capacity for self-renewal. See *Figure 2.5.*

Sticky ends (cohesive termini): short single-stranded projections from a double-stranded DNA molecule, typically formed by digestion with certain restriction enzymes. Molecules with complementary sticky ends can associate, and can then be covalently joined using *DNA ligase* to form recombinant DNA molecules. See *Figures 4.3* and *4.4.*

STS: see *Sequence tagged site.*

STS content mapping: checking a series of clones for the presence or absence of particular sequence tagged sites. Used as a tool in assembling clone contigs (see, for example, *Figure 10.19*).

Suppressor tRNA: a mutant transfer RNA molecule with a nucleotide substitution in the anticocon. Shows an altered coding specificity and is able to translate a nonsense (or missense) codon. See *Box 4.3.*

Synonymous (silent) substitution: a substitution that replaces one codon with another that encodes the same amino acid. See *Box 9.2.*

Synteny: loci are syntenic if on the same chromosome. Syntenic loci are not necessarily linked: if they are sufficiently far apart on the chromosome they will not cosegregate more than by random chance.

Target DNA: (i) template DNA used in a PCR reaction; (ii) DNA to which a known probe DNA is designed to hybridize in a *hybridization assay.*

Telomere: a specialized structure at the tips of chromosomes. It consists of an array of short tandem repeats, $(TTAGGG)_n$ in humans, which form a closed loop and protect the chromosome end.

Template strand: in transcription, the DNA strand that base-pairs with the nascent RNA transcript. See *Figure 1.12.*

Termination codon: a UAG (amber), UAA (ochre), or UGA (opal) codon in a mRNA (and by extension, in a gene) that signals the end of a polypeptide.

Tissue *in situ* hybridization: a form of molecular hybridization in which the target nucleic acid is RNA within cells of tissue sections immobilized on a microscope slide. See *Figure 5.17.*

Tissue-specific transcription factors: proteins that cause a gene to be expressed only in a certain tissue. See *Table 8.2* for examples.

Trans-**acting:** *trans*-acting factors (protein or maybe RNA) affect expression of all copies of a gene, whereas *cis*-acting factors (normally DNA sequences) regulate only the DNA molecule of which they are part.

Transcription: the synthesis of RNA on a DNA template by RNA polymerase. See Section 1.3.3.

Transcription unit: a stretch of DNA that is naturally transcribed in a single operation to produce a single primary transcript. In straightforward cases, a transcription unit is the same thing as a gene.

Transfection: uptake of DNA by eukaryotic cells (includes the exact equivalent of *transformation* in bacteria, but that word has a different meaning for eukaryotic cells). Also uptake of plasmid DNA by bacterial cells. See *Box 20.4.*

Transformation (of a cell): (i) uptake by a *competent* cell of naked high molecular weight DNA from the environment. (ii) Alteration of the growth properties of a normal cell as a step towards evolving into a tumor cell – see, for example, *Figure 18.4.*

Transgenic animal: an animal in which artificially introduced foreign DNA becomes stably incorporated into the germ line. This can be accomplished by pronuclear microinjection or, in the case of transgenic mice, by the use of embryonic stem cells. See *Figures 21.1* and *21.2.*

Transition: G-A (purine for purine) or C-T (pyrimidine for pyrimidine) nucleotide substitution. .

Translocation: transfer of chromosomal regions between nonhomologous chromosomes. See *Figure 2.21.*

Transmission disequilibrium test (**TDT**): a statistical test of allelic association. See *Box 12.1.*

Transposon: a mobile genetic element – see *Figure 7.16.*

Transversion: a nucleotide substitution of purine for pyrimidine or vice versa. See *Figure 9.1.*

Triploid: of a cell, having 3 copies of the genome; of an organism, being made of triploid cells.

Trisomy: having three copies of a particular chromosome, e.g. trisomy 21.

Tumor-suppressor (**TS**) **genes:** genes whose normal function is to inhibit or control cell division. Tumors always have inactivating mutations in TS genes. See Section 18.5.

Two-hit hypothesis: Knudson's theory that hereditary cancers require two successive mutations to affect a single cell. See *Figure 18.8.*

Two-hybrid system: see *Yeast two-hybrid system.*

Unequal crossover (**UEC**): recombination between non-allelic sequences on nonsister chromatids of homologous chromosomes. See *Figure 9.7.*

Unequal sister chromatid exchange (**UESCE**): recombination between nonallelic sequences on sister chromatids of a single chromosome. See *Figure 9.7.*

Uniparental disomy: a cell or organism with normal chromosome numbers, but abnormal parental origin, in that both copies of one particular chromosome pair are derived from one parent. Depending on the chromosome involved, this may or may not be pathogenic. See Section 2.6.4 and *Box 16.6.*

Untranslated regions (**5′UTR, 3′UTR**): regions at the 5′ end of mRNA before the AUG translation start codon, or at the 3′ end after the UAG, UAA or UGA stop codon. See *Figure 1.19.*

Upstream: in the direction of the 5′ end when a gene sequence is written in the conventional way (showing the sense strand, in the 5′ to 3′ direction).

Variable expression: Variable extent and intensity of phenotypic signs among people with a given genotype. See for example, *Figure 3.5C.*

Variable number of tandem repeat (**VNTR**) **polymorphism:** *microsatellites, minisatellites* and megasatellites are arrays of tandemly repeated sequences that often vary between people in the number of repeat units. See *Table 7.11.* The term VNTR is often used to mean specifically minisatellites.

Vector: a nucleic acid that is able to replicate and maintain itself within a host cell and that can be used to confer similar properties on any sequence covalently linked to it.

Vectorette PCR: see *Bubble-linker PCR.*

VNTR: see *Variable number of tandem repeat polymorphisms.*

Western blotting: a process in which proteins are size-fractionated in a polyacrylamide gel, then transferred to a nitrocellulose membrane for probing with an antibody. See *Figure 20.9.*

Whole-genome amplification: a PCR method using highly degenerate primers that can amplify a very large number of random sequences spread across the genome. Can be used to allow repeated testing of DNA from a single cell, for example in typing single sperm (Section 11.5.4).

X-inactivation (**lyonization**): the inactivation of one of the two X chromosomes in the cells of female mammals by a specialized form of genetic *imprinting.* See *Figure 2.6.*

YAC transgenic: a transgenic mouse in which the transgene is a complete *YAC,* allowing studies of the regulatory effects of sequences surrounding the gene involved. See Section 21.2.4.

Yeast artificial chromosome (**YAC**): a vector able to propagate inserts of a megabase or more in yeast cells. See *Figure 4.16.*

Yeast two-hybrid system: a yeast-based system for identifying and purifying proteins that bind to a protein of interest. See *Figure 20.16.*

Zinc finger: a polypeptide motif which is stabilized by binding a zinc atom and confers on proteins an ability to bind specifically to DNA sequences. Commonly found in transcription factors. See *Figure 8.7.*

Zoo blot: a Southern blot containing DNA samples from different species. See *Figure 10.21.*

Zygote: the fertilized egg cell.

Disease index

Index